Table A.3 Standard Normal Curve Areas *(cont.)* $\Phi(z) = P(Z \leq z)$

z	0.00	0.01	0.02	0.03	0.04	0.05	0.06	0.07	0.08	0.09
0.0	0.5000	0.5040	0.5080	0.5120	0.5160	0.5199	0.5239	0.5279	0.5319	0.5359
0.1	0.5398	0.5438	0.5478	0.5517	0.5557	0.5596	0.5636	0.5675	0.5714	0.5753
0.2	0.5793	0.5832	0.5871	0.5910	0.5948	0.5987	0.6026	0.6064	0.6103	0.6141
0.3	0.6179	0.6217	0.6255	0.6293	0.6331	0.6368	0.6406	0.6443	0.6480	0.6517
0.4	0.6554	0.6591	0.6628	0.6664	0.6700	0.6736	0.6772	0.6808	0.6844	0.6879
0.5	0.6915	0.6950	0.6985	0.7019	0.7054	0.7088	0.7123	0.7157	0.7190	0.7224
0.6	0.7257	0.7291	0.7324	0.7357	0.7389	0.7422	0.7454	0.7486	0.7517	0.7549
0.7	0.7580	0.7611	0.7642	0.7673	0.7704	0.7734	0.7764	0.7794	0.7823	0.7852
0.8	0.7881	0.7910	0.7939	0.7967	0.7995	0.8023	0.8051	0.8078	0.8106	0.8133
0.9	0.8159	0.8186	0.8212	0.8238	0.8264	0.8289	0.8315	0.8340	0.8365	0.8389
1.0	0.8413	0.8438	0.8461	0.8485	0.8508	0.8531	0.8554	0.8577	0.8599	0.8621
1.1	0.8643	0.8665	0.8686	0.8708	0.8729	0.8749	0.8770	0.8790	0.8810	0.8830
1.2	0.8849	0.8869	0.8888	0.8907	0.8925	0.8944	0.8962	0.8980	0.8997	0.9015
1.3	0.9032	0.9049	0.9066	0.9082	0.9099	0.9115	0.9131	0.9147	0.9162	0.9177
1.4	0.9192	0.9207	0.9222	0.9236	0.9251	0.9265	0.9278	0.9292	0.9306	0.9319
1.5	0.9332	0.9345	0.9357	0.9370	0.9382	0.9394	0.9406	0.9418	0.9429	0.9441
1.6	0.9452	0.9463	0.9474	0.9484	0.9495	0.9505	0.9515	0.9525	0.9535	0.9545
1.7	0.9554	0.9564	0.9573	0.9582	0.9591	0.9599	0.9608	0.9616	0.9625	0.9633
1.8	0.9641	0.9649	0.9656	0.9664	0.9671	0.9678	0.9686	0.9693	0.9699	0.9706
1.9	0.9713	0.9719	0.9726	0.9732	0.9738	0.9744	0.9750	0.9756	0.9761	0.9767
2.0	0.9772	0.9778	0.9783	0.9788	0.9793	0.9798	0.9803	0.9808	0.9812	0.9817
2.1	0.9821	0.9826	0.9830	0.9834	0.9838	0.9842	0.9846	0.9850	0.9854	0.9857
2.2	0.9861	0.9864	0.9868	0.9871	0.9875	0.9878	0.9881	0.9884	0.9887	0.9890
2.3	0.9893	0.9896	0.9898	0.9901	0.9904	0.9906	0.9909	0.9911	0.9913	0.9916
2.4	0.9918	0.9920	0.9922	0.9925	0.9927	0.9929	0.9931	0.9932	0.9934	0.9936
2.5	0.9938	0.9940	0.9941	0.9943	0.9945	0.9946	0.9948	0.9949	0.9951	0.9952
2.6	0.9953	0.9955	0.9956	0.9957	0.9959	0.9960	0.9961	0.9962	0.9963	0.9964
2.7	0.9965	0.9966	0.9967	0.9968	0.9969	0.9970	0.9971	0.9972	0.9973	0.9974
2.8	0.9974	0.9975	0.9976	0.9977	0.9977	0.9978	0.9979	0.9979	0.9980	0.9981
2.9	0.9981	0.9982	0.9982	0.9983	0.9984	0.9984	0.9985	0.9985	0.9986	0.9986
3.0	0.9987	0.9987	0.9987	0.9988	0.9988	0.9989	0.9989	0.9989	0.9990	0.9990
3.1	0.9990	0.9991	0.9991	0.9991	0.9992	0.9992	0.9992	0.9992	0.9993	0.9993
3.2	0.9993	0.9993	0.9994	0.9994	0.9994	0.9994	0.9994	0.9995	0.9995	0.9995
3.3	0.9995	0.9995	0.9995	0.9996	0.9996	0.9996	0.9996	0.9996	0.9996	0.9997
3.4	0.9997	0.9997	0.9997	0.9997	0.9997	0.9997	0.9997	0.9997	0.9997	0.9998

AT DUXBURY,
we think you know best

Introducing a different kind of learning system . . .
Whether it's interactive software or a self-study lab manual, now you can decide what will help you learn statistics more effectively. All of the books and software below work in tandem with other Duxbury texts to provide you with learning options that offer innovation, flexibility, and choice.

MINITAB HANDBOOK, 3/E
Ryan and Joiner
Ideal for hands-on sessions in the computer lab. Updated to reflect new features of Releases 8 and 9.

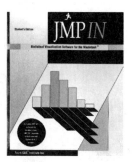

JMP IN,
STUDENT'S EDITION
SAS Institute, Inc
Combines classical statistics with today's most interactive graphics. Completely menu-driven and easy to learn.

STATAQUEST, Release 2
DeLeon and Anagnoson
Powerful, easy-to-use, graphically superior, and inexpensive software with a 250 page user's guide. DOS and Macintosh versions.

STATAQUEST
TEXT COMPANION
Anagnoson and DeLeon
Inexpensive software and an 80 page software manual with essential information to get up and running in StataQuest.

STUDENT EDITION OF
EXECUSTAT 3.0
Strategy Plus, Inc
Exceptional analytical power in an easy-to-use, intuitive format. Menu-driven, with highly interactive graphics. For IBM PC and compatibles.

STUDENT EXECUSTAT 3.0
MINIGUIDE
S. Christian Albright
Offers clear, step-by-step instructions and tutorials for mastering Student Execustat 3.0 software.

LEARNING SAS IN THE
COMPUTER LAB
Rebecca Elliott
An easy introduction to basic data analysis techniques using SAS—with worked examples, exercises, and computer output.

STATISTICS LABORATORY
MANUAL
Gerald Keller
Includes 65 experiments requiring Minitab use that illustrate how basic statistical techniques work.

**Probability and Statistics for
Engineering and the Sciences**

Probability and Statistics for Engineering and the Sciences FOURTH EDITION

Jay L. Devore

California Polytechnic State University
San Luis Obispo

Duxbury Press

An Imprint of Wadsworth Publishing Company

 An International Thomson Publishing Company

Belmont • Albany • Bonn • Boston • Cincinnati • Detroit • London • Madrid • Melbourne
Mexico City • New York • Paris • San Francisco • Singapore • Tokyo • Toronto • Washington

Assistant Editor: Jennifer Burger
Editorial Assistant: Janis Brown
Print Buyer: Randy Hurst
Permissions Editor: Peggy Meehan
Production: Ruth Cottrell
Text Design: Jamie Sue Brooks

Cover Design: Harry Voigt
Copy Editor: Betty Duncan
Illustrations: Graphic World, Inc.
Compositor: Graphic World, Inc.
Printer: R. R. Donnelley/Crawfordsville

Printed in the United States of America
2 3 4 5 6 7 8 9 10—01 00 99 98 97 96 95

For more information, contact Wadsworth Publishing Company:

Wadsworth Publishing Company
10 Davis Drive
Belmont, CA 94002, USA

International Thomson Publishing Europe
Berkshire House 168–173
High Holborn
London WC1V 7AA

Thomas Nelson Australia
102 Dodds Street
South Melbourne 3025
Victoria, Australia

Nelson Canada
1120 Birchmount Road
Scarborough, Ontario
Canada M15 5G4

International Thomson Editores
Campos Eliseos 385, Piso 7
Col. Polanco
11560 México D. F. México

International Thomson Publishing GmbH
Königswinterer 418
53227 Bonn, Germany

International Thomson Publishing Asia
221 Henderson Road
#05-10 Henderson Building
Singapore 0315

International Thomson Publishing Japan
Hirawacho Kyowa Building, 3F
2-2-1 Hirakawacho
Chiyoda-ku, Tokyo 102, Japan

Library of Congress Cataloging-in-Publication Data

Devore, Jay L.
    Probability and statistics for engineering and the
    sciences / Jay L. Devore—4th ed.
        p.   cm.
    Includes bibliographical references and index.
    ISBN 0-534-24264-2
    1. Probabilities.   2. Mathematical statistics.   I. Title.
QA273.D46   1995
519.5—dc20   94-39097

 This book is printed on acid-free recycled paper.

To my wife Carol,
whose dedication to teaching
is a continuing inspiration to me.

Contents

CHAPTER 8   **Tests of Hypotheses Based on a Single Sample   304**

CHAPTER 9   **Inferences Based on Two Samples   346**

CHAPTER 10   **The Analysis of Variance   390**

Preface

Purpose

The use of probability models and statistical methods for analyzing data has become common practice in virtually all scientific disciplines. This book attempts to provide a comprehensive introduction to those models and methods most likely to be encountered and used by students in their careers in engineering and the natural sciences. Although the examples and exercises have been designed with scientists and engineers in mind, most of the methods covered are basic to statistical analyses in many other disciplines, so that students of business and the social sciences will also profit from reading the book.

Approach

Students in a statistics course designed to serve other majors may be initially skeptical of the value and relevance of the subject matter, but my experience is that students *can* be turned on to statistics by the use of good examples and exercises that blend their everyday experiences with their scientific interests. Consequently, I have worked hard to find examples of real, rather than artificial, data—data that someone thought was worth collecting and analyzing. Many of the methods presented, especially in the later chapters on statistical inference, are illustrated by analyzing data taken from a published source, and many of the exercises also involve working with such data. Sometimes the reader may be unfamiliar with the context of a particular problem (as indeed I often was), but I have found that students are more attracted by real problems with a somewhat strange context than by patently artificial problems in a familiar setting.

Mathematical Level

The exposition is relatively modest in terms of mathematical development. Substantial use of the calculus is made only in Chapter 4 and parts of Chapters 5 and 6. In particular, with the exception of an occasional remark or aside, calculus appears in the inference part of the book only in the second section of Chapter 6. Matrix algebra is not used at all. Thus almost all the exposition should be accessible to those whose mathematical background includes one semester or two quarters of differential and integral calculus.

Content

Chapter 1 begins with some basic concepts and terminology—population, sample, descriptive and inferential statistics, enumerative versus analytic studies, and so on—and continues with a survey of important graphical and pictorial descriptive methods. A rather traditional development of probability is given in Chapter 2, followed by probability distributions of discrete and continuous random variables in Chapters 3 and 4, respectively. Joint distributions and their properties are discussed in the first part of Chapter 5. The latter part of this chapter introduces statistics and their sampling distributions, which form the bridge between probability and inference. The next three chapters cover point estimation, statistical intervals, and hypothesis testing based on a single sample. Methods of inference involving two independent samples and paired data are presented in Chapter 9. The analysis of variance is the subject of Chapters 10 and 11 (single-factor and multifactor, respectively). Regression makes its initial appearance in Chapter 12 (the simple linear regression model and correlation) and returns for an extensive encore in Chapter 13. The last three chapters develop chi-squared methods, distribution-free (nonparametric) procedures, and techniques from statistical quality control.

Helping Students Learn

Although the book's mathematical level should give most science and engineering students little difficulty, working toward an understanding of the concepts and gaining an appreciation for the logical development of the methodology may sometimes require substantial effort. To help students gain such an understanding and appreciation, I have provided numerous exercises ranging in difficulty from many that involve routine application of text material to some that ask the reader to extend concepts discussed in the text to somewhat new situations. There are many more exercises than most instructors would want to assign during any particular course, but I recommend that students be required to work a substantial number of them; in a problem-solving discipline, active involvement of this sort is the surest way to identify and close the gaps in understanding that inevitably arise.

The Statistical Tutor

A Statistical Tutor is available with this new edition. It includes complete worked-out solutions to virtually all odd-numbered exercises. The solutions are accompanied by problem-solving hints that relate the solutions to concepts discussed in the text. The immediate feedback provided by such solutions should greatly enhance the learning process.

New for This Edition

- The first section of Chapter 1 has been expanded to include a brief discussion of enumerative versus analytic studies and the collection of data from surveys and designed experiments. Several additional descriptive methods have been added to this chapter.

- The most significant change in this edition appears in the latter part of Chapter 5, which now introduces the concept of a statistic (previously not seen until point estimation in Chapter 6) and its sampling distribution.
- The examples involving the use of probability rules for deriving sampling distributions have been supplemented with a discussion of simulation experiments and the results of several such experiments. I hope that this will give students more help in making the transition from probability to inference. In addition, a general discussion of linear combinations has been put off until the last section, so it can easily be skipped if desired.
- There is increased emphasis on the use of the computer, and substantially more computer output has been included. Correspondingly, less attention is paid to "hand calculations" in ANOVA and regression (for example, hand solutions of the normal equations in the case of quadratic regression have been deleted).
- Finally, numerous examples have been updated, and many new exercises have supplemented or replaced those from previous editions.

Recommended Coverage

There is enough material in this book for a year-long course. Anyone teaching a course of shorter duration will have to be selective in the choice of topics to be included. At Cal Poly, we teach a two-quarter sequence. During the first ten weeks (meeting three hours per week), we cover most of Chapters 1–4, go very lightly over joint distributions while emphasizing the last part of Chapter 5 (sampling distributions), cover only the first section of Chapter 6 (basic concepts of point estimation), and finish with Chapter 7. We used to end the first quarter with several lectures on hypothesis testing, but pushing this into the second course allows for a somewhat more leisurely pace.

With the additional five weeks available in a semester, one could go lightly over the two-sample material and then dabble in ANOVA and/or regression. In our second quarter we go more deeply into these latter topics (with the aid of MINITAB), and we also touch on chi-squared methods and quality control. There is of course never enough time to teach students all that we would like them to know.

Acknowledgments

My colleagues here at Cal Poly have provided me with invaluable support and encouragement over the years. A note of thanks goes to Beth Elbridge and Terri Bittner for their work on the Student Solutions Manual/Statistical Tutor.

I gratefully acknowledge the numerous suggestions and constructive criticisms provided by the following reviewers: Robert L. Armacost, University of Central Florida; Douglas M. Bates, University of Wisconsin—Madison; David M. Cresap, University of Portland; Charles E. Donaghey, University of Houston; Mark Duva, University of Virginia; Nasser S. Fard, Northeastern University; James J. Halavin, Rochester Institute of Technology; Stephen Kokoska, Colgate University; Arnold R.

Miller, University of Denver; Richard M. Soland, The George Washington University; Clifford Speiglman, Texas A&M University; Jery Stedinger, Cornell University; David Steinberg, Tel Aviv University; G. Geoffrey Vining, University of Florida; Bhutan Wadhwa, Cleveland State University; and Michael G. Zabetakis, University of Pittsburgh.

I very much appreciate the editorial and production services rendered by Ruth Cottrell and Betty Duncan (for an excellent job of copy editing), and the staff at Duxbury Press. Finally, words cannot adequately express my gratitude toward my family—Carol, Allie, and Teri—for their ongoing support of my writing efforts.

Jay L. Devore

Probability and Statistics for Engineering and the Sciences

CHAPTER

1

Introduction and Descriptive Statistics

$$\Sigma\Sigma\Sigma\Sigma\Sigma\Sigma\Sigma\Sigma\Sigma\Sigma\Sigma\Sigma\Sigma\Sigma\Sigma\Sigma\Sigma\Sigma$$

1.1 An Overview of Probability and Statistics

In our own work, through conversations with others, and through contact with media of various sorts (books, television, newspapers, etc.), we are continually confronted with collections of facts or **data.** Statistics is the branch of scientific inquiry that provides methods for organizing and summarizing data and for using information in the data to draw various conclusions.

Frequently we wish to acquire information or draw some conclusion about an entire **population** consisting of all individuals or objects of a particular type. The population of interest might consist of all radial tires manufactured by a particular company during the previous calendar year, it might be the collection of all individuals who had been inoculated with a particular flu vaccine, or it might consist of all U.S. colleges and universities. In this last case, we might be interested specifically in the number of students enrolled at each school. If so, rather than think of the schools as the population members, we may speak of the **numerical population** in which each population member is an enrollment figure such as 1536 or 21,311. In the inoculation example, we might wish to focus on whether individuals had subsequently exhibited a certain condition (a rash, dizziness, etc.). We might then visualize the population as consisting of Y's (for yes, the condition was present) and N's (no, the condition was not present). This is an example of a **dichotomous** (two-valued) population. In general, we will define the population to reflect our particular interests at the time of the investigation.

The data at our disposal frequently consists of a portion or subset of the population; any such subset is called a **sample.** If the population is all U.S. colleges and universities, one sample would be {Stanford University, University of Washington, Oberlin College, California Institute of Technology, Iowa State University}. If the population comprises all college and university enrollment figures, a sample might consist of {13,043, 35,234, 2756, 21,831}.

The objectives of organization and summarization of data have been pursued for hundreds of years. The part of statistics that deals with methods for performing these operations is called **descriptive statistics.** Descriptive methods can be used either when we have a list of all population members (a **census**) or when the data consists of a sample.

When the data is a sample and the objective is to go beyond the sample to draw conclusions about the population based on sample information, methods from **inferential statistics** are used. With a few isolated exceptions, the development of inferential statistics has occurred only since the early 1900s, making it of much more recent vintage than descriptive statistics. Yet much of the interest and activity in statistics today, particularly as it relates to scientific activity and experimentation, concerns inferences rather than just description. The psychologist who tries her behavior modification technique on a sample of obese individuals would like to infer something about what the technique would do if applied to all such individuals. The engineer who accumulates data on a sample of computer systems will ultimately wish to draw conclusions about all such systems. The medical team that develops a new vaccine for a disease threatening a particular population is interested in what would happen if the vaccine were administered to all people in the population. The marketing expert may test a product in a few "representative" areas; from the resulting information, he will draw conclusions about what would happen if the product were made available to all potential purchasers.

The main focus of this book is on presenting and illustrating methods of inferential statistics that are useful in scientific work. The three important types of inferential procedures—point estimation, hypothesis testing, and estimation by confidence intervals—are introduced in Chapters 6–8 and then used in more complicated settings in Chapters 9–16. The remainder of this chapter presents methods from descriptive statistics that are most used in the development of inference.

Chapters 2–5 present material from the discipline of probability. This material ultimately forms a bridge between the descriptive and inferential techniques and leads to a better understanding of how inferential procedures are developed and used, how statistical conclusions can be translated into everyday language and interpreted, and when and where pitfalls can occur in applying the methods. Probability and statistics both deal with questions involving populations and samples, but do so in an "inverse manner" to one another.

In a probability problem, properties of the population under study are assumed known (e.g., in a numerical population, some specified distribution of the population values may be assumed), and questions regarding a sample taken from the population are posed and answered. In a statistics problem, characteristics of a sample are available to the experimenter, and this information enables the experimenter to draw conclusions about the population. The relationship between the two disciplines can be summarized by saying that probability reasons from the population to the sample (deductive reasoning), whereas statistics reasons from the sample to the population (inductive reasoning). This is illustrated in Figure 1.1.

Before we can understand what a particular sample can tell us about the population, we should first understand the uncertainty associated with taking a sample from a given population. This is why we study probability before statistics.

Figure 1.1 The relationship between probability and inferential statistics

The following two examples pursue the distinction between the two disciplines and the types of questions asked within each.

Example 1.1 A new type of gasoline pump nozzle has been developed in order to minimize the emission of pollutants into the atmosphere during pumping. One potential design defect is that "topping off" the tanks may cause gasoline to be sucked back into the underground storage tank. (This phenomenon was described in an article that appeared in a May 1981 issue of the *Los Angeles Times*.) Suppose that 10,000 nozzles manufactured by a particular company are currently in use at Southern California gas stations. In probability, we might assume that 10% of them have the defect described above (an assumption about the population of nozzles) and ask, "How likely is it that a sample of 25 nozzles will include at least 5 that are defective?" or "How many defective nozzles can we expect in a sample of 25?" On the other hand, in statistics we might find that there were 5 defectives in a sample of 25 nozzles (sample information) and then ask, "Does this strongly indicate that at least 10% of all nozzles currently in use are defective?" The last question concerns the population as a whole. ∎

Example 1.2 The following measurements (reported in *Science 167,* 1978, pp. 277–279) of the ratio of the mass of the earth to that of the moon were obtained during several different spacecraft flights: 81.3001 (*Mariner 2*), 81.3015 (*Mariner 4*), 81.3006 (*Mariner 5*), 81.3011 (*Mariner 6*), 81.2997 (*Mariner 7*), 81.3005 (*Pioneer 6*), and 81.3021 (*Pioneer 7*). These numbers differ from one another (and presumably from the true ratio) because of measurement error. In probability, we might assume that the distribution of all possible measurements is bell-shaped and centered at the true value of 81.3035, and then ask, "How likely is it that all seven measurements actually made result in observed ratios that are less than the true value?" In statistics, having been given the seven measurements, we might ask, "Does this sample information conclusively demonstrate that the ratio is something other than 81.3035?" or "How confident can we be that the true ratio is in the interval (81.2998, 81.3018)?" ∎

In Example 1.1, the population is a well-defined concrete or existing one—all nozzles presently in use. In Example 1.2, however, while the sample consists of seven observations, the population does not actually exist; instead, the population here is one consisting of all possible measurements that might be made under similar experimental conditions. Such a population is referred to as a **conceptual** or **hypothetical population.** There are a number of problems in which we fit questions into the framework of inferential statistics by conceptualizing a population.

Enumerative Versus Analytic Studies

W. E. Deming, a very influential American statistician who was a moving force in Japan's quality revolution during the 1950s and 1960s, introduced the distinction between *enumerative studies* and *analytic studies*. In the former, interest is focused on a finite, identifiable, unchanging collection of individuals or objects that make up a population. A *sampling frame*—that is, a listing of the individuals or objects to be sampled—is either available to an investigator or else can be constructed. For example, the frame might consist of all signatures on a petition to qualify a certain initiative for the ballot in an upcoming election; a sample is usually selected to ascertain whether the number of *valid* signatures exceeds a specified value. As another example, the frame may contain serial numbers of all furnaces manufactured by a particular company during a certain time period; a sample may be selected to infer something about the average lifetime of these units. The use of inferential methods to be developed in this book is reasonably noncontroversial in such settings (though statisticians may still argue over which particular methods should be used).

An analytic study is broadly defined as one that is not enumerative in nature. Such studies are often carried out with the objective of improving a future product by taking action on a process of some sort (e.g., recalibrating equipment or adjusting the level of some input such as the amount of a catalyst). Data can often be obtained only on an existing process, one that may differ in important respects from the future process. There is thus no sampling frame listing the individuals or objects of interest. For example, a sample of five turbines with a new design may be experimentally manufactured and tested to investigate efficiency. These five could be viewed as a sample from the conceptual population of all prototypes that could be manufactured under similar conditions, but *not* necessarily as representative of the population of units manufactured once regular production gets underway. Methods for using sample information to draw conclusions about future production units may be problematic. Someone with expertise in the area of turbine design and engineering (or whatever other subject area is relevant) should be called upon to judge whether such extrapolation is sensible. A good exposition of these issues is contained in the paper "Assumptions for Statistical Inference" by Gerald Hahn and William Meeker (*The American Statistician,* 1993, pp. 1–11).

Collecting Data

Statistics deals not only with the organization and analysis of data once it has been collected but also with the development of techniques for collecting the data. If data is not properly collected, an investigator may not be able to answer the questions under consideration with a reasonable degree of confidence. One common problem is that the target population—the one about which conclusions are to be drawn—may be different from the population actually sampled. For example, advertisers would like various kinds of information about the television-viewing habits of potential customers. The most systematic information of this sort comes from placing monitoring devices in a small number of homes across the United States. It has been conjectured that placement of such devices in and of itself alters viewing behavior, so that characteristics of the sample may be different from those of the target population.

When data collection entails selecting individuals or objects from a frame, the simplest method for ensuring a representative selection is to take a *simple random sample*. This is one for which any particular subset of the specified size (e.g., a sample of size 100) has the same chance of being selected. For example, if the frame consists of 1,000,000 serial numbers, the numbers 1, 2, . . . , up to 1,000,000 could be placed on identical slips of paper. After placing these slips in a box and thoroughly mixing, slips could be drawn one by one until the requisite sample size has been obtained. Alternatively (and much to be preferred), a table of random numbers or a computer's random number generator could be employed.

Sometimes alternative sampling methods can be used to make the selection process easier, to obtain extra information, or to increase the degree of confidence in conclusions. One such method, *stratified sampling,* entails separating the population units into nonoverlapping groups and taking a sample from each one. For example, a manufacturer of VCRs might want information about customer satisfaction for units produced during the previous year. If three different models were manufactured and sold, a separate sample could be selected from each of the three corresponding strata. This would result in information on all three models and ensure that no one model was over- or underrepresented in the entire sample.

Frequently a "convenience" sample is obtained by selecting individuals or objects without systematic randomization. As an example, a collection of bricks may be stacked in such a way that it is extremely difficult for those in the center to be selected. If the bricks on the top and sides of the stack were somehow different from the others, resulting sample data would not be representative of the population. Often an investigator will assume that such a convenience sample approximates a random sample, in which case a statistician's repertoire of inferential methods can be used; however, this is a judgment call. Most of the methods discussed herein are based on a variation of simple random sampling described in Chapter 5.

Engineers and scientists often collect data by carrying out some sort of designed experiment. This may involve deciding how to allocate several different treatments (such as fertilizers or coatings for corrosion protection) to the various experimental units (plots of land or pieces of pipe). Alternatively, an investigator may systematically vary the levels or categories of certain factors (e.g., pressure or type of insulating material) and observe the effect on some response variable (such as yield from a production process).

Example 1.3   An article in the *New York Times* (Jan. 27, 1987) reported that heart attack risk could be reduced by taking aspirin. This conclusion was based on a designed experiment involving both a control group of individuals who took a placebo having the appearance of aspirin but known to be inert and a treatment group who took aspirin according to a specified regimen. Subjects were randomly assigned to the groups to protect against any biases and so that probability-based methods could be used to analyze the data. Of the 11,034 individuals in the control group, 189 subsequently experienced heart attacks, whereas only 104 of the 11,037 in the aspirin group had a heart attack. The incidence rate of heart attacks in the treatment group was only about half that in the control group. One possible explanation for this result is chance variation—that aspirin really doesn't have the desired effect and the observed difference is just typical variation in the same way that tossing two identical coins would usually produce different numbers of heads. However, in this case, inferential methods suggest that chance variation by itself cannot adequately explain the magnitude of the observed difference.   ■

Example 1.4   An engineer wishes to investigate the effects of both adhesive type and conductor material on bond strength when mounting an integrated circuit (IC) on a certain substrate. Two adhesive types and two conductor materials are under consideration. Two observations are made for each adhesive-type/conductor-material combination, resulting in the accompanying data:

Adhesive Type	Conductor Material	Observed Bond Strength	Average
1	1	82, 77	79.5
1	2	75, 87	81.0
2	1	84, 80	82.0
2	2	78, 90	84.0

The resulting average bond strengths are pictured in Figure 1.2. It appears that adhesive type 2 improves bond strength as compared with type 1 by about the same amount whichever one of the conducting materials is used, with the 2, 2 combination being best. Inferential methods can again be used to judge whether these effects are real or simply due to chance variation.

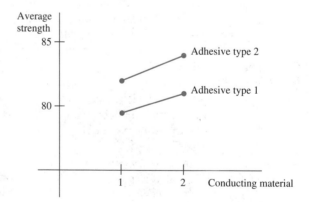

Figure 1.2 Average bond strengths in Example 1.4

Suppose additionally that there are two cure times under consideration and also two types of IC post coating. There are then $2 \cdot 2 \cdot 2 \cdot 2 = 16$ combinations of these four factors, and our engineer may not have enough resources to make even a single observation for each of these combinations. In Chapter 11, we will see how the careful selection of a fraction of these possibilities will usually yield the desired information. ■

Statistics and Computers

The computer has become an almost indispensable tool in any substantial analysis of data. Even when it is feasible to perform an analysis using only an unsophisticated hand-held calculator (this is the case for most methods presented in this book), much time and effort must often be expended doing routine organization and calculation. These latter tasks can be done much more efficiently by a computer than by human beings. This approach frees up time that an investigator can better use to study information provided by a computer as an aid to drawing conclusions.

Most computer analysis is done using a prewritten package of statistical programs. The user has only to enter data into the computer and then request output from various procedures. The most frequently used general-purpose packages are MINITAB, SAS (Statistical Analysis System), BMDP (Biomedical Computer Programs), S-PLUS, SYSTAT, and SPSS (Statistical Package for the Social Sciences). The latter packages are more difficult to learn to use than MINITAB, but they have greater capabilities. MINITAB contains most of the procedures discussed herein, whereas virtually all the methods that we present are embedded in the other five packages. As a way of emphasizing the importance of statistical computing, we will present selected output from these packages throughout the book. Teaching you how to obtain output—for example, building and editing files and using commands—is beyond our scope. For more information on this aspect of statistics, seek expertise at your home institution.

Exercises   Section 1.1 (1–8)

1. List a sample of size four from each of the following populations:

   a. All daily newspapers published in the United States

   b. All U.S. corporations

   c. All students at your college or university

   d. All radio stations

2. For each of the following hypothetical populations, list a plausible sample of size four.

   a. All distances that might result when you throw a football

   b. Page lengths of books you might read during the next year

   c. All possible earthquake-strength measurements (Richter scale) that might be recorded in California during the next year

   d. All pH measurements made on soil samples from a particular region

3. List three different examples of concrete populations and three different examples of hypothetical populations.

4. For one each of your concrete and your conceptual examples (Exercise 3), give an example of a probability question and an example of an inferential statistics question.

5. The California State University system consists of 20 campuses, from San Diego State in the south to Humboldt State near the Oregon border. A CSU

administrator wishes to make an inference about the average distance between the hometowns of students and their campuses. Describe and discuss several different sampling methods that might be employed. Would this be an enumerative or an analytic study? Explain your reasoning.

6. A certain city divides naturally into ten district neighborhoods. How might a real estate appraiser select a sample of single-family homes that could be used as a basis for developing an equation to predict appraised value from characteristics such as age, size, number of bathrooms, distance to the nearest school, and so on? Is the study enumerative or analytic?

7. The amount of flow through a solenoid valve in an automobile's pollution-control system is an important characteristic. An experiment was carried out to study how flow rate depended on three factors: armature length, spring load, and bobbin depth. Two different levels (low and high) of each factor were chosen, and a single observation on flow was made for each combination of levels.

   a. The resulting data set consisted of how many observations?

   b. Is this an enumerative or analytic study? Explain your reasoning.

8. In a famous experiment carried out in 1882, Michelson and Newcomb obtained 66 observations on the time it took for light to travel between two locations in Washington, D.C. A few of the measurements (coded in a certain manner) were 31, 23, 32, 36, −2, 26, 27, and 31.

   a. Why are these measurements not identical?

   b. Is this an enumerative study? Why, or why not?

1.2 Pictorial and Tabular Methods in Descriptive Statistics

Descriptive statistics can be divided into two general subject areas. In this section, we will discuss the first of these areas—representing a data set using visual techniques. In Sections 1.3 and 1.4, we will develop some numerical summary measures for data sets. Many visual techniques may already be familiar to you: frequency tables, tally sheets, histograms, pie charts, bar graphs, scatter diagrams, and the like. Here we focus on a selected few of these techniques that are most useful and relevant to probability and inferential statistics.

Notation

Some general notation will make it easier to apply our methods and formulas to a wide variety of practical problems. The number of observations in a single data set will often be denoted by n, so that $n = 4$ for the sample of universities {Stanford, Iowa State, Wyoming, Rochester} and also for the sample of pH measurements {6.3, 6.2, 5.9, 6.5}. If two data sets are simultaneously under consideration, either m and n or n_1 and n_2 can be used to denote the numbers of observations. Thus, if {29.7, 31.6, 30.9} and {28.7, 29.5, 29.4, 30.3} are thermal-efficiency measurements for two different types of diesel engines, then $m = 3$ and $n = 4$.

   Given a data set consisting of n observations, the observations themselves will be represented by subscripting a selected letter. We will frequently represent the observations by $x_1, x_2, x_3, \ldots, x_n$ (though any other letter could be used in place of x). The subscript bears no relation to the magnitude of a particular observation, so that x_1 will not in general be the smallest observation in the set, nor will x_n typically be the largest. In many applications x_1 will be the first observation gathered by the experimenter, x_2 the second, and so on. The ith observation in the data set will be denoted by x_i.

Stem-and-Leaf Displays

Suppose we have a data set $x_1, x_2, \ldots, x_n$ for which each x_i consists of at least two digits. A quick way to obtain an informative visual representation of the data set is to construct a *stem-and-leaf display*.

Steps for Constructing a Stem-and-Leaf Display

1. Select one or more leading digits for the stem values. The trailing digit or digits become the leaves.
2. List possible stem values in a vertical column.
3. Record the leaf for every observation beside the corresponding stem value.
4. Indicate the units for stems and leaves someplace in the display.

Thus, if the data set consists of exam scores, each between 0 and 100, the score of 83 would have a stem of 8 and a leaf of 3. For a data set of automobile-fuel efficiencies (mpg), all between 8.1 and 47.8, we could use the tens digit as the stem, so 32.6 would then have a leaf of 2.6. In general, a display based on between 5 and 20 stems is recommended.

Example 1.5

The following data on motor octane ratings for various gasoline blends is taken from an article in *Technometrics* (vol. 19, p. 425), a journal devoted to applications of statistics in the physical sciences and engineering:

88.5, 87.7, 83.4, 86.7, 87.5, 91.5, 88.6, 100.3, 95.6, 93.3, 94.7, 91.1, 91.0, 94.2, 87.8, 89.9, 88.3, 87.6, 84.3, 86.7, 88.2, 90.8, 88.3, 98.8, 94.2, 92.7, 93.2, 91.0, 90.3, 93.4, 88.5, 90.1, 89.2, 88.3, 85.3, 87.9, 88.6, 90.9, 89.0, 96.1, 93.3, 91.8, 92.3, 90.4, 90.1, 93.0, 88.7, 89.9, 89.8, 89.6, 87.4, 88.9, 91.2, 89.3, 94.4, 92.7, 91.8, 91.6, 90.4, 91.1, 92.6, 89.8, 90.6, 91.1, 90.4, 89.3, 89.7, 90.3, 91.6, 90.5, 93.7, 92.7, 92.2, 92.2, 91.2, 91.0, 92.2, 90.0, 90.7

Since the smallest observation is 83.4 and the largest is 100.3, we choose as stem values the numbers 83, 84, . . . , 100. The resulting stem-and-leaf display is given in Figure 1.3, which shows immediately that most of the octane ratings lie between 86 and 95, and that the middle value (which divides the data set into two sets of equal size) is someplace between 90 and 92. It also demonstrates that the octane ratings are distributed in a roughly symmetric fashion about the middle value. The display thus allows us to extract some of the salient features of the data set.

Stem: Tens and ones digits
Leaf: Tenths digit

```
 83 ‖ 4
 84 ‖ 3
 85 ‖ 3
 86 ‖ 7  7
 87 ‖ 7  5  8  6  9  4
 88 ‖ 5  6  3  2  3  5  3  6  7  9
 89 ‖ 9  2  0  9  8  6  3  8  3  7
 90 ‖ 8  3  1  9  4  1  4  6  4  3  5  0  7
 91 ‖ 5  1  0  0  8  2  8  6  1  1  6  2  0
 92 ‖ 7  3  7  6  7  2  2  2
 93 ‖ 3  2  4  3  0  7
 94 ‖ 7  2  2  4
 95 ‖ 6
 96 ‖ 1
 97 ‖
 98 ‖ 8
 99 ‖
100 ‖ 3
```

Figure 1.3 Stem-and-leaf display for octane data ■

Example 1.6   As a second example of a stem-and-leaf display, Figure 1.4 presents in this format a random sample of yardages of golf courses that have been designated by *Golf Magazine* as among the most challenging in the United States. Among the sample of 40 yardages, the shortest course is 6433 yards long, and the longest is 7280 yards. The yardages appear to be distributed in a roughly uniform fashion over the range of values in the sample. Notice that a stem choice here of either a single digit (6 or 7) or three digits (643, . . . , 728) would yield an uninformative display, the first because of too few stems and the latter because of too many.

Stem: Thousands and hundreds digits
Leaf: Tens and ones digits

```
64 ‖ 35  64  33  70
65 ‖ 26  27  06  83
66 ‖ 05  94  14
67 ‖ 90  70  00  98  70  45  13
68 ‖ 90  70  73  50
69 ‖ 00  27  36  04
70 ‖ 51  05  11  40  50  22
71 ‖ 31  69  68  05  13  65
72 ‖ 80  09
```

Figure 1.4 Stem-and-leaf display for golf course yardages

Figure 1.5 shows a stem-and-leaf display of the golf course yardage data produced by MINITAB. The computer uses the same stems that we selected for the display of Figure 1.4 but automatically drops (truncates) the second digit in each leaf; for example, the observation 6435 has stem value 64 and leaf 3, the tens digit. Notice also that the leaves in each row of the display are ordered from smallest to largest, something that may be time-consuming and is usually not important when constructing a display by hand.

```
Stem-and-leaf of yardage   N = 40
Leaf Unit = 10

         4       64   3367
         8       65   0228
        11       66   019
        18       67   0147799
        (4)      68   5779
        18       69   0023
        14       70   012455
         8       71   013666
         2       72   08
```

Figure 1.5 A stem-and-leaf display from MINITAB ■

A stem-and-leaf display does not show the order in which observations were obtained, possibly hiding important information about the mechanism that generated the data. For example, during a particular time period, the width of a slot cut in a certain part by a milling machine may tend to increase relative to the target value, indicating an "out-of-control" process. A *digidot plot* combines a picture of the observations over time with a stem-and-leaf display.

Example 1.7   Each observation in the stem-and-leaf display of Figure 1.6 is the value of U.S. beer production (millions of barrels) for a different quarter during the period 1975–1982. The display uses *repeated stems;* for example, the 4L row is for observations with a "low" leaf—0, 1, 2, 3, or 4—and observations with high leaves are placed in the 4H row. The *time-series plot* on the right shows both an increasing trend over time and also higher production in the second and third quarters of any given year than in the other two quarters (a *seasonal* effect).

Figure 1.6 A digidot plot for U.S. beer production ■

Frequency Distributions for Quantitative Data

A *frequency distribution* provides an even more compact summary of a data set than does a stem-and-leaf display. Rather than retaining the entire data set in the display, a frequency distribution essentially provides a count of only the number of observations associated with each stem choice. The first step in constructing a frequency distribution is to divide the relevant measurement scale into a collection of disjoint (nonoverlapping) intervals such that each observation in the data set is contained in one of these intervals. If, for example, the data set consists of compressive strengths of plastic bottles (lb), all between 71 and 128, one choice of intervals would be 70–<75, 75–<80, . . . , 125–<130, for a total of 12 intervals. Each of the resulting intervals is called a **class interval,** or simply a class.

Once the class intervals have been chosen, they are listed along the left-hand margin. The number of observations falling in each class is then determined (usually by keeping a running tally); these are the **frequencies** $f_1, f_2, \ldots$, and are displayed in a column to the right of the classes. Finally, the quantities $f_1/n, f_2/n, \ldots$ are calculated; each of these is a **relative frequency,** the proportion of observations falling in the corresponding class interval.

Example 1.8

The following data consists of lifetimes (in thousands of cycles) of 101 rectangular strips of aluminum subjected to repeated alternating stress at 21,000 psi, 18 cycles per second. The information is taken from an article in the *Journal of the American Statistical Association* (1958, p. 159).

1293, 1567, 1222, 1199, 1782, 1055, 797, 1420, 1016, 2100, 930, 1505, 1310, 1115, 1881, 1238, 370, 1522, 1419, 1262, 1763, 1020, 1102, 1730, 1893, 1594, 716, 1475, 1540, 1203, 988, 2268, 1485, 1792, 1330, 886, 1602, 2023, 2440, 1102, 990, 1390, 1502, 1270, 1140, 1910, 1000, 1450, 1608, 2130, 1290, 706, 1313, 1578, 1478, 1258, 1010, 1018, 1820, 1530, 1420, 2215, 1269, 746, 1134, 1522, 1235, 960, 1768, 1315, 844, 1452, 1940, 1252, 1781, 1108, 785, 1200, 1416, 886, 1750, 1923, 1355, 1085, 858, 1674, 1200, 1890, 1513, 1945, 1120, 1604, 1750, 1481, 855, 1868, 1560, 1300, 1630, 1895, 1642

Since the smallest observation is 370 and the largest is 2440, a reasonable choice of class intervals (one that produces neither too few nor too many intervals) is obtained by starting the first interval at 350 and letting each interval have length 200. The resulting frequency distribution is given in Table 1.1. The frequency distribution

Table I.I A frequency distribution for the lifetimes of Examples 1.8

Class interval	Tally	Frequency $= f_i$	Relative freq. $= f_i/n$
1. 350–<550	I	1	.010
2. 550–<750	III	3	.030
3. 750–<950	ℕℕ III	8	.079
4. 950–<1150	ℕℕ ℕℕ ℕℕ II	17	.168
5. 1150–<1350	ℕℕ ℕℕ ℕℕ IIII	19	.188
6. 1350–<1550	ℕℕ ℕℕ ℕℕ IIII	19	.188
7. 1550–<1750	ℕℕ ℕℕ I	11	.109
8. 1750–<1950	ℕℕ ℕℕ ℕℕ II	17	.168
9. 1950–<2150	III	3	.030
10. 2150–<2350	II	2	.020
11. 2350–<2550	I	1	.010
		$n = 101$	1.000

indicates a small number of relatively small and relatively large lifetimes, with the distribution of values in the middle intervals being relatively flat.

The proportion of observations falling in the 750–<950 class is .079 (7.9%). The proportion of observations less than 950 is .010 + .030 + .079 = .119 (11.9%). ∎

There are no hard-and-fast rules for selecting class intervals. The larger the number of observations available, the more finely the measurement scale can be subdivided to yield an informative summary. Most frequency distributions are based on between 5 and 20 classes. The quantity $\sqrt{n}$ often gives a *rough* idea for an appropriate number of intervals—five intervals when $n = 25$, ten intervals when $n = 100$, and so on. It is not even necessary that all intervals have the same width. Sometimes the data set contains a few observations that are much smaller than or much larger than the bulk of the data; these are often called **outliers.** Figure 1.7 pictures such a situation. A small number of equal-width intervals gives little information about the distribution in the region of high concentration, whereas many short equal-width intervals give a very unwieldy distribution. A reasonable strategy is to use several wide intervals on each end and narrower intervals in the middle.

Figure 1.7 Three different choices of class intervals for a data set with outliers: (a) many short intervals of equal width, (b) a few wide intervals of equal width, and (c) intervals of unequal width

Histograms

A pictorial representation of a frequency distribution can be obtained by constructing a *histogram.* First, draw a horizontal line to represent the measurement axis and then mark the boundaries of adjacent class intervals on the axis. Now above each class interval draw a rectangle whose *area* is proportional to the frequency of that interval. Thus, if interval *i* contains twice as many observations as interval *j*, the rectangle above interval *i* would have twice the area of the rectangle above interval *j*. Figure 1.8 pictures the histogram corresponding to the frequency distribution of Table 1.1. Often relative frequencies (or frequencies) are included either just above each rectangle or inside each rectangle.

A histogram is easiest to construct and gain information from if the class intervals are of equal length, for then we need only construct rectangles whose heights, rather than areas, are proportional to the frequencies. It is much easier to look at two rectangles with the same base length and judge that one is twice as high as the other than it is to compare areas of two rectangles that have different bases and heights.

Figure 1.8 A histogram for the frequency distribution of Table 1.1

To obtain a correct picture in this case, mark a vertical axis with relative frequencies (frequencies themselves can also be used) and draw each rectangle so that its height is the relative frequency of the corresponding interval. This is how Figure 1.8 was in fact obtained.

There is another scaling of the vertical axis that works whether or not the class intervals have equal lengths: First, calculate for each interval the corresponding rectangle height using the formula

$$\text{Rectangle height} = \frac{\text{Relative frequency}}{\text{Base length}}$$

Then mark the vertical axis so that rectangles with these heights can easily be drawn—the resulting vertical axis is often called the *density scale*. Using this method gives a histogram for which the area of a rectangle is equal to the corresponding relative frequency, and thus the total area of all rectangles in the histogram is 1.0. In Figure 1.9, the histogram of Figure 1.8 has been redrawn. Each base length is 200, so the height of the highest rectangle is .188/200 = .00094, and so on.

Figure 1.9 Histogram with area of each rectangle equal to relative frequency

Figure 1.10 shows a histogram of the lifetime data produced by the SAS package. The vertical axis is marked with frequencies, although other scalings can be selected.

Histograms come in a variety of shapes. A **unimodal** histogram is one that rises to a single peak and then declines. A **bimodal** histogram has two different peaks, and a histogram with more than two peaks is said to be **multimodal.** A histogram is **symmetric** if the left half is a mirror image of the right half. A unimodal histogram

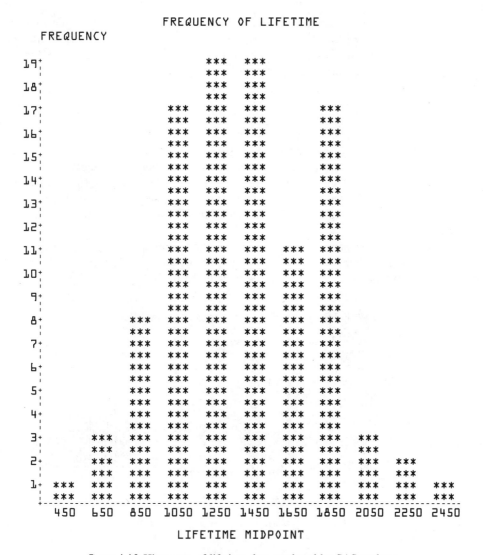

Figure 1.10 Histogram of lifetime data produced by SAS package

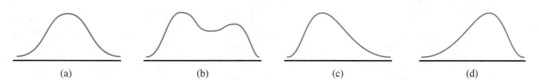

Figure 1.11 Smoothed histograms: (a) symmetric unimodal, (b) bimodal, (c) positively skewed, and (d) negatively skewed

is **positively skewed** if the right or upper tail is stretched out compared with the left or lower tail and **negatively skewed** if the stretching is to the left. Figure 1.11 shows "smoothed" histograms, obtained by superimposing a smooth curve on the rectangles, that illustrate the various possibilities.

Qualitative Data

Both a frequency distribution and a histogram can be constructed when the data set is *qualitative* (categorical) in nature. In some cases, there will be a natural ordering of classes—for example, freshmen, sophomores, juniors, seniors, graduate students—while in other cases the order will be arbitrary—for example, Catholic, Jewish, Protestant, and the like. With such categorical data, the intervals above which rectangles are constructed should have equal length.

Example 1.9   Each member of a sample of 120 individuals owning motorcycles was asked for the name of the manufacturer of his or her bike. The frequency distribution and histogram for the resulting data are given in Table 1.2.

Table 1.2 Frequency distribution of histogram for motorcycle data

	Frequency distribution		Histogram
Manufacturer	Frequency	Relative frequency	
1. Honda	41	.34	
2. Yamaha	27	.23	
3. Kawasaki	20	.17	
4. Suzuki	18	.15	
5. Harley-Davidson	3	.03	
6. Other	11	.09	
	120	1.01	

Multivariate Data

The techniques presented so far have been exclusively for situations in which each observation in a data set is either a single number or a single category. Often, however, the data is *multivariate* in nature. That is, if we obtain a sample of individuals or objects and on each one we make two or more measurements, then each "observation" would consist of several measurements on one individual or object. The sample is bivariate if each observation consists of two measurements or responses, so that the data set can be represented as $(x_1, y_1), \ldots, (x_n, y_n)$. For example, x might refer to engine size and y to miles per gallon, or x might refer to brand of calculator owned and y to academic major. In Chapters 11–14, we will analyze multivariate data sets of this sort, so we will postpone a detailed discussion until that time.

Exercises   Section 1.2 (9–24)

9. The accompanying data set consists of observations on shower-flow rate (L/min) for a sample of $n = 129$ houses in Perth, Australia ("An Application of Bayes Methodology to the Analysis of Diary Records in a Water Use Study," *J. Amer. Stat. Assoc.,* 1987, pp. 705–711):

4.6, 12.3, 7.1, 7.0, 4.0, 9.2, 6.7, 6.9, 11.5, 5.1, 3.8, 11.2, 10.5, 14.3, 8.0, 8.8, 6.4, 5.1, 5.6, 9.6, 7.5, 7.5, 6.2, 5.8, 2.3, 3.4, 10.4 9.8, 6.6, 3.7, 6.4, 6.0, 8.3, 6.5, 7.6, 9.3, 9.2, 7.3, 5.0, 6.3, 13.8, 6.2, 5.4, 4.8, 7.5, 6.0, 6.9, 10.8, 7.5, 6.6, 5.0, 3.3, 7.6, 3.9, 11.9, 2.2, 15.0, 7.2, 6.1, 15.3, 18.9, 7.2, 5.4, 5.5, 4.3, 9.0, 12.7, 11.3, 7.4, 5.0, 3.5, 8.2, 8.4, 7.3, 10.3, 11.9, 6.0, 5.6, 9.5, 9.3, 10.4, 9.7, 5.1, 6.7, 10.2, 6.2, 8.4, 7.0, 4.8, 5.6, 10.5, 14.6, 10.8, 15.5, 7.5, 6.4, 3.4, 5.5, 6.6, 5.9, 15.0, 9.6, 7.8, 7.0, 6.9, 4.1, 3.6, 11.9, 3.7, 5.7, 6.8, 11.3, 9.3, 9.6, 10.4, 9.3, 6.9, 9.8, 9.1, 10.6, 4.5, 6.2, 8.3, 3.2, 4.9, 5.0, 6.0, 8.2, 6.3

a. Construct a stem-and-leaf display of the data.

b. What is a typical, or representative, flow rate?

c. Does the display appear to be highly concentrated or spread out?

d. Does the distribution of values appear to be reasonably symmetric? If not, how would you describe the departure from symmetry?

e. Would you describe any observation as being far from the rest of the data (an outlier)?

10. The concentration of suspended solids in river water is an important environmental characteristic. The paper "Water Quality in Agricultural Watershed: Impact of Riparian Vegetation During Base Flow" (*Water Resources Bull.,* 1981, pp. 233–239) reported on concentration (in parts per million, or ppm) for several different rivers. Suppose the following 50 observations had been obtained for a particular river:

55.8	60.9	37.0	91.3	65.8
42.3	33.8	60.6	76.0	69.0
45.9	39.1	35.5	56.0	44.6
71.7	61.2	61.5	47.2	74.5
83.2	40.0	31.7	36.7	62.3
47.3	94.6	56.3	30.0	68.2
75.3	71.4	65.2	52.6	58.2
48.0	61.8	78.8	39.8	65.0

60.7	77.1	59.1	49.5	69.3
69.8	64.9	27.1	87.1	66.3

a. Construct a stem-and-leaf display.

b. Why can you not base a frequency distribution on the class intervals 0–10, 10–20, 20–30, 30–40, . . . , 90–100?

c. Construct a frequency distribution using class intervals 20–<30, 30–<40, . . . , 90–<100. (The resulting distribution agrees with that for one of the rivers discussed in the paper.)

d. What proportion of the concentration observations were less than 50? At least 60?

11. The accompanying data set consists of observations on shear strengths (lb) of ultrasonic spot welds made on a certain type of alclad sheet. Construct a relative frequency distribution and histogram based on ten class intervals with the first interval having lower limit 4000 and upper limit 4200 (the histogram will agree with the one that appeared in the paper "Comparison of Properties of Joints Prepared by Ultrasonic Welding and Other Means," *J. Aircraft,* 1983, pp. 552–556).

5434, 4948, 4521, 4570, 4990, 5702, 5241, 5112, 5015, 4659, 4806, 4637, 5670, 4381, 4820, 5043, 4886, 4599, 5288, 5299, 4848, 5378, 5260, 5055, 5828, 5218, 4859, 4780, 5027, 5008, 4609, 4772, 5133, 5095, 4618, 4848, 5089, 5518, 5333, 5164, 5342, 5069, 4755, 4925, 5001, 4803, 4951, 5679, 5256, 5207, 5621, 4918, 5138, 4786, 4500, 5461, 5049, 4974, 4592, 4173, 5296, 4965, 5170, 4740, 5173, 4568, 5653, 5078, 4900, 4698, 5248, 5245, 4723, 5275, 5419, 5205, 4452, 5227, 5555, 5388, 5498, 4681, 5076, 4774, 4931, 4493, 5309, 5582, 4308, 4823, 4417, 5364, 5640, 5069, 5188, 5764, 5273, 5042, 5189, 4986

12. A *Consumer Reports* article on peanut butter (Sept. 1990) reported the following scores for various brands:

Creamy	56 44 62 36 39 53 50 65 45 40
	56 68 41 30 40 50 56 30 22
Crunchy	62 53 75 42 47 40 34 62 52
	50 34 42 36 75 80 47 56 62

Construct a *comparative* stem-and-leaf display by listing stems in the middle of your page and then displaying the creamy leaves out to the right and the crunchy leaves out to the left. Describe similarities and differences for the two types.

13. The clearness index was determined for the skies over Baghdad for each of the 365 days during a particular year ("Contribution to the Study of the Solar Radiation Climate of the Baghdad Environment," *Solar Energy,* (1990, pp. 7–12). The accompanying table gives the results.

Class	Frequency
.15–<.25	8
.25–<.35	14
.35–<.45	28
.45–<.50	24
.50–<.55	39
.55–<.60	51
.60–<.65	106
.65–<.70	84
.70–<.75	11

a. Determine relative frequencies and draw the corresponding histogram.

b. Cloudy days are those with a clearness index smaller than .35. What percentage of the days were cloudy?

c. Clear days are those for which the index is at least .65. What percentage of the days were clear?

14. Return to the flow-rate data of Exercise 9 and construct frequencies, relative frequencies, and a histogram based on the classes 2–<3, 3–<4, . . . , 7–<8, 8–<10, 10–<12, 12–<15, and 15–<20.

15. Consider the following data on type of health complaint (*J* = joint swelling, *F* = fatigue, *B* = back pain, *M* = muscle weakness, *C* = coughing, *N* = nose running/ irritation, *O* = other) made by tree planters. Obtain frequencies and relative frequencies for the various categories, and draw a histogram. (The data is consistent with percentages given in the paper "Physiological Effects of Work Stress and Pesticide Exposure in Tree Planting by British Columbia Silviculture Workers," *Ergonomics,* 1993, pp. 951–961):

O O N J C F B B F O J O O M
O F F O O N O N J F J B O C
J O J J F N O B M O J M O B
O F J O O B N C O O O M B F
J O F N

16. Construct a digidot plot for the accompanying time series. The data is monthly and was obtained during the period 1985–1989. Each value is average solar radiation in the band 385–530 nm as a percentage of total radiation ("Global Energy in the Different Spectral Bands at Dhahran, Saudi Arabia," *J. Solar Energy Engr.,* 1991, pp. 290–294). Comment on any interesting features of the data:

20.9	19.6	20.4	20.3	20.8	20.6	20.5	20.4
19.9	19.8	19.5	20.2	16.5	18.3	18.7	19.6
20.0	20.0	19.5	19.6	19.1	18.8	18.3	17.6
17.2	17.8	18.7	19.0	19.0	18.6	18.8	19.0
18.5	18.3	17.5	16.9	17.0	17.8	18.1	18.8
18.9	18.9	19.1	18.8	18.4	17.8	17.0	16.8
17.9	18.4	19.0	19.4	19.7	19.5	19.5	19.5
19.0	18.7	18.1	17.9				

17. The mileage traveled before the first major motor failure for each of 191 buses was reported in an article that appeared in *Technometrics* (1980, p. 588). The frequency distribution appearing in that paper is given here:

Distance traveled (thousands of miles)	Frequency
0–<20	6
20–<40	11
40–<60	16
60–<80	25
80–<100	34
100–<120	46
120–<140	33
140–<160	16
160–<180	2
180–<200	2

a. Draw the corresponding histogram using relative frequencies on the vertical axis.

b. Estimate the proportion of all buses of this type that operate for at least 100,000 miles before the first major motor failure.

c. Estimate the proportion of all such buses that have the first major motor failure after operating for between 50,000 and 125,000 miles.

18. Suppose the information in Exercise 17 had been summarized in the following table:

Distance	Frequency
0–<40	17
40–<60	16
60–<80	25
80–<100	34
100–<120	46
120–<140	33
140–<180	18
180–<220	2

Mark 0, .002, .004, . . . , .014 on a vertical axis and then draw the corresponding histogram.

19. A **Pareto diagram** is a variation of a histogram for categorical data resulting from a quality control study. Each category represents a different type of product nonconformity or production problem. The categories are ordered so that the one with the largest frequency appears on the far left, then the category with the second largest frequency, and so on. Suppose the following information on nonconformities in circuit packs is obtained: failed component, 126; incorrect component, 210; insufficient solder, 67; excess solder, 54; missing component, 131. Construct a Pareto diagram.

20. A transformation of data values by means of some mathematical function, such as $\sqrt{x}$ or $1/x$, can often yield a set of numbers that has "nicer" statistical properties than the original data. In particular, it may be possible to find a function for which the histogram of transformed values is more symmetric (or, even better, more like a bell-shaped curve) than the original data. As an example, the paper "Time Lapse Cinematographic Analysis of Beryllium–Lung Fibroblast Interactions" (*Environ. Research*, 1983, pp. 34–43) reported the results of experiments designed to study the behavior of certain individual cells that had been exposed to beryllium. An important characteristic of such an individual cell is its interdivision time (IDT). IDTs were determined for a large number of cells both in exposed (treatment) and unexposed (control) conditions. The authors of the paper used a logarithmic transformation, that is, transformed value = log(original value). Consider the following representative IDT data:

IDT	$\log_{10}(IDT)$	IDT	$\log_{10}(IDT)$	IDT	$\log_{10}(IDT)$
28.1	1.45	60.1	1.78	21.0	1.32
31.2	1.49	23.7	1.37	22.3	1.35
13.7	1.14	18.6	1.27	15.5	1.19
46.0	1.66	21.4	1.33	36.3	1.56
25.8	1.41	26.6	1.42	19.1	1.28
16.8	1.23	26.2	1.42	38.4	1.58
34.8	1.54	32.0	1.51	72.8	1.86
62.3	1.79	43.5	1.64	48.9	1.69
28.0	1.45	17.4	1.24	21.4	1.33
17.9	1.25	38.8	1.59	20.7	1.32
19.5	1.29	30.6	1.49	57.3	1.76
21.1	1.32	55.6	1.75	40.9	1.61
31.9	1.50	25.5	1.41		
28.9	1.46	52.1	1.72		

Use class intervals 10–<20, 20–<30, . . . to construct a histogram of the original data. Use intervals 1.1–<1.2, 1.2–<1.3, . . . to do the same for the transformed data. What is the effect of the transformation?

21. If in Example 1.8 the relative frequencies are computed to only two decimal places, do they add up to 1?

22. In a study of warp breakage during the weaving of fabric (*Technometrics*, 1982, p. 63), 100 pieces of yarn were tested. The number of cycles of strain to breakage was recorded for each yarn sample. The resulting data was as follows:

86, 146, 251, 653, 98, 249, 400, 292, 131, 169, 175, 176, 76, 264, 15, 364, 195, 262, 88, 264, 157, 220, 42, 321, 180, 198, 38, 20, 61, 121, 282, 224, 149, 180, 325, 250, 196, 90, 229, 166, 38, 337, 65, 151, 341, 40, 40, 135, 597, 246, 211, 180, 93, 315, 353, 571, 124, 279, 81, 186, 497, 182, 423, 185, 229, 400, 338, 290, 398, 71, 246, 185, 188, 568, 55, 55, 61, 244, 20, 284, 393, 396, 203, 829, 239, 236, 286, 194, 277, 143, 198, 264, 105, 203, 124, 137, 135, 350, 193, 188

a. Using class intervals 0–<100, 100–<200, and so on, construct a relative frequency distribution for breaking strength.

b. If weaving specifications require a breaking strength of at least 100 cycles, what proportion of the yarn samples would be considered satisfactory?

23. Every score in the following batch of exam scores is in the 60s, 70s, 80s, or 90s. A stem-and-leaf display with only the four stems 6, 7, 8, and 9 would not give a very detailed description of the distribution of scores. In such situations, it is desirable to use repeated stems. Here we could repeat the stem 6 twice, using 6L for scores in the low 60s (leaves 0, 1, 2, 3, and 4) and 6H for scores in the high 60s (leaves 5, 6, 7, 8, and 9). Similarly, the other stems can be repeated twice to obtain a display consisting of eight rows. Construct such a display for the given scores. What feature of the data is highlighted by this display?

74, 89, 80, 93, 64, 67, 72, 70, 66, 85, 89, 81, 81, 71, 74, 82, 85, 63, 72, 81, 81, 95, 84, 81, 80, 70,

69, 66, 60, 83, 85, 98, 84, 68, 90, 82, 69, 72, 87, 88

24. For quantitative data, the **cumulative frequency** and cumulative relative frequency for a particular class interval are the sum of frequencies and relative frequencies, respectively, for that interval and all intervals lying below it. If, for example, there are four intervals with frequencies 9, 16, 13, and 12, then the cumulative frequencies are 9, 25, 38, and 50, and the cumulative relative frequencies are .18, .50, .76, and 1.00. Compute the cumulative frequencies and cumulative relative frequencies for the data of Exercise 10.

1.3

Measures of Location

Having briefly studied tabular and pictorial methods for organizing and summarizing data, in this section and the next we will focus on numerical summary measures for a given data set. That is, from the data we try to extract several summarizing numbers—numbers that might serve to characterize the data set and convey some of its salient features. Our primary concern will be with numerical data, although some comments regarding categorical data appear at the end of the section.

Suppose, then, that our data set is of the form $x_1, x_2, \ldots, x_n$, where each x_i is a number. What features of such a set of numbers are of most interest and deserve emphasis? One important characteristic of a set of numbers is its location, and in particular its center. This section presents methods for describing the location of a data set, while in Section 1.4 we will turn to methods for measuring the variability in a set of numbers.

The Mean

For a given set of numbers $x_1, x_2, \ldots, x_n$, the most familiar and useful measure of the center is the *mean*, or arithmetic average of the set. Because we will almost always think of the x_i's as constituting a sample, we will often refer to the arithmetic average as the **sample mean** and denote it by $\bar{x}$.

Definition

The **sample mean** $\bar{x}$ of a set of numbers $x_1, x_2, \ldots, x_n$ is given by

$$\bar{x} = \frac{x_1 + x_2 + \cdots + x_n}{n} = \frac{\sum\limits_{i=1}^{n} x_i}{n}$$

The value of $\bar{x}$ is in a sense more precise than the accuracy associated with any single observation. For this reason, we will customarily report the value of $\bar{x}$ using one digit of decimal accuracy beyond what is used in the individual x_i's.

Example 1.10

The amount of light reflectance by leaves has been used for various purposes, including evaluation of turf color, estimation of nitrogen status, and measurement of biomass. The paper "Leaf Reflectance–Nitrogen–Chlorophyll Relations in Buffel-grass" (*Photogrammetric Engineering and Remote Sensing*, 1985, pp. 463–466) gave the following observations, obtained using spectrophotogrammetry, on leaf reflectance under specified experimental conditions:

15.2, 16.8, 12.6, 13.2, 12.8, 13.8, 16.3, 13.0, 12.7, 15.8, 19.2, 12.7, 15.6, 13.5, 12.9

The sum of these 15 x_i's is $\Sigma x_i = 15.2 + 16.8 + \cdots + 12.9 = 216.1$, so the value of the sample mean is

$$\bar{x} = \frac{\sum_{i=1}^{15} x_i}{15} = \frac{216.1}{15} = 14.41$$

∎

A physical interpretation of $\bar{x}$ will demonstrate how it measures the location (center) of a sample. Think of drawing and scaling a horizontal measurement axis, and then represent each sample observation by a 1-lb weight placed at the corresponding point on the axis. The only point at which a fulcrum can be placed to balance the system of weights is the point corresponding to the value of $\bar{x}$ (see Figure 1.12).

Figure 1.12 The mean as the balance point for a system of weights

Just as $\bar{x}$ represents the average value of the observations in a sample, the average of all values in the population can be calculated. This average is called the **population mean** and is denoted by the Greek letter μ. When there are N values in the population (a finite population), then $\mu =$ (sum of the N population values)/N. In Chapters 3 and 4, we will give a more general definition for μ that applies to both finite and (conceptually) infinite populations. Just as $\bar{x}$ is an interesting and important measure of sample location, μ is an interesting and important (often the most important) characteristic of a population. In the chapters on statistical inference, we will present methods based on the sample mean for drawing conclusions about a population mean. For example, we might use the sample mean $\bar{x} = 14.41$ computed in Example 1.10 as a point estimate (a single number which is our "best" guess) of $\mu =$ the true average reflectance for all leaves under the specified conditions.

The sample mean does possess one property that renders it a somewhat unsatisfactory measure of location for some data sets. The computed value of $\bar{x}$ can be greatly influenced by the presence of just one observation that lies very far to one side or the other of the remaining values.

Example 1.11   Suppose we randomly select five recordings of classical music from the *Schwann Record Catalog* (which lists all current recordings of classical and nonclassical music) and determine the listening time for each. If the data values (rounded to the nearest minute) are $x_1 = 37$, $x_2 = 46$, $x_3 = 40$, $x_4 = 57$, and $x_5 = 50$, then $\bar{x} = 46.0$ minutes. However, if the fifth recording selected is not a Tchaikovsky symphony but instead a Wagner opera, so that $x_5 = 200$ (and seems much longer), then $\bar{x} = 76$. Since most of the data values are considerably smaller than 76, many would feel that $\bar{x}$ here is not a reliable measure of location. Of course, this effect would be even more pronounced if x_5 were, for example, 400 (so $\bar{x} = 116$) or 1000.   ■

A sample of incomes often produces a few such outlying values (those lucky few who earn astronomical incomes), and the use of average income as a measure of location will often be misleading. Such examples suggest that we look for a measure that is less sensitive to outlying values than $\bar{x}$, and we will momentarily propose one. However, while $\bar{x}$ does have this potential defect, it is still the most widely used measure, largely because there are many populations for which an extreme outlier in the sample would be highly unlikely. When sampling from such a population (a normal or bell-shaped population being the most important example), the sample mean will be stable and quite representative of the sample.

The Median

The word *median* is synonymous with "middle," and the sample median is indeed the middle value when the observations are ordered from smallest to largest in magnitude. When the observations are denoted by $x_1, \ldots, x_n$, we will use the symbol $\tilde{x}$ to represent the sample median.

Definition

The **sample median** is obtained by first ordering the n observations from smallest to largest (with any repeated values included, so that every sample observation appears in the ordered list). Then,

$$\tilde{x} = \begin{cases} \text{The single} \\ \quad \text{middle} \\ \quad \text{value if } n \quad = \left(\dfrac{n+1}{2}\right)^{\text{th}} \text{ ordered value} \\ \quad \text{is odd} \\ \\ \text{The average} \\ \quad \text{of the two} \\ \quad \text{middle} \quad = \text{average of } \left(\dfrac{n}{2}\right)^{\text{th}} \text{ and } \left(\dfrac{n}{2}+1\right)^{\text{th}} \text{ ordered values} \\ \quad \text{values if } n \\ \quad \text{is even} \end{cases}$$

Example 1.12   The risk of developing iron deficiency is especially high during pregnancy. The problem with detecting such deficiency is that some methods for determining iron status can be affected by the state of pregnancy itself. Consider the following data on transferrin receptor concentration for a sample of women with laboratory evidence of overt iron-deficiency anemia ("Serum Transferrin Receptor for the Detection of Iron Deficiency in Pregnancy," *Amer. J. of Clinical Nutrition* 1991, pp. 1077–1081):

$$x_1 = 15.2 \quad x_2 = 9.3 \quad x_3 = 7.6 \quad x_4 = 11.9 \quad x_5 = 10.4 \quad x_6 = 9.7$$
$$x_7 = 20.4 \quad x_8 = 9.4 \quad x_9 = 11.5 \quad x_{10} = 16.2 \quad x_{11} = 9.4 \quad x_{12} = 8.3$$

The list of ordered values is

7.6   8.3   9.3   9.4   9.4   9.7   10.4   11.5   11.9   15.2   16.2   20.4

Since $n = 12$ is even, we average the $n/2 =$ sixth- and seventh-ordered values:

$$\text{Sample median} = \frac{9.7 + 10.4}{2} = 10.05$$

Notice that if the largest observation, 20.4, had not appeared in the sample, the resulting sample median for the $n = 11$ observations would have been the single middle value, 9.7 (the $(n + 1)/2 =$ sixth-ordered value). The sample mean is $\bar{x} = \Sigma x_i/n = 139.3/12 = 11.61$, which is somewhat larger than the median because of the outliers, 15.2, 16.2, and 20.4.   ■

The data set in Example 1.12 illustrates an important property of $\tilde{x}$ in contrast to $\bar{x}$: The sample median is very insensitive to a number of extremely small or extremely large data values. If, for example, we increased the two largest x_i's from 16.2 and 20.4 to 26.2 and 30.4, $\tilde{x}$ would be unaffected. Thus, in the treatment of outlying data values, $\bar{x}$ and $\tilde{x}$ are at opposite ends of a spectrum: $\bar{x}$ is sensitive to even one such value, while $\tilde{x}$ is insensitive to a large number of outlying values.

Because the large values in the sample of Example 1.12 affect $\bar{x}$ more than $\tilde{x}$, $\tilde{x} < \bar{x}$ for that data. While $\bar{x}$ and $\tilde{x}$ both provide a measure for the center of a data set, they will not in general be equal because they focus on different aspects of the sample.

Analogous to $\tilde{x}$ as the middle value in the sample, there is a middle value in the population, the **population median,** denoted by $\tilde{\mu}$. As with $\bar{x}$ and μ, we can think of using the sample median $\tilde{x}$ to make an inference about $\tilde{\mu}$. In Example 1.12, we might use $\tilde{x} = 10.05$ as an estimate of the median concentration in the entire population from which the sample was selected. A median is often used to describe income or salary data (because it is not greatly influenced by a few large salaries), so if the median salary for a sample of engineers were $\tilde{x} = \$46,416$, we might use this as a basis for concluding that the median salary for all engineers exceeds \$45,000.

The population mean μ and median $\tilde{\mu}$ will not generally be equal to one another. If the population distribution is either positively or negatively skewed, as pictured in Figure 1.13, then $\mu \neq \tilde{\mu}$. When this is the case, in making inferences we must first decide which of the two population characteristics is of greater interest and then proceed accordingly.

(a) Negative skew   (b) Symmetric   (c) Positive skew

Figure 1.13 Three different shapes for a population distribution

Other Measures of Location: Quartiles, Percentiles, and Trimmed Means

The median (population or sample) divides the data set into two parts of equal size. To obtain finer measures of location, we could divide the data into more than two such parts. Roughly speaking, quartiles divide the data set into four equal parts, with the observations above the third quartile constituting the upper quarter of the data set, the second quartile being identical to the median, and the first quartile separating the lower quarter from the upper three-quarters. Similarly, a data set (sample or population) can be divided into 100 equal parts using percentiles; the 99th percentile separates the highest 1% from the bottom 99%, and so on. Unless the number of observations is a multiple of 100, care must be exercised in obtaining percentiles. We will use percentiles in Chapter 4 in connection with certain models for infinite populations, so we will postpone discussion until that point.

The sample mean and sample median are influenced by outlying values in a very different manner—the mean greatly and the median not at all. Since extreme behavior of either type might be undesirable, we briefly consider alternative measures that are neither as sensitive as $\bar{x}$ nor as insensitive as $\tilde{x}$. To motivate these alternatives, note that $\bar{x}$ and $\tilde{x}$ are at opposite extremes of the same "family" of measures. After the data set is ordered, $\tilde{x}$ is computed by throwing away as many values on each end as one can without eliminating everything (leaving just one or two middle values) and averaging what is left, while to compute $\bar{x}$ one throws away nothing before averaging. To paraphrase, the mean involves trimming 0% from each end of the sample, while for the median the maximum possible amount is trimmed from each end. A **trimmed mean** is a compromise between $\bar{x}$ and $\tilde{x}$. A 10% trimmed mean, for example, would be computed by eliminating the smallest 10% and the largest 10% of the sample and then averaging what is left over.

Example 1.13   Consider the following 20 observations, ordered from smallest to largest, each one representing the lifetime (in hours) of a certain type of incandescent lamp:

612, 623, 666, 744, 883, 898, 964, 970, 983, 1003, 1016, 1022, 1029, 1058, 1085, 1088, 1122, 1135, 1197, 1201

The average of all 20 observations is $\bar{x} = 965.0$, while $\tilde{x} = 1009.5$. The 10% trimmed mean is obtained by deleting the smallest two observations (612 and 623) and the largest two (1197 and 1201) and then averaging the remaining 16 to obtain

$\bar{x}_{tr(10)} = 979.1$. The effect of trimming here is to produce a "central value" that is somewhat above the mean ($\bar{x}$ is pulled down by a few small lifetimes) and yet considerably below the median. Similarly, the 20% trimmed mean averages the middle 12 values to obtain $\bar{x}_{tr(20)} = 999.9$, even closer to the median. (See Figure 1.14.)

Figure 1.14 Dot diagram of lifetimes (in hours) of incandescent lamps ■

Generally speaking, using a trimmed mean with a moderate trimming proportion (10% or 20%) will yield a measure that is neither as sensitive to outliers as the mean (since any small number of outliers will be deleted before averaging) nor as insensitive as the median. For this reason, trimmed means have merited increasing attention from statisticians for both descriptive and inferential purposes. More will be said about trimmed means when point estimation is discussed in Chapter 6. As a final point, if the trimming proportion is denoted by α and $n\alpha$ is not an integer, then it is not obvious how the $100\alpha\%$ trimmed mean should be computed. For example, if $\alpha = .10$ (10%) and $n = 22$, then $n\alpha = (22)(.10) = 2.2$, and we cannot trim 2.2 observations from each end of the ordered sample. In this case, the 10% trimmed mean would be obtained by first trimming two observations from each end and calculating $\bar{x}_{tr}$, then trimming three and calculating $\bar{x}_{tr}$, and finally interpolating between the two values to obtain $\bar{x}_{tr(10)}$.

Categorical Data and Sample Proportions

When the data is categorical, a frequency distribution or relative frequency distribution provides an effective tabular summary of the data. The natural numerical summary quantities in this situation are the individual frequencies and the relative frequencies. For example, if a survey of individuals who own stereo receivers is undertaken to study brand preference, then each individual in the sample would identify the brand of receiver that he or she owned, from which we could count the number owning Sony, Marantz, Pioneer, and so on. Consider sampling a dichotomous population—one that consists of only two categories (such as voted or did not vote in the last election, does or does not own a stereo receiver, etc.). If we let x denote the number in the sample falling in category one, then the number in category two is $n - x$. The relative frequency or *sample proportion* in category one is x/n and the sample proportion in category two is $1 - x/n$. To see that x/n *can be regarded as a sample mean,* identify a category-one response by a 1 and a category-two response by a 0. A sample size of $n = 10$ might then yield the responses, 1, 1, 0, 1, 1, 1, 0, 0, 1, 1. The sample mean for this numerical sample is (since number of 1's $= x = 7$)

$$\frac{x_1 + \cdots + x_n}{n} = \frac{1 + 1 + 0 + \cdots + 1 + 1}{10} = \frac{7}{10} = \frac{x}{n} = \begin{matrix} \text{Sample} \\ \text{proportion} \end{matrix}$$

This result can be generalized and summarized as follows: *If in a categorical data situation we focus attention on a particular category and code the sample results so that a 1 is recorded for an individual in the category and a 0 for an individual not in the category, then the sample proportion of individuals in the category is the sample mean of the sequence of 1's and 0's.* Thus, a sample mean can be used to summarize the results of a categorical sample. These remarks also apply to situations in which categories are defined by grouping values in a numerical sample or population (e.g., we might be interested in knowing whether individuals have owned their present automobile for at least 5 years, rather than studying the exact length of ownership).

Analogous to the sample proportion x/n of individuals falling in a particular category, let p represent the proportion of individuals in the entire population falling in the category. As with x/n, p is a quantity between 0 and 1. While x/n is a sample characteristic, p is a characteristic of the population; the relationship between the two parallels the relationship between $\tilde{x}$ and $\tilde{\mu}$ and between $\bar{x}$ and μ. In particular, we will subsequently use x/n to make inferences about p. If, for example, a sample of 100 car owners reveals that 22 owned their car at least 5 years, then we might use $22/100 = .22$ as a point estimate of the proportion of all owners who have owned their car at least 5 years. We will study the properties of x/n as an estimator of p and see how x/n can be used to answer other inferential questions. With k categories $(k > 2)$, we can use the k sample proportions to answer questions about the population proportions $p_1, \ldots, p_k$.

Exercises   Section 1.3 (25–35)

25. The paper "The Pedaling Technique of Elite Endurance Cyclists" (*Int. J. of Sport Biomechanics*, 1991, pp. 29–53) reported the accompanying data on single-leg power at a high workload:

244   191   160   187   180   176   174
205   211   183   211   180   194   200

a. Calculate and interpret the sample mean and median.

b. Suppose that the first observation had been 204 rather than 244. How would the mean and median change?

26. The paper cited in Exercise 25 also reported values of single-leg power for a low workload. The sample mean for $n = 13$ observations was $\bar{x} = 119.8$ (actually 119.7692), and the fourteenth observation, somewhat of an outlier, was 159. What is the value of $\bar{x}$ for the entire sample?

27. Return to the single-leg power data of Exercise 25 and calculate a trimmed mean by eliminating the smallest and largest sample observations. What is the corresponding trimming percentage?

28. The paper "Penicillin in the Treatment of Meningitis" (*J. Amer. Medical Assn.*, 1984, pp. 1870–1873) reported the body temperatures (in degrees Fahrenheit) of patients hospitalized with meningitis. Ten of the observations were as follows:

104.0, 104.8, 101.6, 108.0, 103.8, 100.8, 104.2, 100.2, 102.4, 101.4

a. Compute the sample mean.

b. Do you think the 10% trimmed mean would differ much from the sample mean obtained in part (a)? Answer without doing any calculation and then verify your conjecture by computing this trimmed mean.

29. The paper "Snow Cover and Temperature Relationships in North America and Eurasia" (*J. Climate and Applied Meteorology,* 1983, pp. 460–469) used statistical techniques to relate the amount of snow cover on each continent to average continental temperature. Data presented there included the following ten observations on October snow cover for Eurasia during the years 1970–1979 (in million km^2):

6.5, 12.0, 14.9, 10.0, 10.7, 7.9, 21.9, 12.5, 14.5, 9.2

What would you report as a representative, or typical, value of October snow cover for this period, and what prompted your choice?

30. Blood pressure values are often reported to the nearest 5 mmHg (100, 105, 110, etc.). Suppose the actual blood pressure values for nine randomly selected individuals are

118.6, 127.4, 138.4, 130.0, 113.7, 122.0, 108.3, 131.5, 133.2

a. What is the median of the reported blood pressure values?

b. Suppose the blood pressure of the second individual is 127.6 rather than 127.4 (a small change in a single value). How does this affect the median of the reported values? What does this say about the sensitivity of the median to rounding or grouping in the data?

31. The propagation of fatigue cracks in various aircraft parts has been the subject of extensive study in recent years. The accompanying data consists of propagation lives (flight hours/10^4) to reach a given crack size in fastener holes intended for use in military aircraft ("Statistical Crack Propagation in Fastener Holes under Spectrum Loading," *J. Aircraft,* 1983, pp. 1028–1032):

.736, .863, .865, .913, .915, .937, .983, 1.007, 1.011, 1.064, 1.109, 1.132, 1.140, 1.153, 1.253, 1.394

a. Compute the values of the sample mean and median.

b. By how much could the largest sample observation be decreased without affecting the value of the median?

32. Compute the sample median, 25% trimmed mean, 10% trimmed mean, and sample mean for the concentration data given in Exercise 10 and compare these measures (a stem-and-leaf display will aid in ordering the observations).

33. A sample of $n = 10$ automobiles was selected, and each was subjected to a 5-mph crash test. Denoting a car with no visible damage by *S* (for success) and a car with such damage by *F*, results were as follows:

S, S, F, S, S, S, F, F, S, S

a. What is the value of the sample proportion of successes x/n?

b. Replace each *S* with a 1 and each *F* with a 0. Then calculate $\bar{x}$ for this numerically coded sample. How does $\bar{x}$ compare to x/n?

c. Suppose it is decided to include 15 more cars in the experiment. How many of these would have to be *S*'s to give $x/n = .80$ for the entire sample of 25 cars?

34. a. If a constant *c* is added to each x_i in a sample, yielding $y_i = x_i + c$, how do the sample mean and median of the y_i's relate to the mean and median of the x_i's? Verify your conjectures.

b. If each x_i is multiplied by a constant *c*, yielding $y_i = cx_i$, answer the question of part (a). Again, verify your conjectures.

35. An experiment to study the lifetime (in hours) for a certain type of component involved putting ten components into operation and observing them for 100 hours. Eight of the components failed during that period, and those lifetimes were recorded. Denote the lifetimes of the two components still functioning after 100 hours by 100 +. The resulting sample observations were

48 79 100+ 35 92
86 57 100+ 17 29

Which of the measures of center discussed in this section can be calculated, and what are the values of those measures? (NOTE: The data from this experiment is said to be "censored on the right.")

1.4   Measures of Variability

No single measure of location can give a complete summarization of a data set. Consider the x data set 20, 100, 0, 60, 70 and the y data set 60, 20, 80, 60, 30. Since $\bar{x} = \bar{y} = 50$ and $\tilde{x} = \tilde{y} = 60$, the two standard measures of location by themselves do not distinguish between the two sets. Yet the plots of these two data sets in Figure 1.15 show that the observations in the y data-set cluster more closely about their center than do the observations from the x data set. That is, there is more variability or dispersion in the x's than in the y's.

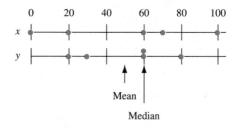

Figure 1.15 Data sets with the same center and differing variability

One simple measure of variability is the **sample range,** defined as the difference between the largest observation and the smallest observation—that is, sample range = $\max(x_i) - \min(x_i)$. We might then say that a small (large) range indicates little (great) variability. A defect of the sample range, though, is that it depends only on the two most extreme observations and disregards the positions of the middle observations. For example, the two samples 0, 5, 5, 5, 10 and 0, 1, 5, 9, 10 both have the same sample range, yet in the first sample there is variability only in the two extreme values, while the second sample has more variability in its middle values. An effective measure depends on more than just a few of the observations.

Deviations from the Mean

The quantity $x_i - \bar{x}$ is called the deviation of the ith observation from the mean, or just the **ith deviation.** A positive deviation indicates an observation to the right of $\bar{x}$ on the measurement axis, whereas a negative deviation indicates an observation to the left of $\bar{x}$. If all the deviations $x_1 - \bar{x}, \ldots, x_n - \bar{x}$ are small in absolute magnitude, then all x_i's are close to $\bar{x}$ and thus to one another, suggesting a relatively small amount of variability in the sample. On the other hand, if some of the $(x_i - \bar{x})$'s are large in absolute magnitude, then some of the x_i's lie far from $\bar{x}$, suggesting large variation. A simple way of combining the n deviations into a single quantity is to

Variability

on

average them (sum them and divide by n). However, since some deviations will be negative and others positive, adding results in cancellation.

$$\sum_{i=1}^{n} (i\text{th deviation}) = \sum_{i=1}^{n} (x_i - \bar{x}) = 0$$

so the average deviation from the mean is always 0.

Proof The verification uses several standard rules of summation and the fact that $\Sigma\bar{x} = \bar{x} + \bar{x} + \cdots + \bar{x} = n\bar{x}$:

$$\Sigma(x_i - \bar{x}) = \Sigma x_i - \Sigma\bar{x} = \Sigma x_i - n\bar{x} = \Sigma x_i - n\left(\frac{1}{n}\Sigma x_i\right) = 0 \qquad \blacksquare$$

To obtain an informative measure of variability, we need to change the deviations to nonnegative quantities before combining. One possibility is to average the absolute deviations $|x_1 - \bar{x}|, \ldots, |x_n - \bar{x}|$. Because this leads to a multitude of theoretical difficulties, consider instead the squared deviations $(x_1 - \bar{x})^2, \ldots, (x_n - \bar{x})^2$. We might now use the average of the squared deviations. There is, however, a technical reason (to be discussed shortly) for dividing the sum of squared deviations by $n - 1$ rather than by n.

The Sample Variance and Standard Deviation

Definition

The **sample variance** of the set $x_1, \ldots, x_n$ of numerical observations, denoted by s^2, is given by

$$s^2 = \frac{\displaystyle\sum_{i=1}^{n} (x_i - \bar{x})^2}{n - 1}$$

The **sample standard deviation,** denoted by s, is the positive square root of the sample variance.

The divisor $n - 1$ in s^2 is smaller than n, so s^2 is somewhat larger than the average squared deviation. Whatever the units in which the x_i's are expressed, s^2 is given in squared units (e.g., in.^{2} when the x_i's are heights in inches or sec^2 when observations are stopping times expressed in seconds). The sample standard deviation has the desirable property of measuring dispersion in units identical to those in which the x_i's themselves are given.

Example 1.14 Strength is an important characteristic of materials used in prefabricated housing. Each of $n = 11$ prefabricated plate elements was subjected to a severe stress test and the maximum width (mm) of the resulting cracks was recorded. The given data (Table 1.3) appeared in the paper "Prefabricated Ferrocement Ribbed Elements for Low-Cost Housing" (*J. Ferrocement,* 1984, pp. 347–364). Effects of rounding account for the sum of deviations not being exactly zero. The numerator of s^2 is 11.9359, so $s^2 = 11.9359/(11 - 1) = 11.9359/10 = 1.19359$ and $s = \sqrt{1.19359} = 1.0925$ mm.

Table 1.3 Data for Example 1.14

x_i	$x_i - \bar{x}$	$(x_i - \bar{x})^2$
.684	−.9841	.9685
2.540	.8719	.7602
.924	−.7441	.5537
3.130	1.4619	2.1372
1.038	−.6301	.3970
.598	−1.0701	1.1451
.483	−1.1851	1.4045
3.520	1.8519	3.4295
1.285	−.3831	.1468
2.650	.9819	.9641
1.497	−.1711	.0293
$\Sigma x_i = 18.349$	$\Sigma(x_i - \bar{x}) = -.0001$	$\Sigma(x_i - \bar{x})^2 = 11.9359$

$$\bar{x} = \frac{18.349}{11} = 1.6681$$

Admittedly, s^2 and s do not lend themselves to the intuitive understanding that measures of location do. Because variability is a less familiar characteristic than location, the usefulness of measures of variability is less obvious. In particular, we are not in a position to say that $s = 1.0925$ indicates a large or a small amount of variability. All you should believe at this point is that if "eyeballing" two different samples suggests that the first clearly has less variability than the second, then s^2 and s for the first sample should be smaller than the corresponding quantities for the second sample.

Motivation for s^2

To explain why s^2 rather than the average squared deviation is used to measure variability, note first that whereas s^2 measures sample variability, there is a measure of variability in the population called the population variance. We will use σ^2 (the square of the lowercase Greek letter sigma) to denote the population variance and σ to denote the population standard deviation (the square root of σ^2). When the population is finite and consists of N values,

$$\sigma^2 = \sum_{i=1}^{N}(x_i - \mu)^2/N$$

which is the average of all squared deviations from the population mean (for the population, the divisor is N and not $N - 1$). More general definitions of σ^2 appear in Chapters 3 and 4.

Just as $\bar{x}$ will be used to make inferences about the population mean μ, we should define the sample variance so that it can be used to make inferences about σ^2. Now note that σ^2 involves squared deviations about the population mean μ. If we actually knew the value of μ, then we could define the sample variance as the average squared deviation of the sample x_i's about μ. However, the value of μ is almost never known, so the sum of squared deviations about $\bar{x}$ must be used. But *the x_i's tend to be closer to their average $\bar{x}$ than to the population average μ, so to compensate for this the divisor $n - 1$ is used rather than n.* Said another way, if we used a divisor n in the sample variance, then the resulting quantity would tend to underestimate σ^2 (produce estimated values that are too small on the average), while dividing by the slightly smaller $n - 1$ corrects this underestimating.

It is customary to refer to s^2 as being based on $n - 1$ **degrees of freedom** (df). This terminology results from the fact that while s^2 is based on the n quantities $x_1 - \bar{x}$, $x_2 - \bar{x}, \ldots, x_n - \bar{x}$, these sum to 0, so specifying the values of any $n - 1$ of the quantities determines the remaining one. For example, if $n = 4$ and $x_1 - \bar{x} = 8$, $x_2 - \bar{x} = -6$, and $x_4 - \bar{x} = -4$, then automatically we have $x_3 - \bar{x} = 2$, so only three of the four values of $x_i - \bar{x}$ are freely determined (3 df).

The Computation of s^2

With n observations, the computation of s^2 from the definition involves n subtractions and n squaring operations. If $\bar{x}$ is not an integer, the subtractions to obtain the deviations can be tedious, and the deviations themselves may be unpleasant to square. There is an equivalent expression for the numerator of s^2 that yields a more efficient method for calculating s^2 when computations are done by hand or hand-held calculator.

Proposition

$$s^2 = \frac{\sum_{i=1}^{n}(x_i - \bar{x})^2}{n-1} = \frac{\sum_{i=1}^{n}x_i^2 - \dfrac{\left(\sum_{i=1}^{n}x_i\right)^2}{n}}{n-1}$$

Proof Because $\bar{x} = \Sigma x_i/n$, $n\bar{x}^2 = (\Sigma x_i)^2/n$. Then,

$$\sum_{i=1}^{n}(x_i - \bar{x})^2 = \sum_{i=1}^{n}(x_i^2 - 2\bar{x} \cdot x_i + \bar{x}^2) = \sum_{i=1}^{n}x_i^2 - 2\bar{x}\sum_{i=1}^{n}x_i + \sum_{i=1}^{n}(\bar{x})^2$$

$$= \sum_{i=1}^{n}x_i^2 - 2\bar{x} \cdot n\bar{x} + n(\bar{x})^2 = \sum_{i=1}^{n}x_i^2 - n(\bar{x})^2$$

$$= \sum_{i=1}^{n} x_i^2 - \frac{\left(\sum_{i=1}^{n} x_i\right)^2}{n}$$ ■

To use this method for computing s^2, square each x_i (before subtraction), then add the squares and subtract $(\Sigma x_i)^2/n$ from Σx_i^2. While this involves squaring $n + 1$ numbers (Σx_i in addition to each x_i), only one subtraction is necessary. We will refer to this formula for computing s^2 as the "shortcut method for s^2." The shortcut for s involves computing s^2 using the shortcut and then taking the square root.

Example 1.15   Table 1.4 displays the $n = 15$ reflectance observations first introduced in Example 1.10.

Table 1.4 Data for Example 1.15

Observation	x_i	x_i^2
1	15.2	231.04
2	16.8	282.24
3	12.6	158.76
4	13.2	174.24
5	12.8	163.84
6	13.8	190.44
7	16.3	265.69
8	13.0	169.00
9	12.7	161.29
10	15.8	249.64
11	19.2	368.64
12	12.7	161.29
13	15.6	243.36
14	13.5	182.25
15	12.9	166.41
	$\Sigma x_i = 216.1$	$\Sigma x_i^2 = 3168.13$

Thus, the shortcut formula gives

$$s^2 = \frac{\Sigma x_i^2 - \dfrac{(\Sigma x_i)^2}{n}}{n - 1} = \frac{3168.13 - \dfrac{(216.1)^2}{15}}{15 - 1} = \frac{54.85}{14} = 3.92$$

and $s = \sqrt{3.92} = 1.98$.   ■

The shortcut method can yield values of s^2 and s that differ from the values computed using the definitions. These differences are due to effects of rounding and will not be important in most problems. To minimize the effects of rounding when using the shortcut formula, particularly when there is little variability in the data, intermediate calculations should be done using several more significant digits than are to be retained in the final answer. Because the numerator of s^2 is the sum of nonnegative quantities (squared deviations), s^2 is guaranteed to be nonnegative. Yet if the shortcut method is used, particularly with data having little variability, a slight numerical error can result in a negative numerator (Σx_i^2 smaller than $(\Sigma x_i)^2/n$). If your value of s^2 is negative, you have made a computational error.

Several other properties of s^2 can sometimes be used to increase computational efficiency. These are summarized in the following proposition.

Proposition

Let $x_1, x_2, \ldots, x_n$ be a sample and c be any nonzero constant.

1. If $y_1 = x_1 + c, y_2 = x_2 + c, \ldots, y_n = x_n + c$, then $s_y^2 = s_x^2$, and
2. If $y_1 = cx_1, \ldots, y_n = cx_n$, then $s_y^2 = c^2 s_x^2$, $s_y = |c| s_x$,

where s_x^2 is the sample variance of the x's and s_y^2 is the sample variance of the y's.

In words, Result 1 says that if a constant c is added to (or subtracted from) each data value, the variance is unchanged. This is intuitive, since adding or subtracting c shifts the location of the data set but leaves distances between data values unchanged. According to Result 2, multiplication of each x_i by c results in s^2 being multiplied by a factor of c^2. These properties can be proved by noting in Result 1 that $\bar{y} = \bar{x} + c$ and in Result 2 that $\bar{y} = c\bar{x}$.

Example 1.16

Recall the spacecraft data (Example 1.2) $x_1 = 81.3001$, $x_2 = 81.3015$, $x_3 = 81.3006$, $x_4 = 81.3011$, $x_5 = 81.2997$, $x_6 = 81.3005$, and $x_7 = 81.3021$. If we subtract the smallest value 81.2997 from all observations, the resulting values are .0004, .0018, .0009, .0014, .0000, .0008, and .0024; the variance of this set is the same as that of the original data and is easier to compute. Now if we multiply each value by 10,000, the variance of the resulting set is $(10,000)^2$ times the original variance. The new data set is 4, 18, 9, 14, 0, 8, and 24 with mean 11.0 and sum of squares 1257, so the variance is $[1257 - (77)^2/7]/6 = 68.33$. The variance of the original data set is therefore $68.33/(10,000)^2 = .0000006833$. ∎

Boxplots

Stem-and-leaf displays and histograms convey rather general impressions about a data set, whereas a single summary such as the mean or standard deviation focuses on just one aspect of the data. In recent years, a pictorial summary called a *boxplot* has been used successfully to describe several of a data set's most prominent features. These features include (1) center, (2) spread, (3) the extent and nature of any departure from symmetry, and (4) identification of "outliers," observations that lie unusually far from the main body of the data. Because even a single outlier can drastically affect the value of some numerical summaries (such as $\bar{x}$ and s), a boxplot is based on measures that are "resistant" to the presence of a few outliers—the median and a measure of spread called the *fourth spread*.

Definition

> After the n observations in a data set are ordered from smallest to largest, the **lower fourth** and **upper fourth** are given by
>
> $$\text{Lower fourth} = \begin{cases} \text{Median of the smallest } n/2 \text{ observations,} & n \text{ even} \\ \text{Median of the smallest } (n + 1)/2 \text{ observations,} & n \text{ odd} \end{cases}$$
>
> $$\text{Upper fourth} = \begin{cases} \text{Median of the largest } n/2 \text{ observations,} & n \text{ even} \\ \text{Median of the largest } (n + 1)/2 \text{ observations,} & n \text{ odd} \end{cases}$$
>
> That is, the lower (upper) fourth is the median of the smallest (largest) half of the data, where the median $\tilde{x}$ is included in both halves if n is odd. A measure of spread that is resistant to outliers is the **fourth spread** f_s, given by
>
> $$f_s = \text{Upper fourth} - \text{Lower fourth}$$

Roughly speaking, the fourth spread is unaffected by the positions of those observations in the smallest 25% or the largest 25% of the data.

A boxplot can now be constructed via the following sequence of steps:

1. Draw and mark a horizontal measurement axis.
2. Construct a rectangle whose left edge lies above the lower fourth and whose right edge lies above the upper fourth.
3. Draw a vertical line segment inside the box above the median.
4. Extend lines from each end of the box out to the farthest observations that are still within $1.5f_s$ of the corresponding edges.
5. Draw an open circle to identify each observation that falls between $1.5f_s$ and $3f_s$ from the edge to which it is closest; these are called **mild outliers.**
6. Draw a solid circle to identify each observation that falls more than $3f_s$ from the closest edge; these are called **extreme outliers.**

Figure 1.16 illustrates the regions for mild and extreme outliers.

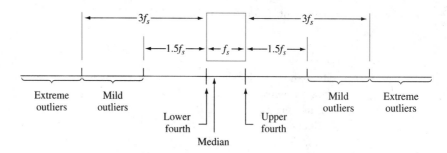

Figure 1.16 Regions for outliers in a boxplot

Example 1.17   The accompanying data on oxygen concentration for a sample of silicon wafers was reported in the paper "Determination of Conversion Factor for Infrared

Measurement of Oxygen in Silicon" (*Solid State Science and Technology,* 1985, pp. 1707–1713). The sample size is $n = 22$, so the smallest (largest) half of the data consists of the 11 smallest (largest) observations:

Smallest: 2.68   3.06   4.31   4.71   5.71   5.99   6.06   7.04   7.17   7.46   7.50

Largest: 8.27   8.42   8.73   8.84   9.14   9.19   9.21   9.39   11.28   15.19   21.06

The median, lower, and upper fourths are

$$\tilde{x} = \frac{7.50 + 8.27}{2} = 7.885, \quad \frac{\text{Lower}}{\text{fourth}} = 5.99, \quad \frac{\text{Upper}}{\text{fourth}} = 9.19$$

Figure 1.17 Boxplots for the data of Example 1.17: (a) hand-drawn and (b) from SAS

Thus, $f_s = 9.19 - 5.99 = 3.20$, $1.5f_s = 4.80$, and $3f_s = 9.60$. Adding 4.80 and 9.60 to the upper fourth gives 13.99 and 18.79, respectively, so a line segment extends from the box's right edge out to 11.28, and 15.19 is a mild outlier and 21.06 an extreme outlier. Because there are no observations more than 4.80 below the lower fourth, the sample contains no outliers on the lower end. The boxplot appears in Figure 1.17. The median is near the middle of the relatively narrow box, giving the impression that the middle 50% of the data is reasonably symmetrically distributed and not too spread out. The upper tail appears much longer than the lower tail because of the two outliers.

Figure 1.18 shows MINITAB output from a request to describe the concentration data. The trimmed mean is the average of the 20 observations that remain after the largest and smallest values are deleted (trimming percentage $\approx 5\%$). Q1 and Q3 are the lower and upper quartiles; these are similar to the fourths but are calculated in a slightly different manner. SEMEAN is $s/\sqrt{n}$; this will be an important quantity in our subsequent work concerning inferences about μ.

	N	MEAN	MEDIAN	TRMEAN	STDEV	SEMEAN
concen	22	8.200	7.885	7.833	3.977	0.848

	MIN	MAX	Q1	Q3
concen	2.680	21.060	5.920	9.195

Figure 1.18 A MINITAB description of the concentration data ■

Example 1.18 In recent years, some evidence suggests that high indoor radon concentration may be linked to the development of childhood cancers, but many health professionals remain unconvinced. A recent article ("Indoor Radon and Childhood Cancer," *The Lancet,* 1991, pp. 1537–1538) presented the accompanying data on radon concentration (Bq/m^3) in two different samples of houses. The first sample consisted of houses in which a child diagnosed with cancer had been residing. Houses in the second sample had no recorded cases of childhood cancer. Figure 1.19 presents a stem-and-leaf display of the data.

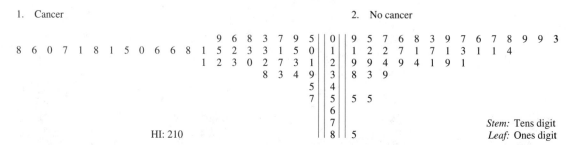

Figure 1.19 Stem-and-leaf display for Example 1.18

Numerical summary quantities are as follows:

	$\bar{x}$	$\tilde{x}$	s	f_s
Cancer	22.8	16.0	31.7	11.0
No cancer	19.2	12.0	17.0	18.0

The values of both the mean and median suggest that the cancer sample is centered somewhat to the right of the no-cancer sample on the measurement scale. The mean, however, exaggerates the magnitude of this shift, largely because of the observation 210 in the cancer sample. The values of s suggest more variability in the cancer sample than in the no-cancer sample, but this impression is contradicted by the fourth spreads. Again, the observation 210, an extreme outlier, is the culprit. Figure 1.20 shows a comparative boxplot from the S-PLUS computer package. The no-cancer box is stretched out compared with the cancer box ($f_s = 18$ vs. $f_s = 11$), and the positions of the median lines in the two boxes show much more skewness in the middle half of the no-cancer sample than the cancer sample. Outliers are represented by horizontal line segments, and there is no distinction between mild and extreme outliers.

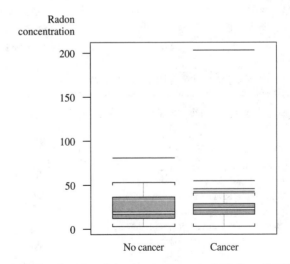

Figure 1.20 A boxplot of the data in Example 1.18, from S-PLUS

Exercises Section 1.4 (36–52)

36. The paper "Oxygen Consumption During Fire Suppression: Error of Heart Rate Estimation" (*Ergo-* *nomics,* 1991, pp. 1469–1474) reported the following data on oxygen consumption (mL/kg/min) for a sample

of ten firefighters performing a fire-suppression simulation:

29.5, 49.3, 30.6, 28.2, 28.0, 26.3, 33.9, 29.4, 23.5, 31.6

Compute the following:

a. The sample range

b. The sample variance s^2 from the definition (i.e., by first computing deviations, then squaring them, etc.)

c. The sample standard deviation

d. s^2 using the shortcut method

37. A sample of eight resistors of a certain type resulted in the sample resistances (ohms) $x_1 = 40$, $x_2 = 43$, $x_3 = 39$, $x_4 = 35$, $x_5 = 37$, $x_6 = 43$, $x_7 = 46$, $x_8 = 37$.

a. Compute s^2 and s directly from the definitions.

b. Compute s^2 and s using the shortcut formula.

c. Subtract 35 from each x_i and then compute s^2.

d. If the resistances were 400, 430, 390, 350, 370, 430, 460, and 370, how would you use the results of parts (a), (b), or (c) to compute s^2 and s?

38. The following observations are weight gains (mg/cm^3) of Ti–Cr alloy samples due to oxidation when exposed to CO_2 for 1 hour at 1000°C:

6.4, 5.9, 6.1, 5.8, 6.6, 6.0

a. Using $\bar{x}$ computed to two decimal places, compute s^2 directly from the definition and then compute s.

b. Repeat part (a) with $\bar{x}$ rounded to three decimal places.

c. Compute s^2 using the shortcut formula.

39. A study of the relationship between age and various visual functions (such as acuity and depth perception) reported the following observations on area of scleral lamina (mm^2) from human optic nerve heads ("Morphometry of Nerve Fiber Bundle Pores in the Optic Nerve Head of the Human," *Experimental Eye Research,* 1988, pp. 559–568):

2.75 2.62 2.74 3.85 2.34 2.74 3.93 4.21 3.88
4.33 3.46 4.52 2.43 3.65 2.78 3.56 3.01

a. Calculate Σx and Σx^2.

b. Use the values calculated in part (a) to compute the sample variance s^2 and then the sample standard deviation s.

40. The paper "Evaluating Variability of Filling Operations" (*Food Technology,* 1984, pp. 51–55) gave data on the actual amount of fluid dispersed by a machine designed to disperse 10 oz. Ten observed values were used to compute $\Sigma x = 100.2$ and $\Sigma x^2 = 1004.4$. Compute the sample variance and standard deviation.

41. Compute s^2 and s for the temperature data from Exercise 28 by first subtracting 100 from each data value. Is there anything special about the choice of 100, or would other choices work just as well?

42. The accompanying data on bearing load-life (million revs.) for bearings of a certain type when subjected to a 9.56 kN load appeared in the paper "The Load-Life Relationship for M50 Bearings with Silicon Nitride Ceramic Balls" (*Lubric. Engr.,* 1984, pp. 153–159):

14.5, 25.6, 52.4, 66.3, 69.3, 69.8, 76.2

a. Calculate the values of the sample mean and median. What do their values relative to one another tell you about the sample?

b. Compute the values of the sample variance and standard deviation.

43. Refer to the fatigue crack propagation data given in Exercise 31.

a. Calculate the values of the sample variance and sample standard deviation.

b. Calculate the value of the fourth spread.

c. Construct a boxplot. Does the data appear to be symmetrically distributed along the measurement axis? Are there any outliers?

44. The paper "A Thin-Film Oxygen Uptake Test for the Evaluation of Automotive Crankcase Lubricants" (*Lubric. Engr.,* 1984, pp. 75–83) reported the following data on oxidation-induction time (min) for various commercial oils:

87, 103, 130, 160, 180, 195, 132, 145, 211, 105, 145, 153, 152, 138, 87, 99, 93, 119, 129

a. Calculate the sample variance and standard deviation.

b. If the observations were reexpressed in hours, what would be the resulting values of the sample variance and sample standard deviation? Answer without actually performing the reexpression.

45. The first four deviations from the mean in a sample of $n = 5$ reaction times were .3, .9, 1.0, and 1.3. What is the fifth deviation from the mean? Give a sample for which these are the five deviations from the mean.

46. Refer to the data in Exercise 39 on scleral lamina areas.

a. Determine the lower and upper fourths.

b. Calculate the value of the fourth spread.

c. If the two largest sample values, 4.33 and 4.52, had instead been 5.33 and 5.52, how would this affect f_s? Explain.

d. By how much could the observation 2.34 be increased without affecting f_s? Explain.

e. If an eighteenth observation, $x_{18} = 4.60$, is added to the sample, what is f_s?

47. Refer to the oxidation-induction time data given in Exercise 44. Summarize the data by constructing a boxplot and comment on its most prominent features.

48. Cardiac output and maximal oxygen uptake typically decrease with age in sedentary individuals, but these decreases are at least partially arrested in middle-aged individuals who engage in a substantial amount of physical exercise. To understand better the effects of exercise and aging on various circulatory functions, the paper "Cardiac Output in Male Middle-Aged Runners" (*J. Sports Medicine*, 1982, pp. 17–22) presented data from a study of 21 male middle-aged runners. A stem-and-leaf display of oxygen uptake values (mL/ kg · min) while pedaling at 100 watts on a bicycle ergometer is given:

```
12 || 81
13 ||
14 || 95
15 || 97  83
16 ||
17 || 90
18 || 34  27
19 || 94  82
20 || 99  93  98  62  88
21 || 15
22 || 16  24          Stem:  Ones
23 || 16  56          Leaf:  Hundredths
HI: 35.78,  36.73
```

Construct a boxplot and discuss its characteristics.

49. In Exercise 28, after subtracting 100 from each data value, multiply each resulting number by 10 and then compute s^2 for the resulting set of numbers. How can you now obtain s^2 for the original data set?

50. The standard deviation and variance are both measures of variability that depend on the units of measurement, with s expressed in the original units of measurement. The **coefficient of variation**, defined by $c.v. = s/\bar{x}$, is a dimensionless quantity that measures the amount of variability relative to the value of the mean. Compute the value of $c.v.$ for the oxygen consumption data of Exercise 36 and the weight-gain data of Exercise 38. Which data set has more relative variation?

51. The accompanying data appeared in an article in *Technometrics* that discussed the analysis of information from weather-modification experiments. Construct side-by-side boxplots and then comment on similarities and differences.

Rainfall from Control Clouds		Rainfall from Seeded Clouds	
1202.6	41.1	2745.6	200.7
830.1	36.6	1697.8	198.6
372.4	29.0	1656.0	129.6
345.5	28.6	978.0	119.0
321.2	26.3	703.4	118.3
244.3	26.1	489.1	115.3
163.0	24.4	430.0	92.4
147.8	21.7	334.1	40.6
95.0	17.3	302.8	32.7
87.0	11.5	274.7	31.4
81.2	4.9	274.7	17.5
68.5	4.9	255.0	7.7
47.3	1.0	242.5	4.1

52. The following inequality is due to the Russian mathematician Chebyshev. Let k be any number satisfying $k \geq 1$. Then for *any* numerical data set, *at least* $100(1 - 1/k^2)\%$ of the observations are within k standard deviations (SDs) of the mean.

a. At least what percentage is guaranteed to be within 2 SDs of the mean? Within 3 SDs? Within 5 SDs?

b. If the mean and standard deviation of a data set consisting of stress-resistance observations are 30 and 2, respectively, what can be said about the percentage of observations between 26 and 34? Outside the interval from 24 to 36?

c. The mean and standard deviation of the shear strength data given in Exercise 11 are 5046.46 and 353.1155, respectively. Compare the actual percentages of observations within 1, 2, and 3 SDs of the mean with what Chebyshev's result says. Comment.

Supplementary Exercises for Chapter I (53–67)

53. The **mode** of a numerical data set is the value that occurs most frequently in the set.

    a. Determine the mode for the data set consisting of scores for crunchy peanut butter given in Exercise 12.

    b. For a categorical sample, how would you define the modal category?

54. The task completion time in a learning experiment was recorded both on the first trial (x) and the tenth trial (y) for six individuals:

Individual i	1	2	3	4	5	6
x_i	23.7	36.9	25.5	30.2	28.0	34.8
y_i	15.0	25.4	21.0	22.3	25.2	28.8

    a. Compute the decrease $d_i = x_i - y_i$ for each individual and then compute the sample average decrease.

    b. Compute the sample average time on the first trial, the sample average time on the tenth trial, and then the difference in the two sample averages.

    c. Using rules of summation, show that, in general, the average of the differences $d_i = x_i - y_i$ equals the difference $\bar{x} - \bar{y}$ of the averages.

55. Fifteen air samples from a certain region were obtained, and for each one the carbon monoxide concentration was determined. The results (in ppm) were

    9.3, 10.7, 8.5, 9.6, 12.2, 15.6, 9.2, 10.5, 9.0, 13.2, 11.0, 8.8, 13.7, 12.1, 9.8

    Using the interpolation method suggested in Section 1.3, compute the 10% trimmed mean.

56. a. For what value of c is the quantity $\Sigma(x_i - c)^2$ minimized? (HINT: Take the derivative with respect to c, set equal to 0, and solve.)

    b. Using the result of part (a), which of the two quantities $\Sigma(x_i - \bar{x})^2$ and $\Sigma(x_i - \mu)^2$ will be smaller than the other (assuming that $\bar{x} \neq \mu$)?

57. a. Let a and b be constants and let $y_i = ax_i + b$ for $i = 1, 2, \ldots, n$. What are the relationships between $\bar{x}$ and $\bar{y}$ and between s_x^2 and s_y^2?

    b. A sample of temperatures for initiating a certain chemical reaction yielded a sample average (°C) of 87.3 and a sample standard deviation of 1.04. What are the sample average and standard deviation measured in °F? (HINT: $F = \frac{9}{5}C + 32$.)

58. The percentages of juice lost after thawing for 19 different strawberry varieties appeared in the paper "Evaluation of Strawberry Cultivars with Different Degrees of Resistance to Red Stele" (*Fruit Varieties J.,* 1991, pp. 12–17):

    46   51   44   50   33   46   60   41   55   46
    53   53   42   44   50   54   46   41   48

    a. Are there any observations that are mild outliers? Extreme outliers?

    b. Construct a boxplot and comment.

59. The article "Can We Really Walk Straight?" (*Amer. J. of Physical Anthropology,* 1992, pp. 19–27) reported on an experiment in which each of 20 healthy men was asked to walk as straight as possible to a target 60 m away at normal speed. Consider the following observations on cadence (number of strides per second):

    .95   .85   .92   .95   .93   .86   1.00   .92   .85   .81
    .78   .93   .93   1.05   .93   1.06   1.06   .96   .81   .96

    Use the methods developed in this chapter to summarize the data; include an interpretation or discussion wherever appropriate. (NOTE: The author of the paper used a rather sophisticated statistical analysis to conclude that people cannot walk in a straight line and suggested several explanations for this.)

60. The amount of radiation received at a greenhouse plays an important role in determining the rate of photosynthesis. The accompanying observations on incoming solar radiation were read from a graph in the paper "Radiation Components over Bare and Planted Soils in a Greenhouse" (*Solar Energy,* 1990, pp. 1–6):

    6.3   6.4   7.1   7.7   8.4   8.5   8.8   8.9   9.0
    9.1   10.0   10.1   10.2   10.6   10.6   10.7   10.7   10.8
    10.9   11.1   11.2   11.2   11.4   11.9   11.9   12.2   13.1

    Use some of the methods discussed in this chapter to describe this data.

61. The accompanying data on milk volume (g/day) was taken from the paper "Smoking During Pregnancy and Lactation and Its Effects on Breast Milk Volume" (*Amer. J. of Clinical Nutrition,* 1991, pp. 1011–1016):

Smoking mothers	621	793	593	545	753	655
	895	767	714	598	693	
Nonsmoking mothers	947	945	1086	1202	973	981
	930	745	903	899	961	

Compare and contrast the two samples.

62. Consider the following data on active repair time (hours) for a sample of $n = 46$ airborne communications receivers:

.2, .3, .5, .5, .5, .6, .6, .7, .7, .7, .8, .8, .8, 1.0,
1.0, 1.0, 1.0, 1.1, 1.3, 1.5, 1.5, 1.5, 1.5, 2.0, 2.0,
2.2, 2.5, 2.7, 3.0, 3.0, 3.3, 3.3, 4.0, 4.0, 4.5, 4.7,
5.0, 5.4, 5.4, 7.0, 7.5, 8.8, 9.0, 10.3, 22.0, 24.5

Construct the following

a. A stem-and-leaf display in which the two largest values are displayed separately in a row labeled HI.

b. A histogram based on six class intervals with 0 as the lower limit of the first interval and interval lengths of 2, 2, 2, 4, 10, and 10, respectively.

63. Consider a sample $x_1, x_2, \ldots, x_n$ and suppose that the values of $\bar{x}, s^2$, and s have been calculated.

a. Let $y_i = x_i - \bar{x}$ for $i = 1, \ldots, n$. How do the values of s^2 and s for the y_i's compare to the corresponding values for the x_i's? Explain.

b. Let $z_i = (x_i - \bar{x})/s$ for $i = 1, \ldots, n$. What are the values of the sample variance and sample standard deviation for the z_i's?

64. Let $\bar{x}_n$ and s_n^2 denote the sample mean and variance for the sample $x_1, \ldots, x_n$ and let $\bar{x}_{n+1}$ and s_{n+1}^2 denote these quantities when an additional observation x_{n+1} is added to the sample.

a. Show how $\bar{x}_{n+1}$ can be computed from $\bar{x}_n$ and x_{n+1}.

b. Show that

$$ns_{n+1}^2 = (n-1)\,s_n^2 + \frac{n}{n+1}\,(x_{n+1} - \bar{x}_n)^2$$

so that s_{n+1}^2 can be computed from x_{n+1}, $\bar{x}_n$, and s_n^2.

c. Use the results of parts (a) and (b) to compute $\bar{x}$ and s^2 recursively (first $\bar{x}_2$ and s_2^2, then $\bar{x}_3$ and s_3^2, etc.) for the data of Exercise 36. (NOTE: These relationships are used by some computer packages to increase computational efficiency and accuracy.)

65. Americium 241 (^{241}Am) is a radioactive material used in the manufacture of smoke detectors. The article "Retention and Dosimetry of Injected ^{241}Am in Beagles" (*Radiation Research*, 1984, pp. 564–575) described a study in which 55 beagles were injected with a dose of ^{241}Am (proportional to the animals' weights). Skeletal retention of ^{241}Am (μCi/kg) was recorded for each beagle, resulting in the given data:

.196	.451	.498	.411	.324
.190	.489	.300	.346	.448
.188	.399	.305	.304	.287
.243	.334	.299	.292	.419
.236	.315	.447	.585	.291
.186	.393	.419	.335	.332
.292	.375	.349	.324	.301
.333	.408	.399	.303	.318
.468	.441	.306	.367	.345
.428	.345	.412	.337	.353
.357	.320	.354	.361	.329

Summarize this data using the techniques presented in this chapter and write a short paragraph discussing the most prominent features of your summaries.

66. The sample data $x_1, x_2, \ldots, x_n$ sometimes represents a **time series,** where x_t = the observed value of a response variable x at time t. Often the observed series shows a great deal of random variation, which makes it difficult to study longer-term behavior. In such situations, it is desirable to produce a smoothed version of the series. One technique for doing so involves **exponential smoothing.** The value of a smoothing constant α is chosen $(0 < \alpha < 1)$. Then with $\bar{x}_t$ = smoothed value at time t, we set $\bar{x}_1 = x_1$, and for $t = 2, 3, \ldots, n$, $\bar{x}_t = \alpha x_t + (1 - \alpha)\bar{x}_{t-1}$.

a. Consider the following time series in which x_t = temperature (°F) of effluent at a sewage treatment plant on day t: 47, 54, 53, 50, 46, 46, 47, 50, 51, 50, 46, 52, 50, 50. Plot each x_t against t on a two-dimensional coordinate system (a time-series plot). Does there appear to be any pattern?

b. Calculate the $\bar{x}_t$'s using $\alpha = .1$. Repeat using $\alpha = .5$. Which value of α gives a smoother $\bar{x}_t$ series?

c. Substitute $\bar{x}_{t-1} = \alpha x_{t-1} + (1 - \alpha)\bar{x}_{t-2}$ on the right-hand side of the expression $\bar{x}_t$, then substitute $\bar{x}_{t-2}$ in terms of x_{t-2} and $\bar{x}_{t-3}$, and so on. On how many of the values $x_t, x_{t-1}, \ldots, x_1$ does $\bar{x}_t$ depend? What happens to the coefficient on x_{t-k} as k increases?

d. Referring to part (c), if t is large, how sensitive is $\bar{x}_t$ to the initialization $\bar{x}_1 = x_1$? Explain.

(NOTE: A relevant reference is the paper "Simple Statistics for Interpreting Environmental Data," *Water Pollution Control Fed. J.,* 1981, pp. 167–175.)

67. Consider numerical observations $x_1, \ldots, x_n$. It is frequently of interest to know whether the x_i's are (at least approximately) symmetrically distributed about some value. If n is at least moderately large, the extent of symmetry can be assessed from a stem-and-leaf display or histogram. However, if n is not very large, such pictures are not particularly informative. Consider the following alternative. Let y_1 denote the smallest x_i, y_2 the second smallest x_i, and so on. Then plot the following pairs as points on a two-dimensional coordinate system: $(y_n - \bar{x}, \bar{x} - y_1)$, $(y_{n-1} - \bar{x}, \bar{x} - y_2)$, $(y_{n-2} - \bar{x}, \bar{x} - y_3)$, ... There are $n/2$ points when n is even and $(n - 1)/2$ when n is odd.

a. What does this plot look like when there is perfect symmetry in the data? What does it look like when observations stretch out more above the median than below it (a long upper tail)?

b. The accompanying data on rainfall (acre-feet) from 26 seeded clouds is taken from the paper "A Bayesian Analysis of a Multiplicative Treatment Effect in Weather Modification" (*Technometrics,* 1975, pp. 161–166). Construct the plot and comment on the extent of symmetry or nature of departure from symmetry.

4.1, 7.7, 17.5, 31.4, 32.7, 40.6, 92.4, 115.3, 118.3, 119.0, 129.6, 198.6, 200.7, 242.5, 255.0, 274.7, 274.7, 302.8, 334.1, 430.0, 489.1, 703.4, 978.0, 1656.0, 1697.8, 2745.6

Bibliography

Chambers, John, William Cleveland, Beat Kleiner, and Paul Tukey, *Graphical Methods for Data Analysis,* Brooks/Cole, Pacific Grove, CA (1983). A highly recommended presentation of both older and more recent graphical and pictorial methodology in statistics.

Devore, Jay, and Roxy Peck, *Statistics: The Exploration and Analysis of Data* (2nd ed.), Duxbury Press, Belmont, CA (1993). The first few chapters give a very nonmathematical survey of methods for describing and summarizing data.

Freedman, David, Robert Pisani, Roger Purves, and Ani Adhikari, *Statistics* (2nd ed.), Norton, New York (1991). An excellent, very nonmathematical survey of basic statistical reasoning and methodology.

Hoaglin, David, Frederick Mosteller, and John Tukey, *Understanding Robust and Exploratory Data Analysis,* Wiley, New York (1983). Discusses why, as well as how, exploratory methods should be employed; it is good on details of stem-and-leaf displays and boxplots.

Hoaglin, David, and Paul Velleman, *Applications, Basics, and Computing of Exploratory Data Analysis,* Duxbury Press, Boston (1980). A good discussion of some basic exploratory methods.

Moore, David, *Statistics: Concepts and Controversies* (3rd ed.), Freeman, San Francisco (1991). An extremely readable and entertaining paperback that contains an intuitive discussion of problems connected with sampling and designed experiments.

Tanur, Judith, et al. (eds.), *Statistics: A Guide to the Unknown* (3rd ed.), Duxbury Press, Belmont, CA (1988). Contains many short nontechnical articles describing various applications of statistics.

CHAPTER

2 Probability

ΣΣΣΣΣΣΣΣΣΣΣΣΣΣΣΣΣΣΣΣ

Introduction

The term **probability** refers to the study of randomness and uncertainty. In any situation in which one of a number of possible outcomes may occur, the theory of probability provides methods for quantifying the chances, or likelihoods, associated with the various outcomes. The language of probability is constantly used in an informal manner in both written and spoken contexts. Examples include such statements as "It is likely that the Dow-Jones average will increase by the end of the year," "There is a 50–50 chance that the incumbent will seek reelection," "There will probably be at least one section of that course offered next year," "The odds favor a quick settlement of the strike," and "It is expected that at least 20,000 concert tickets will be sold." In this chapter, we introduce some elementary probability concepts, indicate how probabilities can be interpreted, and show how the rules of probability can be applied to compute the probabilities of many interesting events. The methodology of probability will then permit us to express in precise language such informal statements as those given above.

The study of probability as a branch of mathematics goes back over 300 years, where it had its genesis in connection with questions involving games of chance. Many books are devoted exclusively to probability, but our objective here is to cover only that part of the subject that has the most direct bearing on problems of statistical inference.

2.1 Sample Spaces and Events

An **experiment** is any action or process that generates observations. Although the word *experiment* generally suggests a planned or carefully controlled laboratory testing situation, we use it here in a much wider sense. Thus, experiments that may be of interest include tossing a coin once or several times, selecting a card or cards from a deck, weighing a loaf of bread, ascertaining the commuting time from home to work on a particular morning, obtaining blood types from a group of individuals, or measuring the compressive strengths of different steel beams.

The Sample Space of an Experiment

Definition

> The **sample space** of an experiment, denoted by $\mathscr{S}$, is the set of all possible outcomes of that experiment.

Example 2.1

The simplest experiment to which probability applies is one with two possible outcomes. One such experiment consists of examining a single fuse to see whether it is defective. The sample space for this experiment can be abbreviated as $\mathscr{S} = \{N, D\}$, where N represents not defective, D represents defective, and the braces are used to enclose the elements of a set. Another such experiment would involve tossing a thumbtack and noting whether it landed point up or point down, with sample space $\mathscr{S} = \{U, D\}$, and yet another would consist of observing the sex of the next child born at the local hospital, with $\mathscr{S} = \{M, F\}$.   ■

Example 2.2

If we examine three fuses in sequence and note the result of each examination, then an outcome for the entire experiment is any sequence of N's and D's of length 3, so

$$\mathscr{S} = \{NNN, NND, NDN, NDD, DNN, DND, DDN, DDD\}$$

If we had tossed a thumbtack three times, the sample space would be obtained by replacing N by U in $\mathscr{S}$ above, with a similar notational change yielding the sample space for the experiment in which the sexes of three newborn children are observed.   ■

Example 2.3

Two gas stations are located at a certain intersection. Each one has six gas pumps. Consider the experiment in which the number of pumps in use at a particular time of day is determined for each of the stations. An experimental outcome specifies how many pumps are in use at the first station and how many are in use at the second one. One possible outcome is (2, 2), another is (4, 1), and yet another is (1, 4). The 49 outcomes in $\mathscr{S}$ are displayed in the accompanying table. The sample space for the experiment in which a six-sided die is thrown twice results from deleting the 0 row and 0 column from the table, giving 36 outcomes.

		Second Station						
		0	*1*	*2*	*3*	*4*	*5*	*6*
	0	(0, 0)	(0, 1)	(0, 2)	(0, 3)	(0, 4)	(0, 5)	(0, 6)
	1	(1, 0)	(1, 1)	(1, 2)	(1, 3)	(1, 4)	(1, 5)	(1, 6)
	2	(2, 0)	(2, 1)	(2, 2)	(2, 3)	(2, 4)	(2, 5)	(2, 6)
First Station	*3*	(3, 0)	(3, 1)	(3, 2)	(3, 3)	(3, 4)	(3, 5)	(3, 6)
	4	(4, 0)	(4, 1)	(4, 2)	(4, 3)	(4, 4)	(4, 5)	(4, 6)
	5	(5, 0)	(5, 1)	(5, 2)	(5, 3)	(5, 4)	(5, 5)	(5, 6)
	6	(6, 0)	(6, 1)	(6, 2)	(6, 3)	(6, 4)	(6, 5)	(6, 6)

■

Example 2.4   If a new type-D flashlight battery has a voltage that is outside certain limits, that battery is characterized as a failure (F); if the battery has a voltage within the prescribed limits, it is a success (S). Suppose an experiment consists of testing each battery as it comes off an assembly line until we first observe a success. Although it may not be very likely, a possible outcome of this experiment is that the first 10 (or 100 or 1000 or . . .) are F's and the next one is an S. That is, for any positive integer n, we may have to examine n batteries before seeing the first S. The sample space is $\mathcal{S} = \{S, FS, FFS, FFFS, \ldots \}$, which contains an infinite number of possible outcomes. The same abbreviated form of the sample space is appropriate for an experiment in which, starting at a specified time, the sex of each newborn infant is recorded until the birth of a male is observed.   ■

Events

In our study of probability, we will be interested not only in the individual outcomes of $\mathcal{S}$ but also in any collection of outcomes from $\mathcal{S}$.

Definition

> An **event** is any collection (subset) of outcomes contained in the sample space $\mathcal{S}$. An event is said to be **simple** if it consists of exactly one outcome and **compound** if it consists of more than one outcome.

When an experiment is performed, a particular event A is said to occur if the resulting experimental outcome is contained in A. In general, exactly one simple event will occur, but many compound events will occur simultaneously.

Example 2.5   Consider an experiment in which each of three automobiles taking a particular freeway exit turns left (L) or right (R) at the end of the exit ramp. The eight possible outcomes that comprise the sample space are *LLL, RLL, LRL, LLR, LRR, RLR, RRL,* and *RRR.* Thus, there are eight simple events, among which are $E_1 = \{LLL\}$ and $E_5 = \{LRR\}$. Some compound events include

$A = \{RLL, LRL, LLR\}$ = the event that exactly one of the three cars turns right.

$B = \{LLL, RLL, LRL, LLR\}$ = the event that at most one of the cars turns right.

$C = \{LLL, RRR\}$ = the event that all three cars turn in the same direction.

Suppose that when the experiment is performed, the outcome is *LLL*. Then the simple event E_1 has occurred and so also have the events *B* and *C* (but not *A*). ■

Example 2.6
(Example 2.3 continued)

When the number of pumps in use at each of two six-pump gas stations is observed, there are 49 possible outcomes, so there are 49 simple events: $E_1 = \{(0, 0)\}$, $E_2 = \{(0, 1)\}, \ldots, E_{49} = \{(6, 6)\}$. Examples of compound events are

$A = \{(0, 0), (1, 1), (2, 2), (3, 3), (4, 4), (5, 5), (6, 6)\}$ = the event that the number of pumps in use is the same for both stations.

$B = \{(0, 4), (1, 3), (2, 2), (3, 1), (4, 0)\}$ = the event that the total number of pumps in use is four.

$C = \{(0, 0), (0, 1), (1, 0), (1, 1)\}$ = the event that at most one pump is in use at each station. ■

Example 2.7
(Example 2.4 continued)

The sample space for the battery examination experiment contains an infinite number of outcomes, so there are an infinite number of simple events. Compound events include

$A = \{S, FS, FFS\}$ = the event that at most three batteries are examined.

$E = \{FS, FFFS, FFFFFS, \ldots \}$ = the event that an even number of batteries are examined. ■

Some Relations from Set Theory

An event is nothing but a set, so that relationships and results from elementary set theory can be used to study events. The following concepts from set theory will be used to construct new events from given events.

Definition

1. The **union** of two events *A* and *B*, denoted by $A \cup B$ and read "*A or B*," is the event consisting of all outcomes that are *either in A or in B or in both events* (so that the union includes outcomes for which both *A* and *B* occur as well as outcomes for which exactly one occurs).
2. The **intersection** of two events *A* and *B*, denoted by $A \cap B$ and read "*A and B*," is the event consisting of all outcomes that are in *both A and B*.
3. The **complement** of an event *A*, denoted by A', is the set of all outcomes in $\mathscr{S}$ that are not contained in *A*.

Example 2.8
(Example 2.3 continued)

For the experiment in which the number of pumps in use at a single six-pump gas station is observed, let $A = \{0, 1, 2, 3, 4\}$, $B = \{3, 4, 5, 6\}$, and $C = \{1, 3, 5\}$. Then

$$A \cup B = \{0, 1, 2, 3, 4, 5, 6\} = \mathscr{S}, \quad A \cup C = \{0, 1, 2, 3, 4, 5\},$$
$$A \cap B = \{3, 4\}, \quad A \cap C = \{1, 3\}, \quad A' = \{5, 6\}, \quad (A \cup C)' = \{6\}. \quad \blacksquare$$

Example 2.9
(Example 2.4 continued)

In the battery experiment, define A, B, and C by

$$A = \{S, FS, FFS\}$$
$$B = \{S, FFS, FFFFS\}$$

and

$$C = \{FS, FFFS, FFFFFS, \dots\}$$

Then

$$A \cup B = \{S, FS, FFS, FFFFS\}$$
$$A \cap B = \{S, FFS\}$$
$$A' = \{FFFS, FFFFS, FFFFFS, \dots\}$$

and

$$C' = \{S, FFS, FFFFS, \dots\} = \{\text{an odd number of batteries are examined}\} \quad \blacksquare$$

Sometimes A and B have no outcomes in common, so that the intersection of A and B contains no outcomes.

Definition

> When A and B have no outcomes in common, they are said to be **mutually exclusive** or **disjoint** events.

Example 2.10

A small city has three automobile dealerships: a GM dealer selling Chevrolets, Pontiacs, and Buicks; a Ford dealer selling Fords and Mercurys; and a Chrysler dealer selling Plymouths and Chryslers. If experiment consists of observing the brand of the next car sold, then the events $A = \{\text{Chevrolet, Pontiac, Buick}\}$ and $B = \{\text{Ford, Mercury}\}$ are mutually exclusive because the next car sold cannot be both a GM product and a Ford product. $\quad \blacksquare$

The operations of union and intersection can be extended to more than two events. For any three events A, B, and C, the event $A \cup B \cup C$ is the set of outcomes contained in at least one of the three events, while $A \cap B \cap C$ is the set of outcomes contained in all three events. Given events $A_1, A_2, A_3, \dots$, these events are said to be mutually exclusive (or pairwise disjoint) if no two events have any outcomes in common.

A pictorial representation of events and manipulations with events is obtained by using Venn diagrams. To construct a Venn diagram, draw a rectangle whose interior will represent the sample space $\mathscr{S}$. Then any event A is represented as the interior of a closed curve (often a circle) contained in $\mathscr{S}$. Examples of Venn diagrams appear in Figure 2.1.

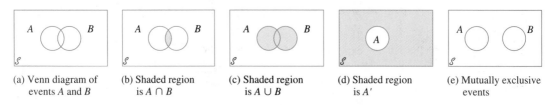

| (a) Venn diagram of events A and B | (b) Shaded region is $A \cap B$ | (c) Shaded region is $A \cup B$ | (d) Shaded region is A' | (e) Mutually exclusive events |

Figure 2.1 Venn diagrams illustrated

Exercises   Section 2.1 (1–10)

1. A family that owns two cars is selected, and for both the older car and the newer car we note whether the car was manufactured in America, Europe, or Asia.

   a. What are the possible outcomes of this experiment?

   b. Which outcomes are contained in the event that one car is American and the other is foreign?

   c. Which outcomes are contained in the event that at least one of the two cars is foreign? What is the complement of this event? Is either of these two events a simple event?

2. Each of a sample of four home mortgages is classified as fixed rate (F) or variable rate (V).

   a. What are the 16 outcomes in $\mathscr{S}$?

   b. Which outcomes are in the event that exactly three of the selected mortgages are a fixed rate?

   c. Which outcomes are in the event that all four mortgages are of the same type?

   d. Which outcomes are in the event that at most one of the four is a variable-rate mortgage?

   e. What is the union of the events in parts (c) and (d), and what is the intersection of these two events?

   f. What are the union and intersection of the two events in parts (b) and (c)?

3. A family consisting of three persons—A, B, and C—belongs to a medical clinic that always has a doctor at each of stations 1, 2, and 3. During a certain week, each member of the family visits the clinic once and is assigned at random to a station. The experiment consists of recording the station number for each member. One outcome is (1, 2, 1) for A to station 1, B to station 2, and C to station 1.

   a. List the 27 outcomes in the sample space.

   b. List all outcomes in the event that all three members go to the same station.

   c. List all outcomes in the event that all members go to different stations.

   d. List all outcomes in the event that no one goes to station 2.

4. A college library has five copies of a certain text on reserve. Two copies (1 and 2) are first printings, and the other three (3, 4, and 5) are second printings. A student examines these books in random order, stopping only when a second printing has been selected. One possible outcome is 5, and another is 213.

   a. List the outcomes in $\mathscr{S}$.

   b. Let A denote the event that exactly one book must be examined. What outcomes are in A?

   c. Let B be the event that book 5 is the one selected. What outcomes are in B?

   d. Let C be the event that book 1 is not examined. What outcomes are in C?

5. To demonstrate my skill at clairvoyance, I shuffle a deck consisting of three cards marked 1, 2, and 3 and place the deck face down on a table. I then guess at the number on the first card and turn it over. If my guess is correct, the experiment terminates. If my first guess is incorrect, I guess again and then turn over the second card. At this point I know what the third card is, so at most the first two guesses and cards need to be reported. An outcome is described by a sequence of either two or four numbers in which a guess and card alternate.

a. List all outcomes in the event that the experiment terminates after just one guess.

b. List all outcomes in the event that the two guesses are both incorrect (I will only guess a number that is possible).

6. Jack Twain has just visited a friend at location *A* in the following figure. He wishes to get to location *I*. He will travel only along indicated straight-line paths and will only go either north (up) or east (to the right). At each intersection at which there is a choice, he will toss a coin to decide in which of the two possible directions he should go. An outcome for Jack's experiment consists of specifying the corners visited on his trip.

a. List all outcomes in the sample space.

b. List all outcomes for which three coin tosses are necessary.

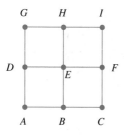

7. An academic department has just completed voting by secret ballot for a department head. The ballot box contains four slips with votes for candidate *A* and three slips with votes for candidate *B*. Suppose these slips are removed from the box one by one.

a. List all possible outcomes.

b. Suppose a running tally is kept as slips are removed. For what outcomes does *A* remain ahead of *B* throughout the tally?

8. An engineering construction firm is currently working on power plants at three different sites. Let A_i denote the event that the plant at site *i* is completed by the contract date. Use the operations of union, intersection, and complementation to describe each of the following events in terms of A_1, A_2, and A_3, draw a Venn diagram, and shade the region corresponding to each one.

a. At least one plant is completed by the contract date.

b. All plants are completed by the contract date.

c. Only the plant at site 1 is completed by the contract date.

d. Exactly one plant is completed by the contract date.

e. Either the plant at site 1 or both of the other two plants are completed by the contract date.

9. Use Venn diagrams to verify the following two relationships for any events *A* and *B* (these are called De Morgan's laws):

a. $(A \cup B)' = A' \cap B'$

b. $(A \cap B)' = A' \cup B'$

10. a. In Example 2.10, identify three events that are mutually exclusive.

b. Suppose there is no outcome common to all three of the events *A*, *B*, and *C*. Are these three events necessarily mutually exclusive? If your answer is yes, explain why; if your answer is no, give a counterexample using the experiment of Example 2.10.

2.2

Axioms, Interpretations, and Properties of Probability

Given an experiment and a sample space $\mathcal{S}$, *the objective of probability is to assign to each event A a number P(A), called the probability of the event A, which will give a precise measure of the chance that A will occur.* To ensure that the probability assignments will be consistent with our intuitive notions of probability, all assignments should satisfy the following axioms (basic properties) of probability.

For any event A, $P(A) \geq 0$.

$P(\mathscr{S}) = 1$.

a. If $A_1, A_2, \ldots A_n$ is a finite collection of mutually exclusive events, then

$$P(A_1 \cup A_2 \cup \cdots \cup A_n) = \sum_{i=1}^{n} P(A_i)$$

b. If $A_1, A_2, A_3, \ldots$ is an infinite collection of mutually exclusive events, then

$$P(A_1 \cup A_2 \cup A_3 \cup \cdots) = \sum_{i=1}^{\infty} P(A_i)$$

Axiom 1 reflects the intuitive notion that the chance of A occurring should be at least 0, so that negative probabilities are not allowed. The sample space is by definition an event that must occur when the experiment is performed ($\mathscr{S}$ contains all possible outcomes), so Axiom 2 says that the maximum possible probability of 1 is assigned to $\mathscr{S}$. The third axiom formalizes the idea that if we wish the probability that at least one of a number of events will occur and no two of the events can occur simultaneously, then the chance of at least one occurring is the sum of the chances of the individual events.

Example 2.11   In the experiment in which a single coin is tossed, the sample space is $\mathscr{S} = \{H, T\}$. The axioms specify $P(\mathscr{S}) = 1$, so to complete the probability assignment, it remains only to determine $P(H)$ and $P(T)$. Since H and T are disjoint events and $H \cup T = \mathscr{S}$, Axiom 3 implies that

$$1 = P(\mathscr{S}) = P(H) + P(T)$$

This implies that $P(T) = 1 - P(H)$. Thus, the only freedom allowed by the axioms in this experiment is the probability assigned to H. One possible assignment of probabilities is $P(H) = .5$, $P(T) = .5$, while another possible assignment is $P(H) = .75, P(T) = .25$. In fact, letting p represent any fixed number between 0 and 1, $P(H) = p$ and $P(T) = 1 - p$ is an assignment consistent with the axioms. ∎

Example 2.12   Consider the experiment in Example 2.4, in which batteries coming off an assembly line are tested one by one until one having a voltage within prescribed limits is found. The simple events are $E_1 = \{S\}$, $E_2 = \{FS\}$, $E_3 = \{FFS\}$, $E_4 = \{FFFS\}, \ldots$. Suppose the probability of any particular battery being satisfactory is .99. Then it can be shown that $P(E_1) = .99$, $P(E_2) = (.01)(.99)$, $P(E_3) = (.01)^2(.99), \ldots$ is an assignment of probabilities to the simple events that satisfies the axioms. In particular, because the E_i's are disjoint and $\mathscr{S} = E_1 \cup E_2 \cup E_3 \cup \cdots$, it must be the case that

$$1 = P(\mathscr{S}) = P(E_1) + P(E_2) + P(E_3) + \cdots$$
$$= .99[1 + .01 + (.01)^2 + (.01)^3 + \cdots]$$

The validity of this equality is a consequence of a mathematical result concerning the sum of a geometric series.

However, another legitimate (according to the axioms) probability assignment of the same "geometric" type is obtained by replacing .99 by any other number p between 0 and 1 (and .01 by $1 - p$). Thus, there are an infinite number of legitimate probability assignments of this type. ∎

Interpreting Probability

Examples 2.11 and 2.12 show that the axioms do not completely determine an assignment of probabilities to events. The axioms only serve to rule out assignments inconsistent with our intuitive notions of probability. In the coin-tossing experiment of Example 2.11, two particular assignments were suggested. The appropriate or correct assignment depends on the manner in which the experiment is carried out and also on one's interpretation of probability. The interpretation that is most frequently used and most easily understood is based on the notion of relative frequencies.

Consider an experiment that can be repeatedly performed in an identical and independent fashion and let A be an event consisting of a fixed set of outcomes of the experiment. Simple examples of such repeatable experiments include the coin-tossing and die-tossing experiments previously discussed. If the experiment is performed n times, on some of the replications the event A will occur (the outcome will be in the set A), and on others, A will not occur. Let $n(A)$ denote the number of replications on which A does occur. Then the ratio $n(A)/n$ is called the *relative frequency* of occurrence of the event A in the sequence of n replications. Empirical evidence, based on the results of many of these sequences of repeatable experiments, indicates that as n grows large, the relative frequency $n(A)/n$ stabilizes, as pictured in Figure 2.2. That is, as n gets arbitrarily large, the relative frequency approaches a limiting value we refer to as the *limiting relative frequency* of the event A. The objective interpretation of probability identifies this limiting relative frequency with $P(A)$.

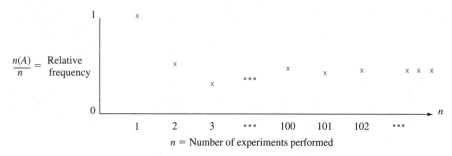

Figure 2.2 Stabilization of relative frequency

If probabilities are assigned to events in accordance with their limiting relative frequencies, then we can interpret a statement such as "The probability of that coin landing with the head facing up when it is tossed is .5" to mean that in a large number of such tosses, a head will appear on approximately half the tosses and a tail on the other half.

This relative frequency interpretation of probability is referred to as an objective interpretation because it rests on a property of the experiment rather than any particular individual concerned with the experiment. For example, two different observers of a sequence of coin tosses should both use the same probability assignments since the observers have nothing to do with limiting relative frequency. In practice, this interpretation is not as objective as it might seem, since the limiting relative frequency of an event will not be known. Thus, we will have to assign probabilities based on our beliefs about the limiting relative frequency of events under study. Fortunately, there are many experiments for which there will be a consensus with respect to probability assignments. When we speak of a fair coin, we shall mean $P(H) = P(T) = .5$, and a fair die is one for which limiting relative frequencies of the six outcomes are all $\frac{1}{6}$, suggesting probability assignments $P(\{1\}) = \cdots = P(\{6\}) = \frac{1}{6}$.

Because the objective interpretation of probability is based on the notion of limiting frequency, its applicability is limited to experimental situations that are repeatable. Yet the language of probability is often used in connection with situations that are inherently unrepeatable. Examples include: "The chances are good for a peace agreement"; "It is likely that our company will be awarded the contract"; and "Because their best quarterback is injured, I expect them to score no more than 10 points against us." In such situations we would like, as before, to assign numerical probabilities to various outcomes and events (e.g., the probability is .9 that we will get the contract). We must therefore adopt an alternative interpretation of these probabilities. Because different observers may have different prior information and opinions concerning such experimental situations, probability assignments may now differ from individual to individual. Interpretations in such situations are thus referred to as *subjective*. The book by Robert Winkler listed in the chapter references gives a very readable survey of several subjective interpretations.

Properties of Probability

Proposition

For any event A, $P(A) = 1 - P(A')$.

Proof   In Axiom 3a, let $n = 2$, $A_1 = A$, and $A_2 = A'$. Since by definition of A', $A \cup A' = \mathcal{S}$ while A and A' are disjoint, $1 = P(\mathcal{S}) = P(A \cup A') = P(A) + P(A')$, from which the desired result follows.   ∎

This proposition is surprisingly useful because there are many situations in which $P(A')$ is more easily obtained by direct methods than is $P(A)$.

Example 2.13   Consider a system of five identical components connected in series as illustrated in Figure 2.3.

Figure 2.3 A system of five components connected in series

Denote a component that fails by F and one that doesn't fail by S (for success). Let A be the event that the *system* fails. For A to occur, at least one of the individual components must fail. Outcomes in A include $SSFSS$ (1, 2, 4, and 5 all work, but 3 does not), $FFSSS$, and so on. There are in fact 31 different outcomes in A. However, A', the event that the system works, consists of the single outcome $SSSSS$. We will see in Section 2.5 that if 90% of all these components do not fail and different components fail independently of one another, then $P(A') = P(SSSSS) = .9^5 = .59$. Thus, $P(A) = 1 - .59 = .41$; so among a large number of such systems, roughly 41% will fail. ■

In general, the foregoing proposition is useful when the event of interest can be expressed as "at least . . . ", since then the complement "less than . . . " is often easier to work with (in some problems, "more than . . . " is easier to deal with than "at most . . . "). When you are having difficulty calculating $P(A)$ directly, think of determining $P(A')$.

Proposition

> If A and B are mutually exclusive, then $P(A \cap B) = 0$.

Proof   Because $A \cap B$ contains no outcomes, $(A \cap B)' = \mathcal{S}$. Thus, we have that $1 = P[(A \cap B)'] = 1 - P(A \cap B)$. ■

When events A and B are mutually exclusive, Axiom 3 gives $P(A \cup B) = P(A) + P(B)$. When A and B are not mutually exclusive, the probability of the union is obtained from the following result.

Proposition

> For any two events A and B,
>
> $P(A \cup B) = P(A) + P(B) - P(A \cap B)$

Notice that the proposition is valid even if A and B are mutually exclusive, since then $P(A \cap B) = 0$. The key idea is that, in adding $P(A)$ and $P(B)$, the probability of the intersection $A \cap B$ is actually counted twice, so $P(A \cap B)$ must be subtracted out.

Proof   Note first that $A \cup B = A \cup (B \cap A')$, as illustrated in Figure 2.4.
Since A and $(B \cap A')$ are mutually exclusive, $P(A \cup B) = P(A) + P(B \cap A')$. But $B = (B \cap A) \cup (B \cap A')$ (the union of that part of B in A and that part of B not

Figure 2.4 $A \cup B = A \cup (B \cap A')$

in A), with $(B \cap A)$ and $(B \cap A')$ mutually exclusive, so that $P(B) = P(B \cap A) + P(B \cap A')$. Combining these results gives

$$P(A \cup B) = P(A) + P(B \cap A') = P(A) + [P(B) - P(A \cap B)]$$
$$= P(A) + P(B) - P(A \cap B) \qquad \blacksquare$$

Example 2.14   In a certain residential suburb, 60% of all households subscribe to the metropolitan newspaper published in a nearby city, 80% subscribe to the local afternoon paper, and 50% of all households subscribe to both papers. If a household is selected at random, what is the probability that it subscribes to (1) at least one of the two newspapers and (2) exactly one of the two newspapers?

With $A = \{$subscribes to the metropolitan paper$\}$ and $B = \{$subscribes to the local paper$\}$, the given information implies that $P(A) = .6$, $P(B) = .8$, and $P(A \cap B) = .5$. The previous proposition then applies to give

P(subscribes to at least one of the two newspapers)

$$= P(A \cup B) = P(A) + P(B) - P(A \cap B) = .6 + .8 - .5 = .9$$

The event that a household subscribes only to the local paper can be written as $A' \cap B$ [(not metropolitan) and local]. Now Figure 2.4 implies that

$$.9 = P(A \cup B) = P(A) + P(A' \cap B) = .6 + P(A' \cap B)$$

from which $P(A' \cap B) = .3$. Similarly, $P(A \cap B') = P(A \cup B) - P(B) = .1$. This is all illustrated in Figure 2.5, from which we see that

$$P(\text{exactly one}) = P(A \cap B') + P(A' \cap B) = .1 + .3 = .4$$

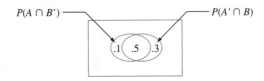

$P(A \cap B')$ ——   —— $P(A' \cap B)$

.1   .5   .3

Figure 2.5 Probabilities for Example 2.14   ∎

The probability of a union of more than two events can be computed analogously. For three events A, B, and C, the result is

$$P(A \cup B \cup C) = P(A) + P(B) + P(C) - P(A \cap B) - P(A \cap C)$$
$$- P(B \cap C) + P(A \cap B \cap C)$$

This can be seen by examining a Venn diagram of $A \cup B \cup C$, which is shown in Figure 2.6. When $P(A)$, $P(B)$, and $P(C)$ are added, certain intersections are counted twice, so must be subtracted out, but this results in $P(A \cap B \cap C)$ being subtracted once too often.

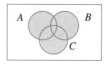

Figure 2.6 $A \cup B \cup C$

Determining Probabilities Systematically

When the number of possible outcomes (simple events) is large, there will be many compound events. A simple way to determine probabilities for these events that avoids violating the axioms and derived properties is to first determine probabilities $P(E_i)$ for all simple events. These should satisfy $P(E_i) \geq 0$ and $\Sigma_{\text{all } i} P(E_i) = 1$. Then the probability of any compound event A is computed by adding together the $P(E_i)$'s for all E_i's in A:

$$P(A) = \sum_{\text{all } E_i's \text{ in } A} P(E_i)$$

Example 2.15   Denote the six elementary events $\{1\}, \ldots, \{6\}$ associated with tossing a six-sided die once by $E_1, \ldots, E_6$. If the die is constructed so that any of the three even outcomes is twice as likely to occur as any of the three odd outcomes, then an appropriate assignment of probabilities to elementary events is $P(E_1) = P(E_3) = P(E_5) = \frac{1}{9}$, $P(E_2) = P(E_4) = P(E_6) = \frac{2}{9}$. Then for the event $A = \{$outcome is even$\} = E_2 \cup E_4 \cup E_6$, $P(A) = P(E_2) + P(E_4) + P(E_6) = \frac{6}{9} = \frac{2}{3}$; for $B = \{$outcome $\leq 3\} = E_1 \cup E_2 \cup E_3$, $P(B) = \frac{1}{9} + \frac{2}{9} + \frac{1}{9} = \frac{4}{9}$.   ∎

Equally Likely Outcomes

In many experiments consisting of N outcomes, it is reasonable to assign equal probabilities to all N simple events. These include such obvious examples as tossing a fair coin or fair die once or twice (or any fixed number of times), or selecting one or several cards from a well-shuffled deck of 52. With $p = P(E_i)$ for every i,

$$1 = \sum_{i=1}^{N} P(E_i) = \sum_{i=1}^{N} p = p \cdot N, \qquad \text{so } p = \frac{1}{N}$$

That is, if there are N possible outcomes, then the probability assigned to each is $1/N$.

Now consider an event A, with $N(A)$ denoting the number of outcomes contained in A. Then

$$P(A) = \sum_{E_i \text{ in } A} P(E_i) = \sum_{E_i \text{ in } A} \frac{1}{N} = \frac{N(A)}{N}$$

Once we have counted the number N of outcomes in the sample space, to compute the probability of any event we must count the number of outcomes contained in that event and take the ratio of the two numbers. Thus, when outcomes are equally likely, computing probabilities reduces to counting.

Example 2.16   When two dice are rolled separately, there are $N = 36$ outcomes (delete the first row and column from the table in Example 2.3). If both the dice are fair, all 36 outcomes are equally likely, so $P(E_i) = \frac{1}{36}$. Then the event $A = \{\text{sum of two numbers} = 7\}$ consists of the six outcomes (1, 6), (2, 5), (3, 4), (4, 3), (5, 2), and (6, 1), so

$$P(A) = \frac{N(A)}{N} = \frac{6}{36} = \frac{1}{6}$$

■

Exercises   Section 2.2 (11–29)

11. A mutual fund company offers its customers several different funds: a money-market fund, three different bond funds (short, intermediate, and long-term), two stock funds (moderate and high-risk), and a balanced fund. Among customers who own shares in just one fund, the percentages of customers in the different funds are as follows:

Money-market	20%	High-risk stock	18%
Short bond	15%	Moderate-risk	25%
Intermediate	10%	stock	
bond		Balanced	7%
Long bond	5%		

A customer who owns shares in just one fund is randomly selected.

a. What is the probability that the selected individual owns shares in the balanced fund?

b. What is the probability that the individual owns shares in a bond fund?

c. What is the probability that the selected individual does not own shares in a stock fund?

12. A family that owns two automobiles is selected at random. Let $A_1 = \{\text{the older car is American}\}$ and $A_2 = \{\text{the newer car is American}\}$. If $P(A_1) = .7, P(A_2) = .5$, and $P(A_1 \cap A_2) = .4$, compute the following:

a. $P(A_1 \cup A_2)$ (the probability that at least one car is American).

b. The probability that neither car is American.

c. The probability that exactly one of the two cars is American.

13. A computer consulting firm presently has bids out on three projects. Let $A_i = \{\text{awarded project } i\}$, for $i = 1, 2, 3$, and suppose that $P(A_1) = .22$, $P(A_2) = .25, P(A_3) = .28, P(A_1 \cap A_2) = .11, P(A_1 \cap A_3) = .05$, $P(A_2 \cap A_3) = .07$, $P(A_1 \cap A_2 \cap A_3) = .01$. Express in words each of the following events and compute the probability of each event.

a. $A_1 \cup A_2$

b. $A_1' \cap A_2'$ [HINT: $(A_1 \cup A_2)' = A_1' \cap A_2'$]

c. $A_1 \cup A_2 \cup A_3$ **d.** $A_1' \cap A_2' \cap A_3'$

e. $A_1' \cap A_2' \cap A_3$ **f.** $(A_1' \cap A_2') \cup A_3$

14. A utility company offers a lifeline rate to any household whose electricity usage falls below 240 kWh during a particular month. Let A denote the event that a randomly selected household in a certain community does not exceed the lifeline usage during January and let B be the analogous event for the month of July (A and B refer to the same household). Suppose $P(A) = .8, P(B) = .7$, and $P(A \cup B) = .9$. Compute the following:

a. $P(A \cap B)$.

b. The probability that the lifeline usage amount is exceeded in exactly one of the two months. Describe this event in terms of A and B.

15. Consider the type of clothes dryer (gas or electric) purchased by each of five different customers at a certain store

a. If the probability that at most one of these purchases an electric dryer is .087, what is the probability that at least two purchase an electric dryer?

b. If P(all five purchase gas) = .0768 and P(all five purchase electric) = .0102, what is the probability that at least one of each type is purchased?

16. A construction firm is working on two different projects. Let A be the event that the first one is completed by the contract date and define B analogously for the second project. If $P(A \cup B) = .9$ and $P(A \cap B) = .5$, what is the probability that exactly one project is completed by the contract date? Explain your reasoning. (HINT: A Venn diagram should help.)

17. In a school machine shop, 60% of all machine breakdowns occur on lathes and 15% on drills. Let $A = \{$the next machine breakdown is a lathe$\}$, and $B = \{$the next machine breakdown is a drill$\}$ (so that A and B are mutually exclusive). With $P(A) = .60$ and $P(B) = .15$, calculate the following:

   a. $P(A')$ b. $P(A \cup B)$ c. $P(A' \cap B')$

18. An individual is presented with three different glasses of cola, labeled C, D, and P. He is asked to taste all three and then list them in order of preference. Suppose the same cola has actually been put into all three glasses.

   a. What are the simple events in this ranking experiment, and what probability would you assign to each one?

   b. What is the probability that C is ranked first?

   c. What is the probability that C is ranked first and D is ranked last?

19. A video store sells two different brands of VCRs, each of which comes with either two heads or four heads. The accompanying table gives the percentages of recent purchasers buying each type of VCR:

Brand	Number of Heads	
	2	4
M	25%	16%
Q	32%	27%

Suppose a recent purchaser is randomly selected and both the brand and the number of heads are determined.

   a. What are the four simple events?

b. What is the probability that the selected purchaser bought brand Q, with two heads?

c. What is the probability that the selected purchaser bought brand M?

20. A library has five copies of a certain text, of which copies 1 and 2 are first printings and copies 3, 4, and 5 are second printings. Two copies are to be randomly selected to be placed on 2-hour reserve (implying 10 equally likely outcomes).

   a. What is the probability that both selected copies are first printings?

   b. What is the probability that both selected copies are second printings?

   c. What is the probability that at least one selected copy is a first printing?

   d. What is the probability that the selected copies are different printings?

21. An insurance company offers four different deductible levels—none, low, medium, and high—for its homeowner's policyholders and three different levels—low, medium, and high—for its automobile policyholders. The accompanying table gives proportions for the various categories of policyholders who have both types of insurance. For example, the proportion of individuals with both low homeowner's deductible and low auto deductible is .06 (6% of all such individuals).

Auto	Homeowner's			
	N	L	M	H
L	.04	.06	.05	.03
M	.07	.10	.20	.10
H	.02	.03	.15	.15

Suppose an individual having both types of policies is randomly selected.

   a. What is the probability that the individual has a medium auto deductible and a high homeowner's deductible?

   b. What is the probability that the individual has a low auto deductible? A low homeowner's deductible?

   c. What is the probability that the individual is in the same category for both auto and homeowner's deductibles?

d. Based on your answer in part (c), what is the probability that the two categories are different?

e. What is the probability that the individual has at least one low deductible level?

f. Using the answer in part (e), what is the probability that neither deductible level is low?

22. The route used by a certain motorist in commuting to work contains two intersections with traffic signals. The probability that he must stop at the first signal is .4, the analogous probability for the second signal is .5, and the probability that he must stop at at least one of the two signals is .6. What is the probability that he must stop

a. At both signals?

b. At the first signal but not at the second one?

c. At exactly one signal?

23. A student has a box containing 25 computer disks, of which 15 are blank and the other 10 are not. If she randomly selects disks one by one, what is the probability that at least two must be selected in order to find one that is blank?

24. When I visit the local library, the probability that someone is reading the current issue of *Sports Illustrated* is .4, the probability that someone is reading *Time* is .3, and the probability that at least one of these two magazines is being read by someone is .5. What is the probability that

a. Both of the magazines are being read?

b. Neither of the two is being read?

c. Exactly one is being read?

25. Use the axioms to show that if one event A is contained in another event B (A is a subset of B), then $P(A) \le P(B)$. [HINT: For such A and B, $B = A \cup (B \cap A')$. Draw a Venn diagram.]

26. If a standard deck of 52 cards (spades, hearts, diamonds, and clubs; with ace, 2, . . . , 10, jack, queen, and king in each suit) is well mixed before a single card is drawn, all 52 outcomes are equally likely. Define events by $A = \{\text{heart}\}$, $B = \{\text{black card}\}$, $C = \{\text{jack, queen, or king}\}$, and $D = \{\text{selected card is at most a 5}\}$ and use elementary counting along with the axioms and properties to compute the following probabilities.

a. $P(A)$, $P(B)$, $P(C)$, and $P(D)$

b. $P(A \cup B)$ and $P(C \cup D)$

c. $P(A \cup C)$ and $P(B \cup D)$

d. $P(A \cup B \cup C \cup D)$

27. The three most popular options on a certain type of new car are automatic transmission (A), power steering (B), and a radio (C). If 70% of all purchasers request A, 80% request B, 75% request C, 85% request A or B, 90% request A or C, 95% request B or C, and 98% request A or B or C, compute the probabilities of the following events.

(NOTE: "A or B" could mean that both were requested. HINT: Try drawing a Venn diagram and labeling all regions.)

a. The next purchaser will select at least one of the three options ($A \cup B \cup C$).

b. The next purchaser will select none of the three options.

c. The next purchaser will select only a radio.

d. The next purchaser will select exactly one of the three options.

28. An academic department with five faculty members—Anderson, Box, Cox, Cramer, and Fisher—must select two of its members to serve on a personnel review committee. Because the work will be time-consuming, no one is anxious to serve, so it is decided that the representative will be selected by putting five slips of paper in a box, mixing them, and selecting two.

a. What is the probability that both Anderson and Box will be selected? (HINT: List the equally likely outcomes.)

b. What is the probability that at least one of the two members whose name begins with C is selected?

c. If the five faculty members have taught for 3, 6, 7, 10, and 14 years, respectively, at the university, what is the probability that the two chosen representatives have at least 15 years' teaching experience at the university?

29. In Exercise 3, suppose that any incoming individual is equally likely to be assigned to any of the three stations irrespective of where other individuals have been assigned. What is the probability that

a. All three family members are assigned to the same station?

b. At most two family members are assigned to the same station?

c. Every family member is assigned to a different station?

Counting Techniques

When the various outcomes of an experiment are equally likely (the same probability is assigned to each simple event), then the task of computing probabilities reduces to counting. In particular, if N is the number of outcomes in a sample space and $N(A)$ is the number of outcomes contained in an event A, then

$$P(A) = \frac{N(A)}{N} \tag{2.1}$$

If a list of the outcomes is available or easy to construct and N is small, then the numerator and denominator of Equation (2.1) can be obtained without the benefit of any general counting principles.

There are, however, many experiments for which the effort involved in constructing such a list is prohibitive because N is quite large. By exploiting some general counting rules, it is possible to compute probabilities of the form (2.1) without a listing of outcomes. These rules are also useful in many problems involving outcomes that are not equally likely. Several of the rules developed here will be used in studying probability distributions in the next chapter.

The Product Rule for Ordered Pairs

Our first counting rule applies to any situation in which a set (event) consists of ordered pairs of objects and we wish to count the number of such pairs. By an ordered pair, we mean that, if O_1 and O_2 are objects, then the pair (O_1, O_2) is different from the pair (O_2, O_1). For example, if an individual selects one airline for a trip from Los Angeles to Chicago and (after transacting business in Chicago) a second one for continuing on to New York, one possibility is (American, United), another is (United, American), and still another is (United, United).

Proposition

> If the first element or object of an ordered pair can be selected in n_1 ways, and for each of these n_1 ways the second element of the pair can be selected in n_2 ways, then the number of pairs is $n_1 n_2$.

Example 2.17   A student wishes to commute first to a junior college for two years and then to a state college campus. Within commuting range there are four junior colleges and three state college campuses. How many choices of junior college and state college are available to her? If junior colleges are denoted by 1, 2, 3, 4 and state colleges by a, b, c, choices are $(1, a)$, $(1, b)$, . . . , $(4, c)$, a total of 12 choices. With $n_1 = 4$ and $n_2 = 3$, $N = n_1 n_2 = 12$ without a list.   ∎

Example 2.18   A homeowner doing some remodeling requires the services of both a plumbing contractor and an electrical contractor. If there are 12 plumbing contractors and 9 electrical contractors available in the area, in how many ways can the contractors be chosen? If we denote the plumbers by $P_1, \ldots, P_{12}$ and the electricians by $Q_1, \ldots, Q_9$, then we wish the number of pairs of the form (P_i, Q_j). With $n_1 = 12$ and $n_2 = 9$, the product rule yields $N = (12)(9) = 108$ possible ways of choosing the two types of contractors. ∎

In both Examples 2.17 and 2.18, the choice of the second element of the pair did not depend on which first element was chosen or occurred. As long as there is the same number of choices of the second element for each first element, the product rule is valid even when the set of possible second elements depends on the first element.

Example 2.19   A family has just moved to a new city and requires the services of both an obstetrician and a pediatrician. There are two easily accessible medical clinics, each having two obstetricians and three pediatricians. The family will obtain maximum health insurance benefits by joining a clinic and selecting both doctors from that clinic. In how many ways can this be done? Denote the obstetricians by $O_1, O_2, O_3,$ and O_4 and the pediatricians by $P_1, \ldots, P_6$. Then we wish the number of pairs (O_i, P_j) for which O_i and P_j are associated with the same clinic. Because there are four obstetricians, $n_1 = 4$, and for each there are three choices of pediatrician, so $n_2 = 3$. Applying the product rule gives $N = n_1 n_2 = 12$ possible choices. ∎

Tree Diagrams

In problems in which the product rule can be applied, a configuration called a *tree diagram* can be used to represent pictorially all the possibilities. The tree diagram associated with Example 2.19 appears in Figure 2.7. Starting from a point on the left side of the diagram, for each possible first element of a pair a straight-line segment emanates rightward. Each of these lines is referred to as a first-generation branch. Now for any given first-generation branch we construct another line segment

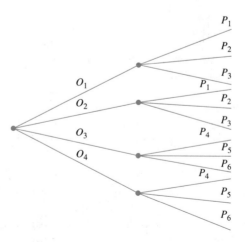

Figure 2.7 Tree diagram for Example 2.19

emanating from the tip of the branch for each possible choice of a second element of the pair. Each such line segment is a second-generation branch. Because there are four obstetricians, there are four first-generation branches, and three pediatricians for each obstetrician yields three second-generation branches emanating from each first-generation branch.

In the general case, there are n_1 first-generation branches, and for each first-generation branch there are n_2 second-generation branches. The total number of second-generation branches is therefore $n_1 n_2$. Since each second-generation branch corresponds to exactly one possible pair (choosing a first element and then a second puts us at the end of exactly one second-generation branch), there are $n_1 n_2$ pairs, so the product rule is verified.

The construction of a tree diagram does not depend on having the same number of second-generation branches emanating from each first-generation branch. If the second clinic had four pediatricians, then there would be only three branches emanating from two of the first-generation branches and four emanating from each of the other two first-generation branches. A tree diagram can thus be used to represent pictorially experiments other than those to which the product rule applies.

A More General Product Rule

If a six-sided die is tossed five times in succession rather than just twice, then each possible outcome is an ordered collection of five numbers such as (1, 3, 1, 2, 4) or (6, 5, 2, 2, 2). We will call an ordered collection of k objects a **k-tuple** (so a pair is a 2-tuple and a triple is a 3-tuple). Each outcome of the die-tossing experiment is then a 5-tuple.

Product Rule for k-Tuples

> Suppose a set consists of ordered collections of k elements (k-tuples) and that there are n_1 possible choices for the first element; for each choice of the first element, there are n_2 possible choices of the second element; . . . ; for each possible choice of the first $k - 1$ elements, there are n_k choices of the kth element. Then there are $n_1 n_2 \cdot \cdots \cdot n_k$ possible k-tuples.

This more general rule can also be illustrated by a tree diagram; simply construct a more elaborate diagram by adding third-generation branches emanating from the tip of each second generation, then fourth-generation branches, and so on, until finally kth-generation branches are added.

Example 2.20
(Example 2.18 continued)

Suppose the home remodeling job involves first purchasing several kitchen appliances. They will all be purchased from the same dealer, and there are five dealers in the area. With the dealers denoted by $D_1, \ldots, D_5$, there are $N = n_1 n_2 n_3 = (5)(12)(9) = 540$ 3-tuples of the form (D_i, P_j, Q_k), so there are 540 ways to choose first an appliance dealer, then a plumbing contractor, and finally an electrical contractor. ∎

Example 2.21
(*Example 2.19*
continued)

If each clinic has both three specialists in internal medicine and two general surgeons, there are $n_1 n_2 n_3 n_4 = (4)(3)(3)(2) = 72$ ways to select one doctor of each type such that all doctors practice at the same clinic. ∎

Permutations

So far the successive elements of a k-tuple were selected from entirely different sets (e.g., appliance dealers, then plumbers, and finally electricians). In several tosses of a die, the set from which successive elements are chosen is always $\{1, 2, 3, 4, 5, 6\}$, but the choices are made "with replacement" so that the same element can appear more than once. We now consider a fixed set consisting of n distinct elements and suppose that a k-tuple is formed by selecting successively from this set *without replacement* so that an element can appear in at most one of the k positions.

Definition

> Any ordered sequence of k objects taken from a set of n distinct objects is called a **permutation** of size k of the objects. The number of permutations of size k that can be constructed from the n objects is denoted by $P_{k,n}$.

The number of permutations of size k is obtained immediately from the general product rule. The first element can be chosen in n ways, for each of these n ways the second element can be chosen in $n - 1$ ways, and so on; finally, for each way of choosing the first $k - 1$ elements, the kth element can be chosen in $n - (k - 1) = n - k + 1$ ways, so

$$P_{k,n} = n(n - 1)(n - 2) \cdots \cdots (n - k + 2)(n - k + 1)$$

Example 2.22

There are eight teaching assistants available for grading papers in a particular course. The first exam consists of four questions, and the professor wishes to select a different assistant to grade each question (only one assistant per question). In how many ways can assistants be chosen to grade the exam? Here $n =$ the number of assistants $= 8$ and $k =$ the number of questions $= 4$. The number of different grading assignments is then $P_{4,8} = (8)(7)(6)(5) = 1680$. ∎

The use of factorial notation allows $P_{k,n}$ to be expressed more compactly.

Definition

> For any positive integer m, $m!$ is read "m factorial" and is defined by $m! = m(m - 1) \cdots \cdots (2)(1)$. Also, $0! = 1$.

Using factorial notation yields

$$P_{k,n} = n(n - 1) \cdots \cdots (n - k + 1)$$

$$= \frac{n(n - 1) \cdots \cdots (n - k + 1)(n - k)(n - k - 1) \cdots \cdots (2)(1)}{(n - k)(n - k - 1) \cdots \cdots (2)(1)}$$

which becomes

$$P_{k,n} = \frac{n!}{(n-k)!}$$

For example, $P_{3,9} = 9!/(9-3)! = 9!/6! = 9 \cdot 8 \cdot 7 \cdot 6!/6! = 9 \cdot 8 \cdot 7$. Note also that because $0! = 1$, $P_{n,n} = n!/(n-n)! = n!/0! = n!/1 = n!$, as it should.

Combinations

There are many counting problems in which one is given a set of n distinct objects and wishes to count the number of *unordered* subsets of size k. For example, in bridge it is only the 13 cards in a hand and not the order in which they are dealt that is important; in the formation of a committee, the order in which committee members are listed is frequently unimportant.

Definition

Given a set of n distinct objects, any unordered subset of size k of the objects is called a **combination.** The number of combinations of size k that can be formed from n distinct objects will be denoted by $\binom{n}{k}$. (This notation is more common in probability than $C_{k,n}$, which would be analogous to notation for permutations.)

The number of combinations of size k from a particular set is smaller than the number of permutations because, when order is disregarded, a number of permutations correspond to the same combination. Consider, for example, the set $\{A, B, C, D, E\}$ consisting of five elements. We know that there are $5!/(5-3)! = 60$ permutations of size 3. There are six permutations of size 3 consisting of the elements A, B, and C since these three can be ordered $3 \cdot 2 \cdot 1 = 3! = 6$ ways: (A, B, C), (A, C, B), (B, A, C), (B, C, A), (C, A, B), and (C, B, A). These six permutations are equivalent to the single combination $\{A, B, C\}$. Similarly, for any other combination of size 3, there are $3!$ permutations, each obtained by ordering the three objects. Thus,

$$60 = P_{3,5} = \binom{5}{3} \cdot 3!; \quad \text{so} \quad \binom{5}{3} = \frac{60}{3!} = 10$$

These ten combinations are

$\{A, B, C\}, \{A, B, D\}, \{A, B, E\}, \{A, C, D\}, \{A, C, E\}, \{A, D, E\}, \{B, C, D\},$
$\{B, C, E\}, \{B, D, E\}, \{C, D, E\}$

When there are n distinct objects, any permutation of size k is obtained by ordering the k unordered objects of a combination in one of $k!$ ways, so the number of permutations is the product of $k!$ and the number of combinations. This gives

$$\binom{n}{k} = \frac{P_{k,n}}{k!} = \frac{n!}{k!(n-k)!}$$

Notice that $\binom{n}{n} = 1$ and $\binom{n}{0} = 1$ since there is only one way to choose a set of (all) n elements or of no elements, and $\binom{n}{1} = n$ since there are n subsets of size 1.

Example 2.23   A bridge hand consists of any 13 cards selected from a 52-card deck without regard to order. There are $\binom{52}{13} = 52!/13!39!$ different bridge hands, which works out to approximately 635 billion. Since there are 13 cards in each suit, the number of hands consisting entirely of clubs and/or spades (no red cards) is $\binom{26}{13} = 26!/13!13! = 10,400,597$. One of these $\binom{26}{13}$ hands consists entirely of spades, and one consists entirely of clubs, so there are $[\binom{26}{13} - 2]$ hands that consist entirely of clubs and spades with both suits represented in the hand. Suppose a bridge hand is dealt from a well-shuffled deck (i.e., 13 cards are randomly selected from among the 52 possibilities) and let

> A = {the hand consists entirely of spades and clubs with
>   both suits represented}
>
> B = {the hand consists of exactly two suits}

The $N = \binom{52}{13}$ possible outcomes are equally likely, so

$$P(A) = \frac{N(A)}{N} = \frac{\binom{26}{13} - 2}{\binom{52}{13}} = .0000164$$

Since there are $\binom{4}{2} = 6$ combinations consisting of two suits, of which spades and clubs is one such combination,

$$P(B) = \frac{6\left[\binom{26}{13} - 2\right]}{\binom{52}{13}} = .0000984$$

That is, a hand consisting entirely of cards from exactly two of the four suits will occur roughly once in every 10,000 hands. If you play bridge only once a month, it is likely that you will never be dealt such a hand.   ∎

Example 2.24   A rental car service facility has 10 foreign cars and 15 domestic cars waiting to be serviced on a particular Saturday morning. Because there are so few mechanics working on Saturday, only 6 can be serviced. If the 6 are chosen at random, what is the probability that 3 of the cars selected are domestic and the other 3 are foreign?

Let D_3 = {exactly 3 of the 6 cars chosen are domestic}. Assuming that any particular set of 6 cars is as likely to be chosen as is any other set of 6, we have equally likely outcomes, so $P(D_3) = N(D_3)/N$, where N is the number of ways of choosing 6 cars from the 25 and $N(D_3)$ is the number of ways of choosing 3 domestic cars and 3 foreign cars. We have immediately that $N = \binom{25}{6}$. To obtain $N(D_3)$, think of first choosing 3 of the 15 domestic cars and then 3 of the foreign cars. There are $\binom{15}{3}$ ways of choosing the 3 domestic cars, and there are $\binom{10}{3}$ ways of choosing the 3 foreign cars; $N(D_3)$ is now the product of these two numbers (visualize a tree diagram—we are really using a product rule argument here), so

$$P(D_3) = \frac{N(D_3)}{N} = \frac{\binom{15}{3}\binom{10}{3}}{\binom{25}{6}} = \frac{\frac{15!}{3!12!} \cdot \frac{10!}{3!7!}}{\frac{25!}{6!19!}} = .3083$$

Let $D_4 = \{$exactly 4 of the 6 cars chosen are domestic$\}$ and define D_5 and D_6 in an analogous manner. Then the probability that at least 3 domestic cars are selected is

$$P(D_3 \cup D_4 \cup D_5 \cup D_6) = P(D_3) + P(D_4) + P(D_5) + P(D_6)$$

$$= \frac{\binom{15}{3}\binom{10}{3}}{\binom{25}{6}} + \frac{\binom{15}{4}\binom{10}{2}}{\binom{25}{6}} + \frac{\binom{15}{5}\binom{10}{1}}{\binom{25}{6}} + \frac{\binom{15}{6}\binom{10}{0}}{\binom{25}{6}} = .8530$$

This is also the probability that at most 3 foreign cars are selected. ∎

Exercises   Section 2.3 (30–45)

30. The Student Engineers Council at a certain college has one student representative from each of the five engineering majors (civil, electrical, industrial, materials, and mechanical). In how many ways can

  a. Both a council president and a vice-president be selected?

  b. A president, a vice-president, and a secretary be selected?

  c. Two members be selected for the President's Council?

31. A real estate agent is showing homes to a prospective buyer. There are ten homes in the desired price range listed in the area. The buyer has time to visit only three of them.

  a. In how many ways could the three homes be chosen if the order of visiting is considered?

  b. In how many ways could the three homes be chosen if the order is unimportant?

  c. If four of the homes are new and six have previously been occupied and if the three homes to visit are randomly chosen, what is the probability that all three are new? (The same answer results whether or not order is considered.)

32. a. Beethoven wrote 9 symphonies and Mozart wrote

27 piano concertos. If a university radio station announcer wishes to play first a Beethoven symphony and then a Mozart concerto, in how many ways can this be done?

  b. The station manager decides that on each successive night (7 days per week), a Beethoven symphony will be played, followed by a Mozart piano concerto, followed by a Schubert string quartet (of which there are 15). For roughly how many years could this policy be continued before exactly the same program would have to be repeated?

33. Television viewers without cable in a certain city can receive seven different channels. From 5:30 to 6:00 P.M. on weekdays, three channels show news; from 6:00 to 6:30, five channels show news; from 6:30 to 7:00, four channels show news; and from 7:00 to 7:30, two channels show news.

  a. If a viewing sequence consists of four half-hour programs between 5:30 and 7:30, how many viewing sequences are there?

  b. If a viewing sequence is chosen at random from the set of all possible sequences, what is the probability that all four programs selected are news programs?

c. Choosing a sequence as in part (b), what is the probability that at least two programs are news programs?

d. Choosing a sequence as in part (b), what is the probability that a news program is selected both at 6:00 P.M. and at 6:30 P.M.?

34. A chain of stereo stores is offering a special price on a complete set of components (receiver, compact disc player, speakers, cassette deck). A purchaser is offered a choice of manufacturer for each component:

Receiver: Kenwood, Onkyo, Pioneer, Sony, Sherwood

Compact disc player: Onkyo, Pioneer, Sony, Technics

Speakers: Boston, Infinity, Polk

Cassette deck: Onkyo, Sony, Teac, Technics

A switchboard display in the store allows a customer to hook together any selection of components (consisting of one of each type). Use the product rules to answer the following questions.

a. In how many ways can one component of each type be selected?

b. In how many ways can components be selected if both the receiver and the compact disc player are to be Sony?

c. In how many ways can components be selected if none is to be Sony?

d. In how many ways can a selection be made if at least one Sony component is to be included?

e. If someone flips switches on the selection in a completely random fashion, what is the probability that the system selected contains at least one Sony component? Exactly one Sony component?

35. Shortly after being put into service, some buses manufactured by a certain company have developed cracks on the underside of the main frame. Suppose a particular city has 20 of these buses, and cracks have actually appeared in 8 of them.

a. How many ways are there to select a sample of 5 buses from the 20 for a thorough inspection?

b. In how many ways can a sample of 5 buses contain exactly 4 with visible cracks?

c. If a sample of 5 buses is chosen at random, what is the probability that exactly 4 of the 5 will have visible cracks?

d. If buses are selected as in part (c), what is the probability that at least 4 of those selected will have visible cracks?

36. A production facility employs 20 workers on the day shift, 15 workers on the swing shift, and 10 workers on the graveyard shift. A quality control consultant is to select 6 of these workers for in-depth interviews. Suppose the selection is made in such a way that any particular group of 6 workers has the same chance of being selected as does any other group (drawing 6 slips without replacement from among 45).

a. How many selections result in all 6 workers coming from the day shift? What is the probability that all 6 selected workers will be from the day shift?

b. What is the probability that all 6 selected workers will be from the same shift?

c. What is the probability that at least two different shifts will be represented among the selected workers?

d. What is the probability that at least one of the shifts will be unrepresented in the sample of workers?

37. An academic department with five faculty members narrowed its choice for department head to either candidate *A* or candidate *B*. Each member then voted on a slip of paper for one of the candidates. Suppose there are actually three votes for *A* and two for *B*. If the slips are selected for tallying in random order, what is the probability that *A* remains ahead of *B* throughout the vote count (e.g., this event occurs if the selected ordering is *AABAB*, but not for *ABBAA*)?

38. An experimenter is studying the effects of temperature, pressure, and type of catalyst on yield from a certain chemical reaction. Three different temperatures, four different pressures, and five different catalysts are under consideration.

a. If any particular experimental run involves the use of a single temperature, pressure, and catalyst, how many experimental runs are possible?

b. How many experimental runs are there that involve use of the lowest temperature and two lowest pressures?

39. Refer to Exercise 38 and suppose that five different experimental runs are to be made on the first day of experimentation. If the five are randomly selected from among all the possibilities, so that any group of five has the same probability of selection, what is the probability that a different catalyst is used on each run?

40. A consumer group is concerned about the possibility that there is systematic underweighing in the meat department of a local supermarket. A representative is sent to purchase five 1-lb packages of ground meat. If there are currently 25 such packages on display and 8 of them are actually underweight, what is the probability that at most 2 of the 5 packages purchased are underweight?

41. Three molecules of type *A*, three of type *B*, three of type *C*, and three of type *D* are to be linked together to form a chain molecule. One such chain molecule is *ABCDABCDABCD*, and another is *BCDDAAABDBCC*.

  a. How many such chain molecules are there? (HINT: If the three *A*'s were distinguishable from one another—A_1, A_2, A_3—and the *B*'s, *C*'s, and *D*'s were also, how many molecules would there be? How is this number reduced when the subscripts are removed from the *A*'s?)

  b. Suppose a chain molecule of the type described is randomly selected. What is the probability that all three molecules of each type end up next to one another (such as in *BBBAAADDDCCC*)?

42. A mathematics professor wishes to schedule an appointment with each of her eight teaching assistants, four men and four women, to discuss her calculus course. Suppose all possible orderings of appointments are equally likely to be selected.

  a. What is the probability that at least one female assistant is among the first three the professor meets with?

  b. What is the probability that after the first five appointments she has met with all female assistants?

  c. Suppose the professor has the same eight assistants the following semester and again schedules appointments without regard to the ordering during the first semester. What is the probability that the orderings of appointments are different?

43. Three married couples have purchased theater tickets and are seated in a row consisting of just six seats. If they take their seats in a completely random fashion (random order), what is the probability that Jim and Paula (husband and wife) sit in the two seats on the far left? What is the probability that Jim and Paula end up sitting next to one another? What is the probability that at least one of the wives ends up sitting next to her husband?

44. In five-card poker, a straight consists of five cards with adjacent denominations (e.g., 9 of clubs, 10 of hearts, jack of hearts, queen of spades, and king of clubs). Assuming that aces can be high or low, if you are dealt a five-card hand, what is the probability that it will be a straight with high card 10? What is the probability that it will be a straight? What is the probability that it will be a straight flush (all cards in the same suit)?

45. Show that $\binom{n}{k} = \binom{n}{n-k}$. Give an interpretation involving subsets.

2.4 Conditional Probability

The probabilities assigned to various events depend on what is known about the experimental situation when the assignment is made. Subsequent to the initial assignment, partial information about or relevant to the outcome of the experiment may become available, and this information may cause us to revise some of our probability assignments. For a particular event *A*, we have used $P(A)$ to represent the probability assigned to *A*; we now think of $P(A)$ as the original or unconditional probability of the event *A*.

In this section, we examine how the information "an event *B* has occurred" affects the probability assigned to *A*. For example, *A* might refer to an individual having a particular disease in the presence of certain symptoms. If a blood test is performed on the individual and the result is negative (*B* = negative blood test), then the probability of having the disease will change (it should decrease, but not usually to zero, since blood tests are not infallible). We will use the notation $P(A|B)$ to represent the **conditional probability of *A* given that the event *B* has occurred.**

Example 2.25   Complex components are assembled in a plant that uses two different assembly lines, *A* and *A'*. Line *A* uses older equipment than *A'*, so it is somewhat slower and less reliable. Suppose on a given day line *A* has assembled 8 components, of which 2 have been identified as defective (*B*) and 6 as nondefective (*B'*), while *A'* has produced 1 defective and 9 nondefective components. This information is summarized in the accompanying table.

		Condition	
		B	*B'*
Line	*A*	2	6
	A'	1	9

Unaware of this information, the sales manager randomly selects 1 of these 18 components for a demonstration. Prior to the demonstration

$$P(\text{line } A \text{ component selected}) = P(A) = \frac{N(A)}{N} = \frac{8}{18} = .44$$

However, if the chosen component turns out to be defective, then the event *B* has occurred, so the component must have been 1 of the 3 in the *B* column of the table. Since these 3 components are equally likely among themselves after *B* has occurred,

$$P(A \mid B) = \frac{2}{3} = \frac{\frac{2}{18}}{\frac{3}{18}} = \frac{P(A \cap B)}{P(B)} \qquad (2.2)$$

∎

In Equation (2.2) the conditional probability is expressed as a ratio of unconditional probabilities: The numerator is the probability of the intersection of the two events, while the denominator is the probability of the conditioning event *B*. A Venn diagram illuminates this relationship (Figure 2.8).

Figure 2.8 Motivating the definition of conditional probability

Given that *B* has occurred, the relevant sample space is no longer 𝒮 but consists of outcomes in *B*; *A* has occurred if and only if one of the outcomes in the intersection occurred, so that the conditional probability of *A* given *B* is proportional to $P(A \cap B)$. The proportionally constant $1/P(B)$ is used to ensure that the probability $P(B|B)$ of the new sample space B equals 1.

The Definition of Conditional Probability

Example 2.25 demonstrates that when outcomes are equally likely, computation of conditional probabilities can be based on intuition. When experiments are more complicated, though, intuition may fail us, so we want to have a general definition of conditional probability that will yield intuitive answers in simple problems. The Venn diagram and Equation (2.2) suggest the appropriate definition.

Definition

> For any two events A and B with $P(B) > 0$, the **conditional probability of A given that B has occurred** is defined by
>
> $$P(A|B) = \frac{P(A \cap B)}{P(B)} \qquad (2.3)$$

Example 2.26

Suppose that of all individuals buying a certain personal computer, 60% include a word processing program in their purchase, 40% include a spreadsheet program, and 30% include both types of programs. Consider randomly selecting a purchaser and let $A = \{$word processing program included$\}$ and $B = \{$spreadsheet program included$\}$. Then $P(A) = .60$, $P(B) = .40$, and $P(\text{both included}) = P(A \cap B) = .30$. Given that the selected individual included a spreadsheet program, the probability that a word processing program was also included is

$$P(A|B) = \frac{P(A \cap B)}{P(B)} = \frac{.30}{.40} = .75$$

That is, of all those purchasing a spreadsheet program, 75% purchased a word processing program. Similarly,

$$P(\text{spreadsheet} \mid \text{word processing}) = P(B|A) = \frac{P(A \cap B)}{P(A)} = \frac{.30}{.60} = .50$$

Notice that $P(A|B) \neq P(A)$ and $P(B|A) \neq P(B)$. ∎

Example 2.27

A news magazine publishes three columns entitled "Art" (A), "Books" (B), and "Cinema" (C). Reading habits of a randomly selected reader with respect to these columns are

Read regularly	A	B	C	$A \cap B$	$A \cap C$	$B \cap C$	$A \cap B \cap C$
Probability	.14	.23	.37	.08	.09	.13	.05

(See Figure 2.9.)

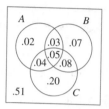

Figure 2.9 Venn diagram for Example 2.27

We thus have

$$P(A \mid B) = \frac{P(A \cap B)}{P(B)} = \frac{.08}{.23} = .348$$

$$P(A \mid B \cup C) = \frac{P(A \cap (B \cup C))}{P(B \cup C)} = \frac{.04 + .05 + .03}{.47} = \frac{.12}{.47} = .255$$

$$P(A \mid \text{reads at least one}) = P(A \mid A \cup B \cup C) = \frac{P(A \cap (A \cup B \cup C))}{P(A \cup B \cup C)}$$

$$= \frac{P(A)}{P(A \cup B \cup C)} = \frac{.14}{.49} = .286$$

and

$$P(A \cup B \mid C) = \frac{P((A \cup B) \cap C)}{P(C)} = \frac{.04 + .05 + .08}{.37} = .459 \qquad ■$$

The Multiplication Rule for $P(A \cap B)$

The definition of conditional probability yields the following result, obtained by multiplying both sides of (2.3) by $P(B)$.

The
Multiplication
Rule

$$P(A \cap B) = P(A \mid B) \cdot P(B)$$

This rule is important because it is often the case that $P(A \cap B)$ is desired while both $P(B)$ and $P(A \mid B)$ can be specified from the problem description. Consideration of $P(B \mid A)$ gives $P(A \cap B) = P(B \mid A) \cdot P(A)$.

Example 2.28   Four individuals have responded to a request by a blood bank for blood donations. None of them has donated before, so their blood types are unknown. Suppose only type A+ is desired and only one of the four actually has this type. If the potential donors are selected in random order for typing, what is the probability that at least three individuals must be typed to obtain the desired type?

Making the identification $B = \{$first type not A+$\}$ and $A = \{$second type not A+$\}$, $P(B) = \frac{3}{4}$. Given that the first type is not A+, two of the three individuals left are not A+, so $P(A \mid B) = \frac{2}{3}$. The multiplication rule now gives

$$P(\text{at least three individuals are typed}) = P(A \cap B)$$
$$= P(A \mid B) \cdot P(B)$$
$$= \frac{2}{3} \cdot \frac{3}{4} = \frac{6}{12}$$
$$= .5 \qquad \blacksquare$$

The multiplication rule is most useful when the experiment consists of several stages in succession. The conditioning event B then describes the outcome of the first stage and A the outcome of the second, so that $P(A \mid B)$—conditioning on what occurs first—will often be known. The rule is easily extended to experiments involving more than two stages. For example,

$$P(A_1 \cap A_2 \cap A_3) = P(A_3 \mid A_1 \cap A_2) \cdot P(A_1 \cap A_2) \qquad (2.4)$$
$$= P(A_3 \mid A_1 \cap A_2) \cdot P(A_2 \mid A_1) \cdot P(A_1)$$

where A_1 occurs first, followed by A_2, and finally A_3.

Example 2.29 For the blood typing experiment of Example 2.28,

$$P(\text{third type is A+}) = P(\text{third is} \mid \text{first isn't} \cap \text{second isn't})$$
$$\cdot P(\text{second isn't} \mid \text{first isn't}) \cdot P(\text{first isn't})$$
$$= \frac{1}{2} \cdot \frac{2}{3} \cdot \frac{3}{4} = \frac{1}{4} = .25 \qquad \blacksquare$$

When the experiment of interest consists of a sequence of several stages, it is convenient to represent these with a tree diagram. Once we have an appropriate tree diagram, probabilities and conditional probabilities can be entered on the various branches; this will make repeated use of the multiplication rule quite straightforward.

Example 2.30 A chain of video stores sells three different brands of VCRs. Of its VCR sales, 50% are brand 1 (the least expensive), 30% are brand 2, and 20% are brand 3. Each manufacturer offers a 1-year warranty on parts and labor. It is known that 25% of brand 1's VCRs require warranty repair work, whereas the corresponding percentages for brands 2 and 3 are 20% and 10%, respectively.

1. What is the probability that a randomly selected purchaser has bought a brand 1 VCR that will need repair while under warranty?

2. What is the probability that a randomly selected purchaser has a VCR that will need repair while under warranty?

3. If a customer returns to the store with a VCR that needs warranty repair work, what is the probability that it is a brand 1 VCR? A brand 2 VCR? A brand 3 VCR?

The first stage of the problem involves a customer selecting one of the three brands of VCR. Let A_i = {brand i is purchased}, for $i = 1$, 2, and 3. Then $P(A_1) = .50$, $P(A_2) = .30$, and $P(A_3) = .20$. Once a brand of VCR is selected, the second stage involves observing whether the selected VCR needs warranty repair. With B = {needs repair} and B' = {doesn't need repair}, the given information implies that $P(B|A_1) = .25$, $P(B|A_2) = .20$, and $P(B|A_3) = .10$.

The tree diagram representing this experimental situation appears in Figure 2.10. The initial branches correspond to different brands of VCRs, and there are two second-generation branches emanating from the tip of each initial branch, one for "needs repair" and the other for "doesn't need repair." The probability $P(A_i)$ appears on the ith initial branch, whereas the conditional probabilities $P(B|A_i)$ and $P(B'|A_i)$ appear on the second-generation branches. To the right of each second-generation branch corresponding to the occurrence of B, we display the product of probabilities on the branches leading out to that point. This is simply the multiplication rule in action. The answer to the question posed in 1 is thus $P(A_1 \cap B) = P(B|A_1) \cdot P(A_1) = .125$. The answer to question 2 is

$$P(B) = P((\text{brand 1 and repair}) \text{ or } (\text{brand 2 and repair}) \text{ or } (\text{brand 3 and repair}))$$
$$= P(A_1 \cap B) + P(A_2 \cap B) + P(A_3 \cap B)$$
$$= .125 + .060 + .020 = .205$$

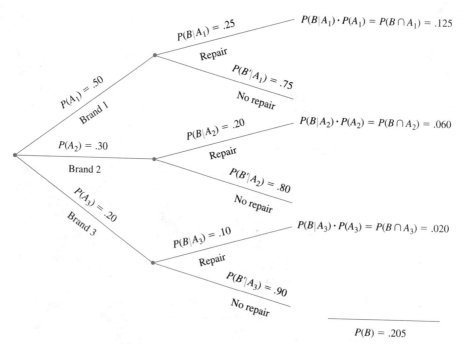

Figure 2.10 Tree diagram for Example 2.30

Finally,

$$P(A_1 | B) = \frac{P(A_1 \cap B)}{P(B)} = \frac{.125}{.205} = .61$$

$$P(A_2 | B) = \frac{P(A_2 \cap B)}{P(B)} = \frac{.060}{.205} = .29$$

and

$$P(A_3 | B) = 1 - P(A_1 | B) - P(A_2 | B) = .10$$

Notice that the initial or *prior probability* of brand 1 is .50, whereas once it is known that the selected VCR needed repair, the *posterior probability* of brand 1 increases to .61. This is because brand 1 VCRs are more likely to need warranty repair than are the other brands. The posterior probability of brand 3 is $P(A_3 | B) = .10$, which is much less than the prior probability $P(A_3) = .20$. ■

Bayes' Theorem

The computation of a posterior probability $P(A_k | B)$ from given prior probabilities $P(A_i)$ and conditional probabilities $P(B | A_i)$ occupies a central position in elementary probability. The general rule for such computations, which is really just a simple application of the multiplication rule, goes back to Reverend Thomas Bayes, who lived in the eighteenth century. To state it we first need another result. Recall that events $A_1, \ldots, A_n$ are mutually exclusive if no two have any common outcomes. The events are *exhaustive* if one A_i must occur, so that $A_1 \cup \cdots \cup A_n = \mathcal{S}$.

The Law
of Total
Probability

> Let $A_1, \ldots, A_n$ be mutually exclusive and exhaustive events. Then for any other event B,
>
> $$P(B) = P(B | A_1)P(A_1) + \cdots + P(B | A_n)P(A_n) \qquad (2.5)$$
>
> $$= \sum_{i=1}^{n} P(B | A_i)P(A_i)$$

Proof Because the A_i's are mutually exclusive and exhaustive, if B occurs it must be in conjunction with exactly one of the A_i's. That is, $B = (A_1$ and $B)$ or $\ldots$ or $(A_n$ and $B) = (A_1 \cap B) \cup \cdots \cup (A_n \cap B)$, where the events $(A_i \cap B)$ are mutually exclusive. This "partitioning of B" is illustrated in Figure 2.11. Thus,

$$P(B) = \sum_{i=1}^{n} P(A_i \cap B) = \sum_{i=1}^{n} P(B | A_i)P(A_i)$$

as desired. ■

Figure 2.11 Partition of B by mutually exclusive and exhaustive A_i's

An example of the use of Equation (2.5) appeared in answering Question (1) of Example 2.30, where $A_1 = \{$brand 1$\}$, $A_2 = \{$brand 2$\}$, $A_3 = \{$brand 3$\}$, and $B = \{$repair$\}$.

Bayes' Theorem

> Let $A_1, A_2, \ldots, A_n$ be a collection of n mutually exclusive and exhaustive events with $P(A_i) > 0$ for $i = 1, \ldots, n$. Then for any other event B for which $P(B) > 0$
>
> $$P(A_k | B) = \frac{P(A_k \cap B)}{P(B)} = \frac{P(B | A_k)P(A_k)}{\sum_{i=1}^{n} P(B | A_i) \cdot P(A_i)} \qquad k = 1, \ldots, n \qquad (2.6)$$

The transition from the second to the third expression in (2.6) rests on using the multiplication rule in the numerator and the law of total probability in the denominator.

While the right-hand side of Expression (2.6) enables one to compute posterior probabilities by substituting prior and conditional probabilities in the appropriate places, the proliferation of events and subscripts can often intimidate a newcomer to probability. Example 2.30 shows that the posterior probabilities can be computed from a tree diagram by using the middle expression in (2.6) without ever referring explicitly to Bayes' theorem.

Example 2.31

Incidence of a rare disease. Only 1 in 1000 adults is afflicted with a rare disease for which a diagnostic test has been developed. The test is such that, when an individual actually has the disease, a positive result will occur 99% of the time, while an individual without the disease will show a positive test result only 2% of the time. If a randomly selected individual is tested and the result is positive, what is the probability that the individual has the disease?

To use Bayes' theorem, let $A_1 = \{$individual has the disease$\}$, $A_2 = \{$individual does not have the disease$\}$, and $B = \{$positive test result$\}$. Then $P(A_1) = .001$, $P(A_2) = .999$, $P(B | A_1) = .99$, and $P(B | A_2) = .02$. The tree diagram for this problem appears in Figure 2.12.

Next to each branch corresponding to a positive test result, the multiplication rule yields the recorded probabilities. Therefore, $P(B) = .00099 + .01998 = .02097$,

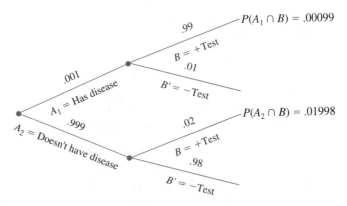

Figure 2.12 Tree diagram for the rare-disease problem

from which we have

$$P(A_1 | B) = \frac{P(A_1 \cap B)}{P(B)} = \frac{.00099}{.02097} = .047$$

This result seems counterintuitive; the diagnostic test appears so accurate we expect someone with a positive test result to be highly likely to have the disease, whereas the computed conditional probability is only .047. The reason for this seemingly paradoxical result is that, because the disease is rare and the test only moderately reliable, most positive test results arise from errors rather than from diseased individuals. The probability of having the disease has increased by a multiplicative factor of 47 (from prior .001 to posterior .047); but to get a further increase in the posterior probability, a diagnostic test with much smaller error rates is needed. If the disease were not so rare (e.g., 25% incidence in the population), then the error rates for the present test would provide good diagnoses. ∎

Exercises   Section 2.4 (46–67)

46. The head size and grip size are determined for a randomly selected tennis racket purchaser at a certain sporting goods store. Relevant probabilities appear in the accompanying table:

		Grip Size		
		$4\frac{3}{8}$ in.	$4\frac{1}{2}$ in.	$4\frac{5}{8}$ in.
Head Size	Midsize	.10	.20	.15
	Oversize	.20	.15	.20

Let A denote the event that a midsize racket was purchased and B denote the event that a racket with a $4\frac{1}{2}$-in. grip was purchased.

a. Determine $P(A)$, $P(B)$, and $P(A \cap B)$.

b. Calculate both $P(A | B)$ and $P(B | A)$ and explain in words what you have calculated in each case.

c. If C denotes the event that grip size is at least $4\frac{1}{2}$ in., calculate and interpret $P(A | C)$.

47. Suppose an individual is randomly selected from the

population of all adult males living in the United States. Let A be the event that the selected individual is over 6 ft in height and let B be the event that the selected individual is a professional basketball player. Which do you think is larger, $P(A|B)$ or $P(B|A)$, and why?

48. A mathematics professor is teaching both a morning and an afternoon section of introductory calculus. Let A = {the professor gives a bad morning lecture} and B = {the professor gives a bad afternoon lecture}. If $P(A) = .3$, $P(B) = .2$, and $P(A \cap B) = .1$, calculate the following probabilities (a Venn diagram might help):

a. $P(B|A)$ **b.** $P(B'|A)$

c. $P(B|A')$ **d.** $P(B'|A')$

e. If at the conclusion of the afternoon class, the professor is heard to mutter "what a rotten lecture," what is the probability that the morning lecture was also bad?

49. When a roulette wheel is spun once, there are 38 possible outcomes—18 red, 18 black, and 2 green (if the outcome is green, the house wins everything). If a wheel is spun twice, all (38)(38) outcomes are equally likely. If you are told that in two spins, at least one resulted in a green outcome, what is the probability that both outcomes were green?

50. Refer to Exercise 12, where A_1 = {older car is American}, A_2 = {newer car is American}, $P(A_1) = .7$, $P(A_2) = .5$, and $P(A_1 \cap A_2) = .4$. Compute the following probabilities:

a. $P(A_2|A_1)$ **b.** $P(A_1|A_2)$ **c.** $P(A_2'|A_1')$

b. $P(A_1 \cap A_2 |$ at least one car is American)

51. A large department store sells sport shirts in three sizes (small, medium, and large), three patterns (plaid, print, and stripe), and two sleeve lengths (long and short). The accompanying tables give the proportions of shirts sold falling in the various category combinations.

Short-sleeved

Size	Pl	Pr	St
	Pattern		
S	.04	.02	.05
M	.08	.07	.12
L	.03	.07	.08

Long-sleeved

Size	Pl	Pr	St
	Pattern		
S	.03	.02	.03
M	.10	.05	.07
L	.04	.02	.08

a. What is the probability that the next shirt sold is a medium, long-sleeved, print shirt?

b. What is the probability that the next shirt sold is a medium print shirt?

c. What is the probability that the next shirt sold is a short-sleeved shirt? A long-sleeved shirt?

d. What is the probability that the size of the next shirt sold is medium? That the pattern of the next shirt sold is a print?

e. Given that the shirt just sold was a short-sleeved plaid, what is the probability that its size was medium?

f. Given that the shirt just sold was a medium plaid, what is the probability that it was short-sleeved? Long-sleeved?

52. One box contains six red balls and four green balls, and a second box contains seven red balls and three green balls. A ball is randomly chosen from the first box and placed in the second box. Then a ball is randomly selected from the second box and placed in the first box.

a. What is the probability that a red ball is selected from the first box and a red ball is selected from the second box?

b. At the conclusion of the selection process, what is the probability that the numbers of red and green balls in the first box are identical to the numbers at the beginning?

53. A pumping system consists of two identical pumps, #1 and #2. If one pump fails, the system will still operate. However, because of the added strain, the extra remaining pump is now more likely to fail than was originally the case. That is, $r = P(\#2 \text{ fails} | \#1 \text{ fails}) > P(\#2 \text{ fails}) = q$. If at least one pump fails by the end of the pump design life in 7% of all systems and both pumps fail during that period in only 1%, what is the probability that pump #1 will fail during the pump design life?

54. A certain shop repairs both audio and video components. Let A denote the event that the next component brought in for repair is an audio component and let B be the event that the next component is a compact disc player (so the event B is contained in A). Suppose that $P(A) = .6$ and $P(B) = .05$. What is $P(B|A)$?

55. In Exercise 13, A_i = {awarded project i}, for $i = 1, 2, 3$. Use the probabilities given there to compute the following probabilities:

   a. $P(A_2|A_1)$ **b.** $P(A_2 \cap A_3|A_1)$

   c. $P(A_2 \cup A_3|A_1)$

   d. $P(A_1 \cap A_2 \cap A_3|A_1 \cup A_2 \cup A_3)$. Express in words the event whose probability you have calculated.

56. In Exercise 43, six people (three married couples) chose seats at random in a row consisting of six seats.

   a. Use the multiplication rule to compute the probability that Jim and Paula sit together on the far left (event A) and that John and Mary Lou (husband and wife) sit together in the middle (event B).

   b. Given that John and Mary Lou sit together in the middle, what is the probability that the two other husbands sit next to their wives?

   c. Given that John and Mary Lou sit together, what is the probability that all husbands sit next to their wives?

57. For any events A and B with $P(B) > 0$, show that $P(A|B) + P(A'|B) = 1$.

58. Show that for any three events A, B, and C with $P(C) > 0$, $P(A \cup B|C) = P(A|C) + P(B|C) - P(A \cap B|C)$.

59. At a certain gas station, 40% of the customers use regular unleaded gas (A_1), 35% use extra unleaded gas (A_2), and 25% use premium unleaded gas (A_3). Of those customers using regular gas, only 30% fill their tanks (event B). Of those customers using extra gas, 60% fill their tanks, while of those using premium, 50% fill their tanks.

   a. What is the probability that the next customer will request extra unleaded gas and fill the tank $(A_2 \cap B)$?

   b. What is the probability that the next customer fills the tank?

   c. If the next customer fills the tank, what is the probability that regular gas is requested? Extra gas? Premium gas?

60. Seventy percent of the light aircraft that disappear while in flight in a certain country are subsequently discovered. Of the aircraft that are discovered, 60% have an emergency locator, whereas 90% of the aircraft not discovered do not have such a locator. Suppose a light aircraft has disappeared.

   a. If it has an emergency locator, what is the probability that it will not be discovered?

   b. If it does not have an emergency locator, what is the probability that it will be discovered?

61. Components of a certain type are shipped to a supplier in batches of ten. Suppose that 50% of all such batches contain no defective components, 30% contain one defective component, and 20% contain two defective components. Two components from a batch are randomly selected and tested. What are the probabilities associated with 0, 1, and 2 defective components being in the batch under each of the following conditions?

   a. Neither tested component is defective.

   b. One of the two tested components is defective. (HINT: Draw a tree diagram with three first-generation branches for the three different types of batches.)

62. A company that manufactures video cameras produces a basic model and a deluxe model. Over the past year, 40% of the cameras sold have been of the basic model. Of those buying the basic model, 30% purchase an extended warranty, whereas 50% of all deluxe purchasers do so. If you learn that a randomly selected purchaser has an extended warranty, how likely is it that he/she has a basic model?

63. For customers purchasing a full set of tires at a particular tire store, consider the events

   A = {tires purchased were made in the United States}

   B = {purchaser has tires balanced immediately}

   C = {purchaser requests front-end alignment}

   along with A', B', and C'. Assume the following unconditional and conditional probabilities:

   $P(A) = .75$, $P(B|A) = .9$, $P(B|A') = .8$,
   $P(C|A \cap B) = .8$, $P(C|A \cap B') = .6$,
   $P(C|A' \cap B) = .7$, $P(C|A' \cap B') = .3$

   a. Construct a tree diagram consisting of first-,

second-, and third-generation branches and place an event label and appropriate probability next to each branch.

b. Compute $P(A \cap B \cap C)$.

c. Compute $P(B \cap C)$.

d. Compute $P(C)$.

e. Compute $P(A \mid B \cap C)$, the probability of a purchase of U.S. tires given that both balancing and an alignment were requested.

64. In Example 2.31, suppose that the incidence rate for the disease is 1 in 25 rather than 1 in 1000. What then is the probability of a positive test result? Given that the test result is positive, what is the probability that the individual has the disease? Given a negative test result, what is the probability that the individual does not have the disease?

65. At a large university, in the never-ending quest for a satisfactory textbook, the Statistics Department has tried a different text during each of the last three quarters. During the fall quarter, 500 students used the text by Professor Mean; during the winter quarter, 300 students used the text by Professor Median; and during the spring quarter, 200 students used the text by Professor Mode. A survey at the end of each quarter showed that 200 students were satisfied with Mean's book, 150 were satisfied with Median's book, and 160 were satisfied with Mode's book. If a student who took statistics during one of these quarters is selected at random and admits to having been satisfied with the text, is the student most likely to have used the book by Mean, Median, or Mode? Who is the least likely author? (HINT: Draw a tree diagram or use Bayes' theorem.)

66. A friend who works in a big city owns two cars, one small and one large. Three-quarters of the time he drives the small car to work, and one-quarter of the time he takes the large car. If he takes the small car, he usually has little trouble parking, and so is at work on time with probability .9. If he takes the large car, he is on time to work with probability .6. Given that he was on time on a particular morning, what is the probability that he drove the small car?

67. In Exercise 59, consider the following additional information on credit card usage:

70% of all regular fill-up customers use a credit card.

50% of all regular non-fill-up customers use a credit card.

60% of all extra fill-up customers use a credit card.

50% of all extra non-fill-up customers use a credit card.

50% of all premium fill-up customers use a credit card.

40% of all premium non-fill-up customers use a credit card.

Compute the probability of each of the following events for the next customer to arrive (a tree diagram might help).

a. {extra and fill-up and credit card}

b. {premium and non-fill-up and credit card}

c. {premium and credit card}

d. {fill-up and credit card}

e. {credit card}

f. If the next customer uses a credit card, what is the probability that premium was requested?

2.5 ## Independence

The definition of conditional probability enables us to revise the probability $P(A)$ originally assigned to A when we are subsequently informed that another event B has occurred; the new probability of A is $P(A \mid B)$. In our examples, it was frequently the case that $P(A \mid B)$ was unequal to the unconditional probability $P(A)$, indicating that the information "B has occurred" resulted in a change in the chance of A occurring. There are other situations, though, in which the chance that A will occur or has occurred is not affected by knowledge that B has occurred, so that $P(A \mid B) = P(A)$. It is then natural to think of A and B as independent events, meaning that the occurrence or nonoccurrence of one event has no bearing on the chance that the other will occur.

Definition

Two events A and B are **independent** if $P(A \mid B) = P(A)$ and are **dependent** otherwise.

The definition of independence might seem "unsymmetric" because we do not demand that $P(B \mid A) = P(B)$ also. However, using the definition of conditional probability and the multiplication rule,

$$P(B \mid A) = \frac{P(A \cap B)}{P(A)} = \frac{P(A \mid B)P(B)}{P(A)} \tag{2.7}$$

The right-hand side of Equation (2.7) is $P(B)$ if and only if $P(A \mid B) = P(A)$ (independence), so the equality in the definition implies the other equality (and vice versa). It is also straightforward to show that if A and B are independent, then so are the following pairs of events: (1) A' and B, (2) A and B', and (3) A' and B'.

Example 2.32

Consider tossing a fair six-sided die once and define events $A = \{2, 4, 6\}$, $B = \{1, 2, 3\}$, and $C = \{1, 2, 3, 4\}$. We then have $P(A) = \frac{1}{2}$, $P(A \mid B) = \frac{1}{3}$, and $P(A \mid C) = \frac{1}{2}$. That is, events A and B are dependent while events A and C are independent. Intuitively, if such a die is tossed and we are informed that the outcome was either 1, 2, 3, or 4 (C has occurred), then the probability that A occurred is $\frac{1}{2}$, as it originally was, since two of the four relevant outcomes are even and the outcomes are still equally likely. ∎

Example 2.33

Let A and B be any two mutually exclusive events with $P(A) > 0$. For example, for a randomly chosen automobile, let $A = \{$the car has four cylinders$\}$ and $B = \{$the car has six cylinders$\}$. Since the events are mutually exclusive, if B occurs, then A cannot possibly have occurred, so $P(A \mid B) = 0 \neq P(A)$. The message here is that *if two events are mutually exclusive, they cannot be independent*. When A and B are mutually exclusive, the information that A occurred says something about B (it cannot have occurred), so independence is precluded. ∎

$P(A \cap B)$ When Events Are Independent

Frequently the nature of an experiment suggests that two events A and B should be assumed independent. This is the case, for example, if a manufacturer receives a circuit board from each of two different suppliers, each board is tested on arrival, and $A = \{$first is defective$\}$ and $B = \{$second is defective$\}$. If $P(A) = .1$, it should also be the case that $P(A \mid B) = .1$; knowing the condition of the second board shouldn't provide information about the condition of the first. Our next result shows how to compute $P(A \cap B)$ when the events are independent.

Proposition

A and B are independent if and only if

$$P(A \cap B) = P(A) \cdot P(B) \tag{2.8}$$

To paraphrase the proposition, A and B are independent events iff* the probability that they both occur $(A \cap B)$ is the product of the two individual probabilities. The verification is as follows:

$$P(A \cap B) = P(A \mid B) \cdot P(B) = P(A) \cdot P(B) \tag{2.9}$$

where the second equality in Equation (2.9) is valid iff A and B are independent. Because of the equivalence of independence with Equation (2.8), the latter can be used as a definition of independence.

Example 2.34   It is known that 30% of a certain company's washing machines require service while under warranty, whereas only 10% of its dryers need such service. If someone purchases both a washer and a dryer made by this company, what is the probability that both machines need warranty service?

Let A denote the event that the washer needs service while under warranty and B be defined analogously for the dryer. Then $P(A) = .30$ and $P(B) = .10$. Assuming that the two machines function independently of one another, the desired probability is

$$P(A \cap B) = P(A) \cdot P(B) = (.30)(.10) = .03$$

The probability that neither machine needs service is

$$P(A' \cap B') = P(A') \cdot P(B') = (.70)(.90) = .63 \qquad \blacksquare$$

Example 2.35   Each day, Monday through Friday, a batch of components sent by a first supplier arrives at a certain inspection facility. Two days a week, a batch also arrives from a second supplier. Eighty percent of all supplier 1's batches pass inspection, and 90% of supplier 2's do likewise. What is the probability that, on a randomly selected day, two batches pass inspection? We will answer this assuming that on days when two batches are tested, whether or not the first passes is independent of whether or not the second does so. Figure 2.13 displays the relevant information.

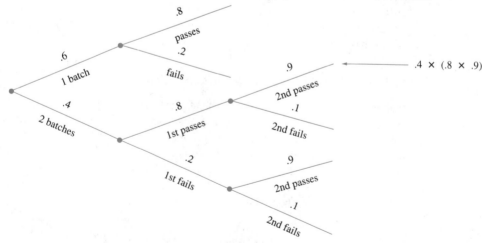

Figure 2.13 Tree diagram for Example 2.35

---

*iff is an abbreviation for "if and only if."

$$P(\text{two pass}) = P(\text{two received} \cap \text{both pass})$$

$$= P(\text{both pass} \mid \text{two received}) \cdot P(\text{two received})$$

$$= [(.8)(.9)](.4) = .288 \qquad \blacksquare$$

Independence of More Than Two Events

The notion of independence of two events can be extended to collections of more than two events. While it is possible to extend the definition for two independent events by working in terms of conditional and unconditional probabilities, it is more direct and less cumbersome to proceed along the lines of the last proposition.

Definition

> Events $A_1, \ldots, A_n$ are **mutually independent** if for every k ($k = 2, 3, \ldots, n$) and every subset of indices $i_1, i_2, \ldots, i_k$,
>
> $$P(A_{i_1} \cap A_{i_2} \cap \cdots \cap A_{i_k}) = P(A_{i_1}) \cdot P(A_{i_2}) \cdot \cdots \cdot P(A_{i_k}).$$

To paraphrase the definition, the events are mutually independent if the probability of the intersection of any subset of the n events is equal to the product of the individual probabilities. As was the case with two events, we frequently specify at the outset of a problem the independence of certain events. The definition can then be used to calculate the probability of an intersection.

Example 2.36

A system consists of four components as illustrated in Figure 2.14. The entire system will work if either the 1–2 subsystem works or if the 3–4 subsystem works (since the two subsystems are connected in parallel). Since the two components in each subsystem are connected in series, a subsystem will work only if both its components work. If components work or fail independently of one another and if each works with probability .9, what is the probability that the entire system will work (the system reliability coefficient)? Letting A_i ($i = 1, 2, 3, 4$) be the event that the ith component works, the A_i's are mutually independent. The event that the 1–2 subsystem works is $A_1 \cap A_2$, and similarly, $A_3 \cap A_4$ denotes the event that the 3–4 subsystem works. The event that the entire system works is $(A_1 \cap A_2) \cup (A_3 \cap A_4)$, so

$$P[(A_1 \cap A_2) \cup (A_3 \cap A_4)]$$

$$= P(A_1 \cap A_2) + P(A_3 \cap A_4) - P[(A_1 \cap A_2) \cap (A_3 \cap A_4)]$$

$$= P(A_1) \cdot P(A_2) + P(A_3) \cdot P(A_4) - P(A_1) \cdot P(A_2) \cdot P(A_3) \cdot P(A_4)$$

$$= (.9)(.9) + (.9)(.9) - (.9)(.9)(.9)(.9) = .9639$$

Figure 2.14 The system of components for Example 2.36

Letting $x = P(A_i)$, for $i = 1, 2, 3, 4$, what value of x would yield a system reliability of .99? Proceeding analogously, $P(\text{system works}) = x^2 + x^2 - x^4 = .99$ or $y^2 - 2y + .99 = 0$ where $y = x^2$. Solving this quadratic gives $y = .9$, so $x = \sqrt{.9} \approx .95$ (recall that $\approx$ means approximately equal). To achieve a system reliability of .99 without introducing more components or changing the system configuration, the reliability of each component would have to be increased from .9 to .95. ∎

Exercises   Section 2.5 (68–84)

68. If $P(A_1) = .7$, $P(A_2) = .5$, and $P(A_1 \cap A_2) = .4$ as in Exercise 12, show that A_1 and A_2 are dependent first by using the definition of independence and then by verifying that the multiplication property does not hold.

69. In Exercise 13, is any A_i independent of any other A_j? Answer using the multiplication property for independent events.

70. If A and B are independent events, show that A' and B are also independent. [HINT: $P(A' \cap B) = P(B) - P(A \cap B)$.]

71. An executive has both a morning and an afternoon meeting on a particular day. Let $A = \{$late to the morning meeting$\}$ and $B = \{$late to the afternoon meeting$\}$.

a. If $P(A) = .4$, $P(B) = .5$, and $P(A \cap B) = .25$, are A and B independent events?

b. If A and B are independent events with $P(A) = .4$ and $P(B) = .5$, what is the probability that the executive is on time to both meetings? To exactly one meeting?

72. The probability that a grader will make a marking error on any particular question of a multiple-choice exam is .1. If there are ten questions and questions are marked independently, what is the probability that no errors are made? That at least one error is made? If there are n questions and the probability of a marking error is p rather than .1, give expressions for these two probabilities.

73. An aircraft seam requires 25 rivets. The seam will have to be reworked if any of these rivets is defective. Suppose rivets are defective independently of one another, each with the same probability.

a. If 14% of all seams need reworking, what is the probability that a rivet is defective?

b. How small should the probability of a defective rivet be to ensure that only 10% of all seams need reworking?

74. A boiler has five identical relief valves. The probability that any particular valve will open on demand is .95. Assuming independent operation of the valves, calculate $P(\text{at least one valve opens})$ and $P(\text{at least one valve fails to open})$.

75. Two pumps connected in parallel fail independently of one another on any given day. The probability that only the older pump will fail is .10, and the probability that only the newer pump will fail is .05. What is the probability that the pumping system will fail on any given day (which happens if both pumps fail)?

76. Consider the system of components connected as in the accompanying picture. Components 1 and 2 are connected in parallel, so that subsystem works iff either 1 or 2 works; since 3 and 4 are connected in series, that subsystem works iff both 3 and 4 work. If components work independently of one another and $P(\text{component works}) = .9$, calculate $P(\text{system works})$.

77. Components arriving at a distributor are checked for defects by two different inspectors (each component is checked by both inspectors). The first inspector detects 90% of all defectives that are present, and the second inspector does likewise. At least one inspector does not detect a defect on 20% of all defective components. What is the probability that

a. A defective component will be detected only by the first inspector? By exactly one of the two inspectors?

b. All three defective components in a batch escape detection by both inspectors (assuming inspections of different components are independent of one another)?

78. Sixty percent of all vehicles examined at a certain emissions inspection station pass the inspection. Assuming that successive vehicles pass or fail independently of one another, calculate the following probabilities.

a. P(all of the next three vehicles inspected pass)

b. P(at least one of the next three inspected fails)

c. P(exactly one of the next three inspected passes) (HINT: With A = {exactly one passes}, $A = E_1 \cup E_2 \cup E_3$ where E_i = {ith passes and the other two do not}.)

d. P(at most one of the next three vehicles inspected passes)

e. Given that at least one of the next three vehicles passes inspection, what is the probability that all three pass (a conditional probability)?

79. A quality control inspector is inspecting newly produced items for faults. The inspector searches an item for faults in a series of independent fixations, each of a fixed duration. Given that a flaw is actually present, let p denote the probability that the flaw is detected during any one fixation (this model is discussed in "Human Performance in Sampling Inspection," *Human Factors,* 1979, pp. 99–105).

a. Assuming that an item has a flaw, what is the probability that it is detected by the end of the second fixation (once a flaw has been detected, the sequence of fixations terminates)? (HINT: {flaw detected in at most two fixations} = {flaw detected on the first fixation} $\cup$ {flaw undetected on the first and detected on the second}.)

b. Give an expression for the probability that a flaw will be detected by the end of the nth fixation.

c. If when a flaw has not been detected in three fixations, the item is passed, what is the probability that a flawed item will pass inspection?

d. Suppose 10% of all items contain a flaw [P(randomly chosen item is flawed) = .1]. With the assumption of part (c), what is the probability that a randomly chosen item will pass inspection (it will automatically pass if it is not flawed, but could also pass if it is flawed)?

e. Given that an item has passed inspection (no flaws in three fixations), what is the probability that it is actually flawed? Calculate for p = .5.

80. a. A lumber company has just taken delivery on a lot of 10,000 2 × 4 boards. Suppose that 20% of these boards (2000) are actually too green to be used in first-quality construction. Two boards are selected at random, one after the other. Let A = {the first board is green} and B = {the second board is green}. Compute $P(A)$, $P(B)$, and $P(A \cap B)$ (a tree diagram might help). Are A and B independent?

b. With A and B independent and $P(A) = P(B) = .2$, what is $P(A \cap B)$? How much difference is there between this answer and $P(A \cap B)$ in part (a)? For purposes of calculating $P(A \cap B)$, can we assume that A and B of part (a) are independent to obtain essentially the correct probability?

c. Suppose the lot consists of ten boards, of which two are green. Does the assumption of independence now yield approximately the correct answer for $P(A \cap B)$? What is the critical difference between the situation here and that of part (a)? When do you think that an independence assumption would be valid in obtaining an approximately correct answer to $P(A \cap B)$?

81. Refer to the assumptions stated in Exercise 76 and answer the question posed there for the system in the accompanying picture. How would the probability change if this were a subsystem connected in parallel to the subsystem pictured in Example 2.36?

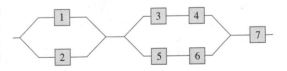

82. Professor Stan der Deviation can take one of two routes on his way home from work. On the first route, there are four railroad crossings. The probability that he will be stopped by a train at any particular one of the crossings is .1, and trains operate independently at the four crossings. The other route is longer but there are only two crossings, independent of one another, with the same stoppage probability for each as on the first route. On a particular day, Professor Deviation has a meeting scheduled at home for a certain time. Whichever route he takes, he calculates that he will be late if he is stopped by trains at at least half the crossings encountered.

a. Which route should he take to minimize the probability of being late to the meeting?

b. If he tosses a fair coin to decide on a route and he is late, what is the probability that he took the four-crossing route?

83. In Exercise 6, suppose Jack Twain is at location *A* on the diagram and that his sister Jill is at location *I*. At exactly the same instant, Jack and Jill decide to leave their respective locations and journey toward one another (Jack traveling north and east only and Jill traveling south and west). If they walk at exactly the same pace and toss a fair coin independently at each corner location at which there is a choice of direction, what is the probability that the Twains meet at *G*? What is the probability that the Twains do not meet?

84. Suppose identical tags are placed on both the left ear and the right ear of a fox. The fox is then let loose for a period of time. Consider the two events C_1 = {left-ear tag is lost} and C_2 = {right-ear tag is lost}. Let $\pi = P(C_1) = P(C_2)$ and assume C_1 and C_2 are independent events. Derive an expression (involving π) for the probability that exactly one tag is lost given that at most one is lost ("Ear Tag Loss in Red Foxes," *J. Wildlife Mgmt.,* 1976, pp. 164–167). (HINT: Draw a tree diagram in which the two initial branches refer to whether or not the left-ear tag was lost.)

Supplementary Exercises for Chapter 2 (85–110)

85. A small manufacturing company will start operating a night shift. There are 20 machinists employed by the company.

a. If a night crew consists of 3 machinists, how many different crews are possible?

b. If the machinists are ranked 1, 2, . . . , 20 in order of competence, how many of these crews would not have the best machinist?

c. How many of the crews would have at least 1 of the 10 best machinists?

d. If one of these crews is selected at random to work on a particular night, what is the probability that the best machinist will not work that night?

86. A certain model of car comes in a two-door version, a four-door version, and a hatchback version. Each version can be equipped with either an automatic transmission or a manual transmission. The accompanying table gives the relevant proportions:

A customer who has purchased one of these cars is randomly selected.

a. What is the probability that this customer purchased a car with an automatic transmission? A four-door car?

b. Given that the customer purchased a four-door car, what is the probability that it has an automatic transmission?

c. Given that the customer did not purchase a hatchback, what is the probability that the car has a manual transmission? How does this conditional probability compare to the (unconditional) probability that the car has a manual transmission?

87. A claims officer at a Social Security office will have time to examine six claims during a particular day. There are ten claims on her desk, of which four concern disability and six concern old-age benefits. If claims are selected at random from the ten, what is the probability that all disability claims will have been examined by the end of the day? What is the probability that by the end of the day, only one of the two types of claims will remain on her desk?

88. Let *A* be the event that a certain book on fluid mechanics is presently checked out of the university

		Version		
		TD	*FD*	*HB*
Transmission type	*A*	.32	.27	.18
	M	.08	.04	.11

library and let B be the event that another book on the same subject is checked out.

a. If $P(A) = .5$, $P(B) = .4$, and $P(A \cup B) = .65$, compute $P(A \cap B)$.

b. Using the probabilities of part (a), compute P(exactly one of the two books is checked out).

c. If $P(A \cup B) = .7$, $P(A \cap B) = .2$, and P(only the first book is checked out) $= .4$, compute $P(A)$ and $P(B)$. (HINT: Venn diagram.)

d. If $P(A \cup B) = .7$, $P(A \cap B) = .2$, and P(exactly one book is checked out) $= .5$, can $P(A)$ and $P(B)$ be determined?

89. Suppose I fly to New York on airline X and return to Los Angeles on airline Y. Let $A = \{X \text{ loses my baggage}\}$ and $B = \{Y \text{ loses my baggage}\}$. If A and B are independent events with $P(A) > P(B)$, $P(A \cap B) = .0002$, and $P(A \cup B) = .03$, determine $P(A)$ and $P(B)$.

90. A transmitter is sending a message by using a binary code, namely, a sequence of 0's and 1's. Each transmitted bit (0 or 1) must pass through three relays in order to reach the receiver. At each relay, the probability is .20 that the bit sent on will be different from the bit received (a reversal). Assume that the relays operate independently of one another.

Transmitter $\rightarrow$ Relay 1 $\rightarrow$ Relay 2 $\rightarrow$ Relay 3
$\rightarrow$ Receiver

a. If a 1 is sent from the transmitter, what is the probability that a 1 is sent on by all three relays?

b. If a 1 is sent from the transmitter, what is the probability that a 1 is received by the receiver? (HINT: The eight experimental outcomes can be displayed on a tree diagram with three generations of branches, one generation for each relay.)

c. Suppose 70% of all bits sent from the transmitter are 1's. If a 1 is received by the receiver, what is the probability that a 1 was sent?

91. Individual A has a circle of five close friends (B, C, D, E, and F). A has heard a certain rumor from outside the circle and has invited the five friends to a party in order to circulate the rumor. To begin, A selects one of the five at random and tells the rumor to the chosen individual. That individual then selects at random one of the four remaining individuals and repeats the rumor. Continuing, a new individual is selected from those not already having heard the rumor by the individual who has just heard it, until everyone has been told.

a. What is the probability that the rumor is repeated in the order B, C, D, E, and F?

b. What is the probability that F is the third person at the party to be told the rumor?

c. What is the probability that F is the last person to hear the rumor?

92. Refer to Exercise 91. If at each stage the person who currently "has" the rumor does not know who has already heard it and selects the next recipient at random from all five possible individuals, what is the probability that F has still not heard the rumor after it has been told ten times at the party?

93. An automobile insurance company classifies each driver as a good risk (A_1), a medium risk (A_2), or a poor risk (A_3). Of those currently insured, 30% are good risks, 50% are medium risks, and 20% are poor risks. In any given year, the probability that a driver will have at least one citation is .1 for a good risk, .3 for a medium risk, and .5 for a poor risk. If a randomly selected driver insured by this company has at least one citation during the next year, what is the probability that the driver was actually a good risk? A medium risk?

94. In Exercise 93, suppose that each insuree has a three-year policy. If accidents are independent from year to year and a randomly selected driver reports no citations during the three years, what is the probability that the driver was actually a good risk?

95. A chemical engineer is interested in determining if a certain trace impurity is present in a product. An experiment has a probability of .80 of detecting the impurity if it is present. The probability of not detecting the impurity if it is absent is .90. The prior probabilities of the impurity being present and being absent are .40 and .60, respectively. Three separate experiments result in only two detections. What is the posterior probability that the impurity is present?

96. Each contestant on a quiz show is asked to specify one of six possible categories from which questions will be asked. Suppose P(contestant requests category i) $= \frac{1}{6}$ and successive contestants choose their categories independently of one another. If there are three contestants on each show and all three contestants on a particular show select different categories, what is the probability that exactly one has selected category 1?

97. One percent of all individuals in a certain population are carriers of a particular disease. A diagnostic test for this disease has a 90% detection rate for carriers and a

5% detection rate for noncarriers. Suppose the test is applied independently to two different blood samples from the same randomly selected individual.

a. What is the probability that both tests yield the same result?

b. If both tests are positive, what is the probability that the selected individual is a carrier?

98. A system consists of two components. The probability that the second component functions in a satisfactory manner during its design life is .9, the probability that at least one of the two components does so is .96, and the probability that both components do so is .75. Given that the first component functions in a satisfactory manner throughout its design life, what is the probability that the second one does also?

99. A certain company sends 40% of its overnight mail parcels via express mail service E_1. Of these parcels, 2% arrive after the guaranteed delivery time (denote by L the event "late delivery"). If a record of an overnight mailing is randomly selected from the company's file, what is the probability that the parcel went via E_1 and was late?

100. Return to Exercise 99 and suppose that 50% of the overnight parcels are sent via express mail service E_2 and the remaining 10% are sent via E_3. Of those sent via E_2, only 1% arrive late, whereas 5% of the parcels handled by E_3 arrive late.

a. What is the probability that a randomly selected parcel arrived late?

b. If a randomly selected parcel has arrived on time, what is the probability that it was not sent via E_1?

101. A company uses three different assembly lines—A_1, A_2, and A_3—to manufacture a particular component. Of those manufactured by line A_1, 5% need rework to remedy a defect, whereas 8% of A_2's components need rework and 10% of A_3's need rework. Suppose that 50% of all components are produced by line A_1, while 30% are produced by line A_2 and 20% come from line A_3. If a randomly selected component needs rework, what is the probability that it came from line A_1? From line A_2? From line A_3?

102. Disregarding the possibility of a February 29 birthday, suppose a randomly selected individual is equally likely to have been born on any one of the other 365 days.

a. If ten people are randomly selected, what is the probability that all have different birthdays? That at least two have the same birthday?

b. With k replacing ten in part (a), what is the smallest k for which there is at least a 50–50 chance that two or more people will have the same birthday?

c. If ten people are randomly selected, what is the probability that either at least two have the same birthday or that at least two have the same last three digits of their Social Security numbers?
[NOTE: The article "Methods for Studying Coincidences" (F. Mosteller and P. Diaconis, *J. Amer. Stat. Assoc.*, 1989, pp. 853–861) discusses problems of this type.]

103. One method used to distinguish between granitic (G) and basaltic (B) rocks is to examine a portion of the infrared spectrum of the sun's energy reflected from the rock surface. Let R_1, R_2, and R_3 denote measured spectrum intensities at three different wavelengths; typically, for granite $R_1 < R_2 < R_3$ whereas for basalt $R_3 < R_1 < R_2$. When measurements are made remotely (using aircraft), various orderings of the R_i's may arise whether the rock is basalt or granite. Flights over regions of known composition have yielded the following information:

	Granite	*Basalt*
$R_1 < R_2 < R_3$	60%	10%
$R_1 < R_3 < R_2$	25%	20%
$R_3 < R_1 < R_2$	15%	70%

Suppose that for a randomly selected rock in a certain region, $P(\text{granite}) = .25$ and $P(\text{basalt}) = .75$.

a. Show that $P(\text{granite} \mid R_1 < R_2 < R_3) > P(\text{basalt} \mid R_1 < R_2 < R_3)$. If measurements yielded $R_1 < R_2 < R_3$, would you classify the rock as granite or basalt?

b. If measurements yielded $R_1 < R_3 < R_2$, how would you classify the rock? Answer the same question for $R_3 < R_1 < R_2$.

c. Using the classification rules indicated in parts (a) and (b), when selecting a rock from this region, what is the probability of an erroneous classification? [HINT: Either G could be classified as B or B as G, and $P(B)$ and $P(G)$ are known.]

d. If $P(\text{granite}) = p$ rather than .25, are there values of p (other than 1) for which one would always classify a rock as granite?

104. A subject is allowed a sequence of glimpses to detect a target. Let $G_i = \{\text{the target is detected on the } i\text{th glimpse}\}$, with $p_i = P(G_i)$. Suppose the G_i's are independent events and write an expression for the probability that the target has been detected by the end of the nth glimpse. (NOTE: This model is discussed in "Predicting Aircraft Detectability," *Human Factors,* 1979, pp. 277–291.)

105. In a Little League baseball game, team A's pitcher throws a strike 50% of the time and a ball 50% of the time, successive pitches are independent of one another, and the pitcher never hits a batter. Knowing this, team B's manager has instructed the first batter not to swing at anything. Calculate the probability that

a. The batter walks on the fourth pitch.

b. The batter walks on the sixth pitch (so two of the first five must be strikes), using a counting argument or constructing a tree diagram.

c. The batter walks.

d. What is the probability that the first batter up scores while no one is out (assuming that each batter pursues a no-swing strategy)?

106. Four engineers, A, B, C, and D, have been scheduled for job interviews at 10 A.M. on Friday, January 13, at Random Sampling, Inc. The personnel manager has scheduled the four for interview rooms 1, 2, 3, and 4, respectively. However, the manager's secretary does not know this, so assigns them to the four rooms in a completely random fashion (what else!). What is the probability that

a. All four end up in the correct rooms?

b. None of the four ends up in the correct room?

107. A particular airline has 10 A.M. flights from Chicago to New York, Atlanta, and Los Angeles. Let A denote the event that the New York flight is full and define events B and C analogously for the other two flights. Suppose $P(A) = .6$, $P(B) = .5$, $P(C) = .4$ and the three events are independent. What is the probability that

a. All three flights are full? That at least one flight is not full?

b. Only the New York flight is full? That exactly one of the three flights is full?

108. A personnel manager is to interview four candidates for a job. These are ranked 1, 2, 3, and 4 in order of preference and will be interviewed in random order. However, at the conclusion of each interview, the manager will know only how the current candidate compares to those previously interviewed. For example, the interview order 3, 4, 1, 2 generates no information after the first interview, shows that the second candidate is worse than the first, and that the third is better than the first two. However, the order 3, 4, 2, 1 would generate the same information after each of the first three interviews. The manager wants to hire the best candidate, but must make an irrevocable hire/no hire decision after each interview. Consider the following strategy: Automatically reject the first s candidates and then hire the first subsequent candidate who is best among those already interviewed (if no such candidate appears, the last one interviewed is hired).

For example, with $s = 2$, the order 3, 4, 1, 2 would result in the best being hired, while 3, 1, 2, 4 would not. Of the four possible s values (0, 1, 2, and 3), which one maximizes $P(\text{best is hired})$? (HINT: Write out the 24 equally likely interview orderings: $s = 0$ means that the first candidate is automatically hired.)

109. Consider four independent events A_1, A_2, A_3, and A_4 and let $p_i = P(A_i)$ for $i = 1, 2, 3, 4$. Express the probability that at least one of these four events occurs in terms of the p_i's and do the same for the probability that at least two of the events occur.

110. A box contains the following four slips of paper, each having exactly the same dimensions: (1) win prize 1; (2) win prize 2; (3) win prize 3; (4) win prizes 1, 2, and 3. One slip will be randomly selected. Let $A_1 = \{\text{win prize 1}\}$, $A_2 = \{\text{win prize 2}\}$, and $A_3 = \{\text{win prize 3}\}$. Show that A_1 and A_2 are independent, that A_1 and A_3 are independent, and that A_2 and A_3 are also independent (this is *pairwise* independence). However, show that $P(A_1 \cap A_2 \cap A_3) \neq P(A_1) \cdot P(A_2) \cdot P(A_3)$, so the three events are *not* mutually independent.

Bibliography

Olkin, Ingram, Cyrus Derman, and Leon Gleser (2nd ed.), *Probability Models and Applications,* Macmillan, New York, 1994. A comprehensive introduction to probability, written at a slightly higher mathematical level than this text but containing many good examples.

Durrett, Richard, *The Essentials of Probability,* Duxbury Press, Belmont, CA, 1993. A concise presentation at a slightly higher level than this text.

Mosteller, Frederick, Robert Rourke, and George Thomas, *Probability with Statistical Applications* (2nd ed.), Addison-Wesley, Reading, MA, 1970. A very good precalculus introduction to probability, with many entertaining examples; especially good on counting rules and their application.

Ross, Sheldon, *A First Course in Probability* (4th ed.), Macmillan, New York, 1994. Rather tightly written and more mathematically sophisticated than this text, but contains a wealth of interesting examples and exercises.

Winkler, Robert, *Introduction to Bayesian Inference and Decision,* Holt, Rinehart & Winston, New York, 1972. A very good introduction to subjective probability.

3 Discrete Random Variables and Probability Distributions

$$\Sigma\Sigma\Sigma\Sigma\Sigma\Sigma\Sigma\Sigma\Sigma\Sigma\Sigma\Sigma\Sigma\Sigma\Sigma\Sigma$$

Introduction

Whether an experiment yields qualitative or quantitative outcomes, methods of statistical analysis require that we focus on certain numerical aspects of the data (such as a sample proportion x/n, mean $\bar{x}$, or standard deviation s). The concept of a random variable allows us to pass from the experimental outcomes themselves to a numerical function of the outcomes. There are two fundamentally different types of random variables—discrete random variables and continuous random variables. In this chapter, we examine the basic properties and discuss the most important examples of discrete variables, and we will study continuous variables in Chapter 4.

3.1 Random Variables

In any experiment, there are numerous characteristics that can be observed or measured, but in most cases an experimenter will focus on some specific aspect or aspects of a sample. For example, in a study of commuting patterns in a metropolitan area, each individual in a sample might be asked about commuting distance and the number of people commuting in the same vehicle, but not about IQ, income, family size, and other such characteristics. Alternatively, a researcher may test a sample of components and record only the number that have failed within 1000 hours, rather than recording the individual failure times.

In general, each outcome of an experiment can be associated with a number by specifying a rule of association (e.g., the number among the sample of ten components that fail to last 1000 hours or the total weight of baggage for a sample of 25 airline passengers). Such a rule of association is called a **random variable**—

a variable because different numerical values are possible and random because the observed value depends on which of the possible experimental outcomes results (Figure 3.1).

Figure 3.1 A random variable

Definition

For a given sample space $\mathscr{S}$ of some experiment, a **random variable** is any rule that associates a number with each outcome in $\mathscr{S}$.

We will often use the abbreviation rv in place of random variable. As is customary in probability and statistics, capital letters such as X or Y will be used to denote random variables. The notation $X(s) = x$ means that x is the number associated with the outcome s by the rv X, so x is called the value of the variable associated with s.

Example 3.1   When a student attempts to log on to a computer time-sharing system, either all ports could be busy (F), in which case the student will fail to obtain access, or else there will be at least one port free (S), in which case the student will be successful in accessing the system. With $\mathscr{S} = \{S, F\}$, define an rv X by

$$X(S) = 1, \qquad X(F) = 0$$

The rv X indicates whether (1) or not (0) the student can log on.   ■

In Example 3.1, the rv X was specified by explicitly listing each element of $\mathscr{S}$ and the associated number. If $\mathscr{S}$ contains more than a few outcomes, such a listing is tedious, but it can frequently be avoided.

Example 3.2   Consider the experiment in which a telephone number in a certain area code is dialed using a random number dialer (such devices are used extensively by polling organizations), and define an rv Y by

$$Y = \begin{cases} 1 & \text{if the selected number is unlisted} \\ 0 & \text{if the selected number is listed in the directory} \end{cases}$$

For example, if 5282966 appears in the telephone directory, then $Y(5282966) = 0$, whereas $Y(7727350) = 1$ tells us that the number 7727350 is unlisted. A word description of this sort is more economical than a complete listing, so we will use such a description whenever possible.   ■

In Examples 3.1 and 3.2, the only possible values of the random variable were 0 and 1. Such a random variable arises frequently enough to be given a special name, after the individual who first studied it.

Definition

> Any random variable whose only possible values are 0 and 1 is called a
> **Bernoulli random variable.**

We will often want to define and study several different random variables from the same sample space.

Example 3.3   Example 2.3 described an experiment in which the number of pumps in use at each of two gas stations was determined. Define rv's X, Y, and U by

X = the total number of pumps in use at the two stations

Y = the difference between the number of pumps in use at station 1 and the number in use at station 2

U = the maximum of the numbers of pumps in use at the two stations

If this experiment is performed and $s = (2, 3)$ results, then $X((2, 3)) = 2 + 3 = 5$, so we say that the observed value of X was $x = 5$. Similarly, the observed value of Y would be $y = 2 - 3 = -1$, and the observed value of U would be $u = \max(2, 3) = 3$. ■

An Infinite Set of Possible X Values

Each of the random variables of Examples 3.1–3.3 can assume only a finite number of possible values. This need not be the case.

Example 3.4   In Example 2.4, we considered the experiment in which batteries coming off an assembly line were examined until a good one (S) was obtained. The sample space was $\mathcal{S} = \{S, FS, FFS, \ldots\}$. Define an rv X by

X = the number of batteries examined before the experiment terminates

Then $X(S) = 1$, $X(FS) = 2$, $X(FFS) = 3, \ldots, X(FFFFFFS) = 7$, and so on. Any positive integer is a possible value of X, so the set of possible values is infinite. ■

Example 3.5   Suppose that in some random fashion, a location (latitude and longitude) in the continental United States is selected and define an rv Y by

Y = the height above sea level at the selected location

For example, if the selected location were (39°50′N, 98°35′W), then we might have $Y((39°50′N, 98°35′W)) = 1748.26$ ft. The largest possible value of Y is 14,494 (Mt. Whitney), and the smallest possible value is −282 (Death Valley). The set of all possible values of Y is the set of all numbers in the interval between −282 and 14,494—that is,

$$\{y: y \text{ is a number}, -282 \le y \le 14{,}494\}$$

and there are an infinite number of numbers in this interval. ■

Discrete Random Variables

Although both X of Example 3.4 and Y of Example 3.5 can assume any one of an infinite number of possible values, the two infinite sets are actually quite different from one another. The variable X is more closely related to the finite-valued variables of Examples 3.1–3.3 than to Y above.

Definition

> A set is **discrete** either if it consists of a finite number of elements or if its elements can be listed in sequence so that there is a first element, a second element, a third element, and so on, in the list.

A discrete set that is infinite is sometimes called countably infinite, since according to the definition we can count (1, 2, 3, . . .) its elements. Most of the infinite discrete sets we will encounter will consist either of the nonnegative integers ($\{0, 1, 2, . . .\}$) or some subset of these integers. The set of possible values of the rv X in Example 3.4 is $D = \{1, 2, 3 . . .\}$, which is clearly discrete.

That there are infinite sets that are not discrete is a profound and important result from pure mathematics. If D is any interval of real numbers, such as $\{y: -282 \leq y \leq 14{,}494\}$ of Example 3.5, then it can be shown that there is no way to list (or count) the elements of this set. More will be said about this at the outset of Chapter 4.

Definition

> A random variable is said to be **discrete** if its set of possible values is a discrete set.

Example 3.6

All random variables of Examples 3.1–3.4 are discrete. As another example, suppose we select married couples at random and do a blood test on each person until we find a husband and wife who both have the same Rh factor. With $X =$ the number of blood tests to be performed, possible values of X are $D = \{2, 4, 6, 8, . . .\}$. Since the possible values have been listed in sequence, D is a discrete set, so X is a discrete rv. ∎

To study basic properties of discrete rv's, only the tools of discrete mathematics—summation and differences—are required. The study of continuous variables requires the continuous mathematics of the calculus—integrals and derivatives. This is the reason for making the distinction between the two types of variables.

Exercises   Section 3.1 (1–10)

1. Three automobiles are selected at random, and each is categorized as having a diesel (S) or nondiesel (F) engine (so outcomes are SSS, SSF, etc.). If $X =$ the number of cars among the three with diesel engines, list each outcome in $\mathscr{S}$ and its associated X value.

2. Give three examples of Bernoulli rv's (other than those in the text).

3. Using the experiment in Example 3.3, define two more random variables and list the possible values of each.

4. Let X = the number of nonzero digits in a randomly selected zip code. What are the possible values of X? Give three possible outcomes and their associated X values.

5. If the sample space $\mathscr{S}$ is an infinite set, does this necessarily imply that any rv X defined from $\mathscr{S}$ will have an infinite set of possible values? If yes, say why. If no, give an example.

6. Starting at a fixed time, each car entering an intersection is observed to see whether it turns left (L), right (R), or goes straight ahead (A). The experiment terminates as soon as a car is observed to turn left. Let X = the number of cars observed. What are possible X values? List five outcomes and their associated X values.

7. For each random variable defined below, describe the set of possible values for the variable and state whether the variable is discrete.

   a. X = the number of unbroken eggs in a randomly chosen standard egg carton

   b. Y = the number of students on a class list for a particular course who are absent on the first day of classes

   c. U = the number of times a duffer has to swing at a golf ball before hitting it

   d. X = the length of a randomly selected rattlesnake

   e. Z = the amount of royalties earned from the sale of a first edition of 10,000 textbooks

   f. Y = the pH of a randomly chosen soil sample

   g. X = the tension (psi) at which a randomly selected tennis racket has been strung

   h. X = the total number of coin tosses required for three individuals to obtain a match (*HHH* or *TTT*)

8. Each time a component is tested, the trial is a success (S) or failure (F). Suppose the component is tested

repeatedly until a success occurs on three *consecutive* trials. Let Y denote the number of trials necessary to achieve this. List all outcomes corresponding to the five smallest possible values of Y and state which Y value is associated with each one.

9. Consider an experiment in which an individual named Claudius is located at the point 0 in the accompanying diagram.

Using an appropriate randomization device (such as a tetrahedral die, one having four sides), Claudius first moves to one of the four locations B_1, B_2, B_3, B_4. Once at one of these locations, another randomization device is used to decide whether Claudius next returns to 0 or next visits one of the other two adjacent points. The experiment then continues in this fashion; after each move, another move to one of the (new) adjacent points is determined by tossing an appropriate die or coin.

   a. Let X = the number of moves that Claudius makes before first returning to 0. What are possible values of X? Is X discrete or continuous?

   b. If moves are allowed also along the diagonal paths connecting 0 to A_1, A_2, A_3, and A_4, respectively, answer the questions in part (a).

10. The number of pumps in use at both a six-pump station and a four-pump station will be determined. Give the possible values for each of the following random variables.

   a. T = the total number of pumps in use

   b. X = the difference between the numbers in use at stations 1 and 2

   c. U = the maximum number of pumps in use at either station

   d. Z = the number of stations having exactly two pumps in use

Probability Distributions for Discrete Random Variables

When probabilities are assigned to various outcomes in $\mathcal{S}$, these in turn determine probabilities associated with the values of any particular rv X. The *probability distribution of X* says how the total probability of 1 is distributed among (allocated to) the various possible X values.

Example 3.7

Six lots of components are ready to be shipped by a certain supplier. The number of defective components in each lot is as follows:

Lot	1 2 3 4 5 6
Number of defectives	0 2 0 1 2 0

One of these lots is to be randomly selected for shipment to a particular customer. Let X be the number of defectives in the selected lot. The three possible X values are 0, 1, and 2. Of the six equally likely simple events, three result in $X = 0$, one in $X = 1$, and the other two in $X = 2$. Let $p(0)$ denote the probability that $X = 0$ and $p(1)$ and $p(2)$ represent the probabilities of the other two possible values of X. Then

$$p(0) = P(X = 0) = P(\text{lot 1 or 3 or 6 is sent}) = \frac{3}{6} = .500$$

$$p(1) = P(X = 1) = P(\text{lot 4 is sent}) = \frac{1}{6} = .167$$

$$p(2) = P(X = 2) = P(\text{lot 2 or 5 is sent}) = \frac{2}{6} = .333$$

That is, a probability of .500 is distributed to the X value 0, a probability of .167 is placed on the X value 1, and the remaining probability, .333, is associated with the X value 2. The values of X along with their probabilities collectively specify the probability distribution or *probability mass function of X*. If this experiment were repeated over and over again, in the long run $X = 0$ would occur one-half of the time, $X = 1$ one-sixth of the time, and $X = 2$ one-third of the time. ∎

Definition

> The **probability distribution** or **probability mass function** (pmf) of a discrete rv is defined for every number x by $p(x) =$ $P(X = x) = P(\text{all } s \in \mathcal{S}: X(s) = x).$*

In words, for every possible value x of the random variable, the pmf specifies the probability of observing that value when the experiment is performed. The conditions $p(x) \geq 0$ and $\Sigma_{\text{all possible } x}\, p(x) = 1$ are required of any pmf.

---

*$P(X = x)$ is read "the probability that the rv X assumes the value x." For example, $P(X = 2)$ denotes the probability that the resulting X value is 2.

Example 3.8

Suppose we go to a university bookstore during the first week of classes and observe whether the next person buying a calculator buys an RPN type or a strictly algebraic type. Let

$$X = \begin{cases} 1 & \text{if the customer purchases an RPN type} \\ 0 & \text{if the customer purchases an algebraic type} \end{cases}$$

If 20% of all purchasers during that week select an RPN type, the pmf for X is

$$p(0) = P(X = 0) = P(\text{next customer purchases algebraic}) = .8$$
$$p(1) = P(X = 1) = P(\text{next customer purchases RPN}) = .2$$
$$p(x) = P(X = x) = 0 \qquad \text{for } x \ne 0 \text{ or } 1$$

An equivalent description is

$$p(x) = \begin{cases} .8 & \text{if } x = 0, \\ .2 & \text{if } x = 1, \\ 0 & \text{if } x \ne 0 \text{ or } 1 \end{cases}$$

A picture of this pmf, called a *line graph,* appears in Figure 3.2.

Figure 3.2 The line graph for the pmf in Example 3.8

Example 3.9

Consider a group of five potential blood donors—A, B, C, D, and E—of whom only A and B have type O+ blood. Five blood samples, one from each individual, will be typed in random order until an O+ individual is identified. Let the rv Y = the number of typings necessary to identify an O+ individual. Then the pmf of Y is

$$p(1) = P(Y = 1) = P(\text{A or B typed first}) = \frac{2}{5} = .4$$

$$p(2) = P(Y = 2) = P(\text{C, D, or E first, and then A or B})$$

$$= P(\text{C, D, or E first}) \cdot P(\text{A or B next} | \text{C, D, or E first}) = \frac{3}{5} \cdot \frac{2}{4} = .3$$

$$p(3) = P(Y = 3) = P(\text{C, D, or E first and second, and then A or B})$$

$$= \left(\frac{3}{5}\right)\left(\frac{2}{4}\right)\left(\frac{2}{3}\right) = .2$$

$$p(4) = P(Y = 4) = P(C, D, \text{ and } E \text{ all done first}) = \left(\frac{3}{5}\right)\left(\frac{2}{4}\right)\left(\frac{1}{3}\right) = .1$$

$$p(y) = 0 \qquad \text{if } y \neq 1, 2, 3, 4$$

The pmf can be presented nicely in tabular form:

y	1	2	3	4
$p(y)$	.4	.3	.2	.1

where any y value not listed receives zero probability. This pmf can also be displayed in a line graph (Figure 3.3).

Figure 3.3 The line graph for the pmf in Example 3.9   ■

The name "probability mass function" is suggested by a model used in physics for a system of "point masses." In this model, masses are distributed at various locations x along a one-dimensional axis. Our pmf describes how the total probability mass of 1 is distributed at various points along the axis of possible values of the random variable (where and how much mass at each x).

Another useful pictorial representation of a pmf, called a **probability histogram,** is similar to histograms discussed in Chapter 1. Above each y with $p(y) > 0$, construct a rectangle centered at y. The height of each rectangle is proportional to $p(y)$, and the base is the same for all rectangles. When possible values are equally spaced, the base is frequently chosen as the distance between successive y values (though it could be smaller). Figure 3.4 shows two probability histograms.

Figure 3.4 Probability histograms for Examples 3.8 and 3.9

A Parameter of a Probability Distribution

In Example 3.8, we had $p(0) = .8$ and $p(1) = .2$ because 20% of all purchasers selected an RPN calculator. At another bookstore, it may be the case that $p(0) = .9$

and $p(1) = .1$. More generally, the pmf of any Bernoulli rv can be expressed in the form $p(1) = \alpha$ and $p(0) = 1 - \alpha$, where $0 < \alpha < 1$. Because the pmf depends on the particular value of α, we often write $p(x;\, \alpha)$ rather than just $p(x)$:

$$p(x;\, \alpha) = \begin{cases} 1 - \alpha & \text{if } x = 0 \\ \alpha & \text{if } x = 1 \\ 0 & \text{otherwise} \end{cases} \qquad (3.1)$$

Then each choice of α in Expression (3.1) yields a different pmf.

Definition

> Suppose $p(x)$ depends on a quantity that can be assigned any one of a number of possible values, with each different value determining a different probability distribution. Such a quantity is called a **parameter** of the distribution. The collection of all probability distributions for different values of the parameter is called a **family** of probability distributions.

The quantity α in Expression (3.1) is a parameter. Each different number α between 0 and 1 determines a different member of a family of distributions; two such members are

$$p(x;\, .6) = \begin{cases} .4 & \text{if } x = 0 \\ .6 & \text{if } x = 1 \\ 0 & \text{otherwise} \end{cases} \qquad \text{and} \qquad p(x;\, .5) = \begin{cases} .5 & \text{if } x = 0 \\ .5 & \text{if } x = 1 \\ 0 & \text{otherwise} \end{cases}$$

Every probability distribution for a Bernoulli rv has the form (3.1), so (3.1) is called the *family of Bernoulli distributions.*

Example 3.10

Starting at a fixed time, we observe the sex of each newborn child at a certain hospital until a boy (B) is born. Let $p = P(B)$, assume that successive births are independent, and define the rv X by $X = $ number of births observed. Then

$$p(1) = P(X = 1) = P(B) = p$$
$$p(2) = P(X = 2) = P(GB) = P(G) \cdot P(B) = (1 - p)p$$

and

$$p(3) = P(X = 3) = P(GGB) = P(G) \cdot P(G) \cdot P(B) = (1 - p)^2 p$$

Continuing in this way, a general formula emerges:

$$p(x) = \begin{cases} (1 - p)^{x-1} p & x = 1, 2, 3, \dots \\ 0 & \text{otherwise} \end{cases} \qquad (3.2)$$

The quantity p in Expression (3.2) represents a number between 0 and 1 and is a parameter of the probability distribution. In the sex example, $p = .51$ might be appropriate, but if we were looking for the first child with Rh-positive blood, then we might let $p = .85$. ∎

The Cumulative Distribution Function

For some fixed value x, we often wish to compute the probability that the observed value of X will be at most x. For example, the pmf in Example 3.7 was

$$p(x) = \begin{cases} .500 & x = 0 \\ .167 & x = 1 \\ .333 & x = 2 \\ 0 & \text{otherwise} \end{cases}$$

The probability that X is at most 1 is then

$$P(X \le 1) = p(0) + p(1) = .500 + .167 = .667$$

In this example, $X \le 1.5$ iff $X \le 1$, so $P(X \le 1.5) = P(X \le 1) = .667$. Similarly, $P(X \le 0) = P(X = 0) = .5$, and $P(X \le .75) = .5$ also. Since 0 is the smallest possible value of X, $P(X \le -1.7) = 0$, $P(X \le -.0001) = 0$, and so on. The largest possible X value is 2, so $P(X \le 2) = 1$, and if x is any number larger than 2, $P(X \le x) = 1$—that is, $P(X \le 5) = 1$, $P(X \le 10.23) = 1$, and so on. Notice that $P(X < 1) = .5 \ne P(X \le 1)$, since the probability of the X value 1 is included in the latter probability but not in the former. When X is a discrete rv and x is a possible value of X, $P(X < x) < P(X \le x)$.

Definition

> The **cumulative distribution function** (cdf) $F(x)$ of a discrete rv variable X with pmf $p(x)$ is defined for every number x by
>
> $$F(x) = P(X \le x) = \sum_{y:\, y \le x} p(y)$$
>
> For any number x, $F(x)$ is the probability that the observed value of X will be at most x.

Example 3.11

The pmf of Y for Example 3.9 was

y	1	2	3	4
$p(y)$	.4	.3	.2	.1

We first determine $F(y)$ for each value in the set $\{1, 2, 3, 4\}$ of possible values:

$$F(1) = P(Y \le 1) = P(Y = 1) = p(1)$$
$$F(2) = P(Y \le 2) = P(Y = 1 \text{ or } 2) = p(1) + p(2) = .7$$
$$F(3) = P(Y \le 3) = P(Y = 1 \text{ or } 2 \text{ or } 3) = p(1) + p(2) + p(3) = .9$$
$$F(4) = P(Y \le 4) = P(Y = 1 \text{ or } 2 \text{ or } 3 \text{ or } 4) = 1$$

Now for any other number y, $F(y)$ will equal the value of F at the closest possible value of Y to the left of y. For example, $F(2.7) = P(Y \le 2.7) = P(Y \le 2) = .7$, and $F(3.999) = F(3) = .9$. The cdf is thus

$$F(y) = \begin{cases} 0 & \text{if } y < 1 \\ .4 & \text{if } 1 \le y < 2 \\ .7 & \text{if } 2 \le y < 3 \\ .9 & \text{if } 3 \le y < 4 \\ 1 & \text{if } 4 \le y \end{cases}$$ (3.3)

A graph of $F(y)$ appears in Figure 3.5.

Figure 3.5 A graph of the cdf of Example 3.11

For X a discrete rv, the graph of $F(x)$ will have a jump at every possible value of X and will be flat between possible values. Such a graph is called a **step function.**

Example 3.12   In Example 3.10, any positive integer was a possible X value, and the pmf was

$$p(x) = \begin{cases} (1 - p)^{x-1}p & x = 1, 2, 3, \ldots \\ 0 & \text{otherwise} \end{cases}$$

For any positive integer x,

$$F(x) = \sum_{y \le x} p(y) = \sum_{y = 1}^{x} (1 - p)^{y-1}p = p \sum_{y = 0}^{x-1} (1 - p)^{y}$$ (3.4)

To evaluate this sum, we use the fact that the partial sum of a geometric series is

$$\sum_{y = 0}^{k} a^{y} = \frac{1 - a^{k+1}}{1 - a}$$

Using this in (3.4), with $a = 1 - p$ and $k = x - 1$, gives

$$F(x) = p \cdot \frac{1 - (1 - p)^{x}}{1 - (1 - p)} = 1 - (1 - p)^{x}, \qquad x \text{ a positive integer}$$

Since F is constant in between positive integers,

$$F(x) = \begin{cases} 0 & x < 1 \\ 1 - (1 - p)^{[x]} & x \ge 1 \end{cases}$$ (3.5)

where $[x]$ is the largest integer $\le x$ (e.g., $[2.7] = 2$). Thus, if $p = .51$ as in the birth example, then the probability of having to examine at most five births to see the first boy is $F(5) = 1 - (.49)^{5} = 1 - .0282 = .9718$, while $F(10) \approx 1.0000$. This cdf is graphed in Figure 3.6.

Figure 3.6 A graph of $F(x)$ for Example 3.12   ■

In examples thus far, the cdf has been derived from the pmf. It is possible to reverse this procedure and obtain the pmf from the cdf whenever the latter function is available. Suppose, for example, that X represents the number of defective components in a shipment consisting of six components, so that possible X values are $0, 1, \ldots, 6$. Then

$$p(3) = P(X = 3)$$
$$= [p(0) + p(1) + p(2) + p(3)] - [p(0) + p(1) + p(2)]$$
$$= P(X \leq 3) - P(X \leq 2)$$
$$= F(3) - F(2)$$

More generally, the probability that X falls in a specified interval is easily obtained from the cdf. For example,

$$P(2 \leq X \leq 4) = p(2) + p(3) + p(4)$$
$$= [p(0) + \cdots + p(4)] - [p(0) + p(1)]$$
$$= P(X \leq 4) - P(X \leq 1)$$
$$= F(4) - F(1)$$

Notice that $P(2 \leq X \leq 4) \neq F(4) - F(2)$. This is because the X value 2 is included in $2 \leq X \leq 4$, so we do not want to subtract out its probability. It *is* true that $P(2 < X \leq 4) = F(4) - F(2)$ because $X = 2$ is not included in the interval $2 < X \leq 4$.

Proposition

For any two numbers a and b with $a \leq b$,

$$P(a \leq X \leq b) = F(b) - F(a-)$$

where "$a-$" represents the largest possible X value that is strictly less than a. In particular, if the only possible values are integers and if a and b are integers, then

$$P(a \leq X \leq b) = P(X = a \text{ or } a + 1 \text{ or } \ldots \text{ or } b)$$
$$= F(b) - F(a - 1)$$

Taking $a = b$ yields $P(X = a) = F(a) - F(a - 1)$ in this case.

Example 3.13   Let X = the number of days of sick leave taken by a randomly selected employee of a large company during a particular year. If the maximum number of allowable sick days per year is 14, possible values of X are $0, 1, \ldots, 14$. With $F(0) = .58$,

$$F(1) = .72, \ F(2) = .76, \ F(3) = .81, \ F(4) = .88, \text{ and } F(5) = .94,$$

$$P(2 \leq X \leq 5) = P(X = 2, 3, 4, \text{ or } 5) = F(5) - F(1) = .22$$

and

$$P(X = 3) = F(3) - F(2) = .05 \qquad \blacksquare$$

The reason for subtracting $F(a-)$ rather than $F(a)$ is that we want to include $P(X = a)$; $F(b) - F(a)$ gives $P(a < X \leq b)$. This proposition will be used extensively when computing binomial and Poisson probabilities in Sections 3.4 and 3.6.

Another View of Probability Mass Functions

It is often helpful to think of a pmf as specifying a mathematical model for a discrete population.

Example 3.14   Consider selecting at random a student who is among the 15,000 registered for the current term at Mega University. Let X = the number of courses for which the selected student is registered and suppose that X has pmf

x	1	2	3	4	5	6	7
$p(x)$	.01	.03	.13	.25	.39	.17	.02

One way to view this situation is to think of the population as consisting of 15,000 individuals, each having his or her own X value; the proportion with each X value is given by $p(x)$ above. An alternative viewpoint is to forget about the students and think of the population itself as consisting of the X values: There are some 1's in the population, some 2's, . . . , and finally some 7's. The population then consists of the numbers 1, 2, . . . , 7 (so is discrete), and $p(x)$ gives a model for the distribution of population values. $\qquad \blacksquare$

Once we have such a mathematical model for a population, we will use it to compute values of population characteristics (such as the mean μ) and make inferences about such characteristics.

Exercises   Section 3.2 (11–27)

11. An automobile service facility specializing in engine tune-ups knows that 45% of all tune-ups are done on four-cylinder automobiles, 40% on six-cylinder automobiles, and 15% on eight-cylinder automobiles. Let X = the number of cylinders on the next car to be tuned.

    a. What is the pmf of X?

    b. Draw both a line graph and a probability histogram for the pmf of part (a).

12. Let X = the number of tires on a randomly selected automobile that are underinflated.

    a. Which of the following three $p(x)$ functions is a legitimate pmf for X, and why are the other two not allowed?

x	0	1	2	3	4
p(x)	.3	.2	.1	.05	.05
p(x)	.4	.1	.1	.1	.3
p(x)	.4	.1	.2	.1	.3

b. For the legitimate pmf of part (a), compute $P(2 \leq X \leq 4)$, $P(X \leq 2)$, and $P(X \neq 0)$.

c. If $p(x) = c \cdot (5 - x)$ for $x = 0, 1, \ldots, 4$, what is the value of c? [HINT: $\sum_{x=0}^{4} p(x) = 1$.]

13. A mail-order computer business has six telephone lines. Let X denote the number of lines in use at a specified time. Suppose the pmf of X is as given in the accompanying table.

x	0	1	2	3	4	5	6
p(x)	.10	.15	.20	.25	.20	.06	.04

Calculate the probability of each of the following events.

a. {At most 3 lines are in use}

b. {Fewer than 3 lines are in use}

c. {At least 3 lines are in use}

d. {Between 2 and 5 lines, inclusive, are in use}

e. {Between 2 and 4 lines, inclusive, are not in use}

f. {At least 4 lines are not in use}

14. A contractor is required by a county planning department to submit one, two, three, four, or five forms (depending on the nature of the project) in applying for a building permit. Let Y = the number of forms required of the next applicant. The probability that y forms are required is known to be proportional to y—that is, $p(y) = ky$ for $y = 1, \ldots, 5$.

a. What is the value of k? [HINT: $\sum_{y=1}^{5} p(y) = 1$.]

b. What is the probability that at most three forms are required?

c. What is the probability that between two and four forms (inclusive) are required?

d. Could $p(y) = y^2/50$ for $y = 1, \ldots, 5$ be the pmf of Y?

15. Many manufacturers have quality control programs that include inspection of incoming materials for defects. Suppose a computer manufacturer receives computer boards in lots of five. Two boards are selected from each lot for inspection. We can represent possible outcomes of the selection process by pairs. For

example, the pair (1, 2) represents the selection of boards 1 and 2 for inspection.

a. List the ten different possible outcomes.

b. Suppose that boards 1 and 2 are the only defective boards in a lot of five. Two boards are to be chosen at random. Define X to be the number of defective boards observed among those inspected. Find the probability distribution of X.

c. Let $F(x)$ denote the cdf of X. First determine $F(0) = P(X \leq 0)$, $F(1)$, and $F(2)$ and then obtain $F(x)$ for all other x.

16. Some parts of California are particularly earthquake-prone. Suppose that in one such area, 20% of all homeowners are insured against earthquake damage. Four homeowners are to be selected at random; let X denote the number among the four who have earthquake insurance.

a. Find the probability distribution of X. [HINT: Let S denote a homeowner who has insurance and F one who does not. Then 1 possible outcome is $SFSS$, with probability $(.2)(.8)(.2)(.2)$ and associated X value 3. There are 15 other outcomes.]

b. Draw the corresponding probability histogram.

c. What is the most likely value for X?

d. What is the probability that at least two of the four selected have earthquake insurance?

17. A new battery's voltage may be acceptable (A) or unacceptable (U). A certain flashlight requires two batteries, so batteries will be independently selected and tested until two acceptable ones have been found. Suppose that 80% of all batteries have acceptable voltages and let Y denote the number of batteries that must be tested.

a. What is $p(2)$, that is, $P(Y = 2)$?

b. What is $p(3)$? (HINT: There are two different outcomes that result in $Y = 3$.)

c. To have $Y = 5$, what must be true of the fifth battery selected? List the four outcomes for which $Y = 5$ and then determine $p(5)$.

d. Use the pattern in your answers for parts (a)–(c) to obtain a general formula for $p(y)$.

18. Two fair six-sided dice are tossed independently. Let M = the maximum of the two tosses (so $M(1, 5) = 5$, $M(3, 3) = 3$, etc.).

a. What is the pmf of M? [HINT: First determine $p(1)$, then $p(2)$, and so on.]

b. Determine the cdf of M and graph it.

19. In Example 3.9 suppose there are only four potential blood donors, of whom only one has type O+ blood. Compute the pmf of Y.

20. A library subscribes to two different weekly news magazines, each of which is supposed to arrive in Wednesday's mail. In actuality, each one may arrive on Wednesday, Thursday, Friday, or Saturday. Suppose the two arrive independently of one another and for each one $P(\text{Wed.}) = .4$, $P(\text{Thurs.}) = .3$, $P(\text{Fri.}) = .2$, and $P(\text{Sat.}) = .1$. Let Y = the number of days beyond Wednesday that it takes for both magazines to arrive (so possible Y values are 0, 1, 2, or 3). Compute the pmf of Y. [HINT: There are 16 possible outcomes; $Y(W, W) = 0$, $Y(F, Th) = 2$, and so on.]

21. Refer to Exercise 13 and calculate and graph the cdf $F(x)$. Then use it to calculate the probabilities of the events given in parts (a)–(d) of that problem.

22. A consumer organization that evaluates new automobiles customarily reports the number of major defects on each car examined. Let X denote the number of major defects on a randomly selected car of a certain type. The cdf of X is as follows:

$$F(x) = \begin{cases} 0 & x < 0 \\ .06 & 0 \le x < 1 \\ .19 & 1 \le x < 2 \\ .39 & 2 \le x < 3 \\ .67 & 3 \le x < 4 \\ .92 & 4 \le x < 5 \\ .97 & 5 \le x < 6 \\ 1 & 6 < x \end{cases}$$

Calculate the following probabilities directly from the cdf.

a. $p(2)$, that is, $P(X = 2)$

b. $P(X > 3)$

c. $P(2 \le X \le 5)$

d. $P(2 < X < 5)$

23. An insurance company offers its policyholders a number of different premium payment options. For a randomly selected policyholder, let X = the number of months between successive payments. The cdf of X is as follows:

$$F(x) = \begin{cases} 0 & x < 1 \\ .30 & 1 \le x < 3 \\ .40 & 3 \le x < 4 \\ .45 & 4 \le x < 6 \\ .60 & 6 \le x < 12 \\ 1 & 12 \le x \end{cases}$$

a. What is the pmf of X?

b. Using just the cdf, compute $P(3 \le X \le 6)$ and $P(4 \le X)$.

24. In Example 3.10, let Y = the number of girls born before the experiment terminates. With $p = P(B)$ and $1 - p = P(G)$, what is the pmf of Y? (HINT: First list the possible values of Y, starting with the smallest, and proceed until you see a general formula.)

25. Alvie Singer lives at 0 in the accompanying diagram and has four friends who live at A, B, C, and D. One day Alvie decides to go visiting, so he tosses a fair coin twice to decide which of the four to visit. Once at a friend's house, he will either return home or else proceed to one of the two adjacent houses (such as 0, A, or C when at B), with each of the three possibilities having probability $\frac{1}{3}$. In this way Alvie continues to visit friends until he returns home.

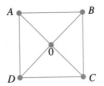

a. Let X = the number of times that Alvie visits a friend. Derive the pmf of X.

b. Let Y = the number of straight-line segments that Alvie traverses (including those leading to and from 0). What is the pmf of Y?

c. Suppose that female friends live at A and C and male friends at B and D. If Z = the number of visits to female friends, what is the pmf of Z?

26. After all students have left the classroom, a statistics professor notices that four copies of the text were left under desks. At the beginning of the next lecture, the professor distributes the four books in a completely random fashion to each of the four students (1, 2, 3, and 4) who claim to have left books. One possible outcome is that 1 receives 2's book, 2 receives 4's book, 3 receives his or her own book, and 4 receives 1's book. This outcome can be abbreviated as (2, 4, 3, 1).

a. List the other 23 possible outcomes.

b. Let X denote the number of students who receive their own book. Determine the pmf of X.

27. Show that the cdf $F(x)$ is a nondecreasing function; that is, $x_1 < x_2$ implies that $F(x_1) \le F(x_2)$. Under what condition will $F(x_1) = F(x_2)$?

3.3   Expected Values of Discrete Random Variables

In Example 3.14, we considered a university having 15,000 students and let X = the number of courses for which a randomly selected student is registered. The pmf of X appears below. Since $p(1) = .01$, we know that $(.01) \cdot (15,000) = 150$ of the students are registered for one course, and similarly for the other x values.

x	1	2	3	4	5	6	7	
$p(x)$	.01	.03	.13	.25	.39	.17	.02	(3.6)
Number registered	150	450	1950	3750	5850	2550	300	

To compute the average number of courses per student, or the average value of X in the population, we should compute the total number of courses and divide by the total number of students. Since each of 150 students is taking one course, these 150 contribute 150 courses to the total. Similarly, 450 students contribute 2(450) courses, and so on. The population average value of X is then

$$\frac{1(150) + 2(450) + 3(1950) + \cdots + 7(300)}{15,000} = 4.57 \qquad (3.7)$$

Since $150/15,000 = .01 = p(1)$, $450/15,000 = .03 = p(2)$, and so on, an alternative expression for (3.7) is

$$1 \cdot p(1) + 2 \cdot p(2) + \cdots + 7 \cdot p(7) \qquad (3.8)$$

Expression (3.8) shows that to compute the population average value of X, we need only the possible values of X along with their probabilities (proportions). In particular, the population size is irrelevant as long as the pmf is given by (3.6). The average or mean value of X is then a *weighted* average of the possible values $1, \ldots, 7$, where the weights are the probabilities of those values.

The Expected Value of X

Definition

> Let X be a discrete rv with set of possible values D and pmf $p(x)$. The **expected value** or **mean value** of X, denoted by $E(X)$ or μ_X, is
>
> $$E(X) = \mu_X = \sum_{x \in D} x \cdot p(x)$$

When it is clear to which X the expected value refers, μ rather than μ_X is often used.

Example 3.15   For the pmf in Expression (3.6),

$$\mu = 1 \cdot p(1) + 2 \cdot p(2) + \cdots + 7 \cdot p(7)$$

$$= (1)(.01) + 2(.03) + \cdots + (7)(.02)$$
$$= .01 + .06 + .39 + 1.00 + 1.95 + 1.02 + .14 = 4.57$$

If we think of the population as consisting of the X values $1, 2, \ldots, 7$, then $\mu = 4.57$ is the population mean. In the sequel, we will often refer to μ as the *population mean* rather than the mean of X in the population. ∎

In Example 3.15, the expected value μ was 4.57, which is not a possible value of X. The word *expected* should be interpreted with caution because one would not expect to see an X value of 4.57 when a single student is selected.

Example 3.16 Just after birth, each newborn child is rated on a scale called the Apgar scale. The possible ratings are $0, 1, \ldots, 10$, with the child's rating determined by color, muscle tone, respiratory effort, heartbeat, and reflex irritability (the best possible score is 10). Let X be the Apgar score of a randomly selected child born at a certain hospital during the next year and suppose that the pmf of X is

x	0	1	2	3	4	5	6	7	8	9	10
$p(x)$	.002	.001	.002	.005	.02	.04	.18	.37	.25	.12	.01

Then the mean value of X is

$$E(X) = \mu = 0(.002) + 1(.001) + 2(.002) + \cdots$$
$$+ 8(.25) + 9(.12) + 10(.01)$$
$$= 7.15$$

Again, μ is not a possible value of the variable X. Also, because the variable refers to a future child, there is no concrete existing population to which μ refers. Instead, we think of the pmf as a model for a conceptual population consisting of the values $0, 1, 2, \ldots, 10$. The mean value of this conceptual population is then $\mu = 7.15$. ∎

Example 3.17 Let X be a Bernoulli rv with pmf

$$p(x) = \begin{cases} 1 - p & x = 0 \\ p & x = 1 \\ 0 & x \neq 0, 1 \end{cases}$$

Then $E(X) = 0 \cdot p(0) + 1 \cdot p(1) = 0(1 - p) + 1(p) = p$. That is, the expected value of X is just the probability that X takes on the value 1. If we conceptualize a population consisting of 0's in proportion $1 - p$ and 1's in proportion p, then the population average is $\mu = p$. ∎

Example 3.18 The general form for the pmf of $X =$ number of children born up to and including the first boy is

$$p(x) = \begin{cases} p(1 - p)^{x-1} & x = 1, 2, 3, \ldots \\ 0 & \text{otherwise} \end{cases}$$

From the definition,

$$E(X) = \sum_{D} x \cdot p(x) = \sum_{x=1}^{\infty} xp(1-p)^{x-1} = p \sum_{x=1}^{\infty} \left[-\frac{d}{dp}(1-p)^x \right] \qquad (3.9)$$

If we interchange the order of taking the derivative and the summation, the sum is that of a geometric series. After the sum is computed, the derivative is taken, and the final result is $E(X) = 1/p$. If p is near 1, we expect to see a boy very soon, while if p is near 0, we expect many births before the first boy. For $p = .5$, $E(X) = 2$. ∎

There is another frequently used interpretation of μ. Consider the pmf

$$p(x) = \begin{cases} (.5) \cdot (.5)^{x-1} & \text{if } x = 1, 2, 3, \ldots \\ 0 & \text{otherwise} \end{cases}$$

This is the pmf of $X =$ the number of tosses of a fair coin necessary to obtain the first H (a special case of Example 3.18). Suppose we observe a value x from this pmf (toss a coin until an H appears), then observe independently another value (keep tossing), then another, and so on. If after observing a very large number of x values, we average them, the resulting sample average will be very near to $\mu = 2$. That is, μ can be interpreted as the long-run average observed value of X when the experiment is performed repeatedly.

Example 3.19   Let X have pmf

$$p(x) = \begin{cases} \dfrac{k}{x^2} & x = 1, 2, 3, \ldots \\ 0 & \text{otherwise} \end{cases}$$

where k is chosen so that $\sum_{x=1}^{\infty} (k/x^2) = 1$ (in a mathematics course on infinite series, it is shown that $\sum_{x=1}^{\infty} (1/x^2) < \infty$, which implies that such a k exists, but its exact value need not concern us). The expected value of X is

$$\mu = E(X) = \sum_{x=1}^{\infty} x \cdot \frac{k}{x^2} = k \sum_{x=1}^{\infty} \frac{1}{x} \qquad (3.10)$$

The sum on the right of Equation (3.10) is the famous harmonic series of mathematics and can be shown to equal ∞. $E(X)$ is not finite here because $p(x)$ does not decrease sufficiently fast as x increases; statisticians say that the probability distribution of X has "a heavy tail." If a sequence of X values is chosen using this distribution, the sample average will not settle down to some finite number but will tend to grow without bound.

Statisticians use the phrase "heavy tails" in connection with any distribution having a large amount of probability far from μ (so heavy tails does not require $\mu = \infty$). Such heavy tails make it difficult to make inferences about μ. ∎

The Expected Value of a Function

Often we will be interested in the expected value of some function $h(X)$ rather than X itself.

Example 3.20 Suppose a bookstore purchases ten copies of a book at $6.00 each, to sell at $12.00 with the understanding that at the end of a 3-month period any unsold copies can be redeemed for $2.00. If $X =$ the number of copies sold, then net revenue $= h(X) = 12X + 2(10 - X) - 60 = 10X - 40$. ∎

An easy way of computing the expected value of $h(X)$ is suggested by the following example.

Example 3.21 Let $X =$ the number of cylinders in the engine of the next car to be tuned up at a certain facility. The cost of a tune-up is related to X by $h(X) = 20 + 3X + .5X^2$. Since X is a random variable, so is $h(X)$; denote this latter rv by Y. The pmf's of X and Y are as follows:

x	4	6	8	y	40	56	76
$p(x)$	.5	.3	.2	$p(y)$	.5	.3	.2

With D^* denoting possible values of Y,

$$E(Y) = E[h(X)] = \sum_{D^*} y \cdot p(y) \tag{3.11}$$

$$= (40)(.5) + (56)(.3) + (76)(.2)$$
$$= h(4) \cdot (.5) + h(6) \cdot (.3) + h(8) \cdot (.2)$$
$$= \sum_{D} h(x) \cdot p(x)$$

According to Equation (3.11), it was not necessary to determine the pmf of Y to obtain $E(Y)$; instead, the desired expected value is a weighted average of the possible $h(x)$ (rather than x) values. ∎

Proposition

> If the rv X has set of possible values D and pmf $p(x)$, then the expected value of any function $h(X)$, denoted by $E[h(X)]$ or $\mu_{h(X)}$, is computed by
>
> $$E[h(X)] = \sum_{D} h(x) \cdot p(x)$$

According to this proposition, $E[h(X)]$ is computed in the same way that $E(X)$ itself is, except that $h(x)$ is substituted in place of x.

Example 3.22 A computer store has purchased three computers of a certain type at $500 apiece. It will sell them for $1000 apiece. The manufacturer has agreed to repurchase any computers still unsold after a specified period at $200 apiece. Let X denote the number of computers sold and suppose that $p(0) = .1$, $p(1) = .2$, $p(2) = .3$, and $p(3) = .4$. With $h(X)$ denoting the profit associated with selling X units, the given information implies that $h(X) =$ revenue $-$ cost $= 1000X + 200(3 - X) -$

$1500 = 800X - 900$. The expected profit is then

$$E[h(X)] = h(0) \cdot p(0) + h(1) \cdot p(1) + h(2) \cdot p(2) + h(3) \cdot p(3)$$
$$= (-900)(.1) + (-100)(.2) + (700)(.3) + (1500)(.4)$$
$$= \$700$$

∎

Rules of Expected Value

The $h(X)$ function of interest is quite frequently a linear function $aX + b$. In this case, $E[h(X)]$ is easily computed from $E(X)$.

Proposition

$$E(aX + b) = a \cdot E(X) + b$$

(or, using alternative notation, $\mu_{aX + b} = a \cdot \mu_X + b$).

To paraphrase, the expected value of a linear function equals the linear function evaluated at the expected value $E(X)$. Since $h(X)$ in Example 3.22 is linear and $E(X) = 2$, $E[h(X)] = 800(2) - 900 = \$700$ as before.

Proof

$$E(aX + b) = \sum_D (ax + b) \cdot p(x) = a \sum_D x \cdot p(x) + b \sum_D p(x)$$
$$= aE(X) + b$$

∎

Two special cases of the proposition yield two important rules of expected value.

> **1.** For any constant a, $E(aX) = a \cdot E(X)$ (take $b = 0$).   (3.12)
> **2.** For any constant b, $E(X + b) = E(X) + b$ (take $a = 1$).

Multiplication of X by a constant a changes the unit of measurement (from dollars to cents, where $a = 100$, inches to cm, where $a = 2.54$, etc.). Rule (1) says that the expected value in the new units equals the expected value in the old units multiplied by the conversion factor a. Similarly, if a constant b is added to each possible value of X, then the expected value will be shifted by that same constant amount.

The Variance of X

The expected value of X measures where the probability distribution is centered. Using the physical analogy of placing point mass $p(x)$ at the value x on a one-dimensional axis, if the axis were then supported by a fulcrum placed at μ, there would be no tendency for the axis to tilt. This is illustrated for two different distributions in Figure 3.7.

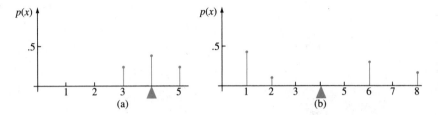

Figure 3.7 Two different probability distributions with $\mu = 4$

Although both distributions pictured in Figure 3.7 have the same center μ, the distribution of Figure 3.7b has greater spread or variability or dispersion than does that of Figure 3.7a. We will use the variance of X to measure the amount of variability in (the distribution of) X, just as s^2 was used in Chapter 1 to measure variability in a sample.

Definition

> Let X have pmf $p(x)$ and expected value μ. Then the **variance** of X, denoted by $V(X)$ or σ_X^2, or just σ^2, is
>
> $$V(X) = \sum_D (x - \mu)^2 \cdot p(x) = E[(X - \mu)^2]$$
>
> The **standard deviation** (SD) of X is
>
> $$\sigma_X = \sqrt{\sigma_X^2}$$

The quantity $h(X) = (X - \mu)^2$ is the squared deviation of X from its mean, and σ^2 is the expected squared deviation. If most of the probability distribution is close to μ, then σ^2 will be relatively small, whereas if there are x values far from μ that have large $p(x)$, then σ^2 will be quite large.

Example 3.23 If X has pmf

x	3	4	5
$p(x)$	.3	.4	.3

(as in Figure 3.7a) with mean $\mu = 4$, then

$$V(X) = \sigma^2 = \sum_{x=3}^{5} (x - 4)^2 \cdot p(x)$$

$$= (3 - 4)^2 \cdot (.3) + (4 - 4)^2 \cdot (.4) + (5 - 4)^2 \cdot (.3) = .6$$

The standard deviation of X is $\sigma = \sqrt{.6} = .77$.

If X has the pmf of Figure 3.7b,

x	1	2	6	8
$p(x)$	.4	.1	.3	.2

then

$$\sigma^2 = (1 - 4)^2 \cdot (.4) + (2 - 4)^2 \cdot (.1) + (6 - 4)^2 \cdot (.3) + (8 - 4)^2 \cdot (.2)$$
$$= 8.4$$

and

$$\sigma = 2.90$$ ∎

When the pmf $p(x)$ specifies a mathematical model for the distribution of population values, both σ^2 and σ measure the spread of values in the population; σ^2 is the population variance, and σ is the population standard deviation.

A Shortcut Formula for σ^2

The number of arithmetic operations necessary to compute σ^2 can be reduced by using an alternative computing formula.

Proposition

$$V(X) = \sigma^2 = \left[\sum_D x^2 \cdot p(x)\right] - \mu^2 = E(X^2) - [E(X)]^2$$

In using this formula, $E(X^2)$ is computed first without any subtraction; then $E(X)$ is computed, squared, and subtracted (once) from $E(X^2)$.

Example 3.24   The pmf of the number of cylinders X on the next car to be tuned at a certain facility was given in Example 3.21 as $p(4) = .5$, $p(6) = .3$, and $p(8) = .2$, from which

$$\mu = E(X) = (4)(.5) + (6)(.3) + (8)(.2) = 5.4$$
$$E(X^2) = (4^2)(.5) + (6^2)(.3) + (8^2)(.2) = 31.6$$
$$\sigma^2 = 31.6 - (5.4)^2 = 2.44$$
$$\sigma = \sqrt{2.44} = 1.56$$ ∎

Proof of the shortcut formula   Expand $(x - \mu)^2$ in the definition of σ^2 to obtain $x^2 - 2\mu x + \mu^2$ and then carry Σ through to each of the three terms:

$$\sigma^2 = \sum_D x^2 \cdot p(x) - 2\mu \cdot \sum_D x \cdot p(x) + \mu^2 \sum_D p(x)$$
$$= E(X^2) - 2\mu \cdot \mu + \mu^2 = E(X^2) - \mu^2$$ ∎

Rules of Variance

The variance of $h(X)$ is the expected value of the squared difference between $h(X)$ and its expected value:

$$V[h(X)] = \sigma^2_{h(X)} = \sum_D \{h(x) - E[h(X)]\}^2 \cdot p(x) \tag{3.13}$$

When $h(X)$ is a linear function, $V[h(X)]$ is easily related to $V(X)$.

Proposition

$$V(aX + b) = \sigma^2_{aX + b} = a^2 \cdot \sigma^2_X \qquad \text{and} \qquad \sigma_{aX + b} = |a| \cdot \sigma_X$$

This result says that the addition of the constant b does not affect the variance, which is intuitive, because the addition of b changes the location (mean value) but not the spread of values. In particular,

$$
\begin{aligned}
&1.\quad \sigma^2_{aX} = a^2 \cdot \sigma^2_X, \quad \sigma_{aX} = |a| \cdot \sigma_X \\
&2.\quad \sigma^2_{X + b} = \sigma^2_X
\end{aligned}
\tag{3.14}
$$

The reason for the absolute value in σ_{aX} is that a may be negative, while a standard deviation cannot be negative; a^2 results when a is brought outside the term being squared in Equation (3.13).

Example 3.25

In the computer sales problem of Example 3.22, $E(X) = 2$ and

$$E(X^2) = (0)^2(.1) + (1)^2(.2) + (2)^2(.3) + (3)^2(.4) = 5$$

so $V(X) = 5 - (2)^2 = 1$. The profit function $h(X) = 800X - 900$ then has variance $(800)^2 \cdot V(X) = (640,000)(1) = 640,000$ and standard deviation 800. ∎

Exercises   Section 3.3 (28–43)

28. The pmf for X = the number of major defects on a randomly selected appliance of a certain type is

x	0	1	2	3	4
$p(x)$	.08	.15	.45	.27	.05

Compute the following:

a. $E(X)$

b. $V(X)$ directly from the definition

c. The standard deviation of X

d. $V(X)$ using the shortcut formula

29. An instructor in a technical writing class has asked that a certain report be turned in the following week, adding the restriction that any report exceeding four pages will not be accepted. Let Y = the number of pages in a randomly chosen student's report and suppose that Y has pmf

y	1	2	3	4
$p(y)$	.01	.19	.35	.45

a. Compute $E(Y)$.

b. Suppose the instructor spends $\sqrt{Y}$ minutes grading

a paper consisting of Y pages. What is the expected amount of time $[E(\sqrt{Y})]$ spent grading a randomly selected paper?

30. Refer to Exercise 29 and calculate $V(Y)$ and σ_Y. Then determine the probability that Y is within 1 SD of its mean value.

31. An appliance dealer sells three different models of upright freezers having 13.5, 15.9, and 19.1 cubic feet of storage space, respectively. Let X = the amount of storage space purchased by the next customer to buy a freezer. Suppose that X has pmf

x	13.5	15.9	19.1
$p(x)$	.2	.5	.3

a. Compute $E(X)$, $E(X^2)$, and $V(X)$.

b. If the price of a freezer having capacity X cubic feet is $25X - 8.5$, what is the expected price paid by the next customer to buy a freezer?

c. What is the variance of the price $25X - 8.5$ paid by the next customer?

d. Suppose that while the rated capacity of a freezer is X, the actual capacity is $h(X) = X - .01X^2$. What is the expected actual capacity of the freezer purchased by the next customer?

32. Let X be a Bernoulli rv with pmf as in Example 3.17.

a. Compute $E(X^2)$.

b. Show that $V(X) = p(1 - p)$.

c. Compute $E(X^{79})$.

33. Suppose that the number of plants of a particular type found in a rectanglar region (called a quadrat by ecologists) in a certain geographic area is an rv X with pmf

$$p(x) = \begin{cases} c/x^3 & x = 1, 2, 3, \ldots \\ 0 & \text{otherwise} \end{cases}$$

Is $E(X)$ finite? Justify your answer (this is another distribution that statisticians would call heavy-tailed).

34. A small drugstore orders copies of a news magazine for its magazine rack each week. Let X = demand for the magazine, with pmf

x	1	2	3	4	5	6
$p(x)$	$\frac{1}{15}$	$\frac{2}{15}$	$\frac{3}{15}$	$\frac{4}{15}$	$\frac{3}{15}$	$\frac{2}{15}$

Suppose the store owner actually pays $.25 for each copy of the news magazine and the price to customers

is $1.00. If magazines left at the end of the week have no salvage value, is it better to order three or four copies of the magazine? (HINT: For both three and four copies ordered, express net revenue as a function of demand X and then compute the expected revenue.)

35. Let X be the damage incurred (in $) in a certain type of accident during a given year. Possible X values are 0, 1000, 5000, and 10000, with probabilities .8, .1, .08, and .02, respectively. A particular company offers a $500 deductible policy. If the company wishes its expected profit to be $100, what premium amount should it charge?

36. The n candidates for a job have been ranked 1, 2, 3, . . . , n. Let X = the rank of a randomly selected candidate, so that X has pmf

$$p(x) = \begin{cases} 1/n & x = 1, 2, 3, \ldots, n \\ 0 & \text{otherwise} \end{cases}$$

(this is called the discrete uniform distribution). Compute $E(X)$ and $V(X)$ using the shortcut formula. [HINT: The sum of the first n positive integers is $n(n + 1)/2$, while the sum of their squares is $n(n + 1)(2n + 1)/6$.]

37. Let X = the outcome when a fair die is rolled once. If before the die is rolled you are offered either (1/3.5) dollars or $h(X) = 1/X$ dollars, would you accept the guaranteed amount or would you gamble? [NOTE: It is not generally true that $1/E(X) = E(1/X)$.]

38. A chemical supply company currently has in stock 100 lb of a certain chemical, which it sells to customers in 5-lb lots. Let X = the number of lots ordered by a randomly chosen customer, and suppose that X has pmf

x	1	2	3	4
$p(x)$	.2	.4	.3	.1

Compute $E(X)$ and $V(X)$. Then compute the expected number of pounds left after the next customer's order is shipped, and the variance of the number of pounds left. (HINT: The number of pounds left is a linear function of X.)

39. a. Draw a line graph of the pmf of X in Exercise 34. Then determine the pmf of $-X$ and draw its line graph. From these two pictures, what can you say about $V(X)$ and $V(-X)$?

b. Use the proposition involving $V(aX + b)$ to establish a general relationship between $V(X)$ and $V(-X)$.

40. Use the definition in Expression (3.13) to prove that $V(aX + b) = a^2 \cdot \sigma_X^2$. [HINT: With $h(X) = aX + b$, $E[h(X)] = a\mu + b$ where $\mu = E(X)$.]

41. Suppose $E(X) = 5$ and $E[X(X - 1)] = 27.5$. What is

    a. $E(X^2)$? [HINT: $E(X(X - 1)) = E[X^2 - X] = E(X^2) - E(X)$.]

    b. $V(X)$?

    c. The general relationship between the quantities $E(X)$, $E[X(X - 1)]$, and $V(X)$?

42. Write a general rule for $E(X - c)$ where c is a constant. What happens when you let $c = \mu$, the expected value of X?

43. A result called **Chebyshev's inequality** states that for any probability distribution of an rv X and any number k that is at least 1, $P(|X - \mu| \geq k\sigma) \leq 1/k^2$. In words, the probability that the value of X lies at least k SDs from its mean is at most $1/k^2$.

    a. What is the value of the upper bound for $k = 2$? $k = 3$? $k = 4$? $k = 5$? $k = 10$?

    b. Compute μ and σ for the distribution of Exercise 13. Then evaluate $P(|X - \mu| \geq k\sigma)$ for the values of k given in part (a). What does this suggest about the upper bound relative to the corresponding probability?

    c. Let X have three possible values, $-1, 0$, and 1, with probabilities $\frac{1}{18}, \frac{8}{9}$, and $\frac{1}{18}$, respectively. What is $P(|X - \mu| \geq 3\sigma)$, and how does it compare to the corresponding bound?

    d. Give a distribution for which $P(|X - \mu| \geq 5\sigma) = .04$.

3.4 The Binomial Probability Distribution

There are many experiments that conform either exactly or approximately to the following list of requirements:

1. The experiment consists of a sequence of n trials, where n is fixed in advance of the experiment.

2. The trials are identical, and each trial can result in one of the same two possible outcomes, which we denote by success (S) or failure (F).

3. The trials are independent, so that the outcome on any particular trial does not influence the outcome on any other trial.

4. The probability of success is constant from trial to trial; we denote this probability by p.

Definition

> An experiment for which Conditions 1–4 are satisfied is called a **binomial experiment.**

Example 3.26

The same coin is tossed successively and independently n times. We arbitrarily use S to denote the outcome H (heads) and F to denote the outcome T (tails). Then this experiment satisfies Conditions 1–4. ■

Many experiments involve a sequence of independent trials for which there are more than two possible outcomes on any one trial. A binomial experiment can then be created by dividing the possible outcomes into two groups.

Example 3.27

The color of pea seeds is determined by a single genetic locus. If the two alleles at this locus are AA or Aa (the genotype), then the pea will be yellow (the phenotype), and if the allele is aa, the pea will be green. Suppose we pair off 20 Aa seeds and cross the two seeds in each of the ten pairs to obtain ten new genotypes. Call each new genotype a success S if it is aa and a failure otherwise. Then with this identification of S and F, the experiment is binomial with $n = 10$ and $p = P(aa$ genotype$)$. If each member of the pair is equally likely to contribute a or A, then $p = P(a) \cdot P(a) = (\frac{1}{2})(\frac{1}{2}) = \frac{1}{4}$.   ∎

Example 3.28

Suppose a certain city has 50 licensed restaurants, of which 15 currently have at least one serious health code violation and the other 35 have no serious violations. There are five inspectors, each of whom will inspect one restaurant during the coming week. The name of each restaurant is written on a different slip of paper, and after the slips are thoroughly mixed, each inspector in turn draws one of the slips *without replacement*. Label the ith trial as a success if the ith restaurant selected ($i = 1, \ldots, 5$) has no serious violations. Then

$$P(S \text{ on first trial}) = \frac{35}{50} = .70$$

and

$$P(S \text{ on second trial}) = P(SS) + P(FS) = P(\text{second } S \mid \text{first } S) \, P(\text{first } S)$$
$$+ \, P(\text{second } S \mid \text{first } F) \, P(\text{first } F)$$
$$= \frac{34}{49} \cdot \frac{35}{50} + \frac{35}{49} \cdot \frac{15}{50} = \frac{35}{50}\left(\frac{34}{49} + \frac{15}{49}\right) = \frac{35}{50} = .70$$

Similarly, it can be shown that $P(S \text{ on } i\text{th trial}) = .70$ for $i = 3, 4, 5$. However,

$$P(S \text{ on fifth trial} \mid SSSS) = \frac{31}{46} = .67$$

whereas

$$P(S \text{ on fifth trial} \mid FFFF) = \frac{35}{46} = .76$$

so the experiment is not binomial because the trials are not independent. In general, if sampling is without replacement, the experiment will not yield independent trials. If each slip had been replaced after being drawn, then trials would have been independent, but this might have resulted in the same restaurant being inspected by more than one inspector.   ∎

Example 3.29

Suppose a certain state has 500,000 registered automobiles, of which 400,000 were manufactured domestically. A sample of ten cars is chosen without replacement. The ith trial is labeled S if the ith car chosen is a domestic model. Although this situation would seem identical to that of Example 3.28, the important difference is that the size of the population being sampled is very large relative to the sample size. In this case

$$P(S \text{ on } 2 \mid S \text{ on } 1) = \frac{399,999}{499,999} = .80000$$

and

$$P(S \text{ on } 10 \mid S \text{ on first } 9) = \frac{399,991}{499,991} = .799996 \approx .80000$$

These calculations suggest that while the trials are not exactly independent, the conditional probabilities differ so slightly from one another that for practical purposes the trials can be regarded as independent with constant $P(S) = .8$. Thus, to a very good approximation, the experiment is binomial with $n = 10$ and $p = .8$.   ■

We will use the following rule of thumb in deciding whether a "without-replacement" experiment can be treated as a binomial experiment.

Rule

> Suppose each trial of an experiment can result in S or F, but the sampling is without replacement from a population of size N. If the sample size (number of trials) n is at most 5% of the population size, the experiment can be analyzed as though it were exactly a binomial experiment.

By "analyzed," we mean that probabilities based on the binomial experiment assumptions will be quite close to the actual "without-replacement" probabilities, which are typically more difficult to calculate. In Example 3.28, $n/N = 5/50 = .1 > .05$, so the binomial experiment is not a good approximation, but in Example 3.29, $n/N = 10/500,000 < .05$.

The Binomial Random Variable and Distribution

In most binomial experiments it is the total number of S's, rather than knowledge of exactly which trials yielded S's, that is of interest.

Definition

> Given a binomial experiment consisting of n trials, the **binomial random variable X** associated with this experiment is defined as
>
> $X =$ the number of S's among the n trials

Suppose, for example, that $n = 3$. Then there are eight possible outcomes for the experiment:

SSS, SSF, SFS, SFF, FSS, FSF, FFS, FFF

From the definition of X, $X(SSF) = 2$, $X(SFF) = 1$, and so on. Possible values for X in an n-trial experiment are $x = 0, 1, 2, \ldots, n$. We will often write $X \sim \text{Bin}(n, p)$ to indicate that X is a binomial rv based on n trials with success probability p.

Notation

> Because the pmf of a binomial rv X depends on the two parameters n and p, we denote the pmf by $b(x; n, p)$.

To gain insight into $b(x; n, p)$ for general n, consider the case $n = 4$ for which each outcome, its probability, and corresponding x value are listed in Table 3.1. For example,

$$P(SSFS) = P(S) \cdot P(S) \cdot P(F) \cdot P(S) \qquad \text{(independent trials)}$$
$$= p \cdot p \cdot (1 - p) \cdot p \qquad \text{(constant } P(S))$$
$$= p^3 \cdot (1 - p)$$

Table 3.1 Outcomes and probabilities for a binomial experiment with four trials

Outcome	x	Probability	Outcome	x	Probability
SSSS	4	p^4	FSSS	3	$p^3(1-p)$
SSSF	3	$p^3(1-p)$	FSSF	2	$p^2(1-p)^2$
SSFS	3	$p^3(1-p)$	FSFS	2	$p^2(1-p)^2$
SSFF	2	$p^2(1-p)^2$	FSFF	1	$p(1-p)^3$
SFSS	3	$p^3(1-p)$	FFSS	2	$p^2(1-p)^2$
SFSF	2	$p^2(1-p)^2$	FFSF	1	$p(1-p)^3$
SFFS	2	$p^2(1-p)^2$	FFFS	1	$p(1-p)^3$
SFFF	1	$p(1-p)^3$	FFFF	0	$(1-p)^4$

In this special case, we wish $b(x; 4, p)$ for $x = 0, 1, 2, 3,$ and 4. For $b(3; 4, p)$, we identify which of the 16 outcomes yield an x value of 3 and add up the probabilities associated with each such outcome:

$$b(3; 4, p) = P(FSSS) + P(SFSS) + P(SSFS) + P(SSSF) = 4p^3(1 - p)$$

Since the number of outcomes with $x = 3$ is four and each of the four has probability $p^3(1 - p)$ (the order of S's and F's is not important, but only the number of S's),

$$b(3; 4, p) = \begin{Bmatrix} \text{number of outcomes} \\ \text{with } X = 3 \end{Bmatrix} \cdot \begin{Bmatrix} \text{probability of any particular} \\ \text{outcome with } X = 3 \end{Bmatrix}$$

Similarly, $b(2; 4, p) = 6p^2(1 - p)^2$, which is also the product of the number of outcomes with $X = 2$ and the probability of any such outcome.

In general,

$$b(x; n, p) = \begin{Bmatrix} \text{number of sequences of} \\ \text{length } n \text{ consisting of } x \text{ } S\text{'s} \end{Bmatrix} \cdot \begin{Bmatrix} \text{probability of any} \\ \text{particular such sequence} \end{Bmatrix}$$

Since the ordering of S's and F's is not important, the second factor above is $p^x(1 - p)^{n-x}$ (e.g., the first x trials resulting in S and the last $n - x$ resulting in F). The first factor is the number of ways of choosing x of the n trials to be S's—that is, the number of combinations of size x that can be constructed from n distinct objects (trials here).

Theorem

$$b(x; n, p) = \begin{cases} \binom{n}{x}p^x(1-p)^{n-x} & x = 0, 1, 2, \ldots \\ 0 & \text{otherwise} \end{cases}$$

Example 3.30   Each of six randomly selected cola drinkers is given a g   g cola S and one containing cola F. The glasses are identical in appear   or a code on the bottom to identify the cola. Suppose there is actually   among cola drinkers to prefer one cola to the other. Then $p = P(a\ s$ dual prefers $S) = .5$, so with $X =$ the number among the six who pref $(6, .5)$.

Thus,

$$P(X = 3) = b(3; 6, .5) = \binom{6}{3}(.5^3(.5)^3 = 20\ (.5)^6 = .31$$

The probability that at least three prefer S is

$$P(3 \le X) = \sum_{x=3}^{6} b(x; 6, .5) = \sum_{x=3}^{6} \binom{6}{x}(.5)^x(.5)^{6-x} = .656$$

while the probability that at most one prefers S is

$$P(X \le 1) = \sum_{x=0}^{1} b(x; 6, .5) = .109$$

∎

Using Binomial Tables

Even for a relatively small value of n, the computation of binomial probabilities can be tedious. Appendix Table A.1 tabulates the cdf $F(x) = P(X \le x)$ for $n = 5, 10, 15, 20, 25$ in combination with selected values of p. Various other probabilities can then be calculated using the proposition on cdf's from Section 3.2.

Notation

For $X \sim \text{Bin}(n, p)$, the cdf will be denoted by

$$P(X \le x) = B(x; n, p) = \sum_{y=0}^{x} b(y; n, p), \qquad x = 0, 1, \ldots, n$$

Example 3.31   Suppose that 20% of all copies of a particular textbook fail a certain binding strength test. Let X denote the number among 15 randomly selected copies that fail the test. Then X has a binomial distribution with $n = 15$ and $p = .2$.

1. The probability that at most 8 fail the test is

$$P(X \le 8) = \sum_{y=0}^{8} b(y; 15, .2) = B(8; 15, .2)$$

which is the entry in the $x = 8$ row and the $p = .2$ column of the $n = 15$ binomial table. From Appendix Table A.1, the probability is $B(8; 15, .2) = .999$.

2. The probability that exactly 8 fail is

$$P(X = 8) = P(X \leq 8) - P(X \leq 7) = B(8; 15, .2) - B(7; 15, .2)$$

which is the difference between two consecutive entries in the $p = .2$ column. The result is $.999 - .996 = .003$.

3. The probability that at least 8 fail is

$$P(X \geq 8) = 1 - P(X \leq 7) = 1 - B(7; 15, .2)$$

$$= 1 - \begin{pmatrix} \text{entry in } x = 7 \\ \text{row of } p = .2 \text{ column} \end{pmatrix}$$

$$= 1 - .996 = .004$$

4. Finally, the probability that between 4 and 7, inclusive, fail is

$$P(4 \leq X \leq 7) = P(X = 4, 5, 6, \text{ or } 7) = P(X \leq 7) - P(X \leq 3)$$
$$= B(7; 15, .2) - B(3; 15, .2) = .996 - .648 = .348$$

Notice that this latter probability is the difference between entries in the $x = 7$ and $x = 3$ rows, *not* the $x = 7$ and $x = 4$ rows.   ■

Example 3.32   An electronics manufacturer claims that at most 10% of its power supply units need service during the warranty period. To investigate this claim, technicians at a testing laboratory purchase 20 units and subject each one to accelerated testing to simulate use during the warranty period. Let p denote the probability that a power supply unit needs repair during the period (the proportion of all such units that need repair). The laboratory technicians must decide whether the data resulting from the experiment supports the claim that $p \leq .10$. Let X denote the number among the 20 sampled that need repair, so $X \sim \text{Bin}(20, p)$. Consider the decision rule

Reject the claim that $p \leq .10$ in favor of the conclusion that $p > .10$ if $x \geq 5$ (where x is the observed value of X) and consider the claim plausible if $x \leq 4$.

The probability that the claim is rejected when $p = .10$ (an incorrect conclusion) is

$P(X \geq 5 \text{ when } p = .10) = 1 - B(4; 20, .1) = 1 - .957 = .043$

The probability that the claim is not rejected when $p = .20$ (a different type of incorrect conclusion) is

$P(X \leq 4 \text{ when } p = .2) = B(4; 20, .2) = .630$

The first probability is rather small, but the second is intolerably large. When $p = .20$, so that the manufacturer has grossly understated the percentage of units that need service, and the stated decision rule is used, 63% of all samples will result in the manufacturer's claim being judged plausible!

One might think that the probability of this second type of erroneous conclusion could be made smaller by changing the cutoff value 5 in the decision rule to something else. However, while replacing 5 by a smaller number would yield a probability smaller than .630, the other probability would then increase. The only way to make both "error probabilities" small is to base the decision rule on an experiment involving many more units. ∎

Note that a table entry of 0 signifies only that a probability is 0 to three significant digits, for all entries in the table are actually positive. Much more extensive tables of binomial probabilities have been published. In Chapter 4, we will present a method for obtaining quick and accurate approximations to binomial probabilities when n is large.

The Mean and Variance of X

For $n = 1$, the binomial distribution becomes the Bernoulli distribution. From Example 3.17 the mean value of a Bernoulli variable is $\mu = p$, so the expected number of S's on any single trial is p. Since a binomial experiment consists of n trials, intuition suggests that for $X \sim \text{Bin}(n, p)$, $E(X) = np$, the product of the number of trials and the probability of success on a single trial. The expression for $V(X)$ is not so intuitive.

Proposition

> If $X \sim \text{Bin}(n, p)$, then $E(X) = np$, $V(X) = np(1 - p) = npq$, and $\sigma_X = \sqrt{npq}$ (where $q = 1 - p$).

Thus, calculating the mean and variance of a binomial rv does not necessitate evaluating summations. The proof of the result for $E(X)$ is sketched in Exercise 58.

Example 3.33

If 75% of all purchases at a certain store are made with a credit card and X is the number among ten randomly selected purchases made with a credit card, then $X \sim \text{Bin}(10, .75)$. Thus $E(X) = np = (10)(.75) = 7.5$, $V(X) = npq = 10(.75)(.25) = 1.875$, and $\sigma = \sqrt{1.875}$. Again, even though X can take on only integer values, $E(X)$ need not be an integer. If we perform a large number of independent binomial experiments, each with $n = 10$ trials and $p = .75$, then the average number of S's per experiment will be close to 7.5. ∎

Exercises Section 3.4* (44–61)

44. Compute the following binomial probabilities directly from the formula for $b(x; n, p)$.

a. $b(3; 8, .6)$

b. $b(5; 8, .6)$

c. $P(3 \le X \le 5)$, when $n = 8$ and $p = .6$

d. $P(1 \le X)$, when $n = 12$ and $p = .1$

45. Use Appendix Table A.1 to obtain the following probabilities:

---

*"Between a and b inclusive" is equivalent to $(a \le X \le b)$.

a. $B(4; 10, .3)$

b. $b(4; 10, .3)$

c. $b(6; 10, .7)$

d. $P(2 \leq X \leq 4)$,   when $X \sim \text{Bin}(10, .3)$

e. $P(2 \leq X)$,   when $X \sim \text{Bin}(10, .3)$

f. $P(X \leq 1)$,   when $X \sim \text{Bin}(10, .7)$

g. $P(2 < X < 6)$,   when $X \sim \text{Bin}(10, .3)$

46. When circuit boards used in the manufacture of compact disc players are tested, the long-run percentage of defectives is 5%. Let $X =$ the number of defective boards in a random sample of size $n = 25$, so $X \sim \text{Bin}(25, .05)$.

a. Determine $P(X \leq 2)$.

b. Determine $P(X \geq 5)$.

c. Determine $P(1 \leq X \leq 4)$.

d. What is the probability that none of the 25 boards are defective?

e. Calculate the expected value and standard deviation of X.

47. A company that produces fine crystal knows from experience that 10% of its goblets have cosmetic flaws and must be classified as "seconds."

a. Among six randomly selected goblets, how likely is it that only one is a second?

b. Among six randomly selected goblets, what is the probability that at least two are seconds?

c. If goblets are examined one-by-one, what is the probability that at most five must be selected to find four that are not seconds?

48. Suppose that only 20% of all drivers come to a complete stop at an intersection having flashing red lights in all directions when no other cars are visible. What is the probability that, of 20 randomly chosen drivers coming to an intersection under these conditions,

a. At most 5 will come to a complete stop?

b. Exactly 5 will come to a complete stop?

c. At least 5 will come to a complete stop?

d. How many of the next 20 drivers do you expect to come to a complete stop?

49. If 90% of all applicants for a certain type of mortgage fail to fill out the application form correctly on first submission, what is the probability that among 15 randomly chosen such applicants,

a. At least 12 fail on first submission?

b. Between 10 and 13 inclusive fail on first submission?

c. At most 2 do fill out their forms correctly prior to initial submission?

50. A particular type of tennis racket comes in a midsize version and an oversize version. Sixty percent of all customers at a certain store want the oversize version.

a. Among ten randomly selected customers who want this type of racket, what is the probability that at least six want the oversize version?

b. Among ten randomly selected customers, what is the probability that the number who want the oversize version is within 1 SD of the mean value?

c. The store currently has seven rackets of each version. What is the probability that all of the next ten customers who want this racket can get the version they want from current stock?

51. Twenty percent of all telephones of a certain type are submitted for service while under warranty. Of these, 60% can be repaired whereas the other 40% must be replaced with new units. If a company purchases ten of these telephones, what is the probability that exactly two will end up being replaced under warranty?

52. A very large batch of components has arrived at a distributor. The batch can be characterized as acceptable only if the proportion of defective components is at most .10. The distributor decides to randomly select 10 components and to accept the batch only if the number of defective components in the sample is at most 2.

a. What is the probability that the batch will be accepted when the actual proportion of defectives is .01? .05? .10? .20? .25?

b. Let p denote the actual proportion of defectives in the batch. A graph of P(lot is accepted) as a function of p with p on the horizontal axis and P(lot is accepted) on the vertical axis is called the *operating characteristic curve* for the lot acceptance sampling plan. Use the results of part (a) to sketch this curve for $0 \leq p \leq 1$.

c. Repeat parts (a) and (b) with "1" replacing "2" in the lot acceptance sampling plan.

d. Repeat parts (a) and (b) with 15 replacing 10 in the lot acceptance sampling plan.

e. Which of the three sampling plans, that of part (a), (c), or (d), appears most satisfactory, and why?

53. An ordinance requiring that a smoke detector be installed in all previously constructed houses has been in effect in a particular city for one year. The fire department is concerned that many houses remain without detectors. Let p = the true proportion of such houses having detectors and suppose that a random sample of 25 homes is inspected. If the sample strongly indicates that fewer than 80% of all houses have a detector, the fire department will campaign for a mandatory inspection program. Because of the costliness of the program, the department prefers not to call for such inspections unless sample evidence strongly argues for their necessity. Let X denote the number of homes with detectors among the 25 sampled. Consider rejecting the claim that $p \geq .8$ if $x \leq 15$, where x is the observed value of X.

a. What is the probability that the claim is rejected when the actual value of p is .8?

b. What is the probability of not rejecting the claim when $p = .7$? When $p = .6$?

c. How do the "error probabilities" of parts (a) and (b) change if the value 15 in the decision rule is replaced by 14?

54. A toll bridge charges $1.00 for passenger cars and $2.50 for other vehicles. Suppose that during daytime hours, 60% of all vehicles are passenger cars. If 25 vehicles cross the bridge during a particular daytime period, what is the resulting expected toll revenue? [HINT: Let X = the number of passenger cars; then the toll revenue $h(X)$ is a linear function of X.]

55. A student who is trying to write a paper for a course has a choice of two topics, A and B. If topic A is chosen, the student will order two books through interlibrary loan, while if topic B is chosen, the student will order four books. The student believes that a good paper necessitates receiving and using at least half the books ordered for either topic chosen. If the probability that a book ordered through interlibrary loan actually arrives in time is .9 and books arrive independently of one another, which topic should the student choose to maximize the probability of writing a good paper? What if the arrival probability is only .5 instead of .9?

56. a. For fixed n, are there values of p ($0 \leq p \leq 1$) for which $V(X) = 0$? Explain why this is so.

b. For what value of p is $V(X)$ maximized? [HINT: Either graph $V(X)$ as a function of p or else take a derivative.]

57. a. Show that $b(x; n, 1 - p) = b(n - x; n, p)$.

b. Show that $B(x; n, 1 - p) = 1 - B(n - x - 1; n, p)$. [HINT: At most x S's is equivalent to at least $(n - x)$ F's.]

c. What do parts (a) and (b) imply about the necessity of including values of p greater than .5 in Appendix Table A.1?

58. Show that $E(X) = np$ when X is a binomial random variable. [HINT: First express $E(X)$ as a sum with lower limit $x = 1$. Then factor out np, let $y = x - 1$ so that the sum is from $y = 0$ to $y = n - 1$, and show that the sum equals 1.]

59. Customers at a gas station select either regular (A), premium (B), or diesel fuel (C). Assume that successive customers make independent choices, with $P(A) = .3$, $P(B) = .2$, and $P(C) = .5$.

a. Among the next 100 customers, what are the mean and variance of the number who select regular fuel? Explain your reasoning.

b. Answer part (a) for the number among the 100 who select a nondiesel fuel.

60. An airport limousine can accommodate up to four passengers on any one trip. The company will accept a maximum of six reservations for a trip, and a passenger must have a reservation. From previous records, 20% of all those making reservations do not appear for the trip. Answer the following questions, assuming independence wherever appropriate.

a. If six reservations are made, what is the probability that at least one individual with a reservation cannot be accommodated on the trip?

b. If six reservations are made, what is the expected number of available places when the limousine departs?

c. Suppose the probability distribution of the number of reservations made is given in the accompanying table.

Number of reservations	3	4	5	6
Probability	.1	.2	.3	.4

Let X denote the number of passengers on a randomly selected trip. Obtain the probability mass function of X.

61. Refer to Chebyshev's inequality given in Exercise 43. Calculate $P(|X - \mu| \geq k\sigma)$ for $k = 2$ and $k = 3$ when $X \sim \text{Bin}(20, .5)$ and compare to the corresponding upper bound. Repeat for $X \sim \text{Bin}(20, .75)$.

3.5   Hypergeometric and Negative Binomial Distributions

The hypergeometric and negative binomial distributions are both closely related to the binomial distribution. Whereas the binomial distribution is the approximate probability model for sampling without replacement from a finite dichotomous (S–F) population, the hypergeometric distribution is the exact probability model for the number of S's in the sample. The binomial rv X is the number of S's when the number n of trials is fixed, whereas the negative binomial distribution arises from fixing the number of S's and letting the number of trials be random.

The Hypergeometric Distribution

The assumptions leading to the hypergeometric distribution are as follows:

1. The population or set to be sampled consists of N individuals, objects, or elements (a *finite* population).

2. Each individual can be characterized as a success (S) or a failure (F), and there are M successes in the population.

3. A sample of n individuals is drawn in such a way that each subset of size n is equally likely to be chosen.

The random variable of interest is $X =$ the number of S's in the sample. The probability distribution of X depends on the parameters n, M, and N, so we wish to obtain $P(X = x) = h(x; n, M, N)$.

Example 3.34

An undergraduate library has 20 copies of a certain introductory economics text, of which 8 are first printings and 12 are second printings (containing corrections of some minor errors that appeared in the first printing). The course instructor has requested that 5 copies be put on 2-hour reserve. If the copies are selected in a completely random fashion, so that every subset of size 5 has the same probability of being selected, what is the probability that x ($x = 0, 1, 2, 3, 4,$ or 5) of those selected are second printings?

In this example, the population size is $N = 20$, the sample size is $n = 5$, and the number of S's (second printing $= S$) and F's in the population are $M = 12$ and $N - M = 8$, respectively. Consider the value $x = 2$. Because all outcomes (each one consisting of five particular books) are equally likely,

$$P(X = 2) = h(2; 5, 12, 20) = \frac{\text{Number of outcomes having } X = 2}{\text{Number of possible outcomes}}$$

The number of possible outcomes in the experiment is the number of ways of selecting 5 from the 20 objects without regard to order—that is, $\binom{20}{5}$. To count the number of outcomes having $X = 2$, note that there are $\binom{12}{2}$ ways of selecting 2 of the second printings, and for each such way there are $\binom{8}{3}$ ways of selecting the 3 first printings to fill out the sample. The product rule from Chapter 2 then gives $\binom{12}{2}\binom{8}{3}$ as the number of outcomes with $X = 2$, so

$$h(2;\ 5,\ 12,\ 20) = \frac{\binom{12}{2}\binom{8}{3}}{\binom{20}{5}} = \frac{77}{323} = .238$$

■

In general, if the sample size n is smaller than the number of successes in the population (M), then the largest possible X value is n. However, if $M < n$ (e.g., a sample size of 25 and only 15 successes in the population), then X can be at most M. Similarly, whenever the number of population failures ($N - M$) exceeds the sample size, the smallest possible X value is 0 (since all sampled individuals might then be failures). However, if $N - M < n$, the smallest possible X value is $n - (N - M)$. Summarizing, the possible values of X satisfy the restriction $\max(0, n - (N - M)) \leq x \leq \min(n, M)$. An argument parallel to that of the previous example gives the pmf of X.

Proposition

> If X is the number of S's in a completely random sample of size n drawn from a population consisting of M S's and $(N - M)$ F's, then the probability distribution of X, called the **hypergeometric distribution,** is given by
>
> $$P(X = x) = h(x;\ n,\ M,\ N) = \frac{\binom{M}{x}\binom{N - M}{n - x}}{\binom{N}{n}} \tag{3.15}$$
>
> for x an integer satisfying $\max(0, n - N + M) \leq x \leq \min(n, M)$.

In Example 3.34, $n = 5$, $M = 12$, and $N = 20$, so $h(x;\ 5,\ 12,\ 20)$ for $x = 0, 1, 2, 3, 4, 5$ can be obtained by substituting these numbers into Equation (3.15).

Example 3.35

Five individuals from an animal population thought to be near extinction in a certain region have been caught, tagged, and released to mix into the population. After they have had an opportunity to mix in, a random sample of 10 of these animals is selected. Let X = the number of tagged animals in the second sample. If there are actually 25 animals of this type in the region, what is the probability that (a) $X = 2$? (b) $X \leq 2$?

The parameter values are $n = 10, M = 5$ (5 tagged animals in the population), and $N = 25$, so

$$h(x;\ 10,\ 5,\ 25) = \frac{\binom{5}{x}\binom{20}{10-x}}{\binom{25}{10}} \qquad x = 0, 1, 2, 3, 4, 5$$

For (a),

$$P(X = 2) = h(2;\ 10,\ 5,\ 25) = \frac{\binom{5}{2}\binom{20}{8}}{\binom{25}{10}} = .385$$

For (b),

$$P(X \leq 2) = P(X = 0, 1, \text{ or } 2) = \sum_{x=0}^{2} h(x;\ 10,\ 5,\ 25)$$
$$= .057 + .257 + .385 = .699 \qquad \blacksquare$$

Comprehensive tables of the hypergeometric distribution are available, but because the distribution has three parameters, these tables require much more space than tables for the binomial distribution.

The Mean and Variance of X

As in the binomial case, there are simple expressions for $E(X)$ and $V(X)$ for hypergeometric rv's.

Proposition

The mean and variance of the hypergeometric rv X having pmf $h(x;\ n,\ M,\ N)$ are

$$E(X) = n \cdot \frac{M}{N}, \qquad V(X) = \left(\frac{N-n}{N-1}\right) \cdot n \cdot \frac{M}{N} \cdot \left(1 - \frac{M}{N}\right)$$

The ratio M/N is the proportion of S's in the population. If we replace M/N by p in $E(X)$ and $V(X)$, we get

$$E(X) = np \tag{3.16}$$
$$V(X) = \left(\frac{N-n}{N-1}\right) \cdot np(1-p)$$

Expression (3.16) shows that the means of the binomial and hypergeometric rv's are equal, while the variances of the two rv's differ by the factor $(N - n)/(N - 1)$, often called the **finite population correction factor.** This factor is less than 1, so the hypergeometric variable has smaller variance than does the binomial rv. The correction factor can be written $(1 - n/N)/(1 - 1/N)$, which is approximately 1 when n is small relative to N.

Example 3.36
(Example 3.35 continued)

In the animal-tagging example, $n = 10$, $M = 5$, and $N = 25$, so $p = \frac{5}{25} = .2$ and

$$E(X) = 10(.2) = 2$$

$$V(X) = \frac{15}{24}(10)(.2)(.8) = (.625)(1.6) = 1$$

If the sampling was carried out with replacement, $V(X) = 1.6$. Suppose the population size N is not actually known, so the value x is observed and we wish to estimate N. It is reasonable to equate the observed sample proportion of S's, x/n, with the population proportion, M/N, giving the estimate

$$\hat{N} = \frac{M \cdot n}{x}$$

If $M = 100$, $n = 40$, and $x = 16$, then $\hat{N} = 250$. ∎

Approximating Hypergeometric Probabilities

Our general rule of thumb in Section 3.4 stated that if sampling was without replacement but n/N was at most .05, then the binomial distribution could be used to compute approximate probabilities involving the number of S's in the sample. A more precise statement is as follows: Let the population size, N, and number of population S's, M, get large with the ratio M/N remaining fixed at p. Then $h(x; n, M, N)$ approaches $b(x; n, p)$, so for n/N small, the two are approximately equal provided that p is not too near either 0 or 1. This is the rationale for our rule of thumb.

The Negative Binomial Distribution

The negative binomial rv and distribution are based on an experiment satisfying the following conditions:

1. The experiment consists of a sequence of independent trials.
2. Each trial can result in either a success (S) or a failure (F).
3. The probability of success is constant from trial to trial, so $P(S$ on trial $i) = p$ for $i = 1, 2, 3 \ldots$.
4. The experiment continues (trials are performed) until a total of r successes have been observed, where r is a specified positive integer.

The random variable of interest is $X = $ the number of failures that precede the rth success; X is called a **negative binomial random variable** because, in contrast to the binomial rv, the number of successes is fixed and the number of trials is random.

Possible values of X are $0, 1, 2, \ldots$. Let $nb(x; r, p)$ denote the pmf of X. The event $\{X = x\}$ is equivalent to $\{r - 1$ S's in the first $(x + r - 1)$ trials and an S on the $(x + r)$th trial} (e.g., if $r = 5$ and $x = 10$, then there must be four S's in the first 14 trials and the trial 15 must be an S). Since trials are independent,

$$nb(x; r, p) = P(X = x) \tag{3.17}$$
$$= P(r - 1 \ S\text{'s on the first } x + r - 1 \text{ trials}) \cdot P(S)$$

The first probability on the far right of Expression (3.17) is the binomial probability

$$\binom{x + r - 1}{r - 1} p^{r-1}(1 - p)^x, \qquad \text{while } P(S) = p$$

Proposition

> The pmf of the negative binomial rv X with parameters r = number of S's and $p = P(S)$ is
>
> $$nb(x; r, p) = \binom{x + r - 1}{r - 1} p^r(1 - p)^x, \qquad x = 0, 1, 2, \ldots$$

Example 3.37   A pediatrician wishes to recruit 5 couples, each of whom is expecting their first child, to participate in a new natural childbirth regimen. Let $p = P$(a randomly selected couple agrees to participate). If $p = .2$, what is the probability that 15 couples must be asked before 5 are found who agree to participate? That is, with $S = \{$agrees to participate$\}$, what is the probability that 10 F's occur before the fifth S? Substituting $r = 5$, $p = .2$, and $x = 10$ into $nb(x; r, p)$ gives

$$nb(10; 5, .2) = \binom{14}{4}(.2)^5(.8)^{10} = .034$$

The probability that at most 10 F's are observed (at most 15 couples are asked) is

$$P(X \leq 10) = \sum_{x=0}^{10} nb(x; 5, .2) = (.2)^5 \sum_{x=0}^{10} \binom{x + 4}{4}(.8)^x = .164$$

■

In some sources, the negative binomial rv is taken to be the number of trials $X + r$ rather than the number of failures.

In the special case $r = 1$, the pmf is

$$nb(x; 1, p) = (1 - p)^x p, \qquad x = 0, 1, 2, \ldots \tag{3.18}$$

In Example 3.10, we derived the pmf for the number of trials necessary to obtain the first S, and the pmf there is similar to Expression (3.18). Both X = number of F's and Y = number of trials ($= 1 + X$) are referred to in the literature as **geometric random variables,** and the pmf (3.18) is called the **geometric distribution.**

In Example 3.18, the expected number of trials until the first S was shown to be $1/p$, so that the expected number of F's until the first S is $(1/p) - 1 = (1 - p)/p$. Intuitively, we would expect to see $r \cdot (1 - p)/p$ F's before the rth S, and this is indeed $E(X)$. There is also a simple formula for $V(X)$.

Proposition

If X is a negative binomial rv with pmf $nb(x; r, p)$, then

$$E(X) = \frac{r(1-p)}{p}, \qquad V(X) = \frac{r(1-p)}{p^2}$$

Finally, by expanding the binomial coefficient in front of $p^r(1-p)^x$ and doing some cancellation, it can be seen that $nb(x; r, p)$ is well defined even when r is not an integer. This *generalized negative binomial distribution* has been found to fit observed data quite well in a wide variety of applications.

Exercises   **Section 3.5** (62–72)

62. A shipment of 15 concrete cylinders has been received by a contractor, 5 for a small project and the other 10 for a larger project. Suppose that 6 of the 15 have a crushing strength below the specified minimum. If the 5 for the smaller project are randomly selected from the 15 and X = the number among the 5 that have a below minimum crushing strength, then X has a hypergeometric distribution with parameters $n = 5$, $M = 6$, and $N = 15$. Compute the following:

a. $P(X = 2)$

b. $P(X \le 2)$

c. $P(X \ge 2)$

d. $E(X)$ and $V(X)$

63. Each of 12 refrigerators of a certain type has been returned to a distributor because of the presence of a high-pitched oscillating noise when the refrigerator is running. Suppose that 4 of these 12 have defective compressors and the other 8 have less serious problems. If they are examined in random order, let X = the number among the first 6 examined that have a defective compressor. Compute the following:

a. $P(X = 1)$

b. $P(X \ge 4)$

c. $P(1 \le X \le 3)$

64. A tennis coach has a basket containing 25 balls; 15 of these are Penn balls and the other 10 are Wilsons. Each of four players randomly selects 3 balls for a match.

a. What is the probability that exactly 8 of the balls selected are Penns?

b. What are the mean value and standard deviation of the number of Penn balls selected?

c. What is the probability that the number of Penn balls selected is more than 1 SD away from its mean value?

d. What are the mean value and standard deviation of the number of Penn balls left in the basket?

65. A geologist has collected 10 specimens of basaltic rock and 10 specimens of granite. If the geologist instructs a laboratory assistant to randomly select 15 of the specimens for analysis, what is the pmf of the number of basalt specimens selected for analysis? What is the probability that all specimens of one of the two types of rock are selected for analysis?

66. A personnel director interviewing 11 senior engineers for four job openings has scheduled six interviews for the first day and five for the second day of interviewing. Assume that the candidates are interviewed in random order.

a. What is the probability that x of the top four candidates are interviewed on the first day?

b. How many of the top four candidates can be expected to be interviewed on the first day?

67. Twenty pairs of individuals playing in a bridge tournament have been seeded 1, . . . , 20. In the first part of the tournament, the 20 are randomly divided into 10 east–west pairs and 10 north–south pairs.

a. What is the probability that x of the top 10 pairs end up playing east–west?

b. What is the probability that all of the top five pairs end up playing the same direction?

c. If there are $2n$ pairs, what is the pmf of X = the number among the top n pairs who end up playing east–west, and what are $E(X)$ and $V(X)$?

68. A second-stage smog alert has been called in a certain area of Los Angeles County in which there are 50 industrial firms. An inspector will visit 10 randomly selected firms to check for violations of regulations.

   a. If 15 of the firms are actually violating at least one regulation, what is the pmf of the number of firms visited by the inspector that are in violation of at least one regulation?

   b. If there are 500 firms in the area, of which 150 are in violation, approximate the pmf of part (a) by a simpler pmf.

   c. For X = the number among the 10 visited that are in violation, compute $E(X)$ and $V(X)$ both for the exact pmf and the approximating pmf in part (b).

69. Suppose that $p = P(\text{male birth}) = .5$. A couple wishes to have exactly two female children in their family. They will have children until this condition is fulfilled.

a. What is the probability that the family has x male children?

b. What is the probability that the family has four children?

c. What is the probability that the family has at most four children?

d. How many male children would you expect this family to have? How many children would you expect this family to have?

70. A family decides to have children until it has three children of the same sex. Assuming $P(B) = P(G) = .5$, what is the pmf of X = the number of children in the family?

71. Three brothers and their wives decide to have children until each family has two female children. What is the pmf of X = the total number of male children born to the brothers? What is $E(X)$, and how does it compare to the expected number of male children born to each brother?

72. Individual A has a red die and B has a green die (both fair). If they each roll until they obtain five "doubles" (1–1, . . . , 6–6), what is the pmf of X = the total number of times a die is rolled? What are $E(X)$ and $V(X)$?

3.6   The Poisson Probability Distribution

The binomial, hypergeometric, and negative binomial distributions were all derived by starting with an experiment consisting of trials or draws and applying the laws of probability to various outcomes of the experiment. There is no simple experiment on which the Poisson distribution is based, though we will shortly describe how it can be obtained by certain limiting operations.

Definition

> An rv X is said to have a **Poisson distribution** if the pmf of X is
>
> $$p(x;\ \lambda) = \frac{e^{-\lambda}\lambda^{x}}{x!}, \qquad x = 0, 1, 2, \ldots$$
>
> for some $\lambda > 0$.

The value of λ is frequently a rate per unit time or per unit area. The letter e in $p(x;\ \lambda)$ represents the base of the natural logarithm system, and its numerical value is approximately 2.71828. Because λ must be positive, $p(x;\ \lambda) > 0$ for all possible *x* values. The fact that $\sum_{x=0}^{\infty} p(x; \lambda) = 1$ is a consequence of the Maclaurin infinite series expansion of e^{λ}, which appears in most calculus texts:

$$e^{\lambda} = 1 + \lambda + \frac{\lambda^2}{2!} + \frac{\lambda^3}{3!} + \cdots = \sum_{x=0}^{\infty} \frac{\lambda^x}{x!} \tag{3.19}$$

If the two extreme terms in Expression (3.19) are multiplied by $e^{-\lambda}$ and then $e^{-\lambda}$ is placed inside the summation, the result is

$$1 = \sum_{x=0}^{\infty} e^{-\lambda} \frac{\lambda^x}{x!}$$

which shows that $p(x; \lambda)$ fulfills the second condition necessary for specifying a pmf.

Example 3.38 Let *X* denote the number of flaws on the surface of a randomly selected boiler of a certain type. Suppose *X* has a Poisson distribution with $\lambda = 5$. Then the probability that a randomly selected boiler has exactly two flaws is

$$P(X = 2) = \frac{e^{-5}(5)^2}{2!} = .084$$

The probability that a boiler contains at most two flaws is

$$P(X \leq 2) = \sum_{x=0}^{2} \frac{e^{-5}(5)^x}{x!} = e^{-5}\left(1 + 5 + \frac{25}{2}\right) = .125 \qquad \blacksquare$$

The Poisson Distribution as a Limit

The rationale for using the Poisson distribution in many situations is provided by the following proposition.

Proposition

> Suppose that in the binomial pmf $b(x; n, p)$, we let $n \to \infty$ and $p \to 0$ in such a way that np remains fixed at value $\lambda > 0$. Then $b(x; n, p) \to p(x; \lambda)$.

According to this proposition, *in any binomial experiment in which n is large and p is small, $b(x; n, p) \approx p(x; \lambda)$ where $\lambda = np$*. As a rule of thumb, this approximation can safely be applied if $n \geq 100$, $p \leq .01$, and $np \leq 20$.

Example 3.39 If a publisher of nontechnical books takes great pains to ensure that its books are free of typographical errors, so that the probability of any given page containing at least one such error is .005 and errors are independent from page to page, what is the probability that one of its 400-page novels will contain exactly 1 page with errors? At most three pages with errors?

With *S* denoting a page containing at least one error and *F* an error-free page, the number *X* of pages containing at least one error is a binomial rv with $n = 400$ and $p = .005$, so $np = 2$. We wish

$$P(X = 1) = b(1; 400, .005) \approx p(1; 2) = \frac{e^{-2}(2)^1}{1!} = .271$$

Similarly,

$$P(X \le 3) \approx \sum_{x=0}^{3} p(x;\, 2) = \sum_{x=0}^{3} e^{-2}\, \frac{2^x}{x!} = .135 + .271 + .271 + .180$$

$$= .857 \qquad \blacksquare$$

Appendix Table A.2 exhibits the cdf $F(x;\, \lambda)$ for $\lambda = .1, .2, \ldots, 1, 2, \ldots, 10$, 15, and 20. For example, if $\lambda = 2$, then $P(X \le 3) = F(3;\, 2) = .857$ as in Example 3.39, while $P(X = 3) = F(3;\, 2) - F(2;\, 2) = .180$.

The Mean and Variance of X

Since $b(x;\, n, p) \to p(x;\, \lambda)$ as $n \to \infty$, $p \to 0$, $np = \lambda$, the mean and variance of a binomial variable should approach those of a Poisson variable. These limits are $np = \lambda$ and $np(1 - p) \to \lambda$.

Proposition

> If X has a Poisson distribution with parameter λ, then $E(X) = V(X) = \lambda$.

These results can also be derived directly from the definitions of mean and variance.

Example 3.40
(Example 3.38 continued)

Both the expected number of flaws and the variance of the number of flaws equal five, and $\sigma_X = \sqrt{\lambda} = \sqrt{5} = 2.24$. $\blacksquare$

The Poisson Process

A very important application of the Poisson distribution arises in connection with the occurrence of events of a particular type over time. As an example, suppose that starting from a time point that we label $t = 0$, we are interested in counting the number of radioactive pulses recorded by a Geiger counter. We make the following assumptions about the way in which pulses occur:

1. There exists a parameter $\alpha > 0$ such that for any short time interval of length Δt, the probability that exactly one pulse is received is $\alpha \cdot \Delta t + o(\Delta t)$.*
2. The probability of more than one pulse being received during Δt is $o(\Delta t)$ [which, along with Assumption 1, implies that the probability of no pulses during Δt is $1 - \alpha \cdot \Delta t - o(\Delta t)$].
3. The number of pulses received during the time interval Δt is independent of the number received prior to this time interval.

Informally, Assumption 1 says that, for a short interval of time, the probability of receiving a single pulse is approximately proportional to the length of the time interval, where α is the constant of proportionality. Now let $P_k(t)$ denote the

---

* A quantity is $o(\Delta t)$ (read "little o of delta t") if as Δt approaches 0, so does $o(\Delta t)/\Delta t$. That is, $o(\Delta t)$ is even more negligible than Δt itself. The quantity $(\Delta t)^2$ has this property, but $\sin(\Delta t)$ does not.

probability that k pulses will be received by the counter during any particular time interval of length t.

Proposition

$P_k(t) = e^{-\alpha t} \cdot (\alpha t)^k/k!$, so that the number of pulses during a time interval of length t is a Poisson rv with parameter $\lambda = \alpha t$. The expected number of pulses during any such time interval is then αt, so the expected number during a unit interval of time is α.

Example 3.41   Suppose pulses arrive at the counter at an average rate of six per minute, so that $\alpha = 6$. To find the probability that in a .5-min interval at least one pulse is received, note that the number of pulses in such an interval has a Poisson distribution with parameter $\alpha t = 6(.5) = 3$ (.5 min is used because α is expressed as a rate per minute). Then with $X =$ the number of pulses received in the 30-sec interval,

$$P(1 \le X) = 1 - P(X = 0) = 1 - \frac{e^{-3}(3)^0}{0!} = .950$$

∎

If in Assumptions 1–3 we replace "pulse" by "event," then the number of events occurring during a fixed time interval of length t has a Poisson distribution with parameter αt. Any process that has this distribution is called a **Poisson process,** and α is called the *rate of the process.* Other examples of situations giving rise to a Poisson process include monitoring the status of a computer system over time, with breakdowns constituting the events of interest, recording the number of accidents in an industrial facility over time, answering calls at a telephone switchboard, and observing the number of cosmic-ray showers from a particular observatory over time.

Instead of observing events over time, consider observing events of some type that occur in a two- or three-dimensional region. For example, we might select on a map a certain region R of a forest, go to that region, and count the number of trees. Each tree would represent an event occurring at a particular point in space. Under assumptions similar to 1–3, it can be shown that the number of events occurring in a region R has a Poisson distribution with parameter $\alpha \cdot a(R)$, where $a(R)$ is the area of R. The quantity α is the expected number of events per unit area or volume.

Exercises   Section 3.6 (73–87)

73. Let X have a Poisson distribution with parameter $\lambda = 5$ and use Appendix Table A.2 to compute the following probabilities.

   a. $P(X \le 8)$

   b. $P(X = 8)$

   c. $P(9 \le X)$

   d. $P(5 \le X \le 8)$

   e. $P(5 < X < 8)$

74. Suppose the number X of tornadoes observed in a particular region during a 1-year period has a Poisson distribution with $\lambda = 8$.

   a. Compute $P(X \le 5)$.

b. Compute $P(6 \leq X \leq 9)$.

c. Compute $P(10 \leq X)$.

d. How many tornadoes can be expected to be observed during the 1-year period, and what is the standard deviation of the number of observed tornadoes?

75. Let X = the number of automobiles of a particular year and model that will at some time in the future suffer a catastrophic failure in the steering mechanism, causing complete loss of control at high speed. Suppose X has a Poisson distribution with parameter $\lambda = 10$.

a. What is the probability that at most 10 cars suffer such a failure?

b. What is the probability that between 10 and 15 (inclusive) cars will suffer such a failure?

c. What are $E(X)$ and σ_X?

76. Consider writing onto a computer disk and then sending it through a certifier that counts the number of missing pulses. Suppose this number X has a Poisson distribution with parameter $\lambda = .2$. (Suggested in "Average Sample Number for Semi-Curtailed Sampling Using the Poisson Distribution," *J. Quality Technology,* 1983, pp. 126–129).

a. What is the probability that a disk has exactly one missing pulse?

b. What is the probability that a disk has at least two missing pulses?

c. If two disks are independently selected, what is the probability that neither contains a missing pulse?

77. An article in the *Los Angeles Times* (Dec. 3, 1993) reports that 1 in 200 people carry the defective gene that causes inherited colon cancer. In a sample of 1000 individuals, what is the approximate distribution of the number who carry this gene? Use this distribution to calculate the approximate probability that

a. Between 4 and 7 (inclusive) carry the gene.

b. At least 8 carry the gene.

78. A notice is sent to all owners of a certain type of automobile, asking them to bring their cars to a dealer to check for the presence of a particular manufacturing defect. Suppose that only .05% of such cars have the defect. Consider a random sample of 10,000 cars.

a. What are the expected value and standard deviation of the number of cars in the sample that have the defect?

b. What is the (approximate) probability that at least 10 sampled cars have the defect?

c. What is the (approximate) probability that no sampled cars have the defect?

79. Suppose small aircraft arrive at a certain airport according to a Poisson process with rate $\alpha = 8$ per hour, so that the number of arrivals during a time period of t hours is a Poisson rv with parameter $\lambda = 8t$.

a. What is the probability that exactly 5 small aircraft arrive during a 1-hour period? At least 5? At least 10?

b. What are the expected value and standard deviation of the number of small aircraft that arrive during a 90-min period?

c. What is the probability that at least 20 small aircraft arrive during a 2½-hour period? That at most 10 arrive during this period?

80. The number of tickets issued by a meter reader for parking-meter violations can be modeled by a Poisson process with a rate parameter of five per hour.

a. What is the probability that exactly five tickets are given out during a particular hour?

b. What is the probability that at least three tickets are given out during a particular hour?

c. How many tickets do you expect to be given during a 45-min period?

81. The number of requests for assistance received by a towing service is a Poisson process with rate $\alpha = 4$ per hour.

a. Compute the probability that exactly ten requests are received during a particular 2-hour period.

b. If the operators of the towing service take a 30-min break for lunch, what is the probability that they do not miss any calls for assistance?

c. How many calls would you expect during their break?

82. In proof testing of circuit boards, the probability that any particular diode will fail is .01. Suppose a circuit board contains 200 diodes.

a. How many diodes would you expect to fail, and what is the standard deviation of the number that are expected to fail?

b. What is the (approximate) probability that at least four diodes will fail on a randomly selected board?

c. If five boards are shipped to a particular customer, how likely is it that at least four of them will work properly? (A board works properly only if all its diodes work.)

83. The paper "Reliability-Based Service-Life Assessment of Aging Concrete Structures" (*J. Structural Engr.*, 1993, pp. 1600–1621) suggests that a Poisson process can be used to represent the occurrence of structural loads over time. Suppose the mean time between occurrences of loads is .5 year.

a. How many loads can be expected to occur during a 2-year period?

b. What is the probability that more than five loads occur during a 2-year period?

c. How long must a time period be so that the probability of no loads occurring during that period is at most .1?

84. Let X have a Poisson distribution with parameter λ. Show that $E(X) = \lambda$ directly from the definition of expected value. (HINT: The first term in the sum equals 0, and then x can be canceled. Now factor out λ and show that what is left sums to 1.)

85. Suppose that trees are distributed in a forest according to a two-dimensional Poisson process with parameter α, the expected number of trees per acre, equal to 80.

a. What is the probability that in a certain quarter-acre plot, there will be at most 16 trees?

b. If the forest covers 85,000 acres, what is the expected number of trees in the forest?

c. Suppose you select a point in the forest and construct a circle of radius .1 mile. Let X = the number of trees within that circular region. What is the pmf of X? (HINT: 1 sq mile = 640 acres.)

86. Automobiles arrive at a vehicle equipment inspection station according to a Poisson process with rate $\alpha = 10$ per hour. Suppose that with probability .5 an arriving vehicle will have no equipment violations.

a. What is the probability that exactly ten arrive during the hour and all ten have no violations?

b. For any fixed $y \geq 10$, what is the probability that y arrive during the hour, of which ten have no violations?

c. What is the probability that ten "no-violation" cars arrive during the next hour? [HINT: Sum the probabilities in part (b) from $y = 10$ to ∞.]

87. a. In a Poisson process, what has to happen in both the time interval $(0, t)$ and the interval $(t, t + \Delta t)$ in order that there be no events occurring in the entire interval $(0, t + \Delta t)$? Use this and Assumptions 1–3 to write a relationship between $P_0(t + \Delta t)$ and $P_0(t)$.

b. Use the result of part (a) to write an expression for the difference $P_0(t + \Delta t) - P_0(t)$. Then divide by Δt and let $\Delta t \to 0$ to obtain an equation involving $(d/dt)P_0(t)$, the derivative of $P_0(t)$ with respect to t.

c. Verify that $P_0(t) = e^{-\alpha t}$ satisfies the equation of part (b).

d. It can be shown in a manner similar to parts (a) and (b) that the $P_k(t)$'s must satisfy the system of differential equations

$$\frac{d}{dt} P_k(t) = \alpha P_{k-1}(t) - \alpha P_k(t)$$

$$k = 1, 2, 3, \ldots$$

Verify that $P_k(t) = e^{-\alpha t}(\alpha t)^k/k!$ satisfies the system. (This is actually the only solution.)

Supplementary Exercises for Chapter 3 (88–114)

88. Consider a deck consisting of seven cards, marked 1, 2, . . . , 7. Three of these cards are selected at random. Define an rv W by W = the sum of the resulting numbers, and compute the pmf of W. Then compute μ and σ^2. [HINT: Consider outcomes as unordered, so that (1, 3, 7) and (3, 1, 7) are not different outcomes. Then there are 35 outcomes, and they can be listed. (This type of rv actually arises in connection with a hypothesis test called Wilcoxon's rank sum test, in which there is an x sample and a y sample and W is the sum of the ranks of the x's in the combined sample.)]

89. After shuffling a deck of 52 cards, a dealer deals out 5. Let X = the number of suits represented in the 5-card hand.

a. Show that the pmf of X is

x	1	2	3	4
$p(x)$	.002	.146	.588	.264

[HINT: $p(1) = 4P$(all are spades), $p(2) = 6P$(only spades and hearts with at least one of each), and $p(4) = 4P(2$ spades $\cap$ one of each other suit).]

b. Compute μ, σ^2, and σ.

90. The negative binomial rv X was defined as the number of F's preceding the rth S. Let Y = the number of trials necessary to obtain the rth S. In the same manner in which the pmf of X was derived, derive the pmf of Y.

91. Of all customers purchasing automatic garage-door openers, 75% purchase a chain-driven model. Let X = the number among the next 15 purchasers who select the chain-driven model.

a. What is the pmf of X?

b. Compute $P(X \leq 12)$.

c. Compute $P(9 \leq X \leq 12)$.

d. Compute μ and σ^2.

e. If the store currently has in stock 10 chain-driven models and 8 shaft-driven models, what is the probability that the requests of these 15 customers can all be met from existing stock?

92. A friend recently planned a camping trip. He had two flashlights, one that required a single 6-V battery and another that used two size-D batteries. He had previously packed two 6-V and four size-D batteries in his camper. Suppose the probability that any particular battery works is p and that batteries work or fail independently of one another. Our friend wants to take just one flashlight. For what values of p should he take the 6-V flashlight?

93. A *k-out-of-n system* is one that will function if and only if at least k of the n individual components in the system function. If individual components function independently of one another, each with probability .9, what is the probability that a 3-out-of-5 system functions?

94. A manufacturer of flashlight batteries wishes to control the quality of its product by rejecting any lot in which the proportion of batteries having unacceptable voltage appears to be too high. To this end, out of each large lot (10,000 batteries), 25 will be selected and tested. If at least 5 of these generate an unacceptable voltage, the entire lot will be rejected. What is the probability that a lot will be rejected if

a. 5% of the batteries in the lot have unacceptable voltage?

b. 10% of the batteries in the lot have unacceptable voltage?

c. 20% of the batteries in the lot have unacceptable voltage?

d. What would happen to the above probabilities if the critical rejection number were increased from 5 to 6?

95. Of the people passing through an airport metal detector, .1% activate it; let X = the number among a randomly selected group of 500 who activate the detector.

a. What is the (approximate) pmf of X?

b. Compute $P(X = 2)$.

c. Compute $P(2 \leq X)$.

96. An educational consulting firm is trying to decide whether high school students who have never before used a hand-held calculator can solve a certain type of problem more easily with a calculator that uses reverse Polish logic or one that does not use this logic. A sample of 25 students is selected and allowed to practice on both calculators. Then each student is asked to work one problem on the reverse Polish calculator and a similar problem on the other. Let $p = P(S)$ where S indicates that a student worked the problem more quickly using reverse Polish logic than without and let X = number of S's.

a. If $p = .5$, what is $P(7 \leq X \leq 18)$?

b. If $p = .8$, what is $P(7 \leq X \leq 18)$?

c. If the claim that $p = .5$ is to be rejected when either $X \leq 7$ or $X \geq 18$, what is the probability of rejecting the claim when it is actually correct?

d. If the decision to reject the claim $p = .5$ is made as in part (c), what is the probability that the claim is not rejected when $p = .6$? When $p = .8$?

e. What decision rule would you choose for rejecting the claim $p = .5$ if you wanted the probability in part (c) to be at most .01?

97. Consider a disease whose presence can be identified by carrying out a blood test. Let p denote the probability that a randomly selected individual has the disease. Suppose n individuals are independently selected for testing. One way to proceed is to carry out a separate test on each of the n blood samples. A potentially more economical approach, group testing, was introduced during World War II to identify syphilitic men among army inductees. First take a part of each blood sample, combine these specimens, and carry out a single test. If no one has the disease, the result will be negative, and only the one test is required. If at least one individual is diseased, the test on the combined sample will yield a positive result, in which case the n individual tests are then carried out. If $p = .1$ and $n = 3$, what is the expected number of tests using this procedure? What is the expected number when $n = 5$? [The paper "Random Multiple-Access Communication and Group Testing" (*IEEE Trans. on Commun.*, 1984, pp. 769–774) applied these ideas to a communication system in which the dichotomy was active/idle user rather than diseased/nondiseased.]

98. Let p_1 denote the probability that any particular code symbol is erroneously transmitted through a communication system. Assume that on different symbols, errors occur independently of one another. Suppose also that with probability p_2 an erroneous symbol is corrected upon receipt. Let X denote the number of correct symbols in a message block consisting of n symbols (after the correction process has ended). What is the probability distribution of X?

99. The purchaser of a power-generating unit requires c consecutive successful start-ups before the unit will be accepted. Assume that the outcomes of individual start-ups are independent of one another and let p denote the probability that any particular start-up is successful. The random variable of interest is $X =$ the number of start-ups that must be made prior to acceptance. Give the pmf of X for the case $c = 2$. If $p = .9$, what is $P(X \le 8)$? [HINT: For $x \ge 5$, express $p(x)$ "recursively" in terms of the pmf evaluated at the smaller values $x - 3, x - 4, \ldots, 2$.] (This problem was suggested by the paper "Evaluation of a Start-Up Demonstration Test," *J. Quality Technology*, 1983, pp. 103–106.)

100. A plan for an executive traveler's club has been developed by an airline on the premise that 10% of its current customers would qualify for membership.

a. Assuming the validity of this premise, among 25 randomly selected current customers, what is the probability that between 2 and 6 (inclusive) qualify for membership?

b. Again assuming the validity of the premise, what are the expected number of customers who qualify and the standard deviation of the number who qualify in a random sample of 100 current customers?

c. Let X denote the number in a random sample of 25 current customers who qualify for membership. Consider rejecting the company's premise in favor of the claim that $p > .10$ if $x \ge 7$. What is the probability that the company's premise is rejected when it is actually valid?

d. Refer to the decision rule introduced in part (c). What is the probability that the company's premise is not rejected even though $p = .20$ (i.e., 20% qualify)?

101. Of all repair work done on TV sets by a certain store, 80% is done on sets that are no longer under warranty [so if $S = \{$not under warranty$\}$, $P(S) = .8$].

a. Among 20 sets brought in for repair work, what is the expected number not under warranty?

b. Among these 20 sets, what is the expected number under warranty?

c. Among these 20 sets, what is the probability that at least 75% are not under warranty?

d. Suppose that there are 12 sets now in the shop of which 4 are under warranty. If the 12 were brought in in completely random order and are serviced in the same order, what is the pmf of $X =$ the number of sets under warranty among the first 5 serviced? What are $E(X)$ and $V(X)$?

102. A trial has just resulted in a hung jury because eight members of the jury were in favor of a guilty verdict while the other four were for acquittal. If the jurors leave the jury room in random order and each of the first four leaving the room is accosted by a reporter in quest of an interview, what is the pmf of $X =$ the number of jurors favoring acquittal among those interviewed? How many of those favoring acquittal do you expect to be interviewed?

103. A telephone company employs five information operators who receive requests for information independently of one another, each according to a Poisson process with rate $\alpha = 2$ per minute.

a. What is the probability that during a given 1-min period, the first operator receives no requests?

b. What is the probability that during a given 1-min period, exactly four of the five operators receive no requests?

c. Write an expression for the probability that during a given 1-min period, all of the operators receive exactly the same number of requests.

104. Grasshoppers are distributed at random in a large field according to a Poisson distribution with parameter $\alpha = 2$ per square yard. How large should the radius R of a circular sampling region be taken so that the probability of finding at least one in the region equals .99?

105. A newsstand has ordered five copies of a certain issue of a photography magazine. Let X = the number of individuals who come in to purchase this magazine. If X has a Poisson distribution with parameter $\lambda = 4$, what is the expected number of copies that are sold?

106. Individuals A and B begin to play a sequence of chess games. Let $S = \{A$ wins a game$\}$ and suppose that outcomes of successive games are independent with $P(S) = p$ and $P(F) = 1 - p$ (they never draw). They will play until one of them wins ten games. Let X = the number of games played (with possible values 10, 11, . . . , 19).

a. For $x = 10, 11, . . . , 19$, obtain an expression for $p(x) = P(X = x)$.

b. If a draw is possible, with $p = P(S)$, $q = P(F)$, $1 - p - q = P(\text{draw})$, what are the possible values of X? What is $P(20 \leq X)$? [HINT: $P(20 \leq X) = 1 - P(X < 20)$.]

107. A test for the presence of a certain disease has probability .20 of giving a false-positive reading (indicating that an individual has the disease when this is not the case) and probability .10 of giving a false-negative result. Suppose that ten individuals are tested, five of whom have the disease and five who do not. Let X = the number of positive readings that result.

a. Does X have a binomial distribution? Explain your reasoning.

b. What is the probability that exactly three of the ten test results are positive?

108. The generalized negative binomial pmf is given by

$$nb(x; r, p) = k(r, x) \cdot p^r(1 - p)^x$$

$$x = 0, 1, 2, . . .$$

Let X, the number of plants of a certain species found in a particular region, have this distribution with $p = .3$ and $r = 2.5$. What is $P(X = 4)$? What is the probability that at least one plant is found?

109. Define a function $p(x; \lambda, \mu)$ by

$p(x; \lambda, \mu)$

$$= \begin{cases} \dfrac{1}{2} e^{-\lambda} \dfrac{\lambda^x}{x!} + \dfrac{1}{2} e^{-\mu} \dfrac{\mu^x}{x!} & x = 0, 1, 2, . . . \\ 0 & \text{otherwise} \end{cases}$$

a. Show that $p(x; \lambda, \mu)$ satisfies the two conditions necessary for specifying a pmf. [NOTE: If a firm employs two typists, one of whom makes typographical errors at the rate of λ per page and the other at rate μ per page and they each do half the firm's typing, then $p(x; \lambda, \mu)$ is the pmf of X = the number of errors on a randomly chosen page.]

b. If the first typist (rate λ) types 60% of all pages, what is the pmf of X of part (a)?

c. What is $E(X)$ for $p(x; \lambda, \mu)$ given by the displayed expression above?

d. What is σ^2 for $p(x; \lambda, \mu)$ given by that expression?

110. The *mode* of a discrete rv X with pmf $p(x)$ is that value x^* for which $p(x)$ is largest (the most probable x value).

a. Let $X \sim \text{Bin}(n, p)$. By considering the ratio $b(x + 1; n, p)/b(x; n, p)$, show that $b(x; n, p)$ increases with x as long as $x < np - (1 - p)$. Conclude that the mode x^* is the integer satisfying $(n + 1)p - 1 \leq x^* \leq (n + 1)p$.

b. Show that if X has a Poisson distribution with parameter λ, the mode is the largest integer less than λ. If λ is an integer, show that both $\lambda - 1$ and λ are modes.

111. A computer disk storage device has ten concentric tracks, numbered 1, 2, . . . , 10 from outermost to innermost, and a single access arm. Let p_i = the probability that any particular requests for data will take the arm to track $i(i = 1, . . . , 10)$. Assume that the tracks accessed in successive seeks are independent and let X = the number of tracks over which the access arm passes during two successive requests (excluding the track that the arm has just left, so possible X values are $x = 0, 1, . . . , 9$). Compute the pmf of X. [HINT: $P(\text{the arm is now on track } i \text{ and } X = j) = P(X = j \mid \text{arm now on } i) \cdot p_i$. After the conditional probability is written in terms of $p_1, . . . , p_{10}$, by the law of total probability, the desired probability is obtained by summing over i.]

112. If X is a hypergeometric rv, show directly from the definition that $E(X) = nM/N$ (consider only the case $n < M$). [HINT: Factor nM/N out of the sum for $E(X)$, and show that the terms inside the sum are of the form $h(y; n - 1, M - 1, N - 1)$ where $y = x - 1$.]

113. Use the fact that

$$\sum_{\text{all } x} (x - \mu)^2 p(x) \geq \sum_{x;|x - \mu| \geq k\sigma} (x - \mu)^2 p(x)$$

to prove Chebyshev's inequality given in Exercise 43.

114. The simple Poisson process of Section 3.6 is characterized by a constant rate α at which events occur per unit time. A generalization of this is to suppose that the probability of exactly one event occurring in the interval $[t, t + \Delta t]$ is $\alpha(t) \cdot \Delta t + o(\Delta t)$. It can then be shown that the number of events occurring during an interval $[t_1, t_2]$ has a Poisson distribution with parameter

$$\lambda = \int_{t_1}^{t_2} \alpha(t) \, dt$$

The occurrence of events over time in this situation is called a *nonhomogeneous Poisson process*. The paper "Inference Based on Retrospective Ascertainment," *J. Amer. Stat. Assoc.*, 1989, pp. 360–372 considers the intensity function

$$\alpha(t) = e^{a + bt}$$

as appropriate for events involving transmission of HIV (the AIDS virus) via blood transfusions. Suppose that $a = 2$ and $b = .6$ (close to values suggested in the paper), with time in years.

a. What is the expected number of events in the interval $[0, 4]$? In $[2, 6]$?

b. What is the probability that at most 15 events occur in the interval $[0, .9907]$?

Bibliography

Olkin, Ingram, Cyrus Derman, and Leon Gleser, *Probability Models and Applications* (2nd ed.), Macmillan, New York, 1994. Contains an in-depth discussion of both general properties of discrete and continuous distributions and results for specific distributions.

Johnson, Norman, Samuel Kotz, and Adrienne Kemp, *Discrete Univariate Distributions,* Wiley, New York, 1992. An encyclopedia of information on discrete distributions.

Ross, Sheldon, *Introduction to Probability Models* (5th ed.), Academic Press, New York, 1993. A good source of material on the Poisson process and generalizations and a nice introduction to other topics in applied probability.

4 Continuous Random Variables and Probability Distributions

Introduction

As mentioned at the beginning of Chapter 3, not all random variables are discrete. In this chapter, we study the second general type of random variable that arises in many applied problems. Sections 4.1 and 4.2 present the basic definitions and properties of continuous random variables and their probability distributions. In Section 4.3, we study in detail the normal random variable and distribution, unquestionably the most important and useful in probability and statistics. Sections 4.4 and 4.5 discuss some other continuous distributions that are often used in applied work. In Section 4.6, we introduce a method for assessing whether given sample data is consistent with a specified distribution.

4.1 Continuous Random Variables and Probability Density Functions

A discrete random variable (rv) is one whose possible values either constitute a finite set or else can be listed in an infinite sequence (a list in which there is a first element, a second element, etc.). A random variable whose set of possible values is an entire interval of numbers is not discrete.

Continuous Random Variables

Definition

> An rv X is said to be **continuous** if its set of possible values is an entire interval of numbers—that is, if for some $A < B$, any number x between A and B is possible.

Example 4.1   If in the study of the ecology of a lake, we make depth measurements at randomly chosen locations, then $X =$ the depth at such a location is a continuous rv. Here A is the minimum depth in the region being sampled, and B is the maximum depth. ■

Example 4.2   If a chemical compound is randomly selected and its pH X is determined, then X is a continuous rv because any pH value between 0 and 14 is possible. If more is known about the compound selected for analysis, then the set of possible values might be a subinterval of [0, 14] such as $5.5 \le x \le 6.5$, but X would still be continuous. ■

If the measurement scale of X can be subdivided to any extent desired, then the variable is continuous; if it cannot, the variable is discrete. For example, if the variable is height or length, then it can be measured in kilometers, meters, centimeters, millimeters, and so on, so the variable is continuous. If, however, $X =$ the billing amount on a randomly chosen monthly natural gas statement, then the smallest unit of measurement is cents, so any value of X is a multiple of \$.01 and X is a discrete variable.

One might argue that although in principle variables such as height, weight, and temperature are continuous, in practice the limitations of our measuring instruments restrict us to a discrete (though sometimes very finely subdivided) world. However, continuous models often approximate real-world situations very well, and continuous mathematics (the calculus) is frequently easier to work with than mathematics of discrete variables and distributions.

Probability Distributions for Continuous Variables

Suppose the variable X of interest is the depth of a lake at a randomly chosen point on the surface. Let $M =$ the maximum depth (in meters), so that any number in the interval $[0, M]$ is a possible value of X. If we "discretize" X by measuring depth to the nearest meter, then possible values are nonnegative integers less than or equal to M. The resulting discrete distribution of depth can be pictured using a probability histogram. If we draw the histogram so that the area of the rectangle above any possible integer k is the proportion of the lake whose depth is (to the nearest meter) k, then the total area of all rectangles is 1. A possible histogram appears in Figure 4.1a.

(a) (b) (c)

Figure 4.1 (a) Probability histogram of depth measured to the nearest meter; (b) probability histogram of depth measured to the nearest centimeter; (c) a limit of a sequence of discrete histograms

If depth is measured to the nearest centimeter and the same measurement axis as in Figure 4.1a is used, each rectangle in the resulting probability histogram is much narrower, though the total area of all rectangles is still 1. A possible histogram is pictured in Figure 4.1b; it has a much smoother appearance than the histogram in Figure 4.1a. If we continue in this way to measure depth more and more finely, the resulting sequence of histograms approaches a smooth curve such as is pictured in Figure 4.1c. Because for each histogram the total area of all rectangles equals 1, the total area under the smooth curve is also 1. The probability that the depth at a randomly chosen point is between a and b is just the area under the smooth curve between a and b. It is exactly a smooth curve of the type pictured in Figure 4.1c that specifies a continuous probability distribution.

Definition

Let X be a continuous rv. Then a **probability distribution** or **probability density function** (pdf) of X is a function $f(x)$ such that for any two numbers a and b with $a \leq b$,

$$P(a \leq X \leq b) = \int_a^b f(x)\,dx$$

That is, the probability that X takes on a value in the interval $[a, b]$ is the area under the graph of the density function, as illustrated in Figure 4.2. The graph of $f(x)$ is often referred to as the *density curve*.

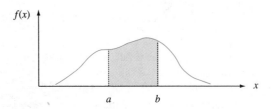

Figure 4.2 $P(a \leq X \leq b)$ = the area under the graph of $f(x)$ between a and b

For $f(x)$ to be a legitimate pdf, it must satisfy the following two conditions:

1. $f(x) \geq 0$ for all x
2. $\int_{-\infty}^{\infty} f(x)\,dx$ = area under the entire graph of $f(x)$
    $= 1$

Example 4.3

Suppose I take a bus to work, and that every 5 minutes a bus arrives at my stop. Because of variation in the time that I leave my house, I don't always arrive at the bus stop at the same time, so my waiting time X for the next bus is a continuous rv. The set of possible values of X is the interval $[0, 5]$. One possible pdf for X is

$$f(x) = \begin{cases} \dfrac{1}{5} & 0 \leq x \leq 5 \\ 0 & \text{otherwise} \end{cases}$$

The pdf $f(x)$ is graphed in Figure 4.3. Clearly $f(x) \geq 0$, and the total area under the graph is $5 \cdot (\frac{1}{5}) = 1$. The probability that I wait between 1 and 3 minutes is

$$P(1 \leq X \leq 3) = \int_1^3 f(x) \, dx = \int_1^3 \frac{1}{5} \, dx = \frac{x}{5} \Big|_{x=1}^{x=3} = \frac{2}{5}$$

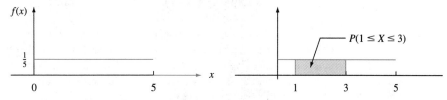

Figure 4.3 The pdf of Example 4.3

Similarly, $P(2 \leq X \leq 4) = \frac{2}{5}$. The probability that I wait at least 4 minutes is

$$P(4 \leq X) = \int_4^\infty f(x) \, dx = \int_4^5 \frac{1}{5} \, dx + \int_5^\infty 0 \, dx = \frac{x}{5} \Big|_{x=4}^{x=5} = \frac{1}{5} \qquad \blacksquare$$

Because whenever $0 \leq a \leq b \leq 5$ in Example 4.3, $P(a \leq X \leq b)$ depends only on the length $b - a$ of the interval, X is said to have a uniform distribution.

Definition

A continuous rv X is said to have a **uniform distribution** on the interval $[A, B]$ if the pdf of X is

$$f(x; A, B) = \begin{cases} \dfrac{1}{B - A} & A \leq x \leq B \\ 0 & \text{otherwise} \end{cases}$$

The graph of any uniform pdf looks like the graph in Figure 4.3 except that the interval of positive density is $[A, B]$ rather than $[0, 5]$.

In the discrete case, a probability mass function (pmf) tells us how little "blobs" of probability mass of various magnitudes are distributed along the measurement axis. In the continuous case, probability density is "smeared" in a continuous fashion along the interval of possible values. When density is smeared uniformly over the interval, a uniform pdf, as in Figure 4.3, results.

When X is a discrete rv, each possible value is assigned positive probability, but this is not the case when X is continuous.

Proposition

If X is a continuous rv, then for any number c, $P(X = c) = 0$. Furthermore, for any two numbers a and b with $a < b$,

$$P(a \leq X \leq b) = P(a < X \leq b) \qquad (4.1)$$
$$= P(a \leq X < b) = P(a < X < b)$$

In words, the probability assigned to any particular value is zero, and the probability of an interval does not depend on whether either of its endpoints is included. These properties follow from the facts that the area under the graph of $f(x)$ and above the single value c is zero and that the area under the graph above an interval is unaffected by exclusion or inclusion of the endpoints of the interval. If X is discrete and there is positive probability mass at both $X = a$ and $X = b$, then all four probabilities in Expression (4.1) will differ from one another.

The fact that a continuous distribution assigns probability zero to each value has a physical analog. Consider a solid circular rod with cross-sectional area = 1 in.². Place the rod alongside a measurement axis and suppose that the density of the rod at any point x is given by the value $f(x)$ of a density function. Then if the rod is sliced at points a and b and this segment is removed, the amount of mass removed is $\int_a^b f(x)\, dx$; if the rod is sliced just at the point c, no mass is removed. Mass is assigned to interval segments of the rod but not to individual points.

Example 4.4   "Time headway" in traffic flow is the elapsed time between the time that one car finishes passing a fixed point and the instant that the next car begins to pass that point. Let X = the time headway for two randomly chosen consecutive cars on a freeway during a period of heavy flow. The following pdf of X is essentially the one suggested in "The Statistical Properties of Freeway Traffic" (*Transportation Research*, vol. 11, pp. 221–228):

$$f(x) = \begin{cases} .15e^{-.15(x - .5)} & x \geq .5 \\ 0 & \text{otherwise} \end{cases}$$

The graph of $f(x)$ appears in Figure 4.4; there is no density associated with headway times less than .5, and headway density decreases rapidly (exponentially fast) as x increases from .5. Clearly $f(x) \geq 0$; to show that $\int_{-\infty}^{\infty} f(x)\, dx = 1$, we use the calculus result $\int_a^\infty e^{-kx}\, dx = (1/k)\, e^{-ka}$. Then

$$\int_{-\infty}^{\infty} f(x)\, dx = \int_{.5}^{\infty} .15e^{-.15(x - .5)}\, dx = .15e^{.075} \int_{.5}^{\infty} e^{-.15x}\, dx$$

$$= .15e^{.075} \cdot \frac{1}{.15}e^{-(.15)(.5)} = 1$$

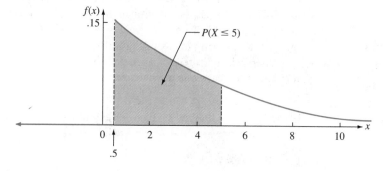

Figure 4.4 The pdf of headway time in Example 4.4

The probability that headway time is at most 5 sec is

$$P(X \leq 5) = \int_{-\infty}^{5} f(x)\, dx = \int_{.5}^{5} .15e^{-.15(x-.5)}\, dx$$

$$= .15e^{.075} \int_{.5}^{5} e^{-.15x}\, dx = .15e^{.075} \cdot -\frac{1}{.15} e^{-.15x} \Big|_{x=.5}^{x=5}$$

$$= e^{.075}(-e^{-.75} + e^{-.075}) = 1.078(-.472 + .928) = .491$$

$$= P(\text{less than 5 sec}) = P(X < 5) \qquad \blacksquare$$

Unlike discrete distributions such as the binomial, hypergeometric, and negative binomial, the distribution of any given continuous rv cannot usually be derived using simple probabilistic arguments. Instead, one must make a judicious choice of pdf based on prior knowledge and available data. Fortunately, there are some general families of pdf's that have been found to fit well in a wide variety of experimental situations; several of these are discussed later in the chapter.

Just as in the discrete case, it is often helpful to think of the population of interest as consisting of X values rather than individuals or objects. The pdf is then a model for the distribution of values in this numerical population, and from this model various population characteristics (such as the mean) can be calculated.

Exercises   Section 4.1 (1–10)

1. Let X denote the amount of time for which a book on 2-hour reserve at a college library is checked out by a randomly selected student and suppose that X has density function

$$f(x) = \begin{cases} .5x & 0 \leq x \leq 2 \\ 0 & \text{otherwise} \end{cases}$$

Calculate the following probabilities:

  a. $P(X \leq 1)$

  b. $P(.5 \leq X \leq 1.5)$

  c. $P(1.5 < X)$

2. Suppose the reaction temperature X (in °C) in a certain chemical process has a uniform distribution with $A = -5$ and $B = 5$.

  a. Compute $P(X < 0)$.

  b. Compute $P(-2 < X < 2)$.

  c. Compute $P(-2 \leq X \leq 3)$.

  d. For k satisfying $-5 < k < k + 4 < 5$, compute $P(k < X < k + 4)$.

3. Suppose the distance X between a point target and a shot aimed at the point in a coin-operated target game is a continuous rv with pdf

$$f(x) = \begin{cases} .75(1 - x^2) & -1 \leq x \leq 1 \\ 0 & \text{otherwise} \end{cases}$$

  a. Sketch the graph of $f(x)$.

  b. Compute $P(X > 0)$.

  c. Compute $P(-.5 < X < .5)$.

  d. Compute $P(X < -.25 \text{ or } X > .25)$.

4. Let X denote the vibratory stress (psi) on a wind turbine blade at a particular wind speed in a wind tunnel. The paper "Blade Fatigue Life Assessment with Application to VAWTS" (*J. Solar Energy Engr.*, 1982, pp. 107–111) proposes the Rayleigh distribution, with pdf

$$f(x;\, \theta) = \begin{cases} \dfrac{x}{\theta^2} \cdot e^{-x^2/(2\theta^2)} & x > 0 \\ 0 & \text{otherwise} \end{cases}$$

as a model for the X distribution.

a. Verify that $f(x; \theta)$ is a legitimate pdf.

b. Suppose $\theta = 100$ (a value suggested by a graph in the paper). What is the probability that X is at most 200? Less than 200? At least 200?

c. What is the probability that X is between 100 and 200?

d. Give an expression for $P(X \leq x)$.

5. A college professor never finishes his lecture before the bell rings to end the period and always finishes his lectures within 1 min after the bell rings. Let $X =$ the time that elapses between the bell and the end of the lecture and suppose the pdf of X is

$$f(x) = \begin{cases} kx^2 & 0 \leq x \leq 1 \\ 0 & \text{otherwise} \end{cases}$$

a. Find the value of k. [HINT: Total area under the graph of $f(x)$ is 1.]

b. What is the probability that the lecture ends within $\frac{1}{2}$ min of the bell ringing?

c. What is the probability that the lecture continues beyond the bell for between 15 and 30 sec?

d. What is the probability that the lecture continues for at least 40 sec beyond the bell?

6. The actual tracking weight of a stereo cartridge that is set to track at 3 g on a particular changer can be regarded as a continuous rv X with pdf

$$f(x) = \begin{cases} k[1 - (x - 3)^2] & 2 \leq x \leq 4 \\ 0 & \text{otherwise} \end{cases}$$

a. Sketch the graph of $f(x)$.

b. Find the value of k.

c. What is the probability that the actual tracking weight is greater than the prescribed weight?

d. What is the probability that the actual weight is within .25 g of the prescribed weight?

e. What is the probability that the actual weight differs from the prescribed weight by more than .5 g?

7. The time X (min) for a lab assistant to prepare the equipment for a certain experiment is believed to have a uniform distribution with $A = 25$ and $B = 35$.

a. Write the pdf of X and sketch its graph.

b. What is the probability that preparation time exceeds 33 min?

c. What is the probability that preparation time is

within 2 min of the mean time? [HINT: Identify μ from the graph of $f(x)$.]

d. For any a such that $25 < a < a + 2 < 35$, what is the probability that preparation time is between a and $a + 2$ min?

8. In commuting to work, I must first get on a bus near my house and then transfer to a second bus. If the waiting time (in minutes) at each stop has a uniform distribution with $A = 0$ and $B = 5$, then it can be shown that my total waiting time Y has the pdf

$$f(y) = \begin{cases} \dfrac{1}{25} y & 0 \leq y < 5 \\ \dfrac{2}{5} - \dfrac{1}{25} y & 5 \leq y \leq 10 \\ 0 & y < 0 \text{ or } y > 10 \end{cases}$$

a. Sketch a graph of the pdf of Y.

b. Verify that $\int_{-\infty}^{\infty} f(y) \, dy = 1$.

c. What is the probability that total waiting time is at most 3 min?

d. What is the probability that total waiting time is at most 8 min?

e. What is the probability that total waiting time is between 3 and 8 min?

f. What is the probability that total waiting time is either less than 2 min or more than 6 min?

9. Consider again the pdf of $X =$ time headway given in Example 4.4. What is the probability that time headway is

a. At most 6 sec?

b. More than 6 sec? At least 6 sec?

c. Between 5 and 6 sec?

10. A family of pdf's that has been used to approximate the distribution of income, city population size, and size of firms is the Pareto family. The family has two parameters, k and θ, both > 0, and the pdf is

$$f(x; k, \theta) = \begin{cases} \dfrac{k \cdot \theta^k}{x^{k+1}} & x \geq \theta \\ 0 & x < \theta \end{cases}$$

a. Sketch the graph of $f(x; k, \theta)$.

b. Verify that the total area under the graph equals 1.

c. If the rv X has pdf $f(x; k, \theta)$, for any fixed $b > \theta$ obtain an expression for $P(X \leq b)$.

d. For $\theta < a < b$, obtain an expression for the probability $P(a \leq X \leq b)$.

Cumulative Distribution Functions and Expected Values

Several of the most important concepts introduced in the study of discrete distributions also play an important role for continuous distributions. Definitions analogous to those in Chapter 3 involve replacing summation by integration.

The Cumulative Distribution Function

The cumulative distribution function (cdf) $F(x)$ for a discrete rv X gives, for any specified number x, the probability $P(X \leq x)$. It is obtained by summing the pmf $p(y)$ over all possible values y satisfying $y \leq x$. The cdf of a continuous rv gives the same probabilities $P(X \leq x)$ and is obtained by integrating the pdf $f(y)$ between the limits $-\infty$ and x.

Definition

> The **cumulative distribution function** $F(x)$ for a continuous rv X is defined for every number x by
>
> $$F(x) = P(X \leq x) = \int_{-\infty}^{x} f(y)\, dy$$
>
> For each x, $F(x)$ is the area under the density curve to the left of x. This is illustrated in Figure 4.5, where $F(x)$ increases smoothly as x increases.

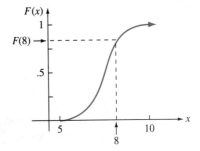

Figure 4.5 A pdf and associated cdf

Example 4.5

Let X have a uniform distribution on $[A, B]$. The density function appears in Figure 4.6. For $x < A$, $F(x) = 0$, since there is no area under the graph of the density function to the left of such an x. For $x \geq B$, $F(x) = 1$, since all the area is accumulated to the left of such an x. Finally, for $A \leq x \leq B$,

$$F(x) = \int_{-\infty}^{x} f(y)\, dy = \int_{A}^{x} \frac{1}{B-A}\, dy = \frac{1}{B-A} \cdot y \Big|_{y=A}^{y=x} = \frac{x-A}{B-A}$$

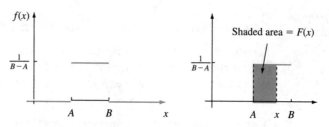

Figure 4.6 The pdf for a uniform distribution

The entire cdf is

$$f(x) = \begin{cases} 0 & x < A \\ \dfrac{x - A}{B - A} & A \le x < B \\ 1 & x \ge B \end{cases}$$

The graph of this cdf appears in Figure 4.7.

Figure 4.7 The cdf for a uniform distribution ∎

Using F(x) to Compute Probabilities

The importance of the cdf here, just as for discrete rv's, is that probabilities of various intervals can be computed from a formula for or table of $F(x)$.

Proposition

> Let X be a continuous rv with pdf $f(x)$ and cdf $F(x)$. Then for any two numbers a, b with $a < b$,
>
> $$P(a \le X \le b) = F(b) - F(a)$$

Figure 4.8 illustrates this proposition; the desired probability is the shaded area under the density curve between a and b and equals the difference between the two shaded cumulative areas.

Example 4.6

Suppose the pdf of the magnitude X of a dynamic load on a bridge (in newtons) is given by

$$f(x) = \begin{cases} \dfrac{1}{8} + \dfrac{3}{8}x & 0 \le x \le 2 \\ 0 & \text{otherwise} \end{cases}$$

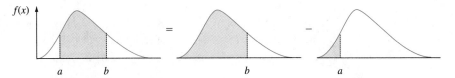

Figure 4.8 Computing $P(a \leq X \leq b)$ from cumulative probabilities

The cdf $F(x)$ is 0 for $x < 0$, 1 for $x > 2$, and

$$F(x) = \int_{-\infty}^{x} f(y) \, dy = \int_0^x \left(\frac{1}{8} + \frac{3}{8} y \right) dy = \frac{x}{8} + \frac{3}{16} x^2, \qquad 0 \leq x \leq 2$$

The graphs of $f(x)$ and $F(x)$ appear in Figure 4.9. The probability that the load is between 1 and 1.5 is

$$P(1 \leq X \leq 1.5) = F(1.5) - F(1)$$

$$= \left[\frac{1}{8} (1.5) + \frac{3}{16} (1.5)^2 \right] - \left[\frac{1}{8} (1) + \frac{3}{16} (1)^2 \right]$$

$$= \frac{19}{64} = .297$$

The probability that the load exceeds 1 is

$$P(X > 1) = 1 - P(X \leq 1) = 1 - F(1) = 1 - \left[\frac{1}{8} (1) + \frac{3}{16} (1)^2 \right]$$

$$= \frac{11}{16} = .688$$

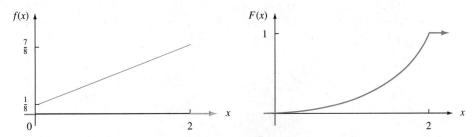

Figure 4.9 The pdf and cdf for Example 4.6

In the previous example, $P(1 \leq X \leq 1.5)$ could just as easily have been obtained by integrating $f(x)$, but when the integration must be done numerically (as with the normal and gamma distributions to be encountered shortly), tabulating $F(x)$ enables probabilities to be computed quickly.

Obtaining $f(x)$ from $F(x)$

For X discrete, the pmf is obtained from the cdf by taking the difference between two $F(x)$ values. The continuous analog of a difference is a derivative. The following result is a consequence of the Fundamental Theorem of Calculus.

Proposition

> If X is a continuous rv with pdf $f(x)$ and cdf $F(x)$, then at every x at which the derivative $F'(x)$ exists, $F'(x) = f(x)$.

Example 4.7
(Example 4.5 continued)

When X has a uniform distribution, $F(x)$ is differentiable except at $x = A$ and $x = B$, where the graph of $F(x)$ has sharp corners. Since $F(x) = 0$ for $x < A$ and $F(x) = 1$ for $x > B$, $F'(x) = 0 = f(x)$ for such x. For $A < x < B$,

$$F'(x) = \frac{d}{dx}\left(\frac{x - A}{B - A}\right) = \frac{1}{B - A} = f(x)$$

∎

Percentiles of a Continuous Distribution

When we say that an individual's test score was at the 85th percentile of the population, we mean that 85% of all population scores were below that score and 15% were above. Similarly, the 40th percentile is the score that exceeds 40% of all scores and is exceeded by 60% of all scores.

Definition

> Let p be a number between 0 and 1. The **(100p)th percentile** of the distribution of a continuous rv X, denoted by $\eta(p)$, is defined by
>
> $$p = F(\eta(p)) = \int_{-\infty}^{\eta(p)} f(y)\, dy \tag{4.2}$$

According to Expression (4.2), $\eta(p)$ is that value on the measurement axis such that $100p\%$ of the area under the graph of $f(x)$ lies to the left of $\eta(p)$ and $100(1 - p)\%$ lies to the right. Thus $\eta(.75)$, the 75th percentile, is such that the area under the graph of $f(x)$ to the left of $\eta(.75)$ is .75. Figure 4.10 illustrates the definition.

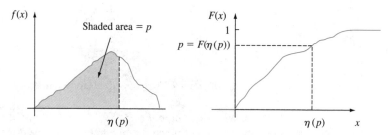

Figure 4.10 The (100p)th percentile of a continuous distribution

Example 4.8

The distribution of the amount of gravel (in tons) sold by a particular construction supply company in a given week is a continuous rv X with pdf

$$f(x) = \begin{cases} \dfrac{3}{2}(1 - x^2) & 0 \le x \le 1 \\ 0 & \text{otherwise} \end{cases}$$

The cdf of sales is then, for $0 < x < 1$,

$$F(x) = \int_0^x \frac{3}{2}(1 - y^2)\,dy = \frac{3}{2}\left(y - \frac{y^3}{3}\right)\Bigg|_{y=0}^{y=x} = \frac{3}{2}\left(x - \frac{x^3}{3}\right)$$

The graphs of both $f(x)$ and $F(x)$ appear in Figure 4.11. The $(100p)$th percentile of this distribution satisfies the equation

$$p = F(\eta(p)) = \frac{3}{2}\left[\eta(p) - \frac{(\eta(p))^3}{3}\right]$$

that is,

$$(\eta(p))^3 - 3\eta(p) + 2p = 0$$

For the 50th percentile, $p = .5$, and the equation to be solved is $\eta^3 - 3\eta + 1 = 0$; the solution is $\eta = \eta(.5) = .347$. If the distribution remains the same from week to week, then in the long run 50% of all weeks will result in sales of less than .347 ton and 50% in more than .347 ton.

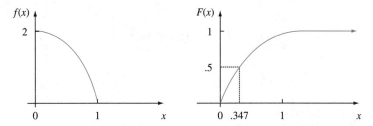

Figure 4.11 The pdf and cdf for Example 4.8   ■

Definition

> The **median** of a continuous distribution, denoted by $\tilde{\mu}$, is the 50th percentile, so $\tilde{\mu}$ satisfies $.5 = F(\tilde{\mu})$. That is, half the area under the density curve is to the left of $\tilde{\mu}$ and half is to the right of $\tilde{\mu}$.

Example 4.9   A continuous distribution whose pdf is **symmetric**—which means that the graph of the pdf to the left of some point is a mirror image of the graph to the right of that point—has median $\tilde{\mu}$ equal to the point of symmetry, since half the area under the curve lies to either side of this point. Figure 4.12 gives several examples. The amount of error in a measurement of a physical quantity is often assumed to have a symmetric distribution.

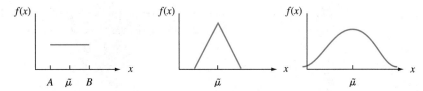

Figure 4.12 Medians of symmetric distributions   ■

Expected Values for Continuous Random Variables

For a discrete rv X, $E(X)$ was obtained by summing $x \cdot p(x)$ over possible X values. Here we replace summation by integration and the pmf by the pdf to get a continuous weighted average.

Definition

> The **expected** or **mean value** of a continuous rv X with pdf $f(x)$ is
>
> $$\mu_X = E(X) = \int_{-\infty}^{\infty} x \cdot f(x) \, dx$$

Example 4.10
(Example 4.8 continued)

The pdf of weekly gravel sales X was

$$f(x) = \begin{cases} \dfrac{3}{2}(1 - x^2) & 0 \le x \le 1 \\ 0 & \text{otherwise} \end{cases}$$

so

$$E(X) = \int_{-\infty}^{\infty} x \cdot f(x) \, dx = \int_{0}^{1} x \cdot \frac{3}{2}(1 - x^2) \, dx$$

$$= \frac{3}{2} \int_{0}^{1} (x - x^3) \, dx = \frac{3}{2} \left(\frac{x^2}{2} - \frac{x^4}{4} \right) \Bigg|_{x=0}^{x=1} = \frac{3}{8} \qquad \blacksquare$$

When the pdf $f(x)$ specifies a model for the distribution of values in a numerical population, then μ is the population mean, which is the most frequently used measure of population location or center.

Often we wish to compute the expected value of some function $h(X)$ of the rv X. If we think of $h(X)$ as a new rv Y, techniques from mathematical statistics can be used to derive the pdf of Y, and $E(Y)$ can be computed from the definition. Fortunately, as in the discrete case, there is an easier way to compute $E[h(X)]$.

Proposition

> If X is a continuous rv with pdf $f(x)$ and $h(X)$ is any function of X, then
>
> $$E[h(X)] = \mu_{h(X)} = \int_{-\infty}^{\infty} h(x) \cdot f(x) \, dx$$

Example 4.11

Two species are competing in a region for control of a limited amount of a certain resource. Let X = the proportion of the resource controlled by species 1 and suppose X has pdf

$$f(x) = \begin{cases} 1 & 0 \le x \le 1 \\ 0 & \text{otherwise} \end{cases}$$

which is a uniform distribution on [0, 1] (in her book *Ecological Diversity,* E. C. Pielou calls this the "broken-stick" model for resource allocation, since it is analogous to breaking a stick at a randomly chosen point). Then the species that controls the majority of this resource controls the amount

$$h(X) = \max(X,\ 1 - X) = \begin{cases} 1 - X & \text{if } 0 \le X < \dfrac{1}{2} \\[2mm] X & \text{if } \dfrac{1}{2} \le X \le 1 \end{cases}$$

The expected amount controlled by the species having majority control is then

$$E[h(X)] = \int_{-\infty}^{\infty} \max(x,\ 1 - x) \cdot f(x)\ dx = \int_{0}^{1} \max(x,\ 1 - x) \cdot 1\ dx$$

$$= \int_{0}^{1/2} (1 - x) \cdot 1\ dx + \int_{1/2}^{1} x \cdot 1\ dx = \frac{3}{4} \qquad \blacksquare$$

For $h(X)$ a linear function, $E[h(X)] = E(aX + b) = aE(X) + b$.

The Variance of a Continuous Random Variable

Definition

> The **variance** of a continuous rv X with pdf $f(x)$ and mean value μ is
>
> $$\sigma_X^2 = V(X) = \int_{-\infty}^{\infty} (x - \mu)^2 \cdot f(x)\ dx = E[(X - \mu)^2]$$
>
> The **standard deviation** (SD) of X is $\sigma_X = \sqrt{V(X)}$.

As in the discrete case, σ_X^2 is the expected or average squared deviation about the mean μ and gives a measure of how much spread there is in the distribution or population of x values. The easiest way to compute σ^2 is to again use a shortcut formula.

Proposition

> $$V(X) = E(X^2) - [E(X)]^2$$

Example 4.12
(Example 4.10 continued)

For X = weekly gravel sales, we computed $E(X) = \frac{3}{8}$. Since

$$E(X^2) = \int_{-\infty}^{\infty} x^2 \cdot f(x)\ dx = \int_{0}^{1} x^2 \cdot \frac{3}{2} (1 - x^2)\ dx$$

$$= \int_{0}^{1} \frac{3}{2} (x^2 - x^4)\ dx = \frac{1}{5}$$

$$V(X) = \frac{1}{5} - \left(\frac{3}{8}\right)^2 = \frac{19}{320} = .059 \qquad \text{and} \qquad \sigma_X = .244 \qquad \blacksquare$$

When $h(X)$ is a linear function and $V(X) = \sigma^2$, $V[h(X)] = V(aX + b) = a^2 \cdot \sigma^2$ and $\sigma_{aX + b} = |a| \cdot \sigma$.

Exercises   Section 4.2 (11–25)

11. The cdf of checkout duration X as described in Exercise 1 is

$$F(x) = \begin{cases} 0 & x < 0 \\ \dfrac{x^2}{4} & 0 \le x < 2 \\ 1 & 2 \le x \end{cases}$$

Use this to compute the following:

a. $P(X \le 1)$ b. $P(.5 \le X \le 1)$

c. $P(X > .5)$

d. The median checkout duration $\tilde{\mu}$ [solve $.5 = F(\tilde{\mu})$]

e. $F'(x)$ to obtain the density function $f(x)$

12. The cdf for $X(=$ shot-to-target distance) of Exercise 3 is

$$F(x) = \begin{cases} 0 & x < -1 \\ \dfrac{1}{2} + \dfrac{3}{4}\left(x - \dfrac{x^3}{3}\right) & -1 \le x < 1 \\ 1 & 1 \le x \end{cases}$$

a. Compute $P(X < 0)$.

b. Compute $P(-.5 < X < .5)$.

c. Compute $P(.25 < X)$.

d. Verify that $f(x)$ is as given in Exercise 3 by obtaining $F'(x)$.

e. Verify that $\tilde{\mu} = 0$.

13. Let X denote checkout time duration with pdf given in Exercise 1.

a. Compute $E(X)$.   b. Compute $V(X)$ and σ_X.

c. If the borrower is charged an amount $h(X) = X^2$ when checkout duration is X, compute the expected charge $E[h(X)]$.

14. The article "Modeling Sediment and Water Column Interactions for Hydrophobic Pollutants" (*Water Research*, 1984, pp. 1169–1174) suggests the uniform distribution on the interval (7.5, 20) as a model for depth (cm) of the bioturbation layer in sediment in a certain region.

a. What are the mean and variance of depth?

b. What is the cdf of depth?

c. What is the probability that observed depth is at most 10? Between 10 and 15?

d. What is the probability that the observed depth is within 1 SD of the mean value? Within 2 SDs?

15. Suppose the pdf of weekly gravel sales X (in tons) is

$$f(x) = \begin{cases} 2(1 - x) & 0 \le x \le 1 \\ 0 & \text{otherwise} \end{cases}$$

a. Obtain the cdf of X and graph it.

b. What is $P(X \le .5)$ [i.e., $F(.5)$]?

c. Using part (a), what is $P(.25 < X \le .5)$? What is $P(.25 \le X \le .5)$?

d. What is the 75th percentile of the sales distribution?

e. What is the median $\tilde{\mu}$ of the sales distribution?

f. Compute $E(X)$ and σ_X.

16. Answer parts (a)–(f) of Exercise 15 with $X =$ lecture time past the bell given in Exercise 5.

17. Consider the pdf of $X =$ actual tracking weight given in Exercise 6.

a. Obtain and graph the cdf of X.

b. From the graph of $f(x)$, what is $\tilde{\mu}$?

c. Compute $E(X)$ and $V(X)$.

18. Let X have a uniform distribution on the interval $[A, B]$.

a. Obtain an expression for the $(100p)$th percentile.

b. Compute $E(X)$, $V(X)$, and σ_X.

c. For n a positive integer, compute $E(X^n)$.

19. Let X be a continuous rv with cdf

$$F(x) = \begin{cases} 0 & x \le 0 \\ \dfrac{x}{4}\left[1 + \ln\left(\dfrac{4}{x}\right)\right] & 0 < x \le 4 \\ 1 & x > 4 \end{cases}$$

[This type of cdf is suggested in the paper "Variability in Measured Bedload-Transport Rates" (*Water Resources Bull.*, 1985, pp. 39–48) as a model for a certain hydrologic variable.] What is

a. $P(X \le 1)$?

b. $P(1 \le X \le 3)$?

c. The pdf of X?

20. Consider the pdf for total waiting time Y for two buses

$$f(y) = \begin{cases} \dfrac{1}{25} y & 0 \le y < 5 \\ \dfrac{2}{5} - \dfrac{1}{25} y & 5 \le y \le 10 \\ 0 & \text{otherwise} \end{cases}$$

introduced in Exercise 8.

a. Compute and sketch the cdf of Y. [HINT: Consider separately $0 \le y < 5$ and $5 \le y \le 10$ in computing $F(y)$. A picture of the pdf should be helpful.]

b. Obtain an expression for the $(100p)$th percentile. (HINT: Consider separately $0 < p < .5$ and $.5 < p < 1$.)

c. Compute $E(Y)$ and $V(Y)$. How do these compare with the expected waiting time and variance for a single bus when the time is uniformly distributed on $[0, 5]$?

21. An ecologist wishes to mark off a circular sampling region having radius 10 m. However, the radius of the resulting region is actually a random variable R with pdf

$$f(r) = \begin{cases} \dfrac{3}{4} [1 - (10 - r)^2] & 9 \le r \le 11 \\ 0 & \text{otherwise} \end{cases}$$

What is the expected area of the resulting circular region?

22. The weekly demand for propane gas (in 1000s of gallons) from a particular facility is an rv X with pdf

$$f(x) = \begin{cases} 2\left(1 - \dfrac{1}{x^2}\right) & 1 \le x \le 2 \\ 0 & \text{otherwise} \end{cases}$$

a. Compute the cdf of X.

b. Obtain an expression for the $(100p)$th percentile. What is the value of $\tilde{\mu}$?

c. Compute $E(X)$ and $V(X)$.

d. If 1.5 thousand gallons is in stock at the beginning of the week and no new supply is due in during the week, how much of the 1.5 thousand gallons is expected to be left at the end of the week?

23. If the temperature at which a certain compound melts is a random variable with mean value 120°C and standard deviation 2°C, what are the mean temperature and standard deviation measured in °F? (HINT: °F = 1.8°C + 32.)

24. Let X have the Pareto pdf

$$f(x; k, \theta) = \begin{cases} \dfrac{k \cdot \theta^k}{x^{k+1}} & x \ge \theta \\ 0 & x < \theta \end{cases}$$

introduced in Exercise 10.

a. If $k > 1$, compute $E(X)$.

b. What can you say about $E(X)$ if $k = 1$?

c. If $k > 2$, show that $V(X) = k\theta^2(k-1)^{-2}(k-2)^{-1}$.

d. If $k = 2$, what can you say about $V(X)$?

e. What conditions on k are necessary to ensure that $E(X^n)$ is finite?

25. Let X be the temperature in °C at which a certain chemical reaction takes place and let Y be the temperature in °F (so $Y = 1.8X + 32$).

a. If the median of the X distribution is $\tilde{\mu}$, show that $1.8\tilde{\mu} + 32$ is the median of the Y distribution.

b. How is the 90th percentile of the Y distribution related to the 90th percentile of the X distribution? Verify your conjecture.

c. More generally, if $Y = aX + b$, how is any particular percentile of the Y distribution related to the corresponding percentile of the X distribution?

The Normal Distribution

The normal distribution is the most important one in all of probability and statistics. Many numerical populations have distributions that can be fit very closely by an appropriate normal curve. Examples include heights, weights, and other physical characteristics (the famous 1903 *Biometrika* paper "On the Laws of Inheritance in Man" discussed many examples of this sort), measurement errors in scientific experiments, anthropometric measurements on fossils, reaction times in psychological experiments, measurements of intelligence and aptitude, scores on various tests, and numerous economic measures and indicators. Even when the underlying distribution is discrete, the normal curve often gives an excellent approximation. In addition, even when individual variables themselves are not normally distributed, sums and averages of the variables will under suitable conditions have approximately a normal distribution; this is the content of the Central Limit Theorem discussed in this next chapter.

Definition

> A continuous rv X is said to have a **normal distribution with parameters** μ **and** σ **(or** μ **and** σ^2**),** where $-\infty < \mu < \infty$ and $0 < \sigma$, if the pdf of X is
>
> $$f(x;\ \mu,\ \sigma) = \frac{1}{\sqrt{2\pi}\ \sigma}\ e^{-(x-\mu)^2/(2\sigma^2)} \qquad -\infty < x < \infty \qquad (4.3)$$

Again e denotes the base of the natural logarithm system and equals approximately 2.71828, while π represents the familiar mathematical constant with approximate value 3.14159. The statement that X is normally distributed with parameters μ and σ^2 is often abbreviated $X \sim N(\mu, \sigma^2)$.

Clearly $f(x;\ \mu,\ \sigma) \geq 0$ for any number x, but techniques from multivariate calculus must be used to show that $\int_{-\infty}^{\infty} f(x;\ \mu,\ \sigma)\ dx = 1$. It can be shown that $E(X) = \mu$ and $V(X) = \sigma^2$, so the parameters are the mean and the standard deviation of X. Figure 4.13 presents graphs of $f(x;\ \mu,\ \sigma)$ for several different (μ, σ) pairs. Each graph is symmetric about μ and bell-shaped, so the center of the bell (point of symmetry) is both the mean of the distribution and the median. The value of σ is the distance from μ to the inflection points of the curve (the points at which the curve changes from turning downward to turning upward). Large values of σ yield graphs that are quite spread out about μ, whereas small values of σ yield graphs with a high peak above μ and most of the area under the graph quite close to μ. Thus, a large σ implies that a value of X far from μ may well be observed, while such a value is quite unlikely when σ is small.

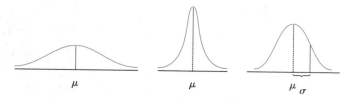

Figure 4.13 Graphs of normal pdf's

The Standard Normal Distribution

To compute $P(a \leq X \leq b)$ when X is a normal rv with parameters μ and σ, we must evaluate

$$\int_a^b \frac{1}{\sqrt{2\pi}\,\sigma}\, e^{-(x-\mu)^2/(2\sigma^2)}\, dx \tag{4.4}$$

None of the standard integration techniques can be used to evaluate Expression (4.4). Instead, for $\mu = 0$ and $\sigma = 1$, Expression (4.4) has been numerically evaluated and tabulated for certain values of a and b. This table can also be used to compute probabilities for any other values of μ and σ under consideration.

Definition

> The normal distribution with parameter values $\mu = 0$ and $\sigma = 1$ is called a **standard normal distribution.** A random variable that has a standard normal distribution is called a **standard normal random variable** and will be denoted by Z. The pdf of Z is
>
> $$f(z; 0, 1) = \frac{1}{\sqrt{2\pi}}\, e^{-z^2/2} \qquad -\infty < z < \infty$$
>
> The cdf of Z is $P(Z \leq z) = \int_{-\infty}^z f(y; 0, 1)\, dy$, which we will denote by $\Phi(z)$.

The standard normal distribution does not frequently serve as a model for a naturally arising population. Instead, it is a reference distribution from which information about other normal distributions can be obtained. Appendix Table A.3 gives $\Phi(z) = P(Z \leq z)$, the area under the graph of the standard normal pdf to the left of z, for $z = -3.49, -3.48, \ldots, 3.48, 3.49$. Figure 4.14 illustrates the type of cumulative area (probability) tabulated in Table A.3. From this table, various other probabilities involving Z can be calculated.

Shaded area $= \Phi(z)$

Standard normal (z) curve

$0 \quad z$

Figure 4.14 Standard normal curve areas tabulated in Appendix Table A.3

Example 4.13

Compute the following probabilities: (a) $P(Z \leq 1.25)$,   (b) $P(Z > 1.25)$, (c) $P(Z \leq -1.25)$, and (d) $P(-.38 \leq Z \leq 1.25)$.

a. $P(Z \leq 1.25) = \Phi(1.25)$, a probability that is tabulated in Appendix Table A.3 at the intersection of the row marked 1.2 and the column marked .05. The number there is .8944, so $P(Z \leq 1.25) = .8944$. (See Figure 4.15a.)

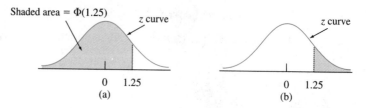

Figure 4.15 Normal curve areas (probabilities) for Example 4.13

b. $P(Z > 1.25) = 1 - P(Z \leq 1.25) = 1 - \Phi(1.25)$, the area under the standard normal curve to the right of 1.25 (an upper-tail area). Since $\Phi(1.25) = .8944$, $P(Z > 1.25) = .1056$. Since Z is a continuous rv, $P(Z \geq 1.25)$ also equals .1056. (See Figure 4.15b.)

c. $P(Z \leq -1.25) = \Phi(-1.25)$, a lower-tail area. Directly from Table A.3, $\Phi(-1.25) = .1056$. By symmetry of the normal curve, this is the same answer as in (b).

d. $P(-.38 \leq Z \leq 1.25)$ is the area under the standard normal curve above the interval whose left endpoint is $-.38$ and whose right endpoint is 1.25. From Section 4.2 if X is a continuous rv with cdf $F(x)$, then $P(a \leq X \leq b) = F(b) - F(a)$. This gives $P(-.38 \leq Z \leq 1.25) = \Phi(1.25) - \Phi(-.38) = .8944 - .3520 = .5424$. (See Figure 4.16.)

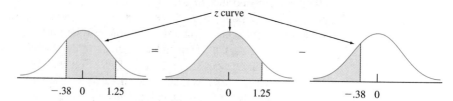

Figure 4.16 $P(-.38 \leq Z \leq 1.25)$ as the difference between two cumulative areas ∎

Percentiles of the Standard Normal Distribution

For any p between 0 and 1, Appendix Table A.3 can be used to obtain the $(100p)$th percentile of the standard normal distribution.

Example 4.14   The 99th percentile of the standard normal distribution is that point on the axis such that the area under the curve to the left of the point is .9900. Now Appendix Table A.3 gives for fixed z the area under the standard normal curve to the left of z, whereas here we have the area and want the value of z. This is the "inverse" problem to "$P(Z \leq z) = ?$" so the table is used in an inverse fashion: Find in the middle of the table .9900; the row and column in which it lies identify the 99th z percentile. Here .9901 lies in the row marked 2.3 and column marked .03, so the 99th percentile is (approximately) $z = 2.33$. (See Figure 4.17.) By symmetry, the first percentile is the negative of the 99th percentile, so equals -2.33 (1% lies below the first and above the 99th). (See Figure 4.18.)

Figure 4.17 Finding the 99th percentile   Figure 4.18 The relationship between the 1st
                                              and 99th percentiles

In general, the (100p)th percentile is identified by the row and column of Appendix Table A.3 in which the entry p is found (e.g., the 67th percentile is obtained by finding .6700 in the body of the table, which gives $z = .44$). If p does not appear, the number closest to it is often used, although linear interpolation gives a more accurate answer. For example, to find the 95th percentile, we look for .9500 inside the table. While .9500 does not appear, both .9495 and .9505 do, corresponding to $z = 1.64$ and 1.65, respectively. Since .9500 is halfway between the two probabilities that do appear, we will use 1.645 as the 95th percentile and -1.645 as the 5th percentile.

z_α Notation

In statistical inference, we will need the values on the measurement axis that capture certain small tail areas under the standard normal curve.

Notation

> z_α will denote the value on the measurement axis for which α of the area under the z curve lies to the right of z_α. (See Figure 4.19.)

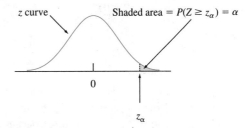

Figure 4.19 z_α notation illustrated

Since α of the area under the standard normal curve lies to the right of z_α, $1 - \alpha$ of the area lies to the left of z_α. Thus, z_α *is the* $100(1 - \alpha)$*th percentile of the standard normal distribution.* Further, by symmetry the area under the standard normal curve to the left of $-z_\alpha$ is also α. The z_α's are usually referred to as **z critical values.** Table 4.1 lists the most useful standard normal percentiles and z_α values.

Table 4.1 Standard normal percentiles and critical values

Percentile	90	95	97.5	99	99.5	99.9	99.95
α (tail area)	.1	.05	.025	.01	.005	.001	.0005
$z_\alpha = 100(1 - \alpha)th$ *percentile*	1.28	1.645	1.96	2.33	2.58	3.08	3.27

Example 4.15 $z_{.05}$ is the $100(1 - .05)$th = 95th percentile of the standard normal distribution, so $z_{.05} = 1.645$. The area under the standard normal curve to the left of $-z_{.05}$ is also .05. (See Figure 4.20.)

Figure 4.20 Finding $z_{.05}$ ■

Nonstandard Normal Distributions

When $X \sim N(\mu, \sigma^2)$, probabilities involving X are computed by "standardizing." The **standardized variable** is $(X - \mu)/\sigma$. Subtracting μ shifts the mean from μ to zero, and then dividing by σ scales the variable so that the standard deviation is 1 rather than σ.

Proposition

If X has a normal distribution with mean μ and standard deviation σ, then

$$Z = \frac{X - \mu}{\sigma}$$

is a standard normal rv.

The key idea of the proposition is that by standardizing, any probability involving X can be expressed as a probability involving a standard normal rv Z, so that Appendix Table A.3 can be used. In particular,

$$P(X \leq x) = P[Z \leq (x - \mu)/\sigma] = \Phi[(x - \mu)/\sigma]$$

as illustrated in Figure 4.21. The proposition can be proved by writing the cdf of $Z = (X - \mu)/\sigma$ as

$$P(Z \leq z) = P(X \leq \sigma z + \mu) = \int_{-\infty}^{\sigma z + \mu} f(x; \mu, \sigma) \, dx$$

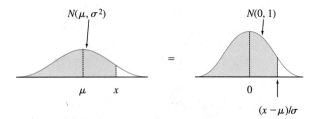

Figure 4.21 Equality of nonstandard and standard normal curve areas

Using a result from calculus, this integral can be differentiated with respect to z to yield the desired pdf $f(z; 0, 1)$.

Example 4.16

The time that it takes a driver to react to the brake lights on a decelerating vehicle is critical in helping to avoid rear-end collisions. The paper "Fast-Rise Brake Lamp as a Collision-Prevention Device" (*Ergonomics*, 1993, pp. 391–395) suggests that reaction time for an in-traffic response to a brake signal from standard brake lights can be modeled with a normal distribution having mean value 1.25 sec and standard deviation of .46 sec. What is the probability that reaction time is between 1.00 sec and 1.75 sec? If we let X denote reaction time, then standardizing gives

$$1.00 \le X \le 1.75$$

if and only if

$$\frac{1.00 - 1.25}{.46} \le \frac{X - 1.25}{.46} \le \frac{1.75 - 1.25}{.46}$$

Thus,

$$P(1.00 \le X \le 1.75) = P\left(\frac{1.00 - 1.25}{.46} \le Z \le \frac{1.75 - 1.25}{.46}\right)$$

$$= P(-.54 \le Z \le 1.09) = \Phi(1.09) - \Phi(-.54)$$

$$\doteq .8621 - .2946 = .5675$$

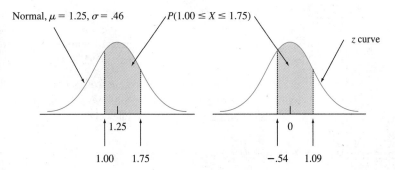

Figure 4.22 Normal curves for Example 4.16

This is illustrated in Figure 4.22. Similarly, if we view 2 sec as a critically long reaction time, the probability that actual reaction time will exceed this value is

$$P(X > 2) = P\left(Z > \frac{2 - 1.25}{.46}\right) = P(Z > 1.63) = 1 - \Phi(1.63) = .0516$$ ■

Standardizing amounts to nothing more than calculating a distance from the mean value and then reexpressing the distance as some number of standard deviations. For example, if $\mu = 100$ and $\sigma = 15$, then $x = 130$ corresponds to $z = (130 - 100)/15 = 30/15 = 2.00$. That is, 130 is 2 SDs above (to the right of) the mean value. Similarly, standardizing 85 gives $(85 - 100)/15 = -1.00$, so 85 is 1 SD below the mean. The z table applies to *any* normal distribution provided that we think in terms of number of standard deviations away from the mean value.

Example 4.17   The breakdown voltage of a randomly chosen diode of a particular type is known to be normally distributed. What is the probability that a diode's breakdown voltage is within 1 SD of its mean value? This question can be answered without knowing either μ or σ, as long as the distribution is known to be normal; put another way, the answer is the same for *any* normal distribution:

$$P(X \text{ is within 1 SD of its mean}) = P(\mu - \sigma \leq X \leq \mu + \sigma)$$

$$= P\left(\frac{\mu - \sigma - \mu}{\sigma} \leq Z \leq \frac{\mu + \sigma - \mu}{\sigma}\right)$$

$$= P(-1.00 \leq Z \leq 1.00)$$

$$= \Phi(1.00) - \Phi(-1.00) = .6826$$

The probability that X is observed to be within 2 SDs is $P(-2.00 \leq Z \leq 2.00) = .9544$ and within 3 SDs is $P(-3.00 \leq Z \leq 3.00) = .9974$. ■

The results of Example 4.17 are often reported in percentage form and referred to as the *empirical rule* (because empirical evidence has shown that histograms of real data can very frequently be approximated by normal curves).

If the population distribution of a variable is (approximately) normal, then

1. Roughly 68% of the values are within 1 SD of the mean.
2. Roughly 95% of the values are within 2 SDs of the mean.
3. Roughly 99.7% of the values are within 3 SDs of the mean.

It is indeed unusual to observe a value from a normal population that is much farther than 2 SDs from μ. These results will be important in the development of hypothesis-testing procedures in later chapters.

Percentiles of an Arbitrary Normal Distribution

The $(100p)$th percentile of a normal distribution with mean μ and standard deviation σ is easily related to the $(100p)$th percentile of the standard normal distribution.

Proposition

$$\begin{array}{l} (100p)\text{th percentile} \\ \text{for normal } (\mu,\, \sigma) \end{array} = \mu + \left[\begin{array}{l} (100p)\text{th for} \\ \text{standard normal} \end{array}\right] \cdot \sigma$$

Another way of saying this is that if z is the desired percentile for the standard normal distribution, then the desired percentile for the normal (μ, σ) distribution is z standard deviations from μ.

Example 4.18
The amount of distilled water dispensed by a certain machine is normally distributed with mean value 64 oz and standard deviation .78 oz. What container size c will ensure that overflow occurs only .5% of the time? If X denotes the amount dispensed, the desired condition is that $P(X > c) = .005$, or, equivalently, that $P(X \le c) = .995$. Thus c is the 99.5th percentile of the normal distribution with $\mu = 64$ and $\sigma = .78$. The 99.5th percentile of the standard normal distribution is 2.58, so

$$c = \eta(.995) = 64 + (2.58)(.78) = 64 + 2.0 = 66 \text{ oz}$$

This is illustrated in Figure 4.23.

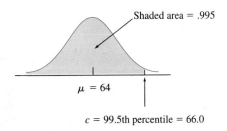

Shaded area = .995

$\mu = 64$

$c = 99.5\text{th percentile} = 66.0$

Figure 4.23 Distribution of amount dispensed for Example 4.18 ∎

The Normal Distribution and Discrete Populations

The normal distribution is often used as an approximation to the distribution of values in a discrete population. In such situations, extra care must be taken to ensure that probabilities are computed in an accurate manner.

Example 4.19
IQ in a particular population (as measured by a standard test) is known to be approximately normally distributed with $\mu = 100$ and $\sigma = 15$. What is the probability that a randomly selected individual has an IQ of at least 125? Letting X = the IQ of a randomly chosen person, we wish $P(X \ge 125)$. The temptation here is to standardize $X \ge 125$ immediately as in previous examples. However, the IQ population is actually discrete, since IQs are integer-valued, so the normal curve is an approximation to a discrete probability histogram as pictured in Figure 4.24.

The rectangles of the histogram are *centered* at integers, so IQs of at least 125 correspond to rectangles beginning at 124.5, as shaded in Figure 4.24. Thus, we really want $P(X \ge 124.5)$, which can now be standardized to yield $P(Z \ge 1.63) = .0516$. If we had standardized $X \ge 125$, we would have obtained $P(Z \ge 1.67) = .0475$. The difference is not great, but the answer .0516 is more accurate. Similarly, $P(X = 125)$ would be approximated by the area between 124.5 and 125.5, since the area under the normal curve above the single value 125 is zero.

125

Figure 4.24 A normal approximation to a discrete distribution   ■

The correction for discreteness of the underlying distribution in Example 4.19 is often called a **continuity correction.** It is useful in the following application of the normal distribution to the computation of binomial probabilities.

The Normal Approximation to the Binomial Distribution

Recall that the mean value and standard deviation of a binomial random variable X are $\mu_X = np$ and $\sigma_X = \sqrt{npq}$, respectively. Figure 4.25 displays a binomial probability histogram for the binomial distribution with $n = 20$, $p = .6$ [so $\mu = 20(.6) = 12$ and $\sigma = \sqrt{20(.6)(.4)} = 2.19$]. A normal curve with mean value and standard deviation equal to the corresponding values for the binomial distribution has been superimposed on the probability histogram. Although the probability histogram is a bit skewed (because $p \neq .5$), the normal curve gives a very good approximation, especially in the middle part of the picture. The area of any rectangle (probability of any particular X value) except those in the extreme tails can be accurately approximated by the corresponding normal curve area. For example, $P(X = 10) = B(10; 20, .6) - B(9; 20, .6) = .117$, whereas the area under the normal curve between 9.5 and 10.5 is $P(-1.14 \leq Z \leq -.68) = .1212$.

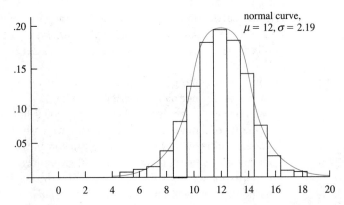

Figure 4.25 Binomial probability histogram for $n = 20$, $p = .6$ with normal approximation curve superimposed

More gene~~r~~ ~~Binomial~~ s the binomial probability histogram is not too skewed,
binomial prob ~~Dist.~~ e well approximated by normal curve areas. It is then
customary to ; approximately a normal distribution.

Proposition

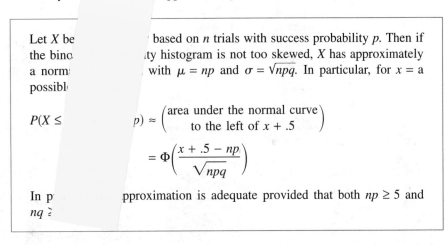

Let X be   based on n trials with success probability p. Then if
the bino.   ty histogram is not too skewed, X has approximately
a norm: with $\mu = np$ and $\sigma = \sqrt{npq}$. In particular, for $x = a$
possibl

$$P(X \le \qquad\qquad p) \approx \left(\begin{array}{c}\text{area under the normal curve}\\ \text{to the left of } x + .5\end{array}\right)$$

$$= \Phi\left(\frac{x + .5 - np}{\sqrt{npq}}\right)$$

In p'   pproximation is adequate provided that both $np \ge 5$ and
$nq \ge$

If either $np < 5$ or $nq < 5$, the binomial distribution is too skewed for the (symmetric)
normal curve to give accurate approximations.

Example 4.20   *The Los Angeles Times* (Dec. 11, 1983) reports that of roughly 3 million gasoline and
petroleum product storage tanks buried below ground, 1 in 4 is leaking. Suppose
$n = 50$ tanks are randomly selected for inspection. Assuming that the 1-in-4 figure
is correct, calculate the probability that at most 10 of the selected tanks leak. Let a
success be identified with a leaking tank, so that $p = .25$. Then $\mu = 12.5$ and
$\sigma = 3.06$. Since $np = 50(.25) = 12.5 \ge 5$ and $nq = 37.5 \ge 5$, the approximation can
safely be applied:

$$P(X \le 10) = B(10; 50, .25) \approx \Phi\left(\frac{10 + .5 - 12.5}{3.06}\right)$$

$$= \Phi(-.65) = .2578$$

Similarly, the probability that between 5 and 15 (inclusive) of the selected tanks are
leaking is

$$P(5 \le X \le 15) = B(15; 50, .25) - B(4; 50, .25)$$

$$\approx \Phi\left(\frac{15.5 - 12.5}{3.06}\right) - \Phi\left(\frac{4.5 - 12.5}{3.06}\right) = .8320$$

The exact probabilities are .2622 and .8348, respectively, so the approximations are
quite good. In the last calculation, the probability $P(5 \le X \le 15)$ is being
approximated by the area under the normal curve between 4.5 and 15.5—the
continuity correction is used for both the upper and lower limits.   ∎

When the objective of our investigation is to make an inference about a
population proportion p, interest will focus on the sample proportion of successes X/n

rather than on X itself. Because this proportion is just X multiplied by the constant $1/n$, it will also have approximately a normal distribution (with mean $\mu = p$ and standard deviation $\sigma = \sqrt{pq/n}$) provided that both $np \geq 5$ and $nq \geq 5$. This normal approximation is the basis for several large-sample inferential procedures to be discussed in later chapters.

It is quite difficult to give a direct proof of the validity of this normal approximation (the first one goes back over 150 years to Laplace). In the next chapter, we'll show that it is a consequence of an important general result called the Central Limit Theorem.

Exercises   Section 4.3 (26–52)

26. Let Z be a standard normal rv and calculate the following probabilities, drawing pictures wherever appropriate.

   a. $P(0 \leq Z \leq 2.17)$ **b.** $P(0 \leq Z \leq 1)$

   c. $P(-2.50 \leq Z \leq 0)$

   d. $P(-2.50 \leq Z \leq 2.50)$

   e. $P(Z \leq 1.37)$ **f.** $P(-1.75 \leq Z)$

   g. $P(-1.50 \leq Z \leq 2.00)$

   h. $P(1.37 \leq Z \leq 2.50)$ **i.** $P(1.50 \leq Z)$

   j. $P(|Z| \leq 2.50)$

27. In each case below, determine the value of the constant c that makes the probability statement correct.

   a. $\Phi(c) = .9838$ **b.** $P(0 \leq Z \leq c) = .291$

   c. $P(c \leq Z) = .121$

   d. $P(-c \leq Z \leq c) = .668$

   e. $P(c \leq |Z|) = .016$

28. Find the following percentiles for the standard normal distribution. Interpolate where appropriate.

   a. 91st   **b.** 9th   **c.** 75th
   d. 25th   **e.** 6th

29. Determine z_α for the following:

   a. $\alpha = .0055$ **b.** $\alpha = .09$ **c.** $\alpha = .663$

30. If X is a normal rv with mean 80 and standard deviation 10, compute the following probabilities by standardizing.

   a. $P(X \leq 100)$ **b.** $P(X \leq 80)$

   c. $P(65 \leq X \leq 100)$ **d.** $P(70 \leq X)$

   e. $P(85 \leq X \leq 95)$ **f.** $P(|X - 80| \leq 10)$

31. Suppose the force acting on a column that helps to support a building is normally distributed with mean 15.0 kips and standard deviation 1.25 kips. What is the probability that the force

   a. Is at most 17 kips?

   b. Is between 12 and 17 kips?

   c. Differs from 15.0 kips by at most 2 SDs?

32. Assume that development time for a particular type of photographic printing paper when it is exposed to a light source for 5 sec is normally distributed with a mean of 25 sec and a standard deviation of 1.3 sec. What is the probability that

   a. A particular print will require more than 26.5 sec to develop?

   b. Development time is at least 23 sec?

   c. Development time differs from the expected time by more than 2.5 sec?

33. A particular type of gasoline tank for a compact car is designed to hold 15 gal. Suppose the actual capacity X of a randomly chosen tank of this type is normally distributed with mean 15 gal and standard deviation .2 gal.

   a. What is the probability that a randomly selected tank will hold at most 14.8 gal?

   b. What is the probability that a randomly selected tank will hold between 14.7 and 15.1 gal?

   c. If the car on which a randomly chosen tank is mounted gets exactly 25 mpg, what is the probability that the car can travel 370 miles without refueling?

34. There are two machines available for cutting corks intended for use in wine bottles. The first produces corks with diameters that are normally distributed with mean 3 cm and standard deviation .1 cm. The second machine produces corks with diameters that have a normal distribution with mean 3.04 cm and standard deviation .02 cm. Acceptable corks have diameters between 2.9 cm and 3.1 cm. Which machine is more likely to produce an acceptable cork?

35. a. If a normal distribution has $\mu = 25$ and $\sigma = 5$, what is the 91st percentile of the distribution?

  b. What is the 6th percentile of the distribution of part (a)?

  c. The width of a line etched on an integrated circuit chip is normally distributed with mean 3.000 μm and standard deviation .150. What width value separates the widest 10% of all such lines from the other 90 percent?

36. The paper "Monte Carlo Simulation—Tool for Better Understanding of LRFD" (*J. Structural Engr.*, 1993, pp. 1586–1599) suggests that yield strength (ksi) for A36 steel grade is normally distributed with $\mu = 43$ and $\sigma = 4.5$.

  a. What is the probability that yield strength is at most 40? Greater than 60?

  b. What yield strength value separates the strongest 75% from the others?

37. The automatic opening device of a military cargo parachute has been designed to open when the parachute is 200 m above the ground. Suppose opening altitude actually has a normal distribution with mean value 200 m and standard deviation 30 m. Equipment damage will occur if the parachute opens at an altitude of less than 100 m. What is the probability that there is equipment damage to the payload of at least one of five independently dropped parachutes?

38. The temperature reading from a thermocouple placed in a constant-temperature medium is normally distributed with mean μ, the actual temperature of the medium, and standard deviation σ. What would the value of σ have to be to ensure that 95% of all readings are within .1° of μ?

39. The distribution of resistance for resistors of a certain type is known to be normal, 10% of all resistors have a resistance exceeding 10.256 ohms, and 5% have resistance smaller than 9.671 ohms. What are the mean value and standard deviation of the resistance distribution?

40. If bearing diameter is normally distributed, what is the probability that the diameter of a randomly selected bearing is

  a. Within 1.5 SDs of its mean value?

  b. Farther than 2.5 SDs from its mean value?

  c. Between 1 and 2 SDs from its mean value?

41. The air pressure in a randomly selected tire put on a certain model new car is normally distributed with mean value 31 psi and standard deviation .2 psi.

  a. What is the probability that the pressure for a randomly selected tire exceeds 30.5 psi?

  b. What is the probability that the pressure for a randomly selected tire is between 30.5 and 31.5 psi? Between 30 and 32 psi?

  c. Suppose a tire is classed as underinflated if its pressure is less than 30.4 psi. What is the probability that at least one of the four tires on a car is underinflated? (HINT: If $A = \{$at least 1 tire is underinflated$\}$, what is the complement of A?)

42. The Rockwell hardness of a metal is determined by impressing a hardened point into the surface of the metal and then measuring the depth of penetration of the point. Suppose the Rockwell hardness of a particular alloy is normally distributed with mean 70 and standard deviation 3 (assume that Rockwell hardness is measured on a continuous scale).

  a. If a specimen is acceptable only if its hardness is between 67 and 75, what is the probability that a randomly chosen specimen has an acceptable hardness?

  b. If the acceptable range of hardness was $(70 - c, 70 + c)$, for what value of c would 95% of all specimens have acceptable hardness?

  c. If the acceptable range is as in part (a) and the hardness of each of ten randomly selected specimens is independently determined, what is the expected number of acceptable specimens among the ten?

  d. What is the probability that at most eight of ten independently selected specimens have a hardness of less than 73.84? (HINT: $Y = $ the number among the ten specimens with hardness less than 73.84 is a binomial variable; what is p?)

43. The weight distribution of parcels sent in a certain manner is normal with mean value 10 lb and standard deviation 2 lb. The parcel service wishes to establish a weight value c beyond which there will be a surcharge. What value of c is such that 99% of all parcels are at least 1 lb under the surcharge weight?

44. Suppose Appendix Table A.3 contained $\Phi(z)$ only for $z \geq 0$. Explain how you could still compute

    a. $P(-1.72 \leq Z \leq -.55)$.

    b. $P(-1.72) \leq Z \leq .55)$.

Is it necessary to table $\Phi(z)$ for z negative? What property of the standard normal curve justifies your answer?

45. Chebyshev's inequality, introduced in Exercise 43 (Chapter 3) is valid for continuous as well as discrete distributions. It states that for any number k satisfying $k \geq 1$, $P(|X - \mu| \geq k\sigma) \leq 1/k^2$ (see Exercise 43 in Chapter 3 for an interpretation). Obtain this probability in the case of a normal distribution for $k = 1, 2$, and 3 and compare to the upper bound.

46. Let X represent the number of pages of text in a randomly chosen mathematics Ph.D. thesis. Although X can assume only positive integer values, suppose it is approximately normally distributed with expected value 90 and standard deviation 15. What is the probability that a randomly chosen thesis contains

    a. At most 100 pages (using the continuity correction)?

    b. Between 80 and 110 pages (using the continuity correction)?

47. Let X have a binomial distribution with parameters $n = 25$ and p. Calculate each of the following probabilities using the normal approximation (with the continuity correction) for the cases $p = .5, .6$, and $.8$ and compare to the exact probabilities calculated from Appendix Table A.1.

    a. $P(15 \leq X \leq 20)$ **b.** $P(X \leq 15)$

    c. $P(20 \leq X)$

48. Suppose that 10% of all steel shafts produced by a certain process are nonconforming but can be reworked (rather than having to be scrapped). Consider a random sample of 200 shafts and let X denote the number among these that are nonconforming and can be reworked. What is the (approximate) probability that X is

    a. At most 30?   **b.** Less than 30?

    c. Between 15 and 25 (inclusive)?

49. Suppose only 40% of all drivers in a certain state regularly wear a seatbelt. A random sample of 500 drivers is selected. What is the probability that

    a. Between 180 and 230 (inclusive) of the drivers in the sample regularly wear a seatbelt?

    b. Fewer than 175 of those in the sample regularly wear a seatbelt? Fewer than 150?

50. Show that the relationship between a general normal percentile and the corresponding z percentile is as stated in this section.

51. a. Show that if X has a normal distribution with parameters μ and σ, then $Y = aX + b$ (a linear function of X) also has a normal distribution. What are the parameters of the distribution of Y [i.e., $E(Y)$ and $V(Y)$]? [HINT: Write the cdf of Y, $P(Y \leq y)$, as an integral involving the pdf of X, and then differentiate with respect to y to get the pdf of Y.]

    b. If when measured in °C, temperature is normally distributed with mean 115 and standard deviation 2, what can be said about the distribution of temperature measured in °F?

52. There is no nice formula for the standard normal cdf $\Phi(z)$, but several good approximations have appeared in published papers. The following one comes from "Approximations for Hand Calculators Using Small Integer Coefficients" (*Mathematics of Computation*, 1977, pp. 214–222). For $0 \leq z \leq 5.5$,

$$P(Z \geq z) = 1 - \Phi(z)$$

$$\approx .5 \exp\left\{-\left[\frac{(83z + 351)z + 562}{703/z + 165}\right]\right\}$$

The relative error of this approximation is less than .042%. Use this to calculate approximations to the following probabilities and compare whenever possible to the probabilities obtained from Appendix Table A.3.

    a. $P(Z \geq 1)$ **b.** $P(Z < -3)$

    c. $P(-4 < Z < 4)$ **d.** $P(Z > 5)$

4.4 The Gamma Distribution and Its Relatives

The graph of any normal pdf is bell-shaped and thus symmetric. There are many practical situations in which the variable of interest to the experimenter might have a skewed distribution. A family of pdf's that yields a wide variety of skewed distributional shapes is the gamma family. To define the family of gamma distributions, we first need to introduce a function that plays an important role in many branches of mathematics.

Definition

For $\alpha > 0$, the **gamma function** $\Gamma(\alpha)$ is defined by

$$\Gamma(\alpha) = \int_0^\infty x^{\alpha - 1} e^{-x} \, dx \tag{4.5}$$

The most important properties of the gamma function are the following:

1. For any $\alpha > 1$, $\Gamma(\alpha) = (\alpha - 1) \cdot \Gamma(\alpha - 1)$ [via integration by parts]
2. For any positive integer n, $\Gamma(n) = (n - 1)!$
3. $\Gamma\left(\frac{1}{2}\right) = \sqrt{\pi}$

By Expression (4.5) if we let

$$f(x; \alpha) = \begin{cases} \dfrac{x^{\alpha - 1} e^{-x}}{\Gamma(\alpha)} & x \geq 0 \\ 0 & \text{otherwise} \end{cases} \tag{4.6}$$

then $f(x; \alpha) \geq 0$ and $\int_0^\infty f(x; \alpha) \, dx = \Gamma(\alpha)/\Gamma(\alpha) = 1$, so $f(x; \alpha)$ satisfies the two basic properties of a pdf.

The Family of Gamma Distributions

Definition

A continuous rv X is said to have a **gamma distribution** if the pdf of X is

$$f(x; \alpha, \beta) = \begin{cases} \dfrac{1}{\beta^\alpha \Gamma(\alpha)} x^{\alpha - 1} e^{-x/\beta} & x \geq 0 \\ 0 & \text{otherwise} \end{cases} \tag{4.7}$$

where the parameters α and β satisfy $\alpha > 0$, $\beta > 0$. The **standard gamma distribution** has $\beta = 1$, so the pdf of a standard gamma rv is given by (4.6).

Figure 4.26a illustrates the graphs of the gamma pdf $f(x; \alpha, \beta)$ (4.7) for several (α, β) pairs, while Figure 4.26b presents graphs of the standard gamma pdf. For the standard pdf, when $\alpha \leq 1$, $f(x; \alpha)$ is strictly decreasing as x increases from 0; when $\alpha > 1$, $f(x; \alpha)$ rises from 0 at $x = 0$ to a maximum and then decreases. The parameter

β in (4.7) is called the *scale parameter* because values other than 1 either stretch or compress the pdf in the *x* direction.

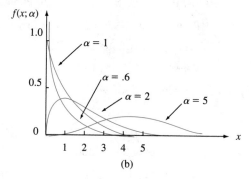

Figure 4.26 (a) Gamma density functions; (b) standard gamma density functions

$E(X)$ and $E(X^2)$ can be obtained from a reasonably straightforward integration, and then $V(X) = E(X^2) - [E(X)]^2$.

Proposition

> The mean and variance of an rv X having the gamma distribution $f(x;\ \alpha,\ \beta)$ are
>
> $$E(X) = \mu = \alpha\beta, \quad V(X) = \sigma^2 = \alpha\beta^2$$

Computing Probabilities from the Gamma Distribution

When X is a standard gamma rv, the cdf of X,

$$F(x;\ \alpha) = \int_0^x \frac{y^{\alpha - 1}e^{-y}}{\Gamma(\alpha)}\ dy, \qquad x > 0 \tag{4.8}$$

is called the **incomplete gamma function** [sometimes the incomplete gamma function refers to Expression (4.8) without the denominator $\Gamma(\alpha)$ in the integrand].

There are extensive tables of $F(x; \alpha)$ available; in Appendix Table A.4, we present a small tabulation for $\alpha = 1, 2, \ldots, 10$ and $x = 1, 2, \ldots, 15$.

Example 4.21 Suppose the reaction time X of a randomly selected individual to a certain stimulus has a standard gamma distribution with $\alpha = 2$ sec. Since

$$P(a \leq X \leq b) = F(b) - F(a)$$

when X is continuous,

$$P(3 \leq X \leq 5) = F(5; 2) - F(3; 2) = .960 - .801 = .159$$

The probability that the reaction time is more than 4 sec is

$$P(X > 4) = 1 - P(X \leq 4) = 1 - F(4; 2) = 1 - .908 = .092 \qquad \blacksquare$$

The incomplete gamma function can also be used to compute probabilities involving nonstandard gamma distributions.

Proposition

> Let X have a gamma distribution with parameters α and β. Then for any $x > 0$, the cdf of X is given by
>
> $$P(X \leq x) = F(x; \alpha, \beta) = F\left(\frac{x}{\beta}; \alpha\right)$$
>
> where $F(\cdot \,; \alpha)$ is the incomplete gamma function.*

Example 4.22 Suppose the survival time X in weeks of a randomly selected male mouse exposed to 240 rads of gamma radiation has a gamma distribution with $\alpha = 8$ and $\beta = 15$ (data in *Survival Distributions: Reliability Applications in the Biomedical Services,* by A. J. Gross and V. Clark, suggests $\alpha \approx 8.5$ and $\beta \approx 13.3$). The expected survival time is $E(X) = (8)(15) = 120$ weeks, while $V(X) = (8)(15)^2 = 1800$ and $\sigma_X = \sqrt{1800} = 42.43$ weeks. The probability that a mouse survives between 60 and 120 weeks is

$$P(60 \leq X \leq 120) = P(X \leq 120) - P(X \leq 60)$$
$$= F(120/15; 8) - F(60/15; 8)$$
$$= F(8; 8) - F(4; 8) = .547 - .051 = .496$$

The probability that a mouse survives at least 30 weeks is

$$P(X \geq 30) = 1 - P(X < 30) = 1 - P(X \leq 30)$$
$$= 1 - F(30/15; 8) = .999 \qquad \blacksquare$$

---

*MINITAB and other statistical packages will calculate $F(x; \alpha, \beta)$ once values of x, α, and β have been specified.

The Exponential Distribution

The family of exponential distributions provides probability models that are very widely used in engineering and science disciplines.

Definition

> X is said to have an **exponential distribution** if the pdf of X is
>
> $$f(x; \lambda) = \begin{cases} \lambda e^{-\lambda x} & x \geq 0 \\ 0 & \text{otherwise} \end{cases} \qquad \text{where } \lambda > 0 \qquad\qquad (4.9)$$

The exponential pdf is a special case of the general gamma pdf Expression (4.7) in which $\alpha = 1$ and β has been replaced by $1/\lambda$ [some authors use the form $(1/\beta)e^{-x/\beta}$]. The mean and variance of X are then

> $$\mu = \alpha\beta = \frac{1}{\lambda}, \qquad \sigma^2 = \alpha\beta^2 = \frac{1}{\lambda^2}$$

Both the mean and standard deviation of the exponential distribution equal $1/\lambda$. Graphs of several exponential pdf's appear in Figure 4.27.

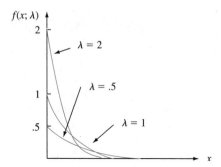

Figure 4.27 Exponential density functions

Unlike the general gamma pdf, the exponential pdf can be easily integrated. In particular, the cdf of X is

> $$F(x; \lambda) = \begin{cases} 0 & x < 0 \\ 1 - e^{-\lambda x} & x \geq 0 \end{cases}$$

Example 4.23

Suppose the response time X at a certain on-line computer terminal (the elapsed time between the end of a user's inquiry and the beginning of the system's response to that inquiry) has an exponential distribution with expected response time equal to 5 sec. Then $E(X) = 1/\lambda = 5$, so $\lambda = .2$. The probability that the response time is at most 10 sec is

$$P(X \le 10) = F(10; .2) = 1 - e^{-(.2)(10)} = 1 - e^{-2} = 1 - .135 = \quad \text{Exp. Dist.}$$

The probability that response time is between 5 and 10 sec is

$$P(5 \le X \le 10) = F(10; .2) - F(5; .2)$$

$$= (1 - e^{-2}) - (1 - e^{-1}) = .233 \qquad \blacksquare$$

Applications of the Exponential Distribution

The exponential distribution is frequently used as a model for the distr   es between the occurrence of successive events such as customers arriv   e facility or calls coming in to a switchboard. The reason for this is that t   al distribution is closely related to the Poisson process discussed in Cl

Proposition

> Suppose the number of events occurring in any time interval of le a Poisson distribution with parameter αt (where α, the rate of process, is the expected number of events occurring in 1 unit of tim , ...u uiaι numbers of occurrences in nonoverlapping intervals are independent of one another. Then the distribution of elapsed time between the occurrence of two successive events is exponential with parameter $\lambda = \alpha$.

While a complete proof is beyond the scope of the text, the result is easily verified for the time X_1 until the first event occurs:

$$P(X_1 \le t) = 1 - P(X_1 > t) = 1 - P[\text{no events in } (0, t)]$$

$$= 1 - \frac{e^{-\alpha t} \cdot (\alpha t)^0}{0!} = 1 - e^{-\alpha t}$$

which is exactly the cdf of the exponential distribution.

Example 4.24

Suppose calls are received at a 24-hour "suicide hotline" according to a Poisson process with rate $\alpha = .5$ calls per day. Then the number of days X between successive calls has an exponential distribution with parameter value .5, so the probability that more than 2 days elapse between calls is

$$P(X > 2) = 1 - P(X \le 2) = 1 - F(2; .5) = e^{-(.5)(2)} = .368$$

The expected time between successive calls is $1/.5 = 2$ days. $\qquad \blacksquare$

Another important application of the exponential distribution is to model the distribution of component lifetime. A partial reason for the popularity of such applications is the **"memoryless" property** of the exponential distribution. Suppose component lifetime is exponentially distributed with parameter λ. After putting the component into service, we leave for a period of t_0 hours and then return to find the component still working; what now is the probability that it lasts at least an additional t hours? In symbols we wish $P(X \geq t + t_0 | X \geq t_0)$. By the definition of conditional probability,

$$P(X \geq t + t_0 | X \geq t_0) = \frac{P[(X \geq t + t_0) \cap (X \geq t_0)]}{P(X \geq t_0)}$$

But the event $X \geq t_0$ in the numerator is redundant, since both events can occur if and only if $X \geq t + t_0$. Therefore,

$$P(X \geq t + t_0 | X \geq t_0) = \frac{P(X \geq t + t_0)}{P(X \geq t_0)} = \frac{1 - F(t + t_0; \lambda)}{1 - F(t_0; \lambda)} = e^{-\lambda t}$$

This conditional probability is identical to the original probability $P(X \geq t)$ that the component lasted t hours. Thus, *the distribution of additional lifetime is exactly the same as the original distribution of lifetime,* so at each point in time the component shows no effect of wear. A way to paraphrase this is to say that the distribution of remaining lifetime is independent of current age.

While the memoryless property can be justified at least approximately in many applied problems, in other situations components deteriorate with age or occasionally improve with age (at least up to a certain point). More general lifetime models are then furnished by the gamma, Weibull, or lognormal distributions (the latter two are discussed in the next section).

The Chi-Squared Distribution

Definition

> Let v be a positive integer. Then an rv X is said to have a **chi-squared distribution** with parameter v if the pdf of X is the gamma density with $\alpha = v/2$ and $\beta = 2$. The pdf of a chi-squared rv is thus
>
> $$f(x; v) = \begin{cases} \dfrac{1}{2^{v/2}\Gamma(v/2)} x^{(v/2) - 1} e^{-x/2} & x \geq 0 \\ 0 & x < 0 \end{cases} \qquad (4.10)$$
>
> The parameter v is called the **number of degrees of freedom** (df) of X. The symbol χ^2 is often used in place of "chi-squared."

The chi-squared distribution is important because it is the basis for a number of procedures in statistical inference. The reason for this is that chi-squared distributions are intimately related to normal distributions (see Exercise 65). We will discuss the chi-squared distribution in more detail in the chapters on inference.

Exercises   Section 4.4 (53–65)

53. Evaluate the following:

a. $\Gamma(6)$ **b.** $\Gamma(5/2)$

c. $F(4; 5)$ (the incomplete gamma function)

d. $F(5; 4)$ **e.** $F(0; 4)$

54. Let X have a standard gamma distribution with $\alpha = 7$. Evaluate the following:

a. $P(X \le 5)$ **b.** $P(X < 5)$

c. $P(X > 8)$ **d.** $P(3 \le X \le 8)$

e. $P(3 < X < 8)$ **f.** $P(X < 4 \text{ or } X > 6)$

55. Suppose the time (in hours) taken by a homeowner to mow his lawn is an rv X having a gamma distribution with parameters $\alpha = 2$ and $\beta = \frac{1}{2}$. What is the probability that it takes:

a. At most 1 hour to mow the lawn?

b. At least 2 hours to mow the lawn?

c. Between .5 and 1.5 hours to mow the lawn?

56. Suppose the time spent by a randomly selected student who uses a terminal connected to a local time-sharing computer facility has a gamma distribution with mean 20 min and variance 80 min^2.

a. What are the values of α and β?

b. What is the probability that a student uses the terminal for at most 24 min?

c. What is the probability that a student spends between 20 and 40 min using the terminal?

57. Suppose that when a transistor of a certain type is subjected to an accelerated life test, the lifetime X (in weeks) has a gamma distribution with mean 24 weeks and standard deviation 12 weeks.

a. What is the probability that a transistor will last between 12 and 24 weeks?

b. What is the probability that a transistor will last at most 24 weeks? Is the median of the lifetime distribution less than 24? Why, or why not?

c. What is the 99th percentile of the lifetime distribution?

d. Suppose the test will actually be terminated after t weeks. What value of t is such that only one-half of 1% of all transistors would still be operating at termination?

58. Let X = the time between two successive arrivals at the drive-up window of a local bank. If X has an exponential distribution with $\lambda = 1$ (which is identical to a standard gamma distribution with $\alpha = 1$), compute the following:

a. The expected time between two successive arrivals

b. The standard deviation of the time between successive arrivals

c. $P(X \le 4)$ **d.** $P(2 \le X \le 5)$

59. The time X (in seconds) that it takes a librarian to locate an entry in a file of records on checked-out books has an exponential distribution with expected time 20 sec. Calculate the following probabilities:

a. $P(X \le 30)$ **b.** $P(X \ge 20)$

c. $P(20 \le X \le 30)$

d. For what value of t is $P(X \le t) = .5$ (t is the 50th percentile of the distribution)?

60. Extensive experience with fans of a certain type used in diesel engines has suggested that the exponential distribution provides a good model for time until failure. Suppose the mean time until failure is 25,000 hours. What is the probability that

a. A randomly selected fan will last at least 20,000 hours? At most 30,000 hours? Between 20,000 and 30,000 hours?

b. The lifetime of a fan exceeds the mean value by more than 2 SDs? More than 3 SDs?

61. The special case of the gamma distribution in which α is a positive integer n is called an Erlang distribution. If we replace β by $1/\lambda$ in Expression (4.7), the Erlang pdf is

$$f(x; \lambda, n) = \begin{cases} \dfrac{\lambda(\lambda x)^{n-1} e^{-\lambda x}}{(n-1)!} & x \ge 0 \\ 0 & x < 0 \end{cases}$$

It can be shown that if the times between successive events are independent, each with an exponential distribution with parameter λ, then the total time X that elapses before all of the next n events occur has pdf $f(x; \lambda, n)$.

a. What is the expected value of X? If the time (in minutes) between arrivals of successive customers is exponentially distributed with $\lambda = .5$, how much time can be expected to elapse before the tenth customer arrives?

b. If customer interarrival time is exponentially distributed with $\lambda = .5$, what is the probability that the tenth customer (after the one who has just arrived) will arrive within the next 30 min?

c. The event $\{X \leq t\}$ occurs iff at least n events occur in the next t units of time. Use the fact that the number of events occurring in an interval of length t has a Poisson distribution with parameter λt to write down an expression (involving Poisson probabilities) for the Erlang cdf $F(t; \lambda, n) = P(X \leq t)$.

62. A system consists of five identical components connected in series as shown:

As soon as one component fails, the entire system will fail. Suppose each component has a lifetime that is exponentially distributed with $\lambda = .01$ and that components fail independently of one another. Define events $A_i = \{i\text{th component lasts at least } t \text{ hours}\}$, $i = 1, \ldots, 5$, so that the A_i's are independent events. Let $X =$ the time at which the system fails—that is, the shortest (minimum) lifetime among the five components.

a. The event $\{X \geq t\}$ is equivalent to what event involving $A_1, \ldots, A_5$?

b. Using the independence of the A_i's, compute $P(X \geq t)$. Then obtain $F(t) = P(X \leq t)$ and the pdf of X. What type of distribution does X have?

c. Suppose there are n components, each having exponential lifetime with parameter λ. What type of distribution does X have?

63. If X has an exponential distribution with parameter λ, derive a general expression for the $(100p)$th percentile of the distribution. Then specialize to obtain the median.

64. Show that $\int_0^\infty f(x; \alpha, \beta)\, dx = 1$ for the gamma pdf of (4.7). [HINT: Let $y = x/\beta$ to obtain an integrand like that of Expression (4.6).]

65. a. The event $\{X^2 \leq y\}$ is equivalent to what event involving X itself?

b. If X has a standard normal distribution, use part (a) to write the integral that equals $P(X^2 \leq y)$. Then differentiate this with respect to y to obtain the pdf of X^2 [the square of a $N(0, 1)$ variable]. Finally, show that X^2 has a chi-squared distribution with $v = 1$ df [see (4.10)]. (HINT: Use the following identity.)

$$\frac{d}{dy}\left\{\int_{a(y)}^{b(y)} f(x)\, dx\right\}$$
$$= f[b(y)] \cdot b'(y) - f[a(y)] \cdot a'(y)$$

Other Continuous Distributions

The normal, gamma (including exponential), and uniform families of distributions provide a wide variety of probability models for continuous variables, but there are many practical situations in which no member of these families fits a set of observed data very well. Statisticians and other investigators have developed other families of distributions that are often appropriate in practice.

The Weibull Distribution

The family of Weibull distributions was introduced by the Swedish physicist Waloddi Weibull in 1939; his 1951 paper "A Statistical Distribution Function of Wide Applicability" (*J. Applied Mechanics,* vol. 18, pp. 293–297) discusses a number of applications.

Definition

An rv X is said to have a **Weibull distribution** with parameters α and β $(\alpha > 0, \beta > 0)$ if the pdf of X is

$$f(x; \alpha, \beta) = \begin{cases} \dfrac{\alpha}{\beta^\alpha} x^{a-1} e^{-(x/\beta)^\alpha} & x \geq 0 \\ 0 & x < 0 \end{cases} \qquad (4.11)$$

In some situations, there are theoretical justifications for the appropriateness of the Weibull distribution, but in many applications $f(x; \alpha, \beta)$ simply provides a good fit to observed data for particular values of α and β. When $\alpha = 1$, the pdf reduces to the exponential distribution (with $\lambda = 1/\beta$), so the exponential distribution is a special case of both the gamma and Weibull distributions. However, there are gamma distributions that are not Weibull distributions and vice versa, so one family is not a subset of the other. Both α and β can be varied to obtain a number of different distributional shapes, as illustrated in Figure 4.28. β is a scale parameter, so different values stretch or compress the graph in the x direction.

Integrating to obtain $E(X)$ and $E(X^2)$ yields

$$\mu = \beta\Gamma\left(1 + \frac{1}{\alpha}\right), \quad \sigma^2 = \beta^2\left\{\Gamma\left(1 + \frac{2}{\alpha}\right) - \left[\Gamma\left(1 + \frac{1}{\alpha}\right)\right]^2\right\}$$

The computation of μ and σ^2 thus necessitates using a table of the gamma function. The integration $\int_0^x f(y; \alpha, \beta)\, dy$ is easily carried out to obtain the cdf of X.

Proposition

The cdf of a Weibull rv having parameters α and β is

$$F(x; \alpha, \beta) = \begin{cases} 0 & x < 0 \\ 1 - e^{-(x/\beta)^\alpha} & x \geq 0 \end{cases} \qquad (4.12)$$

Example 4.25   Let X = the ultimate tensile strength (ksi) at $-200°F$ of a type of steel that exhibits "cold brittleness" at low temperatures. Suppose X has a Weibull distribution with parameters $\alpha = 20$ and $\beta = 100$. Then

$$P(X \leq 105) = F(105; 20, 100) = 1 - e^{-(105/100)^{20}} = 1 - .070 = .930$$

and

$$P(98 \leq X \leq 102) = F(102; 20, 100) - F(98; 20, 100)$$
$$= e^{-(.98)^{20}} - e^{-(1.02)^{20}} = .513 - .226 = .287 \qquad \blacksquare$$

Frequently, in practical situations, a Weibull model may be reasonable except that the smallest possible X value may be some value γ not assumed to be zero (this would also apply to a gamma model). The quantity γ can then be regarded as a third parameter of the distribution, which is what Weibull did in his original work. For,

Figure 4.28 Graphs of Weibull pdf's

say, $\gamma = 3$, all curves in Figure 4.28 would be shifted 3 units to the right. This is equivalent to saying that $X - \gamma$ has the pdf (4.11), so that the cdf of X is obtained by replacing x in (4.12) by $x - \gamma$.

Example 4.26   Let X = the corrosion weight loss for a small square magnesium alloy plate immersed for 7 days in an inhibited aqueous 20% solution of $MgBr_2$. Suppose the minimum possible weight loss is $\gamma = 3$ and that the excess $X - 3$ over this minimum has a

Weibull distribution with $\alpha = 2$ and $\beta = 4$. (This example was considered in "Practical Applications of the Weibull Distribution," *Industrial Quality Control,* Aug. 1964, pp. 71–78; values for α and β were taken to be 1.8 and 3.67, respectively, though a slightly different choice of parameters was used in the article.) The cdf of X is then

$$F(x;\ \alpha,\ \beta,\ \gamma) = F(x;\ 2,\ 4,\ 3) = \begin{cases} 0 & x < 3 \\ 1 - e^{-\ [(x\ -\ 3)/4]^2} & x \geq 3 \end{cases}$$

Therefore,

$$P(X > 3.5) = 1 - F(3.5;\ 2,\ 4,\ 3) = e^{-.0156} = .985$$

and

$$P(7 \leq X \leq 9) = 1 - e^{-2.25} - (1 - e^{-1}) = .895 - .632 = .263 \qquad \blacksquare$$

The Lognormal Distribution

Definition

> A nonnegative rv X is said to have a **lognormal distribution** if the rv $Y = \ln(X)$ has a normal distribution. The resulting pdf of a lognormal rv when $\ln(X)$ is normally distributed with parameters μ and σ is
>
> $$f(x;\ \mu,\ \sigma) = \begin{cases} \dfrac{1}{\sqrt{2\pi}\ \sigma x}\ e^{-\ [\ln(x)\ -\ \mu]^2/(2\sigma^2)} & x \geq 0 \\ 0 & x < 0 \end{cases}$$

Be careful here; μ and σ are not the mean and standard deviation of X but of $\ln(X)$. The mean and variance of X can be shown to be

> $$E(X) = e^{\mu\ +\ \sigma^2/2}, \qquad V(X) = e^{2\mu\ +\ \sigma^2} \cdot (e^{\sigma^2} - 1)$$

In the next chapter, we will present a theoretical justification for this distribution in connection with the Central Limit Theorem, but as with other distributions, the lognormal can be used as a model even in the absence of such justification. Figure 4.29 illustrates the graphs of the lognormal pdf; although the normal curve is symmetric, a lognormal curve has a positive skew.

Because $\ln(X)$ has a normal distribution, the cdf of X can be expressed in terms of the cdf $\Phi(z)$ of a standard normal rv Z. For $x \geq 0$,

> $$F(x;\ \mu,\ \sigma) = P(X \leq x) = P[\ln(X) \leq \ln(x)]$$
>
> $$= P\left(Z \leq \frac{\ln(x) - \mu}{\sigma}\right) = \Phi\left(\frac{\ln(x) - \mu}{\sigma}\right) \qquad (4.13)$$

Figure 4.29 Graphs of lognormal pdf's

Example 4.27   Let X = the hourly median power (in decibels) of received radio signals transmitted between two cities. The authors of the paper "Families of Distributions for Hourly Median Power and Instantaneous Power of Received Radio Signals" (*J. Research National Bureau of Standards*, 1963, vol. 67D, pp. 753–762) argue that the lognormal distribution provides a reasonable probability model for X. If the parameter values are $\mu = 3.5$ and $\sigma = 1.2$, then

$$E(X) = e^{3.5 + .72} = 68.0, \quad V(X) = e^{8.44}(e^{1.44} - 1) = 14{,}907.2$$

The probability that received power is between 50 and 250 dB is

$$P(50 \leq X \leq 250) = F(250; 3.5, 1.2) - F(50; 3.5, 1.2)$$

$$= \Phi\left(\frac{\ln(250) - 3.5}{1.2}\right) - \Phi\left(\frac{\ln(50) - 3.5}{1.2}\right)$$

$$= \Phi(1.68) - \Phi(.34)$$

$$= .9535 - .6331 = .3204 \qquad \text{(Appendix Table A.3)}$$

The probability that X does not exceed its mean is

$$P(X \leq 68.0) = \Phi\left(\frac{\ln(68.0) - 3.5}{1.2}\right) = \Phi(.60) = .7257$$

If the distribution were symmetric, this probability would equal .5; it is much larger because of the positive skew (long upper tail) of the distribution, which pulls μ outward past the median. ∎

The Beta Distribution

All families of continuous distributions discussed so far except for the uniform distribution have positive density over an infinite interval (though typically the density function decreases rapidly to zero beyond a few standard deviations from the mean). The beta distribution provides positive density only for X in an interval of finite length.

Definition

An rv X is said to have a **beta distribution** with parameters α, β (both positive), A, and B if the pdf of X is

$f(x; \alpha, \beta, A, B)$

$$
= \begin{cases} \dfrac{1}{B - A} \cdot \dfrac{\Gamma(\alpha + \beta)}{\Gamma(\alpha) \cdot \Gamma(\beta)} \left(\dfrac{x - A}{B - A} \right)^{\alpha - 1} \left(\dfrac{B - x}{B - A} \right)^{\beta - 1} & A \leq x \leq B \\ 0 & \text{otherwise} \end{cases}
$$

The case $A = 0$, $B = 1$ gives the **standard beta distribution.**

Figure 4.30 illustrates several standard beta pdf's. Graphs of the general pdf are similar, except they are shifted and then stretched or compressed to fit over $[A, B]$. Unless α and β are integers, integration of the pdf to calculate probabilities is difficult, so a table of the incomplete beta function is generally used. The mean and variance of X are

$$
\mu = A + (B - A) \cdot \frac{\alpha}{\alpha + \beta}, \qquad \sigma^2 = \frac{(B - A)^2 \alpha \beta}{(\alpha + \beta)^2 (\alpha + \beta + 1)}
$$

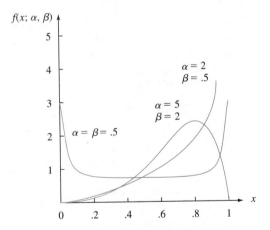

Figure 4.30 Graphs of standard beta pdf's

Example 4.28

Project managers often use a method labeled PERT—for program evaluation and review technique—to coordinate the various activities making up a large project (one successful application was in the construction of the *Apollo* spacecraft). A standard assumption in PERT analysis is that the time necessary to complete any particular activity once it has been started has a beta distribution with A = the optimistic time (if everything goes well) and B = the pessimistic time (if everything goes badly). Suppose that in constructing a single-family house, the time X (in days) necessary for laying the foundation has a beta distribution with $A = 2$, $B = 5$, $\alpha = 2$, and $\beta = 3$.

Then $\alpha/(\alpha + \beta) = .4$, so $E(X) = 2 + (3)(.4) = 3.2$. For these values of α and β, the pdf of X is a simple polynomial function. The probability that it takes at most 3 days to lay the foundation is

$$P(X \le 3) = \int_2^3 \frac{1}{3} \cdot \frac{4!}{1!2!} \left(\frac{x-2}{3}\right)\left(\frac{5-x}{3}\right)^2 dx$$

$$= \frac{4}{27} \int_2^3 (x-2)(5-x)^2 \, dx = \frac{4}{27} \cdot \frac{11}{4} = \frac{11}{27} = .407$$

      ■

The standard beta distribution is commonly used to model variation in the proportion or percentage of a quantity occurring in different samples, such as the proportion of a 24-hour day that an individual is asleep or the proportion of a certain element in a chemical compound.

Exercises   Section 4.5 (66–78)

66. The lifetime X (in hundreds of hours) of a certain type of vacuum tube has a Weibull distribution with parameters $\alpha = 2$ and $\beta = 3$. Compute the following:

   a. $E(X)$ and $V(X)$ **b.** $P(X \le 6)$

   c. $P(1.5 \le X \le 6)$

(This Weibull distribution is suggested as a model for time in service in "On the Assessment of Equipment Reliability: Trading Data Collection Costs for Precision," *J. Engr. Manuf.*, 1991, pp. 105–109).

67. The authors of the paper "A Probabilistic Insulation Life Model for Combined Thermal-Electrical Stresses" (*IEEE Trans. on Elect. Insulation*, 1985, pp. 519–522) state that "the Weibull distribution is widely used in statistical problems relating to aging of solid insulating materials subjected to aging and stress." They propose the use of the distribution as a model for time (in hours) to failure of solid insulating specimens subjected to AC voltage. The values of the parameters depend on the voltage and temperature; suppose $\alpha = 2.5$ and $\beta = 200$ (values suggested by data in the paper).

   a. What is the probability that a specimen's lifetime is at most 200? Less than 200? More than 300?

   b. What is the probability that a specimen's lifetime is between 100 and 200?

   c. What value is such that exactly 50% of all specimens have lifetimes exceeding that value?

68. Let X = the time (in 10^{-1} weeks) from shipment of a defective product until the customer returns the product. Suppose that the minimum return time is $\gamma = 3.5$ and that the excess $X - 3.5$ over the minimum has a Weibull distribution with parameters $\alpha = 2$ and $\beta = 1.5$ (see the *Industrial Quality Control* article referenced in Example 4.26).

   a. What is the cdf of X?

   b. What are the expected return time and variance of return time? [HINT: First obtain $E(X - 3.5)$ and $V(X - 3.5)$.]

   c. Compute $P(X > 5)$.

   d. Compute $P(5 \le X \le 8)$.

69. Let X have a Weibull distribution with the pdf from Expression (4.11). Verify that $\mu = \beta\Gamma(1 + 1/\alpha)$. (HINT: In the integral for $E(X)$, make the change of variable $y = (x/\beta)^{\alpha}$, so that $x = \beta y^{1/\alpha}$.)

70. a. In Exercise 66, what is the median lifetime of such tubes? (HINT: Use Expression (4.12).)

   b. In Exercise 68, what is the median return time?

   c. If X has a Weibull distribution with the cdf from Expression (4.12), obtain a general expression for the $(100p)$th percentile of the distribution.

   d. In Exercise 68, the company wants to refuse to accept returns after t weeks. For what value of t will only 10% of all returns be refused?

71. In Example 4.27, in which X has a lognormal distribution with parameters $\mu = 3.5$ and $\sigma = 1.2$, compute the following:

   a. $P(X \le 100)$

   b. $P(100 \le X \le 200)$

72. a. Use Equation (4.13) to write a formula for the median $\tilde{\mu}$ of the lognormal distribution. What is the median for the power distribution of Example 4.27?

   b. Recalling that z_α is our notation for the $100(1 - \alpha)$ percentile of the standard normal distribution, write an expression for the $100(1 - \alpha)$ percentile of the lognormal distribution. In Example 4.27, above what value will median power be only 5% of the time?

73. A theoretical justification based on a certain material failure mechanism underlies the assumption that ductile strength X of a material has a lognormal distribution. Suppose the parameters are $\mu = 5$ and $\sigma = .1$.

   a. Compute $E(X)$ and $V(X)$.

   b. Compute $P(X > 120)$.

   c. Compute $P(110 \le X \le 130)$.

   d. What is the value of median ductile strength?

   e. If ten different samples of an alloy steel of this type were subjected to a strength test, how many would you expect to have strength at least 120?

   f. If the smallest 5% of strength values were unacceptable, what would the minimum acceptable strength be?

74. The paper "The Statistics of Phytotoxic Air Pollutants" (*J. Royal Stat. Soc.*, 1989, pp. 183–198) suggests the lognormal distribution as a model for SO_2 concentration above a certain forest. Suppose the parameter values are $\mu = 1.9$ and $\sigma = .9$.

   a. What are the mean value and standard deviation of concentration?

   b. What is the probability that concentration is at most 10? Between 5 and 10?

75. What condition on α and β is necessary for the standard beta pdf to be symmetric?

76. Suppose the proportion X of surface area in a randomly selected quadrat that is covered by a certain plant has a standard beta distribution with $\alpha = 5$ and $\beta = 2$.

   a. Compute $E(X)$ and $V(X)$.

   b. Compute $P(X \le .2)$.

   c. Compute $P(.2 \le X \le .4)$.

   d. What is the expected proportion of the sampling region not covered by the plant?

77. Let X have a standard beta density with parameters α and β.

   a. Verify the formula for $E(X)$ given in the section.

   b. Compute $E[(1 - X)^m]$. If X represents the proportion of a substance consisting of a particular ingredient, what is the expected proportion that does not consist of this ingredient?

78. Stress is applied to a 20-in. steel bar that is clamped in a fixed position at each end. Let $Y = $ the distance from the left end at which the bar snaps. Suppose $Y/20$ has a standard beta distribution with $E(Y) = 10$ and $V(Y) = \frac{100}{7}$.

   a. What are the parameters of the relevant standard beta distribution?

   b. Compute $P(8 \le Y \le 12)$.

   c. Compute the probability that the bar snaps more than 2 in. from where you expect it to.

4.6 Probability Plots

An investigator will often have obtained a numerical sample $x_1, x_2, \ldots, x_n$ and wish to know whether it is plausible that it came from a population distribution of some particular type (e.g., from a normal distribution). For one thing, many formal procedures from statistical inference are based on the assumption that the population distribution is of a specified type. The use of such a procedure is inappropriate if the actual underlying probability distribution differs greatly from the assumed type.

Additionally, understanding the underlying distribution can sometimes give insight into the physical mechanisms involved in generating the data. An effective way to check a distributional assumption is to construct what is called a **probability plot.** The essence of such a plot is that if the distribution on which the plot is based is correct, the points in the plot will fall close to a straight line. If the actual distribution is quite different from the one used to construct the plot, the points should depart substantially from a linear pattern.

Sample Percentiles

The details involved in constructing probability plots differ a bit from source to source. The basis for our construction is a comparison between percentiles of the sample data and the corresponding percentiles of the distribution under consideration. Recall that the $(100p)$th percentile of a continuous distribution with cdf $F(\cdot)$ is the number $\eta(p)$ that satisfies $F(\eta(p)) = p$. That is, $\eta(p)$ is the number on the measurement scale such that the area under the density curve to the left of $\eta(p)$ is p. Thus, the 50th percentile $\eta(.5)$ satisfies $F(\eta(.5)) = .5$, and the 90th percentile satisfies $F(\eta(.9)) = .9$. Consider as an example the standard normal distribution, for which we have denoted the cdf by $\Phi(\cdot)$. From Appendix Table A.3, the 20th percentile is found by locating the row and column in which .2000 (or a number as close to it as possible) appears inside the table. Since .2005 appears at the intersection of the $-.8$ row and the .04 column, the 20th percentile is approximately $-.84$. Similarly, the 25th percentile of the standard normal distribution is (using linear interpolation) approximately $-.675$.

Roughly speaking, sample percentiles are defined in the same way that percentiles of a population distribution are defined. The fiftieth-sample percentile should separate the smallest 50% of the sample from the largest 50%, the 90th percentile should be such that 90% of the sample lies below that value and 10% lies above, and so on. Unfortunately, we run into problems when we actually try to compute the sample percentiles for a particular sample of n observations. If, for example, $n = 10$, we can split off 20% of these values or 30% of the data, but there is no value that will split off exactly 23% of these ten observations. To proceed further, we need an operational definition of sample percentiles (this is one place where different people do slightly different things). Recall that when n is odd, the sample median or 50th sample percentile is the middle value in the ordered list—for example, the sixth largest value when $n = 11$. This amounts to regarding the middle observation as being half in the lower half of the data and half in the upper half. Similarly, suppose $n = 10$. Then if we call the third smallest value the 25th percentile, we are regarding that value as being half in the lower group (consisting of the two smallest observations) and half in the upper group (the seven largest observations). This leads to the following general definition of sample percentiles.

Definition

> Order the n-sample observations from smallest to largest. Then the ith smallest observation in the list is taken to be the **$[100(i - .5)/n]$th sample percentile.**

Once the percentage values $100(i - .5)/n$ $(i = 1, 2, \ldots, n)$ have been calculated, sample percentiles corresponding to intermediate percentages can be obtained by linear interpolation. For example, if $n = 10$, the percentages corresponding to the ordered sample observations are $100(1 - .5)/10 = 5\%$, $100(2 - .5)/10 = 15\%$, $25\%, \ldots$, and $100(10 - .5)/10 = 95\%$. The 10th percentile is then halfway between the 5th percentile (smallest sample observation) and the 15th percentile (second smallest observation). For our purposes such interpolation is not necessary because a probability plot will be based only on the percentages $100(i - .5)/n$ corresponding to the n sample observations.

A Probability Plot

Suppose now that for percentages $100(i - .5)/n$ $(i = 1, \ldots, n)$ the percentiles are determined for a specified population distribution whose plausibility is being investigated. If the sample was actually selected from the specified distribution, the sample percentiles (ordered sample observations) should be reasonably close to the corresponding population distribution percentiles. That is, for $i = 1, 2, \ldots, n$ there should be reasonable agreement between the ith smallest sample observation and the $[100(i - .5)/n]$th percentile for the specified distribution. Consider the (population percentile, sample percentile) pairs—that is, the pairs

$$\left(\begin{array}{ll} [100(i - .5)/n]\text{th percentile} & i\text{th smallest} \\ \text{of the distribution} & , \quad \text{sample observation} \end{array} \right)$$

for $i = 1, \ldots, n$. Each such pair can be plotted as a point on a two-dimensional coordinate system. If the sample percentiles are close to the corresponding population distribution percentiles, the first number in each pair will be roughly equal to the second number. The plotted points will then fall close to a 45° line. Substantial deviations of the plotted points from a 45° line cast doubt on the assumption that the distribution under consideration is the correct one.

Example 4.29   The value of a certain physical constant is known to an experimenter. The experimenter makes $n = 10$ independent measurements of this value using a particular measurement device and records the resulting measurement errors (error = observed value − true value). These observations appear in the accompanying table.

Percentage	5	15	25	35	45
z percentile	−1.645	−1.037	−.675	−.385	−.126
Sample observation	−1.91	−1.25	−.75	−.53	.20

Percentage	55	65	75	85	95
z percentile	.126	.385	.675	1.037	1.645
Sample observation	.35	.72	.87	1.40	1.56

Is it plausible that the random variable *measurement error* has a standard normal distribution? The needed standard normal (z) percentiles are also displayed in the table. Thus, the points in the probability plot are $(-1.645, -1.91)$, $(-1.037, -1.25)$, . . . , and $(1.645, 1.56)$. Figure 4.31 shows the resulting plot. Although the points deviate a bit from the 45° line, the predominant impression is that this line fits the points very well. The plot suggests that the standard normal distribution is a reasonable probability model for measurement error.

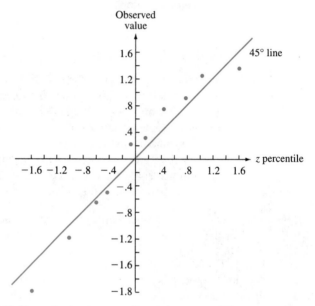

Figure 4.31 Plots of pairs (z percentile, observed value) for the data of Example 4.29: first sample

Figure 4.32 shows a plot of pairs (z percentile, observation) for a second sample of ten observations. The 45° line gives a good fit to the middle part of the sample but not to the extremes. The plot has a well-defined S-shaped appearance. The two smallest sample observations are considerably larger than the corresponding z percentiles (the points on the far left of the plot are well above the 45° line). Similarly, the two largest sample observations are much smaller than the associated z percentiles. This plot indicates that the standard normal distribution would not be a plausible choice for the probability model that gave rise to these observed measurement errors.   ■

An investigator is typically not interested in knowing just whether a specified probability distribution, such as the standard normal distribution (normal with $\mu = 0$ and $\sigma = 1$) or the exponential distribution with $\lambda = .1$, is a plausible model for the population distribution from which the sample was selected. Instead, the investigator will want to know whether *some* member of a family of probability distributions

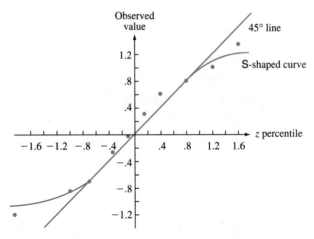

Figure 4.32 Plots of pairs (z percentile, observed value) for the data of Example 4.29: second sample

specifies a plausible model—the family of normal distributions, the family of exponential distributions, the family of Weibull distributions, and so on. The values of the parameters of a distribution are usually not specified at the outset. If the family of Weibull distributions is under consideration as a model for lifetime data, the issue is whether there are *any* values of the parameters α and β for which the corresponding Weibull distribution gives a good fit to the data. Fortunately, it is almost always the case that just one probability plot will suffice for assessing the plausibility of an entire family. If the plot deviates substantially from a straight line, no member of the family is plausible. When the plot is quite straight, further work is necessary to estimate values of the parameters (e.g., find values for μ and σ) that yield the most reasonable distribution of the specified type.

Let's focus on a plot for checking normality. Such a plot can be very useful in applied work because many formal statistical procedures are appropriate (give accurate inferences) only when the population distribution is at least approximately normal. These procedures should generally not be used if the normal probability plot shows a very pronounced departure from linearity. The key to constructing an omnibus normal probability plot is the relationship between standard normal (z) percentiles and those for any other normal distribution:

$$\begin{matrix} \text{percentile for a normal} \\ (\mu,\ \sigma)\ \text{distribution} \end{matrix} = \mu + \sigma \cdot (\text{corresponding } z \text{ percentile})$$

Consider first the case $\mu = 0$. Then if each observation is exactly equal to the corresponding normal percentile for a particular value of σ, the pairs ($\sigma \cdot [z$ percentile], observation) fall on a 45° line, which has slope 1. This implies that the pairs (z percentile, observation) fall on a line passing through (0, 0) (i.e., one with y intercept 0) but having slope σ rather than 1. The effect of a nonzero value of μ is simply to change the y intercept from 0 to μ.

> A plot of the n pairs
>
> ($[100(i - .5)/n]$th z percentile, corresponding observation)
>
> on a two-dimensional coordinate system is called a **normal probability plot.** If the sample observations are in fact drawn from a normal distribution with mean value μ and standard deviation σ, the points should fall close to a straight line with slope σ and intercept μ. Thus, a plot for which the points fall close to some straight line suggests that the assumption of a normal population distribution is plausible.

Example 4.30   The article "Nonbloated Burned Clay Aggregate Concrete" (*J. Materials*, 1972, pp. 555–563) reports 7-day compressive strength observations for $n = 30$ concrete specimens. The data appears in the accompanying stem-and-leaf display (a stem with an L includes observations with "low leaves," those between 000 and 499, whereas an H stem is for leaves between 500 and 999).

```
1L ||  400
1H ||  932
2L ||  000   200   200
2H ||  530   630   665   735   735   800   935
3L ||  000   000   030   065   065   065   170   200   235   260   335   365   465
3H ||  500   600   600   835
4L ||  460
```

A normal probability plot of the data is given in Figure 4.33. Each 2 in the plot denotes two points that are very close to one another, and the 3 plays a similar role. The plot is clearly very straight, providing substantial support for the conclusion that compressive strength is a normally distributed variable.   ■

A nonnormal population distribution can often be placed in one of the following three categories:

1. It is symmetric and has "lighter tails" than does a normal distribution—that is, the density curve declines more rapidly out in the tails than does a normal curve.

2. It is symmetric and heavy-tailed compared to a normal distribution.

3. It is skewed.

A uniform distribution is light-tailed, since its density function drops to zero outside a finite interval. The density function $f(x) = 1/[\pi(1 + x^2)]$ for $-\infty < x < \infty$ is one example of a heavy-tailed distribution, since $1/(1 + x^2)$ declines much less rapidly than does $e^{-x^2/2}$. Lognormal and Weibull distributions are among those that are skewed. When the points in a normal probability plot do not adhere to a straight line, the pattern will frequently suggest that the population distribution is in a particular one of these three categories.

Figure 4.33 A normal probability plot of the compressive strength data

When the distribution from which the sample is selected is light-tailed, the largest and smallest observations are usually not as extreme as would be expected from a normal random sample. Visualize a straight line drawn through the middle part of the plot; points on the far right tend to be below the line (observed value < z percentile), whereas points on the left end of the plot tend to fall above the straight line (observed value > z percentile). The result is an S-shaped pattern of the type pictured in Figure 4.32.

A sample from a heavy-tailed distribution also tends to produce an S-shaped plot. However, in contrast to the light-tailed case, the left end of the plot curves downward (observed < z percentile). An example appears in Figure 4.34a. If the underlying distribution is positively skewed (a short left tail and a long right tail), the smallest sample observations will be larger than expected from a normal sample and so will the largest observations. In this case, points on both ends of the plot will fall above a straight line through the middle part, yielding a curved pattern as illustrated in Figure 4.34b. A sample from a lognormal distribution will usually produce such a pattern. A plot of (z percentile, $\ln(x)$) pairs should then resemble a straight line.

Even when the population distribution is normal, the sample percentiles will not coincide exactly with the theoretical percentiles because of sampling variability. How much can the points in the probability plot deviate from a straight-line pattern before the assumption of population normality is no longer plausible? This is not an easy question to answer. Generally speaking, a small sample from a normal distribution is more likely to yield a plot with a nonlinear pattern than is a large sample. The book *Fitting Equations to Data* (see Chapter 13 bibliography) presents the results of a simulation study in which numerous samples of different sizes were selected from normal distributions. The authors concluded that there is typically greater variation

(a)

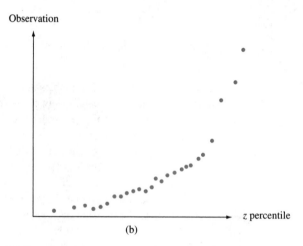

(b)

Figure 4.34 Probability plots that suggest a nonnormal distribution: (a) a plot consistent with a heavy-tailed distribution; (b) a plot consistent with a positively skewed distribution

in the appearance of the probability plot for sample sizes smaller than 30, and only for much larger sample sizes does a linear pattern generally predominate. When a plot is based on a small sample size, only a very substantial departure from linearity should be taken as conclusive evidence of nonnormality. A similar comment applies to probability plots for checking the plausibility of other types of distributions.

Beyond Normality

Consider a family of probability distributions involving two parameters, θ_1 and θ_2, and let $F(x; \theta_1, \theta_2)$ denote the corresponding cdf's. The family of normal distributions is one such family, with $\theta_1 = \mu$, $\theta_2 = \sigma$, and $F(x; \mu, \sigma) = \Phi[(x - \mu)/\sigma]$. Another example is the Weibull family, with $\theta_1 = \alpha$, $\theta_2 = \beta$, and

$$F(x; \alpha, \beta) = 1 - e^{-(x/\beta)^\alpha}$$

Still another family of this type is the gamma family, for which the cdf is an integral involving the incomplete gamma function that cannot be expressed in any simpler form.

The parameters θ_1 and θ_2 are said to be **location** and **scale parameters,** respectively, if $F(x; \theta_1, \theta_2)$ is a function of $(x - \theta_1)/\theta_2$. The parameters μ and σ of the normal family are location and scale parameters, respectively. Changing μ shifts the location of the bell-shaped density curve to the right or left, and changing σ amounts to stretching or compressing the measurement scale (the scale on the horizontal axis when the density function is graphed). Another example is given by the cdf

$$F(x; \theta_1, \theta_2) = 1 - e^{-e^{(x - \theta_1)/\theta_2}} \qquad -\infty < x < \infty$$

A random variable with this cdf is said to have an *extreme value distribution.* It is used in applications involving component lifetime and material strength.

Although the form of the extreme value cdf might at first glance suggest that θ_1 is the point of symmetry for the density function, and therefore the mean and median, this is not the case. Instead, $P(X \leq \theta_1) = F(\theta_1; \theta_1, \theta_2) = 1 - e^{-1} = .632$, and the density function $f(x; \theta_1, \theta_2) = F'(x; \theta_1, \theta_2)$ is negatively skewed (a long lower tail). Similarly, the scale parameter θ_2 is not the standard deviation ($\mu = \theta_1 - .5772\theta_2$ and $\sigma = 1.283\theta_2$). However, changing the value of θ_1 does change the location of the density curve, whereas a change in θ_2 rescales the measurement axis.

The parameter β of the Weibull distribution is a scale parameter, but α is not a location parameter. The parameter α is usually referred to as a **shape parameter.** A similar comment applies to the parameters α and β of the gamma distribution. In the usual form, the density function for any member of either the gamma or Weibull distribution is positive for $x > 0$ and zero otherwise. A location parameter can be introduced as a third parameter γ (we did this for the Weibull distribution) to shift the density function so that it is positive if $x > \gamma$ and zero otherwise.

When the family under consideration has only location and scale parameters, the issue of whether any member of the family is a plausible population distribution can be addressed via a single easily constructed probability plot. One first obtains the percentiles of the *standard distribution,* the one with $\theta_1 = 0$ and $\theta_2 = 1$, for percentages $100(i - .5)/n$ ($i = 1, \ldots, n$). The n (standardized percentile, observation) pairs give the points in the plot. This is of course exactly what we did to obtain an omnibus normal probability plot. Somewhat surprisingly, this methodology can be applied to yield an omnibus Weibull probability plot. The key result is that if X has a Weibull distribution with shape parameter α and scale parameter β, then the transformed variable $\ln(X)$ has an extreme value distribution with location parameter $\theta_1 = \ln(\beta)$ and scale parameter α. Thus a plot of the (extreme value standardized percentile, $\ln(x)$) pairs that shows a strong linear pattern provides support for choosing the Weibull distribution as a population model.

Example 4.31

The accompanying observations are on lifetime (in hours) of power apparatus insulation when thermal and electrical stress acceleration were fixed at particular values ("On the Estimation of Life of Power Apparatus Insulation Under Combined Electrical and Thermal Stress," *IEEE Trans. on Electrical Insulation,* 1985, pp. 70–78). A Weibull probability plot necessitates first computing the 5th, 15th, . . . , and 95th percentiles of the standard extreme value distribution. The $(100p)$th percentile $\eta(p)$ satisfies

$$p - F(\eta(p)) = 1 - e^{-e^{\eta(p)}}$$

from which $\eta(p) = \ln[-\ln(1-p)]$.

Percentile	−2.97	−1.82	−1.25	−.84	−.51
x	282	501	741	851	1072
$\ln(x)$	5.64	6.22	6.61	6.75	6.98

Percentile	−.23	.05	.33	.64	1.10
x	1122	1202	1585	1905	2138
$\ln(x)$	7.02	7.09	7.37	7.55	7.67

The pairs (−2.97, 5.64), (−1.82, 6.22), . . . , (1.10, 7.67) are plotted as points in Figure 4.35. The straightness of the plot argues strongly for using the Weibull distribution as a model for insulation life, a conclusion also reached by the author of the cited paper.

Figure 4.35 A Weibull probability plot of the insulation lifetime data ■

The gamma distribution is an example of a family involving a shape parameter for which there is no transformation $h(\cdot)$ such that $h(X)$ has a distribution depending only on location and scale parameters. Construction of a probability plot necessitates first estimating the shape parameter from sample data (some methods for doing this are described in Chapter 6). Sometimes an investigator wishes to know whether the transformed variable X^θ has a normal distribution for some value of θ (by convention, $\theta = 0$ is identified with the logarithmic transformation, in which case X has a lognormal distribution). The book *Graphical Methods for Data Analysis,* listed in the Chapter 1 bibliography, discusses this type of problem as well as other refinements of probability plotting.

Finally, construction of various probability plots can be facilitated by the use of commercially available special probability paper. The axes of such paper are scaled so that calculation of percentiles is unnecessary. The book *Applied Life Data Analysis,* listed in the bibliography of this chapter, presents examples and related discussion. However, the ability of statistical computer packages to construct probability plots will eventually make probability paper as obsolete as the slide rule.

Exercises   Section 4.6 (79–85)

79. Consider the following ten observations on bearing lifetime (in hours):

152.7, 172.0, 172.5, 173.3, 193.0,
204.7, 216.5, 234.9, 262.6, 422.6

Construct a normal probability plot and comment on the plausibility of the normal distribution as a model for bearing lifetime (data from "Modified Moment Estimation for the Three-Parameter Lognormal Distribution," *J. Quality Technology,* 1985, pp. 92–99).

80. Construct a normal probability plot for the fatigue-crack propagation data given in Exercise 31 (Chapter 1). Does it appear plausible that propagation life has a normal distribution? Explain.

81. The paper "The Load-Life Relationship for M50 Bearings with Silicon Nitride Ceramic Balls" (*Lubrication Engr.,* 1984, pp. 153–159) reports the accompanying data on bearing load life (million revs.) for bearings tested at a 6.45 kN load.

47.1	68.1	68.1	90.8	103.6	106.0	115.0
126.0	146.6	229.0	240.0	240.0	278.0	278.0
289.0	289.0	367.0	385.9	392.0	505.0	

a. Construct a normal probability plot. Is normality plausible?

b. Construct a Weibull probability plot. Is the Weibull distribution family plausible?

82. If you have access to a statistical computer package, use it to construct a normal probability plot for the shear strength data of Exercise 11 (Chapter 1). (If not,

construct a plot using just the first 25 observations in the data set.) Comment.

83. Construct a probability plot that will allow you to assess the plausibility of the lognormal distribution as a model for the rainfall data of Exercise 67 (Chapter 1).

84. The accompanying observations are precipitation values during March over a 30-year period in Minneapolis–St. Paul.

.77	1.20	3.00	1.62	2.81	2.48
1.74	.47	3.09	1.31	1.87	.96
.81	1.43	1.51	.32	1.18	1.89
1.20	3.37	2.10	.59	1.35	.90
1.95	2.20	.52	.81	4.75	2.05

a. Construct and interpret a normal probability plot for this data set.

b. Calculate the square root of each value and then construct a normal probability plot based on this transformed data. Does it seem plausible that the square root of precipitation is normally distributed?

c. Repeat part (b) after transforming by cube roots.

85. The following failure time observations (1000s of hours) resulted from accelerated life testing of 16 integrated circuit chips of a certain type.

82.8	11.6	359.5	502.5	307.8	179.7
242.0	26.5	244.8	304.3	379.1	212.6
229.9	558.9	366.7	204.6		

Use the corresponding percentiles of the exponential distribution with $\lambda = 1$ to construct a probability plot. Then explain why the plot assesses the plausibility of the sample having been generated from *any* exponential distribution.

Supplementary Exercises for Chapter 4 (86–109)

86. Let X = the time it takes a read/write head to locate a desired record on a computer disk memory device once the head has been positioned over the correct track. If the disks rotate once every 25 millisec, a reasonable assumption is that X is uniformly distributed on the interval $[0, 25]$.

 a. Compute $P(10 \leq X \leq 20)$.

 b. Compute $P(X \geq 10)$.

 c. Obtain the cdf $F(X)$.

 d. Compute $E(X)$ and σ_X.

87. A 12-in. bar that is clamped at both ends is to be subjected to an increasing amount of stress until it snaps. Let Y = the distance from the left end at which the break occurs and suppose Y has pdf

$$f(y) = \begin{cases} \left(\dfrac{1}{24}\right)y\left(1 - \dfrac{y}{12}\right) & 0 \leq y \leq 12 \\ 0 & \text{otherwise} \end{cases}$$

Compute the following:

 a. The cdf of Y and graph it

 b. $P(Y \leq 4)$, $P(Y > 6)$, and $P(4 \leq Y \leq 6)$

 c. $E(Y)$, $E(Y^2)$, and $V(Y)$

 d. The probability that the break point occurs more than 2 in. from the expected break point

 e. The expected length of the shorter segment when the break occurs

88. Let X denote the time to failure (in years) of a certain hydraulic component. Suppose the pdf of X is $f(x) = 32/(x + 4)^3$ for $x > 0$.

 a. Verify that $f(x)$ is a legitimate pdf.

 b. Determine the cdf.

 c. Use the result of part (b) to calculate the probability that time to failure is between 2 and 5 years.

 d. What is the expected time to failure?

 e. If the component has a salvage value equal to

$100/(4 + x)$ when its time to failure is x, what is the expected salvage value?

89. The completion time X for a certain task has cdf $F(x)$ given by

$$\begin{cases} 0 & x < 0 \\ \dfrac{x^3}{3} & 0 \leq x < 1 \\ 1 - \dfrac{1}{2}\left(\dfrac{7}{3} - x\right)\left(\dfrac{7}{4} - \dfrac{3}{4}x\right) & 1 \leq x < \dfrac{7}{3} \\ 1 & x \geq \dfrac{7}{3} \end{cases}$$

 a. Obtain the pdf $f(x)$ and sketch its graph.

 b. Compute $P(.5 \leq X \leq 2)$.

 c. Compute $E(X)$.

90. The breakdown voltage of a randomly chosen diode of a certain type is known to be normally distributed with mean value 40 V and standard deviation 1.5 V.

 a. What is the probability that the voltage of a single diode is between 39 and 42?

 b. What value is such that only 15% of all diodes have voltages exceeding that value?

 c. If four diodes are independently selected, what is the probability that at least one has a voltage exceeding 42?

91. The paper "Computer Assisted Net Weight Control" (*Quality Progress*, 1983, pp. 22–25) suggests a normal distribution with mean 137.2 oz and standard deviation 1.6 oz. for the actual contents of jars of a certain type. The stated contents was 135 oz.

 a. What is the probability that a single jar contains more than the stated contents?

 b. Among ten randomly selected jars, what is the probability that at least eight contain more than the stated contents?

 c. Assuming that the mean remains at 137.2, to what value would the standard deviation have to be changed so that 95% of all jars contain more than the stated contents?

92. The reaction time (in seconds) to a certain stimulus is a continuous random variable with pdf

$$f(x) = \begin{cases} \dfrac{3}{2} \cdot \dfrac{1}{x^2} & 1 \le x \le 3 \\ 0 & \text{otherwise} \end{cases}$$

   a. Obtain the cdf.

   b. What is the probability that reaction time is at most 2.5 sec? Between 1.5 and 2.5 sec?

   c. Compute the expected reaction time.

   d. Compute the standard deviation of reaction time.

   e. If an individual takes more than 1.5 sec to react, a light comes on and stays on either until one further second has elapsed or until the person reacts (whichever happens first). Determine the expected amount of time that the light remains lit. [HINT: Let $h(X)$ = the time that the light is on as a function of reaction time X.]

93. Let X denote the temperature at which a certain chemical reaction takes place and suppose that X has pdf

$$f(x) = \begin{cases} \dfrac{1}{9}(4 - x^2) & -1 \le x \le 2 \\ 0 & \text{otherwise} \end{cases}$$

   a. Sketch the graph of $f(x)$.

   b. Determine the cdf and sketch it.

   c. Is 0 the median temperature at which the reaction takes place? If not, is the median temperature smaller or larger than 0?

   d. Suppose this reaction is independently carried out once in each of ten different labs and that the pdf of reaction time in each lab is as given. Let Y = the number among the ten labs at which the temperature exceeds 1. What kind of distribution does Y have? (Give the name and values of any parameters.)

94. The paper "Determination of the MTF of Positive Photoresists Using the Monte Carlo Method" (*Photographic Sci. and Engr.*, 1983, pp. 254–260) proposes the exponential distribution with parameter $\lambda = .93$ as a model for the distribution of a photon's free path length (μm) under certain circumstances. Suppose this is the correct model.

   a. What is the expected path length, and what is the standard deviation of path length?

   b. What is the probability that path length exceeds 3.0? What is the probability that path length is between 1.0 and 3.0?

   c. What value is exceeded by only 10% of all path lengths?

95. The paper "The Prediction of Corrosion by Statistical Analysis of Corrosion Profiles" (*Corrosion Science,* 1985, pp. 305–315) suggests the following cdf for the depth X of the deepest pit in an experiment involving the exposure of carbon manganese steel to acidified seawater:

$$F(x;\ \alpha,\ \beta) = e^{-e^{-(x-\alpha)/\beta}}, \qquad -\infty < x < \infty$$

The authors propose the values $\alpha = 150$, $\beta = 90$. Assume this to be the correct model.

   a. What is the probability that the depth of the deepest pit is at most 150? At most 300? Between 150 and 300?

   b. Below what value will the depth of the maximum pit be observed in 90% of all such experiments?

   c. What is the density function of X?

   d. The density function can be shown to be unimodal (a single peak). Above what value on the measurement axis does this peak occur? (This value is the mode.)

   e. It can be shown that $E(X) \approx .5772\beta + \alpha$. What is the mean for the given values of α and β, and how does it compare to the median and mode? Sketch the graph of the density function. (NOTE: This is called the *largest extreme value distribution*.)

96. A component has lifetime X that is exponentially distributed with parameter λ.

   a. If the cost of operation per unit time is c, what is the expected cost of operating this component over its lifetime?

   b. Instead of a constant cost rate c as in part (a), suppose the cost rate is $c(1 - .5e^{ax})$ with $a < 0$, so that the cost per unit time is less than c when the component is new and gets more expensive as the component ages. Now compute the expected cost of operation over the lifetime of the component.

97. The *mode* of a continuous distribution is the value x^* that maximizes $f(x)$.

a. What is the mode of a normal distribution with parameters μ and σ?

b. Does the uniform distribution with parameters A and B have a single mode? Why, or why not?

c. What is the mode of an exponential distribution with parameter λ? (Draw a picture here.)

d. If X has a gamma distribution with parameters α and β, and $\alpha > 1$, find the mode. [HINT: $\ln[f(x)]$ will be maximized iff $f(x)$ is, and it may be simpler to take the derivative of $\ln[f(x)]$.]

e. What is the mode of a chi-squared distribution having v degrees of freedom?

98. The article "Error Distribution in Navigation" (*J. Institute of Navigation,* 1971, pp. 429–442) suggests that the frequency distribution of positive errors (magnitudes of errors) is well approximated by an exponential distribution. Let X = the lateral position error (nautical miles), which can be either negative or positive, and suppose the pdf of X is

$$f(x) = \begin{cases} (.1)e^{-.2\,|x|} & -\infty < x < \infty \\ 0 & \text{otherwise} \end{cases}$$

a. Sketch a graph of $f(x)$ and verify that $f(x)$ is a legitimate pdf (show that it integrates to 1).

b. Obtain the cdf of X and sketch it.

c. Compute $P(X \le 0)$, $P(X \le 2)$, $P(-1 \le X \le 2)$, and the probability that an error of more than 2 miles is made.

99. In some systems, a customer is allocated to one of two service facilities. If the service time for a customer served by facility i has an exponential distribution with parameter $\lambda_i (i = 1, 2)$ and p is the proportion of all customers served by facility 1, then the pdf of X = the service time of a randomly selected customer is

$$f(x; \lambda_1, \lambda_2, p)$$
$$= \begin{cases} p\lambda_1 e^{-\lambda_1 x} + (1 - p)\lambda_2 e^{-\lambda_2 x} & x \ge 0 \\ 0 & \text{otherwise} \end{cases}$$

This is often called the hyperexponential or mixed exponential distribution. As an example, many computer systems process some programs using both a fast in-core compiler (FORTRAN, WATFIV) and a slower compiler. This distribution is also proposed as a model for rainfall amount in "Modeling Monsoon Affected Rainfall of Pakistan by Point Processes" (*J. Water Resources Planning and Mgmnt.,* 1992, pp. 671–688).

a. Verify that $f(x; \lambda_1, \lambda_2, p)$ is indeed a pdf.

b. What is the cdf $F(x; \lambda_1, \lambda_2, p)$?

c. If X has $f(x; \lambda_1, \lambda_2, p)$ as its pdf, what is $E(X)$?

d. Using the fact that $E(X^2) = 2/\lambda^2$ when X has an exponential distribution with parameter λ, compute $E(X^2)$ when X has pdf $f(x; \lambda_1, \lambda_2, p)$. Then compute $V(X)$.

e. The coefficient of variation of a random variable (or distribution) is $CV = \sigma/\mu$. What is CV for an exponential rv? What can you say about the value of CV when X has a hyperexponential distribution?

f. What is CV for an Erlang distribution with parameters λ and n as defined in Exercise 61? (NOTE: In applied work, the sample CV is used to decide which of the three distributions might be appropriate.)

100. Suppose a particular state allows individuals filing tax returns to itemize deductions only if the total of all itemized deductions is at least \$5000 and let X (in 1000s of dollars) be the total of itemized deductions on a randomly chosen form. Assume that X has the pdf

$$f(x; \alpha) = \begin{cases} \dfrac{k}{x^\alpha} & x \ge 5 \\ 0 & \text{otherwise} \end{cases}$$

a. Find the value of k. What restriction on α is necessary?

b. What is the cdf of X?

c. What is the expected total deduction on a randomly chosen form? What restriction on α is necessary for $E(X)$ to be finite?

d. Show that $\ln(X/5)$ has an exponential distribution with parameter $\alpha - 1$.

101. Let I_i be the input current to a transistor and I_0 be the output current. Then the current gain is proportional to $\ln(I_0/I_i)$. Suppose the constant of proportionality is 1 (which amounts to choosing a particular unit of measurement), so that current gain = $X = \ln(I_0/I_i)$. Assume X is normally distributed with $\mu = 1$ and $\sigma = .05$.

a. What type of distribution does the ratio I_0/I_i have?

b. What is the probability that the output current is more than twice the input current?

c. What are the expected value and variance of the ratio of output to input current?

102. Let Z have a standard normal distribution and define a new rv Y by $Y = \sigma Z + \mu$. Show that Y has a normal distribution with parameters μ and σ. (HINT: $Y \le y$ iff $Z \le$? Use this to find the cdf of Y and then differentiate it with respect to y.)

103. a. Suppose the lifetime X of a component, when measured in hours, has a gamma distribution with parameters α and β. Let $Y =$ the lifetime measured in minutes. Derive the pdf of Y. (HINT: $Y \le y$ iff $X \le y/60$. Use this to obtain the cdf of Y and then differentiate to obtain the pdf.)

   b. If X has a gamma distribution with parameters α and β, what is the probability distribution of $Y = cX$?

104. In Exercises 102 and 103, as well as many other situations, one has the pdf $f(x)$ of X and wishes the pdf of $Y = h(X)$. Assume that $h(\cdot)$ is an invertible function, so that $y = h(x)$ can be solved for x to yield $x = k(y)$. Then it can be shown that the pdf of Y is

$$g(y) = f[k(y)] \cdot |k'(y)|$$

   a. If X has a uniform distribution with $A = 0$ and $B = 1$, derive the pdf of $Y = -\ln(X)$.

   b. Work Exercise 102, using this result.

   c. Work Exercise 103(b), using this result.

105. Let X denote the lifetime of a component, with $f(x)$ and $F(x)$ the pdf and cdf of X. The probability that the component fails in the interval $(x, x + \Delta x)$ is approximately $f(x) \cdot \Delta x$. The conditional probability that it fails in $(x, x + \Delta x)$ given that it has lasted at least x is $f(x) \cdot \Delta x/[1 - F(x)]$. Dividing this by Δx produces the **failure rate function:**

$$r(x) = \frac{f(x)}{1 - F(x)}$$

An increasing failure rate function indicates that older components are increasingly likely to wear out, while a decreasing failure rate is evidence of increasing reliability with age. In practice a "bathtub-shaped" failure is often assumed.

   a. If X is exponentially distributed, what is $r(x)$?

   b. If X has a Weibull distribution with parameters α and β, what is $r(x)$? For what parameter values will $r(x)$ be increasing? For what parameter values will $r(x)$ decrease with x?

c. Since $r(x) = -(d/dx)\ln[1 - F(x)]$, $\ln[1 - F(x)] = -\int r(x)\,dx$. Suppose

$$r(x) = \begin{cases} \alpha\left(1 - \dfrac{x}{\beta}\right) & 0 \le x \le \beta \\ 0 & \text{otherwise} \end{cases}$$

so that if a component lasts β hours, it will last forever (while seemingly unreasonable, this model can be used to study just "initial wearout"). What are the cdf and pdf of X?

106. Let U have a uniform distribution on the interval $[0, 1]$. Then observed values having this distribution can be obtained from a computer's random number generator. Let $X = -(1/\lambda) \ln(1 - U)$.

   a. Show that X has an exponential distribution with parameter λ. [HINT: The cdf of X is $F(x) = P(X \le x)$; $X \le x$ is equivalent to $U \le$?]

   b. How would you use part (a) and a random number generator to obtain observed values from an exponential distribution with parameter $\lambda = 10$?

107. Consider an rv X with mean μ and standard deviation σ, and let $g(X)$ be a specified function of X. The first-order Taylor series approximation to $g(X)$ in the neighborhood of μ is

$$g(X) \approx g(\mu) + g'(\mu) \cdot (X - \mu)$$

The right-hand side of this equation is a linear function of X. If the distribution of X is concentrated in an interval over which $g(\cdot)$ is approximately linear [e.g., $\sqrt{x}$ is approximately linear in $(1, 2)$], then the equation yields approximations to $E(g(X))$ and $V(g(X))$.

   a. Give expressions for these approximations. (HINT: Use rules of expected value and variance for a linear function $aX + b$.)

   b. If the voltage v across a medium is fixed but current I is random, then resistance will also be a random variable related to I by $R = v/I$. If $\mu_I = 20$ and $\sigma_I = .5$, calculate approximations to μ_R and σ_R.

108. A function $g(x)$ is *convex* if the chord connecting any two points on the function's graph lies above the graph. When $g(x)$ is differentiable, an equivalent condition is that for every x, the tangent line at x lies entirely on or below the graph. (See the accompanying figures.)

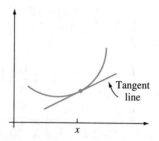

How does $g(\mu) = g(E(X))$ compare to $E(g(X))$? [HINT: The equation of the tangent line at $x = \mu$ is $y = g(\mu) + g'(\mu) \cdot (x - \mu)$. Use the condition of convexity, substitute X for x, and take expected values. NOTE: Unless $g(x)$ is linear, the resulting inequality (usually called Jensen's inequality) is strict ($<$ rather than $\le$); it is valid for both continuous and discrete rv's.]

109. Let X have a Weibull distribution with parameters $\alpha = 2$ and β. Show that $Y = 2X^2/\beta^2$ has a chi-squared distribution with $v = 2$. [HINT: The cdf of Y is $P(Y \le y)$; express this probability in the form $P(X \le g(y))$, use the fact that X has cdf of the form (4.12), and differentiate with respect to y to obtain the pdf of Y.]

Bibliography

Bury, Karl, *Statistical Models in Applied Science,* Wiley, New York, 1975. Somewhat difficult to read, but the author presents a number of good engineering and science examples involving continuous probability models.

Olkin, Ingram, Cyrus Derman, and Leon Gleser, *Probability Models and Applications,* (2nd ed.), Macmillan, New York, 1994. Good coverage of general properties and specific distributions.

Johnson, Norman, and Samuel Kotz, *Distributions in Statistics: Continuous Distributions 1 and 2,* Houghton Mifflin, Boston, 1970. These two volumes together present an exhaustive survey of various continuous distributions.

Nelson, Wayne, *Applied Life Data Analysis,* Wiley, New York, 1982. Gives a comprehensive discussion of distributions and methods that are used in the analysis of lifetime data.

5 Joint Probability Distributions and Random Samples

ΣΣΣΣΣΣΣΣΣΣΣΣΣΣΣΣΣΣΣ

Introduction

In Chapters 3 and 4, we studied probability models for a single random variable. Many problems in probability and statistics lead to models involving several random variables simultaneously. In this chapter, we first discuss probability models for the joint behavior of several random variables, putting special emphasis on the case in which the variables are independent of one another. We then study expected values of functions of several random variables, including covariance and correlation as measures of the degree of association between two variables.

The last three sections of the chapter consider functions of n random variables $X_1, X_2, \ldots, X_n$, focusing especially on their average $(X_1 + \cdots + X_n)/n$. We call any such function, itself a random variable, a *statistic*. Results from probability are used to obtain information about the distribution of a statistic. The premier result of this type is the Central Limit Theorem (CLT), the basis for many inferential procedures involving large sample sizes.

5.1 Jointly Distributed Random Variables

There are many experimental situations in which more than one random variable (rv) will be of interest to an investigator. We shall first consider joint probability distributions for two discrete rv's, then for two continuous variables, and finally for more than two variables.

The Joint Probability Mass Function for Two Discrete Random Variables

The probability mass function (pmf) of a single discrete rv X specifies how much probability mass is placed on each possible X value. The joint pmf of two discrete rv's X and Y describes how much probability mass is placed on each possible pair of values (x, y).

Definition

> Let X and Y be two discrete rv's defined on the sample space S of an experiment. The **joint probability mass function** $p(x, y)$ is defined for each pair of numbers (x, y) by
>
> $$p(x, y) = P(X = x \text{ and } Y = y)$$
>
> Let A be any set consisting of pairs of (x, y) values. Then the probability $P[(X, Y) \in A]$ is obtained by summing the joint pmf over pairs in A:
>
> $$P[(X, Y) \in A] = \sum_{(x,\, y)\, \in\, A} \sum p(x, y)$$

Example 5.1

A large insurance agency services a number of customers who have purchased both a homeowner's policy and an automobile policy from the agency. For each type of policy, a deductible amount must be specified. For an automobile policy, the choices are \$100 and \$250, while for a homeowner's policy the choices are 0, \$100, and \$200. Suppose an individual with both types of policy is selected at random from the agency's files and let $X =$ the deductible amount on the auto policy and $Y =$ the deductible amount on the homeowner's policy. Possible (X, Y) pairs are then (100, 0), (100, 100), (100, 200), (250, 0), (250, 100), and (250, 200); the joint pmf specifies the probability associated with each one of these pairs, with any other pair having probability zero. Suppose the joint pmf is given in the accompanying **joint probability table:**

$p(x, y)$		0	y 100	200
x	100	.20	.10	.20
	250	.05	.15	.30

Then $p(100, 100) = P(X = 100 \text{ and } Y = 100) = P(\$100 \text{ deductible on both poli-cies}) = .10$. The probability $P(Y \geq 100)$ is computed by summing probabilities of all (x, y) pairs for which $y \geq 100$:

$$P(Y \geq 100) = p(100, 100) + p(250, 100) + p(100, 200) + p(250, 200)$$
$$= .75$$ ∎

A function $p(x, y)$ can be used as a joint pmf provided that $p(x, y) \geq 0$ for all x and y and $\sum_x \sum_y p(x, y) = 1$.

The pmf of one of the variables alone is obtained by summing $p(x, y)$ over values of the other variable. The result is called a *marginal pmf* because when the $p(x, y)$'s appear in a rectangular table, the sums are just marginal (row or column) totals.

Definition

> The **marginal probability mass functions** of X and of Y, denoted by $p_X(x)$ and $p_Y(y)$, respectively, are given by
>
> $$p_X(x) = \sum_y p(x, y), \qquad p_Y(y) = \sum_x p(x, y)$$

Thus, to obtain the marginal pmf of X evaluated at, say, $x = 100$, the probabilities $p(100, y)$ are added over all possible y values. Doing this for each possible X value gives the marginal pmf of X alone (without reference to Y). From the marginal pmf's, probabilities of events involving only X or only Y can be computed.

Example 5.2
(Example 5.1 continued)

The possible X values are $x = 100$ and $x = 250$, so computing row totals in the joint probability table yields

$$p_X(100) = p(100, 0) + p(100, 100) + p(100, 200) = .50$$

and

$$p_X(250) = p(250, 0) + p(250, 100) + p(250, 200) = .50$$

The marginal pmf of X is then

$$p_X(x) = \begin{cases} .5 & x = 100, 250 \\ 0 & \text{otherwise} \end{cases}$$

Similarly, the marginal pmf of Y is obtained from column totals as

$$p_Y(y) = \begin{cases} .25 & y = 0, 100 \\ .50 & y = 200 \\ 0 & \text{otherwise} \end{cases}$$

so $P(Y \geq 100) = p_Y(100) + p_Y(200) = .75$ as before. ∎

The Joint Probability Density Function for Two Continuous Random Variables

The probability that the observed value of a continuous rv X lies in a one-dimensional set A (such as an interval) is obtained by integrating the pdf $f(x)$ over the set A. Similarly, the probability that the pair (X, Y) of continuous rv's falls in a two-dimensional set A (such as a rectangle) is obtained by integrating a function called the *joint density function*.

Definition

> Let X and Y be continuous rv's. Then $f(x, y)$ is the **joint probability density function** for X and Y if for any two-dimensional set A
>
> $$P[(X, Y) \in A] = \iint_A f(x, y) \, dx \, dy$$

In particular, if A is the two-dimensional rectangle $\{(x, y): a \leq x \leq b,$ $c \leq y \leq d\}$, then

$$P[(X, Y) \in A] = P(a \leq X \leq b, c \leq Y \leq d) = \int_a^b \int_c^d f(x, y) \, dy \, dx$$

For $f(x, y)$ to be a candidate for a joint pdf, it must satisfy $f(x, y) \geq 0$ and $\int_{-\infty}^{\infty} \int_{-\infty}^{\infty} f(x, y) \, dx \, dy = 1$. We can think of $f(x, y)$ as specifying a surface at height $f(x, y)$ above the point (x, y) in a three-dimensional coordinate system. Then $P[(X, Y) \in A]$ is the volume underneath this surface and above the region A, analogous to the area under a curve in the one-dimensional case. This is illustrated in Figure 5.1.

Figure 5.1 $P[(X, Y) \in A] = $ volume under density surface above A

Example 5.3   A bank operates both a drive-up facility and a walk-up window. On a randomly selected day, let $X = $ the proportion of time that the drive-up facility is in use (at least one customer is being served or waiting to be served) and $Y = $ the proportion of time that the walk-up window is in use. Then the set of possible values for (X, Y) is the rectangle $D = \{(x, y): 0 \leq x \leq 1, 0 \leq y \leq 1\}$. Suppose the joint pdf of (X, Y) is given by

$$f(x, y) = \begin{cases} \dfrac{6}{5}(x + y^2) & 0 \leq x \leq 1, 0 \leq y \leq 1 \\ 0 & \text{otherwise} \end{cases}$$

To verify that this is a legitimate pdf, note that $f(x, y) \geq 0$ and

$$\int_{-\infty}^{\infty} \int_{-\infty}^{\infty} f(x, y) \, dx \, dy = \int_0^1 \int_0^1 \frac{6}{5}(x + y^2) \, dx \, dy$$

$$= \int_0^1 \int_0^1 \frac{6}{5}x \, dx \, dy + \int_0^1 \int_0^1 \frac{6}{5}y^2 \, dx \, dy$$

$$= \int_0^1 \frac{6}{5}x \, dx + \int_0^1 \frac{6}{5}y^2 \, dy = \frac{6}{10} + \frac{6}{15} = 1$$

The probability that neither facility is busy more than one-quarter of the time is

$$P\left(0 \le X \le \frac{1}{4}, 0 \le Y \le \frac{1}{4}\right) = \int_0^{1/4} \int_0^{1/4} \frac{6}{5}(x + y^2) \, dx \, dy$$

$$= \frac{6}{5} \int_0^{1/4} \int_0^{1/4} x \, dx \, dy + \frac{6}{5} \int_0^{1/4} \int_0^{1/4} y^2 \, dx \, dy$$

$$= \frac{6}{20} \cdot \frac{x^2}{2}\Big|_{x=0}^{x=1/4} + \frac{6}{20} \cdot \frac{y^3}{3}\Big|_{y=0}^{y=1/4} = \frac{7}{640}$$

$$= .0109$$

As with joint pmf's, from the joint pdf of X and Y, each of the two marginal density functions can be computed.

Definition

The **marginal probability density functions** of X and Y, denoted by $f_X(x)$ and $f_Y(y)$, respectively, are given by

$$f_X(x) = \int_{-\infty}^{\infty} f(x, y) \, dy, \qquad \text{for} -\infty < x < \infty$$

$$f_Y(y) = \int_{-\infty}^{\infty} f(x, y) \, dx, \qquad \text{for} -\infty < y < \infty$$

Example 5.4
(Example 5.3 continued)

The marginal pdf of X, which gives the probability distribution of busy time for the drive-up facility without reference to the walk-up window, is

$$f_X(x) = \int_{-\infty}^{\infty} f(x, y) \, dy = \int_0^1 \frac{6}{5}(x + y^2) \, dy = \frac{6}{5}x + \frac{2}{5}$$

for $0 \le x \le 1$ and 0 otherwise. The marginal pdf of Y is

$$f_Y(y) = \begin{cases} \frac{6}{5}y^2 + \frac{3}{5} & 0 \le y \le 1 \\ 0 & \text{otherwise} \end{cases}$$

Then

$$P\left(\frac{1}{4} \le Y \le \frac{3}{4}\right) = \int_{1/4}^{3/4} f_Y(y) \, dy = \frac{37}{80} = .4625$$

In Example 5.3, the region of positive joint density was a rectangle, which made computation of the marginal pdf's relatively easy. Consider now an example in which the region of positive density is a more complicated figure.

Example 5.5

A nut company markets cans of deluxe mixed nuts containing almonds, cashews, and peanuts. Suppose the net weight of each can is exactly 1 lb, but the weight contribution of each type of nut is random. Because the three weights sum to 1, a joint probability model for any two gives all necessary information about the weight of the third type. Let X = the weight of almonds in a selected can and Y = the weight of cashews. Then the region of positive density is $D = \{(x, y): 0 \leq x \leq 1, 0 \leq y \leq 1, x + y \leq 1\}$, the shaded region pictured in Figure 5.2.

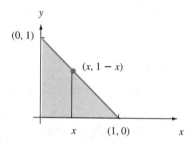

Figure 5.2 Region of positive density for Example 5.5

Now let the joint pdf for (X, Y) be

$$f(x, y) = \begin{cases} 24xy & 0 \leq x \leq 1, 0 \leq y \leq 1, x + y \leq 1 \\ 0 & \text{otherwise} \end{cases}$$

For any fixed x, $f(x, y)$ increases with y; for fixed y, $f(x, y)$ increases with x. This is appropriate because the word *deluxe* implies that most of the can should consist of almonds and cashews rather than peanuts, so that the density function should be large near the upper boundary and small near the origin. The surface determined by $f(x, y)$ slopes upward from zero as (x, y) moves away from either axis.

Clearly $f(x, y) \geq 0$. To verify the second condition on a joint pdf, recall that a double integral is computed as an iterated integral by holding one variable fixed (such as x as in Figure 5.2), integrating over values of the other variable lying along the straight line passing through the value of the fixed variable, and finally integrating over all possible values of the fixed variable. Thus,

$$\int_{-\infty}^{\infty} \int_{-\infty}^{\infty} f(x, y)\, dy\, dx = \int\int_D f(x, y)\, dy\, dx = \int_0^1 \left\{ \int_0^{1-x} 24xy\, dy \right\} dx$$

$$= \int_0^1 24x \left\{ \frac{y^2}{2} \Big|_{y=0}^{y=1-x} \right\} dx = \int_0^1 12x(1-x)^2\, dx = 1$$

To compute the probability that the two types of nuts together make up at most 50% of the can, let $A = \{(x, y): 0 \leq x \leq 1, 0 \leq y \leq 1, \text{ and } x + y \leq .5\}$ as shown in Figure 5.3. Then

$$P((X, Y) \in A) = \int\int_A f(x, y)\, dx\, dy = \int_0^{.5} \int_0^{.5-x} 24xy\, dy\, dx = .0625$$

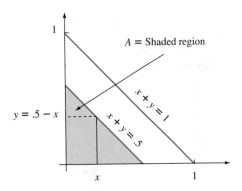

Figure 5.3 Computing $P[(X, Y) \in A]$ for Example 5.5

The marginal pdf for almonds is obtained by holding X fixed at x and integrating $f(x, y)$ along the vertical line through x:

$$f_X(x) = \int_{-\infty}^{\infty} f(x, y) \, dy = \begin{cases} \int_0^{1-x} 24xy \, dy = 12x(1-x)^2 & 0 \le x \le 1 \\ 0 & \text{otherwise} \end{cases}$$

By symmetry of $f(x, y)$ and the region D, the marginal pdf of Y is obtained by replacing x and X in $f_X(x)$ by y and Y, respectively. ∎

Independent Random Variables

In many situations, information about the observed value of one of the two variables X and Y gives information about the value of the other variable. In Example 5.1, the marginal probability of X at $x = 250$ was .5, as was the probability that $X = 100$. If, however, we are told that the selected individual had $Y = 0$, then $X = 100$ is four times as likely as $X = 250$. Thus, there is a dependence between the two variables.

In Chapter 2 we pointed out that one way of defining independence of two events is to say that A and B are independent if $P(A \cap B) = P(A) \cdot P(B)$. We now use an analogous definition for the independence of two rv's.

Definition

> Two rv's X and Y are said to be **independent** if for every pair of x and y values,
>
> $p(x, y) = p_X(x) \cdot p_Y(y)$,   when X and Y are discrete
>
> or   (5.1)
>
> $f(x, y) = f_X(x) \cdot f_Y(y)$,   when X and Y are continuous
>
> If (5.1) is not satisfied for all (x, y), then X and Y are said to be **dependent**.

The definition says that two variables are independent if their joint pmf or pdf is the product of the two marginal pmf's or pdf's.

Example 5.6

In the insurance problem of Example 5.1, $p(100, 100) = .10$, while $p_X(100) = .5$ and $p_Y(100) = .25$. Since $.10 \neq (.5)(.25)$, X and Y are not independent. Independence of X and Y requires that *every* entry in the joint probability table be the product of the corresponding row and column marginal probabilities. ∎

Example 5.7

(Example 5.5 continued)

Because $f(x, y)$ has the form of a product, X and Y would appear to be independent. However, while $f_X(\frac{3}{4}) = f_Y(\frac{3}{4}) = \frac{9}{16}$, $f(\frac{3}{4}, \frac{3}{4}) = 0 \neq \frac{9}{16} \cdot \frac{9}{16}$, so the variables are not in fact independent. To be independent, $f(x, y)$ must have the form $g(x) \cdot h(y)$ *and the* region of positive density must be a rectangle whose sides are parallel to the coordinate axes. ∎

Independence of two random variables is most useful when the description of the experiment under study tells us that X and Y have no effect on one another. Then once the marginal pmf's or pdf's have been specified, the joint pmf or pdf is simply the product of the two marginal functions. It follows that

$$P(a \leq X \leq b, c \leq Y \leq d) = P(a \leq X \leq b) \cdot P(c \leq Y \leq d)$$

Example 5.8

Suppose that the lifetimes of two components are independent of one another and that the first lifetime X_1 has an exponential distribution with parameter λ_1 while the second, X_2, has an exponential distribution with parameter λ_2. Then the joint pdf is

$$f(x_1, x_2) = f_{X_1}(x_1) \cdot f_{X_2}(x_2)$$

$$= \begin{cases} \lambda_1 e^{-\lambda_1 x_1} \cdot \lambda_2 e^{-\lambda_2 x_2} = \lambda_1 \lambda_2 e^{-\lambda_1 x_1 - \lambda_2 x_2} & x_1 > 0, x_2 > 0 \\ 0 & \text{otherwise} \end{cases}$$

Let $\lambda_1 = 1/1000$ and $\lambda_2 = 1/1200$, so that the expected lifetimes are 1000 hours and 1200 hours, respectively. The probability that both component lifetimes are at least 1500 hours is

$$P(1500 \leq X_1, 1500 \leq X_2) = P(1500 \leq X_1) \cdot P(1500 \leq X_2)$$
$$= e^{-\lambda_1(1500)} \cdot e^{-\lambda_2(1500)}$$
$$= (.2231)(.2865) = .0639$$ ∎

More Than Two Random Variables

To model the joint behavior of more than two random variables, we extend the concept of a joint distribution of two variables.

Definition

If $X_1, X_2, \ldots, X_n$ are all discrete rv's, the joint pmf of the variables is the function

$$p(x_1, x_2, \ldots, x_n) = P(X_1 = x_1, X_2 = x_2, \ldots, X_n = x_n)$$

If the variables are continuous, the joint pdf of $X_1, \ldots, X_n$ is the function $f(x_1, x_2, \ldots, x_n)$ such that for any n intervals $[a_1, b_1], \ldots, [a_n, b_n]$,

$$P(a_1 \le X_1 \le b_1, \ldots, a_n \le X_n \le b_n)$$

$$= \int_{a_1}^{b_1} \cdots \int_{a_n}^{b_n} f(x_1, \ldots, x_n) \, dx_n \cdots dx_1$$

Example 5.9

In a binomial experiment, each trial could result in one of only two possible outcomes. Consider now an experiment consisting of n independent and identical trials, in which each trial can result in any one of r possible outcomes. Let $p_i = P(\text{outcome } i \text{ on any particular trial})$ and define random variables by $X_i = $ the number of trials resulting in outcome i $(i = 1, \ldots, r)$. Such an experiment is called a multinomial experiment, and the joint pmf of $X_1, \ldots, X_r$ is called the multinomial distribution. By using a counting argument analogous to the one used in deriving the binomial distribution, the joint pmf of $X_1, \ldots, X_r$ can be shown to be

$$p(x_1, \ldots, x_r)$$

$$= \begin{cases} \dfrac{n!}{(x_1!)(x_2!) \cdot \ldots \cdot (x_r!)} \, p_1^{x_1} \cdot \ldots \cdot p_r^{x_r} & x_i = 0, 1, 2, \ldots, \text{ with } x_1 + \cdots + x_r = n \\ 0 & \text{otherwise} \end{cases}$$

As an example, if the allele of each of ten independently obtained pea sections is determined and $p_1 = P(AA)$, $p_2 = P(Aa)$, $p_3 = P(aa)$, $X_1 = $ number of AA's, $X_2 = $ number of Aa's, and $X_3 = $ number of aa's, then

$$p(x_1, x_2, x_3) = \frac{10!}{(x_1!)(x_2!)(x_3!)} \, p_1^{x_1} p_2^{x_2} p_3^{x_3}, \qquad \begin{array}{l} x_i = 0, 1, \ldots \quad \text{and} \\ x_1 + x_2 + x_3 = 10 \end{array}$$

If $p_1 = p_3 = .25$, $p_2 = .5$, then

$$P(X_1 = 2, X_2 = 5, X_3 = 3) = p(2, 5, 3)$$

$$= \frac{10!}{2! \, 5! \, 3!} (.25)^2 (.5)^5 (.25)^3 = .0769$$

The case $r = 2$ gives the binomial distribution, with $X_1 = $ number of successes and $X_2 = n - X_1 = $ number of failures. ∎

Example 5.10

When a certain method is used to collect a fixed volume of rock samples in a region, there are four resulting rock types. Let X_1, X_2, and X_3 denote the proportion by volume of rock types 1, 2, and 3 in a randomly selected sample (the proportion of rock type 4 is $1 - X_1 - X_2 - X_3$, so a variable X_4 would be redundant). If the joint pdf of X_1, X_2, X_3 is

$$f(x_1, x_2, x_3)$$

$$= \begin{cases} kx_1x_2(1 - x_3) & 0 \le x_1 \le 1, 0 \le x_2 \le 1, 0 \le x_3 \le 1, x_1 + x_2 + x_3 \le 1 \\ 0 & \text{otherwise} \end{cases}$$

then k is determined by

$$1 = \int_{-\infty}^{\infty} \int_{-\infty}^{\infty} \int_{-\infty}^{\infty} f(x_1, x_2, x_3) \, dx_3 \, dx_2 \, dx_1$$

$$= \int_0^1 \left\{ \int_0^{1-x_1} \left[\int_0^{1-x_1-x_2} kx_1x_2(1 - x_3) \, dx_3 \right] dx_2 \right\} dx_1$$

This iterated integral has value $k/144$, so $k = 144$. The probability that rocks of type 1 and 2 together account for at most 50% of the sample is

$$P(X_1 + X_2 \le .5) = \iiint\limits_{\substack{\{0 \le x_i \le 1 \text{ for } i = 1, 2, 3 \\ x_1 + x_2 + x_3 \le 1, \, x_1 + x_2 \le .5\}}} f(x_1, x_2, x_3) \, dx_3 \, dx_2 \, dx_1$$

$$= \int_0^{.5} \left\{ \int_0^{.5 - x_1} \left[\int_0^{1-x_1-x_2} 144x_1x_2(1 - x_3) \, dx_3 \right] dx_2 \right\} dx_1$$

$$= .6066 \qquad \blacksquare$$

The notion of independence of more than two random variables is similar to the notion of independence of more than two events.

Definition

> The rv's X_1, X_2, . . . , X_n are said to be **independent** if for *every* subset X_{i_1}, X_{i_2}, . . . , X_{i_k} of the variables (each pair, each triple, and so on), the joint pmf or pdf of the subset is equal to the product of the marginal pmf's or pdf's.

Thus, if the variables are independent with $n = 4$, then the joint pmf or pdf of any two variables is the product of the two marginals, and similarly for any three variables and all four variables together. Most important, once we are told that n variables are independent, then the joint pmf or pdf is the product of the n marginals.

Example 5.11

If X_1, . . . , X_n represent the lifetimes of n components, the components operate independently of one another, and each lifetime is exponentially distributed with parameter λ, then

$$f(x_1, x_2, \ldots, x_n) = (\lambda e^{-\lambda x_1}) \cdot (\lambda e^{-\lambda x_2}) \cdots \cdot (\lambda e^{-\lambda x_n})$$

$$= \begin{cases} \lambda^n e^{-\lambda \Sigma x_i} & x_1 \ge 0, x_2 \ge 0, \ldots, x_n \ge 0 \\ 0 & \text{otherwise} \end{cases}$$

If these n components constitute a system that will fail as soon as a single component fails, then the probability that the system lasts past time t is

$$P(X_1 > t, \ldots, X_n > t) = \int_t^\infty \cdots \int_t^\infty f(x_1, \ldots, x_n) \, dx_1 \ldots dx_n$$

$$= \left(\int_t^\infty \lambda e^{-\lambda x_1} \, dx_1 \right) \cdots \left(\int_t^\infty \lambda e^{-\lambda x_n} \, dx_n \right)$$

$$= (e^{-\lambda t})^n = e^{-n\lambda t}$$

Therefore,

$$P(\text{system lifetime} \le t) = 1 - e^{-n\lambda t}, \qquad \text{for } t \ge 0$$

which shows that system lifetime has an exponential distribution with parameter $n\lambda$, so the expected lifetime is $1/n\lambda$. ∎

In many experimental situations to be considered in this book, independence is a reasonable assumption, so that specifying the joint distribution reduces to deciding on appropriate marginal distributions.

Conditional Distributions

Suppose X = the number of major defects on a randomly selected new automobile and Y = the number of minor defects on that same auto. If we learn that the number of major defects on the selected car is one, what now is the probability that the car has at most three minor defects—that is, what is $P(Y \le 3 \mid X = 1)$? Similarly, if X and Y denote the lifetimes of two components in a system and it happens that $X = 100$, what is the probability that $Y \ge 200$, and what is the expected lifetime of the second component "conditional on" this value of X? Questions of this sort can be answered by studying conditional probability distributions.

Definition

> Let X and Y be two continuous rv's with joint pdf $f(x, y)$ and marginal X pdf $f_X(x)$. Then for any X value x for which $f_X(x) > 0$, the **conditional probability density function of Y given that $X = x$** is
>
> $$f_{Y|X}(y|x) = \frac{f(x, y)}{f_X(x)} \qquad -\infty < y < \infty$$
>
> If X and Y are discrete, replacing pdf's by pmf's in this definition gives the **conditional probability mass function of Y when $X = x$.**

Notice that the definition of $f_{Y|X}(y|x)$ parallels that of $P(B|A)$, the conditional probability that B will occur, given that A has occurred. Once the conditional pdf or pmf has been determined, questions of the type posed at the outset of this subsection can be answered by integrating or summing over an appropriate set of Y values.

Example 5.12   Reconsider the situation of Examples 5.3 and 5.4 involving X = the proportion of time that a bank's drive-up facility is busy and Y = the analogous proportion for the walk-up window. The conditional pdf of Y given that $X = .8$ is

$$f_{Y|X}(y|.8) = \frac{f(.8, y)}{f_X(.8)} = \frac{1.2(.8 + y^2)}{1.2(.8) + .4} = \frac{1}{34}(24 + 30y^2) \qquad 0 < y < 1$$

The probability that the walk-in facility is busy at most half the time given that $X = .8$ is then

$$P(Y \le .5 | X = .8) = \int_{-\infty}^{.5} f_{Y|X}(y|.8) \, dy = \int_0^{.5} \frac{1}{34}(24 + 30y^2) \, dy = .390$$

Using the marginal pdf of Y gives $P(Y \le .5) = .350$. Also $E(Y) = .6$, whereas the expected proportion of time that the walk-in facility is busy given that $X = .8$ (a *conditional* expectation) is

$$E(Y|X = .8) = \int_{-\infty}^{\infty} y \cdot f_{Y|X}(y|.8) \, dy = \frac{1}{34}\int_0^1 y(24 + 30y^2) \, dy = .574 \qquad \blacksquare$$

Exercises   Section 5.1 (1–21)

1. A service station has both self-service and full-service islands. On each island, there is a single regular unleaded pump with two hoses. Let X denote the number of hoses being used on the self-service island at a particular time and let Y denote the number of hoses on the full-service island in use at that time. The joint pmf of X and Y appears in the accompanying tabulation.

$p(x, y)$		y	
	0	1	2
0	.10	.04	.02
x 1	.08	.20	.06
2	.06	.14	.30

   a. What is $P(X = 1$ and $Y = 1)$?

   b. Compute $P(X \le 1$ and $Y \le 1)$.

   c. Give a word description of the event $\{X \ne 0$ and $Y \ne 0\}$ and compute the probability of this event.

   d. Compute the marginal pmf of X and of Y. Using $p_X(x)$, what is $P(X \le 1)$?

   e. Are X and Y independent rv's? Explain.

2. When an automobile is stopped by a roving safety patrol, each tire is checked for tire wear, and each headlight is checked to see whether it is properly aimed. Let X denote the number of headlights that need adjustment and let Y denote the number of defective tires.

   a. If X and Y are independent with $p_X(0) = .5$, $p_X(1) = .3, p_X(2) = .2$, and $p_Y(0) = .6, p_Y(1) = .1$, $p_Y(2) = p_Y(3) = .05, p_Y(4) = .2$, display the joint pmf of (X, Y) in a joint probability table.

   b. Compute $P(X \le 1$ and $Y \le 1)$ from the joint probability table and verify that it equals the product $P(X \le 1) \cdot P(Y \le 1)$.

   c. What is $P(X + Y = 0)$ (the probability of no violations)?

   d. Compute $P(X + Y \le 1)$.

3. A certain market has both an express checkout line and a superexpress checkout line. Let X_1 denote the number of customers in line at the express checkout at a particular time of day and let X_2 denote the number of customers in line at the superexpress checkout at the same time. Suppose the joint pmf of X_1 and X_2 is as given in the accompanying table.

			x_2		
		0	1	2	3
	0	.08	.07	.04	.00
	1	.06	.15	.05	.04
x_1	2	.05	.04	.10	.06
	3	.00	.03	.04	.07
	4	.00	.01	.05	.06

a. What is $P(X_1 = 1, X_2 = 1)$, that is, the probability that there is exactly one customer in each line?

b. What is $P(X_1 = X_2)$, that is, the probability that the numbers of customers in the two lines are identical?

c. Let A denote the event that there are at least two more customers in one line than in the other line. Express A in terms of X_1 and X_2 and calculate the probability of this event.

d. What is the probability that the total number of customers in the two lines is exactly four? At least four?

4. Return to the situation described in Exercise 3.

a. Determine the marginal pmf of X_1 and then calculate the expected number of customers in line at the express checkout.

b. Determine the marginal pmf of X_2.

c. By inspection of the probabilities $P(X_1 = 4)$, $P(X_2 = 0)$, and $P(X_1 = 4, X_2 = 0)$, are X_1 and X_2 independent random variables? Explain.

5. The number of customers waiting for gift-wrap service at a department store is an rv X with possible values 0, 1, 2, 3, 4 and corresponding probabilities .1, .2, .3, .25, .15. A randomly selected customer will have 1, 2, or 3 packages for wrapping with probabilities .6, .3, and .1, respectively. Let Y = the total number of packages to be wrapped for the customers waiting in line (assume that the number of packages submitted by one customer is independent of the number submitted by any other customer).

a. Determine $P(X = 3, Y = 3)$.

b. Determine $P(X = 4, Y = 11)$.

6. Let X denote the number of brand X VCRs sold during a particular week by a certain store. The pmf of X is

x	0	1	2	3	4
$p_X(x)$	.1	.2	.3	.25	.15

Sixty percent of all customers who purchase brand X VCRs also buy an extended warranty. Let Y denote the number of purchasers during this week who buy an extended warranty.

a. What is $P(X = 4, Y = 2)$? [HINT: This probability equals $P(Y = 2 \mid X = 4) \cdot P(X = 4)$; now think of the four purchases as four trials of a binomial experiment, with success on a trial corresponding to buying an extended warranty.]

b. Calculate $P(X = Y)$.

c. Determine the joint pmf of X and Y and then the marginal pmf of Y.

7. The joint probability distribution of the number X of cars and the number Y of buses per signal cycle at a proposed left turn lane is displayed in the accompanying joint probability table.

$p(x, y)$	y 0	1	2
0	.025	.015	.010
1	.050	.030	.020
2	.125	.075	.050
x 3	.150	.090	.060
4	.100	.060	.040
5	.050	.030	.020

a. What is the probability that there is exactly one car and exactly one bus during a cycle?

b. What is the probability that there is at most one car and at most one bus during a cycle?

c. What is the probability that there is exactly one car during a cycle? Exactly one bus?

d. Suppose the left turn lane is to have a capacity of five cars and one bus is equivalent to three cars. What is the probability of an overflow during a cycle?

e. Are X and Y independent rv's? Explain.

8. A stockroom currently has 30 components of a certain type, of which 8 were provided by supplier 1, 10 by supplier 2, and 12 by supplier 3. Six of these are to be randomly selected for a particular assembly. Let X = the number of supplier 1's components selected, Y = the number of supplier 2's components selected, and $p(x, y)$ denote the joint pmf of X and Y.

a. What is $p(3, 2)$? [HINT: Each sample of size six is equally likely to be selected. Therefore, $p(3, 2)$ = (number of outcomes with $X = 3$ and $Y = 2$)/ (total number of outcomes). Now use the product rule for counting to obtain the numerator and denominator.]

b. Using the logic of part (a), obtain $p(x, y)$. (This can be thought of as a multivariate hypergeometric distribution—sampling without replacement from a finite population consisting of more than two categories.)

9. Each front tire on a particular type of automobile is supposed to be filled to a pressure of 26 psi. Suppose the actual air pressure in each tire is a random variable—X for the right tire and Y for the left tire, with joint pdf

$$f(x, y)$$
$$= \begin{cases} K(x^2 + y^2) & 20 \le x \le 30,\ 20 \le y \le 30 \\ 0 & \text{otherwise} \end{cases}$$

   a. What is the value of K?

   b. What is the probability that both tires are under-filled?

   c. What is the probability that the difference in air pressure between the two tires is at most 2 psi?

   d. Determine the (marginal) distribution of air pressure in the right tire alone.

   e. Are X and Y independent rv's?

10. Annie and Alvie have agreed to meet between 5:00 P.M. and 6:00 P.M. for dinner at a local health-food restaurant. Let X = Annie's arrival time and Y = Alvie's arrival time and suppose X and Y are independent with each uniformly distributed on the interval [5, 6].

   a. What is the joint pdf of X and Y?

   b. What is the probability that they both arrive between 5:15 and 5:45?

   c. If the first one to arrive will wait only 10 min before leaving to eat elsewhere, what is the probability that they have dinner at the health-food restaurant? [HINT: The event of interest is $A = \{(x,\ y):\ |x - y| \le \frac{1}{6}\}$.]

11. A professor has just given one long technical paper to one typist and a somewhat shorter paper to another typist. Let X = the number of typing errors on the first paper and Y the number of typing errors on the second paper. Suppose X has a Poisson distribution with parameter λ, Y has a Poisson distribution with parameter μ, and X and Y are independent.

   a. What is the joint pmf of X and Y?

   b. What is the probability that at most one error is made on both papers combined?

   c. Obtain a general expression for the probability that the total number of errors in the two papers is m (where m is a nonnegative integer). [HINT: $A = \{(x,\ y):\ x + y = m\} = \{(m,\ 0),\ (m - 1,\ 1),\ \ldots,\ (1,\ m - 1),\ (0,\ m)\}$. Now sum the joint pmf over $(x,\ y) \in A$ and use the binomial theorem, which

says that

$$\sum_{k=0}^{m} \binom{m}{k} a^k b^{m-k} = (a + b)^m$$

for any a, b.]

12. Two components of a minicomputer have the following joint pdf for their useful lifetimes X and Y:

$$f(x, y) = \begin{cases} xe^{-x(1 + y)} & x \ge 0 \text{ and } y \ge 0 \\ 0 & \text{otherwise} \end{cases}$$

   a. What is the probability that the lifetime X of the first component exceeds 3?

   b. What are the marginal pdf's of X and Y? Are the two lifetimes independent? Explain.

   c. What is the probability that the lifetime of at least one component exceeds 3?

13. You have two light bulbs for a particular lamp. Let X = the lifetime of the first bulb and Y = the lifetime of the second bulb (both in 1000s of hours). Suppose that X and Y are independent and that each has an exponential distribution with parameter $\lambda = 1$.

   a. What is the joint pdf of X and Y?

   b. What is the probability that each bulb lasts at most 1000 hours (i.e., $X \le 1$ and $Y \le 1$)?

   c. What is the probability that the total lifetime of the two bulbs is at most 2? [HINT: Draw a picture of the region $A = \{(x,\ y): x \ge 0,\ y \ge 0,\ x + y \le 2\}$ before integrating.]

   d. What is the probability that the total lifetime is between 1 and 2?

14. Suppose that you have ten light bulbs, that the lifetime of each is independent of all the other lifetimes, and that each lifetime has an exponential distribution with parameter λ.

   a. What is the probability that all ten bulbs fail before time t?

   b. What is the probability that exactly k of the ten bulbs fail before time t?

   c. Suppose nine of the bulbs have lifetimes that are exponentially distributed with parameter λ and the remaining bulb has a lifetime that is exponentially distributed with parameter θ (it is made by another manufacturer). What is the probability that exactly five of the ten bulbs fail before time t?

15. Consider a system consisting of three components as pictured. The system will continue to function as long

as the first component functions and either component 2 or component 3 functions. Let X_1, X_2, and X_3 denote the lifetimes of components 1, 2, and 3, respectively. Suppose the X_i's are independent of one another and each X_i has an exponential distribution with parameter λ.

a. Let Y denote the system lifetime. Obtain the cumulative distribution function of Y and differentiate it to obtain the pdf. [HINT: $F(y) = P(Y \le y)$; express the event $\{Y \le y\}$ in terms of unions and/or intersections of the three events $\{X_1 \le y\}$, $\{X_2 \le y\}$, and $\{X_3 \le y\}$.]

b. Compute the expected system lifetime.

16. a. For $f(x_1, x_2, x_3)$ as given in Example 5.10, compute the **joint marginal density function** of X_1 and X_3 alone (by integrating over x_2).

b. What is the probability that rocks of types 1 and 3 together make up at most 50% of the sample? [HINT: Use the result of part (a).]

c. Compute the marginal pdf of X_1 alone. [HINT: Use the result of part (a).]

17. An ecologist wishes to select a point inside a circular sampling region according to a uniform distribution (in practice this could be done by first selecting a direction and then a distance from the center in that direction). Let X = the x coordinate of the point selected and Y = the y coordinate of the point selected. If the circle is centered at $(0, 0)$ and has radius R, then the joint pdf of X and Y is

$$f(x, y) = \begin{cases} \dfrac{1}{\pi R^2} & x^2 + y^2 \le R^2 \\ 0 & \text{otherwise} \end{cases}$$

a. What is the probability that the selected point is within $R/2$ of the center of the circular region? [HINT: Draw a picture of the region of positive density D.]

Because $f(x, y)$ is constant on D, computing a probability reduces to computing an area.]

b. What is the probability that both X and Y differ from 0 by at most $R/2$?

c. Answer part (b) for $R/\sqrt{2}$ replacing $R/2$.

d. What is the marginal pdf of X? Of Y? Are X and Y independent?

18. Refer to Exercise 1 and answer the following questions.

a. Given that $X = 1$, determine the conditional pmf of Y—that is, $p_{Y|X}(0\,|\,1)$, $p_{Y|X}(1\,|\,1)$, and $p_{Y|X}(2\,|\,1)$.

b. Given that two hoses are in use at the self-service island, what is the conditional pmf of the number of hoses in use on the full-service island?

c. Use the result of part (b) to calculate the conditional probability $P(Y \le 1\,|\,X = 2)$.

d. Given that two hoses are in use at the full-service island, what is the conditional pmf of the number in use at the self-service island?

19. The joint pdf of pressures for right and left front tires is given in Exercise 9.

a. Determine the conditional pdf of Y given that $X = x$ and the conditional pdf of X given that $Y = y$.

b. If the pressure in the right tire is found to be 22 psi, what is the probability that the left tire has a pressure of at least 25 psi? Compare this to $P(Y \ge 25)$.

c. If the pressure in the right tire is found to be 22 psi, what is the expected pressure in the left tire, and what is the standard deviation of pressure in this tire?

20. Let X_1, X_2, and X_3 be the lifetimes of components 1, 2, and 3 in a three-component system.

a. How would you define the conditional pdf of X_3 given that both $X_1 = x_1$ and $X_2 = x_2$?

b. How would you define the conditional joint pdf of X_2 and X_3 given that $X_1 = x_1$?

21. What condition on $f_{Y|X}(y\,|\,x)$ is equivalent to the independence of X and Y?

5.2

Expected Values, Covariance, and Correlation

We previously saw that any function $h(X)$ of a single rv X is itself a random variable. However, to compute $E[h(X)]$, it was not necessary to obtain the probability distribution of $h(X)$; instead, $E[h(X)]$ was computed as a weighted average of $h(x)$ values, where the weight function was the pmf $p(x)$ or pdf $f(x)$ of X. A similar result holds for a function $h(X, Y)$ of two jointly distributed random variables.

Proposition

> Let X and Y be jointly distributed rv's with pmf $p(x, y)$ or pdf $f(x, y)$ according to whether the variables are discrete or continuous. Then the expected value of a function $h(X, Y)$, denoted by $E[h(X, Y)]$ or $\mu_{h(X,Y)}$, is given by
>
> $$E[h(X, Y)] = \begin{cases} \sum_x \sum_y h(x, y) \cdot p(x, y) & \text{if } X \text{ and } Y \text{ are discrete} \\ \int_{-\infty}^{\infty} \int_{-\infty}^{\infty} h(x, y) \cdot f(x, y) \, dx \, dy & \text{if } X \text{ and } Y \text{ are continuous} \end{cases}$$

Example 5.13

Five friends have purchased tickets to a certain concert. If the tickets are for seats 1–5 in a particular row and the tickets are randomly distributed among the five, what is the expected number of seats separating any particular two of the five? Let X and Y denote the seat number of the first and second individuals, respectively. Possible (X, Y) pairs are $\{(1, 2), (1, 3), \ldots, (5, 4)\}$, and the joint pmf of (X, Y) is

$$p(x, y) = \begin{cases} \dfrac{1}{20} & x = 1, \ldots, 5; \, y = 1, \ldots, 5; \, x \neq y \\ 0 & \text{otherwise} \end{cases}$$

The number of seats separating the two individuals is $h(X, Y) = |X - Y| - 1$. The accompanying table gives $h(x, y)$ for each possible (x, y) pair.

$h(x, y)$		1	2	x 3	4	5
	1	—	0	1	2	3
	2	0	—	0	1	2
y	3	1	0	—	0	1
	4	2	1	0	—	0
	5	3	2	1	0	—

Thus,

$$E[h(X, Y)] = \sum\sum_{(x,y)} h(x, y) \cdot p(x, y) = \sum_{\substack{x=1 \\ x \neq y}}^{5} \sum_{y=1}^{5} (|x - y| - 1) \cdot \frac{1}{20} = 1 \qquad \blacksquare$$

Example 5.14

In Example 5.5, the joint pdf of the amount X of almonds and amount Y of cashews in a 1-lb can of nuts was

$$f(x, y) = \begin{cases} 24xy & 0 \leq x \leq 1, \, 0 \leq y \leq 1, \, x + y \leq 1 \\ 0 & \text{otherwise} \end{cases}$$

If 1 lb of almonds costs the company $1.00, 1 lb of cashews costs $1.50, and 1 lb of peanuts costs $.50, then the total cost of the contents of a can is

$$h(X, Y) = (1)X + (1.5)Y + (.5)(1 - X - Y) = .5 + .5X + Y$$

(since $1 - X - Y$ of the weight consists of peanuts). The expected total cost is

$$E[h(X, Y)] = \int_{-\infty}^{\infty} \int_{-\infty}^{\infty} h(x, y) \cdot f(x, y) \, dx \, dy$$

$$= \int_{0}^{1} \int_{0}^{1-x} (.5 + .5x + y) \cdot 24xy \, dy \, dx = \$1.10 \qquad \blacksquare$$

The method of computing the expected value of a function $h(X_1, \ldots, X_n)$ of n random variables is similar to that for two random variables. If the X_i's are discrete, $E[h(X_1, \ldots, X_n)]$ is an n-dimensional sum; if the X_i's are continuous, it is an n-dimensional integral.

Covariance

When two random variables X and Y are not independent, it is frequently of interest to measure how strongly they are related to one another.

Definition

> The **covariance** between two rv's X and Y is
>
> $$\text{Cov}(X, Y) = E[(X - \mu_X)(Y - \mu_Y)]$$
>
> $$= \begin{cases} \sum_x \sum_y (x - \mu_X)(y - \mu_Y)p(x, y) & X, Y \text{ discrete} \\ \int_{-\infty}^{\infty} \int_{-\infty}^{\infty} (x - \mu_X)(y - \mu_Y)f(x, y) \, dx \, dy & X, Y \text{ continuous} \end{cases}$$

The rationale for the definition is as follows. Suppose X and Y have a strong positive relationship to one another, by which we mean that large values of X tend to occur with large values of Y and small values of X with small values of Y. Then most of the probability mass or density will be associated with $(x - \mu_X)$ and $(y - \mu_Y)$ either both positive (both X and Y above their respective means) or both negative, so the product $(x - \mu_X)(y - \mu_Y)$ will tend to be positive. Thus, for a strong positive relationship, $\text{Cov}(X, Y)$ should be quite positive. For a strong negative relationship, the signs of $(x - \mu_X)$ and $(y - \mu_Y)$ will tend to be opposite, yielding a negative product. Thus, for a strong negative relationship, $\text{Cov}(X, Y)$ should be quite negative. If X and Y are not strongly related, positive and negative products will tend to cancel one another, yielding a covariance near 0. Figure 5.4 illustrates the different possibilities. The covariance depends on *both* the set of possible pairs and the probabilities. In Figure 5.4 the probabilities could be changed without altering the set of possible pairs, and this could drastically change the value of $\text{Cov}(X, Y)$.

Figure 5.4 $p(x, y) = \frac{1}{10}$ for each of ten pairs corresponding to indicated points; (a) positive covariance; (b) negative covariance; (c) covariance near zero

Example 5.15   The joint and marginal pmf's for X = automobile policy deductible amount and Y = homeowner policy deductible amount in Example 5.1 were

p(x, y)		0	y 100	200
x	100	.20	.10	.20
	250	.05	.15	.30

x	100	250
$p_X(x)$	.5	.5

y	0	100	200
$p_Y(y)$	.25	.25	.5

from which $\mu_X = \Sigma x p_X(x) = 175$ and $\mu_Y = 125$. Therefore,

$$
\begin{aligned}
\text{Cov}(X, Y) &= \sum_{(x,y)}\sum (x - 175)(y - 125)p(x, y) \\
&= (100 - 175)(0 - 125)(.20) + \cdots \\
&\quad + (250 - 175)(200 - 125)(.30) \\
&= 1875
\end{aligned}
$$

∎

The following shortcut formula for $\text{Cov}(X, Y)$ simplifies the computations.

Proposition

$$\text{Cov}(X, Y) = E(XY) - \mu_X \cdot \mu_Y$$

According to this formula, no intermediate subtractions are necessary; only at the end of the computation is $\mu_X \cdot \mu_Y$ subtracted from $E(XY)$. The proof involves expanding $(X - \mu_X)(Y - \mu_Y)$ and then taking the expected value of each term separately. Note that $\text{Cov}(X, X) = E(X^2) - \mu_X^2 = V(X)$.

Example 5.16
(Example 5.5 continued)

The joint and marginal pmf's of X = amount of almonds and Y = amount of cashews were

$$f(x, y) = \begin{cases} 24xy & 0 \le x \le 1, 0 \le y \le 1, x + y \le 1 \\ 0 & \text{otherwise} \end{cases}$$

$$f_X(x) = \begin{cases} 12x(1-x)^2 & 0 \leq x \leq 1 \\ 0 & \text{otherwise} \end{cases}$$

with $f_Y(y)$ obtained by replacing x by y in $f_X(x)$. It is easily verified that $\mu_X = \mu_Y = \frac{2}{5}$, and

$$E(XY) = \int_{-\infty}^{\infty} \int_{-\infty}^{\infty} xy \, f(x, y) \, dx \, dy = \int_0^1 \int_0^{1-x} xy \cdot 24xy \, dy \, dx$$

$$= 8 \int_0^1 x^2 (1-x)^3 \, dx = \frac{2}{15}$$

Thus $\text{Cov}(X, Y) = \frac{2}{15} - \left(\frac{2}{5}\right)\left(\frac{2}{5}\right) = \frac{2}{15} - \frac{4}{25} = -\frac{2}{75}$. A negative covariance is reasonable here because more almonds in the can implies fewer cashews. ∎

It would appear that the relationship in the insurance example is quite strong since $\text{Cov}(X, Y) = 1875$, whereas $\text{Cov}(X, Y) = -\frac{2}{75}$ in the nut example would seem to imply quite a weak relationship. Unfortunately, the covariance has a serious defect that makes it impossible to interpret a computed value of the covariance. In the insurance example, suppose we had expressed deductible amount in cents rather than in dollars. Then $100X$ would replace X, $100Y$ would replace Y, and the resulting covariance would be $\text{Cov}(100X, 100Y) = (100)(100)\text{Cov}(X, Y) = 18,750,000$. If, on the other hand, the deductible amount had been expressed in hundreds of dollars, the computed covariance would have been $(.01)(.01)(1875) = .1875$. *The defect of covariance is that its computed value depends critically on the units of measurement.* Ideally the choice of units should have no effect on a measure of strength of relationship. This is achieved by scaling the covariance.

Correlation

Definition

> The **correlation coefficient** of X and Y, denoted by $\text{Corr}(X, Y)$, $\rho_{X,Y}$, or just ρ, is defined by
>
> $$\rho_{X,Y} = \frac{\text{Cov}(X, Y)}{\sigma_X \cdot \sigma_Y}$$

Example 5.17

It is easily verified that in the insurance problem of Example 5.15, $E(X^2) = 36,250$, $\sigma_X^2 = 36,250 - (175)^2 = 5625$, $\sigma_X = 75$, $E(Y^2) = 22,500$, $\sigma_Y^2 = 6875$, and $\sigma_Y = 82.92$. This gives

$$\rho = \frac{1875}{(75)(82.92)} = .301$$

∎

The following proposition shows that ρ remedies the defect of $\text{Cov}(X, Y)$ and also suggests how to recognize the existence of a strong (linear) relationship.

Proposition

> 1. If a and c are either both positive or both negative
>
> $\text{Corr}(aX + b, cY + d) = \text{Corr}(X, Y)$
>
> 2. For any two rv's X and Y, $-1 \le \text{Corr}(X, Y) \le 1$.

Statement 1 says precisely that the correlation coefficient is not affected by a linear change in the units of measurement (if, say, X = temperature in °C, then $9X/5 + 32$ = temperature in °F). According to Statement 2, the strongest possible positive relationship is evidenced by $\rho = +1$, while the strongest possible negative relationship corresponds to $\rho = -1$. The proof of the first statement is sketched in Exercise 35, while that of the second appears in a supplementary exercise at the end of the chapter. For descriptive purposes, the relationship will be described as strong if $|\rho| \ge .8$, moderate if $.5 < |\rho| < .8$, and weak if $|\rho| \le .5$.

If we think of $p(x, y)$ or $f(x, y)$ as prescribing a mathematical model for how the two numerical variables X and Y are distributed in some population (height and weight, verbal SAT score and quantitative SAT score, etc.), then ρ is a population characteristic or parameter that measures how strongly X and Y are related in the population. In Chapter 12, we will consider taking a sample of pairs $(x_1, y_1), \ldots,$ (x_n, y_n) from the population. The sample correlation coefficient r will then be defined and used to make inferences about ρ.

The correlation coefficient ρ is actually not a completely general measure of the strength of a relationship.

Proposition

> 1. If X and Y are independent, then $\rho = 0$, but $\rho = 0$ does not imply independence.
> 2. $\rho = 1$ or -1 iff $Y = aX + b$ for some numbers a and b with $a \ne 0$.

This proposition says that ρ is a measure of the degree of **linear** relationship between X and Y, and only when the two variables are perfectly related in a linear manner will ρ be as positive or negative as it can be. A ρ less than 1 in absolute value indicates only that the relationship is not completely linear, but there may still be a very strong nonlinear relation. Also, $\rho = 0$ does not imply that X and Y are independent, but only that there is complete absence of a linear relationship. When $\rho = 0$, X and Y are said to be **uncorrelated.** Two variables could be uncorrelated yet highly dependent because there is a strong nonlinear relationship, so be careful not to conclude too much from knowing that $\rho = 0$.

Example 5.18   Let X and Y be discrete rv's with joint pmf.

$$p(x, y) = \begin{cases} \dfrac{1}{4} & (x, y) = (-4, 1), (4, -1), (2, 2), (-2, -2) \\ 0 & \text{otherwise} \end{cases}$$

The points that receive positive probability mass are identified on the (x, y) coordinate system in Figure 5.5. It is evident from the figure that the value of X is completely determined by the value of Y and vice versa, so the two variables are completely dependent. However, by symmetry $\mu_X = \mu_Y = 0$ and $E(XY) = (-4)\frac{1}{4} + (-4)\frac{1}{4} + (4)\frac{1}{4} + (4)\frac{1}{4} = 0$, so $\text{Cov}(X, Y) = E(XY) - \mu_X \cdot \mu_Y = 0$ and thus $\rho_{X,Y} = 0$. Although there is perfect dependence, there is also complete absence of any linear relationship!

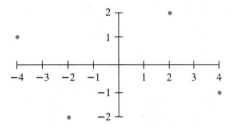

Figure 5.5 The population of pairs for Example 5.18 ∎

A value of ρ near 1 does not necessarily imply that increasing the value of X *causes* Y to increase. It implies only that large X values are *associated* with large Y values. For example, in the population of children, vocabulary size and number of cavities are quite positively correlated, but it is certainly not true that cavities cause vocabulary to grow. Instead, the values of both these variables tend to increase as the value of age, a third variable, increases. For children of a fixed age, there is probably a very low correlation between number of cavities and vocabulary size. In summary, association (a high correlation) is not the same as causation.

Exercises   Section 5.2 (22–36)

22. An instructor has given a short quiz consisting of two parts. For a randomly selected student, let $X =$ the number of points earned on the first part, $Y =$ the number of points earned on the second part, and suppose that the joint pmf of X and Y is given in the accompanying tabulation.

$p(x, y)$		y		
	0	5	10	15
0	.02	.06	.02	.10
x **5**	.04	.15	.20	.10
10	.01	.15	.14	.01

a. If the score recorded in the grade book is the total number of points earned on the two parts, what is the expected recorded score $E(X + Y)$?

b. If the maximum of the two scores is recorded, what is the expected recorded score?

23. The difference between the number of customers in line at the express checkout and the number in line at the superexpress checkout in Exercise 3 is $X_1 - X_2$. Calculate the expected difference.

24. Six individuals, including A and B, take seats around a circular table in a completely random fashion. Suppose the seats are numbered 1, . . . , 6 and let $X =$ A's seat number and $Y =$ B's seat number. If A sends a written message around the table to B in the direction in which they are closest, how many individuals (including A and B) would you expect to handle the message?

25. A surveyor wishes to lay out a square region with each side having length L. However, because of measurement error, he instead lays out a rectangle in which the north–south sides both have length X and the east–west sides both have length Y. Suppose that X and Y are independent and that each is uniformly distributed on the interval $[L - A, L + A]$ (where $0 < A < L$). What is the expected area of the resulting rectangle?

26. Consider a small ferry that can accommodate cars and buses. The toll for cars is $3, and the toll for buses is $10. Let X and Y denote the number of cars and buses, respectively, carried on a single trip, and suppose the joint distribution of X and Y is as given in the table of Exercise 7. Compute the expected revenue from a single trip.

27. Annie and Alvie have agreed to meet for lunch between noon (0:00 P.M.) and 1:00 P.M. Denote Annie's arrival time by X, Alvie's by Y, and suppose X and Y are independent with pdf's

$$f_X(x) = \begin{cases} 3x^2 & 0 \le x \le 1 \\ 0 & \text{otherwise} \end{cases}$$

$$f_Y(y) = \begin{cases} 2y & 0 \le y \le 1 \\ 0 & \text{otherwise} \end{cases}$$

What is the expected amount of time that the one who arrives first must wait for the other person? [HINT: $h(X, Y) = |X - Y|$.]

28. Show that if X and Y are independent rv's, then $E(XY) = E(X) \cdot E(Y)$. Then apply this in Exercise 25. [HINT: Consider the continuous case with $f(x, y) = f_X(x) \cdot f_Y(y)$.]

29. Compute the correlation coefficient ρ for X and Y of Example 5.16 (the covariance has already been computed).

30. a. Compute the covariance for X and Y in Exercise 22.
    b. Compute ρ for X and Y in the same exercise.

31. a. Compute the covariance between X and Y in Exercise 9.
    b. Compute the correlation coefficient ρ for this X and Y.

32. Reconsider the minicomputer component lifetimes X and Y as described in Exercise 12. Determine $E(XY)$. What can be said about Cov(X, Y) and ρ?

33. Use the result of Exercise 28 to show that when X and Y are independent, Cov(X, Y) = Corr(X, Y) = 0.

34. a. Recalling the definition of σ^2 for a single rv X, write a formula that would be appropriate for computing the variance of a function $h(X, Y)$ of two random variables. [HINT: Remember that variance is just a special expected value.]
    b. Use this formula to compute the variance of the recorded score $h(X, Y)$ [= max(X, Y)] in part (b) of Exercise 22.

35. a. Use the rules of expected value to show that Cov($aX + b$, $cY + d$) = ac Cov(X, Y).
    b. Use part (a) along with the rules of variance and standard deviation to show that Corr($aX + b$, $cY + d$) = Corr(X, Y) when a and c have the same sign.
    c. What happens if a and c have opposite signs?

36. Show that if $Y = aX + b$ ($a \ne 0$), then Corr(X, Y) = +1 or −1. Under what conditions will $\rho = +1$?

5.3 Statistics and Their Distributions

The observations in a single sample were denoted in Chapter 1 by $x_1, x_2, \ldots,$ x_n. Prior to sampling or experimentation, any particular observation is subject to uncertainty. For example, the fuel efficiency (mpg) of a first automobile selected might be $x_1 = 30.7$ or $x_1 = 29.4$, and there are many other possible values. Because of this uncertainty, before data is available, we regard the observations as random variables and denote them by uppercase letters $X_1, X_2, \ldots, X_n$. This in turn implies that until data has been obtained, any function of the observations—the sample median, sample standard deviation, sample range, and so on—should also be viewed as a random variable. Consider as an example the sample mean. If we take one sample of n ~~Stat~~ d obtain fuel efficiencies $x_1 = 29.2$, $x_2 = 31.0$, and $x_3 = 27.6$, then $\bar{x} = $ 27. Alternatively, we might have $x_1 = 30.3$, $x_2 = 28.2$, and $x_3 = 32.1$, $\bar{x} = 30.20$; these are, of course, not the only possibilities for $\bar{x}$.

Definition

> A **statistic** is any quan___lue can be calculated from sample data. Prior to obtaining data, ___rtainty as to what value of any particular statistic will result. Th___tistic is a random variable and will be denoted by an upperca___wercase letter is used to represent the calculated or observed ___statistic.

Thus, the sample mean, rega___istic (before a sample has been selected or an experiment carried out), i___$\overline{X}$; the calculated value of this statistic is $\bar{x}$. Similarly, S represents the ___ard deviation thought of as a statistic, and its computed value is s. If sa___different types of bricks are selected and the individual compressive st___noted by $X_1, \ldots, X_m$ and $Y_1, \ldots, Y_n$, respectively, then the statistic $X - \overline{Y}$, the difference between the two sample mean compressive strengths, is often of great interest.

Any statistic, being a random variable, has a probability distribution. In particular, the sample mean $\overline{X}$ has a probability distribution. Suppose, for example, that $n = 2$ components are randomly selected and the number of breakdowns while under warranty is determined for each one. Possible values for the sample mean number of breakdowns $\overline{X}$ are 0 (if $X_1 = X_2 = 0$), .5 (if either $X_1 = 0$ and $X_2 = 1$ or $X_1 = 1$ and $X_2 = 0$), 1, 1.5, ... The probability distribution of $\overline{X}$ specifies $P(\overline{X} = 0)$, $P(\overline{X} = .5)$, and so on, from which other probabilities such as $P(1 \leq \overline{X} \leq 3)$ and $P(\overline{X} \geq 2.5)$ can be calculated. Similarly, if for a sample of size $n = 2$, the only possible values of the sample variance are 0, 12.5, and 50 (which is the case if X_1 and X_2 can each take on only the values 40, 45, and 50), then the probability distribution of S^2 gives $P(S^2 = 0)$, $P(S^2 = 12.5)$, and $P(S^2 = 50)$. The probability distribution of a statistic is sometimes referred to as its **sampling distribution** to emphasize that it describes how the statistic varies in value across all samples that might be selected.

Random Samples

The probability distribution of any particular statistic depends not only on the population distribution (normal, uniform, etc.) and the sample size n but also on the method of sampling. Consider selecting a sample of size $n = 2$ from a population consisting of just the three values 1, 5, and 10 and suppose the statistic of interest is the sample variance. If sampling is done "with replacement," then $S^2 = 0$ will result if $X_1 = X_2$. However, S^2 cannot equal 0 if sampling is "without replacement." So $P(S^2 = 0) = 0$ for one sampling method, and this probability is positive for the other method. Our next definition describes a sampling method often encountered (at least approximately) in practice.

Definition

> The rv's $X_1, X_2, \ldots, X_n$ are said to form a (simple) **random sample** of size n if
>
> 1. The X_i's are independent rv's.
> 2. Every X_i has the same probability distribution.

Conditions 1 and 2 can be paraphrased by saying that the X_i's are *independent* and *identically distributed* (iid). If sampling is either with replacement or from an infinite (conceptual) population, Conditions 1 and 2 are satisfied exactly. These conditions will be approximately satisfied if sampling is without replacement, yet the sample size n is much smaller than the population size N. In practice, if $n/N \le .05$ (at most 5% of the population is sampled), we can pretend that the X_i's form a random sample. The virtue of this sampling method is that the probability distribution of any statistic can be more easily obtained than for any other sampling method.

There are two general methods for obtaining information about a statistic's sampling distribution. One method involves calculations based on probability rules, and the other involves carrying out a simulation experiment.

Deriving the Sampling Distribution of a Statistic

Probability rules can be used to obtain the distribution of a statistic provided that it is a "fairly simple" function of the X_i's and either there are relatively few different X values in the population or else the population distribution has a "nice" form. Our next two examples illustrate such situations.

Example 5.19

A large automobile service center charges \$40, \$45, and \$50 for a tune-up of four-, six-, and eight-cylinder cars, respectively. If 20% of its tune-ups are done on four-cylinder cars, 30% on six-cylinder cars, and 50% on eight-cylinder cars, then the probability distribution of revenue from a single randomly selected tune-up is given by

x	40	45	50
$p(x)$	.2	.3	.5

with $\mu = 46.5$, $\sigma^2 = 15.25$ (5.2)

Suppose on a particular day only two servicing jobs involve tune-ups. Let $X_1 =$ the revenue from the first tune-up, $X_2 =$ the revenue from the second, and suppose that X_1 and X_2 are independent, each having probability distribution shown in (5.2) [so that X_1 and X_2 constitute a random sample from the distribution (5.2)]. Table 5.1 lists possible (x_1, x_2) pairs, the probability of each (computed using (5.2) and the assumption of independence), and the resulting $\bar{x}$ and s^2 values. Now to obtain the probability distribution of $\bar{X}$, the sample average revenue per tune-up, we must consider each possible value $\bar{x}$ and compute its probability. For example, $\bar{x} = 45$ occurs three times in the table with probabilities .10, .09, and .10, so

$$p_{\bar{X}}(45) = P(\bar{X} = 45) = .10 + .09 + .10 = .29$$

Table 5.1 Outcomes, probabilities, and values of $\bar{x}$ and s^2 For Example 5.19

x_1	x_2	$p(x_1, x_2)$	$\bar{x}$	s^2
40	40	.04	40	0
40	45	.06	42.5	12.5
40	50	.10	45	50
45	40	.06	42.5	12.5
45	45	.09	45	0
45	50	.15	47.5	12.5
50	40	.10	45	50
50	45	.15	47.5	12.5
50	50	.25	50	0

Similarly,

$$p_{S^2}(50) = P(S^2 = 50) = P(X_1 = 40, X_2 = 50 \text{ or } X_1 = 50, X_2 = 40)$$
$$= .10 + .10 = .20$$

The complete sampling distributions of $\bar{X}$ and S^2 appear in (5.3) and (5.4).

$\bar{x}$	40	42.5	45	47.5	50
$p_{\bar{X}}(\bar{x})$	.04	.12	.29	.30	.25

(5.3)

s^2	0	12.5	50
$p_{S^2}(s^2)$	.38	.42	.20

(5.4)

Figure 5.6 pictures a probability histogram for both the original distribution (5.2) and the $\bar{X}$ distribution (5.3). Figure 5.6 suggests first that the mean (expected value) of the $\bar{X}$ distribution is equal to the mean 46.5 of the original distribution, since both histograms appear to be centered at the same place.

Figure 5.6 Probability histograms for the underlying distribution and $\bar{X}$ distribution in Example 5.19

From (5.3),

$$\mu_{\bar{X}} = E(\bar{X}) = \Sigma \bar{x} p_{\bar{X}}(\bar{x}) = (40)(.04) + \cdots + (50)(.25) = 46.5 = \mu$$

Second, it appears that the $\bar{X}$ distribution has smaller spread (variability) than the original distribution, since probability mass has moved in toward the mean. Again from (5.3),

$$\sigma_{\bar{X}}^2 = V(\bar{X}) = \Sigma \bar{x}^2 \cdot p_{\bar{X}}(\bar{x}) - \mu_{\bar{X}}^2$$
$$= (40)^2(.04) + \cdots + (50)^2(.25) - (46.5)^2$$
$$= 7.625 = \frac{(15.25)}{2} = \frac{\sigma^2}{2}$$

The variance of $\bar{X}$ is precisely half that of the original variance.

The mean value of S^2 is

$$\mu_{S^2} = E(S^2) = \Sigma s^2 \cdot p_{S^2}(s^2)$$
$$= (0)(.38) + (12.5)(.42) + (50)(.20) = 15.25 = \sigma^2$$

That is, the $\bar{X}$ sampling distribution is centered at the population mean μ and the S^2 sampling distribution is centered at the population variance σ^2.

If four tune-ups had been done on the day of interest, the sample average revenue $\bar{X}$ would be based on a random sample of four X_i's, each having distribution (5.2). More calculation eventually yields the pmf of $\bar{X}$ for $n = 4$ as

$\bar{x}$	40	41.25	42.5	43.75	45	46.25	47.5	48.75	50
$p_{\bar{X}}(\bar{x})$	.0016	.0096	.0376	.0936	.1761	.2340	.2350	.1500	.0625

From this, for $n = 4$, $\mu_{\bar{X}} = 46.50 = \mu$ and $\sigma_{\bar{X}}^2 = 3.8125 = \sigma^2/4$. A probability histogram of this pmf appears in Figure 5.7.

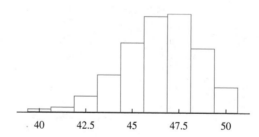

Figure 5.7 Probability histogram for $\bar{X}$ based on $n = 4$ in Example 5.19   ■

Example 5.19 should suggest first of all that the computation of $p_{\bar{X}}(\bar{x})$ and $p_{S^2}(s^2)$ can be tedious. If the original distribution (5.2) had allowed for more than three possible values 40, 45, and 50, then even for $n = 2$ the computations would have been more involved. The example should also suggest, however, that there are some general relationships between $E(\bar{X})$, $V(\bar{X})$, $E(S^2)$, and the mean μ and variance σ^2 of the original distribution. These are stated in the next section. Now consider an example in which the random sample is drawn from a continuous distribution.

Example 5.20   The time that it takes to serve a customer at the cash register in a minimarket is a random variable having an exponential distribution with parameter λ. Suppose X_1 and X_2 are service times for two different customers, assumed independent of each other. Consider the total service time $T_o = X_1 + X_2$ for the two customers, also a statistic. The cdf of T_o is, for $t \geq 0$,

$$F_{T_o}(t) = P(X_1 + X_2 \leq t) = \iint\limits_{\{(x_1, x_2) : x_1 + x_2 \leq t\}} f(x_1, x_2) \, dx_1 \, dx_2$$

$$= \int_0^t \int_0^{t-x_1} \lambda e^{-\lambda x_1} \cdot \lambda e^{-\lambda x_2} \, dx_2 \, dx_1 = \int_0^t [\lambda e^{-\lambda x_1} - \lambda e^{-\lambda t}] \, dx_1$$

$$= 1 - e^{-\lambda t} - \lambda t e^{-\lambda t}$$

The region of integration is pictured in Figure 5.8.

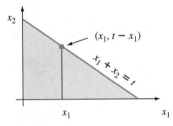

Figure 5.8 Region of integration to obtain cdf of T_o in Example 5.20 ■

The *pdf* of T_o is obtained by differentiating $F_{T_o}(t)$:

$$f_{T_o}(t) = \begin{cases} \lambda^2 t e^{-\lambda t} & t \geq 0 \\ 0 & t < 0 \end{cases} \tag{5.5}$$

This is a gamma pdf ($\alpha = 2$ and $\beta = 1/\lambda$). The pdf of $\overline{X} = T_o/2$ is obtained from the relation $\{\overline{X} \leq \bar{x}\}$ iff $\{T_o \leq 2\bar{x}\}$ as

$$f_{\overline{X}}(\bar{x}) \begin{cases} 4\lambda^2 \bar{x} e^{-2\lambda\bar{x}} & \bar{x} \geq 0 \\ 0 & \bar{x} < 0 \end{cases} \tag{5.6}$$

The mean and variance of the underlying exponential distribution are $\mu = 1/\lambda$ and $\sigma^2 = 1/\lambda^2$. From Expressions (5.5) and (5.6), it can be verified that $E(\overline{X}) = 1/\lambda$, $V(\overline{X}) = 1/(2\lambda^2)$, $E(T_o) = 2/\lambda$, and $V(T_o) = 2/\lambda^2$. These results again suggest some general relationships between means and variances of $\overline{X}$, T_o, and the underlying distribution.

Simulation Experiments

The second method of obtaining information about a statistic's sampling distribution is to perform a simulation experiment. This method is usually used when a derivation via probability rules is too difficult or complicated to be carried out. Such an experiment is virtually always done with the aid of a computer. The following characteristics of an experiment must be specified:

1. The statistic of interest ($\overline{X}$, S, a particular trimmed mean, etc.)
2. The population distribution (normal with $\mu = 100$ and $\sigma = 15$, uniform with lower limit $A = 5$ and upper limit $B = 10$, etc.)
3. The sample size n (e.g., $n = 10$ or $n = 50$)
4. The number of replications k (e.g., $k = 500$)

Then use a computer to obtain k different random samples, each of size n, from the designated population distribution. For each such sample, calculate the value of the statistic and construct a histogram of the k calculated values. This histogram gives the *approximate* sampling distribution of the statistic. The larger the value of k, the better the approximation will tend to be (the actual sampling distribution emerges as $k \to \infty$). In practice, $k = 500$ or 1000 is usually enough if the statistic is "fairly simple."

Example 5.21   The population distribution for our first simulation study is normal with $\mu = 8.25$ and $\sigma = .75$ as pictured in Figure 5.9.

Figure 5.9 Normal distribution, with $\mu = 8.25$ and $\sigma = .75$

We actually performed four different experiments, with 500 replications for each one. In the first experiment, 500 samples of $n = 5$ observations each were generated using MINITAB, and the sample sizes for the other three were $n = 10$, $n = 20$, and $n = 30$, respectively. The sample mean was calculated for each sample, and the resulting histograms of $\bar{x}$ values appear in Figure 5.10.

The first thing to notice about the histograms is their shape. To a reasonable approximation, each of the four looks like a normal curve. The resemblance would be even more striking if each histogram had been based on many more than 500 $\bar{x}$ values. Second, each histogram is centered approximately at 8.25, the mean of the population being sampled. Had the histograms been based on an unending sequence of $\bar{x}$ values, their centers would have been exactly the population mean, 8.25.

The final aspect of the histograms to note is their spread relative to one another. The smaller the value of n, the greater the extent to which the sampling distribution spreads out about the mean value. This is why the histograms for $n = 20$ and $n = 30$

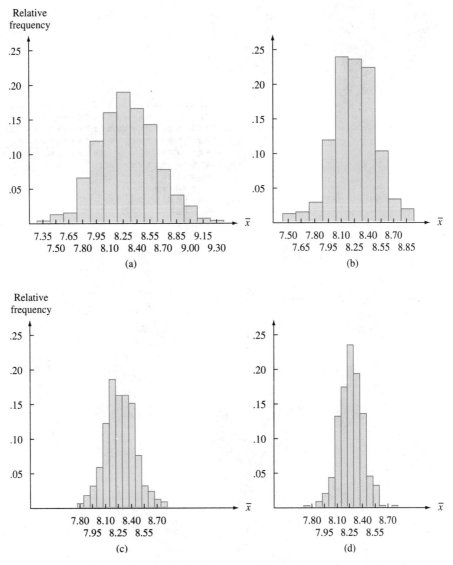

Figure 5.10 Sample histograms for $\bar{x}$ based on 500 samples, each consisting of n observations: (a) $n = 5$, (b) $n = 10$, (c) $n = 20$, and (d) $n = 30$

are based on narrower class intervals than those for the two smaller sample sizes. For the larger sample sizes, most of the $\bar{x}$ values are quite close to 8.25. This is the effect of averaging. When n is small, a single unusual x value can result in an $\bar{x}$ value far from the center. With a larger sample size, any unusual x values, when averaged in with the other sample values, still tend to yield an $\bar{x}$ value close to μ. Combining these insights yields a result that should appeal to your intuition: **$\overline{X}$ based on a large n tends to be closer to μ than does $\overline{X}$ based on a small n.**   ■

Example 5.22   Now consider a simulation experiment in which the population distribution is highly skewed as shown in Figure 5.11.

```
MIDDLE OF  NUMBER OF
INTERVAL   OBSERVATIONS
  1625.0     6    ******
  1675.0    49    *************************************************
  1725.0    24    ************************
  1775.0    30    ******************************
  1825.0    48    ************************************************
  1875.0    29    *****************************
  1925.0    15    ***************
  1975.0     8    ********
  2025.0    11    ***********
  2075.0     9    *********
  2125.0     6    ******
  2175.0     6    ******
  2225.0     3    ***
  2275.0     2    **
  2325.0     0
  2375.0     0
  2425.0     1    *
  2475.0     1    *
  2525.0     1    *
  2575.0     0
  2625.0     1    *
  2675.0     0
  2725.0     1    *
  2775.0     1    *
  2825.0     0
  2875.0     0
  2925.0     1    *
  2975.0     0
  3025.0     0
  3075.0     0
  3125.0     0
  3175.0     0
  3225.0     0
  3275.0     0
  3325.0     0
  3375.0     0
  3425.0     0
  3475.0     0
  3525.0     0
  3575.0     0
  3625.0     0
  3675.0     0
  3725.0     0
  3775.0     1    *
```

Figure 5.11 The population distribution for Example 5.22 ($\mu = 1864$)

Again, the statistic of interest is the sample mean $\overline{X}$, and the same four sample sizes as in the previous example were utilized here. The resulting histograms of $\bar{x}$ values appear in Figure 5.12.

Unlike the normal population case, these histograms all differ in shape. In particular, the histograms become progressively less skewed as the sample size n

```
MIDDLE OF  NUMBER OF
INTERVAL   OBSERVATIONS
  1685.0    0
  1700.0    3   ***
  1715.0    5   *****
  1730.0   17   *****************
  1745.0   24   ************************
  1760.0   29   *****************************
  1775.0   31   *******************************
  1790.0   22   **********************
  1805.0   41   *****************************************
  1820.0   38   **************************************
  1835.0   27   ***************************
  1850.0   32   ********************************
  1865.0   34   **********************************
  1880.0   34   **********************************
  1895.0   28   ****************************
  1910.0   20   ********************
  1925.0   22   **********************
  1940.0   15   ***************
  1955.0   15   ***************
  1970.0   10   **********
  1985.0    8   ********
  2000.0    8   ********
  2015.0    4   ****
  2030.0    6   ******
  2045.0    4   ****
  2060.0    2   **
  2075.0    0
  2090.0    4   ****
  2105.0    2   **
  2120.0    2   **
  2135.0    4   ****
  2150.0    0
  2165.0    0
  2180.0    0
  2195.0    0
  2210.0    0
  2225.0    1   *
  2240.0    1   *
  2255.0    3   ***
  2270.0    2   **
  2285.0    0
  2300.0    0
  2315.0    1   *
  2330.0    0

            (a)
```

EACH * REPRESENTS 2 OBSERVATIONS

```
MIDDLE OF  NUMBER OF
INTERVAL   OBSERVATIONS
  1685.0    0
  1700.0    1   *
  1715.0    1   *
  1730.0    3   **
  1745.0    4   **
  1760.0    8   ****
  1775.0   18   *********
  1790.0   31   ***************
  1805.0   49   *************************
  1820.0   50   *************************
  1835.0   53   **************************
  1850.0   43   *********************
  1865.0   49   *************************
  1880.0   44   **********************
  1895.0   25   ************
  1910.0   25   ************
  1925.0   20   **********
  1940.0   20   **********
  1955.0    9   *****
  1970.0    8   ****
  1985.0    8   ****
  2000.0    7   ****
  2015.0    6   ***
  2030.0    2   *
  2045.0    6   ***
  2060.0    1   *
  2075.0    2   *
  2090.0    2   *
  2105.0    1   *
  2120.0    1   *
  2135.0    1   *
  2150.0    0
  2165.0    0
  2180.0    1   *
  2195.0    0
  2210.0    0
  2225.0    0
  2240.0    0
  2255.0    1   *

            (b)
```

Figure 5.12 MINITAB histograms of 500 $\bar{x}$ values, each based on n observations:
(a) $n = 5$, and (b) $n = 10$

```
EACH * REPRESENTS 2 OBSERVATIONS

MIDDLE OF   NUMBER OF
INTERVAL    OBSERVATIONS
  1685.0        0
  1700.0        0
  1715.0        0
  1730.0        0
  1745.0        1    *
  1760.0        3    **
  1775.0       12    ******
  1790.0       22    ***********
  1805.0       30    ***************
  1820.0       52    **************************
  1835.0       67    **********************************
  1850.0       65    *********************************
  1865.0       58    *****************************
  1880.0       38    *******************
  1895.0       34    *****************
  1910.0       44    **********************
  1925.0       20    **********
  1940.0       14    *******
  1955.0       15    ********
  1970.0       13    *******
  1985.0        3    **
  2000.0        3    **
  2015.0        2    *
  2030.0        2    *
  2045.0        1    *
  2060.0        1    *
```

(c)

```
EACH * REPRESENTS 2 OBSERVATIONS

MIDDLE OF   NUMBER OF
INTERVAL    OBSERVATIONS
  1685.0        0
  1700.0        0
  1715.0        0
  1730.0        0
  1745.0        0
  1760.0        1    *
  1775.0        1    *
  1780.0       12    ******
  1805.0       33    *****************
  1820.0       44    **********************
  1835.0       77    **************************************
  1850.0       77    **************************************
  1865.0       80    ****************************************
  1880.0       53    **************************
  1895.0       45    ***********************
  1910.0       30    ***************
  1925.0       22    ***********
  1940.0       13    *******
  1955.0        3    **
  1970.0        4    **
  1985.0        4    **
  2000.0        1    *
```

(d)

Figure 5.12 (continued) MINITAB histograms of 500 $\bar{x}$ values, each based on n observations: (c) $n = 20$, and (d) $n = 30$

increases. The averages of the 500 $\bar{x}$ values for the five different sample sizes are all quite close to the population mean $\mu = 1864$. If each histogram had been based on an unending sequence of $\bar{x}$ values rather than just 500 values, each histogram would have been centered at exactly 1864. Thus, different values of n change the shape but not the center of $\bar{x}$'s sampling distribution. Comparison of the four $\bar{x}$ histograms in Figure 5.12 also shows that as n increases, the extent to which the histogram spreads out about its center decreases markedly. Increasing n both changes the shape of the distribution and squeezes it in toward the center, so that $\bar{X}$ based on a large n is a less variable quantity than is $\bar{X}$ based on a small value of n. ∎

Exercises Section 5.3 (37–43)

37. A convenience store sells three different brands of yogurt in 8-oz containers. Of all customers purchasing a single container, 50% purchase one containing 160 calories, 30% purchase the 200-calorie yogurt, and the other 20% buy the container having 250 calories. Let X_1 and X_2 denote the numbers of calories in containers purchased by two independently selected customers.

a. Determine the sampling distribution of $\bar{X}$, calculate $E(\bar{X})$, and compare to μ.

b. Determine the sampling distribution of the sample variance S^2, calculate $E(S^2)$, and compare to σ^2.

38. There are two traffic lights on my way to work. Let X_1 be the number of lights at which I must stop and suppose that the distribution of X_1 is as follows:

x_1	0	1	2
$p(x_1)$	.2	.5	.3

$\mu = 1.1, \sigma^2 = .49$

Let X_2 be the number of lights at which I must stop on the way home, clearly independent of X_1. Assume that X_2 has the same distribution as X_1, so that X_1, X_2 is a random sample of size $n = 2$.

a. Let $T_o = X_1 + X_2$ and determine the probability distribution of T_o.

b. Calculate μ_{T_o}. How does it relate to μ, the population mean?

c. Calculate $\sigma_{T_o}^2$. How does it relate to σ^2, the population variance?

39. It is known that 80% of all brand X disk drives work in a satisfactory manner throughout the warranty period (are "successes"). Suppose that $n = 10$ drives are randomly selected and let $X =$ the number of successes

in the sample. The statistic X/n is the sample proportion (fraction) of successes. Obtain the sampling distribution of this statistic. [HINT: One possible value of X/n is .3, corresponding to $X = 3$. What is the probability of this value (what kind of random variable is X)?]

40. A box contains ten sealed envelopes numbered 1, . . . , 10. The first five contain no money, the next three each contain \$5, and there is a \$10 bill in each of the last two. A sample of size 3 is selected *with* replacement (so we have a random sample), and you get the largest amount in any of the envelopes selected. If X_1, X_2, and X_3 denote the amounts in the selected envelopes, the statistic of interest is $M =$ the maximum of X_1, X_2, and X_3.

a. Obtain the probability distribution of this statistic.

b. Describe how you would carry out a simulation experiment to compare the distributions of M for various sample sizes. How would you guess the distribution would change as n increases?

41. Let X be the number of packages being mailed by a randomly selected customer at a certain shipping facility. Suppose the distribution of X is as follows:

x	1	2	3	4
$p(x)$	.4	.3	.2	.1

a. Consider a random sample of size $n = 2$ (two customers) and let $\bar{X}$ be the sample mean number of packages shipped. Obtain the probability distribution of $\bar{X}$.

b. Refer to part (a) and calculate $P(\bar{X} \leq 2.5)$.

c. Again consider a random sample of size $n = 2$, but now focus on the statistic $R =$ the sample range (difference between the largest and smallest values in the sample). Obtain the distribution of R. [HINT: Calculate the value of R for each outcome and use the probabilities from part (a).]

d. If a random sample of size $n = 4$ is selected, what is $P(\overline{X} \leq 1.5)$? (HINT: You should not have to list all possible outcomes, only those for which $\bar{x} \leq 1.5$.)

42. A company maintains three offices in a certain region, each staffed by two employees. Information concerning yearly salaries (1000s of dollars) is as follows:

Office	1	1	2	2	3	3
Employee	1	2	3	4	5	6
Salary	19.7	23.6	20.2	23.6	15.8	19.7

a. Suppose two of these employees are randomly selected from among the six (without replacement).

Determine the sampling distribution of the sample mean salary $\overline{X}$.

b. Suppose one of the three offices is randomly selected and let X_1 and X_2 denote the salaries of the two employees. Determine the sampling distribution of $\overline{X}$.

c. How does $E(\overline{X})$ from parts (a) and (b) compare to the population mean salary μ?

43. Suppose the amount of liquid dispensed by a certain machine is uniformly distributed with lower limit $A = 8$ oz and upper limit $B = 10$ oz. Describe how you would carry out simulation experiments to compare the sampling distribution of the (sample) fourth spread for sample sizes $n = 5$, 10, 20, and 30.

5.4 The Distribution of the Sample Mean

The importance of the sample mean $\overline{X}$ springs from its use in drawing conclusions about the population mean μ. Some of the most frequently used inferential procedures are based on properties of $\overline{X}$'s sampling distribution. A preview of these properties appeared in the calculations and simulation experiments of the previous section, where we noted relationships between $E(\overline{X})$ and μ and also between $V(\overline{X})$, σ^2, and n.

Proposition

Let $X_1, X_2, \ldots, X_n$ be a random sample from a distribution with mean value μ and standard deviation σ. Then

1. $E(\overline{X}) = \mu_{\overline{X}} = \mu$
2. $V(\overline{X}) = \sigma_{\overline{X}}^2 = \sigma^2/n$ and $\sigma_{\overline{X}} = \sigma/\sqrt{n}$

In addition, with $T_o = X_1 + \cdots + X_n$ (the sample total), $E(T_o) = n\mu$, $V(T_o) = n\sigma^2$, and $\sigma_{T_o} = \sqrt{n}\,\sigma$.

Proofs of these results are given in the next section. According to Result 1, the sampling (i.e., probability) distribution of $\overline{X}$ is centered precisely at the mean of the population from which the sample has been selected. Result 2 shows that the $\overline{X}$ distribution becomes more concentrated about μ as the sample size n increases. In marked contrast, the distribution of T_o becomes more spread out as n increases. Averaging moves probability in toward the middle, whereas totaling spreads probability out over a wider and wider range of values.

Example 5.23

In a notched tensile fatigue test on a titanium specimen, the expected number of cycles to first acoustic emission (used to indicate crack initiation) is $\mu = 28{,}000$, and the standard deviation of the number of cycles is $\sigma = 5000$. Let $X_1, X_2, \ldots, X_{25}$ be a random sample of size 25, where each X_i is the number of cycles on a different randomly selected specimen. Then the expected value of the sample mean number of cycles until first emission is $E(\overline{X}) = \mu = 28{,}000$, and the expected total number of cycles for the 25 specimens is $E(T_o) = n\mu = 25(28{,}000) = 700{,}000$. The standard deviations of $\overline{X}$ and T_o are

$$\sigma_{\overline{X}} = \sigma/\sqrt{n} = \frac{5000}{\sqrt{25}} = 1000$$

$$\sigma_{T_o} = \sqrt{n}\sigma = \sqrt{25}(5000) = 25{,}000$$

If the sample size increases to $n = 100$, $E(\overline{X})$ is unchanged, but $\sigma_{\overline{X}} = 500$, half of its previous value (the sample size must be quadrupled to halve the standard deviation of $\overline{X}$). ∎

The Case of a Normal Population Distribution

Looking back to the simulation experiment of Example 5.21, we see that when the population distribution is normal, each histogram of $\bar{x}$ values is well approximated by a normal curve. Here is the precise result.

Proposition

> Let $X_1, X_2, \ldots, X_n$ be a random sample from a *normal* distribution with mean μ and standard deviation σ. Then for *any n*, $\overline{X}$ is normally distributed (with mean μ and standard deviation $\sigma/\sqrt{n}$), as is T_o (with mean $n\mu$ and standard deviation $\sqrt{n}\sigma$).*

We know everything there is to know about the $\overline{X}$ and T_o distributions when the population distribution is normal. In particular, probabilities such as $P(a \le \overline{X} \le b)$ and $P(c \le T_o \le d)$ can be obtained simply by standardizing. Figure 5.13 illustrates the proposition.

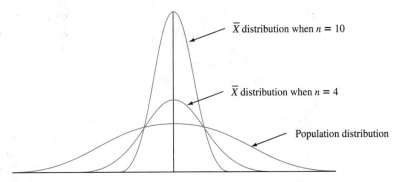

Figure 5.13 A normal population distribution and $\overline{X}$ sampling distributions

---

*A proof of the result for T_o when $n = 2$ is possible using the method in Example 5.20, but the details are messy. The general result is usually proved using a theoretical tool called a *moment generating function*. One of the chapter references can be consulted for more information.

Example 5.24   The time that it takes a randomly selected rat of a certain subspecies to find its way through a maze is a normally distributed rv with $\mu = 1.5$ min and $\sigma = .35$ min. Suppose five rats are selected and let $X_1, \ldots, X_5$ denote their times in the maze. Assuming the X_i's to be a random sample from this normal distribution, what is the probability that the total time $T_o = X_1 + \cdots + X_5$ for the five is between 6 and 8 min? By the proposition, T_o has a normal distribution with $\mu_{T_o} = n\mu = 5(1.5) = 7.5$ and variance $\sigma^2_{T_o} = n\sigma^2 = 5(.1225) = .6125$, so $\sigma_{T_o} = .783$. To standardize T_o, subtract μ_{T_o} and divide by σ_{T_o}:

$$P(6 \leq T_o \leq 8) = P\left(\frac{6 - 7.5}{.783} \leq Z \leq \frac{8 - 7.5}{.783}\right)$$

$$= P(-1.92 \leq Z \leq .64) = \Phi(.64) - \Phi(-1.92) = .7115$$

Determination of the probability that the sample average time $\bar{X}$ (a normally distributed variable) is at most 2.0 min requires $\mu_{\bar{X}} = \mu = 1.5$ and $\sigma_{\bar{X}} = \sigma/\sqrt{n} = .35/\sqrt{5} = .1565$. Then

$$P(\bar{X} \leq 2.0) = P\left(Z \leq \frac{2.0 - 1.5}{.1565}\right) = P(Z \leq 3.19) = \Phi(3.19) = .9993 \quad \blacksquare$$

The Central Limit Theorem

When the X_i's are normally distributed, so is $\bar{X}$ for every sample size n. The simulation experiment of Example 5.22 suggests that even when the population distribution is highly nonnormal, averaging produces a distribution more bell-shaped than the one being sampled. A reasonable conjecture is that if n is large, a suitable normal curve will approximate the actual distribution of $\bar{X}$. The formal statement of this result is the most important theorem of probability.

Theorem

> **The Central Limit Theorem (CLT)**
>
> Let $X_1, X_2, \ldots, X_n$ be a random sample from a distribution with mean μ and variance σ^2. Then if n is sufficiently large, $\bar{X}$ has approximately a normal distribution with $\mu_{\bar{X}} = \mu$ and $\sigma^2_{\bar{X}} = \sigma^2/n$, and T_o also has approximately a normal distribution with $\mu_{T_o} = n\mu$, $\sigma^2_{T_o} = n\sigma^2$. The larger the value of n, the better the approximation.

Figure 5.14 gives an illustration of the Central Limit Theorem. According to the CLT, when n is large and we wish to calculate a probability such as $P(a \leq \bar{X} \leq b)$, we need only "pretend" that $\bar{X}$ is normal, standardize it, and use the normal table. The resulting answer will be approximately correct. The exact answer could be obtained only by first finding the distribution of $\bar{X}$, so the CLT provides a truly impressive shortcut. The proof of the theorem involves much advanced mathematics.

Example 5.25   The nicotine content in a single cigarette of a particular brand is a random variable with mean $\mu = .8$ mg and standard deviation .1 mg. If an individual smokes five

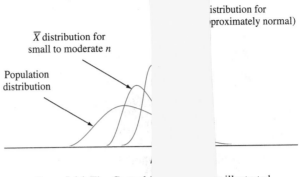

$\overline{X}$ distribution for
small to moderate n

Population
distribution

istribution for
proximately normal)

Figure 5.14 The Central   illustrated

packs of these cigarettes per week, what   bility that the total amount of
nicotine consumed in a week is at least   .

Notice that there is no assumption of normality here, but five packs consist
of $n = 100$ cigarettes. Letting $X_1, \ldots, X_{100}$ be the nicotine contents of the 100
cigarettes, the X_i's constitute a random sample from a distribution with $\mu = .8$
and $\sigma = .1$. Assuming that $n = 100$ is sufficiently large for the CLT to apply, T_o
is approximately normal with mean $(100)(.8) = 80$ and standard deviation
$(.1) \cdot \sqrt{100} = 1$. Therefore, $(T_o - 80)/1$ is *approximately* standard normal, so that

$$P(T_o \geq 82) \approx P[Z \geq (82 - 80)/1] = P(Z \geq 2) = .0228$$

Thus, the probability that at least 82 mg is consumed is approximately .0228. ■

Example 5.26 When a batch of a certain chemical product is prepared, the amount of a particular
impurity in the batch is a random variable with mean value 4.0 g and standard
deviation 1.5 g. If 50 batches are independently prepared, what is the (approximate)
probability that the sample average amount of impurity $\overline{X}$ is between 3.5 and 3.8 g?
According to the rule of thumb to be stated shortly, $n = 50$ is large enough for the
CLT to be applicable. $\overline{X}$ then has approximately a normal distribution with mean
value $\mu_{\overline{X}} = 4.0$ and $\sigma_{\overline{X}} = 1.5/\sqrt{50} = .2121$, so

$$P(3.5 \leq \overline{X} \leq 3.8) \approx P\left(\frac{3.5 - 4.0}{.2121} \leq Z \leq \frac{3.8 - 4.0}{.2121}\right)$$

$$= \Phi(-.94) - \Phi(-2.36) = .1645$$ ■

Example 5.27 A certain consumer organization customarily reports the number of major defects for
each new automobile that it tests. Suppose the number of such defects for a certain
model is a random variable with mean value 3.2 and standard deviation 2.4. Among
100 randomly selected cars of this model, how likely is it that the sample average
number of major defects exceeds 4? Let X_i denote the number of major defects for
the ith car in the random sample. Notice that X_i is a discrete rv, but the CLT is
applicable whether the variable of interest is discrete or continuous. Also, although
the fact that the standard deviation of this nonnegative variable is quite large relative
to the mean value suggests that its distribution is positively skewed, the large sample

size implies that $\overline{X}$ does have approximately a normal distribution. Using $\mu_{\overline{X}} = 3.2$ and $\sigma_{\overline{X}} = .24$,

$$P(\overline{X} > 4) \approx P\!\left(Z > \frac{4 - 3.2}{.24}\right) = 1 - \Phi(3.33) = .0004 \qquad \blacksquare$$

The CLT provides insight into why many random variables have probability distributions that are approximately normal. For example, the measurement error in a scientific experiment can be thought of as the sum of a number of underlying perturbations and errors of small magnitude.

Although the usefulness of the CLT for inference will soon be apparent, the intuitive content of the result gives many beginning students difficulty. Again looking back to Figure 5.6, the probability histogram on the left is a picture of the distribution being sampled. It is discrete and quite skewed, so does not look at all like a normal distribution. The distribution of $\overline{X}$ for $n = 2$ starts to exhibit some symmetry, and this is even more pronounced for $n = 4$ in Figure 5.7. Figure 5.15 contains the probability distribution of $\overline{X}$ for $n = 8$, as well as a probability histogram for this distribution. With $\mu_{\overline{X}} = \mu = 46.5$ and $\sigma_{\overline{X}} = \sigma/\sqrt{n} = 3.905/\sqrt{8} = 1.38$, if we fit a normal curve with this mean and standard deviation through the histogram of $\overline{X}$, the areas of rectangles in the probability histogram are reasonably well approximated by the normal curve

$\overline{x}$	40	40.625	41.25	41.875	42.5	43.125
$p(\overline{x})$	.0000	.0000	.0003	.0012	.0038	.0112
$\overline{x}$	43.75	44.375	45	45.625	46.25	46.875
$p(\overline{x})$	.0274	.0556	.0954	.1378	.1704	.1746
$\overline{x}$	47.5	48.125	48.75	49.375	50	
$p(\overline{x})$	.1474	.0998	.0519	.0188	.0039	

(a)

(b)

Figure 5.15 (a) Probability distribution of $\overline{X}$ for $n = 8$, and (b) probability histogram and normal approximation to the distribution of $\overline{X}$, when the original distribution is as in Example 5.19

areas, at least in the central part of the distribution. The picture for T_o is similar except that the horizontal scale is much more spread out, with T_o ranging between 320 ($\bar{x} = 40$) and 400 ($\bar{x} = 50$).

A practical difficulty in applying the CLT is in knowing when n is sufficiently large. The problem is that the accuracy of the approximation for a particular n depends on the shape of the original underlying distribution being sampled. If the underlying distribution is close to bell-shaped, then the approximation will be good even for a small n, while if it is far from bell-shaped then a large n will be required. We will use the following rule of thumb, which is frequently somewhat conservative.

Rule of Thumb

> If $n > 30$, the CLT can be used.

Other Applications of the Central Limit Theorem

The CLT can be used to justify the normal approximation to the binomial distribution discussed in Chapter 4. Recall that a binomial variable X is the number of successes in a binomial experiment consisting of n independent success/failure trials with $p = P(S)$ for any particular trial. Define new rv's $X_1, X_2, \ldots, X_n$ by

$$X_i = \begin{cases} 1 & \text{if the } i\text{th trial results in a success} \\ 0 & \text{if the } i\text{th trial results in a failure} \end{cases} \quad (i = 1, \ldots, n)$$

Because the trials are independent and $P(S)$ is constant from trial to trial, the X_i's are iid (a random sample from a Bernoulli distribution). The CLT then implies that if n is sufficiently large, both the sum and the average of the X_i's have approximately normal distributions. When the X_i's are summed, a 1 is added in for every S that occurs and a 0 for every F, so $X_1 + \cdots + X_n = X$. The sample mean of the X_i's is X/n, the sample proportion of successes. That is, both X and X/n are approximately normal when n is large. The necessary sample size for this approximation depends on the value of p: When p is close to .5, the distribution of each X_i is reasonably symmetric (Figure 5.16), whereas the distribution is quite skewed when p is near 0 or 1. Using the approximation only if both $np \geq 5$ and $n(1 - p) \geq 5$ ensures that n is large enough to overcome any skewness in the underlying Bernoulli distribution.

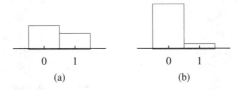

Figure 5.16 Two Bernoulli distributions: (a) $p = .4$ (reasonably symmetric); (b) $p = .1$ (very skewed)

Recall from Section 4.5 that an rv X has a lognormal distribution if $\ln(X)$ has a normal distribution.

Proposition

> Let $X_1, X_2, \ldots, X_n$ be a random sample from a distribution for which only positive values are possible [$P(X_i > 0) = 1$]. Then if n is sufficiently large, the product $Y = X_1 X_2 \cdots \cdots X_n$ has approximately a lognormal distribution.

To verify this, note that

$$\ln(Y) = \ln(X_1) + \ln(X_2) + \cdots + \ln(X_n)$$

Since $\ln(Y)$ is a sum of independent and identically distributed rv's [the $\ln(X_i)$'s], it is approximately normal when n is large, so Y itself has approximately a lognormal distribution. As an example of the applicability of this result, Bury (*Statistical Models in Applied Science,* Wiley, p. 590) argues that the damage process in plastic flow and crack propagation is a multiplicative process, so that variables such as percentage elongation and rupture strength have approximately lognormal distributions.

Exercises   Section 5.4 (44–55)

44. The inside diameter of a randomly selected piston ring is a random variable with mean value 12 cm and standard deviation .04 cm.

 a. If $\bar{X}$ is the sample mean diameter for a random sample of $n = 16$ rings, where is the sampling distribution of $\bar{X}$ centered, and what is the standard deviation of the $\bar{X}$ distribution?

 b. Answer the questions posed in part (a) for a sample size of $n = 64$ rings.

 c. For which of the two random samples, the one of part (a) or the one of part (b), is $\bar{X}$ more likely to be within .01 cm of 12 cm? Explain your reasoning.

45. Refer to Exercise 44 and suppose the distribution of diameter is normal.

 a. Calculate $P(11.99 \le \bar{X} \le 12.01)$ when $n = 16$.

 b. How likely is it that the sample mean diameter exceeds 12.01 when $n = 25$?

46. Let $X_1, X_2, \ldots, X_{100}$ denote the actual net weights of 100 randomly selected 50-lb bags of fertilizer.

 a. If the expected weight of each bag is 50 and the variance is 1, calculate $P(49.75 \le \bar{X} \le 50.25)$ (approximately) using the CLT.

 b. If the expected weight is 49.8 lb rather than 50 lb so that on average bags are underfilled, calculate $P(49.75 \le \bar{X} \le 50.25)$.

47. There are 40 students in an elementary statistics class. On the basis of years of experience, the instructor knows that the time needed to grade a randomly chosen first examination paper is a random variable with an expected value of 6 min and a standard deviation of 6 min.

 a. If grading times are independent and the instructor begins grading at 6:50 P.M. and grades continuously, what is the (approximate) probability that he is through grading before the 11:00 P.M. TV news begins?

 b. If the sports report begins at 11:10, what is the probability that he misses part of the report if he waits until grading is done before turning on the TV?

48. The breaking strength of a rivet has a mean value of 10,000 psi and a standard deviation of 500 psi.

 a. What is the probability that the sample mean breaking strength for a random sample of 40 rivets is between 9900 and 10,200?

 b. If the sample size had been 15 rather than 40, could the probability requested in part (a) be calculated from the given information?

49. The time taken by a randomly selected applicant for a mortgage to fill out a certain form has a normal distribution with mean value 10 min and standard

deviation 2 min. If five individuals fill out a form on one day and six on another, what is the probability that the sample average amount of time taken on each day is at most 11 min?

50. The lifetime of a certain type of battery is normally distributed with mean value 8 hours and standard deviation 1 hour. There are four batteries in a package. What lifetime value is such that the total lifetime of all batteries in a package exceeds that value for only 5% of all packages?

51. Rockwell hardness of pins of a certain type is known to have a mean value of 50 and a standard deviation of 1.5.

   a. If the distribution is normal, what is the probability that the sample mean hardness for a random sample of 9 pins is at least 52?

   b. What is the (approximate) probability that the sample mean hardness for a random sample of 40 pins is at least 52?

52. Suppose the sediment density (g/cm) of a randomly selected specimen from a certain region is normally distributed with mean 2.65 and standard deviation .85 (suggested in "Modeling Sediment and Water Column Interactions for Hydrophobic Pollutants," *Water Research*, 1984, pp. 1169–1174).

   a. If a random sample of 25 specimens is selected, what is the probability that the sample average sediment density is at most 3.00? Between 2.65 and 3.00?

   b. How large a sample size would be required to ensure that the first probability in part (a) is at least .99?

53. The first assignment in an introductory computer programming class involves running a short program. If past experience indicates that 40% of all beginning students will make no typographical errors, compute the (approximate) probability that in a class of 50 students

   a. At least 25 will make no errors. (HINT: Normal approximation to the binomial.)

   b. Between 15 and 25 (inclusive) will make no errors.

54. The number of parking tickets issued in a certain city on any given weekday has a Poisson distribution with parameter $\lambda = 50$. What is the approximate probability that

   a. Between 35 and 70 tickets are given out on a particular day? (HINT: When λ is large, a Poisson rv has approximately a normal distribution.)

   b. The total number of tickets given out during a 5-day week is between 225 and 275?

55. Suppose the distribution of the time X (in hours) spent by students at a certain university on their senior projects is gamma with parameters $\alpha = 50$ and $\beta = 2$. Because α is large, it can be shown that X has approximately a normal distribution. Use this fact to compute the probability that a randomly selected student spends at most 125 hours on the senior project.

5.5   The Distribution of a Linear Combination

The sample mean $\overline{X}$ and sample total T_o are special cases of a type of random variable that arises very frequently in statistical applications.

Definition

> Given a collection of n rv's $X_1, \ldots, X_n$ and n numerical constants $a_1, \ldots, a_n$, the rv
>
> $$Y = a_1 X_1 + \cdots + a_n X_n = \sum_{i=1}^{n} a_i X_i \qquad (5.7)$$
>
> is called a **linear combination** of the X_i's.

Taking $a_1 = a_2 = \cdots = a_n = 1$ gives $Y = X_1 + \cdots + X_n = T_o$, and $a_1 = a_2 = \cdots = a_n = 1/n$ yields $Y = \frac{1}{n}X_1 + \cdots + \frac{1}{n}X_n = \frac{1}{n}(X_1 + \cdots + X_n) = \frac{1}{n}T_o = \overline{X}$. Notice that we are not requiring the X_i's to be independent or identically distributed. All the X_i's could have different distributions and therefore different mean values and variances. We first consider the expected value and variance of a linear combination.

Proposition

Let $X_1, X_2, \ldots, X_n$ have mean values $\mu_1, \ldots, \mu_n$, respectively, and variances $\sigma_1^2, \ldots, \sigma_n^2$, respectively.

1. Whether or not the X_i's are independent,

$$E(a_1X_1 + a_2X_2 + \cdots + a_nX_n) = a_1E(X_1) + a_2E(X_2) + \cdots + a_nE(X_n)$$
$$= a_1\mu_1 + \cdots + a_n\mu_n \qquad (5.8)$$

2. If $X_1, \ldots, X_n$ are independent,

$$V(a_1X_1 + a_2X_2 + \cdots + a_nX_n) = a_1^2V(X_1) + a_2^2 V(X_2) + \cdots + a_n^2V(X_n)$$
$$= a_1^2\sigma_1^2 + \cdots + a_n^2\sigma_n^2 \qquad (5.9)$$

and

$$\sigma_{a_1X_1+ \cdots + a_nX_n} = \sqrt{a_1^2\sigma_1^2 + \cdots + a_n^2\sigma_n^2} \qquad (5.10)$$

3. For any $X_1, \ldots, X_n$,

$$V(a_1X_1 + \cdots + a_nX_n) = \sum_{i=1}^{n}\sum_{j=1}^{n} a_ia_j\mathrm{Cov}(X_i, X_j) \qquad (5.11)$$

Proofs are sketched out at the end of the section. A paraphrase of (5.8) is that the expected value of a linear combination is the same linear combination of the expected values—for example, $E(2X_1 + 5X_2) = 2\mu_1 + 5\mu_2$. The result (5.9) in Statement 2 is a special case of (5.11) in Statement 3; when the X_i's are independent, $\mathrm{Cov}(X_i, X_j) = 0$ for $i \neq j$ and $= V(X_i)$ for $i = j$ (this simplification actually occurs when the X_i's are uncorrelated, a weaker condition than independence). Specializing to the case of a random sample (X_i's iid) with $a_i = 1/n$ for every i gives $E(\overline{X}) = \mu$ and $V(\overline{X}) = \sigma^2/n$ as discussed in Section 5.4. A similar comment applies to the rules for T_o.

Example 5.28

A gas station sells three grades of gasoline: regular unleaded, extra unleaded, and super unleaded. These are priced at \$1.20, \$1.35, and \$1.50 per gallon, respectively. Let X_1, X_2, and X_3 denote the amounts of these grades purchased (gallons) on a particular day. Suppose the X_i's are independent with $\mu_1 = 1000$, $\mu_2 = 500$, $\mu_3 = 300$, $\sigma_1 = 100$, $\sigma_2 = 80$, and $\sigma_3 = 50$. The revenue from sales is $Y = 1.2X_1 + 1.35X_2 + 1.5X_3$, and

$$E(Y) = 1.2\mu_1 + 1.35\mu_2 + 1.5\mu_3 = \$2325$$
$$V(Y) = (1.2)^2\sigma_1^2 + (1.35)^2\sigma_2^2 + (1.5)^2\sigma_3^2 = 31{,}689$$
$$\sigma_Y = \sqrt{31{,}689} = \$178.01$$

∎

The Difference Between Two Random Variables

An important special case of a linear combination results from taking $n = 2$, $a_1 = 1$, and $a_2 = -1$:

$$Y = a_1X_1 + a_2X_2 = X_1 - X_2$$

We then have the following corollary to the proposition.

Corollary

> $E(X_1 - X_2) = E(X_1) - E(X_2)$ and, if X_1 and X_2 are independent, $V(X_1 - X_2) = V(X_1) + V(X_2)$

The expected value of a difference is the difference of the two expected values, but the variance of a difference between two independent variables is the *sum, not* the difference, of the two variances. There is just as much variability in $X_1 - X_2$ as in $X_1 + X_2$ (writing $X_1 - X_2 = X_1 + (-1)X_2$, $(-1)X_2$ has the same amount of variability as X_2 itself).

Example 5.29

A certain automobile manufacturer equips a particular model with either a six-cylinder engine or a four-cylinder engine. Let X_1 and X_2 be fuel efficiencies for independently and randomly selected six-cylinder and four-cylinder cars, respectively. With $\mu_1 = 22$, $\mu_2 = 26$, $\sigma_1 = 1.2$, and $\sigma_2 = 1.5$,

$$E(X_1 - X_2) = \mu_1 - \mu_2 = 22 - 26 = -4$$
$$V(X_1 - X_2) = \sigma_1^2 + \sigma_2^2 = (1.2)^2 + (1.5)^2 = 3.69$$
$$\sigma_{X_1 - X_2} = \sqrt{3.69} = 1.92$$

If we relabel so that X_1 refers to the four-cylinder car, then $E(X_1 - X_2) = 4$, but the variance of the difference is still 3.69. ∎

The Case of Normal Random Variables

When the X_i's form a random sample from a normal distribution, $\overline{X}$ and T_o are both normally distributed. Here is a more general result concerning linear combinations.

Proposition

> If $X_1, X_2, \ldots, X_n$ are independent, normally distributed rv's (with possibly different means and/or variances), then any linear combination of the X_i's also has a normal distribution. In particular, the difference $X_1 - X_2$ between two independent, normally distributed variables is itself normally distributed.

Example 5.30
(Example 5.28 continued)

The total revenue from the sale of the three grades of gasoline on a particular day was $Y = 1.2X_1 + 1.35X_2 + 1.5X_3$, and we calculated $\mu_Y = 2325$ and (assuming independence) $\sigma_Y = 178.01$. If the X_i's are normally distributed, the probability that revenue exceeds 2500 is

$$P(Y > 2500) = P(Z > \frac{2500 - 2325}{178.01})$$

$$= P(Z > .98) = 1 - \Phi(.98) = .1635 \qquad \blacksquare$$

The CLT can also be generalized so it applies to certain linear combinations. Roughly speaking, if n is large and no individual term is likely to contribute too much to the overall value, then Y has approximately a normal distribution.

Proofs for the case $n = 2$ For the result concerning expected values, suppose that X_1 and X_2 are continuous with joint pdf $f(x_1, x_2)$. Then

$$E(a_1X_1 + a_2X_2) = \int_{-\infty}^{\infty} \int_{-\infty}^{\infty} (a_1x_1 + a_2x_2)f(x_1, x_2) \, dx_1 \, dx_2$$

$$= a_1 \int_{-\infty}^{\infty} \int_{-\infty}^{\infty} x_1 f(x_1, x_2) \, dx_2 \, dx_1$$

$$+ a_2 \int_{-\infty}^{\infty} \int_{-\infty}^{\infty} x_2 f(x_1, x_2) \, dx_1 \, dx_2$$

$$= a_1 \int_{-\infty}^{\infty} x_1 f_{X_1}(x_1) \, dx_1 + a_2 \int_{-\infty}^{\infty} x_2 f_{X_2}(x_2) \, dx_2$$

$$= a_1 E(X_1) + a_2 E(X_2)$$

Summation replaces integration in the discrete case. The argument for the variance result does not require specifying whether either variable is discrete or continuous. Recalling that $V(Y) = E[(Y - \mu_Y)^2]$,

$$V(a_1X_1 + a_2X_2) = E\{[a_1X_1 + a_2X_2 - (a_1\mu_1 + a_2\mu_2)]^2\}$$
$$= E\{a_1^2(X_1 - \mu_1)^2 + a_2^2(X_2 - \mu_2)^2 + 2a_1a_2(X_1 - \mu_1)(X_2 - \mu_2)\}$$

The expression inside the braces is a linear combination of the variables $Y_1 = (X_1 - \mu_1)^2$, $Y_2 = (X_2 - \mu_2)^2$, and $Y_3 = (X_1 - \mu_1)(X_2 - \mu_2)$, so carrying the E operation through to the three terms gives $a_1^2V(X_1) + a_2^2V(X_2) + 2a_1a_2\text{Cov}(X_1, X_2)$ as required. $\blacksquare$

Exercises   Section 5.5 (56–72)

56. A shipping company handles containers in three different sizes: (1) 27 ft³ (3 × 3 × 3), (2) 125 ft³, and (3) 512 ft³. Let X_i ($i = 1, 2, 3$) denote the number of type i containers shipped during a given week. With $\mu_i = E(X_i)$ and $\sigma_i^2 = V(X_i)$, suppose that the mean values and standard deviations are as follows:

$\mu_1 = 200 \qquad \mu_2 = 250 \qquad \mu_3 = 100$

$\sigma_1 = 10 \qquad \sigma_2 = 12 \qquad \sigma_3 = 8$

a. Assuming that X_1, X_2, X_3 are independent, calculate the expected value and variance of the total volume shipped. [HINT: Volume = $27X_1 + 125X_2 + 512X_3$.]

b. Would your calculations necessarily be correct if the X_i's were not independent? Explain.

57. Let X_1, X_2, and X_3 be three independent normal rv's with expected values μ_1, μ_2, and μ_3 and variances σ_1^2, σ_2^2, and σ_3^2, respectively.

a. If $\mu_1 = \mu_2 = \mu_3 = 100$ and $\sigma_1^2 = \sigma_2^2 = \sigma_3^2 = 12$, calculate $P(X_1 + X_2 + X_3 \le 309)$ and $P(288 \le X_1 + X_2 + X_3 \le 312)$.

b. Using the μ_i's and σ_i's given in part (a), calculate $P(105 \le \bar{X})$ and $P(98 \le \bar{X} \le 102)$.

c. Using the μ_i's and σ_i's given in part (a), calculate $P(-10 \le X_1 - .5X_2 - .5X_3 \le 5)$.

d. If $\mu_1 = 90$, $\mu_2 = 100$, $\mu_3 = 110$, $\sigma_1^2 = 10$, $\sigma_2^2 = 12$, and $\sigma_3^2 = 14$, calculate $P(X_1 + X_2 + X_3 \le 306)$ and $P(98 \le \bar{X} \le 102)$.

58. Five automobiles of the same type are to be driven on a 300-mile trip. The first two will use an economy brand of gasoline, and the other three will use a name brand. Let $X_1, X_2, X_3, X_4,$ and X_5 be the observed numbers of miles per gallon for the five cars and suppose these variables are independent and normally distributed with $\mu_1 = \mu_2 = 20$, $\mu_3 = \mu_4 = \mu_5 = 21$, and $\sigma^2 = 4$ for all five variables. Define an rv Y by

$$Y = \frac{X_1 + X_2}{2} - \frac{X_3 + X_4 + X_5}{3}$$

so that Y is a measure of the difference in efficiency between economy gas and name-brand gas. Compute $P(0 \le Y)$ and $P(-1 \le Y \le 1)$. [HINT: $Y = a_1X_1 + \cdots + a_5X_5$, with $a_1 = \frac{1}{2}, \ldots, a_5 = -\frac{1}{3}$.]

59. Exercise 26 introduced random variables X and Y, the number of cars and buses, respectively, carried by a ferry on a single trip. The joint pmf of X and Y is given in the table in Exercise 7. It is readily verified that X and Y are independent.

a. Compute the expected value, variance, and standard deviation of the total number of vehicles on a single trip.

b. If each car is charged $3 and each bus $10, compute the expected value, variance, and standard deviation of the revenue resulting from a single trip.

60. Manufacture of a certain component requires three different machining operations. Machining time for each operation has a normal distribution, and the three times are independent of one another. The mean values are 15, 30, and 20 min, respectively, and the standard deviations are 1, 2, and 1.5 min, respectively. What is the probability that it takes at most 1 hour of machining time to produce a randomly selected component?

61. Refer to Exercise 3.

a. Calculate the covariance between X_1 = the number of customers in the express checkout and X_2 = the number of customers in the superexpress checkout.

b. Calculate $V(X_1 + X_2)$. How does this compare to $V(X_1) + V(X_2)$?

62. Suppose your waiting time for a bus in the morning is uniformly distributed on $[0, 5]$, while waiting time in the evening is uniformly distributed on $[0, 10]$ independent of morning waiting time.

a. If you take the bus each morning and evening for a week, what is your total expected waiting time? (HINT: Define rv's $X_1, \ldots, X_{10}$ and use a rule of expected value.)

b. What is the variance of your total waiting time?

c. What are the expected value and variance of the difference between morning and evening waiting time on a given day?

d. What are the expected value and variance of the difference between total morning waiting time and total evening waiting time for a particular week?

63. Suppose that when the pH of a certain chemical compound is 5.00, the pH measured by a randomly selected beginning chemistry student is a random variable with mean 5.00 and standard deviation .3. A large batch of the compound is subdivided and a sample given to each student in a morning lab and each student in an afternoon lab. Let $\bar{X}$ = the average pH as determined by the morning students and $\bar{Y}$ = the average pH as determined by the afternoon students.

a. If pH is a normal variable and there are 25 students in each lab, compute $P(-.1 \le \bar{X} - \bar{Y} \le .1)$. (HINT: $\bar{X} - \bar{Y}$ is a linear combination of normal variables, so is normally distributed. Compute $\mu_{\bar{X} - \bar{Y}}$ and $\sigma_{\bar{X} - \bar{Y}}$.)

b. If there are 36 students in each lab, but pH determinations are not assumed normal, calculate (approximately) $P(-.1 \le \bar{X} - \bar{Y} \le .1)$.

64. If two loads are applied to a cantilever beam as shown in the accompanying drawing, the bending moment at 0 due to the loads is $a_1X_1 + a_2X_2$.

0

a. Suppose that X_1 and X_2 are independent rv's with means 2 and 4 kips, respectively, and standard deviations .5 and 1.0 kip, respectively. If $a_1 = 5$ ft and $a_2 = 10$ ft, what is the expected bending moment and what is the standard deviation of the bending moment?

b. If X_1 and X_2 are normally distributed, what is the probability that the bending moment will exceed 75 kip-ft?

c. Suppose the positions of the two loads are random variables. Denoting them by A_1 and A_2, assume that these variables have means of 5 and 10 ft, respectively, that each has a standard deviation of .5, and that all A_i's and X_i's are independent of one another. What is the expected moment now?

d. For the situation of part (c), what is the variance of the bending moment?

e. If the situation is as described in part (a) except that $\text{Corr}(X_1, X_2) = .5$ (so that the two loads are not independent), what is the variance of the bending moment?

65. One piece of PVC pipe is to be inserted inside another piece. The length of the first piece is normally distributed with mean value 20 in. and standard deviation .5 in. The length of the second piece is a normal rv with mean and standard deviation 15 in. and .4 in., respectively. The amount of overlap is normally distributed with mean value 1 in. and standard deviation .1 in. Assuming that the lengths and amount of overlap are independent of one another, what is the probability that the total length after insertion is between 34.5 in. and 35 in.?

66. Two airplanes are flying in the same direction in adjacent parallel corridors. At time $t = 0$, the first airplane is 10 km ahead of the second one. Suppose the speed of the first plane (km/hr) is normally distributed with mean 520 and standard deviation 10 and the second plane's speed is also normally distributed with mean and standard deviation 500 and 10, respectively.

a. What is the probability that after 2 hours of flying, the second plane has not caught up to the first plane?

b. Determine the probability that the planes are separated by at most 10 km after 2 hours.

67. Three different roads feed into a particular freeway entrance. Suppose that during a fixed time period, the number of cars coming from each road onto the freeway

is a random variable, with expected value and standard deviation as given in the table.

	Road 1	Road 2	Road 3
Expected value	800	1000	600
Standard deviation	16	25	18

a. What is the expected total number of cars entering the freeway at this point during the period? (HINT: Let X_i = the number from road i.)

b. What is the variance of the total number of entering cars? Have you made any assumptions about the relationship between the numbers of cars on the different roads?

c. With X_i denoting the number of cars entering from road i during the period, suppose that $\text{Cov}(X_1, X_2) = 80$, $\text{Cov}(X_1, X_3) = 90$, and $\text{Cov}(X_2, X_3) = 100$ (so that the three streams of traffic are not independent). Compute the expected total number of entering cars and the standard deviation of the total.

68. Suppose we take a random sample of size n from a continuous distribution having median 0, so that the probability of any one observation being positive is .5. We now disregard the signs of the observations, rank them from smallest to largest in absolute value, and then let W = the sum of the ranks of the observations having positive signs. For example, if the observations are $-.3$, $+.7$, $+2.1$, and -2.5, then the ranks of positive observations are 2 and 3, so $W = 5$. In Chapter 15, W will be called *Wilcoxon's signed rank statistic*. W can be represented as follows:

$$W = 1 \cdot Y_1 + 2 \cdot Y_2 + 3 \cdot Y_3 + \cdots + n \cdot Y_n$$

$$= \sum_{i=1}^{n} i \cdot Y_i$$

where the Y_i's are independent Bernoulli rv's, each with $p = .5$ ($Y_i = 1$ corresponds to the observation with rank i being positive). Compute the following:

a. $E(Y_i)$ and then $E(W)$ using the equation for W [HINT: The first n positive integers sum to $n(n+1)/2$.]

b. $V(Y_i)$ and then $V(W)$ [HINT: The sum of the squares of the first n positive integers is $n(n+1)(2n+1)/6$.]

69. In Exercise 64, the weight of the beam itself contributes to the bending moment. Assume that the beam is of uniform thickness and density, so that the resulting load is uniformly distributed on the beam. If the weight of the beam is random, the resulting load from the weight is also random; denote this load by W (kips/ft).

a. If the beam is 12 ft long, W has mean 1.5 and standard deviation .25, and the fixed loads are as described in part (a) of Exercise 64, what are the expected value and variance of the bending moment? (HINT: If the load due to the beam were w kips/ft, the contribution to the bending moment would be $w \int_0^{12} x \, dx$.)

b. If all three variables (X_1, X_2, and W) are normally distributed, what is the probability that the bending moment will be at most 200 kip-ft?

70. I have three errands to take care of in the Administration Building. Let X_i = the time that it takes for the ith errand (i = 1, 2, 3) and let X_4 = the total time that I spend walking to and from the building and between each errand. Suppose the X_i's are independent, normally distributed, with the following means and standard deviations: $\mu_1 = 15$ min, $\sigma_1 = 4$, $\mu_2 = 5$, $\sigma_2 = 1$, $\mu_3 = 8$, $\sigma_3 = 2$, $\mu_4 = 12$, $\sigma_4 = 3$. I plan to leave my office at precisely 10:00 A.M. and wish to post a note on my door that reads, "I will return by t A.M." What time t should I write down if I want the probability of my arriving after t to be .01?

71. Suppose the expected tensile strength of type-A steel is 106 ksi and the standard deviation of tensile strength is 8 ksi. For type-B steel, suppose the expected tensile strength and standard deviation of tensile strength are 104 ksi and 6 ksi, respectively. Let $\bar{X}$ = the sample average tensile strength of a random sample of 40 type-A specimens and let $\bar{Y}$ = the sample average tensile strength of a random sample of 35 type-B specimens.

a. What is the approximate distribution of $\bar{X}$? Of $\bar{Y}$?

b. What is the approximate distribution of $\bar{X} - \bar{Y}$? Justify your answer.

c. Calculate (approximately) $P(-1 \le \bar{X} - \bar{Y} \le 1)$.

d. Calculate $P(\bar{X} - \bar{Y} \ge 6)$. If you actually observed $\bar{X} - \bar{Y} \ge 6$, would you doubt that $\mu_1 - \mu_2 = 2$?

72. In an area having sandy soil, 50 small trees of a certain type were planted, and another 50 trees were planted in an area having clay soil. Let X = the number of trees planted in sandy soil that survive 1 year and Y = the number of trees planted in clay soil that survive 1 year. If the probability that a tree planted in sandy soil will survive 1 year is .7 and the probability of 1-year survival in clay soil is .6, compute an approximation to $P(-5 \le X - Y \le 5)$ (do not bother with the continuity correction).

Supplementary Exercises for Chapter 5 (73–92)

73. A restaurant serves three fixed-price dinners costing $7, $9, and $10. For a randomly selected couple dining at this restaurant, let X = the cost of the man's dinner and Y = the cost of the woman's dinner. The joint pmf of X and Y is given in the following table:

$p(x, y)$		7	9	10
	7	.05	.05	.10
x	9	.05	.10	.35
	10	0	.20	.10

a. Compute the marginal pmf's of X and Y.

b. What is the probability that the man's and the woman's dinner cost at most $9 each?

c. Are X and Y independent? Justify your answer.

d. What is the expected total cost of the dinner for the two people?

e. Suppose that when a couple opens fortune cookies at the conclusion of the meal, they find the message "You will receive as a refund the difference between the cost of the more expensive and the less expensive meal that you have chosen." How much does the restaurant expect to refund?

74. In cost estimation, the total cost of a project is the sum of component task costs. Each of these costs is a random variable with a probability distribution. It is customary to obtain information about the total cost distribution by adding together characteristics of the individual component cost distributions—this is called the "roll-up" procedure. For example, $E(X_1 + \cdots + X_n) = E(X_1) + \cdots + E(X_n)$, so the roll-

up procedure is valid for mean cost. Suppose there are two component tasks and that X_1 and X_2 are independent normally distributed random variables. Is the roll-up procedure valid for the 75th percentile? That is, is the 75th percentile of the distribution of $X_1 + X_2$ the same as the sum of the 75th percentiles of the two individual distributions? If not, what is the relationship between the percentile of the sum and the sum of percentiles? For what percentiles is the roll-up procedure valid in this case?

75. A health-food store stocks two different brands of a certain type of grain. Let X = the amount (lb) of brand A on hand, and Y = the amount of brand B on hand. Suppose the joint pdf of X and Y is

$$f(x, y) = \begin{cases} kxy & x \geq 0,\ y \geq 0,\ 20 \leq x + y \leq 30 \\ 0 & \text{otherwise} \end{cases}$$

a. Draw the region of positive density and determine the value of k.

b. Are X and Y independent? Answer by first deriving the marginal pdf of each variable.

c. Compute $P(X + Y \leq 25)$.

d. What is the expected total amount of this grain on hand?

e. Compute $\text{Cov}(X, Y)$ and $\text{Corr}(X, Y)$.

f. What is the variance of the total amount of grain on hand?

76. Let $X_1, X_2, \ldots, X_n$ be random variables denoting n independent bids for an item that is for sale. Suppose each X_i is uniformly distributed on the interval [100, 200]. If the seller sells to the highest bidder, how much can he expect to earn on the sale? [HINT: Let $Y = \max(X_1, X_2, \ldots, X_n)$. First find $F_Y(y)$ by noting that $Y \leq y$ iff each X_i is $\leq y$. Then obtain the pdf and $E(Y)$.]

77. Suppose my calorie intake at breakfast is a random variable with expected value 500 and standard deviation 50, my calorie intake at lunch is random with expected value 800 and standard deviation 100, and my calorie intake at dinner is a random variable with expected value 1700 and standard deviation 200. Assuming that intakes at different meals are independent of one another and that I eat three meals per day each day, what is the probability that my average calorie intake per day over the next (365-day) year is between 2950 and 3050? [HINT: Let X_i, Y_i, and Z_i denote the three calorie intakes on day i. Then total intake is given by $\Sigma(X_i + Y_i + Z_i)$.]

78. The mean weight of luggage submitted by a randomly selected tourist-class passenger flying between two cities on a certain airline is 40 lb, and the standard deviation is 10 lb. The mean and standard deviation for a business-class passenger are 25 lb and 6 lb, respectively.

a. If there are 12 business-class passengers and 50 tourist-class passengers on a particular flight, what are the expected value of total luggage weight and the standard deviation of total luggage weight?

b. If individual luggage weights are independent normally distributed rv's, what is the probability that total luggage weight is at most 2500 lb?

79. We have seen that if $E(X_1) = E(X_2) = \cdots = E(X_n) = \mu$, then $E(X_1 + \cdots + X_n) = n\mu$. In some applications, the number of X_i's under consideration is not a fixed number n but instead is an rv N. For example, let N = the number of components that are brought into a repair shop on a particular day and let X_i denote the repair shop time for the ith component. Then the total repair time is $X_1 + X_2 + \cdots + X_N$, the sum of a *random* number of random variables. When N is independent of the X_i's, it can be shown that

$$E(X_1 + \cdots + X_N) = E(N) \cdot \mu$$

a. If the expected number of components brought in on a particular day is 10 and expected repair time for a randomly submitted component is 40 min, what is the expected total repair time for components submitted on any particular day?

b. Suppose components of a certain type come in for repair according to a Poisson process with rate 5 per hour. The expected number of defects per component is 3.5. What is the expected value of the total number of defects on components submitted for repair during a 4-hour period? Be sure to indicate how your answer follows from the general result just given.

80. Suppose the proportion of rural voters in a certain state who favor a particular gubernatorial candidate is .45 and the proportion of suburban and urban voters favoring the candidate is .60. If a sample of 200 rural voters and 300 urban and suburban voters is obtained, what is the approximate probability that at least 250 of these voters favor this candidate?

81. Let μ denote the true pH of a chemical compound. A sequence of n independent sample pH determinations will be made. Suppose each sample pH is a random variable with expected value μ and standard deviation

.1. How many determinations are required if we wish the probability that the sample average is within .02 of the true pH to be at least .95? What theorem justifies your probability calculation?

82. If the amount of soft drink that I consume on any given day is independent of consumption on any other day and is normally distributed with $\mu = 13$ oz and $\sigma = 2$ and if I currently have two six-packs of 16-oz bottles, what is the probability that I still have some soft drink left at the end of 2 weeks (14 days)?

83. Refer to Exercise 56, and suppose that the X_i's are independent with each one having a normal distribution. What is the probability that the total volume shipped is at most 100,000 ft^3?

84. A student has a class that is supposed to end at 9:00 A.M. and another that is supposed to begin at 9:10 A.M. Suppose the actual ending time of the 9 A.M. class is a normally distributed rv X_1 with mean 9:02 and standard deviation 1.5 min and that the starting time of the next class is also a normally distributed rv X_2 with mean 9:10 and standard deviation 1 min. Suppose also that the time necessary to get from one classroom to the other is a normally distributed rv X_3 with mean 6 min and standard deviation 1 min. What is the probability that the student makes it to the second class before the lecture starts? (Assume independence of X_1, X_2, and X_3, which is reasonable if the student pays no attention to the finishing time of the first class.)

85. a. Use the general formula for the variance of a linear combination to write an expression for $V(aX + Y)$. Then let $a = \sigma_Y/\sigma_X$ and show that $\rho \geq -1$. [HINT: Variance is always ≥ 0, and Cov(X, Y) = $\sigma_X \cdot \sigma_Y \cdot \rho$.]

b. By considering $V(aX - Y)$, conclude that $\rho \leq 1$.

c. Use the fact that $V(W) = 0$ only if W is a constant to show that $\rho = 1$ only if $Y = aX + b$.

86. Suppose a randomly chosen individual's verbal score X and quantitative score Y on a nationally administered aptitude examination have joint pdf

$$f(x, y) = \begin{cases} \frac{2}{5}(2x + 3y) & 0 \leq x \leq 1, 0 \leq y \leq 1 \\ 0 & \text{otherwise} \end{cases}$$

You are asked to provide a prediction t of the individual's total score $X + Y$. The error of prediction is the mean squared error $E[(X + Y - t)^2]$. What value of t minimizes the error of prediction?

87. a. Let X_1 have a chi-squared distribution with parameter v_1 (see Section 4.4) and let X_2 be independent of X_1 and have a chi-squared distribution with parameter v_2. Use the technique of Example 5.20 to show that $X_1 + X_2$ has a chi-squared distribution with parameter $v_1 + v_2$.

b. In Exercise 65 of Chapter 4, you were asked to show that if Z is a standard normal rv, then Z^2 has a chi-squared distribution with $v = 1$. Let Z_1, Z_2, . . . , Z_n be n independent standard normal rv's. What is the distribution of $Z_1^2 + \cdots + Z_n^2$? Justify your answer.

c. Let X_1, . . . , X_n be a random sample from a normal distribution with mean μ and variance σ^2. What is the distribution of the sum $Y = \sum_{i=1}^{n} [(X_i - \mu)/\sigma]^2$? Justify your answer.

88. a. Show that Cov(X, $Y + Z$) = Cov(X, Y) + Cov(X, Z).

b. Let X_1 and X_2 be quantitative and verbal scores on one aptitude exam and Y_1 and Y_2 be corresponding scores on another exam. If Cov(X_1, Y_1) = 5, Cov(X_1, Y_2) = 1, Cov(X_2, Y_1) = 2, and Cov(X_2, Y_2) = 8, what is the covariance between the two total scores $X_1 + X_2$ and $Y_1 + Y_2$?

89. A rock specimen from a particular area is randomly selected and weighed two different times. Let W denote the actual weight and X_1 and X_2 the two measured weights. Then $X_1 = W + E_1$ and $X_2 = W + E_2$, where E_1 and E_2 are the two measurement errors. Suppose that the E_i's are independent of one another and of W and that $V(E_1) = V(E_2) = \sigma_E^2$.

a. Express ρ, the correlation coefficient between the two measured weights X_1 and X_2, in terms of σ_W^2, the variance of actual weight, and σ_X^2, the variance of measured weight.

b. Compute ρ when $\sigma_W = 1$ kg and $\sigma_E = .01$ kg.

90. Let A denote the percentage of one constituent in a randomly selected rock specimen and let B denote the percentage of a second constituent in that same specimen. Suppose D and E are measurement errors in determining the values of A and B, so that measured values are $X = A + D$ and $Y = B + E$, respectively. Assume that measurement errors are independent of one another and of actual values.

a. Show that

Corr(X, Y)

$= $ Corr(A, B) $\cdot \sqrt{\text{Corr}(X_1, X_2)} \cdot \sqrt{\text{Corr}(Y_1, Y_2)}$

where X_1 and X_2 are replicate measurements on the value of A, and Y_1 and Y_2 are defined analogously with respect to B. What effect does the presence of measurement error have on the correlation?

b. What is the maximum value of Corr(X, Y) when Corr(X_1, X_2) = .8100 and Corr(Y_1, Y_2) = .9025? Is this disturbing?

91. Let $X_1, \ldots, X_n$ be independent rv's with mean values $\mu_1, \ldots, \mu_n$ and variances $\sigma_1^2, \ldots, \sigma_n^2$. Consider a function $h(x_1, \ldots, x_n)$ and use it to define a new rv $Y = h(X_1, \ldots, X_n)$. Under rather general conditions on the h function, if the σ_i's are all small relative to the corresponding μ_i's, it can be shown that $E(Y) \approx h(\mu_1, \ldots, \mu_n)$ and

$$V(Y) \approx \left(\frac{\partial h}{\partial x_1}\right)^2 \cdot \sigma_1^2 + \cdots + \left(\frac{\partial h}{\partial x_n}\right)^2 \cdot \sigma_n^2$$

where each partial derivative is evaluated at $(x_1, \ldots, x_n) = (\mu_1, \ldots, \mu_n)$. Suppose three resistors with resistances X_1, X_2, X_3 are connected in parallel across a battery with voltage X_4. Then by Ohm's law, the current is

$$Y = X_4 \left[\frac{1}{X_1} + \frac{1}{X_2} + \frac{1}{X_3}\right]$$

Let $\mu_1 = 10$ ohms, $\sigma_1 = 1.0$ ohm, $\mu_2 = 15$ ohms, $\sigma_2 = 1.0$ ohm, $\mu_3 = 20$ ohms, $\sigma_3 = 1.5$ ohms, $\mu_4 = 120$ V, $\sigma_4 = 4.0$ V. Calculate the (approximate) expected value and standard deviation of the current (suggested by "Random Samplings," *CHEMTECH*, 1984, pp. 696–697).

92. A more accurate approximation to $E[h(X_1, \ldots, X_n)]$ in Exercise 91 is

$$h(\mu_1, \ldots, \mu_n) + \frac{1}{2}\sigma_1^2\left(\frac{\partial^2 h}{\partial x_1^2}\right) + \cdots + \frac{1}{2}\sigma_n^2\left(\frac{\partial^2 h}{\partial x_n^2}\right)$$

Compute this for $Y = h(X_1, X_2, X_3, X_4)$ given in Exercise 91 and compare it to the leading term $h(\mu_1, \ldots, \mu_n)$.

Bibliography

Olkin, Ingram, Cyrus Derman, and Leon Gleser, *Probability Models and Applications,* (2nd ed.), Macmillan, New York, 1994. Contains a careful and comprehensive exposition of joint distributions, rules of expectation, and limit theorems.

Larsen, Richard, and Morris Marx, *Introduction to Mathematical Statistics* (2nd ed.), Prentice Hall, Englewood Cliffs, NJ, 1986. More limited coverage than in the book by Olkin et al., but well written and readable.

6 Point Esti

ΣΣΣΣΣΣΣΣΣΣΣΣΣΣΣΣΣΣΣ

Introduction

Given a parameter of interest, such as a population mean μ or population proportion p, the objective of point estimation is to use a sample to compute a number that represents in some sense a good guess for the true value of the parameter. The resulting number is called a *point estimate*. In Section 6.1, we present some general concepts of point estimation. In Section 6.2, we describe and illustrate two important methods for obtaining point estimates—the method of moments and the method of maximum likelihood.

6.1 Some General Concepts of Point Estimation

Statistical inference is almost always directed toward drawing some type of conclusion about one or more parameters (population characteristics). To do so requires that an investigator obtain sample data from each of the populations under study. Conclusions can then be based on the computed values of various sample quantities. For example, let μ (a parameter) denote the true average breaking strength of wire connections used in bonding semiconductor wafers. A random sample of $n = 10$ connections might be made, and the breaking strength of each one determined, resulting in observed strengths $x_1, x_2, \ldots, x_{10}$. The sample mean breaking strength $\bar{x}$ could then be used to draw a conclusion about the value of μ. Similarly, if σ^2 is the variance of the breaking strength distribution (population variance, another parameter), the value of the sample variance s^2 can be used to infer something about σ^2.

When discussing general concepts and methods of inference, it is convenient to have a generic symbol for the parameter of interest. We will use the Greek letter θ for this purpose. The objective of point estimation is to select a single number, based on sample data, that represents the most plausible value of θ. Suppose, for example, that the parameter of interest is μ, the true average lifetime of calculator batteries of a certain type. A random sample of $n = 3$ batteries might yield observed lifetimes (hours) $x_1 = 5.0, x_2 = 6.4, x_3 = 5.9$. The computed value of the sample mean lifetime is $\bar{x} = 5.77$, and it is reasonable to regard 5.77 as the most plausible value of μ— our "best guess" for the value of μ based on the available sample information.

Suppose we want to estimate a parameter of a single population (e.g., μ or σ) based on a random sample of size n. Recall from the previous chapter that before data is available, the sample observations must be considered random variables (rv's) X_1, $X_2, \ldots, X_n$. It follows that any function of the X_i's—that is, any statistic—such as the sample mean $\bar{X}$ or sample standard deviation S is also a random variable. The same is true if available data consists of more than one sample. For example, we can represent tensile strengths of m type 1 specimens and n type 2 specimens by $X_1, \ldots, X_m$ and $Y_1, \ldots, Y_n$, respectively. The difference between the two sample mean strengths is $\bar{X} - \bar{Y}$, the natural statistic for making inferences about $\mu_1 - \mu_2$, the difference between the population mean strengths.

<table>
<tr><td>Definition</td><td>

A **point estimate** of a parameter θ is a single number that can be regarded as the most plausible value of θ. A point estimate is obtained by selecting a suitable statistic and computing its value from the given sample data. The selected statistic is called the **point estimator** of θ.

</td></tr>
</table>

In the calculator battery example just given, the estimator used to obtain the point estimate of μ was $\bar{X}$, and the point estimate of μ was 5.77. If the three observed lifetimes had instead been $x_1 = 5.6$, $x_2 = 4.5$, and $x_3 = 6.1$, use of the estimator $\bar{X}$ would have resulted in the estimate $\bar{x} = (5.6 + 4.5 + 6.1)/3 = 5.40$. The symbol $\hat{\theta}$ ("theta hat") is customarily used to denote both the estimator of θ and the point estimate resulting from a given sample.* Thus $\hat{\mu} = \bar{X}$ is read as "the point estimator of μ is the sample mean $\bar{X}$," and the statement "the point estimate of μ is 5.77" can be written concisely as $\hat{\mu} = 5.77$. Notice that in writing $\hat{\theta} = 72.5$, there is no indication of how this point estimate was obtained (what statistic was used). It is recommended that both the estimator and the resulting estimate be reported.

Example 6.1

An automobile manufacturer has developed a new type of bumper, which is supposed to absorb impacts with less damage than previous bumpers. The manufacturer has used this bumper in a sequence of 25 controlled crashes against a wall, each at 10 mph, using one of its compact car models. Let X = the number of crashes that result in no visible damage to the automobile. The parameter to be estimated is p = the proportion of all such crashes that result in no damage [alternatively, $p = P(\text{no damage in a single crash})$]. If X is observed to be $x = 15$, the most reasonable

---

*Following earlier notation, we could use $\hat{\Theta}$ (an uppercase theta) for the estimator, but this is cumbersome to write.

estimator and estimate are

$$\text{estimator } \hat{p} = \frac{X}{n}, \quad \text{Estimate} = \frac{x}{n} = \frac{15}{25} = .60 \qquad \blacksquare$$

If for each parameter of interest there were only one reasonable point estimator, there would not be much to point estimation. In most problems, though, there will be more than one reasonable estimator.

Example 6.2

3 FEB 97

A consumer organization wishes to obtain a point estimate for the true average lifetime (hours of writing) for an inexpensive brand of ballpoint pen. Suppose the distribution of lifetime is normal with parameters μ and σ, so that μ is both the expected lifetime and median lifetime [since the normal probability distribution function (pdf) is symmetric about μ]. The organization purchases $n = 10$ such pens and inserts each in a specially constructed machine that will cause the pen to write continuously until the ink runs out. Let $X_1, X_2, \ldots, X_{10}$ denote the lifetimes, assumed to be a random sample from the normal distribution with parameters μ and σ. Suppose the observed lifetimes are $x_1 = 26.3$, $x_2 = 35.1$, $x_3 = 23.0$, $x_4 = 28.4$, $x_5 = 31.6$, $x_6 = 30.9$, $x_7 = 25.2$, $x_8 = 28.0$, $x_9 = 27.3$, and $x_{10} = 29.2$. Consider the following estimators and resulting estimates for μ:

a. Estimator $= \bar{X}$, estimate $= \bar{x} = \Sigma x_i/10 = 28.50$ sample mean

b. Estimator $= \tilde{X}$, estimate $= \tilde{x} = (28.0 + 28.4)/2 = 28.20$ median

c. Estimator $= [\min(X_i) + \max(X_i)]/2 =$ the average of the two extreme lifetimes, estimate $= [\min(x_i) + \max(x_i)]/2 = (23.0 + 35.1)/2 = 29.05$

d. Estimator $= \bar{X}_{\text{tr}(10)}$, the 10% trimmed mean (discard the smallest and largest 10% of the sample and then average),   Discard maximum

estimate $= \bar{x}_{\text{tr}(10)}$

$$= \frac{25.2 + 26.3 + 27.3 + 28.0 + 28.4 + 29.2 + 30.9 + 31.6}{8}$$

$$= 28.36$$

Each one of the estimators (a)–(d) uses a different measure of the center of the sample to estimate μ. Which of the estimates is closest to the true value? We cannot answer this without knowing the true value. A question that can be answered is, "Which estimator, when used on other samples of X_i's, will tend to produce estimates closest to the true value?" We will shortly consider this type of question. $\blacksquare$

Example 6.3

3 FEB 97

Officials of a paint company have been concerned about the variability in drying time of the company's top-quality interior latex paint. Let X be the drying time of a paint sample on a test board and let $\sigma^2 = V(X)$ (the variance of the population of all such drying times). If n test boards are set up and the drying times are $X_1, X_2, \ldots, X_n$, then one estimator of σ^2 is the sample variance:

$$\hat{\sigma}^2 = S^2 = \frac{\Sigma(X_i - \bar{X})^2}{n - 1} = \frac{\Sigma X_i^2 - (\Sigma X_i)^2/n}{n - 1} \qquad \text{unbiased}$$

If $n = 10$ and the observed x_i's are as in Example 6.2 (with minutes replacing hours), then the corresponding estimate is

$$\hat{\sigma}^2 = s^2 = \frac{\Sigma x_i^2 - (\Sigma x_i)^2/10}{9} = \frac{(26.3)^2 + \cdots + (29.2)^2 - (285)^2/10}{9}$$

$$= 11.90$$

The estimate of σ would then be $\hat{\sigma} = s = \sqrt{11.90} = 3.45$.

An alternative estimator would result from using divisor n instead of $n - 1$ (i.e., the average squared deviation):

$$\hat{\sigma}^2 = \frac{\Sigma(X_i - \bar{X})^2}{n}, \quad \text{estimate} = \frac{107.10}{10} = 10.71 \qquad \textit{biased}$$

We will shortly indicate why many statisticians prefer S^2 to the estimator with divisor n. ∎

In the best of all possible worlds, we could find an estimator $\hat{\theta}$ for which $\hat{\theta} = \theta$ always. However, $\hat{\theta}$ is a function of the sample X_i's, so is itself a random variable. For some samples $\hat{\theta}$ will yield a value larger than θ, while for other samples $\hat{\theta}$ will underestimate θ. If we write

$$\hat{\theta} = \theta + \text{error of estimation}$$

then an accurate estimator would be one resulting in small estimation errors, so that estimated values will be near the true value. An estimator that has the properties of unbiasedness and minimum variance will often be accurate in this sense.

Unbiased Estimators

Suppose we have two measuring instruments; one instrument has been accurately calibrated, but the other systematically gives readings smaller than the true value being measured. When each instrument is used repeatedly on the same object, because of measurement error, the observed measurements will not be identical. However, the measurements produced by the first instrument will be distributed about the true value in such a way that on average this instrument measures what it purports to measure, so is called an unbiased instrument. The second instrument yields observations that have a systematic error component or bias.

Definition

> A point estimator $\hat{\theta}$ is said to be an **unbiased estimator** of θ if $E(\hat{\theta}) = \theta$ for every possible value of θ. If $\hat{\theta}$ is not unbiased, the difference $E(\hat{\theta}) - \theta$ is called the **bias** of $\hat{\theta}$.

Thus, $\hat{\theta}$ is unbiased if its probability (i.e., sampling) distribution is always "centered" at the true value of the parameter. Figure 6.1 pictures the distributions of several biased and unbiased estimators. Note that "centered" here means that the expected value, not the median, of the distribution of $\hat{\theta}$ is equal to θ.

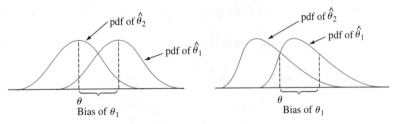

Figure 6.1 The pdf's of a biased estimator $\hat{\theta}_1$ and an unbiased estimator $\hat{\theta}_2$ for a parameter θ

It may seem as though in order to see whether $\hat{\theta}$ is unbiased, it is necessary to know the value of θ (in which case estimation is unnecessary). This is not usually the case because often a general expected value argument can be used to verify unbiasedness.

In Example 6.1, the sample proportion X/n was used as an estimator of p, where X, the number of sample successes, had a binomial distribution with parameters n and p. Thus,

$$E(\hat{p}) = E\left(\frac{X}{n}\right) = \frac{1}{n}\,E(X) = \frac{1}{n}\,(np) = p$$

Proposition

> When X is a binomial rv with parameters n and p, the sample proportion $\hat{p} = X/n$ is an unbiased estimator of p.

No matter what the true value of p is, the distribution of the estimator $\hat{p}$ will be centered at the true value.

Example 6.4   Suppose that X, the reaction time to a certain stimulus, has a uniform distribution on the interval from 0 to an unknown upper limit θ (so the density function of X is rectangular in shape with height $1/\theta$ for $0 \le x \le \theta$). It is desired to estimate θ on the basis of a random sample $X_1, X_2, \ldots, X_n$ of reaction times. Since θ is the largest possible time in the entire population of reaction times, consider as a first estimator the largest sample reaction time: $\hat{\theta}_1 = \max(X_1, \ldots, X_n)$. If $n = 5$ and $x_1 = 4.2$, $x_2 = 1.7$, $x_3 = 2.4$, $x_4 = 3.9$, $x_5 = 1.3$, the point estimate of θ is $\hat{\theta}_1 = \max(4.2, 1.7, 2.4, 3.9, 1.3) = 4.2$.

Unbiasedness implies that some samples will yield estimates that exceed θ and other samples will yield estimates smaller than θ—otherwise θ could not possibly be the center (balance point) of $\hat{\theta}_1$'s distribution. However, our proposed estimator will never overestimate θ (the largest sample value cannot exceed the largest population value) and will underestimate θ unless the largest sample value equals θ. This intuitive argument shows that $\hat{\theta}_1$ is a biased estimator. More precisely, it can be shown (see Exercise 32) that

$$E(\hat{\theta}_1) = \frac{n}{n+1} \cdot \theta < \theta \qquad \left(\text{since } \frac{n}{n+1} < 1\right)$$

The bias of $\hat{\theta}_1$ is given by $n\theta/(n + 1) - \theta = -\theta/(n + 1)$, which approaches 0 as n gets large.

It is easy to modify $\hat{\theta}_1$ to obtain an unbiased estimator of θ. Consider the estimator

$$\hat{\theta}_2 = \frac{n + 1}{n} \cdot \max(X_1, \ldots, X_n)$$

Using this estimator on the above data gives the estimate $(6/5)(4.2) = 5.04$. The fact that $(n + 1)/n > 1$ implies that $\hat{\theta}_2$ will overestimate θ for some samples and underestimate it for others. The mean value of this estimator is

$$E(\hat{\theta}_2) = E\left[\frac{n + 1}{n} \max(X_1, \ldots, X_n)\right] = \frac{n + 1}{n} \cdot E[\max(X_1, \ldots, X_n)]$$

$$= \frac{n + 1}{n} \cdot \frac{n}{n + 1} \theta = \theta$$

If $\hat{\theta}_2$ is used repeatedly on different samples to estimate θ, some estimates will be too large and others will be too small, but in the long run there will be no systematic tendency to underestimate or overestimate θ.   ■

Principle of Unbiased Estimation

When choosing among several different estimators of θ, select one that is unbiased.

According to this principle, the unbiased estimator $\hat{\theta}_2$ in Example 6.4 should be preferred to the biased estimator $\hat{\theta}_1$. Consider now the problem of estimating σ^2.

Proposition

Let $X_1, X_2, \ldots, X_n$ be a random sample from a distribution with mean μ and variance σ^2. Then the estimator

$$\hat{\sigma}^2 = S^2 = \frac{\Sigma(X_i - \bar{X})^2}{n - 1}$$

is an unbiased estimator of σ^2.

Proof   For any rv Y, $V(Y) = E(Y^2) - [E(Y)]^2$, so $E(Y^2) = V(Y) + [E(Y)]^2$. Applying this to

$$S^2 = \frac{1}{n - 1}\left[\Sigma X_i^2 - \frac{(\Sigma X_i)^2}{n}\right]$$

gives

$$E(S^2) = \frac{1}{n - 1}\left\{\Sigma E(X_i^2) - \frac{1}{n} E[(\Sigma X_i)^2]\right\}$$

$$= \frac{1}{n - 1} \left\{ \Sigma(\sigma^2 + \mu^2) - \frac{1}{n} \{ V(\Sigma X_i) + [E(\Sigma X_i)]^2 \} \right\}$$

$$= \frac{1}{n - 1} \left\{ n\sigma^2 + n\mu^2 - \frac{1}{n} n\sigma^2 - \frac{1}{n} (n\mu)^2 \right\}$$

$$= \frac{1}{n - 1} \{ n\sigma^2 - \sigma^2 \} = \sigma^2 \qquad \text{(as desired)} \qquad \blacksquare$$

The estimator that uses divisor n can be expressed as $(n - 1)S^2/n$, so

$$E\left[\frac{(n - 1)S^2}{n} \right] = \frac{n - 1}{n} E(S^2) = \frac{n - 1}{n} \sigma^2$$

This estimator is therefore not unbiased. The bias is $(n - 1)\sigma^2/n - \sigma^2 = -\sigma^2/n$. Because the bias is negative, the estimator with divisor n tends to underestimate σ^2, and this is why the divisor $n - 1$ is preferred by many statisticians (though when n is large, the bias is small and there is little difference between the two).

Although S^2 is unbiased for σ^2, S is a biased estimator of σ (its bias is small unless n is quite small). However, there are other good reasons to use S as an estimator, especially when the population distribution is normal. These will become more apparent when we discuss confidence intervals and hypothesis testing in the next several chapters.

In Example 6.2, we proposed several different estimators for the mean μ of a normal distribution. If there were a unique unbiased estimator for μ, the estimation problem would be resolved by using that estimator. Unfortunately, this is not the case.

Proposition

> If $X_1, X_2, \ldots, X_n$ is a random sample from a distribution with mean μ, then $\bar{X}$ is an unbiased estimator of μ. If in addition the distribution is continuous and symmetric, then $\tilde{X}$ and any trimmed mean are also unbiased estimators of μ.

The fact that $\bar{X}$ is unbiased is just a restatement of one of our rules of expected value: $E(\bar{X}) = \mu$ for every possible value of μ (for discrete as well as continuous distributions). The unbiasedness of the other estimators is more difficult to verify.

According to this proposition, the principle of unbiasedness by itself does not always allow us to select a single estimator. When the underlying population is normal, even the third estimator in Example 6.2 is unbiased, and there are many other unbiased estimators. What we now need is a way of selecting among unbiased estimators.

Estimators with Minimum Variance

Suppose $\hat{\theta}_1$ and $\hat{\theta}_2$ are two estimators of θ that are both unbiased. Then, although the distribution of each estimator is centered at the true value of θ, the spreads of the distributions about the true value may be different.

> **Principle of Minimum Variance Unbiased Estimation**
>
> Among all estimators of θ that are unbiased, choose the one that has minimum variance. The resulting $\hat{\theta}$ is called the **minimum variance unbiased estimator (MVUE)** of θ.

Figure 6.2 pictures the pdf's of two unbiased estimators, with $\hat{\theta}_1$ having smaller variance than $\hat{\theta}_2$. Then $\hat{\theta}_1$ is more likely than $\hat{\theta}_2$ to produce an estimate close to the true θ. The MVUE is in a certain sense the most likely among all unbiased estimators to produce an estimate close to the true θ.

Figure 6.2 Graphs of the pdf's of two different unbiased estimators

Example 6.5   We argued in Example 6.4 that when $X_1, \ldots, X_n$ is a random sample from a uniform distribution on $[0, \theta]$, the estimator

$$\hat{\theta}_1 = \frac{n+1}{n} \cdot \max(X_1, \ldots, X_n)$$

is unbiased for θ (we previously denoted this estimator by $\hat{\theta}_2$). This is not the only unbiased estimator of θ. The expected value of a uniformly distributed rv is just the midpoint of the interval of positive density, so $E(X_i) = \theta/2$. This implies that $E(\bar{X}) = \theta/2$, from which $E(2\bar{X}) = \theta$. That is, the estimator $\hat{\theta}_2 = 2\bar{X}$ is unbiased for θ.

If X is uniformly distributed on the interval $[A, B]$, then $V(X) = \sigma^2 = (B - A)^2/12$. Thus, in our situation, $V(X_i) = \theta^2/12$, $V(\bar{X}) = \sigma^2/n = \theta^2/(12n)$, and $V(\hat{\theta}_2) = V(2\bar{X}) = 4V(\bar{X}) = \theta^2/(3n)$. The results of Exercise 32 can be used to show that $V(\hat{\theta}_1) = \theta^2/[n(n + 2)]$. The estimator $\hat{\theta}_1$ has smaller variance than does $\hat{\theta}_2$ if $3n < n(n + 2)$—that is, if $0 < n^2 - n = n(n - 1)$. As long as $n > 1$, $V(\hat{\theta}_1) < V(\hat{\theta}_2)$, so $\hat{\theta}_1$ is a better estimator than $\hat{\theta}_2$. More advanced methods can be used to show that $\hat{\theta}_1$ is the MVUE of θ—every other unbiased estimator of θ has variance that exceeds $\theta^2/[n(n + 2)]$. ∎

One of the triumphs of mathematical statistics has been the development of methodology for identifying the MVUE in a wide variety of situations. The most important result of this type for our purposes concerns estimating the mean μ of a normal distribution.

Theorem

> Let $X_1, \ldots, X_n$ be a random sample from a normal distribution with parameters μ and σ. Then the estimator $\hat{\mu} = \bar{X}$ is the MVUE for μ.

Whenever we are convinced that the population being sampled is normal, the result says that $\bar{X}$ should be used to estimate μ. In Example 6.2, then, our estimate would be $\bar{x} = 28.50$.

In some situations, it is possible to obtain an estimator with small bias that would be preferred to the best unbiased estimator. This is illustrated in Figure 6.3. However, MVUEs are often easier to obtain than the type of biased estimator whose distribution is pictured.

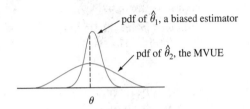

Figure 6.3 A biased estimator that is preferable to the MVUE

Some Complications

The last theorem does not say that in estimating a population mean μ, the estimator $\bar{X}$ should be used irrespective of the distribution being sampled.

Example 6.6

Suppose we wish to estimate the thermal conductivity μ of a certain material. Using standard measurement techniques, we will obtain a random sample $X_1, \ldots, X_n$ of n thermal conductivity measurements. Let's assume that the population distribution is a member of one of the following three families:

$$f(x) = \frac{1}{\sqrt{2\pi\sigma^2}} e^{-(x - \mu)^2/(2\sigma^2)} \qquad -\infty < x < \infty \qquad (6.1)$$

$$f(x) = \frac{1}{\pi[1 + (x - \mu)^2]} \qquad -\infty < x < \infty \qquad (6.2)$$

$$f(x) = \begin{cases} \dfrac{1}{2c} & -c \leq x - \mu \leq c \\ 0 & \text{otherwise} \end{cases} \qquad (6.3)$$

The pdf (6.1) is the normal distribution, (6.2) is called the Cauchy distribution, and (6.3) is a uniform distribution. All three distributions are symmetric about μ, and in fact the Cauchy distribution is bell-shaped but with much heavier tails (more probability farther out) than the normal curve. The uniform distribution has no tails.

The four estimators for μ considered earlier are $\bar{X}$, $\tilde{X}$, $\bar{X}_e$ (the average of the two extreme observations), and $\bar{X}_{\text{tr}(10)}$, a trimmed mean.

The very important moral here is that the best estimator for μ depends crucially on which distribution is being sampled. In particular,

1. If the random sample comes from a normal distribution, then $\bar{X}$ is the best of the four estimators, since it has minimum variance among all unbiased estimators.

2. If the random sample comes from a Cauchy distribution, then $\bar{X}$ and $\bar{X}_e$ are terrible estimators for μ, whereas $\tilde{X}$ is quite good (the MVUE is not known); $\bar{X}$ is bad because it is very sensitive to outlying observations, and the heavy tails of the Cauchy distribution make a few such observations likely to appear in any sample.

3. If the underlying distribution is uniform, the best estimator is $\bar{X}_e$; this estimator is greatly influenced by outlying observations, but the lack of tails makes such observations impossible.

4. *The trimmed mean is best in none of these three situations, but works reasonably well in all three.* That is, $\bar{X}_{\text{tr}(10)}$ does not suffer too much in comparison with the best procedure in any of the three situations.   ∎

More generally, recent research in statistics has established that when estimating a point of symmetry μ of a continuous probability distribution, a trimmed mean with trimming proportion 10% or 20% (from each end of the sample) produces reasonably behaved estimates over a very wide range of possible models. For this reason, a trimmed mean with small trimming percentage is said to be a **robust estimator**.

In some situations, the choice is not between two different estimators constructed from the same sample, but instead between estimators based on two different experiments.

Example 6.7

Suppose a certain type of component has a lifetime distribution that is exponential with parameter λ, so that expected lifetime is $\mu = 1/\lambda$. A sample of n such components is selected, and each is put into operation. If the experiment is continued until all n lifetimes, $X_1, \ldots, X_n$, have been observed, then $\bar{X}$ is an unbiased estimator of μ.

In some experiments, though, the components are left in operation only until the time of the rth failure, where $r < n$. This procedure is referred to as **censoring**. Let Y_1 denote the time of the first failure (the minimum lifetime among the n components), Y_2 denote the time at which the second failure occurs (the second smallest lifetime), and so on. Since the experiment terminates at time Y_r, the total accumulated lifetime at termination is

$$T_r = \sum_{i=1}^{r} Y_i + (n - r)Y_r$$

We now demonstrate that $\hat{\mu} = T_r/r$ is an unbiased estimator for μ. To do so, we need two properties of exponential variables:

1. The memoryless property (see Section 4.4), which says that at any time point, remaining lifetime has the same exponential distribution as original lifetime.

2. If $X_1, \ldots, X_k$ are independent, each exponential with parameter λ, then $\min(X_1, \ldots, X_k)$ is exponential with parameter $k\lambda$ and has expected value $1/(k\lambda)$.

Since all n components last until Y_1, $n - 1$ last an additional $Y_2 - Y_1$, $n - 2$ an additional $Y_3 - Y_2$ amount of time, and so on, another expression for T_r is

$$T_r = nY_1 + (n - 1)(Y_2 - Y_1) + (n - 2)(Y_3 - Y_2) + \cdots$$
$$+ (n - r + 1)(Y_r - Y_{r-1})$$

But Y_1 is the minimum of n exponential variables, so $E(Y_1) = 1(n\lambda)$. Similarly, $Y_2 - Y_1$ is the smallest of the $n - 1$ remaining lifetimes, each exponential with parameter λ (by the memoryless property), so $E(Y_2 - Y_1) = 1/[(n - 1)\lambda]$. Continuing, $E(Y_{i+1} - Y_i) = 1/[(n - i)\lambda]$, so

$$E(T_r) = nE(Y_1) + (n - 1)E(Y_2 - Y_1) + \cdots + (n - r + 1)E(Y_r - Y_{r-1})$$

$$= n \cdot \frac{1}{n\lambda} + (n - 1) \cdot \frac{1}{(n-1)\lambda} + \cdots + (n - r + 1) \cdot \frac{1}{(n-r+1)\lambda}$$

$$= \frac{r}{\lambda}$$

Therefore, $E(T_r/r) = (1/r)E(T_r) = (1/r) \cdot (r/\lambda) = 1/\lambda = \mu$ as claimed.

As an example, suppose 20 components are put on test and $r = 10$. Then if the first ten failure times are 11, 15, 29, 33, 35, 40, 47, 55, 58, and 72, the estimate of μ is

$$\hat{\mu} = \frac{11 + 15 + \cdots + 72 + (10)(72)}{10} = 111.5$$

The advantage of the experiment with censoring is that it terminates more quickly than the uncensored experiment. However, it can be shown that $V(T_r/r) = 1/(\lambda^2 r)$, which is larger than $1/(\lambda^2 n)$, the variance of $\bar{X}$ in the uncensored experiment. ■

Reporting a Point Estimate: The Standard Error

Besides reporting the value of a point estimate, some indication of its precision should be given. The usual measure of precision is the standard error of the estimator used.

Definition

> The **standard error** of an estimator $\hat{\theta}$ is its standard deviation $\sigma_{\hat{\theta}} = \sqrt{V(\hat{\theta})}$. If the standard error itself involves unknown parameters whose values can be estimated, substitution of these estimates into $\sigma_{\hat{\theta}}$ yields the **estimated standard error** (estimated standard deviation) of the estimator. The estimated standard error can be denoted either by $\hat{\sigma}_{\hat{\theta}}$ (the ^ over σ emphasizes that $\sigma_{\hat{\theta}}$ is being estimated) or by $s_{\hat{\theta}}$.

Example 6.8
(Example 6.2 continued) ~~SFEB97~~

Assuming that writing lifetime is normally distributed, $\hat{\mu} = \bar{X}$ is the best estimator of μ. If the value of σ is known to be 3.5, the standard error of $\bar{X}$ is $\sigma_{\bar{X}} = \sigma/\sqrt{n} = 3.5/\sqrt{10} = 1.11$. If, as is usually the case, the value of σ is unknown, the estimate $\hat{\sigma} = s = 3.45$ is substituted into $\sigma_{\bar{X}}$ to obtain the estimated standard error $\hat{\sigma}_{\bar{X}} = s_{\bar{X}} = s/\sqrt{n} = 3.45/\sqrt{10} = 1.09$. ■

Example 6.9
(Example 6.1 continued) ~~SFEB97~~

The standard error of $\hat{p} = X/n$ is

$$\sigma_{\hat{p}} = \sqrt{V(X/n)} = \sqrt{\frac{V(X)}{n^2}} = \sqrt{\frac{npq}{n^2}} = \sqrt{\frac{pq}{n}}$$

Since p (and $q = 1 - p$) are unknown (else why estimate?), we substitute $\hat{p} = x/n$ and $\hat{q} = 1 - x/n$ into $\sigma_{\hat{p}}$, yielding the estimated standard error $\hat{\sigma}_{\hat{p}} = \sqrt{\hat{p}\hat{q}/n} = \sqrt{(.6)(.4)/25} = .098$. Alternatively, since the largest value of pq is attained when $p = q = .5$, an upper bound on the standard error is $\sqrt{1/(4n)} = .10$. ■

When the point estimator $\hat{\theta}$ has approximately a normal distribution, which will often be the case when n is large, then we can be reasonably confident that the true value of θ lies within approximately two standard errors (standard deviations) of $\hat{\theta}$. Thus, in the pen-lifetime example, if $n = 36$, $\hat{\mu} = \bar{x} = 28.50$, and $s = 3.60$, then $s/\sqrt{n} = .60$, so within two estimated standard errors of $\hat{\mu}$ translates to the interval $28.50 \pm (2)(.60) = (27.30, 29.70)$.

If $\hat{\theta}$ is not necessarily approximately normal but is unbiased, then it can be shown that the estimate will deviate from θ by as much as four standard errors at most 6% of the time. We would then expect the true value to lie within four standard errors of $\hat{\theta}$ (and this is a very conservative statement, since it applies to *any* unbiased $\hat{\theta}$). Summarizing, the standard error tells us roughly within what distance of $\hat{\theta}$ we can expect the true value of θ to lie.

The form of the estimator $\hat{\theta}$ may be sufficiently complicated so that standard statistical theory cannot be applied to obtain an expression for $\sigma_{\hat{\theta}}$. This is true, for example, in the case $\theta = \sigma$, $\hat{\theta} = S$; the standard deviation of the statistic S, σ_S, cannot in general be determined. In recent years, a new computer-intensive method called the **bootstrap** has been introduced to address this problem. Suppose the population pdf is $f(x; \theta)$, a member of a particular parametric family, and that data $x_1, x_2, \ldots, x_n$ gives $\hat{\theta} = 21.7$. We now use the computer to obtain "bootstrap samples" from the pdf $f(x; 21.7)$, and for each sample we calculate a "bootstrap estimate" $\hat{\theta}^*$:

First bootstrap sample: $x_1^*, x_2^*, \ldots, x_n^*$; estimate $= \hat{\theta}_1^*$

Second bootstrap sample: $x_1^*, x_2^*, \ldots, x_n^*$; estimate $= \hat{\theta}_2^*$

$\vdots$

Bth bootstrap sample: $x_1^*, x_2^*, \ldots, x_n^*$; estimate $= \hat{\theta}_B^*$

$B = 100$ or 200 is often used. Now let $\bar{\theta}^* = \Sigma\hat{\theta}_i^*/B$, the sample mean of the bootstrap estimates. The **bootstrap estimate** of $\hat{\theta}$'s standard error is now just the sample standard deviation of the $\hat{\theta}_i^*$'s:

$$S_{\hat{\theta}} = \sqrt{\frac{1}{B-1} \sum (\hat{\theta}_i^* - \bar{\theta}^*)^2}$$

(In the bootstrap literature, B is often used in place of $B - 1$; for typical values of B, there is usually little difference between the resulting estimates.)

Example 6.10   A theoretical model suggests that X, the time to breakdown of an insulating fluid between electrodes at a particular voltage, has $f(x; \lambda) = \lambda e^{-\lambda x}$, an exponential distribution. A random sample of $n = 10$ breakdown times (min) gives the following data:

41.53, 18.73, 2.99, 30.34, 12.33, 117.52, 73.02, 223.63, 4.00, 26.78

Since $E(X) = 1/\lambda$, $E(\bar{X}) = 1/\lambda$, so a reasonable estimate of λ is $\hat{\lambda} = 1/\bar{x} = 1/55.087 = .018153$. We then used a statistical computer package to obtain $B = 100$ bootstrap samples, each of size 10, from $f(x; .018153)$. The first such sample was 41.00, 109.70, 16.78, 6.31, 6.76, 5.62, 60.96, 78.81, 192.25, 27.61, from which $\Sigma x_i^* = 545.8$ and $\hat{\lambda}_1^* = 1/54.58 = .01832$. The average of the 100 bootstrap estimates is $\bar{\lambda}^* = .02153$, and the sample standard deviation of these 100 estimates is $s_{\hat{\lambda}} = .0091$, the bootstrap estimate of $\hat{\lambda}$'s standard error. A histogram of the 100 $\hat{\lambda}_i^*$'s was somewhat positively skewed, suggesting that $\hat{\lambda}$'s sampling distribution also has this property. ■

Sometimes an investigator wishes to estimate a population characteristic without assuming that the population distribution belongs to a particular parametric family. An instance of this occurred in Example 6.6, where a 10% trimmed mean was proposed for estimating a symmetric population distribution's center θ. The data of Example 6.2 gave $\hat{\theta} = \bar{x}_{tr(10)} = 28.36$, but now there is no assumed $f(x; \theta)$, so how can we obtain a bootstrap sample? The answer is to regard the sample itself as constituting the population (the $n = 10$ observations in Example 6.2) and take B different samples, each of size n, *with* replacement from this population. The book by Bradley Efron and Robert Tibshirani or the one by John Rice listed in the chapter bibliography provide more information.

Exercises   Section 6.1 (1–19)

1. The interpupillary distance was measured for each of ten randomly selected adult males, yielding the following measurements (mm): 64.9, 62.1, 70.8, 65.3, 68.7, 57.2, 62.5, 69.8, 64.3, 71.6.

a. Assuming that interpupillary distance is normally distributed, use the sample mean $\bar{X}$ and the sample standard deviation S to obtain estimates of μ and σ.

b. Supposing only that the distribution is continuous, use the sample median $\tilde{X}$ to obtain an estimate of the population median $\tilde{\mu}$. Why in this situation would you not use $\bar{X}$ to estimate $\tilde{\mu}$?

c. Assuming that interpupillary distance has a symmetric but not necessarily normal distribution, use a 10% trimmed mean to obtain an estimate of μ.

d. Let p denote the proportion of all men having interpupillary distance between 60 and 70. Determine the number of sample x_i's that satisfy $60 \le x_i \le 70$ and use it to estimate the population proportion p.

2. A sample of 20 students who had recently taken elementary statistics yielded the following information on brand of calculator owned (T = Texas Instruments, H = Hewlett Packard, C = Casio, S = Sharp):

T, T, H, T, C, T, T, S, C, H,
S, S, T, H, C, T, T, T, H, T

a. Estimate the true proportion of all such students who own a Texas Instruments calculator.

b. Among the brands owned, only Hewlett Packard uses reverse Polish logic. Estimate the proportion of all such students who own a calculator that does not use reverse Polish logic.

c. Three of the four Hewlett Packard calculators in the sample were programmable. Estimate the proportion of all students who own a calculator that is both made by HP and is programmable.

3. Twelve samples of a certain brand of white bread were analyzed and the carbohydrate content (percentage of nitrogen-free extract) of each sample was determined, yielding the following data:

76.93, 76.88, 77.07, 76.68, 76.39, 75.09,
76.88, 77.67, 78.15, 76.50, 77.16, 76.42

a. Assuming that the distribution of carbohydrate content is normal, estimate the true average carbohydrate content for this brand. What estimator did you use?

b. Making no assumption whatsoever about the distribution of carbohydrate content, estimate the value below which (and above which) half of all carbohydrate contents will fall. What estimator did you use?

c. Again making no assumption about the underlying distribution, estimate the true proportion of samples whose carbohydrate content would not exceed 76%.

d. What is the estimated standard error of the estimator that you used in part (a)?

4. As an example of a situation in which several different statistics could reasonably be used to calculate a point estimate, consider a population of N invoices. Associated with each invoice is its "book value," the recorded amount of that invoice. Let T denote the total book value, a known amount. Some of these book values are erroneous. An audit will be carried out by randomly selecting n invoices and determining the audited (correct) value for each one. Suppose that the sample gives the following results (in dollars).

	Invoice				
	1	2	3	4	5
Book value	300	720	526	200	127
Audited value	300	520	526	200	157
Error	0	200	0	0	−30

Let

$\bar{Y}$ = sample mean book value

$\bar{X}$ = sample mean audited value

$\bar{D}$ = sample mean error

Several different statistics for estimating the total audited (correct) value have been proposed (see "Statistical Models and Analysis in Auditing," *Statistical Science*, 1989, pp. 2–33). These include

Mean per unit statistic = $N\bar{X}$

Difference statistic = $T - N\bar{D}$

Ratio statistic = $T \cdot (\bar{X}/\bar{Y})$

If $N = 5000$ and $T = 1,761,300$, calculate the three corresponding point estimates. (The cited paper discusses properties of these estimators.)

5. In Exercise 3, let $X_1, \ldots, X_m$ denote the random variables observed, assumed to be a random sample from a distribution with mean μ_1 and variance σ_1^2. For n samples of a second brand, let $Y_1, \ldots, Y_n$ denote the carbohydrate contents, assumed to be a random sample from a distribution with mean μ_2 and variance σ_2^2.

a. Using rules of expected value, show that $\bar{X} - \bar{Y}$ is an unbiased estimator of $\mu_1 - \mu_2$. If the observed $\bar{Y}$ is $\bar{y} = 74.28$, for $n = 14$ samples, what is the point estimate?

b. Use the rules of variance from Chapter 5 to obtain an expression for the variance and standard deviation (standard error) of $\bar{X} - \bar{Y}$.

c. If σ_1 and σ_2 are unknown, but $s_1 = .75$ and $s_2 = .70$, compute the estimated standard error of $\bar{X} - \bar{Y}$.

6. Consider the accompanying observations on streamflow (1000s of acre-feet) recorded at a station in Colorado for the period April 1–August 31 over a 31-year span (from an article in the 1974 volume of *Water Resources Research*).

127.96 210.07 203.24 108.91 178.21
285.37 100.85 89.59 185.36 126.94
200.19 66.24 247.11 299.87 109.64
125.86 114.79 109.11 330.33 85.54
117.64 302.74 280.55 145.11 95.36
204.91 311.13 150.58 262.09 477.08
94.33

An appropriate probability plot supports the use of the lognormal distribution (see Section 4.5) as a reasonable model for streamflow.

a. Estimate the parameters of the distribution. [HINT: Remember that X has a lognormal distribution with parameters μ and σ^2 if $\ln(X)$ is normally distributed with mean μ and variance σ^2.]

b. Use the estimates of part (a) to calculate an estimate of the expected value of streamflow. [HINT: What is $E(X)$?]

7. a. A random sample of 10 houses in a particular area, each of which is heated with natural gas, is selected and the amount of gas (therms) used during the month of January is determined for each house. The resulting observations are 103, 156, 118, 89, 125, 147, 122, 109, 138, 99. Let μ denote the average gas usage during January by all houses in this area. Compute a point estimate of μ.

b. Suppose there are 10,000 houses in this area that use natural gas for heating. Let τ denote the total amount of gas used by all of these houses during January. Estimate τ using the data of part (a). What estimator did you use in computing your estimate?

c. Use the data in part (a) to estimate p, the proportion of all houses that used at least 100 therms.

d. Give a point estimate of the population median usage (the middle value in the population of all houses) based on the sample of part (a). What estimator did you use?

8. In a random sample of 80 components of a certain type, 12 are found to be defective.

a. Give a point estimate of the proportion of all such components that are *not* defective.

b. A system is to be constructed by randomly selecting two of these components and connecting them in series, as shown here.

The series connection implies that the system will function if and only if neither component is defective (i.e., both components work properly). Estimate the proportion of all such systems that work properly. [HINT: If p denotes the probability that a component works properly, how can P(system works) be expressed in terms of p?]

9. Each of 150 newly manufactured items is examined and the number of scratches per item is recorded (the items are supposed to be free of scratches), yielding the following data:

Number of scratches per item	0	1	2	3	4	5	6	7
Observed frequency	18	37	42	30	13	7	2	1

Let X = the number of scratches on a randomly chosen item and assume that X has a Poisson distribution with parameter λ.

a. Find an unbiased estimator of λ and compute the estimate for the above data. [HINT: $E(X) = \lambda$ for X Poisson, so $E(\bar{X}) = ?$]

b. What is the standard deviation (standard error) of your estimator? Compute the estimated standard error. (HINT: $\sigma_X^2 = \lambda$ for X Poisson.)

10. Using a long rod that has length μ, you are going to lay out a square plot in which the length of each side is μ. Thus, the area of the plot will be μ^2. However, you do not know the value of μ, so you decide to make n independent measurements $X_1, X_2, \ldots, X_n$ of the length. Assume that each X_i has mean μ (unbiased measurements) and variance σ^2.

a. Show that $\bar{X}^2$ is not an unbiased estimator for μ^2. (HINT: For any rv Y, $E(Y^2) = V(Y) + [E(Y)]^2$. Apply this with $Y = \bar{X}$.)]

b. For what value of k is the estimator $\bar{X}^2 - kS^2$ unbiased for μ^2? [HINT: Compute $E(\bar{X}^2 - kS^2)$.]

11. Of n_1 randomly selected male smokers, X_1 smoked filter cigarettes, while of n_2 randomly selected female smokers, X_2 smoked filter cigarettes. Let p_1 and p_2 denote the probabilities that a randomly selected male and female, respectively, smoke filter cigarettes.

a. Show that $(X_1/n_1) - (X_2/n_2)$ is an unbiased estimator for $p_1 - p_2$. [HINT: $E(X_i) = n_i p_i$ for $i = 1, 2$.]

b. What is the standard error of the estimator in part (a)?

c. How would you use the observed values x_1 and x_2 to estimate the standard error of your estimator?

d. If $n_1 = n_2 = 200$, $x_1 = 127$, and $x_2 = 176$, use the estimator of part (a) to obtain an estimate of $p_1 - p_2$.

e. Use the result of part (c) and the data of part (d) to estimate the standard error of the estimator.

12. Suppose a certain type of fertilizer has an expected yield per acre of μ_1 with variance σ^2, while the expected yield for a second type of fertilizer is μ_2 with the same variance σ^2. Let S_1^2 and S_2^2 denote the sample variances of yields based on sample sizes n_1 and n_2, respectively, of the two fertilizers. Show that the pooled (combined) estimator

$$\hat{\sigma}^2 = \frac{(n_1 - 1)S_1^2 + (n_2 - 1)S_2^2}{n_1 + n_2 - 2}$$

is unbiased for σ^2 (this estimator will be used in Chapter 9).

13. Consider a random sample $X_1, \ldots, X_n$ from the pdf

$$f(x; \theta) = .5(1 + \theta x) \qquad -1 \leq x \leq 1$$

where $-1 \leq \theta \leq 1$ (this distribution arises in particle physics). Show that $\hat{\theta} = 3\bar{X}$ is an unbiased estimator of θ. [HINT: First determine $\mu = E(X) = E(\bar{X})$.]

14. A sample of n captured Pandemonium jet fighters results in serial numbers $x_1, x_2, x_3, \ldots, x_n$. The CIA knows that the aircraft were numbered consecutively at the factory starting with α and ending with β, so that the total number of planes manufactured is $\beta - \alpha + 1$ (e.g., if $\alpha = 17$ and $\beta = 29$, then $29 - 17 + 1 = 13$ planes having serial numbers 17, 18, 19, . . . , 28, 29 were manufactured). However, the CIA does not know the values of α or β. A CIA statistician suggests using the estimator $\max(X_i) - \min(X_i) + 1$ to estimate the total number of planes manufactured.

a. If $n = 5$, $x_1 = 237$, $x_2 = 375$, $x_3 = 202$, $x_4 = 525$, and $x_5 = 418$, what is the corresponding estimate?

b. Under what conditions on the sample will the value of the estimate be exactly equal to the true total number of planes? Will the estimate ever be larger than the true total? Do you think the estimator is

unbiased for estimating $\beta - \alpha + 1$? Explain in one or two sentences.

15. Let $X_1, X_2, \ldots, X_n$ represent a random sample from a Rayleigh distribution with pdf

$$f(x; \theta) = \frac{x}{\theta} e^{-x^2/(2\theta)} \qquad x > 0$$

a. It can be shown that $E(X^2) = 2\theta$. Use this fact to construct an unbiased estimator of θ based on ΣX_i^2 (and use rules of expected value to show that it is unbiased).

b. Estimate θ from the following $n = 10$ observations on vibratory stress of a turbine blade under specified conditions:

16.88, 10.23, 4.59, 6.66, 13.68,
14.23, 19.87, 9.40, 6.51, 10.95.

16. Suppose the true average growth μ of one type of plant during a 1-year period is identical to that of a second type, but the variance of growth for the first type is σ^2 while for the second type the variance is $4\sigma^2$. Let $X_1, \ldots, X_m$ be m independent growth observations on the first type [so $E(X_i) = \mu$, $V(X_i) = \sigma^2$] and let $Y_1, \ldots, Y_n$ be n independent growth observations on the second type [$E(Y_i) = \mu$, $V(Y_i) = 4\sigma^2$].

a. Show that for any δ between 0 and 1, the estimator $\hat{\mu} = \delta\bar{X} + (1 - \delta)\bar{Y}$ is unbiased for μ.

b. For fixed m and n, compute $V(\hat{\mu})$ and then find the value of δ that minimizes $V(\hat{\mu})$. [HINT: Differentiate $V(\hat{\mu})$ with respect to δ.]

17. In Chapter 3, we defined a negative binomial rv as the number of failures that occur before the rth success in a sequence of independent and identical success/failure trials. The probability mass function (pmf) of X is

$$nb(x; r, p) = \begin{cases} \binom{x + r - 1}{x} p^r(1 - p)^x & x = 0, 1, 2, \ldots \\ 0 & \text{otherwise} \end{cases}$$

a. Suppose that $r \geq 2$. Show that

$$\hat{p} = (r - 1)/(X + r - 1)$$

is an unbiased estimator for p. [HINT: Write out $E(\hat{p})$ and cancel $x + r - 1$ inside the sum.]

b. A reporter wishes to interview five individuals who support a certain candidate, so begins asking people whether (S) or not (F) they support the candidate. If the sequence of responses is $SFFSFFFSS$, estimate p = the true proportion who support the candidate.

18. Let $X_1, X_2, \ldots, X_n$ be a random sample from a pdf $f(x)$ that is symmetric about μ, so that $\tilde{X}$ is an unbiased estimator of μ. If n is large, it can be shown that $V(\tilde{X}) \approx 1/(4n[f(\mu)]^2)$.

   a. Compare $V(\tilde{X})$ to $V(\bar{X})$ when the underlying distribution is normal.

   b. When the underlying pdf is Cauchy (see Example 6.6), $V(\bar{X}) = \infty$, so $\bar{X}$ is a terrible estimator. What is $V(\tilde{X})$ in this case when n is large?

19. An investigator wishes to estimate the proportion of students at a certain university who have violated the honor code. Having obtained a random sample of n students, she realizes that asking each, "Have you violated the honor code?" will probably result in some untruthful responses. Consider the following scheme, called a **randomized response** technique. The investigator makes up a deck of 100 cards, of which 50 are of type I and 50 are of type II.

Type I: Have you violated the honor code (yes or no)?

Type II: Is the last digit of your telephone number a 0, 1, or 2 (yes or no)?

Each student in the random sample is asked to mix the deck, draw a card, and answer the resulting question truthfully. Because of the irrelevant question on type II cards, a yes response no longer stigmatizes the respondent, so we assume that responses are truthful. Let p denote the proportion of honor-code violators (i.e., the probability of a randomly selected student being a violator), and let $\lambda = P(\text{yes response})$. Then λ and p are related by $\lambda = .5p + (.5)(.3)$.

   a. Let Y denote the number of yes responses, so $Y \sim \text{Bin}(n, \lambda)$. Thus, Y/n is an unbiased estimator of λ. Derive an estimator for p based on Y. If $n = 80$ and $y = 20$, what is your estimate? (HINT: Solve $\lambda = .5p + .15$ for p and then substitute Y/n in place of λ.)

   b. Use the fact that $E(Y/n) = \lambda$ to show that your estimator $\hat{p}$ is unbiased.

   c. If there were 70 type I and 30 type II cards, what would your estimator for p be?

6.2 Methods of Point Estimation

The definition of unbiasedness does not in general indicate how unbiased estimators can be derived. We now discuss two "constructive" methods for obtaining point estimators—the method of moments and the method of maximum likelihood. By constructive we mean that the general definition of each type of estimator suggests explicitly how to obtain the estimator in any specific problem. While maximum likelihood estimators are generally preferable to moment estimators because of certain efficiency properties, they often require significantly more computation than do moment estimators. It is sometimes the case that these methods yield unbiased estimators.

The Method of Moments

The basic idea of this method is to equate certain characteristics, such as the mean, to the corresponding population expected values. Then solving these equations for unknown parameter values yields the estimators.

Definition

> Let $X_1, \ldots, X_n$ be a random sample from a pmf or pdf $f(x)$. For $k = 1, 2, 3, \ldots$, the **kth population moment**, or **kth moment of the distribution** $f(x)$, is $E(X^k)$. The **kth sample moment** is $(1/n)\Sigma_{i=1}^n X_i^k$.

Thus, the first population moment is $E(X) = \mu$ and the first sample moment is $\Sigma X_i/n = \bar{X}$. The second population and sample moments are $E(X^2)$ and $\Sigma X_i^2/n$, respectively. The population moments will be functions of any unknown parameters $\theta_1, \theta_2, \ldots$.

Definition

> Let $X_1, X_2, \ldots, X_n$ be a random sample from a distribution with pmf or pdf $f(x;\ \theta_1, \ldots, \theta_m)$, where $\theta_1, \ldots, \theta_m$ are parameters whose values are unknown. Then the **moment estimators** $\hat{\theta}_1, \ldots, \hat{\theta}_m$ are obtained by equating the first m sample moments to the corresponding first m population moments and solving for $\theta_1, \ldots, \theta_m$.

If, for example, $m = 2$, $E(X)$ and $E(X^2)$ will be functions of θ_1 and θ_2. Setting $E(X) = (1/n) \Sigma X_i (= \bar{X})$ and $E(X^2) = (1/n) \Sigma X_i^2$ gives two equations in θ_1 and θ_2. The solution then defines the estimators. For estimating a population mean μ, the method gives $\mu = \bar{X}$, so the estimator is the sample mean.

Example 6.11

Let $X_1, X_2, \ldots, X_n$ represent a random sample of service times of n customers at a certain facility, where the underlying distribution is assumed exponential with parameter λ. Since there is only one parameter to be estimated, the estimator is obtained by equating $E(X)$ to $\bar{X}$. Since $E(X) = 1/\lambda$ for an exponential distribution, this gives $1/\lambda = \bar{X}$ or $\lambda = 1/\bar{X}$. The moment estimator of λ is then $\hat{\lambda} = 1/\bar{X}$.   ∎

Example 6.12

Let $X_1, \ldots, X_n$ be a random sample from a gamma distribution with parameters α and β. From Section 4.4, $E(X) = \alpha\beta$ and $E(X^2) = \beta^2\Gamma(\alpha + 2)/\Gamma(\alpha) = \beta^2(\alpha + 1)\alpha$. The moment estimators of α and β are obtained by solving

$$\bar{X} = \alpha\beta, \qquad \frac{1}{n} \Sigma X_i^2 = \alpha(\alpha + 1)\beta^2$$

Since $\alpha(\alpha + 1)\beta^2 = \alpha^2\beta^2 + \alpha\beta^2$ and the first equation implies $\alpha^2\beta^2 = \bar{X}^2$, the second equation becomes

$$\frac{1}{n} \Sigma X_i^2 = \bar{X}^2 = \alpha\beta^2$$

Now dividing each side of this second equation by the corresponding side of the first equation and substituting back gives the estimators

$$\hat{\alpha} = \frac{\bar{X}^2}{(1/n) \Sigma X_i^2 - \bar{X}^2}, \qquad \hat{\beta} = \frac{(1/n) \Sigma X_i^2 - \bar{X}^2}{\bar{X}}$$

To illustrate, the survival time data mentioned in Example 4.22 is

152, 115, 109, 94, 88, 137, 152, 77, 160, 165,
125, 40, 128, 123, 136, 101, 62, 153, 83, 69

with $\bar{x} = 113.5$ and $(1/20)\Sigma x_i^2 = 14{,}087.8$. The estimates are

$$\hat{\alpha} = \frac{(113.5)^2}{14{,}087.8 - (113.5)^2} = 10.7, \qquad \hat{\beta} = \frac{14{,}087.8 - (113.5)^2}{113.5} = 10.6$$

These estimates of α and β differ from the values suggested by Gross and Clark because they used a different estimation technique. ∎

Example 6.13

Let $X_1, \ldots, X_n$ be a random sample from a generalized negative binomial distribution with parameters r and p (Section 3.5). Since $E(X) = r(1 - p)/p$ and $V(X) = r(1 - p)/p^2$, $E(X^2) = V(X) + [E(X)]^2 = r(1 - p)(r - rp + 1)/p^2$. Equating $E(X)$ to $\bar{X}$ and $E(X^2)$ to $(1/n)\Sigma X_i^2$ eventually gives

$$\hat{p} = \frac{\bar{X}}{(1/n)\ \Sigma X_i^2 - \bar{X}^2}, \qquad \hat{r} = \frac{\bar{X}^2}{(1/n)\Sigma X_i^2 - \bar{X}^2 - \bar{X}}$$

As an illustration, Reep, Pollard, and Benjamin ("Skill and Chance in Ball Games," *J. Royal Stat. Soc.*, 1971, pp. 623–629) consider the negative binomial distribution as a model for the number of goals per game scored by National Hockey League teams. The data for 1966–1967 follows (420 games):

Goals	0	1	2	3	4	5	6	7	8	9	10
Freq.	29	71	82	89	65	45	24	7	4	1	3

Then,

$$\bar{x} = \Sigma x_i/420 = [(0)(29) + (1)(71) + \cdots + (10)(3)]/420 = 2.98$$

and

$$\Sigma x_i^2/420 = [(0)^2(29) + (1)^2(71) + \cdots + (10)^2(3)]/420 = 12.40$$

Thus,

$$\hat{p} = \frac{2.98}{12.40 - (2.98)^2} = .85, \qquad \hat{r} = \frac{(2.98)^2}{12.40 - (2.98)^2 - 2.98} = 16.5$$

Although r by definition must be positive, the denominator of $\hat{r}$ could be negative, indicating that the negative binomial distribution is not appropriate (or that the moment estimator is flawed). ∎

Maximum Likelihood Estimation

The method of maximum likelihood was first introduced by R. A. Fisher, a geneticist and statistician, in the 1920s. Most statisticians recommend this method, at least when the sample size is large, since the resulting estimators have certain desirable efficiency properties (see the proposition on page 270).

Example 6.14

A sample of ten new bike helmets manufactured by a certain company is obtained. Upon testing, it is found that the first, third, and tenth helmets are flawed while the others are not. Let $p = P(\text{flawed helmet})$ and define $X_1, \ldots, X_{10}$ by $X_i = 1$ if the ith helmet is flawed and zero otherwise. Then the observed x_i's are 1, 0, 1, 0, 0, 0, 0, 0, 0, 1, so the joint pmf of the sample is

$$f(x_1, x_2, \ldots, x_{10}; p) = p(1 - p)p \cdots p = p^3(1 - p)^7 \qquad (6.4)$$

We now ask "For what value of p is the observed sample most likely to have occurred?" That is, we wish to find the value of p that maximizes pmf (6.4), or, equivalently, maximizes the natural log of (6.4).* Since

$$\ln[f(x_1, \ldots, x_{10}; p)] = 3\ln(p) + 7\ln(1-p) \tag{6.5}$$

which is a differentiable function of p; equating the derivative of (6.5) to zero gives the maximizing value†

$$\frac{d}{dp}\ln[f(x_1, \ldots, x_{10}; p)] = \frac{3}{p} - \frac{7}{1-p} = 0 \Rightarrow p = \frac{3}{10} = \frac{x}{n}$$

where x is the observed number of successes (flawed helmets). The estimate of p is now $\hat{p} = \frac{3}{10}$. It is called the maximum likelihood estimate because for fixed $x_1, \ldots, x_{10}$, it is the parameter value that maximizes the likelihood (joint pmf) of the observed sample.

Note that if we had been told only that among the ten helmets there were three that were flawed, (6.4) would be replaced by the binomial pmf $\binom{10}{3}p^3(1-p)^7$, which is also maximized for $\hat{p} = \frac{3}{10}$. ∎

Definition

> Let $X_1, X_2, \ldots, X_n$ have joint pmf or pdf
>
> $$f(x_1, x_2, \ldots, x_n; \theta_1, \ldots, \theta_m) \tag{6.6}$$
>
> where the parameters $\theta_1, \ldots, \theta_m$ have unknown values. When $x_1, \ldots, x_n$ are the observed sample values and (6.6) is regarded as a function of $\theta_1, \ldots, \theta_m$, it is called the **likelihood function**. The maximum likelihood estimates (mle's) $\hat{\theta}_1, \ldots, \hat{\theta}_m$ are those values of the θ_i's that maximize the likelihood function, so that
>
> $$f(x_1, \ldots, x_n; \hat{\theta}_1, \ldots, \hat{\theta}_m)$$
> $$\geq f(x_1, \ldots, x_n; \theta_1, \ldots, \theta_m) \text{ for all } \theta_1, \ldots, \theta_m$$
>
> When the X_i's are substituted in place of the x_i's, the **maximum likelihood estimators** result.

The likelihood function tells us how likely the observed sample is as a function of the possible parameter values. Maximizing the likelihood gives the parameter values for which the observed sample is most likely to have been generated—that is, the parameter values that "agree most closely" with the observed data.

Example 6.15   Suppose $X_1, X_2, \ldots, X_n$ is a random sample from an exponential distribution with parameter λ. Because of independence, the likelihood function is a product of the

---

*Since $\ln[g(x)]$ is a monotonic function of $g(x)$, finding x to maximize $\ln[g(x)]$ is equivalent to maximizing $g(x)$ itself. In statistics, taking the logarithm frequently changes a product to a sum, which is easier to work with.
†This conclusion requires checking the second derivative, but the details are omitted.

individual pdf's:

$$f(x_1, \ldots, x_n; \lambda) = (\lambda e^{-\lambda x_1}) \cdot \ldots \cdot (\lambda e^{-\lambda x_n}) = \lambda^n e^{-\lambda \Sigma x_i}$$

The ln(likelihood) is

$$\ln[f(x_1, \ldots, x_n; \lambda)] = n \ln(\lambda) - \lambda \Sigma x_i$$

Equating $(d/d\lambda)$ [ln(likelihood)] to zero results in $n/\lambda - \Sigma x_i = 0$, or $\lambda = n/\Sigma x_i = 1/\bar{x}$. Thus, the mle is $\hat{\lambda} = 1/\bar{X}$; it is identical to the method of moments estimator [but is not an unbiased estimator, since $E(1/\bar{X}) \neq 1/E(\bar{X})$]. ∎

Example 6.16

7(B97

Let $X_1, \ldots, X_n$ be a random sample from a normal distribution. The likelihood function is

$$f(x_1, \ldots, x_n; \mu, \sigma^2) = \frac{1}{\sqrt{2\pi\sigma^2}} e^{-(x_1 - \mu)^2/(2\sigma^2)} \cdot \ldots \cdot \frac{1}{\sqrt{2\pi\sigma^2}} e^{-(x_n - \mu)^2/(2\sigma^2)}$$

$$= \left(\frac{1}{2\pi\sigma^2}\right)^{n/2} e^{-\Sigma(x_i - \mu)^2/(2\sigma^2)}$$

so

$$\ln[f(x_1, \ldots, x_n; \mu, \sigma^2)] = -\frac{n}{2} \ln(2\pi\sigma^2) - \frac{1}{2\sigma^2} \Sigma(x_i - \mu)^2$$

To find the maximizing values of μ and σ^2, we must take the partial derivatives of $\ln(f)$ with respect to μ and σ^2, equate them to zero, and solve the resulting two equations. Omitting the details, the resulting mle's are

$$\hat{\mu} = \bar{X}, \quad \hat{\sigma}^2 = \frac{\Sigma(X_i - \bar{X})^2}{n}$$

The mle of σ^2 is not the unbiased estimator, so two different principles of estimation (unbiasedness and maximum likelihood) yield two different estimators. ∎

Example 6.17

In Chapter 3, we discussed the use of the Poisson distribution for modeling the number of "events" that occur in a two-dimensional region. Assume that when the region R being sampled has area $a(R)$, the number X of events occurring in R has a Poisson distribution with parameter $\lambda a(R)$ (where λ is the expected number of events per unit area), and that nonoverlapping regions yield independent X's.

Suppose an ecologist selects n nonoverlapping regions $R_1, \ldots, R_n$ and counts the number of plants of a certain species found in each region. The joint pmf (likelihood) is then

$$p(x_1, \ldots, x_n; \lambda) = \frac{[\lambda \cdot a(R_1)]^{x_1} e^{-\lambda \cdot a(R_1)}}{x_1!} \cdot \ldots \cdot \frac{[\lambda \cdot a(R_n)]^{x_n} e^{-\lambda \cdot a(R_n)}}{x_n!}$$

$$= \frac{[a(R_1)]^{x_1} \cdot \ldots \cdot [a(R_n)]^{x_n} \cdot \lambda^{\Sigma x_i} \cdot e^{-\lambda \Sigma a(R_i)}}{x_1! \cdot \ldots \cdot x_n!}$$

The ln(likelihood) is

$$\ln[p(x_1, \ldots, x_n; \lambda)] = \Sigma x_i \cdot \ln[a(R_i)] + \ln(\lambda) \cdot \Sigma x_i - \lambda \Sigma a(R_i) - \Sigma \ln(x_i!)$$

Taking $d/d\lambda \ln(p)$ and equating it to zero yields

$$\frac{\Sigma x_i}{\lambda} - \Sigma a(R_i) = 0$$

so

$$\lambda = \frac{\Sigma x_i}{\Sigma a(R_i)}$$

The mle is then $\hat{\lambda} = \Sigma X_i / \Sigma a(R_i)$. This is intuitively reasonable because λ is the true density (plants per unit area), while $\hat{\lambda}$ is the sample density since $\Sigma a(R_i)$ is just the total area sampled. Because $E(X_i) = \lambda \cdot a(R_i)$, the estimator is unbiased.

Sometimes an alternative sampling procedure is used. Instead of fixing regions to be sampled, the ecologist will select n points in the entire region of interest and let $y_i = $ the distance from the ith point to the nearest plant. The cumulative distribution function (cdf) of $Y = $ distance to the nearest plant is

$$F_Y(y) = P(Y \le y) = 1 - P(Y > y) = 1 - P \begin{pmatrix} \text{no plants in a} \\ \text{circle of radius } y \end{pmatrix}$$

$$= 1 - \frac{e^{-\lambda \pi y^2}(\lambda \pi y^2)^0}{0!} = 1 - e^{-\lambda \cdot \pi y^2}$$

Taking the derivative of $F_Y(y)$ with respect to y yields

$$f_Y(y; \lambda) = \begin{cases} 2\pi\lambda y e^{-\lambda \pi y^2} & y \ge 0 \\ 0 & \text{otherwise} \end{cases}$$

If we now form the likelihood $f_Y(y_1; \lambda) \cdot \cdots \cdot f_Y(y_n; \lambda)$, differentiate ln(likelihood), and so on, the resulting mle is

$$\hat{\lambda} = \frac{n}{\pi \Sigma Y_i^2} = \frac{\text{Number of plants observed}}{\text{Total area sampled}}$$

which is also a sample density. It can be shown that in a sparse environment (small λ), the distance method is in a certain sense better, while in a dense environment the first sampling method is better.   ∎

Example 6.18   Let $X_1, \ldots, X_n$ be a random sample from a Weibull pdf

$$f(x; \alpha, \beta) = \begin{cases} \dfrac{\alpha}{\beta^\alpha} \cdot x^{\alpha - 1} \cdot e^{-(x/\beta)^\alpha} & x \ge 0 \\ 0 & \text{otherwise} \end{cases}$$

Writing the likelihood and ln(likelihood), then setting both $(\partial/\partial\alpha)\,[\ln(f)] = 0$ and $(\partial/\partial\beta)\,[\ln(f)] = 0$ yields the equations

$$\alpha = \left[\frac{\Sigma x_i^\alpha \cdot \ln(x_i)}{\Sigma x_i^\alpha} - \frac{\Sigma \ln(x_i)}{n}\right]^{-1}, \quad \beta = \left(\frac{\Sigma x_i^\alpha}{n}\right)^{1/\alpha}$$

These two equations cannot be solved explicitly to give general formulas for the mle's $\hat{\alpha}$ and $\hat{\beta}$. Instead, for each sample $x_1, \ldots, x_n$, the equations must be solved using an iterative numerical procedure. Even moment estimators of α and β are somewhat complicated (see Exercise 21). ∎

Estimating Functions of Parameters

In Example 6.16, we obtained the mle of σ^2 when the underlying distribution is normal. The mle of $\sigma = \sqrt{\sigma^2}$, as well as many other mle's, can be easily derived using the following proposition.

Proposition

> The Invariance Principle
>
> Let $\hat{\theta}_1, \hat{\theta}_2, \ldots, \hat{\theta}_m$ be the mle's of the parameters $\theta_1, \theta_2, \ldots, \theta_m$. Then the mle of any function $h(\theta_1, \theta_2, \ldots, \theta_m)$ of these parameters is the function $h(\hat{\theta}_1, \hat{\theta}_2, \ldots, \hat{\theta}_m)$ of the mle's.

Example 6.19
(Example 6.16 continued)
7 FEB 97

In the normal case, the mle's of μ and σ^2 are $\hat{\mu} = \bar{X}$ and $\hat{\sigma}^2 = \Sigma(X_i - \bar{X})^2/n$. To obtain the mle of the function $h(\mu, \sigma^2) = \sqrt{\sigma^2} = \sigma$, substitute the mle's into the function:

$$\hat{\sigma} = \sqrt{\hat{\sigma}^2} = \left[\frac{1}{n}\Sigma(X_i - \bar{X})^2\right]^{1/2}$$

The mle of σ is not the sample standard deviation S, though they are close unless n is quite small. ∎

Example 6.20
(Example 6.18 continued)

The mean value of an rv X that has a Weibull distribution is

$$\mu = \beta \cdot \Gamma(1 + 1/\alpha)$$

The mle of μ is therefore $\hat{\mu} = \hat{\beta}\Gamma(1 + 1/\hat{\alpha})$, where $\hat{\alpha}$ and $\hat{\beta}$ are the mle's of α and β. In particular, $\bar{X}$ is not the mle of μ, though it is an unbiased estimator. At least for large n, $\hat{\mu}$ is a better estimator than $\bar{X}$. ∎

A Desirable Property of the Maximum Likelihood Estimate

While the principle of maximum likelihood estimation has considerable intuitive appeal, the following proposition provides additional rationale for the use of mle's.

Proposition

> Under very general conditions on the joint distribution of the sample, when the sample size n is large, the maximum likelihood estimator of any parameter θ is approximately unbiased $[E(\hat{\theta}) \approx \theta]$ and has variance that is nearly as small as can be achieved by any estimator. Stated another way, the mle $\hat{\theta}$ is approximately the MVUE of θ.

Because of this result and the fact that calculus-based techniques can usually be used to derive the mle's (though often numerical methods, such as Newton's method, are necessary), maximum likelihood estimation is the most widely used estimation technique among statisticians. Many of the estimators used in the remainder of the book are mle's. Obtaining an mle, however, does require that the underlying distribution be specified.

Some Complications

Sometimes calculus cannot be used to obtain mle's.

Example 6.21   Suppose my waiting time for a bus is uniformly distributed on $[0, \theta]$ and the results $x_1, \ldots, x_n$ of a random sample from this distribution have been observed. Since $f(x;\ \theta) = 1/\theta$ for $0 \le x \le \theta$ and 0 otherwise,

$$f(x_1, \ldots, x_n;\ \theta) = \begin{cases} \dfrac{1}{\theta^n} & 0 \le x_1 \le \theta, \ldots, 0 \le x_n \le \theta \\ 0 & \text{otherwise} \end{cases}$$

As long as $\max(x_i) \le \theta$, the likelihood is $1/\theta^n$, which is positive, but as soon as $\theta < \max(x_i)$, the likelihood drops to 0. This is illustrated in Figure 6.4. Calculus will not work because the maximum of the likelihood occurs at a point of discontinuity, but the figure shows that $\hat{\theta} = \max(X_i)$. Thus, if my waiting times are 2.3, 3.7, 1.5, .4, and 3.2, then the mle is $\hat{\theta} = 3.7$.

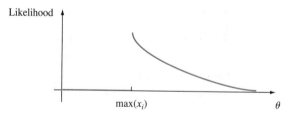

Figure 6.4 The likelihood function for Example 6.21   ∎

Example 6.22   A method that is often used to estimate the size of a wildlife population involves performing a capture/recapture experiment. In this experiment, an initial sample of

M animals is captured, each of these animals is tagged, and the animals are then returned to the population. After allowing enough time for the tagged individuals to mix into the population, another sample of size n is captured. With X = the number of tagged animals in the second sample, the objective is to use the observed x to estimate the population size N.

The parameter of interest is $\theta = N$, which can assume only integer values, so even after determining the likelihood function (pmf of X here), using calculus to obtain N would present difficulties. If we think of a success as a previously tagged animal being recaptured, then sampling is without replacement from a population containing M successes and $N - M$ failures, so that X is a hypergeometric rv and the likelihood function is

$$p(x; N) = h(x; n, M, N) = \frac{\binom{M}{x} \cdot \binom{N - M}{n - x}}{\binom{N}{n}}$$

The integer-valued nature of N notwithstanding, it would be difficult to take the derivative of $p(x; N)$. However, if we consider the ratio of $p(x; N)$ to $p(x; N - 1)$, we have

$$\frac{p(x; N)}{p(x; N - 1)} = \frac{(N - M) \cdot (N - n)}{N(N - M - n + x)}$$

This ratio is larger than 1 if and only if (iff) $N < Mn/x$. The value of N for which $p(x; N)$ is maximized is therefore the largest integer less than Mn/x. If we use standard mathematical notation $[r]$ for the largest integer less than or equal to r, the mle of N is $\hat{N} = [Mn/x]$. As an illustration, if $M = 200$ fish are taken from a lake and tagged, subsequently $n = 100$ fish are recaptured, and among the 100 there are $x = 11$ tagged fish, then $\hat{N} = [(200)(100)/11] = [1818.18] = 1818$. The estimate is actually rather intuitive; x/n is the proportion of the recaptured sample that is tagged, whereas M/N is the proportion of the entire population that is tagged. The estimate is obtained by equating these two proportions (estimating a population proportion by a sample proportion). ■

Suppose $X_1, X_2, \ldots, X_n$ is a random sample from a pdf $f(x; \theta)$ that is symmetric about θ, but that the investigator is unsure of the form of the f function. It is then desirable to use an estimator $\hat{\theta}$ that is *robust*—that is, one that performs well for a wide variety of underlying pdf's. One such estimator is a trimmed mean. In recent years, statisticians have proposed another type of estimator, called an *M-estimator*, based on a generalization of maximum likelihood estimation. Instead of maximizing the log likelihood $\Sigma\ln[f(x_i; \theta)]$ for a specified f, one maximizes $\Sigma\rho(x_i; \theta)$. The "objective function" ρ is selected to yield an estimator with good robustness properties. The book by David Hoaglin et al. (see the bibliography) contains a good exposition.

Exercises   Section 6.2 (20–30)

20. A random sample of n bike helmets manufactured by a certain company is selected. Let $X =$ the number among the n that are flawed and let $p = P(\text{flawed})$. Assume that only X is observed, rather than the sequence of S's and F's.

  a. Derive the maximum likelihood estimator of p. If $n = 20$ and $x = 3$, what is the estimate?

  b. Is the estimator of part (a) unbiased?

  c. If $n = 20$ and $x = 3$, what is the mle of the probability $(1 - p)^5$ that none of the next five helmets examined is flawed.

21. Let X have a Weibull distribution with parameters α and β, so

$$E(X) = \beta \cdot \Gamma(1 + 1/\alpha)$$

$$V(X) = \beta^2\{\Gamma(1 + 2/\alpha) - [\Gamma(1 + 1/\alpha)]^2\}$$

  a. Based on a random sample $X_1, \ldots, X_n$, write equations for the method of moments estimators of β and α. Show that, once the estimate of α has been obtained, the estimate of β can be found from a table of the gamma function and that the estimate of α is the solution to a complicated equation involving the gamma function.

  b. If $n = 20$, $\bar{x} = 28.0$, and $\Sigma x_i^2 = 16,500$, compute the estimates. [HINT: $[\Gamma(1.2)]^2/\Gamma(1.4) = .95$.]

22. Let X denote the proportion of allotted time that a randomly selected student spends working on a certain aptitude test and suppose the pdf of X is

$$f(x;\ \theta) = \begin{cases} (\theta + 1)x^\theta & 0 \le x \le 1 \\ 0 & \text{otherwise} \end{cases}$$

where $-1 < \theta$. A random sample of ten students yields data $x_1 = .92$, $x_2 = .79$, $x_3 = .90$, $x_4 = .65$, $x_5 = .86$, $x_6 = .47$, $x_7 = .73$, $x_8 = .97$, $x_9 = .94$, $x_{10} = .77$.

  a. Use the method of moments to obtain an estimator of θ and then compute the estimate for this data.

  b. Obtain the maximum likelihood estimator of θ and then compute the estimate for the given data.

23. Two different computer systems are monitored for a total of n weeks. Let X_i denote the number of breakdowns of the first system during the ith week and suppose the X_i's are independent and drawn from a Poisson distribution with parameter λ_1. Similarly, let Y_i denote the number of breakdowns of the second

system during the ith week and assume independence with each Y_i Poisson with parameter λ_2. Derive the mle's of λ_1, λ_2, and $\lambda_1 - \lambda_2$. [HINT: Using independence, write the joint pmf (likelihood) of the X_i's and Y_i's together.]

24. Refer to Exercise 20. Instead of selecting $n = 20$ helmets to examine, suppose I examine helmets in succession until I have found $r = 3$ flawed ones. If the twentieth helmet is the third flawed one (so that the number of helmets examined that were not flawed is $x = 17$), what is the mle of p? Is this the same as the estimate in Exercise 20? Why, or why not? Is it the same as the estimate computed from the unbiased estimator of Exercise 17?

25. The shear strength of each of ten test spot welds is determined, yielding the following data (psi):

392, 376, 401, 367, 389, 362, 409, 415, 358, 375.

  a. Assuming that shear strength is normally distributed, estimate the true average shear strength and standard deviation of shear strength using the method of maximum likelihood.

  b. Again assuming a normal distribution, estimate the strength value below which 95% of all welds will have their strengths. (HINT: What is the 95th percentile in terms of μ and σ? Now use the invariance principle.]

26. Refer to Exercise 25. Suppose we decide to examine another test spot weld. Let $X =$ shear strength of the weld. Use the given data to obtain the mle of $P(X \le 400)$. [HINT: $P(X \le 400) = \Phi((400 - \mu)/\sigma)$.]

27. Let $X_1, \ldots, X_n$ be a random sample from a gamma distribution with parameters α and β.

  a. Derive the equations whose solution yields the maximum likelihood estimators of α and β. Do you think they can be solved explicitly?

  b. Show that the mle of $\mu = \alpha\beta$ is $\hat{\mu} = \bar{X}$.

28. Let $X_1, X_2, \ldots, X_n$ represent a random sample from the Rayleigh distribution with density function given in Exercise 15. Determine

  a. The maximum likelihood estimator of θ and then calculate the estimate for the vibratory stress data given in that exercise. Is this estimator the same as the unbiased estimator suggested in Exercise 15?

b. The mle of the median of the vibratory stress distribution. (HINT: First express the median in terms of θ.)

29. Consider a random sample $X_1, X_2, \ldots, X_n$ from the shifted exponential pdf

$$f(x; \lambda, \theta) = \begin{cases} \lambda e^{-\lambda(x - \theta)} & x \geq \theta \\ 0 & \text{otherwise} \end{cases}$$

Taking $\theta = 0$ gives the pdf of the exponential distribution considered previously (with positive density to the right of zero). An example of the shifted exponential distribution appeared in Example 4.4, in which the variable of interest was time headway in traffic flow and $\theta = .5$ was the minimum possible time headway.

a. Obtain the maximum likelihood estimators of θ and λ.

b. If $n = 10$ time headway observations are made, resulting in the values 3.11, .64, 2.55, 2.20, 5.44, 3.42, 10.39, 8.93, 17.82, and 1.30, calculate the estimates of θ and λ.

30. At time $t = 0$, 20 identical components are put on test. The lifetime distribution of each is exponential with parameter λ. The experimenter then leaves the test facility unmonitored. On his return 24 hours later, the experimenter immediately terminates the test after noticing that $y = 15$ of the 20 components are still in operation (so 5 have failed). Derive the mle of λ. [HINT: Let $Y =$ the number that survive 24 hours. Then $Y \sim \text{Bin}(n, p)$. What is the mle of p? Now notice that $p = P(X_i \geq 24)$ where X_i is exponentially distributed. This relates λ to p, so the former can be estimated once the latter has been.]

Supplementary Exercises Chapter 6 (31–38)

31. An estimator $\hat{\theta}$ is said to be **consistent** if for any $\epsilon > 0$, $P(|\hat{\theta} - \theta| \geq \epsilon) \to 0$ as $n \to \infty$. That is, $\hat{\theta}$ is consistent if, as the sample size gets larger, it is less and less likely that $\hat{\theta}$ will be further than ϵ from the true value of θ. Show that $\bar{X}$ is a consistent estimator of μ when $\sigma^2 < \infty$ by using Chebyshev's inequality from Exercise 43 of Chapter 3. (HINT: The inequality can be rewritten in the form

$$P(|Y - \mu_Y| \geq \epsilon) \leq \sigma_Y^2/\epsilon$$

Now identify Y with $\bar{X}$.)

32. a. Let $X_1, \ldots, X_n$ be a random sample from a uniform distribution on $[0, \theta]$. Then the mle of θ is $\hat{\theta} = Y = \max(X_i)$. Use the fact that $Y \leq y$ iff each $X_i \leq y$ to derive the cdf of Y. Then show that the pdf of $Y = \max(X_i)$ is

$$f_Y(y) = \begin{cases} \dfrac{ny^{n-1}}{\theta^n} & 0 \leq y \leq \theta \\ 0 & \text{otherwise} \end{cases}$$

b. Use the result of part (a) to show that the mle is biased but that $(n + 1) \max(X_i)/n$ is unbiased.

33. At time $t = 0$, there is one individual alive in a certain population. A **pure birth process** then unfolds as follows. The time until the first birth is exponentially distributed with parameter λ. After the first birth, there are two individuals alive. The time until the first gives

birth again is exponential with parameter λ, and similarly for the second individual. Therefore, the time until the next birth is the minimum of two exponential (λ) variables, which is exponential with parameter 2λ. Similarly, once the second birth has occurred, there are three individuals alive, so the time until the next birth is an exponential rv with parameter 3λ, and so on (the memoryless property of the exponential distribution is being used here). Suppose the process is observed until the sixth birth has occurred and the successive birth times are 25.2, 41.7, 51.2, 55.5, 59.5, 61.8 (from which you should calculate the times between successive births). Derive the mle of λ. (HINT: The likelihood is a product of exponential terms.)

34. The **mean squared error** of an estimator $\hat{\theta}$ is $\text{MSE}(\hat{\theta}) = E(\hat{\theta} - \theta)^2$. If $\hat{\theta}$ is unbiased, then $\text{MSE}(\hat{\theta}) = V(\hat{\theta})$, but in general $\text{MSE}(\hat{\theta}) = V(\hat{\theta}) + (\text{bias})^2$. Consider the estimator $\hat{\sigma}^2 = KS^2$, where $S^2 =$ sample variance. What value of K minimizes the mean squared error of this estimator when the population distribution is normal? [HINT: It can be shown that

$$E[(S^2)^2] = (n + 1)\sigma^4/(n - 1)$$

In general, it is difficult to find $\hat{\theta}$ to minimize $\text{MSE}(\hat{\theta})$, which is why we look only at unbiased estimators and minimize $V(\hat{\theta})$.]

35. Let $X_1, \ldots, X_n$ be a random sample from a pdf that is symmetric about μ. An estimator for μ that has been found to perform well for a variety of underlying distributions is the *Hodges–Lehmann estimator*. To define it, first compute for each $i \le j$ and each $j = 1, 2, \ldots, n$ the pairwise average $\overline{X}_{i,j} = (X_i + X_j)/2$. Then the estimator is $\hat{\mu}$ = the median of the $\overline{X}_{i,j}$'s. Compute the value of this estimate using the data of Exercise 36 of Chapter 1. (HINT: Construct a square table with the x_i's listed on the left margin and on top. Then compute averages on and above the diagonal.)

36. When the population distribution is normal, the statistic median $\{\, |X_1 - \overline{X}|, \ldots, |X_n - \overline{X}| \,\}/.6745$ can be used to estimate σ. This estimator is more resistant to the effects of outliers (observations far from the bulk of the data) than is the sample standard deviation. Compute both the corresponding point estimate and s for the data of Example 6.2.

37. When the sample standard deviation S is based on a random sample from a normal population distribution, it can be shown that

$$E(S) = \sqrt{2/(n-1)}\,\Gamma(n/2)\sigma/\Gamma((n-1)/2)$$

Use this to obtain an unbiased estimator for σ of the form cS. What is c when $n = 20$?

38. Each of n specimens is to be weighed twice on the same scale. Let X_i and Y_i denote the two observed weights for the ith specimen. Suppose X_i and Y_i are independent of one another, each normally distributed with mean value μ_i (the true weight of specimen i) and variance σ^2.

a. Show that the maximum likelihood estimator of σ^2 is $\hat{\sigma}^2 = \Sigma(X_i - Y_i)^2/(4n)$. [HINT: If $\overline{z} = (z_1 + z_2)/2$, then $\Sigma(z_i - \overline{z})^2 = (z_1 - z_2)^2/2$.]

b. Is the mle $\hat{\sigma}^2$ an unbiased estimator of σ^2? Find an unbiased estimator of σ^2. [HINT: For any rv Z, $E(Z^2) = V(Z) + [E(Z)]^2$. Apply this to $Z = X_i - Y_i$.]

Bibliography

DeGroot, Morris, *Probability and Statistics* (2nd ed.), Addison-Wesley, Reading, MA, 1986. Includes an excellent discussion of both general properties and methods of point estimation; of particular interest are examples showing how general principles and methods can yield unsatisfactory estimators in particular situations.

Efron, Bradley, and Robert Tibshirani, *An Introduction to the Bootstrap*, Chapman and Hall, New York, 1993. The bible of the bootstrap.

Hoaglin, David, Frederick Mosteller, and John Tukey, *Understanding Robust and Exploratory Data Analysis*, Wiley, New York, 1983. Contains several good chapters on robust point estimation, including one on *M*-estimation.

Hogg, Robert, and Allen Craig, *Introduction to Mathematical Statistics* (4th ed.), Macmillan, New York, 1978. A good discussion of unbiasedness.

Larsen, Richard, and Morris Marx, *Introduction to Mathematical Statistics* (2nd ed.), Prentice Hall, Englewood Cliffs, NJ, 1985. A very good discussion of point estimation from a slightly more mathematical perspective than the present text.

Rice, John, *Mathematical Statistics and Data Analysis* (2nd ed.), Duxbury Press, Belmont, CA, 1994. A nice blending of statistical theory and data.

Statistical Intervals Based on a Single Sample

ΣΣΣΣΣΣΣΣΣΣΣΣΣΣΣΣΣΣ

Introduction

Almost any parameter that we might wish to estimate has as its set of possible values an entire interval of numbers. If, for example, we wish to estimate the true average net weight μ of fertilizer bags having nominal weight 50 lb, then μ might be any number between, say, 45 and 55. Even if the variable of interest is discrete, as with X = the number of movies rented by a randomly chosen customer leaving a certain video store, the true average number of movies rented per customer (μ) could be any number between, say, .5 and 4.0, and the true proportion p of customers renting at least one movie might be any number between 0 and 1.

The fact that the set of possible parameter values is a continuum implies that a point estimate, though it will represent our best guess for the true value of the parameter, may be close to that true value but will virtually never actually equal it. Because of this, reporting only the estimated value is generally unsatisfactory; some measure of how close the point estimate is likely to be to the true value is required. One way to do this, as suggested in Section 6.1, is to report both the estimate and its standard deviation (standard error). Then, if the estimator has at least approximately a normal distribution, we can be quite confident that the true value lies within 2 or 3 standard deviations (SDs) of the estimated value. This amounts to replacing the point estimate, a single number, by an entire interval of plausible values, and that is exactly what an interval estimate or confidence interval is—an interval of plausible values for the parameter being estimated. The degree of plausibility will be specified by a confidence level, so that we will speak of a 95% confidence interval (confidence level 95%) or a 99% interval. If the confidence level is high and the resulting interval is narrow, the investigator has reasonably precise knowledge of the parameter's value.

Basic Properties of Confidence Intervals

The basic concepts and properties of confidence intervals (CIs) are most easily introduced by first focusing on a simple, albeit somewhat unrealistic, problem situation. Suppose that the parameter of interest is a population mean μ and that

1. The population distribution is normal.
2. The value of the population standard deviation σ is known.

Normality of the population distribution is often a reasonable assumption. However, if the value of μ is unknown, it is implausible that the value of σ would be available (knowledge of a population's center typically precedes information concerning spread). In later sections, we will develop methods based on less restrictive assumptions.

Example 7.1   Industrial engineers who specialize in ergonomics are concerned with designing workspace and devices operated by workers so as to achieve high productivity and comfort. The paper "Studies on Ergonomically Designed Alphanumeric Keyboards" (*Human Factors*, 1985, pp. 175–187) reports on a study of preferred height for an experimental keyboard with large forearm–wrist support. A sample of $n = 31$ trained typists was selected, and the preferred keyboard height was determined for each typist. The resulting sample average preferred height was $\bar{x} = 80.0$ cm. Assuming that preferred height is normally distributed with $\sigma = 2.0$ cm (a value suggested by data in the paper), obtain a CI for μ, the true average preferred height for the population of all experienced typists.   ■

The actual sample observations $x_1, x_2, \ldots, x_n$ are assumed to be the result of a random sample $X_1, \ldots, X_n$ from a normal distribution with mean value μ and standard deviation σ. The results of Chapter 5 then imply that irrespective of the sample size n, the sample mean $\bar{X}$ is normally distributed with expected value μ and standard deviation $\sigma/\sqrt{n}$. Standardizing $\bar{X}$ by first subtracting its expected value and then dividing by its standard deviation yields the variable

$$Z = \frac{\bar{X} - \mu}{\sigma/\sqrt{n}} \tag{7.1}$$

which has a standard normal distribution. Because the area under the standard normal curve between -1.96 and 1.96 is .95,

$$P\left(-1.96 < \frac{\bar{X} - \mu}{\sigma/\sqrt{n}} < 1.96\right) = .95 \tag{7.2}$$

The next step in the development is to manipulate the inequalities inside the parentheses in (7.2) so that they appear in the equivalent form $l < \mu < u$, where the endpoints l and u involve $\bar{X}$ and $\sigma/\sqrt{n}$. This is achieved through the following sequence of operations, each one yielding inequalities equivalent to those we started with:

1. Multiply through by $\sigma/\sqrt{n}$ to obtain

$$-1.96 \cdot \frac{\sigma}{\sqrt{n}} < \bar{X} - \mu < 1.96 \cdot \frac{\sigma}{\sqrt{n}}$$

2. Subtract $\bar{X}$ from each term to obtain

$$-\bar{X} - 1.96 \cdot \frac{\sigma}{\sqrt{n}} < -\mu < -\bar{X} + 1.96 \cdot \frac{\sigma}{\sqrt{n}}$$

3. Multiply through by -1 to eliminate the minus sign in front of μ (which reverses the direction of each inequality) to obtain

$$\bar{X} + 1.96 \cdot \frac{\sigma}{\sqrt{n}} > \mu > \bar{X} - 1.96 \cdot \frac{\sigma}{\sqrt{n}}$$

that is,

$$\bar{X} - 1.96 \cdot \frac{\sigma}{\sqrt{n}} < \mu < \bar{X} + 1.96 \cdot \frac{\sigma}{\sqrt{n}}$$

Because each set of inequalities in the sequence is equivalent to the original one, the probability associated with each is .95. In particular,

$$P\left(\bar{X} - 1.96 \frac{\sigma}{\sqrt{n}} < \mu < \bar{X} + 1.96 \frac{\sigma}{\sqrt{n}} \right) = .95 \qquad (7.3)$$

The event inside the parentheses in (7.3) has a somewhat unfamiliar appearance; always before the random quantity has appeared in the middle with constants on both ends, as in $a \le Y \le b$. In (7.3) the random quantity appears on the two ends while the unknown constant μ appears in the middle. To interpret (7.3), think of a **random interval** having left endpoint $\bar{X} - 1.96 \cdot \sigma/\sqrt{n}$ and right endpoint $\bar{X} + 1.96 \cdot \sigma/\sqrt{n}$, which in interval notation is

$$\left(\bar{X} - 1.96 \cdot \frac{\sigma}{\sqrt{n}}, \quad \bar{X} + 1.96 \cdot \frac{\sigma}{\sqrt{n}} \right) \qquad (7.4)$$

The interval (7.4) is random because the two endpoints of the interval involve a random variable (rv). Note that the interval is centered at the sample mean $\bar{X}$, and extends $1.96\sigma/\sqrt{n}$ to each side of $\bar{X}$. Thus, the interval's length is $2 \cdot (1.96) \cdot \sigma/\sqrt{n}$, which is not random; only the location of the interval (its midpoint $\bar{X}$) is random (Figure 7.1). Now (7.3) can be paraphrased as "*the probability is .95 that the random interval (7.4) includes or covers the true value*

Figure 7.1 The random interval (7.4) centered at $\bar{X}$

of μ." Before any experiment is performed and any data is gathered, it is quite likely (probability .95) that μ will lie inside (7.4).

Definition

If after observing $X_1 = x_1, X_2 = x_2, \ldots, X_n = x_n$, we compute the observed sample mean $\bar{x}$ and then substitute $\bar{x}$ into (7.4) in place of $\bar{X}$, the resulting fixed interval is called a **95% confidence interval for** μ. This CI can be expressed either as

$$\left(\bar{x} - 1.96 \cdot \frac{\sigma}{\sqrt{n}}, \quad \bar{x} + 1.96 \cdot \frac{\sigma}{\sqrt{n}} \right) \quad \text{is a 95\% CI for } \mu$$

or as

$$\bar{x} - 1.96 \cdot \frac{\sigma}{\sqrt{n}} < \mu < \bar{x} + 1.96 \cdot \frac{\sigma}{\sqrt{n}} \quad \text{with 95\% confidence}$$

A concise expression for the interval is $\bar{x} \pm 1.96 \cdot \sigma/\sqrt{n}$, where $-$ gives the left endpoint (lower limit) and $+$ gives the right endpoint (upper limit).

Example 7.2
(Example 7.1 continued)

The quantities needed for computation of the 95% CI for true average preferred height are $\sigma = 2.0$, $n = 31$, and $\bar{x} = 80.0$. The resulting interval is

$$\bar{x} \pm 1.96 \cdot \frac{\sigma}{\sqrt{n}} = 80.0 \pm (1.96) \frac{2.0}{\sqrt{31}} = 80.0 \pm .7 = (79.3, 80.7)$$

That is, we can be highly confident that $79.3 < \mu < 80.7$. This interval is relatively narrow, indicating that μ has been rather precisely estimated. ∎

Interpreting a Confidence Interval

The confidence level 95% for the interval defined above was inherited from the probability .95 for the random interval (7.4). Intervals having other levels of confidence will be introduced shortly. For now, though, consider how 95% confidence can be interpreted.

Because we started with an event whose probability was .95—that the random interval (7.4) would capture the true value of μ—and then used the data in Example 7.1 to compute the fixed interval (79.3, 80.7), it is tempting to conclude that μ is within this fixed interval with probability .95. But by substituting $\bar{x} = 80.0$ in for $\bar{X}$, all randomness disappears; the interval (79.3, 80.7) is not a random interval, and μ is a constant (unfortunately unknown to us), so it is *incorrect* to write the statement $P(\mu$ lies in (79.3, 80.7)$) = .95$.

A correct interpretation of "95% confidence" relies on the long-run frequency interpretation of probability: To say that an event A has probability .95 is to say that if the experiment on which A is defined is performed over and over again, in the long run A will occur 95% of the time. Suppose we obtain another sample of typists' preferred heights and compute another 95% interval. Then suppose we consider repeating this for a third sample, a fourth sample, and so on. Let A be the event that $\bar{X} - 1.96 \cdot \sigma/\sqrt{n} < \mu < \bar{X} + 1.96 \cdot \sigma/\sqrt{n}$. Since $P(A) = .95$, in the long run 95% of our computed CIs will contain μ. This is illustrated in Figure 7.2, where the vertical line cuts the measurement axis at the true (but unknown) value of μ. Notice that of the 11 intervals pictured, only intervals 3 and 11 fail to contain μ. In the long run, only 5% of the intervals so constructed would fail to contain μ.

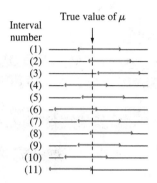

Figure 7.2 Repeated construction of 95% CIs

According to this interpretation, the confidence level 95% is not so much a statement about any particular interval such as (79.3, 80.7) but pertains to what would happen if a very large number of like intervals were to be constructed. Although this may seem unsatisfactory, the root of the difficulty lies with our interpretation of probability—it applies to a long sequence of replications of an experiment rather than just a single replication. There is another approach to the construction of and interpretation of CIs that uses the notion of subjective probability and Bayes' theorem, but the technical details are beyond the scope of this text; the book by Winkler (see the Chapter 2 bibliography) is a good source. The interval presented here (as well as each interval presented subsequently) is called a "classical" CI because its interpretation rests on the classical notion of probability (though the main ideas were developed as recently as the 1930s).

Other Levels of Confidence

Suppose we want a 99% CI rather than a 95% interval. Then rather than starting with a probability of .95, we must begin with a probability of .99. Since the area under the standard normal curve between -2.58 and 2.58 equals .99, replacing 1.96 with 2.58 in the definition yields a 99% interval.

This suggests that any desired level of confidence can be achieved by replacing 1.96 or 2.58 with the appropriate standard normal critical value. As Figure 7.3 shows, a probability of $1 - \alpha$ is achieved by using $z_{\alpha/2}$ in place of 1.96.

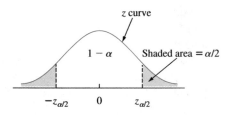

Figure 7.3 $P(-z_{\alpha/2} \le Z < z_{\alpha/2}) = 1 - \alpha$

Definition

> A **100(1 − α)% confidence interval** for the mean μ of a normal population when the value of σ is known is given by
>
> $$\left(\bar{x} - z_{\alpha/2} \cdot \frac{\sigma}{\sqrt{n}}, \quad \bar{x} + z_{\alpha/2} \cdot \frac{\sigma}{\sqrt{n}} \right) \tag{7.5}$$
>
> or, equivalently, by $\bar{x} \pm z_{\alpha/2} \cdot \sigma/\sqrt{n}$.

Example 7.3   A savings and loan association that finances a large number of home purchases in a particular region wanted information on the extent to which mortgage payments reduced the amount of disposable income during the initial year of occupancy for first-time home buyers. A sample of 35 recently granted mortgage applications was obtained, and for each the amount of the monthly mortgage payment as a percentage of take-home income (100 · payment/income) was computed. The sample average percentage was found to be 24.7. Assuming that σ is known to be 3.0 and that percentage is a normally distributed variable (among all successful applicants), compute a 90% CI for the true average percentage μ.

From the definition above, a 90% interval requires that $100(1 - \alpha) = 90$, so that $\alpha = .10$; then $z_{\alpha/2} = z_{.05} = 1.645$. Substituting this into (7.5) along with $\bar{x} = 24.7$, $\sigma = 3.0$, and $n = 35$ yields the interval

$$24.7 \pm (1.645) \frac{(3.0)}{\sqrt{35}} = 24.7 \pm .8 = (23.9, 25.5)$$ ∎

Although (7.5) can be used to obtain an interval with any desired degree of confidence, there are only three levels of confidence that appear at all frequently in the literature. These are 99%, 95%, and 90%, and the intervals are obtained by using 2.58, 1.96, and 1.645, respectively, in place of $z_{\alpha/2}$.

Confidence Level, Precision, and Choice of Sample Size

Why settle for a confidence level of 95% when a level of 99% is achievable? Because the price paid for the higher confidence level is a wider interval. Because the

95% interval extends $1.96 \cdot \sigma/\sqrt{n}$ to each side of $\bar{x}$, the length of the interval is $2(1.96) \cdot \sigma/\sqrt{n} = 3.92 \cdot \sigma/\sqrt{n}$. Similarly, the length of the 99% interval is $2(2.58) \cdot \sigma/\sqrt{n} = 5.16 \cdot \sigma/\sqrt{n}$. That is, we have more confidence in the 99% interval precisely because it is longer. The higher the desired degree of confidence, the longer the resulting interval. In fact, the only 100% CI for μ is $(-\infty, \infty)$, which is not terribly informative because, even before sampling, we knew that this interval covers μ.

If we think of the length of the interval as specifying its precision or accuracy, then the confidence level (or reliability) of the interval is inversely related to its precision. A highly reliable interval estimate may be imprecise in that the endpoints of the interval may be far apart, while a precise interval may entail relatively low reliability. Thus, it cannot be said unequivocally that a 99% interval is to be preferred to a 95% interval; the gain in reliability entails a loss in precision.

An appealing strategy is to specify both the desired confidence level and interval length and then determine the necessary sample size.

Example 7.4

Extensive monitoring of a computer time-sharing system has suggested that response time to a particular editing command is normally distributed with standard deviation 25 millisec. A new operating system has been installed, and it is desired to estimate the true average response time μ for the new environment. Assuming that response times are still normally distributed with $\sigma = 25$, what sample size is necessary to ensure that the resulting 95% CI has a length of (at most) 10? The sample size n must satisfy

$$10 = 2 \cdot (1.96)(25/\sqrt{n})$$

Rearranging this equation gives

$$\sqrt{n} = 2 \cdot (1.96)(25)/10 = 9.80$$

so

$$n = (9.80)^2 = 96.04$$

Since n must be an integer, a sample size of 97 is required. ∎

The general formula for the sample size n necessary to ensure an interval length L is obtained from $L = 2 \cdot z_{\alpha/2} \cdot \sigma/\sqrt{n}$ as

$$n = \left(2z_{\alpha/2} \cdot \frac{\sigma}{L} \right)^2$$

The smaller the desired length L, the larger n must be. In addition, n is an increasing function of σ (more population variability necessitates a larger sample size) and of the confidence level $100(1 - \alpha)$ (as α decreases, $z_{\alpha/2}$ increases).

Deriving a Confidence Interval

Let $X_1, X_2, \ldots, X_n$ denote the sample on which the CI for a parameter θ is to be based. Suppose a random variable satisfying the following two properties can be found:

1. The variable depends functionally on both $X_1, \ldots, X_n$ and θ.

2. The probability distribution of the variable does not depend on θ or on any other unknown parameters.

Let $h(X_1, X_2, \ldots, X_n; \theta)$ denote this random variable. For example, if the population distribution is normal with known σ and $\theta = \mu$, then $h(X_1, \ldots, X_n; \mu) = (\bar{X} - \mu)/(\sigma/\sqrt{n})$ satisfies both properties; it clearly depends functionally on μ, yet has the standard normal probability distribution, which does not depend on μ. In general, the form of the h function is usually suggested by examining the distribution of an appropriate estimator $\hat{\theta}$.

For any α between 0 and 1, constants a and b can be found to satisfy

$$P(a < h(X_1, \ldots, X_n; \theta) < b) = 1 - \alpha \tag{7.6}$$

Because of the second property, a and b do not depend on θ. In the normal example, $a = -z_{\alpha/2}$ and $b = z_{\alpha/2}$. Now suppose that the inequalities in (7.6) can be manipulated to isolate θ, giving the equivalent probability statement

$$P(l(X_1, X_2, \ldots, X_n) < \theta < u(X_1, X_2, \ldots, X_n)) = 1 - \alpha$$

Then $l(x_1, x_2, \ldots, x_n)$ and $u(x_1, \ldots, x_n)$ are the lower and upper confidence limits, respectively, for a $100(1 - \alpha)\%$ CI. In the normal example, we saw that $l(X_1, \ldots, X_n) = \bar{X} - z_{\alpha/2} \cdot \sigma/\sqrt{n}$ and $u(X_1, \ldots, X_n) = \bar{X} + z_{\alpha/2} \cdot \sigma/\sqrt{n}$.

Example 7.5   A theoretical model suggests that the time to breakdown of an insulating fluid between electrodes at a particular voltage has an exponential distribution with parameter λ (see Section 4.4). A random sample of $n = 10$ breakdown times yields the following sample data (in min): $x_1 = 41.53$, $x_2 = 18.73$, $x_3 = 2.99$, $x_4 = 30.34$, $x_5 = 12.33$, $x_6 = 117.52$, $x_7 = 73.02$, $x_8 = 223.63$, $x_9 = 4.00$, $x_{10} = 26.78$. A 95% CI for λ and for the true average breakdown time are desired.

Let $h(X_1, X_2, \ldots, X_n; \lambda) = 2\lambda\Sigma X_i$. It can be shown that this random variable has a probability distribution called a chi-squared distribution with $2n$ degrees of freedom (df) ($\nu = 2n$, where ν is the parameter of a chi-squared distribution as discussed in Section 4.4). Appendix Table A.6 pictures a typical chi-squared density curve and tabulates critical values that capture specified tail areas. The relevant number of degrees of freedom here is $2(10) = 20$. The $\nu = 20$ row of the table shows that 34.170 captures upper-tail area .025 and 9.591 captures lower-tail area .025 (upper-tail area .975). Thus, for $n = 10$,

$$P(9.591 < 2\lambda\Sigma X_i < 34.170) = .95$$

Division by $2\Sigma X_i$ isolates λ, yielding

$$P(9.591/(2\Sigma X_i) < \lambda < (34.170/(2\Sigma X_i)) = .95$$

The lower limit of the 95% CI for λ is $9.591/(2\Sigma x_i)$, and the upper limit is $34.170/(2\Sigma x_i)$. For the given data, $\Sigma x_i = 550.87$, giving the interval $(.00871, .03101)$.

The expected value of an exponential rv is $\mu = 1/\lambda$. Since

$$P(2\Sigma X_i/34.170 < 1/\lambda < 2\Sigma X_i/9.591) = .95$$

the 95% CI for true average breakdown time is $(2\Sigma x_i/34.170, 2\Sigma x_i/9.591) = (32.24, 114.87)$. This interval is obviously quite wide, reflecting substantial variability in breakdown times and a small-sample size.   ■

In general, the upper and lower confidence limits result from replacing each $<$ in (7.6) by $=$ and solving for θ. In the insulating fluid example just considered, $2\lambda\Sigma x_i = 34.170$ gives $\lambda = 34.170/(2\Sigma x_i)$ as the upper confidence limit, and the lower limit is obtained from the other equation. Notice that the two interval limits are not equidistant from the point estimate, since the interval is not of the form $\hat{\theta} \pm c$.

Bootstrap Confidence Intervals

The bootstrap technique was introduced in Chapter 6 as a way of estimating $\sigma_{\hat{\theta}}$. It can also be applied to obtain a CI for θ. Consider again estimating the mean μ of a normal distribution when σ is known. Let's replace μ by θ and use $\hat{\theta} = \overline{X}$ as the point estimator. Notice that $1.96\ \sigma/\sqrt{n}$ is the 97.5th percentile of the distribution of $\hat{\theta} - \theta$ $[P(\overline{X} - \mu < 1.96\sigma/\sqrt{n}) = P(Z < 1.96) = .9750]$. Similarly, $-1.96\ \sigma/\sqrt{n}$ is the 2.5th percentile, so

$$.95 = P(\text{2.5th percentile} < \hat{\theta} - \theta < \text{97.5th percentile})$$
$$= P(\hat{\theta} - \text{2.5th percentile} > \theta > \hat{\theta} - \text{97.5th percentile})$$

That is, with

$$l = \hat{\theta} - \text{97.5th percentile of } \hat{\theta} - \theta$$
$$u = \hat{\theta} - \text{2.5th percentile of } \hat{\theta} - \theta$$

(7.7)

the CI for θ is (l, u). In many cases, the percentiles in (7.7) cannot be calculated, but they *can* be estimated from bootstrap samples. Suppose we obtain $B = 1000$ bootstrap samples and calculate $\hat{\theta}_1^*, \ldots, \hat{\theta}_{1000}^*$, and $\overline{\theta}^*$ and then the 1000 differences $\hat{\theta}_1^* - \overline{\theta}^*, \ldots, \hat{\theta}_{1000}^* - \overline{\theta}^*$. The twenty-fifth largest and twenty-fifth smallest of these differences are estimates of the unknown percentiles in (7.7). Consult the Rice or Efron books cited in Chapter 6 for more information.

Exercises   Section 7.1 (1–9)

1. Consider a normal population distribution with the value of σ known.

    a. What is the confidence level for the interval $\bar{x} \pm 2.81\sigma/\sqrt{n}$?

    b. What is the confidence level for the interval $\bar{x} \pm 1.44\sigma/\sqrt{n}$?

    c. What value of $z_{\alpha/2}$ in the CI formula (7.5) results in a confidence level of 99.7%?

d. Answer the question posed in part (c) for a confidence level of 75%.

2. A CI is desired for the true average stray-load loss μ (watts) for a certain type of induction motor when the line current is held at 10 amps for a speed of 1500 rpm. Assume that stray-load loss is normally distributed with $\sigma = 3.0$.

a. Compute a 95% CI for μ when $n = 25$ and $\bar{x} = 58.3$.

b. Compute a 95% CI for μ when $n = 100$ and $\bar{x} = 58.3$.

c. Compute a 99% CI for μ when $n = 100$ and $\bar{x} = 58.3$.

d. Compute an 82% CI for μ when $n = 100$ and $\bar{x} = 58.3$.

e. How large must n be if the length of the 99% interval for μ is to be 1.0?

3. Assume that the helium porosity (in percentage) of coal samples taken from any particular seam is normally distributed with true standard deviation .75.

a. Compute a 95% CI for the true average porosity of a certain seam if the average porosity for 20 specimens from the seam was 4.85.

b. Compute a 98% CI for true average porosity of another seam based on 16 specimens with a sample average porosity of 4.56.

c. How large a sample size is necessary if the length of the 95% interval is to be .40?

4. On the basis of extensive tests, the yield point of a particular type of mild steel-reinforcing bar is known to be normally distributed with $\sigma = 100$. The composition of the bar has been slightly modified, but the modification is not believed to have affected either the normality or the value of σ.

a. Assuming this to be the case, if a sample of 25 modified bars resulted in a sample average yield point of 8439 lb, compute a 90% CI for the true average yield point of the modified bar.

b. How would you modify the interval in part (a) to obtain a confidence level of 92%?

5. By how much must the sample size n be increased if the length of the CI (7.5) is to be halved? If the sample size is increased by a factor of 25, what effect will this have on the length of the interval? Justify your assertions.

6. Let $\alpha_1 > 0$, $\alpha_2 > 0$, with $\alpha_1 + \alpha_2 = \alpha$. Then

$$P\left(-z_{\alpha_1} < \frac{\bar{X} - \mu}{\sigma/\sqrt{n}} < z_{\alpha_2}\right) = 1 - \alpha$$

a. Use this equation to derive a more general expression for a $100(1 - \alpha)\%$ CI for μ of which the interval (7.5) is a special case.

b. Let $\alpha = .05$ and $\alpha_1 = \alpha/4$, $\alpha_2 = 3\alpha/4$. Does this result in a shorter or longer interval than the interval (7.5)?

7. a. Under the same conditions as those leading to the interval (7.5), $P[(\bar{X} - \mu)/(\sigma/\sqrt{n}) < 1.645] = .95$. Use this to derive a one-sided interval for μ that has infinite length and provides a lower confidence bound on μ. What is this interval for the data in Exercise 3a?

b. Generalize the result of part (a) to obtain a lower bound with confidence level $100(1 - \alpha)\%$.

c. What is an analogous interval to that of part (a) that provides an upper bound on μ? Compute this 99% interval for the data of Exercise 2a.

8. A random sample of $n = 15$ heat pumps of a certain type yielded the following observations on lifetime (in years):

2.0, 1.3, 6.0, 1.9, 5.1, .4, 1.0, 5.3,
15.7, .7, 4.8, .9, 12.2, 5.3, .6

a. Assume that the lifetime distribution is exponential and use an argument parallel to that of Example 7.5 to obtain a 95% CI for expected (true average) lifetime.

b. How should the interval of part (a) be altered to achieve a confidence level of 99%?

c. What is a 95% CI for the standard deviation of the lifetime distribution? (HINT: What is the standard deviation of an exponential random variable?)

9. Consider the next 1000 95% CIs for μ that a statistical consultant will obtain for various clients. Suppose the data sets on which the intervals are based are selected independently of one another. How many of these 1000 intervals do you expect to capture the corresponding value of μ? What is the probability that between 940 and 960 of these intervals contain the corresponding value of μ? (HINT: Let $Y =$ the number among the 1000 intervals that contain μ. What kind of random variable is Y?)

Large-Sample Confidence Intervals for a Population Mean and Proportion

Large Sample CI

The CI for μ given in the previous section assumed that the population distribution is normal and that the value of σ is known. We now present a large-sample CI whose validity does not require these assumptions. After showing how the argument leading to this interval generalizes to yield other large-sample intervals, we focus on an interval for a population proportion p.

A Large-Sample Interval for μ

Let $X_1, X_2, \ldots, X_n$ be a random sample from a population having a mean μ and standard deviation σ. Provided that n is large, the Central Limit Theorem (CLT) implies that $\bar{X}$ has approximately a normal distribution whatever the nature of the population distribution. It then follows that $Z = (\bar{X} - \mu)/(\sigma/\sqrt{n})$ has approximately a standard normal distribution, so that

$$P\left(-z_{\alpha/2} < \frac{\bar{X} - \mu}{\sigma/\sqrt{n}} < z_{\alpha/2}\right) \approx 1 - \alpha$$

An argument parallel to that given in Section 7.1 yields $\bar{x} \pm z_{\alpha/2} \cdot \sigma/\sqrt{n}$ as a large-sample CI for μ with a confidence level of *approximately* $100(1 - \alpha)\%$. That is, when n is large, the CI for μ given earlier remains valid whatever the population distribution, provided that the qualifier "approximately" is inserted in front of the confidence level.

A practical difficulty with the above development is that computation of the interval requires the value of σ, which will almost never be known. Consider the standardized variable

$$Z = \frac{\bar{X} - \mu}{S/\sqrt{n}}$$

in which the sample standard deviation S replaces σ. Previously there was randomness only in the numerator of Z (by virtue of $\bar{X}$). Now there is randomness in both the numerator and the denominator—the values of both $\bar{X}$ and S vary from sample to sample. However, when n is large, the use of S rather than σ adds very little extra variability to Z. This is because the computed value s will be quite close to σ for almost all samples. More specifically, in this case the new Z also has approximately a standard normal distribution. Manipulation of the inequalities in a probability statement involving this new Z yields a general large-sample interval for μ.

Proposition

If n is sufficiently large, the standardized variable

$$Z = \frac{\bar{X} - \mu}{S/\sqrt{n}}$$

has approximately a standard normal distribution. This implies that

$$\bar{x} \pm z_{\alpha/2} \cdot \frac{s}{\sqrt{n}} \tag{7.8}$$

is a **large-sample confidence interval for** μ with confidence level approximately $100(1 - \alpha)\%$.

Generally speaking, $n > 30$ will be sufficient to justify the use of this interval.

Example 7.6   Heavy-metal pollution of various ecosystems is a serious environmental threat, in part because of the potential transference of hazardous substances to humans via food. The paper "Cadmium, Zinc, and Total Mercury Levels in the Tissues of Several Fish Species from La Plata River Estuary, Argentina" (*Environmental Monitoring and Assessment*, 1993, pp. 119–130) reports that, for a sample of $n = 56$ fish of the *Mugil liza* species, the sample mean zinc concentration in the liver was 9.15 $\mu g/g$ and the sample standard deviation was 1.27 $\mu g/g$. A CI for μ, the true average liver concentration of zinc for all fish of this species at the specified location, using a 95% confidence level, is

$$\bar{x} \pm 1.96 \cdot \frac{s}{\sqrt{n}} = 9.15 \pm (1.96) \cdot \frac{1.27}{\sqrt{56}} = 9.15 \pm .33$$

$$= (8.82, \ 9.48)$$

We can be highly confident, at the 95% confidence level, that $8.82 < \mu < 9.48$.

The paper also reports that for the *Pogonias cromis* species, $n = 61$, $\bar{x} = 3.08$, and $s = 1.71$, so a 99% CI for true average concentration is

$$3.08 \pm (2.58) \cdot \frac{1.71}{\sqrt{61}} = 3.08 \pm .56 = (2.52, \ 3.64)$$

Even though the sample size on which this interval is based is slightly larger than for the *Mugil liza* species, it is substantially wider both because of the higher confidence level and the larger value of s. Notice that if the population mean and standard deviation were $\mu = 3.08$ and $\sigma = 1.71$ and the distribution of concentration were normal, then $P(X < 0) = P(Z < -1.80) = .0359$. Because concentration cannot be negative, the magnitude of s relative to $\bar{x}$ suggests that the population distribution is not in fact normal but perhaps positively skewed. However, the validity of the CIs just calculated does not depend on the shape of the population distribution as long as n is large enough for the CLT to hold.   ∎

Unfortunately, the choice of sample size to yield a desired interval length is not as straightforward here as it was for the case of known σ. This is because the length of (7.8) is $2 \cdot z_{\alpha/2} \cdot s/\sqrt{n}$; since s is not known before the data has been gathered, the length of the interval cannot be determined solely by choice of n. The only option for an investigator who wishes to specify a desired length is to make a guess as to what the value of s (or σ) might be—by being conservative and guessing at a large value of s, an n larger than necessary will be chosen.

A General Large-Sample Confidence Interval

The large-sample intervals $\bar{x} \pm z_{\alpha/2} \cdot \sigma/\sqrt{n}$ and $\bar{x} \pm z_{\alpha/2} \cdot s/\sqrt{n}$ are special cases of a general large-sample CI for a parameter θ. Suppose that $\hat{\theta}$ is an estimator satisfying the following properties: (1) It has approximately a normal distribution, (2) it is (at least approximately) unbiased, and (3) an expression for $\sigma_{\hat{\theta}}$, the standard deviation of $\hat{\theta}$, is available. For example, in the case $\theta = \mu$, $\hat{\mu} = \bar{X}$ is an unbiased estimator whose distribution is approximately normal when n is large and $\sigma_{\hat{\mu}} = \sigma_{\bar{X}} = \sigma/\sqrt{n}$. Standardizing $\hat{\theta}$ yields the rv $Z = (\hat{\theta} - \theta)/\sigma_{\hat{\theta}}$, which has approximately a standard normal distribution. This justifies the probability statement

$$P\left(-z_{\alpha/2} < \frac{\hat{\theta} - \theta}{\sigma_{\hat{\theta}}} < z_{\alpha/2}\right) \approx 1 - \alpha \tag{7.9}$$

Suppose, first, that $\sigma_{\hat{\theta}}$ does not involve any unknown parameters (e.g., known σ in the case $\theta = \mu$). Then replacing each $<$ in (7.9) by $=$ results in $\theta = \hat{\theta} \pm z_{\alpha/2} \cdot \sigma_{\hat{\theta}}$, so the lower and upper confidence limits are $\hat{\theta} - z_{\alpha/2} \cdot \sigma_{\hat{\theta}}$ and $\hat{\theta} + z_{\alpha/2} \cdot \sigma_{\hat{\theta}}$, respectively. Now suppose that $\sigma_{\hat{\theta}}$ does not involve θ but does involve at least one other unknown parameter. Let $s_{\hat{\theta}}$ be the estimate of $\sigma_{\hat{\theta}}$ obtained by using estimates in place of the unknown parameters (e.g., $s/\sqrt{n}$ estimates $\sigma/\sqrt{n}$). Under general conditions (essentially that $s_{\hat{\theta}}$ be close to $\sigma_{\hat{\theta}}$ for most samples), a valid CI is $\hat{\theta} \pm z_{\alpha/2} \cdot s_{\hat{\theta}}$. The interval $\bar{x} \pm z_{\alpha/2} \cdot s/\sqrt{n}$ is an example.

Finally, suppose that $\sigma_{\hat{\theta}}$ does involve the unknown θ. This is the case, for example, when $\theta = p$, a population proportion. Then $(\hat{\theta} - \theta)/\sigma_{\hat{\theta}} = z_{\alpha/2}$ can be difficult to solve. An approximate solution can often be obtained by replacing θ in $\sigma_{\hat{\theta}}$ by its estimate $\hat{\theta}$. This results in an estimated standard deviation $s_{\hat{\theta}}$, and the corresponding interval is again $\hat{\theta} \pm z_{\alpha/2} \cdot s_{\hat{\theta}}$. We now apply this in the case $\theta = p$.

A Large-Sample Confidence Interval for a Population Proportion

Let p denote the proportion of "successes" in a population, where *success* identifies an individual or object that has a specified property. A random sample of n individuals is to be selected, and X is the number of successes in the sample. Provided that n is small compared to the population size, X can be regarded as a binomial rv with $E(X) = np$ and $\sigma_X = \sqrt{np(1 - p)}$. Furthermore, if n is large, X has approximately a normal distribution.

The natural estimator of p is $\hat{p} = X/n$, the sample fraction of successes. Since $\hat{p}$ is just X multiplied by the constant $1/n$, $\hat{p}$ also has approximately a normal distribution. As shown in Section 6.1, $E(\hat{p}) = p$ (unbiasedness) and $\sigma_{\hat{p}} = \sqrt{p(1 - p)/n}$. The standard deviation $\sigma_{\hat{p}}$ does involve the unknown parameter p. Proceeding as indicated in the previous subsection on a general CI, estimate $\sigma_{\hat{p}}$ by $\sqrt{\hat{p}(1 - \hat{p})/n}$. This estimated standard deviation then appears in both the lower and upper confidence limits.

Proposition

> A large-sample $100(1 - \alpha)\%$ confidence interval for a population proportion p is
>
> $$\hat{p} \pm z_{\alpha/2} \sqrt{\hat{p}\hat{q}/n} \qquad\qquad (7.10)$$
>
> where $\hat{p} = x/n$, n is the sample size, x is the observed number of successes, and $\hat{q} = 1 - \hat{p}$. This interval can be used whenever $n\hat{p} \geq 5$ and $n\hat{q} \geq 5$.

An alternative derivation of the CI starts with the probability statement

$$P\left(-z_{\alpha/2} < \frac{\hat{p} - p}{\sqrt{p(1 - p)/n}} < z_{\alpha/2}\right) \approx 1 - \alpha$$

Replacing each $<$ by $=$ yields a quadratic equation in p whose solutions are

$$p = \frac{\hat{p} + \dfrac{z_{\alpha/2}^2}{2n} \pm z_{\alpha/2} \sqrt{\dfrac{\hat{p}\hat{q}}{n} + \dfrac{z_{\alpha/2}^2}{4n^2}}}{1 + (z_{\alpha/2}^2)/n}$$

The $+$ sign goes with the upper endpoint of the interval and the $-$ sign with the lower endpoint. Since n is large, the three terms involving $z_{\alpha/2}^2$ are negligible compared to other terms. Ignoring them gives (7.10).

Example 7.7

The 1983 Tylenol poisoning episode and other similar incidents have focused attention on the desirability of packaging various commodities in a tamper-resistant manner. The article "Tamper-Resistant Packaging: Is It Really?" (*Package Engr.*, June 1983, pp. 96–104) reports the results of a survey of consumer attitudes toward such packaging. Of the 270 consumers surveyed, 189 indicated that they would be willing to pay extra for tamper-resistant packaging. Let p denote the proportion of all consumers who, at the time of the survey, would pay extra for such packaging. Then $\hat{p} = x/n = 189/270 = .700$, and the estimated standard deviation (standard error) of $\hat{p}$ is

$$\sqrt{\hat{p}\hat{q}/n} = \sqrt{(.700)(.300)/270} = .0279$$

A 95% CI for p is

$$\hat{p} \pm z_{.025} \sqrt{\hat{p}\hat{q}/n} = .700 \pm (1.96)(.0279) = .700 \pm .055 = (.645, .755)$$

We can be highly confident (at the 95% confidence level) that between 64.5% and 75.5% of all consumers would be willing to absorb the extra cost of such packaging, a conclusion that should be encouraging to manufacturers. ∎

It is also possible to obtain a confidence interval for p when n is small. The interval is based directly on the binomial distribution.

Precision and Sample Size

The interval (7.10) extends a distance $z_{\alpha/2} \sqrt{\hat{p}\hat{q}/n}$ to both the left and the right of the point estimate $\hat{p}$, so the length of the interval is

$$L = 2z_{\alpha/2} \sqrt{\frac{\hat{p}\hat{q}}{n}} \tag{7.11}$$

Because the right side of (7.11) involves the unknown $\hat{p}$, the length is not determined solely by specification of n. Solving for n yields

$$n = \frac{4z_{\alpha/2}^2 \cdot \hat{p}\hat{q}}{L^2} = \frac{4z_{\alpha/2}^2 \hat{p}(1 - \hat{p})}{L^2} \tag{7.12}$$

There are now several possible approaches to a choice of n. One approach is to specify approximately what you expect the value of $\hat{p}$ to be, and use that in (7.12). Alternatively, a choice of n can be made by taking advantage of the fact that $\hat{p}(1 - \hat{p})$ is maximized for $\hat{p} = \frac{1}{2}$ and decreases as $\hat{p}$ moves away from $\frac{1}{2}$ in either direction. Suppose, for example, that p_0 represents an upper bound on what we believe the value of $\hat{p}$ will be and that $p_0 \leq \frac{1}{2}$. Then use of p_0 in (7.12) in place of $\hat{p}$ will give the maximum value of n necessary to arrive at length L. *The most conservative approach is to use $\hat{p} = \frac{1}{2}$, for then the length will be $\leq L$ no matter what $\hat{p}$ is actually observed.*

Example 7.8 The article "An Evaluation of Football Helmets Under Impact Conditions" (*Amer. J. Sports Medicine*, 1984, pp. 233–237) reports that when each football helmet in a random sample of 37 suspension-type helmets was subjected to a certain impact test, 24 showed damage. Let p denote the proportion of all helmets of this type that would show damage when tested in the prescribed manner. With $\hat{p} = 24/37 = .649$ and $\sqrt{\hat{p}\hat{q}/n} = .0785$, a 95% CI for p based on this data has length $2(1.96)(.0785) \approx .31$. This length indicates that p would be rather imprecisely estimated. The sample size n required to yield a 95% CI whose length is at most .10, whatever the resulting value of $\hat{p}$, is

$$n = \frac{4(1.96)^2(.5)^2}{(.10)^2} = 384.16. \qquad n = \left(\frac{z_{\alpha/2}}{L}\right)^2$$

It would be necessary to test 385 helmets in order to fulfill the requirement. ∎

Exercises   Section 7.2 (10–19)

10. A random sample of 110 lightning flashes in a certain region resulted in a sample average radar echo duration of .81 sec and a sample standard deviation of .34 sec ("Lightning Strikes to an Airplane in a Thunderstorm," *J. Aircraft*, 1984, pp. 607–611). Compute a 99% CI for the true average echo duration μ.

11. A sample of 56 research cotton samples resulted in a sample average percentage elongation of 8.17 and a sample standard deviation of 1.42 ("An Apparent Relation Between the Spiral Angle ϕ, the Percent Elongation E_1, and the Dimensions of the Cotton Fiber," *Textile Research J.*, 1978, pp. 407–410). Calculate a 95% large-sample CI for the true average percentage elongation μ. What assumptions are you making about the distribution of percentage elongation?

12. The paper "Ultimate Load Capacities of Expansion Anchor Bolts" (*J. Engr. Energy*, 1993, pp. 139–158) reports the following summary data on shear strength for a sample of 3/8 in. anchor bolts: $n = 78$, $\bar{x} = 4.25$ kip, and $s = 1.30$.

   a. Calculate a CI for true average shear strength using a confidence level of 90%.

   b. What is $P(\bar{X} - 1.28S/\sqrt{n} < \mu)$ when n is large? Use this result to calculate a 90% lower confidence bound for true average shear strength.

13. The sample average ultimate tensile strength for a sample of 35 high-strength magnetic alloy steel rings used in turbine generators was 152.3 ksi, while the sample standard deviation was 4.8 ksi. Obtain a 99% CI for the true average ultimate tensile strength of such rings.

14. The paper "Extravisual Damage Detection? Defining the Standard Normal Tree" (*Photogrammetric Engineering and Remote Sensing*, 1981, pp. 515–522) discusses the use of color infrared photography in identification of normal trees in Douglas fir stands. Among data reported were summary statistics for green-filter analytic optical densitometric measurements on samples of both healthy and diseased trees. For a sample of 69 healthy trees, the sample average dye-layer density was 1.028, and the sample standard deviation was .163. Compute a 95% CI for the true average dye-layer density μ for all such trees.

15. In a survey of 277 randomly selected adult female shoppers, 69 stated that whenever an advertised item is unavailable at their local supermarket, they request a raincheck ("Consumer Attitudes Toward Unavailability and Mispricing of Advertised Items by Grocery Stores," *J. Consumer Affairs*, 1977, no. 1, pp. 158–166). Obtain a 99% CI for the true proportion p of adult female shoppers who request a raincheck in such situations.

16. An article in the *Los Angeles Times* (Dec. 12, 1977) reports that the technique graphic stress telethermometry (GST) accurately detected 23 out of 29 known breast cancer cases. Construct a 90% CI for the true proportion of breast cancers that would be detected by the GST technique (because n is small, the interval will be quite wide).

17. A state legislator wishes to survey residents of her district to see what proportion of the electorate is aware of her position on using state funds to pay for abortions.

   a. What sample size is necessary if the 95% CI for p is to have length at most .10 irrespective of p (i.e., what n is necessary for the legislator to be 95% confident that $\hat{p}$ is within .05 of the true proportion)?

   b. If the legislator has strong reason to believe that at least $\frac{2}{3}$ of the electorate know of her position, how large a sample size would you recommend?

18. A manufacturer of VCR tapes sells tapes labeled as giving 6 hours of playing time. Sixty-four of these tapes are selected, and the actual playing time for each is determined. If the mean and standard deviation of the 64 observed playing times are 352 min and 8 min, respectively, construct a 99% CI for μ, the true average playing time for 6-hour tapes made by this manufacturer. Based on your interval, do you think the manufacturer could be accused of false advertising? Explain.

19. The superintendent of a large school district, having once had a course in probability and statistics, believes that the number of teachers absent on any given day has a Poisson distribution with parameter λ. Use the accompanying data on absences for 50 days to derive a large-sample CI for λ. [HINT: The mean and variance of a Poisson variable both equal λ, so

$$Z = \frac{\bar{X} - \lambda}{\sqrt{\lambda/n}}$$

has approximately a standard normal distribution. Now proceed as in the derivation of the interval for p by making a probability statement (with probability $1 - \alpha$) and solving the resulting inequalities for λ (see the argument just after (7.10)).]

Number of absences	0	1	2	3	4	5	6	7	8	9	10
Frequency	1	4	8	10	8	7	5	3	2	1	1

7.3 Intervals Based on a Normal Population Distribution

The CI for μ presented in Section 7.2 is valid provided that n is large enough for the CLT to apply. The resulting interval can be used whatever the nature of the population distribution. The CLT cannot be invoked, however, when n is small. In this case, one way to proceed is to make a specific assumption about the form of the population distribution and then derive a CI tailored to that assumption. For example, we could develop a CI for μ when the population is ———— nma distribution, another interval for the case of a Weibull popul $T-Dist$ Statisticians have indeed carried out this program for a number o ———— itional families. Because the normal distribution is more frequentl———— population model than is any other type of distribution, we will f————————for this situation.

Assumption

> The population of interest is normal, ———— constitutes a random sample from a normal distribution ———— unknown.

The key result underlying the interv———— as that for large n, the rv $Z = (\bar{X} - \mu)/(S/\sqrt{n})$ has approximately ———— distribution. When n is small, S is no longer likely to be close ———— lity in the distribution of Z arises from randomness in both the n————————lenominator. This implies that the probability distribution of $(\bar{X} - \mu)/(S/\sqrt{n})$ ———— more spread out than the standard normal distribution. The result ————————ces are based introduces a new family of probability distributions ———— of t distributions.

Theorem

> When $\bar{X}$ is the mean of a random sample of size n from a normal distribution with mean μ, the rv
>
> $$T = \frac{\bar{X} - \mu}{S/\sqrt{n}} \qquad (7.13)$$
>
> has a probability distribution called a t distribution with $n - 1$ degrees of freedom (df).

Properties of t Distributions

Before applying this theorem, we must first discuss properties of t distributions. While the variable of interest is still $(\bar{X} - \mu)/(S/\sqrt{n})$, we now denote it by T to emphasize that it does not have a standard normal distribution when n is small. Recall that a normal distribution is governed by two parameters, the mean μ and the standard deviation σ. A t distribution is governed by only one parameter, called the **number of degrees of freedom** of the distribution, abbreviated by df. We denote this parameter by the Greek letter ν. Possible values of ν are the positive integers 1, 2, 3, . . . , Each different value of ν corresponds to a different t distribution.

For any fixed value of the parameter ν, the density function that specifies the associated t curve has an even more complicated appearance than the normal density function. Fortunately, we need concern ourselves only with several of the more important features of these curves.

Properties of t Distributions

Let t_ν denote the density function curve for ν df.

1. Each t_ν curve is bell-shaped and centered at 0.
2. Each t_ν curve is more spread out than the standard normal (z) curve.
3. As ν increases, the spread of the corresponding t_ν curve decreases.
4. As $\nu \to \infty$, the sequence of t_ν curves approaches the standard normal curve (so the z curve is often called the t curve with df $= \infty$).

Figure 7.4 illustrates several of these properties for selected values of ν.

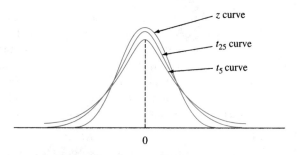

Figure 7.4 t_ν and z curves

The number of df for T in (7.13) is $n - 1$ because, although S is based on the n deviations $X_1 - \bar{X}, \ldots, X_n - \bar{X}, \Sigma(X_i - \bar{X}) = 0$ implies that only $n - 1$ of these are "freely determined." In general, the number of df for a t variable is the number of freely determined deviations on which the estimated standard deviation in the denominator of T is based.

Since we want to use T to obtain a CI in the same way that Z was previously used, it is necessary to establish notation analogous to z_α for the t distribution.

Notation

Let $t_{\alpha,\nu}$ = the number on the measurement axis for which the area under the t curve with ν df to the right of $t_{\alpha,\nu}$ is α; $t_{\alpha,\nu}$ is called a t **critical value**.

This notation is illustrated in Figure 7.5. Table A.5 in the appendix gives $t_{\alpha,\nu}$ for selected values of α and ν. This table also appears inside the back cover. The columns of the table correspond to different values of α. To obtain $t_{.05,15}$, go to the $\alpha = .05$ column, look down to the $\nu = 15$ row, and read $t_{.05,15} = 1.753$. Similarly, $t_{.05,22} = 1.717$ (.05 column, $\nu = 22$ row), and $t_{.01,22} = 2.508$.

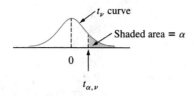

Figure 7.5 A pictorial definition of $t_{\alpha,\nu}$

The values of $t_{\alpha,\nu}$ exhibit regular behavior as we move across a row or down a column. For fixed ν, $t_{\alpha,\nu}$ increases as α decreases, since we must move farther to the right of zero to capture area α in the tail. For fixed α, as ν is increased (i.e., as we look down any particular column of the t table) the value of $t_{\alpha,\nu}$ decreases. This is because a larger value of ν implies a t distribution with smaller spread, so it is not necessary to go so far from zero to capture tail area α. Furthermore, $t_{\alpha,\nu}$ decreases more slowly as ν increases. Consequently, the table skips from $\nu = 30$ to 40 to 60 to 120 to ∞; many tables go directly from $\nu = 30$ to ∞. Because t_∞ is the standard normal curve, the familiar z_α values appear in the last row of the table. The rule of thumb suggested earlier for deciding whether the CLT could be used (if $n > 30$) comes from the approximate equality of the standard normal and t distributions for $\nu \geq 30$.

The One-Sample t Confidence Interval

The standardized variable T has a t distribution with $n - 1$ df, and the area under the corresponding t density curve between $-t_{\alpha/2,n-1}$ and $t_{\alpha/2,n-1}$ is $1 - \alpha$ (area $\alpha/2$ lies in each tail), so

$$P(-t_{\alpha/2,n-1} < T < t_{\alpha/2,n-1}) = 1 - \alpha \qquad (7.14)$$

Expression (7.14) differs from expressions in earlier sections in that T and $t_{\alpha/2,n-1}$ are used in place of Z and $z_{\alpha/2}$, but it can be manipulated in the same manner to obtain a confidence interval for μ.

Proposition

> Let $\bar{x}$ and s be the sample mean and sample standard deviation computed from the results of a random sample from a normal population with mean μ. Then a **$100(1 - \alpha)\%$ confidence interval for μ** is
>
> $$\left(\bar{x} - t_{\alpha/2,n-1} \cdot \frac{s}{\sqrt{n}}, \quad \bar{x} + t_{\alpha/2,n-1} \cdot \frac{s}{\sqrt{n}} \right) \qquad (7.15)$$
>
> or, more compactly, $\bar{x} \pm t_{\alpha/2,n-1} \cdot s/\sqrt{n}$.

Example 7.9   The results of a Wagner turbidity test performed on 15 samples of standard Ottawa testing sand were (in microamperes)

26.7, 25.8, 24.0, 24.9, 26.4, 25.9, 24.4, 21.7,
24.1, 25.9, 27.3, 26.9, 27.3, 24.8, 23.6

("Influence of Washing on Properties of Standard Ottawa Testing Sand," *J. Materials*, 1971, pp. 218–233). Figure 7.6 displays a normal probability plot of the data. Only the lowest point in the plot departs at all from linearity. In our view, the magnitude of the departure is not substantial enough to cast doubt on the assumption that turbidity has a normal distribution. To obtain a 95% CI for μ = the true average Wagner turbidity of all such sand samples, we first compute $\bar{x} = 25.31$ and $s = 1.58$. With $n = 15$ and $\alpha = .05$, $t_{\alpha/2,n-1} = t_{.025,14} = 2.145$. The desired interval is then

$$\bar{x} \pm t_{\alpha/2,n-1} \frac{s}{\sqrt{n}} = 25.31 \pm \frac{(2.145)(1.58)}{\sqrt{15}} = 25.31 \pm .88$$

$$= (24.43, \ 26.19)$$

Figure 7.6 Normal probability plot of the turbidity data

Here is MINITAB output resulting from a request for a t CI (SE MEAN = $s/\sqrt{n}$); the user specifies the confidence level:

```
            N    MEAN   STDEV   SE MEAN   95.0 PERCENT C.I.
turbid     15   25.313  1.579    0.408    ( 24.439, 26.188)
```
■

Unfortunately, it is not easy to select n to control the length of the t interval. This is because the length involves the unknown (before the data is collected) s and because n enters not only through $1/\sqrt{n}$ but also through $t_{\alpha/2,n-1}$. As a result, an appropriate n can be obtained only by trial and error.

In Chapter 15, we will discuss a small-sample CI for μ that is valid provided only that the population distribution is symmetric, a weaker assumption than normality.

PI
future value

However, when the population distribution is normal, the ⬚⬚⬚⬚⬚⬚ shorter than would be any other interval with the same confide⬚⬚⬚⬚ ⬚⬚all and the investigator believes that the population distribution i⬚ ⬚⬚⬚⬚⬚⬚ ⬚⬚istician should be consulted.

A Prediction Interval for a Single Future Val⬚⬚

In many applications, an investigator wishes to *predict* ⬚⬚⬚⬚⬚⬚⬚⬚ variable to be observed at some future time, rather than to *estin*⬚⬚ ⬚⬚⬚ ⬚⬚⬚⬚ of that variable.

Example 7.10

21 FEB 97

Consider the following sample of fat content (in percei⬚⬚⬚⬚⬚⬚ ⬚⬚ndomly selected hot dogs ("Sensory and Mechanical Asses⬚⬚⬚⬚⬚ ⬚⬚ ⬚⬚⬚ ⬚⬚ality of Frankfurters," *J. Texture Studies*, 1990, pp. 395–409):

25.2   21.3   22.8   17.0   29.8   21.0   25.5   16.0   20.9   19.5

Assuming that these were selected from a normal population distribution, a 95% CI for (interval estimate of) the population mean fat content is

$$\bar{x} \pm t_{.025,9} \cdot \frac{s}{\sqrt{n}} = 21.90 \pm 2.262 \cdot \frac{4.134}{\sqrt{10}} = 21.90 \pm 2.96$$

$$= (18.94, 24.86)$$

Suppose, however, you are going to eat a single hot dog of this type and want a *prediction* for the resulting fat content. A *point* prediction, analogous to a *point* estimate, is just $\bar{x} = 21.90$. This prediction unfortunately gives no information about reliability or precision. ∎

The general setup is as follows: We will have available a random sample X_1, $X_2, \ldots, X_n$ from a normal population distribution and wish to predict the value of X_{n+1}, a single future observation. A point predictor is $\bar{X}$, and the resulting prediction

21 FEB 97

error is $\bar{X} - X_{n+1}$. The expected value of the prediction error is

$$E(\bar{X} - X_{n+1}) = E(\bar{X}) - E(X_{n+1}) = \mu - \mu = 0$$

Since X_{n+1} is independent of $X_1, \ldots, X_n$, it is independent of $\bar{X}$, so the variance of the prediction error is

$$V(\bar{X} - X_{n+1}) = V(\bar{X}) + V(X_{n+1}) = \frac{\sigma^2}{n} + \sigma^2 = \sigma^2 \left(1 + \frac{1}{n}\right)$$

The prediction error is a linear combination of independent normally distributed rv's, so itself is normally distributed. Thus,

$$Z = \frac{(\bar{X} - X_{n+1}) - 0}{\sqrt{\sigma^2\left(1 + \frac{1}{n}\right)}} = \frac{\bar{X} - X_{n+1}}{\sqrt{\sigma^2\left(1 + \frac{1}{n}\right)}}$$

has a standard normal distribution. It can be shown that replacing σ by the sample standard deviation S (of $X_1, \ldots, X_n$) results in

$$T = \frac{\bar{X} - X_{n+1}}{S\sqrt{1 + \dfrac{1}{n}}} \sim t \text{ distribution with } n - 1 \text{ df}$$

Manipulating this T variable as $T = (\bar{X} - \mu)/(S/\sqrt{n})$ was manipulated in the development of a CI gives the following result.

Proposition

> A **prediction interval** (PI) for a single observation to be selected from a normal population distribution is
>
> $$\bar{x} \pm t_{\alpha/2,\, n-1} \cdot s\sqrt{1 + \frac{1}{n}} \qquad\qquad (7.16)$$
>
> The *prediction level* is $100(1 - \alpha)\%$.

The interpretation of a 95% prediction level is similar to that of a 95% confidence level; if the interval (7.16) is calculated for sample after sample, in the long run 95% of these intervals will include the corresponding future values of X.

Example 7.11
(Example 7.10 continued)

21 FEB 97

With $n = 10$, $\bar{x} = 21.90$, $s = 4.134$, and $t_{.025,9} = 2.262$, a 95% PI for the fat content of a single hot dog is

$$21.90 \pm (2.262)(4.134)\sqrt{1 + \frac{1}{10}} = 21.90 \pm 9.81$$

$$= (12.09,\ 31.71)$$

This interval is quite wide, indicating substantial uncertainty about fat content. Notice that the width of the PI is more than three times that of the CI. ∎

The error of prediction is $\bar{X} - X_{n+1}$, a difference between two random variables, whereas the estimation error is $\bar{X} - \mu$, the difference between a random variable and a fixed (but unknown) value. The PI is wider than the CI because there is more variability in the prediction error (due to X_{n+1}) than in the estimation error. In fact, as n gets arbitrarily large, the CI shrinks to the single value μ, and the PI approaches $\mu \pm z_{\alpha/2} \cdot \sigma$. There is uncertainty about a single X value even when there is no need to estimate.

Exercises   Section 7.3 (20–29)

20. Determine the values of the following quantities:

    a. $t_{.1,15}$ **b.** $t_{.05,15}$ **c.** $t_{.05,25}$ **d.** $t_{.05,40}$ **e.** $t_{.005,40}$

21. a. What is the 90th percentile of the t distribution with 22 df?

b. What is the 10th percentile of the t distribution with 22 df?

c. What is the 5th percentile of the t distribution with 16 df?

d. What is $P(T \leq 3.365)$ when T has a t distribution with 5 df?

e. What is $P(-2.306 < T < 2.306)$ when T has a t distribution with 8 df?

f. What is $P(|T| \geq 3.435)$ when T has a t distribution with 26 df?

22. A random sample of $n = 8$ E-glass fiber test specimens of a certain type yielded a sample average interfacial shear yield stress of 30.2 and a sample standard deviation of 3.1 ("On Interfacial Failure in Notched Unidirectional Glass/Epoxy Composites," *J. Composite Materials*, 1985, pp. 276–286). Assuming that interfacial shear yield stress is normally distributed, compute a 95% CI for true average stress (as did the authors of the paper).

23. In a study of the flammability of material used in children's sleepwear, the char length (inches) for five samples of washed acetate/nylon brushed tricot fabric was measured. The resulting observations were 8.1, 10.4, 9.5, 8.9, 10.7.

a. Assuming that char length is a normally distributed variable, compute a 95% CI for the true average char length for this fabric.

b. Use the fact that $P(T \geq -t_{\alpha,n-1}) = 1 - \alpha$ to obtain a $100(1 - \alpha)\%$ upper confidence bound for the mean μ of a normal population distribution. Calculate a 95% bound for true average char length.

24. A triathlon consisting of swimming, cycling, and running is one of the more strenuous amateur sporting events. The paper "Cardiovascular and Thermal Response of Triathlon Performance" (*Medicine and Science in Sports and Exercise*, 1988, pp. 385–389) reports on a research study involving nine male triathletes. Maximum heart rate (beats/min) was recorded during performance of each of the three events. For swimming, the sample mean and sample standard deviation were 188.0 and 7.2, respectively. Assuming that the heart rate distribution is (approximately) normal, construct a 98% CI for true mean heart rate of triathletes while swimming.

25. Refer to Exercise 24 and calculate a prediction interval with a 95% prediction level for the maximum heart rate of a single triathlete.

26. Trisodium carboxymethylaxysuccinate (NaCMOS) is a chemical compound being considered for inclusion in various detergent products. Because it would eventually become a component of wastewater, an experiment was performed to assess its effects on aquatic life. Following exposure to 100 mg NaCMOS/L for a 16-hour period, the NaCMOS content was determined for each of six different goldfish. The resulting sample mean and sample standard deviation were 9.1 and 3.2, respectively. Compute a 90% CI for the true average NaCMOS content for goldfish undergoing such exposure ("Acute Fish Toxicity and Absorption Tests of an Experimental Detergent Builder, Trisodium Carboxymethylaxysuccinate," *J. Testing and Evaluation*, 1979, pp. 16–17).

27. For each of 18 preserved cores from oil-wet carbonate reservoirs, the amount of residual gas saturation after a solvent injection was measured at water flood-out. Observations, in percentage of pore volume, were

23.5, 31.5, 34.0, 46.7, 45.6, 32.5,
41.4, 37.2, 42.5, 46.9, 51.5, 36.4,
44.5, 35.7, 33.5, 39.3, 22.0, 51.2

Compute a 98% CI for the true average amount of residual gas saturation ("Relative Permeability Studies of Gas-Water Flow Following Solvent Injection in Carbonate Rocks," *Soc. Petroleum Engineers J.*, 1976, pp. 23–30).

28. A more extensive tabulation of t critical values than that given in Appendix Table A.5 shows that $t_{.25,20} = .687$, $t_{.20,20} = .860$, and $t_{.15,20} = 1.064$. What is the confidence level for each of the following three CIs for the mean μ of a normal population distribution? Which of the three intervals would you recommend be used, and why?

a. $(\bar{x} - .687s/\sqrt{21}, \bar{x} + 1.725s/\sqrt{21})$

b. $(\bar{x} - .860s/\sqrt{21}, \bar{x} + 1.325s/\sqrt{21})$

c. $(\bar{x} - 1.064s/\sqrt{21}, \bar{x} + 1.064s/\sqrt{21})$

29. Refer to Example 7.10, in which the fat content was given for each of ten hot dogs in a sample. Suppose a friend of yours is going to eat two hot dogs of this type. Denote their fat contents by X_{11} and X_{12} and let $\bar{X}^* = (X_{11} + X_{12})/2$. Obtain a 95% PI for $\bar{X}^*$. (HINT: Consider the prediction error $\bar{X} - \bar{X}^*$ and proceed as we did with $\bar{X} - X_{11}$.)

7.4   Confidence Intervals for the Variance and Standard Deviation of a Normal Population

Although inferences concerning a population variance σ^2 or standard deviation σ are usually of less interest than those about a mean or proportion, there are occasions when such procedures are needed. In the case of a normal population distribution, inferences are based on the following result concerning the sample variance S^2.

Theorem

Let $X_1, X_2, \ldots, X_n$ be a random sample from a normal distribution with parameters μ and σ^2. Then the rv

$$\frac{(n-1)S^2}{\sigma^2} = \frac{\Sigma(X_i - \bar{X})^2}{\sigma^2}$$

has a chi-squared (χ^2) probability distribution with $n - 1$ df.

As discussed in Sections 4.4 and 7.1, the chi-squared distribution is a continuous probability distribution with a single parameter ν, called the number of degrees of freedom, with possible values 1, 2, 3, ... The graphs of several χ^2 probability distribution functions (pdf's) are illustrated in Figure 7.7. Each pdf $f(x; \nu)$ is positive only for $x > 0$, and each has a positive skew (long upper tail), though the distribution moves rightward and becomes more symmetric as ν increases. To specify inferential procedures that use the chi-squared distribution, we need notation analogous to that for a t critical value $t_{\alpha,\nu}$.

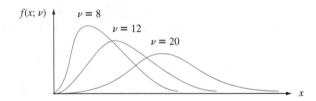

Figure 7.7 Graphs of chi-squared density functions

Notation

Let $\chi^2_{\alpha,\nu}$, called a **chi-squared critical value**, denote the number on the measurement axis such that α of the area under the chi-squared curve with ν df lies to the right of $\chi^2_{\alpha,\nu}$.

Because the t distribution is symmetric, it was necessary to tabulate only upper-tail critical values ($t_{\alpha,\nu}$ for small values of α). The chi-squared distribution is not symmetric, so Appendix Table A.6 contains values of $\chi^2_{\alpha,\nu}$ both for α near 0 and near 1, as illustrated in Figure 7.8b. For example, $\chi^2_{.025,14} = 26.119$ and $\chi^2_{.95,20}$ (the 5th percentile) = 10.851.

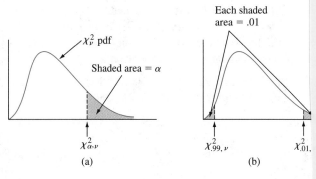

Figure 7.8 $\chi^2_{\alpha,\nu}$ notation illustrated

The rv $(n-1)S^2/\sigma^2$ satisfies the two properties on which the general method for obtaining a CI is based: It is a function of the parameter of interest σ^2, yet its probability distribution (chi-squared) does not depend on this parameter. The area under a chi-squared curve with ν df to the right of $\chi^2_{\alpha/2,\nu}$ is $\alpha/2$, as is the area to the left of $\chi^2_{1-\alpha/2,\nu}$. Thus, the area captured between these two critical values is $1-\alpha$. As a consequence of this and the theorem just stated,

$$P\left(\chi^2_{1-\alpha/2,n-1} < \frac{(n-1)S^2}{\sigma^2} < \chi^2_{\alpha/2,n-1}\right) = 1 - \alpha \tag{7.17}$$

The inequalities in (7.17) are equivalent to

$$\frac{(n-1)S^2}{\chi^2_{\alpha/2,n-1}} < \sigma^2 < \frac{(n-1)S^2}{\chi^2_{1-\alpha/2,n-1}}$$

Substitution of the computed value s^2 into the limits gives a CI for σ^2, and taking square roots gives an interval for σ.

A **100$(1-\alpha)$% confidence interval for the variance** σ^2 **of a normal population** has lower limit

$(n-1)s^2/\chi^2_{\alpha/2,n-1}$

and upper limit

$(n-1)s^2/\chi^2_{1-\alpha/2,n-1}$

A **confidence interval for** σ has lower and upper limits that are the square roots of the corresponding limits in the interval for σ^2.

Example 7.12   The accompanying data on breakdown voltage of electrically stressed circuits was read from a normal probability plot that appeared in the paper "Damage of Flexible

Printed Wiring Boards Associated with Lightning-Induced Voltage Surges" (*IEEE Transactions on Components, Hybrids, and Manuf. Tech.*, 1985, pp. 214–220). The straightness of the plot gave strong support to the assumption that breakdown voltage is approximately normally distributed.

1470, 1510, 1690, 1740, 1900, 2000, 2030, 2100, 2190, 2200, 2290, 2380, 2390, 2480, 2500, 2580, 2700

Let σ^2 denote the variance of the breakdown voltage distribution. The computed value of the sample variance is $s^2 = 137{,}324.3$, the point estimate of σ^2. With df $= n - 1 = 16$, a 95% CI requires $\chi^2_{.975,16} = 6.908$ and $\chi^2_{.025,16} = 28.845$. The interval is

$$\left(\frac{16(137{,}324.3)}{28.845}, \frac{16(137{,}324.3)}{6.908} \right) = (76{,}172.3, 318{,}064.4)$$

Taking the square root of each endpoint yields (276.0, 564.0) as the 95% CI for σ. These intervals are quite wide, reflecting substantial variability in breakdown voltage in combination with a small sample size.   ∎

CIs for σ^2 and σ when the population distribution is not normal can be difficult to obtain, even when the sample size is large. For such cases, consult a knowledgeable statistician.

Exercises   Section 7.4 (30–33)

30. Determine the values of the following quantities:

a. $\chi^2_{.1,15}$

b. $\chi^2_{.1,25}$

c. $\chi^2_{.01,25}$

d. $\chi^2_{.005,25}$

e. $\chi^2_{.99,25}$

f. $\chi^2_{.995,25}$

31. Determine the following:

a. The 95th percentile of the chi-squared distribution with $\nu = 10$.

b. The 5th percentile of the chi-squared distribution with $\nu = 10$.

c. $P(10.98 \le \chi^2 \le 36.78)$ where χ^2 is a chi-squared rv with $\nu = 22$.

d. $P(\chi^2 < 14.611$ or $\chi^2 > 37.652)$ where χ^2 is a chi-squared rv with $\nu = 25$.

32. The amount of lateral expansion (mils) was determined for a sample of $n = 9$ pulsed-power gas metal arc welds used in LNG ship containment tanks. The resulting sample standard deviation was $s = 2.81$ mils. Assuming normality, derive a 95% CI for σ^2 and for σ.

33. The following observations were made on fracture toughness of base plate of 18% nickel maraging steel ["Fracture Testing of Weldments," *ASTM Special Publ. No. 381*, 1965, pp. 328–356 (in ksi $\sqrt{\text{in.}}$, given in increasing order)]:

69.5, 71.9, 72.6, 73.1, 73.3, 73.5, 75.5, 75.7, 75.8, 76.1, 76.2, 76.2, 77.0, 77.9, 78.1, 79.6, 79.7, 79.9, 80.1, 82.2, 83.7, 93.7

Calculate a 99% CI for the standard deviation of the fracture toughness distribution. Is this interval valid whatever the nature of the distribution? Explain.

Supplementary Exercises Chapter 7 (34–49)

34. Television advertisers are becoming concerned over the use of VCRs to tape television shows because many viewers fast-forward through the commercials when viewing the taped shows. A survey conducted by A. C. Nielsen Co. of 1100 VCR owners found that 715 used the fast-forward feature to avoid commercials on taped programs (*Los Angeles Times*, Sept. 2, 1984). Construct and interpret a 95% CI for the proportion of all VCR owners who use the fast-forward feature to avoid advertisements.

35. Analysis of the venom of seven 8-day-old worker bees yielded the following observations on histamine content (nanograms):

649, 832, 418, 530, 384, 899, 755

Compute a 90% CI for the true average histamine content for all worker bees of this age. Were any assumptions necessary to justify your computation?

36. A sample of 40 transistors of a certain type yielded a sample average saturation current of 9.46 milliamps and a sample standard deviation of .58. Compute a 95% CI for the true average saturation current for transistors of this type. What assumptions, if any, are you making?

37. The financial manager of a large department-store chain selected a random sample of 200 of its credit card customers and found that 136 had incurred an interest charge during the previous year because of an unpaid balance.

   a. Compute a 90% CI for the true proportion of credit card customers who incurred an interest charge during the previous year.

   b. If the desired length of the 90% interval is .05, what sample size is necessary to ensure this?

   c. Compute an 82% CI for the true proportion.

38. The reaction time (RT) to a stimulus is the interval of time commencing with stimulus presentation and ending with the first discernible movement of a certain type. The paper "Relationship of Reaction Time and Movement Time in a Gross Motor Skill" (*Perceptual and Motor Skills*, 1973, pp. 453–454) reports that the sample average RT for 16 experienced swimmers to a pistol start was .214 sec and the sample standard deviation was .036 sec. Making any necessary assumptions, derive a 90% CI for true average RT for all experienced swimmers.

39. Aphid infestation of fruit trees can be controlled either by spraying with pesticide or by inundation with ladybugs. In a particular area, four different groves of fruit trees are selected for experimentation. The first three groves are sprayed with pesticides 1, 2, and 3, respectively, and the fourth is treated with ladybugs, with the following results on yield:

Treatment	$n_i =$ number of Trees	$\bar{x}_i$ (bushels/tree)	s_i
1	100	10.5	1.5
2	90	10.0	1.3
3	100	10.1	1.8
4	120	10.7	1.6

Let $\mu_i =$ the true average yield (bushels/tree) after receiving the ith treatment. Then

$$\theta = \tfrac{1}{3}(\mu_1 + \mu_2 + \mu_3) - \mu_4$$

measures the difference in true average yields between treatment with pesticides and treatment with ladybugs. When n_1, n_2, n_3, and n_4 are all large, the estimator $\hat{\theta}$ obtained by replacing each μ_i by $\bar{X}_i$ is approximately normal. Use this to derive a large-sample $100(1 - \alpha)\%$ CI for θ and compute the 95% interval for the given data.

40. It is important that face masks used by firefighters be able to withstand high temperatures because firefighters commonly work in temperatures of 200–500°F. In a test of one type of mask, 11 of 35 had their lenses pop out at 250°. Construct a 90% CI for the true proportion of masks of this type whose lenses would pop out at 250°.

41. A manufacturer of college textbooks is interested in estimating the strength of the bindings produced by a particular binding machine. Strength can be measured by recording the force required to pull the pages from the binding. If this force is measured in pounds, how many books should be tested in order to estimate the average force required to break the binding to within .1 lb with 95% confidence? Assume that σ is known to be .8.

42. Chronic exposure to asbestos fiber is a well-known health hazard. The paper "The Acute Effects of Chrysotile Asbestos Exposure on Lung Function" (*Environ. Research*, 1978, pp. 360–372) reports results of a study based on a sample of construction workers who had been exposed to asbestos over a prolonged period. Among the data given in the article were the following (ordered) values of pulmonary compliance ($cm^3/cm\ H_2O$) for each of 16 subjects 8 months after the exposure period (pulmonary compliance is a measure of lung elasticity, or how effectively the lungs are able to inhale and exhale):

167.9 180.8 184.8 189.8 194.8 200.2
201.9 206.9 207.2 208.4 226.3 227.7
228.5 232.4 239.8 258.6

a. Is it plausible that the population distribution is normal?

b. Compute a 95% CI for the true average pulmonary compliance after such exposure.

43. In Example 6.7, we introduced the concept of a censored experiment in which n components are put on test and the experiment terminates as soon as r of the components have failed. Suppose component lifetimes are independent, each having an exponential distribution with parameter λ. Let Y_1 denote the time at which the first failure occurs, Y_2 the time at which the second failure occurs , and so on, so that $T_r = Y_1 + \cdots + Y_r + (n - r)Y_r$ is the total accumulated lifetime at termination. Then it can be shown that $2\lambda T_r$ has a chi-squared distribution with $2r$ df. Use this fact to develop a $100(1 - \alpha)\%$ CI formula for true average lifetime $1/\lambda$. Compute a 95% CI from the data in the aforementioned example.

44. Let $X_1, X_2, \ldots, X_n$ be a random sample from a continuous probability distribution having median $\tilde{\mu}$ (so that $P(X_i \leq \tilde{\mu}) = P(X_i \geq \tilde{\mu}) = .5$).

a. Show that

$$P(\min(X_i) < \tilde{\mu} < \max(X_i)) = 1 - (\tfrac{1}{2})^{n-1}$$

so that $(\min(x_i), \max(x_i))$ is a $100(1 - \alpha)\%$ confidence interval for $\tilde{\mu}$ with $\alpha = (\tfrac{1}{2})^{n-1}$. [HINT: The complement of the event $\{\min(X_i) < \tilde{\mu} < \max(X_i)\}$ is $\{\max(X_i) \leq \tilde{\mu}\} \cup \{\min(X_i) \geq \tilde{\mu}\}$. But $\max(X_i) \leq \tilde{\mu}$ iff $X_i \leq \tilde{\mu}$ for all i.]

b. For each of six normal male infants, the amount of the amino acid alanine (mg/100 mL) was determined while on an isoleucine-free diet, resulting in the following data:

2.84 3.54 2.80 1.44 2.94 2.70

Compute a 97% CI for the true median amount of alanine for infants on such a diet ("The Essential Amino Acid Requirements of Infants," *Amer. J. Nutrition*, 1964, pp. 322–330).

c. Let $x_{(2)}$ denote the second smallest of the x_i's and $x_{(n-1)}$ denote the second largest of the x_i's. What is the confidence coefficient of the interval $(x_{(2)}, x_{(n-1)})$ for $\tilde{\mu}$?

45. Let $X_1, X_2, \ldots, X_n$ be a random sample from a uniform distribution on the interval $[0, \theta]$, so that

$$f(x) = \begin{cases} \dfrac{1}{\theta} & 0 \leq x \leq \theta \\ 0 & \text{otherwise} \end{cases}$$

Then if $Y = \max(X_i)$, it can be shown that the rv $U = Y/\theta$ has density function

$$f_U(u) = \begin{cases} nu^{n-1} & 0 \leq u \leq 1 \\ 0 & \text{otherwise} \end{cases}$$

a. Use $f_U(u)$ to verify that

$$P\left((\alpha/2)^{1/n} \leq \frac{Y}{\theta} \leq (1 - \alpha/2)^{1/n}\right) = 1 - \alpha$$

and use this to derive a $100(1 - \alpha)\%$ CI for θ.

b. Verify that $P(\alpha^{1/n} \leq Y/\theta \leq 1) = 1 - \alpha$ and derive a $100(1 - \alpha)\%$ CI for θ based on this probability statement.

c. Which of the two intervals derived above is shorter? If my waiting time for a morning bus is uniformly distributed and observed waiting times are $x_1 = 4.2$, $x_2 = 3.5$, $x_3 = 1.7$, $x_4 = 1.2$, and $x_5 = 2.4$, derive a 95% CI for θ by using the shorter of the two intervals.

46. Let $0 \leq \gamma \leq \alpha$. Then a $100(1 - \alpha)\%$ CI for μ when n is large is

$$\left(\bar{x} - z_\gamma \cdot \frac{s}{\sqrt{n}}, \quad \bar{x} + z_{\alpha - \gamma} \cdot \frac{s}{\sqrt{n}}\right)$$

The choice $\gamma = \alpha/2$ yields the usual interval derived in Section 7.2; if $\gamma \neq \alpha/2$, the above interval is not symmetric about $\bar{x}$. The length of this interval is $L = s(z_\gamma + z_{\alpha - \gamma})/\sqrt{n}$. Show that L is minimized for the choice $\gamma = \alpha/2$, so that the symmetric interval is the shortest. [HINTS: (a) By definition of z_α, $\Phi(z_\alpha) = 1 - \alpha$, so that $z_\alpha = \Phi^{-1}(1 - \alpha)$; (b) the relationship between the derivative of a function $y = f(x)$ and the inverse function $x = f^{-1}(y)$ is $(d/dy) f^{-1}(y) = 1/f'(x)$.]

47. Suppose $x_1, x_2, \ldots, x_n$ are observed values resulting from a random sample from a symmetric but possibly heavy-tailed distribution. Let $\bar{x}$ and f_s denote the sample median and fourth spread, respectively. Chapter 11 of *Understanding Robust and Exploratory Data Analysis* (see the bibliography in Chapter 6) suggests the following robust 95% CI for the population mean (point of symmetry):

$$\bar{x} \pm \left(\frac{\text{conservative } t \text{ critical value}}{1.075}\right) \cdot \frac{f_s}{\sqrt{n}}$$

The value of the quantity in parentheses is 2.10 for $n = 10$, 1.94 for $n = 20$, and 1.91 for $n = 30$. Compute this CI for the data of Exercise 33 and compare to the t CI appropriate for a normal population distribution.

48. If a statistic $l(X_1, X_2, \ldots, X_n)$ can be found for which $P(l(X_1, \ldots, X_n) < \theta) = 1 - \alpha$, then $l(x_1, x_2, \ldots, x_n)$ is called a **100(1 − α)% lower confidence bound** for θ.

  a. Use the results of Example 7.5 to obtain a 95% lower confidence bound for the parameter λ of an exponential distribution and calculate the bound based on the data given in the example.

  b. If lifetime X has an exponential distribution, the probability that lifetime exceeds t is $P(X > t) = e^{-\lambda t}$. Use the result of part (a) to obtain a 95% lower confidence bound for the probability that breakdown time exceeds 100 min.

49. If bearing diameter is normally distributed, the interval $\mu \pm 1.645\sigma$ will include 90% of all diameters (an interval for a designated percentage of individual measurements). Since μ and σ are typically unknown, consider the interval $\bar{x} \pm ks$. No matter how large a value of k is selected, we cannot be *sure* that the interval will include at least 90% of all individual values. We can, however, select k so that we are 95% confident that at least 90% of all diameters lie in the resulting **tolerance interval**. The value of k depends on n as well as the confidence level 95% and coverage percentage 90% (there are extensive tabulations of k for various n, confidence levels, and coverage percentages).

  a. Reconsider the breakdown voltage data of Example 7.12; for $n = 17$, $k = 2.400$. Calculate a 95% tolerance interval for capturing at least 90% of all breakdown voltages in the population distribution.

  b. Explain what 95% confidence means in this context.

  c. How does the width of the tolerance interval calculated in part (a) compare to the width of a 95% CI for μ?

  d. What happens to k and the width of the tolerance interval as $n \to \infty$ (note that $k = 1.874$ when $n = 100$ and 1.709 when $n = 1000$)? What happens to the CI width as $n \to \infty$?

Bibliography

DeGroot, Morris, *Probability and Statistics* (2nd ed.), Addison-Wesley, Reading, MA, 1986. A very good exposition of the general principles of statistical inference.

Hahn, Gerald, and William Meeker, *Statistical Intervals*, Wiley, New York, 1991. Everything you ever wanted to know about statistical intervals (confidence, prediction, tolerance, and others).

Larsen, Richard, and Morris Marx, *Introduction to Mathematical Statistics* (2nd ed.), Prentice Hall, Englewood Cliffs, NJ, 1986. Similar to DeGroot's presentation, but slightly less mathematical.

CHAPTER

8

Tests of Hypotheses Based on a Single Sample

$$\Sigma\Sigma\Sigma\Sigma\Sigma\Sigma\Sigma\Sigma\Sigma\Sigma\Sigma\Sigma\Sigma\Sigma\Sigma\Sigma\Sigma\Sigma\Sigma$$

Introduction

A parameter can be estimated from sample data either by a single number (a point estimate) or an entire interval of plausible values (a confidence interval). Frequently, however, the objective of an investigation is not to estimate a parameter but to decide which of two contradictory claims about the parameter is correct. Methods for accomplishing this comprise the part of statistical inference called *hypothesis testing*. In this chapter, we first discuss some of the basic concepts and terminology in hypothesis testing and then develop decision procedures for the most frequently encountered testing problems based on a sample from a single population.

8.1 Hypotheses and Test Procedures

A **statistical hypothesis**, or hypothesis, is a claim either about the value of a single population characteristic or about the values of several population characteristics. One example of a hypothesis is the claim $\mu = .75$, where μ is the true average inside diameter of a certain type of PVC pipe. Another example is the statement $p < .10$, where p is the proportion of defective circuit boards among all circuit boards produced by a certain manufacturer. If μ_1 and μ_2 denote the true average breaking strengths of two different types of twine, one hypothesis is the assertion that $\mu_1 - \mu_2 = 0$, and another is the statement $\mu_1 - \mu_2 > 5$.

In any hypothesis-testing problem, there are two contradictory hypotheses under consideration. One hypothesis might be the claim $\mu = .75$ and the other $\mu \neq .75$, or the two contradictory statements might be $p \geq .10$ and $p < .10$. The objective is to decide, based on sample information, which of the two hypotheses is correct. There is a familiar analogy to this in a criminal trial. One claim is the assertion that the accused individual is innocent. In the U.S. judicial system, this is the claim that is initially believed to be true. Only in the face of strong evidence to the contrary should the jury reject this claim in favor of the alternative assertion that the accused is guilty. In this sense, the claim of innocence is the favored or protected hypothesis, and the burden of proof is placed on those who believe in the alternative claim.

Similarly, in testing statistical hypotheses, the problem will be formulated so that one of the claims is initially favored. This initially favored claim will not be rejected in favor of the alternative claim unless sample evidence contradicts it and provides strong support for the alternative assertion. The claim initially favored or believed to be true is called the **null hypothesis** and is denoted by H_0. The other claim in a hypothesis-testing problem is called the **alternative hypothesis** and is denoted by H_a. Thus, we might test $H_0: \mu = .75$ against the alternative $H_a: \mu \neq .75$. Only if sample data strongly suggests that μ is something other than .75 should the null hypothesis be rejected. In the absence of such evidence, H_0 should not be rejected, since it is still quite plausible.

Sometimes an investigator does not want to accept a particular assertion unless and until data can provide strong support for the assertion. As an example, suppose a company is considering putting a new type of coating on bearings that it produces. The true average wear life with the current coating is known to be 1000 hours. With μ denoting the true average life for the new coating, the company would not want to make a change unless evidence strongly suggested that μ exceeds 1000. An appropriate problem formulation would involve testing $H_0: \mu = 1000$ against $H_a: \mu > 1000$. The conclusion that a change is justified is identified with H_a, and it would take conclusive evidence to justify rejecting H_0 and switching to the new coating.

Scientific research often involves trying to decide whether a current theory should be replaced by a more plausible and satisfactory explanation of the phenomenon under investigation. A conservative approach is to identify the current theory with H_0 and the researcher's alternative explanation with H_a. Rejection of the current theory will then occur only when evidence is much more consistent with the new theory. In many situations, H_a is referred to as the "researcher's hypothesis," since it is the claim that the researcher would really like to validate. The word *null* means "of no value, effect, or consequence," which suggests that H_0 should be identified with the hypothesis of no change (from current opinion), no difference, no improvement, and so on. Suppose, for example, that 10% of all circuit boards produced by a certain manufacturer during a recent period were defective. An engineer has suggested a change in the production process in the belief that it will result in a reduced defective rate. Let p denote the true proportion of defective boards resulting from the changed process. Then the research hypothesis, on which the burden of proof is placed, is the assertion that $p < .10$. Thus the alternative hypothesis is $H_a: p < .10$.

In our treatment of hypothesis testing, H_0 will always be stated as an equality claim. If θ denotes the parameter of interest, the null hypothesis will have the form $H_0: \theta = \theta_0$, where θ_0 is a specified number called the *null value* of the parameter (value claimed for θ by the null hypothesis). As an example, consider the circuit board example just discussed. The suggested alternative hypothesis was $H_a: p < .10$, the claim that the defective rate is reduced by the process modification. A natural choice of H_0 in this situation is the claim that $p \geq .10$, according to which the new process is either no better *or* worse than the one currently used. We will instead consider H_0:

$p = .10$ versus H_a: $p < .10$. The rationale for using this simplified null hypothesis is that any reasonable decision procedure for deciding between H_0: $p = .10$ and H_a: $p < .10$ will also be reasonable for deciding between the claim that $p \geq .10$ and H_a. The use of a simplified H_0 is preferred because it has certain technical benefits, which will be apparent shortly.

The alternative to the null hypothesis H_0: $\theta = \theta_0$ will look like one of the following three assertions: (1) H_a: $\theta > \theta_0$ (in which case the implicit null hypothesis is $\theta \leq \theta_0$), (2) H_a: $\theta < \theta_0$ (so the implicit null hypothesis states that $\theta \geq \theta_0$), or (3) H_a: $\theta \neq \theta_0$. For example, let σ denote the standard deviation of the distribution of inside diameters (inches) for a certain type of metal sleeve. If the decision was made to use the sleeve unless sample evidence conclusively demonstrated that $\sigma > .001$, the appropriate hypotheses would be H_0: $\sigma = .001$ versus H_a: $\sigma > .001$.

Test Procedures

A test procedure is a rule, based on sample data, for deciding whether to reject H_0. A test of H_0: $p = .10$ versus H_a: $p < .10$ in the circuit board problem might be based on examining a random sample of $n = 200$ boards. Let X denote the number of defective boards in the sample, a binomial random variable; x represents the observed value of X. If H_0 is true, $E(X) = np = 200(.10) = 20$, whereas we can expect fewer than 20 defective boards if H_a is true. A value x just a bit below 20 does not strongly contradict H_0, so it is reasonable to reject H_0 only if x is substantially less than 20. One such test procedure is to reject H_0 if $x \leq 15$ and not reject H_0 otherwise. This procedure has two constituents: (1) a *test statistic* or function of the sample data used to make a decision and (2) a *rejection region* consisting of those x values for which H_0 will be rejected in favor of H_a. For the rule just suggested, the rejection region consists of $x = 0, 1, 2, \ldots$, and 15. H_0 will not be rejected if $x = 16, 17, \ldots, 199$, or 200.

A test procedure is specified by the following:

1. A **test statistic**, a function of the sample data on which the decision (reject H_0 or do not reject H_0) is to be based
2. A **rejection region**, the set of all test statistic values for which H_0 will be rejected

The null hypothesis will then be rejected if and only if the observed or computed test statistic value falls in the rejection region.

As another example, suppose a cigarette manufacturer claims that the average nicotine content μ of brand B cigarettes is (at most) 1.5 mg. It would be unwise to reject the manufacturer's claim without strong contradictory evidence, so an appropriate problem formulation is to test H_0: $\mu = 1.5$ versus H_a: $\mu > 1.5$. Consider a decision rule based on analyzing a random sample of 32 cigarettes. Let $\bar{X}$ denote the sample average nicotine content. If H_0 is true, $E(\bar{X}) = \mu = 1.5$, whereas if H_0 is false, we expect $\bar{X}$ to exceed 1.5. Strong evidence against H_0 is provided by a value

$\bar{x}$ that considerably exceeds 1.5. Thus, we might use $\bar{X}$ as a test statistic along with the rejection region $\bar{x} \geq 1.6$.

In both the circuit board and nicotine examples, the choice of test statistic and form of the rejection region make sense intuitively. However, the choice of cutoff value used to specify the rejection region is somewhat arbitrary. Instead of rejecting $H_0: p = .10$ in favor of $H_a: p < .10$ when $x \leq 15$, we could use the rejection region $x \leq 14$. For this region, H_0 would not be rejected if 15 defective boards are observed, whereas this occurrence would lead to rejection of H_0 if the initially suggested region is employed. Similarly, the rejection region $\bar{x} \geq 1.55$ might be used in the nicotine problem in place of the region $\bar{x} \geq 1.60$.

Errors in Hypothesis Testing

The basis for choosing a particular rejection region lies in an understanding of the errors that one might be faced with in drawing a conclusion. Consider the rejection region $x \leq 15$ in the circuit board problem. Even when $H_0: p = .10$ is true, it might happen that an unusual sample results in $x = 13$, so that H_0 is erroneously rejected. On the other hand, even when $H_a: p < .10$ is true, an unusual sample might yield $x = 20$, in which case H_0 would not be rejected, again an incorrect conclusion. Thus, it is possible that H_0 may be rejected when it is true or that H_0 may not be rejected when it is false. These possible errors are not consequences of a foolishly chosen rejection region. Either one of these two errors might result when the region $x \leq 14$ is employed, or indeed when any other region is used.

Definition

> A **type I error** consists of rejecting the null hypothesis H_0 when it is true. A **type II error** involves not rejecting H_0 when H_0 is false.

In the nicotine problem, a type I error consists of rejecting the manufacturer's claim that $\mu = 1.5$ when it is actually true. If the rejection region $\bar{x} \geq 1.6$ is employed, it might happen that $\bar{x} = 1.63$ even when $\mu = 1.5$, resulting in a type I error. Alternatively, it may be that H_0 is false and yet $\bar{x} = 1.52$ is observed, leading to H_0 not being rejected (a type II error).

In the best of all possible worlds, test procedures for which neither type of error is possible could be developed. However, this ideal can be achieved only by basing a decision on an examination of the entire population, which is almost always impractical. The difficulty with using a procedure based on sample data is that because of sampling variability, an unrepresentative sample may result. Even though $E(\bar{X}) = \mu$, the observed value $\bar{x}$ may differ substantially from μ (at least if n is small). Thus, when $\mu = 1.5$ in the nicotine situation, $\bar{x}$ may be much larger than 1.5, resulting in erroneous rejection of H_0. Alternatively, it may be that $\mu = 1.6$ yet an $\bar{x}$ much smaller than this is observed, leading to a type II error.

Instead of demanding error-free procedures, we must look for procedures for which either type of error is unlikely to occur. That is, a good procedure is one for which the probability of making either type of error is small. The choice of a particular rejection region cutoff value fixes the probabilities of type I and type II errors. These error probabilities are traditionally denoted by α and β, respectively. Because H_0 specifies a unique value of the parameter, there is a single value of α. However, there is a different value of β for each value of the parameter consistent with H_a.

Example 8.1

A certain type of automobile is known to sustain no visible damage 25% of the time in 10-mph crash tests. A modified bumper design has been proposed in an effort to increase this percentage. Let p denote the proportion of all 10-mph crashes with this new bumper that result in no visible damage. The hypotheses to be tested are H_0: $p = .25$ (no improvement) versus H_a: $p > .25$. The test will be based on an experiment involving $n = 20$ independent crashes with prototypes of the new design. Intuitively, H_0 should be rejected if a substantial number of the crashes show no damage. Consider the following test procedure:

Test statistic: $X =$ the number of crashes with no visible damage.

Rejection region: $R_8 = \{8, 9, 10, \ldots, 19, 20\}$; that is, reject H_0 if $x \geq 8$, where x is the observed value of the test statistic.

This rejection region is called *upper-tailed* because it consists only of large values of the test statistic.

When H_0 is true, X has a binomial probability distribution with $n = 20$ and $p = .25$. Thus,

$$\alpha = P(\text{type I error}) = P(H_0 \text{ is rejected when it is true})$$
$$= P(X \geq 8 \text{ when } X \sim \text{Bin}(20, .25)) = 1 - B(7; 20, .25)$$
$$= 1 - .898 = .102$$

That is, when H_0 is actually true, roughly 10% of all experiments consisting of 20 crashes would result in H_0 being incorrectly rejected (a type I error).

In contrast to α, there is not a single β. Instead, there is a different β for each different p that exceeds .25. Thus there is a value of β for $p = .3$ (in which case $X \sim \text{Bin}(20, .3)$), another value of β for $p = .5$, and so on. For example,

$$\beta(.3) = P(\text{type II error when } p = .3)$$
$$= P(H_0 \text{ is not rejected when it is false because } p = .3)$$
$$= P(X \leq 7 \text{ when } X \sim \text{Bin}(20, .3)) = B(7; 20, .3) = .772$$

When p is actually .3 rather than .25 (a "small" departure from H_0), roughly 77% of all experiments of this type would result in H_0 being incorrectly not rejected!

The accompanying table displays β for selected values of p (each calculated for the rejection region R_8). Clearly, β decreases as the value of p moves further to the right of the null value .25. Intuitively, the greater the departure from H_0, the less likely it is that such a departure will not be detected.

p	.3	.4	.5	.6	.7	.8
$\beta(p)$	.772	.416	.132	.021	.001	.000

The proposed test procedure is still reasonable for testing the more realistic null hypothesis that $p \leq .25$. In this case, there is no longer a single α, but instead an α for each p that is at most .25—$\alpha(.25)$, $\alpha(.23)$, $\alpha(.20)$, $\alpha(.15)$, and so on. It is easily verified, though, that $\alpha(p) < \alpha(.25) = .102$ if $p < .25$. That is, the largest value of α occurs for the boundary value .25 between H_0 and H_a. Thus, if α is small for the simplified null hypothesis, it will also be as small as or smaller for the more realistic H_0.   ∎

Example 8.2

The drying time of a certain type of paint under specified test conditions is known to be normally distributed with mean value 75 min and standard deviation 9 min. Chemists have proposed a new additive designed to decrease average drying time. It is believed that drying times with this additive will remain normally distributed with $\sigma = 9$. Because of the expense associated with the additive, evidence should strongly suggest an improvement in average drying time before such a conclusion is adopted. Let μ denote the true average drying time when the additive is used. The appropriate hypotheses are $H_0: \mu = 75$ versus $H_a: \mu < 75$. Only if H_0 can be rejected will the additive be declared successful and used.

Experimental data is to consist of drying times from $n = 25$ test specimens. Let $X_1, \ldots, X_{25}$ denote the 25 drying times—a random sample of size 25 from a normal distribution with mean value μ and standard deviation $\sigma = 9$. The sample mean drying time $\bar{X}$ then has a normal distribution with expected value $\mu_{\bar{X}} = \mu$ and standard deviation $\sigma_{\bar{X}} = \sigma/\sqrt{n} = 9/\sqrt{25} = 1.80$. When H_0 is true, $\mu_{\bar{X}} = 75$, so an $\bar{x}$ value somewhat less than 75 would not strongly contradict H_0. A reasonable rejection region has the form $\bar{x} \leq c$, where the cutoff value c is suitably chosen. Consider the choice $c = 70.8$, so that the test procedure consists of test statistic $\bar{X}$ and rejection region $\bar{x} \leq 70.8$. Because the rejection region consists only of small values of the test statistic, the test is said to be *lower-tailed*. Calculation of α and β now involves a routine standardization of $\bar{X}$ followed by reference to the standard normal probabilities of Appendix Table A.3:

$$\alpha = P(\text{type I error}) = P(H_0 \text{ is rejected when it is true})$$
$$= P(\bar{X} \leq 70.8 \text{ when } \bar{X} \sim \text{normal with } \mu_{\bar{X}} = 75,\ \sigma_{\bar{X}} = 1.8)$$
$$= \Phi\left(\frac{70.8 - 75}{1.8}\right) = \Phi(-2.33) = .01$$

$$\beta(72) = P(\text{type II error when } \mu = 72)$$
$$= P(H_0 \text{ is not rejected when it is false because } \mu = 72)$$
$$= P(\bar{X} > 70.8 \text{ when } \bar{X} \sim \text{normal with } \mu_{\bar{X}} = 72 \text{ and } \sigma_{\bar{X}} = 1.8)$$
$$= 1 - \Phi\left(\frac{70.8 - 72}{1.8}\right) = 1 - \Phi(-.67) = 1 - .2514 = .7486$$

$$\beta(70) = 1 - \Phi\left(\frac{70.8 - 70}{1.8}\right) = .3300, \quad \beta(67) = .0174$$

For the specified test procedure, only 1% of all experiments carried out as described will result in H_0 being rejected when it is actually true. However, the chance of a type

II error is very large if $\mu = 72$ (only a small departure from H_0), somewhat less if $\mu = 70$, and quite small when $\mu = 67$ (a very substantial departure from H_0). These error probabilities are illustrated in Figure 8.1. Notice that α is computed using the probability distribution of the test statistic when H_0 is true, whereas determination of β requires knowing the test statistic's distribution when H_0 is false.

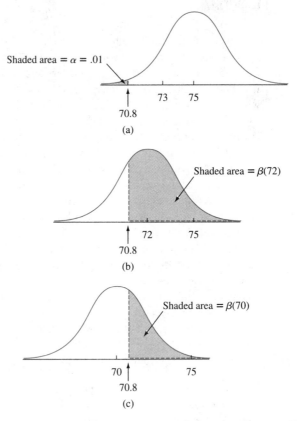

Figure 8.1 α and β illustrated for Example 8.2: (a) the distribution of $\bar{X}$ when $\mu = 75$ (H_0 true); (b) the distribution of $\bar{X}$ when $\mu = 72$ (H_0 false); (c) the distribution of $\bar{X}$ when $\mu = 70$ (H_0 false)

As in Example 8.1, if the more realistic null hypothesis $\mu \geq 75$ is considered, there is an α for each parameter value for which H_0 is true—$\alpha(75)$, $\alpha(75.8)$, $\alpha(76.5)$, and so on. It is easily verified, though, that $\alpha(75)$ is the largest of all these type I error probabilities. Focusing on the boundary value amounts to working explicitly with the "worst case."   ■

The specification of a cutoff value for the rejection region in the examples just considered was somewhat arbitrary. Use of the rejection region $R_8 = \{8, 9, \ldots, 20\}$ in Example 8.1 resulted in $\alpha = .102$, $\beta(.3) = .772$, and $\beta(.5) = .132$. Many would think these error probabilities intolerably large. Perhaps they can be decreased by changing the cutoff value.

Example 8.3
(Example 8.1 continued)

Let us use the same experiment and test statistic X as previously described in the automobile bumper problem, but now consider the rejection region $R_9 = \{9, 10, \ldots, 20\}$. Since X still has a binomial distribution with parameters $n = 20$ and p,

$$\alpha = P(H_0 \text{ is rejected when } p = .25)$$
$$= P(X \geq 9 \text{ when } X \sim \text{Bin}(20, .25)) = 1 - B(8; 20, .25) = .041$$

The type I error probability has been decreased by using the new rejection region. However, a price has been paid for this decrease:

$$\beta(.3) = P(H_0 \text{ is not rejected when } p = .3)$$
$$= P(X \leq 8 \text{ when } X \sim \text{Bin}(20, .3)) = B(8; 20, .3) = .887$$
$$\beta(.5) = B(8; 20, .5) = .252$$

Both these β's are larger than the corresponding error probabilities .772 and .132 for the region R_8. In retrospect, this is not surprising; α is computed by summing over probabilities of test statistic values *in the rejection region*, whereas β is the probability that X falls in the *complement* of the rejection region. Making the rejection region smaller must therefore decrease α while increasing β for any fixed alternative value of the parameter.   ■

Example 8.4
(Example 8.2 continued)

The use of cutoff value $c = 70.8$ in the paint-drying example resulted in a very small value of α (.01) but rather large β's. Consider the same experiment and test statistic $\bar{X}$ with the new rejection region $\bar{x} \leq 72$. Because $\bar{X}$ is still normally distributed with mean value $\mu_{\bar{X}} = \mu$ and $\sigma_{\bar{X}} = 1.8$,

$$\alpha = P(H_0 \text{ is rejected when it is true})$$
$$= P(\bar{X} \leq 72 \text{ when } \bar{X} \sim N(75, 1.8^2))$$
$$= \Phi\left(\frac{72 - 75}{1.8}\right) = \Phi(-1.67) = .0475 \approx .05$$

$$\beta(72) = P(H_0 \text{ is not rejected when } \mu = 72)$$
$$= P(\bar{X} > 72 \text{ when } \bar{X} \text{ is a normal rv with mean 72 and}$$
$$\text{standard deviation 1.8})$$
$$= 1 - \Phi\left(\frac{72 - 72}{1.8}\right) = 1 - \Phi(0) = .5$$

$$\beta(70) = 1 - \Phi\left(\frac{72 - 70}{1.8}\right) = .1335, \quad \beta(67) = .0027$$

The change in cutoff value has made the rejection region larger (it includes more $\bar{x}$ values), resulting in a decrease in β for each fixed μ less than 75. However, α for this new region has increased from the previous value .01 to approximately .05. If a type I error probability this large can be tolerated, though, the second region ($c = 72$) is preferable to the first ($c = 70.8$) because of the smaller β's.   ■

The results of these examples can be generalized in the following manner.

Proposition

> Suppose an experiment and a sample size are fixed, and a test statistic is chosen. Then decreasing the size of the rejection region to obtain a smaller value of α results in a larger value of β for any particular parameter value consistent with H_a.

This proposition says that once the test statistic and n are fixed, there is no rejection region that will simultaneously make both α and all β's small. A region must be chosen to effect a compromise between α and β.

Because of the suggested guidelines for specifying H_0 and H_a, a type I error is usually more serious than a type II error (this can always be achieved by proper choice of the hypotheses). The approach adhered to by most statistical practitioners is then to specify the largest value of α that can be tolerated and find a rejection region having that value of α rather than anything smaller. This makes β as small as possible subject to the bound on α. The resulting value of α is often referred to as the **significance level** of the test. Traditional levels of significance are .10, .05, and .01, though the level in any particular problem will depend on the seriousness of a type I error—the more serious this error, the smaller should be the significance level. The corresponding test procedure is called a **level α test** (e.g., a level .05 test or a level .01 test). A test with significance level α is one for which the type I error probability is controlled at the specified level.

Example 8.5

26 FEB 97

Consider the situation mentioned earlier in which μ was the true average nicotine content of brand B cigarettes. The objective is to test $H_0\!: \mu = 1.5$ versus $H_a\!: \mu > 1.5$ based on a random sample $X_1, X_2, \ldots, X_{32}$ of nicotine contents. Suppose the distribution of nicotine content is known to be normal with $\sigma = .20$. Then $\bar{X}$ is normally distributed with mean value $\mu_{\bar{X}} = \mu$ and standard deviation $\sigma_{\bar{X}} = .20/\sqrt{32} = .0354$.

Rather than use $\bar{X}$ itself as the test statistic, let's standardize $\bar{X}$ assuming that H_0 is true.

Test statistic: $Z = \dfrac{\bar{X} - 1.5}{\sigma/\sqrt{n}} = \dfrac{\bar{X} - 1.5}{.0354}$

Z expresses the distance between $\bar{X}$ and its expected value when H_0 is true as some number of standard deviations. For example, $z = 3$ results from an $\bar{x}$ that is 3 SDs larger than we would have expected it to be were H_0 true.

Rejecting H_0 when $\bar{x}$ "considerably" exceeds 1.5 is equivalent to rejecting H_0 when z "considerably" exceeds 0. That is, the form of the rejection region is $z \geq c$. Let's now determine c so that $\alpha = .05$. When H_0 is true, Z has a standard normal distribution. Thus,

$\alpha = P(\text{type I error}) = P(\text{rejecting } H_0 \text{ when } H_0 \text{ is true})$

$= P(Z \geq c \text{ when } Z \sim N(0, 1))$

The value c must capture upper-tail area .05 under the z curve. Either from Section 4.3 or directly from Appendix Table A.3, $c = z_{.05} = 1.645$.

Notice that $z \geq 1.645$ is the same as $\bar{x} - 1.5 \geq (.0354)(1.645)$, that is, $\bar{x} \geq 1.56$. Then β is the probability that $\bar{X} < 1.56$ and can be calculated for any μ greater than 1.5. ∎

Exercises   Section 8.1 (1–13)

1. For each of the following assertions, say whether it is a legitimate statistical hypothesis and why.

  a. $H: \sigma > 100$ **b.** $H: \bar{x} = 45$

  c. $H: s \leq .20$ **d.** $H: \sigma_1/\sigma_2 < 1$

  e. $H: \bar{X} - \bar{Y} = 5$

  f. $H: \lambda \leq .01$, where λ is the parameter of an exponential distribution used to model component lifetime

2. For the following pairs of assertions, indicate which do not comply with our rules for setting up hypotheses and why (the subscripts 1 and 2 differentiate between quantities for two different populations or samples).

  a. $H_0: \mu = 100, H_a: \mu > 100$

  b. $H_0: \sigma = 20, H_a: \sigma \leq 20$

  c. $H_0: p \neq .25, H_a: p = .25$

  d. $H_0: \mu_1 - \mu_2 = 25, H_a: \mu_1 - \mu_2 > 100$

  e. $H_0: S_1^2 = S_2^2, H_a: S_1^2 \neq S_2^2$

  f. $H_0: \mu = 120, H_a: \mu = 150$

  g. $H_0: \sigma_1/\sigma_2 = 1, H_a: \sigma_1/\sigma_2 \neq 1$

  h. $H_0: p_1 - p_2 = -.1, H_a: p_1 - p_2 < -.1$

3. To determine whether the pipe welds in a nuclear power plant meet specifications, a random sample of welds is selected, and tests are conducted on each weld in the sample. Weld strength is measured as the force required to break the weld. Suppose the specifications state that mean strength of welds should exceed 100 lb/in.²; the inspection team decides to test $H_0: \mu = 100$ versus $H_a: \mu > 100$. Explain why it might be preferable to use this H_a rather than $\mu < 100$.

4. Let μ denote the true average radioactivity level (picocuries per liter). The value 5 pCi/L is considered the dividing line between safe and unsafe water. Would you recommend testing $H_0: \mu = 5$ versus $H_a: \mu > 5$ or $H_0: \mu = 5$ versus $H_a: \mu < 5$? Explain your reasoning.

(HINT: Think about the consequences of a type I and type II error for each possibility.)

5. Before agreeing to purchase a large order of polyethylene sheaths for a particular type of high-pressure oil-filled submarine power cable, a company wants to see conclusive evidence that the true standard deviation of sheath thickness is less than .05 mm. What hypotheses should be tested, and why? In this context, what are the type I and type II errors?

6. Many older homes have electrical systems that use fuses rather than circuit breakers. A manufacturer of 40-amp fuses wants to make sure that the mean amperage at which its fuses burn out is in fact 40. If the mean amperage is lower than 40, customers will complain because the fuses require replacement too often. If the mean amperage is higher than 40, the manufacturer might be liable for damage to an electrical system due to fuse malfunction. To verify the amperage of the fuses, a sample of fuses is to be selected and inspected. If a hypothesis test were to be performed on the resulting data, what null and alternative hypotheses would be of interest to the manufacturer? Describe type I and type II errors in the context of this problem situation.

7. Water samples are taken from water used for cooling as it is being discharged from a power plant into a river. It has been determined that as long as the mean temperature of the discharged water is at most 150°F, there will be no negative effects on the river's ecosystem. To investigate whether the plant is in compliance with regulations that prohibit a mean discharge-water temperature above 150°, 50 water samples will be taken at randomly selected times, and the temperature of each sample recorded. The resulting data will be used to test the hypotheses $H_0: \mu = 150°$ versus $H_a: \mu > 150°$. In the context of this situation, describe type I and type II errors. Which type of error would you consider more serious? Explain.

8. Let p denote the proportion of all registered voters in a certain area who favor candidate A over candidate B in the race for county supervisor. Consider testing H_0: $p = .5$ versus H_a: $p \neq .5$ based on a random sample of 25 registered voters. Let X denote the number in the sample who favor A and x represent the observed value of X.

a. Which of the following rejection regions is most appropriate and why?

$R_1 = \{x: x \leq 7 \text{ or } x \geq 18\}$, $R_2 = \{x: x \leq 8\}$,

$R_3 = \{x: x \geq 17\}$

b. In the context of this problem situation, describe what type I and type II errors are.

c. What is the probability distribution of the test statistic X when H_0 is true? Use it to compute the probability of a type I error.

d. Compute the probability of a type II error for the selected region when $p = .3$, again when $p = .4$, and also for both $p = .6$ and $p = .7$.

e. Using the selected region, what would you conclude if 6 of the 25 queried favored A?

9. A mixture of pulverized fuel ash and Portland cement to be used for grouting should have a compressive strength of more than 1300 KN/m². The mixture will not be used unless experimental evidence indicates conclusively that the strength specification has been met. Suppose compressive strength for specimens of this mixture is normally distributed with $\sigma = 60$. Let μ denote the true average compressive strength.

a. What are the appropriate null and alternative hypotheses?

b. Let $\bar{X}$ denote the sample average compressive strength for $n = 20$ randomly selected specimens. Consider the test procedure with test statistic $\bar{X}$ and rejection region $\bar{x} \geq 1331.26$. What is the probability distribution of the test statistic when H_0 is true? What is the probability of a type I error for the test procedure?

c. What is the probability distribution of the test statistic when $\mu = 1350$? Using the test procedure of part (b), what is the probability that the mixture will be judged unsatisfactory when in fact $\mu = 1350$ (a type II error)?

d. How would you change the test procedure of part (b) to obtain a test with significance level .05? What impact would this change have on the error probability of part (c)?

e. Consider the standardized test statistic $Z = (\bar{X} - 1300)/(\sigma/\sqrt{n}) = (\bar{X} - 1300)/13.42$. What are the values of Z corresponding to the rejection region of part (b)?

10. The calibration of a scale is to be checked by weighing a 10-kg test specimen 25 times. Suppose that the results of different weighings are independent of one another and that the weight on each trial is normally distributed with $\sigma = .200$ kg. Let μ denote the true average weight reading on the scale.

a. What hypotheses should be tested?

b. Suppose the scale is to be recalibrated if either $\bar{x} \geq 10.1032$ or $\bar{x} \leq 9.8968$. What is the probability that recalibration is carried out when it is actually unnecessary?

c. What is the probability that recalibration is judged unnecessary when in fact $\mu = 10.1$? When $\mu = 9.8$?

d. Let $z = (\bar{x} - 10)/(\sigma/\sqrt{n})$. For what value c is the rejection region of part (b) equivalent to the "two-tailed" region *either $z \geq c$ or $z \leq -c$?*

e. If the sample size were only 10 rather than 25, how should the procedure of part (d) be altered so that $\alpha = .05$?

f. Using the test of part (e), what would you conclude from the following sample data:

9.981, 10.006, 9.857, 10.107, 9.888, 9.793, 9.728, 10.439, 10.214, 10.190

g. Reexpress the test procedure of part (b) in terms of the standardized test statistic $Z = (\bar{X} - 10)/(\sigma/\sqrt{n})$.

11. A new design for the braking system on a certain type of car has been proposed. For the current system, the true average braking distance at 40 mph under specified conditions is known to be 120 ft. It is proposed that the new design be implemented only if sample data strongly indicates a reduction in true average braking distance for the new design.

a. Define the parameter of interest and state the relevant hypotheses.

b. Suppose braking distance for the new system is normally distributed with $\sigma = 10$. Let $\bar{X}$ denote the sample average braking distance for a random sample of 36 observations. Which of the following rejection regions is appropriate: $R_1 = \{\bar{x}: \bar{x} \geq 124.80\}$, $R_2 = \{\bar{x}: \bar{x} \leq 115.20\}$, $R_3 = \{\bar{x}: \text{either } \bar{x} \geq 125.13 \text{ or } \bar{x} \leq 114.87\}$?

c. What is the significance level for the appropriate region of part (b)? How would you change the region to obtain a test with $\alpha = .001$?

d. What is the probability that the new design is not implemented when its true average braking distance is actually 115 ft and the appropriate region from part (b) is used?

e. Let $Z = (\bar{X} - 120)/(\sigma/\sqrt{n})$. What is the significance level for the rejection region $\{z: z \le -2.33\}$? For the region $\{z: z \le -2.88\}$?

12. Let $X_1, \ldots, X_n$ denote a random sample from a normal population distribution with a known value of σ.

a. For testing the hypotheses $H_0: \mu = \mu_0$ versus $H_a: \mu > \mu_0$ (where μ_0 is a fixed number), show that the

test with test statistic $\bar{X}$ and rejection region $\bar{x} \ge \mu_0 + 2.33\sigma/\sqrt{n}$ has significance level .01.

b. Suppose the procedure of part (a) is used to test $H_0: \mu \le \mu_0$ versus $H_a: \mu > \mu_0$. If $\mu_0 = 100, n = 25$, and $\sigma = 5$, what is the probability of committing a type I error when $\mu = 99$? When $\mu = 98$? In general, what can be said about the probability of a type I error when the actual value of μ is less than μ_0? Verify your assertion.

13. Reconsider the situation of Exercise 10 and suppose the rejection region is $\{\bar{x}: \bar{x} \ge 10.1004 \text{ or } \bar{x} \le 9.8940\} = \{z: z \ge 2.51 \text{ or } z \le -2.65\}$.

a. What is α for this procedure?

b. What is β when $\mu = 10.1$? When $\mu = 9.9$? Is this desirable?

8.2 Tests About a Population Mean

The general discussion in Chapter 7 of confidence intervals for a population mean μ focused on three different cases. We now develop test procedures for these same three cases.

Case I: A Normal Population with Known σ

Although the assumption that the value of σ is known is rarely met in practice, this case provides a good starting point because of the ease with which general procedures and their properties can be developed. The null hypothesis in all three cases will state that μ has a particular numerical value, the *null value*, which we will denote by μ_0. Let $X_1, \ldots, X_n$ represent a random sample of size n from the normal population. Then the sample mean $\bar{X}$ has a normal distribution with expected value $\mu_{\bar{X}} = \mu$ and standard deviation $\sigma_{\bar{X}} = \sigma/\sqrt{n}$. When H_0 is true, $\mu_{\bar{X}} = \mu_0$. Consider now the statistic Z obtained by standardizing $\bar{X}$ under the assumption that H_0 is true:

$$Z = \frac{\bar{X} - \mu_0}{\sigma/\sqrt{n}}$$

Substitution of the computed sample mean $\bar{x}$ gives z, the distance between $\bar{x}$ and μ_0 expressed in "standard deviation units." For example, if the null hypothesis is $H_0: \mu = 100$, $\sigma_{\bar{X}} = \sigma/\sqrt{n} = 10/\sqrt{25} = 2.0$ and $\bar{x} = 103$, then the test statistic value is $z = (103 - 100)/2.0 = 1.5$. That is, the observed value of $\bar{x}$ is 1.5 standard deviations (of $\bar{X}$) above what we expect it to be when H_0 is true. The statistic Z is a natural measure of the distance between $\bar{X}$, the estimator of μ, and its expected value when

H_0 is true. If this distance is too great in a direction consistent with H_a, the null hypothesis should be rejected.

Suppose first that the alternative hypothesis has the form $H_a: \mu > \mu_0$. Then an $\bar{x}$ value less than μ_0 certainly does not provide support for H_a. Such an $\bar{x}$ corresponds to a negative value of z (since $\bar{x} - \mu_0$ is negative and the divisor $\sigma/\sqrt{n}$ is positive). Similarly, an $\bar{x}$ value that exceeds μ_0 by only a small amount (corresponding to z which is positive but small) does not suggest that H_0 should be rejected in favor of H_a. The rejection of H_0 is appropriate only when $\bar{x}$ considerably exceeds μ_0—that is, when the z value is positive and large. In summary, the appropriate rejection region, based on the test statistic Z rather than $\bar{X}$, has the form $z \geq c$.

As discussed in Section 8.1, the cutoff value c should be chosen to control the probability of a type I error at the desired level α. This is easily accomplished because the distribution of the test statistic Z when H_0 is true is the standard normal distribution (that's why μ_0 was subtracted in standardizing). The required cutoff c is the z critical value that captures upper-tail area α under the standard normal curve. As an example, let $c = 1.645$, the value that captures tail area .05 ($z_{.05} = 1.645$). Then,

$$\alpha = P(\text{type I error}) = P(H_0 \text{ is rejected when } H_0 \text{ is true})$$
$$= P(Z \geq 1.645 \text{ when } Z \sim N(0, 1)) = 1 - \Phi(1.645) = .05$$

More generally, the rejection region $z \geq z_\alpha$ has type I error probability α. The test procedure is *upper-tailed* because the rejection region consists only of large values of the test statistic.

Analogous reasoning for the alternative hypothesis $H_a: \mu < \mu_0$ suggests a rejection region of the form $z \leq c$, where c is a suitably chosen negative number ($\bar{x}$ is far below μ_0 if and only if z is quite negative). Because Z has a standard normal distribution when H_0 is true, taking $c = -z_\alpha$ yields $P(\text{type I error}) = \alpha$. This is a *lower-tailed* test. For example, $z_{.10} = 1.28$ implies that the rejection region $z \leq -1.28$ specifies a test with significance level .10.

Finally, when the alternative hypothesis is $H_a: \mu \neq \mu_0$, H_0 should be rejected if $\bar{x}$ is too far to either side of μ_0. This is equivalent to rejecting H_0 either if $z \geq c$ or if $z \leq -c$. Suppose we desire $\alpha = .05$. Then,

$$.05 = P(Z \geq c \text{ or } Z \leq -c \text{ when } Z \text{ has a standard normal distribution})$$
$$= \Phi(-c) + 1 - \Phi(c) = 2[1 - \Phi(c)]$$

Thus, c is such that $1 - \Phi(c)$, the area under the standard normal curve to the right of c, is .025 (and not .05!). From Section 4.3 or Appendix Table A.3, $c = 1.96$, and the rejection region is $z \geq 1.96$ *or* $z \leq -1.96$. For any α, the *two-tailed* rejection region $z \geq z_{\alpha/2}$ *or* $z \leq -z_{\alpha/2}$ has type I error probability α (since area $\alpha/2$ is captured under each of the two tails of the z curve). Again, the key reason for using the standardized test statistic Z is that, because Z has a known distribution when H_0 is true (standard normal), a rejection region with desired type I error probability is easily obtained by using an appropriate critical value.

The test procedure for case I is summarized in the accompanying box, and the corresponding rejection regions are illustrated in Figure 8.2.

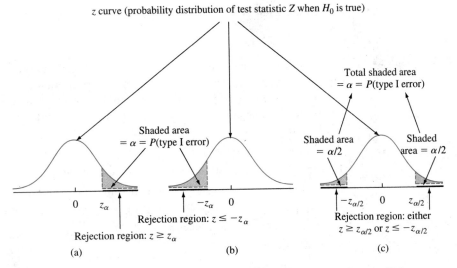

Figure 8.2 Rejection regions for z tests: (a) upper-tailed test; (b) lower-tailed test; (c) two-tailed test

Null hypothesis: H_0: $\mu = \mu_0$

Test statistic value: $z = \dfrac{\bar{x} - \mu_0}{\sigma/\sqrt{n}}$

Alternative hypothesis	Rejection region for level α test
H_a: $\mu > \mu_0$	$z \geq z_\alpha$
H_a: $\mu < \mu_0$	$z \leq -z_\alpha$
H_a: $\mu \neq \mu_0$	either $z \geq z_{\alpha/2}$ or $z \leq -z_{\alpha/2}$

Use of the following sequence of steps is recommended in any hypothesis testing analysis.

1. Identify the parameter of interest and describe it in the context of the problem situation.
2. Determine the null value and state the null hypothesis.
3. State the appropriate alternative hypothesis.
4. Give the formula for the computed value of the test statistic (substituting the null value and the known values of any other parameters, but *not* those of any sample-based quantities).
5. State the rejection region for the selected significance level α.
6. Compute any necessary sample quantities, substitute into the formula for the test statistic value, and compute that value.
7. Decide whether H_0 should be rejected and state this conclusion in the problem context.

The formulation of hypotheses (Steps 2 and 3) should be done prior to examining the data.

Example 8.6

A manufacturer of sprinkler systems used for fire protection in office buildings claims that the true average system-activation temperature is 130°. A sample of $n = 9$ systems, when tested, yields a sample average activation temperature of 131.08°F. If the distribution of activation times is normal with standard deviation 1.5°F, does the data contradict the manufacturer's claim at significance level $\alpha = .01$?

1. **Parameter of interest:** μ = true average activation temperature.
2. **Null hypothesis:** H_0: $\mu = 130$ (null value = $\mu_0 = 130$).
3. **Alternative hypothesis:** H_a: $\mu \neq 130$ (a departure from the claimed value in *either* direction is of concern.
4. **Test statistic value:**

$$z = \frac{\bar{x} - \mu_0}{\sigma/\sqrt{n}} = \frac{\bar{x} - 130}{1.5/\sqrt{n}}$$

5. **Rejection region:**   The form of H_a implies use of a two-tailed test with rejection region *either* $z \geq z_{.005}$ or $z \leq -z_{.005}$. From Section 4.3 or Appendix Table A.3, $z_{.005} = 2.58$, so we reject H_0 if either $z \geq 2.58$ or $z \leq -2.58$.
6. **Substituting** $n = 9$ and $\bar{x} = 131.08$,

$$z = \frac{131.08 - 130}{1.5/\sqrt{9}} = \frac{1.08}{.5} = 2.16$$

That is, the observed sample mean is a bit more than 2 SDs above what would have been expected were H_0 true.

7. The computed value $z = 2.16$ does not fall in the rejection region ($-2.58 < 2.16 < 2.58$), so H_0 cannot be rejected at significance level .01. The data does not give strong support to the claim that true average differs from the design value of 130.   ∎

β and Sample Size Determination

The z tests for case I are among the few in statistics for which there are simple formulas available for β, the probability of a type II error. Consider first the upper-tailed test with rejection region $z \geq z_\alpha$. This is equivalent to $\bar{x} \geq \mu_0 + z_\alpha \cdot \sigma/\sqrt{n}$, so H_0 will not be rejected if $\bar{x} < \mu_0 + z_\alpha \cdot \sigma/\sqrt{n}$. Now let μ' denote a particular value of μ that exceeds the null value μ_0. Then

$$\beta(\mu') = P(H_0 \text{ is not rejected when } \mu = \mu')$$
$$= P(\bar{X} < \mu_0 + z_\alpha \cdot \sigma/\sqrt{n} \text{ when } \mu = \mu')$$
$$= P\left(\frac{\bar{X} - \mu'}{\sigma/\sqrt{n}} < z_\alpha + \frac{\mu_0 - \mu'}{\sigma/\sqrt{n}} \text{ when } \mu = \mu'\right)$$

$$= \Phi\left(z_\alpha + \frac{\mu_0 - \mu'}{\sigma/\sqrt{n}}\right)$$

As μ' increases, $\mu_0 - \mu'$ becomes more negative, so $\beta(\mu')$ will be small when μ' greatly exceeds μ_0 (because the value at which Φ is evaluated will then be quite negative). Error probabilities for the lower-tailed and two-tailed tests are derived in an analogous manner.

If σ is large, the probability of a type II error can be large at an alternative value μ' that is of particular concern to an investigator. Suppose we fix α and also specify β for such an alternative value. In the sprinkler example, company officials might view $\mu' = 132$ as a very substantial departure from H_0: $\mu = 130$, and therefore wish $\beta(132) = .10$ in addition to $\alpha = .01$. More generally, consider the two restrictions $P(\text{type I error}) = \alpha$ and $\beta(\mu') = \beta$ for specified α, μ', and β. Then for an upper-tailed test, the sample size n should be chosen to satisfy

$$\Phi\left(z_\alpha + \frac{\mu_0 - \mu'}{\sigma/\sqrt{n}}\right) = \beta$$

This implies that

$$-z_\beta = \begin{array}{c} z \text{ critical value that} \\ \text{captures lower-tail area } \beta \end{array} = z_\alpha + \frac{\mu_0 - \mu'}{\sigma/\sqrt{n}}$$

It is easy to solve this equation for the desired n. A parallel argument yields the necessary sample size for lower- and two-tailed tests as summarized below.

Alternative hypothesis	Type II error probability $\beta(\mu')$ for a level α test
H_a: $\mu > \mu_0$	$\Phi\left(z_\alpha + \dfrac{\mu_0 - \mu'}{\sigma/\sqrt{n}}\right)$
H_a: $\mu < \mu_0$	$1 - \Phi\left(-z_\alpha + \dfrac{\mu_0 - \mu'}{\sigma/\sqrt{n}}\right)$
H_a: $\mu \neq \mu_0$	$\Phi\left(z_{\alpha/2} + \dfrac{\mu_0 - \mu'}{\sigma/\sqrt{n}}\right) - \Phi\left(-z_{\alpha/2} + \dfrac{\mu_0 - \mu'}{\sigma/\sqrt{n}}\right)$

where $\Phi(z) =$ the standard normal cdf.

The sample size n for which a level α test also has $\beta(\mu') = \beta$ at the alternative value μ' is

$$n = \begin{cases} \left[\dfrac{\sigma(z_\alpha + z_\beta)}{\mu_0 - \mu'}\right]^2 & \begin{array}{l}\text{for a one-tailed} \\ \text{(upper or lower) test}\end{array} \\[4mm] \left[\dfrac{\sigma(z_{\alpha/2} + z_\beta)}{\mu_0 - \mu'}\right]^2 & \begin{array}{l}\text{for a two-tailed test} \\ \text{(an approximate solution)}\end{array} \end{cases}$$

Example 8.7

Let μ denote the true average tread life of a certain type of tire. Consider testing H_0: $\mu = 20,000$ versus H_a: $\mu > 20,000$ based on a sample of size $n = 16$ from a normal population distribution with $\sigma = 1500$. A test with $\alpha = .01$ requires $z_\alpha = z_{.01} = 2.33$. The probability of making a type II error when $\mu = 21,000$ is

$$\beta(21,000) = \Phi\left(2.33 + \frac{20,000 - 21,000}{1500/\sqrt{16}}\right) = \Phi(-.34) = .3669$$

Since $z_{.1} = 1.28$, the requirement that the level .01 test also have $\beta(21,000) = .1$ necessitates

$$n = \left[\frac{1500(2.33 + 1.28)}{20,000 - 21,000}\right]^2 = (-5.42)^2 = 29.32$$

The sample size must be an integer, so $n = 30$ tires should be used.   ■

Case II: Large-Sample Tests

When the sample size is large, the z tests for case I are easily modified to yield valid test procedures without requiring either a normal population distribution or known σ. The key result was used in Chapter 7 to justify large-sample confidence intervals: A large n implies that the sample standard deviation S will be close to σ for most samples, so that the standardized variable

$$Z = \frac{\bar{X} - \mu}{S/\sqrt{n}}$$

has *approximately* a standard normal distribution. Substitution of the null value μ_0 in place of μ yields the test statistic

$$Z = \frac{\bar{X} - \mu_0}{S/\sqrt{n}}$$

which has approximately a standard normal distribution when H_0 is true. The use of rejection regions given earlier for case I (e.g., $z \geq z_\alpha$ when the alternative hypothesis is H_a: $\mu > \mu_0$) then results in test procedures for which the significance level is approximately (rather than exactly) α. The rule of thumb $n > 30$ will again be used to characterize a large sample size.

Example 8.8

28 FEB 97

A certain type of brick is being considered for use in a particular construction project. The brick will be used unless sample evidence strongly suggests that the true average compressive strength is below 3200 psi. A random sample of 36 bricks is selected, and each is subjected to a compressive strength test. The resulting sample average compressive strength and sample standard deviation of compressive strength are 3109 psi and 156 psi, respectively. State the relevant hypotheses and carry out a test to reach a decision using level of significance .05.

1. μ = true average compressive strength for this type of brick.
2. H_0: $\mu = 3200$.

3. H_a: $\mu < 3200$ (so that the brick will be used unless H_0 is rejected in favor of H_a).

4. Test statistic value:

$$z = \frac{\bar{x} - 3200}{s/\sqrt{n}}$$

5. Because $<$ appears in H_a, a lower-tailed test is used. For $\alpha = .05$, the appropriate one-tailed critical value is $z_\alpha = z_{.05} = 1.645$. H_0 will now be rejected if $z \le -1.645$.

6. $n = 36$, $\bar{x} = 3109$, $s = 156$, so

$$z = \frac{3109 - 3200}{156/\sqrt{36}} = \frac{-91}{26} = -3.50$$

That is, $\bar{X}$ has been observed to fall 3.5 estimated standard deviations (of $\bar{X}$) below what would have been expected were H_0 true.

7. Since $-3.50 \le -1.645$, H_0 is rejected at the 5% level of significance. We conclude that true average compressive strength is below 3200, so the brick should not be used. Notice that in reaching this conclusion, we may have made a type I error (rejecting H_0 when it is true), but the low level of significance (.05) has made that unlikely. ∎

Determination of β and the necessary sample size for these large-sample tests can be based either on specifying a plausible value of σ and using the case I formulas (even though s is used in the test) or on using the curves to be introduced shortly in connection with case III.

Case III: A Normal Population Distribution

When n is small, the Central Limit Theorem (CLT) can no longer be invoked to justify the use of a large-sample test. We faced this same difficulty in obtaining a small-sample confidence interval (CI) for μ in Chapter 7. Our approach here will be the same one used there: We will assume that the population distribution is at least approximately normal and describe test procedures whose validity rests on this assumption. If an investigator has good reason to believe that the population distribution is quite nonnormal, a distribution-free test from Chapter 15 can be used. Alternatively, a statistician can be consulted regarding procedures valid for specific families of population distributions other than the normal family.

The key result on which tests for a normal population mean are based was used in Chapter 7 to derive the t CI: If $X_1, X_2, \ldots, X_n$ is a random sample from a normal distribution, the standardized variable

$$T = \frac{\bar{X} - \mu}{S/\sqrt{n}}$$

has a t distribution with $n - 1$ degrees of freedom (df). Consider testing H_0: $\mu = \mu_0$ against H_a: $\mu > \mu_0$ by using the test statistic $T = (\bar{X} - \mu_0)/(S/\sqrt{n})$. That is, the test statistic results from standardizing $\bar{X}$ under the assumption that H_0 is true (using

$S/\sqrt{n}$, the estimated standard deviation of $\bar{X}$, rather than $\sigma/\sqrt{n}$). When H_0 is true, the test statistic has a t distribution with $n - 1$ df. Knowledge of the test statistic's distribution when H_0 is true (the "null distribution") allows us to construct a rejection region for which the type I error probability is controlled at the desired level. In particular, use of the upper-tail t critical value $t_{\alpha,n-1}$ to specify the rejection region $t \geq t_{\alpha,n-1}$ implies that

$$P(\text{type I error}) = P(H_0 \text{ is rejected when it is true})$$
$$= P(T \geq t_{\alpha,n-1} \text{ when } T \text{ has a } t \text{ distribution with } n - 1 \text{ df})$$
$$= \alpha$$

The test statistic is really the same here as in the large-sample case but is labeled T to emphasize that its null distribution is a t distribution with $n - 1$ df rather than the standard normal (z) distribution. The rejection region for the t test differs from that for the z test only in that a t critical value $t_{\alpha,n-1}$ replaces the z critical value z_α. Similar comments apply to alternatives for which a lower-tailed or two-tailed test is appropriate.

The One-Sample t Test

Null hypothesis: $H_0: \mu = \mu_0$

Test statistic value: $t = \dfrac{\bar{x} - \mu_0}{s/\sqrt{n}}$

Alternative hypothesis	Rejection region for a level α test
$H_a: \mu > \mu_0$	$t \geq t_{\alpha,n-1}$
$H_a: \mu < \mu_0$	$t \leq -t_{\alpha,n-1}$
$H_a: \mu \neq \mu_0$	either $t \geq t_{\alpha/2,n-1}$ or $t \leq -t_{\alpha/2,n-1}$

Example 8.9

The accompanying data on maximum weight of lift (MAWL, in kg) for a frequency of four lifts/min was reported in the paper "The Effects of Speed, Frequency, and Load on Measured Hand Forces for a Floor-to-Knuckle Lifting Task" (*Ergonomics*, 1992, pp. 833–843); subjects were randomly selected from the population of healthy males aged 18–30. Assuming that MAWL is normally distributed, does the data suggest that the population mean MAWL exceeds 25? Let's carry out a test using a significance level of .05

25.8   36.6   26.3   21.8   27.2

1. μ = population mean MAWL.

2. $H_0: \mu = 25$

3. $H_a: \mu > 25$

4. $t = \dfrac{\bar{x} - 25}{s/\sqrt{n}}$

5. Reject H_0 if $t \geq t_{\alpha,n-1} = t_{.05,4} = 2.132$

6. $\Sigma x_i = 137.7$ and $\Sigma x_i^2 = 3911.97$, from which $\bar{x} = 27.54$, $s = 5.47$, and

$$t = \frac{27.54 - 25}{5.47/\sqrt{5}} = 1.04$$

7. Since 1.04 does not fall in the rejection region $(1.04 < 2.132)$, H_0 cannot be rejected at significance level .05. It is still plausible that μ is (at most) 25. ∎

β and Sample Size Determination

The calculation of β at the alternative value μ' in case I was carried out by expressing the rejection region in terms of $\bar{x}$ (e.g., $\bar{x} \geq \mu_0 + z_\alpha \cdot \sigma/\sqrt{n}$) and then subtracting μ' to standardize correctly. An equivalent approach involves noting that when $\mu = \mu'$, the test statistic $Z = (\bar{X} - \mu_0)/(\sigma/\sqrt{n})$ still has a normal distribution with variance 1, but now the mean value is given by $(\mu' - \mu_0)/(\sigma/\sqrt{n})$. That is, when $\mu = \mu'$, the test statistic still has a normal distribution though not the standard normal distribution. Because of this, $\beta(\mu')$ is an area under the normal curve with mean value $(\mu' - \mu_0)(\sigma/\sqrt{n})$ and variance 1. Both α and β involve working with normally distributed variables.

The calculation of $\beta(\mu')$ for the t test is much less straightforward. This is because the distribution of the test statistic $T = (\bar{X} - \mu_0)/(S/\sqrt{n})$ is quite complicated when H_0 is false and H_a is true. Thus, for an upper-tailed test, determining

$$\beta(\mu') = P(T < t_{\alpha,n-1} \text{ when } \mu = \mu' \text{ rather than } \mu_0)$$

involves integrating a very unpleasant density function. This must be done numerically, but fortunately it has been done by research statisticians for both one- and two-tailed t tests. The results are summarized in graphs of β that appear in Appendix Table A.13. There are four sets of graphs, corresponding to one-tailed tests at level .05 and level .01 and two-tailed tests at the same levels.

To understand how these graphs are used, note first that both β and the necessary sample size n in case I are functions not just of the absolute difference $|\mu_0 - \mu'|$ but of $d = |\mu_0 - \mu'|/\sigma$. Suppose, for example, that $|\mu_0 - \mu'| = 10$. This departure from H_0 will be much easier to detect (smaller β) when $\sigma = 2$, in which case μ_0 and μ' are 5 population standard deviations apart, than when $\sigma = 10$. The fact that β for the t test depends on d rather than just $|\mu_0 - \mu'|$ is unfortunate, since to use the graphs one must have some idea of the true value of σ. A conservative (large) guess for σ will yield a conservative (large) value of $\beta(\mu')$ and a conservative estimate of the sample size necessary for prescribed α and $\beta(\mu')$.

Once the alternative μ' and value of σ are selected, d is calculated and its value located on the horizontal axis of the relevant set of curves. The value of β is the height of the $n - 1$ df curve above the value of d (visual interpolation is necessary if $n - 1$ is not a value for which the corresponding curve appears). An illustration appears in Figure 8.3.

Rather than fixing n (i.e., $n - 1$, and thus the particular curve from which β is read), one might prescribe both α (.05 or .01 here) and a value of β for the chosen μ' and σ. After computing d, the point (d, β) is located on the relevant set of graphs.

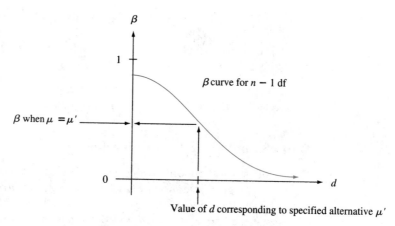

Figure 8.3 A typical β curve for the t test

The curve below and closest to this point gives $n - 1$ and thus n (again, interpolation is often necessary).

Example 8.10 An engineering company has run out of space at its present facility, so is interested in obtaining space at a nearby location. The company is considering several existing buildings. To accommodate interaction between employees, a shuttle bus is to be operated between the two locations during working hours. It is decided that some trial runs will be carried out to obtain data, and then any location for which there is strong evidence that μ, the true average midday travel time, exceeds 20 minutes should be eliminated from consideration. The hypotheses to be tested are then $H_0: \mu = 20$ versus $H_a: \mu > 20$, and the selected significance level is $\alpha = .05$.

Suppose prior evidence suggests $\sigma = 5$ as a plausible value and $n = 10$ runs are to be made. What is the chance of a type II error when $\mu = 25$ (the value 25 represents a substantial departure from H_0, so not rejecting H_0 when $\mu = 25$ is a serious error)? With $d = |20 - 25|/5 = 1$, the point on the $n - 1 = 9$ df curve above $d = 1$ has a height of approximately .1, so $\beta \approx .1$. The company might think $\beta = .1$ too large for such an extreme departure from H_0, so might wish to use $\beta = .05$. Since $d = 1$, the point $(d, \beta) = (1, .05)$ must be located. This point is very close to the 14 df curve, so using $n = 15$ will achieve $\alpha = .05$ and $\beta = .05$ when $\mu' = 25$, $\sigma = 5$. Of course, if σ is actually larger than 5, this value of n will not suffice. ∎

Exercises   Section 8.2 (14–32)

 14. Let the test statistic Z have a standard normal distribution when H_0 is true. Give the significance level for each of the following situations.

  a. $H_a: \mu > \mu_0$, rejection region $z \geq 1.88$

  b. $H_a: \mu < \mu_0$, rejection region $z \leq -2.75$

  c. $H_a: \mu \neq \mu_0$, rejection region $z \geq 2.88$ *or* $z \leq -2.88$

15. Let the test statistic T have a t distribution when H_0 is true. Give the significance level for each of the following situations:

a. H_a: $\mu > \mu_0$, df = 15,
rejection region $t \geq 3.733$

b. H_a: $\mu < \mu_0$, $n = 24$,
rejection region $t \leq -2.500$

c. H_a: $\mu \neq \mu_0$, $n = 31$,
rejection region $t \geq 1.697$ or $t \leq -1.697$

16. Answer the following questions for the tire problem in Example 8.7.

a. If $\bar{x} = 20{,}960$ and a level $\alpha = .01$ test is used, what is the decision?

b. If a level .01 test is used, what is $\beta(20{,}500)$?

c. If a level .01 test is used and it is also required that $\beta(20{,}500) = .05$, what sample size n is necessary?

d. If $\bar{x} = 20{,}960$, what is the smallest α at which H_0 can be rejected (based on $n = 16$)?

17. Reconsider the paint-drying situation of Example 8.2, in which drying time for a test specimen is normally distributed with $\sigma = 9$. The hypotheses H_0: $\mu = 75$ versus H_a: $\mu < 75$ are to be tested using a random sample of $n = 25$ observations.

a. How many standard deviations (of $\bar{X}$) below the null value is $\bar{x} = 72.3$?

b. If $\bar{x} = 72.3$, what is the conclusion using $\alpha = .01$?

c. What is α for the test procedure that rejects H_0 when $z \leq -2.88$?

d. For the test procedure of part (c), what is $\beta(70)$?

e. If the test procedure of part (c) is used, what n is necessary to ensure that $\beta(70) = .01$?

f. If a level .01 test is used with $n = 100$, what is the probability of a type I error when $\mu = 76$?

18. The melting point of each of 16 samples of a certain brand of hydrogenated vegetable oil was determined, resulting in $\bar{x} = 94.32$. Assume that the distribution of melting point is normal with $\sigma = 1.20$.

a. Test H_0: $\mu = 95$ versus H_a: $\mu \neq 95$ using a two-tailed level .01 test.

b. If a level .01 test is used, what is $\beta(94)$, the probability of a type II error when $\mu = 94$?

c. What value of n is necessary to ensure that $\beta(94) = .1$ when $\alpha = .01$?

19. A television manufacturer claims that at most 250 microamperes (μA) of current are needed to attain a certain brightness level with a particular type of set. A

sample of 20 sets yields a sample average current of $\bar{x} = 257.3$. Let μ denote the true average current necessary to achieve the desired brightness with sets of this type, and assume that μ is the mean of a normal population with $\sigma = 15$.

a. Test at level .05 the null hypothesis that μ is at most 250 against the appropriate alternative.

b. If $\mu = 260$, what is the probability of a type II error?

c. Suppose it was decided that the manufacturer's claim should not be accepted unless the data strongly supports it. Would you choose as your null hypothesis the H_0 suggested in part (a), or would you select a different H_0? Explain. (NOTE: From a practical viewpoint, the claim that at most 250 μA are required is the same as the claim that fewer than 250 μA are required.)

20. The desired percentage of SiO_2 in a certain type of aluminous cement is 5.5. To test whether the true average percentage is 5.5 for a particular production facility, 16 independently obtained samples are analyzed. Suppose the percentage of SiO_2 in a sample is normally distributed with $\sigma = .3$, and that $\bar{x} = 5.25$.

a. Does this indicate conclusively that the true average percentage differs from 5.5? Carry out the analysis using the sequence of steps suggested in the text.

b. If the true average percentage is $\mu = 5.6$ and a level $\alpha = .01$ test based on $n = 16$ is used, what is the probability of detecting this departure from H_0?

c. What value of n is required to satisfy $\alpha = .01$ and $\beta(5.6) = .01$?

21. To obtain information on the corrosion-resistance properties of a certain type of steel conduit, 35 specimens are buried in soil for a 2-year period. The maximum penetration (in mils) for each specimen is then measured, yielding a sample average penetration of $\bar{x} = 52.7$ and a sample standard deviation of $s = 4.8$. The conduits were manufactured with the specification that true average penetration be at most 50 mils. To see whether experimental data indicates that specifications have not been met, test H_0: $\mu = 50$ versus H_a: $\mu > 50$ using a large-sample test with $\alpha = .05$.

22. A sample of 40 speedometers of a particular brand is obtained, and each is calibrated to check for accuracy at 55 mph. The resulting sample average and sample standard deviation are $\bar{x} = 53.8$ and $s = 1.3$, respectively. Let $\mu =$ the true average reading when actual speed is 55 mph. Does the sample evidence suggest strongly that $\mu \neq 55$? Use a level .01 test.

23. The Charpy V-notch impact test is the basis for studying many material toughness criteria. This test was applied to 32 samples of a particular alloy at 110°F. The sample average amount of transverse lateral expansion was computed to be 73.1 mils, and the sample standard deviation was $s = 5.9$ mils. To be suitable for a particular application, the true average amount of expansion should be less than 75 mils. The alloy will not be used unless the sample provides strong evidence that this criterion has been met. Test the relevant hypotheses using $\alpha = .01$ to decide whether the alloy is suitable.

24. The relative conductivity of a semiconductor device is determined by the amount of impurity "doped" into the device during its manufacture. A silicon diode to be used for a specific purpose requires a cut-on voltage of .60 V, and if this is not achieved then the mechanism governing the amount of impurity must be adjusted. A sample of 120 such diodes yielded a sample average voltage of .62 V and sample standard deviation of .11. At level .001, does the data indicate that the true average cut-on voltage is something other than .60?

25. Minor surgery on horses under field conditions requires a reliable short-term anesthetic producing good muscle relaxation, minimal cardiovascular and respiratory changes, and a quick, smooth recovery with minimal aftereffects so that horses can be left unattended. The article "A Field Trial of Ketamine Anesthesia in the Horse" (*Equine Vet. J.*, 1984, pp. 176–179) reports that for a sample of $n = 73$ horses to which ketamine was administered under certain conditions, the sample average lateral recumbency (lying-down) time was 18.86 min and the standard deviation was 8.6 min. Does this data suggest that true average lateral recumbency time under these conditions is less than 20 min? Test the appropriate hypotheses at level of significance .10.

26. The amount of shaft wear (.0001 in.) after a fixed mileage was determined for each of $n = 8$ internal combustion engines having copper lead as a bearing material, resulting in $\bar{x} = 3.72$ and $s = 1.25$.

a. Assuming that the distribution of shaft wear is normal with mean μ, use the t test at level .05 to test $H_0: \mu = 3.50$ versus $H_a: \mu > 3.50$.

b. Using $\sigma = 1.25$, what is the type II error probability $\beta(\mu')$ of the test for the alternative $\mu' = 4.00$?

27. The recommended daily dietary allowance for zinc among males older than age 50 years is 15 mg/day. The article "Nutrient Intakes and Dietary Patterns of Older Americans: A National Study" (*J. Gerontology*, 1992, pp. M145–150) reports the following summary data on intake for a sample of males aged 65–74 years: $n = 115$, $\bar{x} = 11.3$, and $s = 6.43$. Does this data indicate that average daily zinc intake in the population of all males aged 65–74 falls below the recommended allowance?

28. In an experiment designed to measure the time necessary for an inspector's eyes to become used to the reduced amount of light necessary for penetrant inspection, the sample average time for $n = 9$ inspectors was 6.32 sec and the sample standard deviation was 1.65 sec. It has previously been assumed that the average adaptation time was at least 7 sec. Assuming adaptation time to be normally distributed, does the data contradict prior belief? Use the t test with $\alpha = .1$.

29. A certain type of soil was determined to have a natural mean pH value of 8.75. The authors of the paper "Effects of Brewery Effluent on Agricultural Soil and Crop Plants" (*Environ. Pollution*, 1984, pp. 341–351) treated soil samples with various dilutions of an acidic effluent. Five soil samples were treated with a solution of 25% water and 75% effluent. The mean and standard deviation of the 5 pH measurements were 8.00 and .05, respectively. Does this data indicate that at this concentration the effluent results in a mean pH that differs from the natural pH of the soil? Use a level .01 test.

30. A sample of 12 radon detectors of a certain type was selected, and each was exposed to 100 pCi/L of radon. The resulting readings were as follows:

105.6, 90.9, 91.2, 96.9, 96.5, 91.3,
100.1, 105.0, 99.6, 107.7, 103.3, 92.4

a. Does this data suggest that the population mean reading under these conditions differs from 100? State and test the appropriate hypotheses using $\alpha = .05$.

b. Suppose that prior to the experiment, a value of $\sigma = 7.5$ had been assumed. How many determinations would then have been appropriate to obtain $\beta = .10$ for the alternative $\mu = 95$?

31. Show that for any $\Delta > 0$, when the population distribution is normal and σ is known, the two-tailed test satisfies $\beta(\mu_0 - \Delta) = \beta(\mu_0 + \Delta)$, so that $\beta(\mu')$ is symmetric about μ_0.

32. For a fixed alternative value μ', show t̶   Teston   in the case of a normal population distribution with
as $n \to \infty$ for either a one-tailed or a tw̶   P   known σ.

Tests Conce̶rning a Popu̶lation Proportion

Let p denote the p̶roportion of in̶dividuals or objects in a population who possess a
specified property̶ (e.g., cars with̶ manual transmissions or smokers who smoke a
filter cigarette). If̶ an individual or o̶bject with the property is labeled a success (S),
then p is the popu̶lation proportion o̶f successes. Tests concerning p will be based on
a random sample o̶f size n from the̶ population. Provided that n is small relative to
the population si̶ze, X (the numbe̶r of S's in the sample) has (approximately) a
binomial distribu̶tion. Furthermore̶, if n itself is large, both X and the estimator
$\hat{p} = X/n$ are appro̶ximately normally̶ distributed. We first consider large-sample tests
based on this latte̶r fact and then tur̶n to the small-sample case that directly uses the
binomial distribut̶ion.

Large-Sample Tests

Large-sample tests concerning p are a special case of the more general large-sample
procedures for a parameter θ. Let $\hat{\theta}$ be an estimator of θ that is (at least approximately) unbiased and has approximately a normal distribution. The null
hypothesis has the form $H_0: \theta = \theta_0$, where θ_0 denotes a number (the null value)
appropriate to the problem context. Suppose that when H_0 is true, the standard
deviation of $\hat{\theta}$, $\sigma_{\hat{\theta}}$, involves no unknown parameters. For example, if $\theta = \mu$ and
$\hat{\theta} = \bar{X}$, $\sigma_{\hat{\theta}} = \sigma_{\bar{X}} = \sigma/\sqrt{n}$, which involves no unknown parameters only if the value
of σ is known. A large-sample test statistic results from standardizing $\hat{\theta}$ under the
assumption that H_0 is true (so that $E(\hat{\theta}) = \theta_0$):

$$\text{Test statistic} = Z = \frac{\hat{\theta} - \theta_0}{\sigma_{\hat{\theta}}}$$

If the alternative hypothesis is $H_a: \theta > \theta_0$, an upper-tailed test whose significance
level is approximately α is specified by the rejection region $z \geq z_\alpha$. The other two
alternatives, $H_a: \theta < \theta_0$ and $H_a: \theta \neq \theta_0$, are tested using a lower-tailed z test and a
two-tailed z test, respectively.

In the case $\theta = p$, $\sigma_{\hat{\theta}}$ will not involve any unknown parameters when H_0 is true,
but this is atypical. When $\sigma_{\hat{\theta}}$ does involve unknown parameters, it is often possible
to use an estimated standard deviation $S_{\hat{\theta}}$ in place of $\sigma_{\hat{\theta}}$ and still have Z approximately
normally distributed when H_0 is true (because when n is large, $s_{\hat{\theta}} \approx \sigma_{\hat{\theta}}$ for most
samples). The large-sample test of the previous section furnishes an example of this:
Because σ is usually unknown, we use $s_{\hat{\theta}} = s_{\bar{X}} = s/\sqrt{n}$ in place of $\sigma/\sqrt{n}$ in the
denominator of z.

The estimator $\hat{p} = X/n$ is unbiased ($E(\hat{p}) = p$), has approximately a normal
distribution, and its standard deviation is $\sigma_{\hat{p}} = \sqrt{p(1-p)/n}$. These facts were used
in Section 7.2 to obtain a large-sample confidence interval for p. When H_0 is true,

$E(\hat{p}) = p_0$ and $\sigma_{\hat{p}} = \sqrt{p_0(1 - p_0)/n}$, so $\sigma_{\hat{p}}$ does not involve any unknown parameters. It then follows that when n is large and H_0 is true, the test statistic

$$Z = \frac{\hat{p} - p_0}{\sqrt{p_0(1 - p_0)/n}}$$

has approximately a standard normal distribution. If the alternative hypothesis is H_a: $p > p_0$ and the upper-tailed rejection region $z \geq z_\alpha$ is used, then

$P(\text{type I error}) = P(H_0 \text{ is rejected when it is true})$

$= P(Z \geq z_\alpha \text{ when } Z \text{ has approximately a standard normal distribution}) \approx \alpha$

Thus, the desired level of significance α is attained by using the critical value that captures area α in the upper tail of the z curve. Rejection regions for the other two alternative hypotheses, lower-tailed for H_a: $p < p_0$ and two-tailed for H_a: $p \neq p_0$, are justified in an analogous manner.

Null hypothesis: H_0: $p = p_0$

Test statistic value: $z = \dfrac{\hat{p} - p_0}{\sqrt{p_0(1 - p_0)/n}}$

Alternative hypothesis	Rejection region
H_a: $p > p_0$	$z \geq z_\alpha$
H_a: $p < p_0$	$z \leq -z_\alpha$
H_a: $p \neq p_0$	either $z \geq z_{\alpha/2}$ or $z \leq -z_{\alpha/2}$

These test procedures are valid provided that both $np_0 \geq 5$ and $n(1 - p_0) \geq 5$.

Notice that the standard deviation in the denominator of z is $\sqrt{p_0(1 - p_0)/n}$, whereas the estimated standard deviation $\sqrt{\hat{p}(1 - \hat{p})/n}$ was used in the large-sample confidence interval for p.

Example 8.11

Many consumers are turning to generics as a way of reducing the cost of prescription medications. The paper "Commercial Information on Drugs: Confusing to the Physician?" (*J. of Drug Issues*, 1988, pp. 245–257) gives the results of a survey of 102 doctors. Only 47 of those surveyed knew the generic name for the drug methadone. Does this provide strong evidence for concluding that fewer than half of all physicians know the generic name for methadone? Let's carry out a test of hypotheses using a significance level of .01.

1. p = the proportion of all physicians who know the generic name for methadone.

2. H_0: $p = .5$

3. H_a: $p < .5$

4. Since $102(.5) \geq 5$, the z test can be used. The test statistic value is

$$z = (\hat{p} - .5)/\sqrt{(.5)(.5)/n}$$

5. The form of H_a implies that a lower-tailed test is appropriate: Reject H_0 if $z \le -z_{.01} = -2.33$.

6. $\hat{p} = 47/102 = .461$, so $z = (.461 - .5)/\sqrt{(.5)(.5)/102} = -.039/.0495 = -.79$.

7. Since $-.79$ is not in the rejection region, H_0 cannot be rejected at level .01. There is not conclusive evidence that fewer than 50% of all physicians know the generic name for methadone. ∎

β and Sample Size Determination

omit

When H_0 is true, the test statistic Z has approximately a standard normal distribution. Now suppose that H_0 is *not* true and that $p = p'$. Then Z still has approximately a normal distribution (because it is a linear function of $\hat{p}$), but its mean value and variance are no longer 0 and 1, respectively. Instead,

$$E(Z) = \frac{p' - p_0}{\sqrt{p_0(1 - p_0)/n}}, \quad V(Z) = \frac{p'(1 - p')/n}{p_0(1 - p_0)/n}$$

The probability of a type II error for an upper-tailed test is $\beta(p') = P(Z < z_\alpha$ when $p = p')$. This can be computed by using the above mean and variance to standardize and then referring to the standard normal cdf. In addition, if it is desired that the level α test also have $\beta(p') = \beta$ for a specified value of β, this equation can be solved for the necessary n as in Section 8.2. General expressions for $\beta(p')$ and n are given in the accompanying box.

Alternative hypothesis	$\beta(p')$
$H_a: p > p_0$	$\Phi\left[\dfrac{p_0 - p' + z_\alpha\sqrt{p_0(1 - p_0)/n}}{\sqrt{p'(1 - p')/n}}\right]$
$H_a: p < p_0$	$1 - \Phi\left[\dfrac{p_0 - p' - z_\alpha\sqrt{p_0(1 - p_0)/n}}{\sqrt{p'(1 - p')/n}}\right]$
$H_a: p \ne p_0$	$\Phi\left[\dfrac{p_0 - \cdots}{\cdots}\right]$
	$-\Phi\left[\dfrac{p_0 - \cdots}{\cdots}\right]$

Large Sample Tests

The sample size n for which the level α test also satisfies $\beta(p') = \beta$ is

$$n = \begin{cases} \left[\dfrac{z_\alpha\sqrt{p_0(1 - p_0)} + z_\beta\sqrt{p'(1 - p')}}{p' - p_0}\right]^2 & \text{one–tailed test} \\[3ex] \left[\dfrac{z_{\alpha/2}\sqrt{p_0(1 - p_0)} + z_\beta\sqrt{p'(1 - p')}}{p' - p_0}\right]^2 & \text{two–tailed test (an approximate solution)} \end{cases}$$

Example 8.12   A package-delivery service advertises that at least 90% of all packages brought to its office by 9 A.M. for delivery in the same city are delivered by noon that day. Let p denote the true proportion of such packages that are delivered as advertised and consider the hypotheses $H_0: p = .9$ versus $H_a: p < .9$. If only 80% of the packages are delivered as advertised, how likely is it that a level .01 test based on $n = 225$ packages will detect such a departure from H_0? What should the sample size be in order to ensure that $\beta(.8) = .01$? With $\alpha = .01$, $p_0 = .9$, $p' = .8$, and $n = 225$,

$$\beta(.8) = 1 - \Phi\left(\frac{.9 - .8 - 2.33\sqrt{(.9)(.1)/225}}{\sqrt{(.8)(.2)/225}}\right)$$

$$= 1 - \Phi(2.00) = .0228$$

Thus, the probability that H_0 will be rejected using the test when $p = .8$ is .9772—roughly 98% of all samples will result in correct rejection of H_0.

Using $z_\alpha = z_\beta = 2.33$ in the sample size formula yields

$$n = \left[\frac{2.33\sqrt{(.9)(.1)} + 2.33\sqrt{(.8)(.2)}}{.8 - .9}\right]^2 \approx 266$$ ∎

Small-Sample Tests

Test procedures when the sample size n is small are based directly on the binomial distribution rather than the normal approximation. Consider the alternative hypothesis $H_a: p > p_0$ and again let X be the number of successes in the sample. Then X is the test statistic, and the upper-tailed rejection region has the form $x \geq c$. When H_0 is true, X has a binomial distribution with parameters n and p_0, so

$$P(\text{type I error}) = P(H_0 \text{ is rejected when it is true})$$
$$= P(X \geq c \text{ when } X \sim \text{Bin}(n, p_0))$$
$$= 1 - P(X \leq c - 1 \text{ when } X \sim \text{Bin}(n, p_0))$$
$$= 1 - B(c - 1; n, p_0)$$

As the critical value c decreases, more x values are included in the rejection region and $P(\text{type I error})$ increases. Because X has a discrete probability distribution, it is usually not possible to find a value of c for which $P(\text{type I error})$ is exactly the desired significance level α (e.g., .05 or .01). Instead, the largest rejection region of the form $\{c, c + 1, \ldots, n\}$ satisfying $1 - B(c - 1; n, p_0) \leq \alpha$ is used.

Let p' denote an alternative value of p ($p' > p_0$). Then when $p = p'$, $X \sim \text{Bin}(n, p')$, so

$$\beta(p') = P(\text{type II error when } p = p')$$
$$= P(X < c \text{ when } X \sim \text{Bin}(n, p')) = B(c - 1; n, p')$$

That is, $\beta(p')$ is the result of a straightforward binomial probability calculation. The sample size n necessary to ensure that a level α test also has specified β at a particular alternative value p' must be determined by trial and error using the binomial cdf.

Test procedures for H_a: $p < p_0$ and for H_a: $p \neq p_0$ are constructed in a similar manner. In the former case, the appropriate rejection region has the form $x \leq c$ (a lower-tailed test). The critical value c is the largest number satisfying $B(c; n, p_0) \leq \alpha$. The rejection region when the alternative hypothesis is H_a: $p \neq p_0$ consists of both large and small x values.

Example 8.13   A plastics manufacturer has developed a new type of plastic trash can and proposes to sell them with an unconditional 6-year warranty. To see whether this is economically feasible, 20 prototype cans are subjected to an accelerated life test to simulate 6 years of use. The proposed warranty will be modified only if the sample data strongly suggests that fewer than 90% of such cans would survive the 6-year period. Let p denote the proportion of all cans that survive the accelerated test. The relevant hypotheses are H_0: $p = .9$ versus H_a: $p < .9$. A decision will be based on the test statistic X, the number among the 20 that survive. If the desired significance level is $\alpha = .05$, c must satisfy $B(c; 20, .9) \leq .05$. From Appendix Table A.1, $B(15; 20, .9) = .043$, whereas $B(16; 20, .9) = .133$. The appropriate rejection region is therefore $x \leq 15$. If the accelerated test results in $x = 14$, H_0 would be rejected in favor of H_a, necessitating a modification of the proposed warranty. The probability of a type II error for the alternative value $p' = .8$ is

$$\beta(.8) = P(H_0 \text{ is not rejected when } X \sim \text{Bin}(20, .8))$$
$$= P(X \geq 16 \text{ when } X \sim \text{Bin}(20, .8))$$
$$= 1 - B(15; 20, .8) = 1 - .370 = .630$$

That is, when $p = .8$, 63% of all samples consisting of $n = 20$ cans would result in H_0 being incorrectly not rejected. This error probability is high because 20 is a small sample size and $p' = .8$ is close to the null value $p_0 = .9$. ∎

Exercises   Section 8.3 (33–41)

33. State DMV records indicate that of all vehicles undergoing emissions testing during the previous year, 70% passed on the first try. A random sample of 200 cars tested in a particular county during the current year yields 156 that passed on the initial test. Does this suggest that the true proportion for this county during the current year differs from the previous statewide proportion? Test the relevant hypotheses using $\alpha = .05$.

34. The city council in a large city has become concerned about the trend toward exclusion of renters with children in apartments within the city. The housing coordinator has decided to select a random sample of 125 apartments and determine for each whether children would be permitted. Let p = the true proportion of apartments that

prohibit children. If $p > .75$, the council will consider appropriate legislation.

a. If 102 of the 125 sampled exclude renters with children, test H_0: $p = .75$ versus H_a: $p > .75$ using a level .05 test.

b. What is the probability of a type II error when $p = .8$ (not rejecting H_0 when $p = .8$)?

c. How many apartments would have to be sampled to ensure that $\beta(.8) = .1$?

35. A telephone company is trying to decide whether some new lines in a large community should be installed underground. Because a small surcharge will be added to telephone bills to pay for the extra installation costs, the company has decided to survey customers and proceed only if the survey strongly indicates that more than 60% of all customers favor underground installation. If 118 of 160 customers surveyed favor underground installation despite the surcharge, what should the company do? Test the relevant hypotheses using $\alpha = .05$.

36. A university library ordinarily has a complete shelf inventory done once every year. Because of new shelving rules instituted the previous year, the head librarian believes it may be possible to save money by postponing the inventory. The librarian decides to select at random 800 books from the library's collection and have them searched in a preliminary manner. If evidence indicates strongly that the true proportion of misshelved or unlocatable books is less than .02, then the inventory will be postponed.

    a. Among the 800 books searched, 12 were misshelved or unlocatable. Test the relevant hypotheses and advise the librarian what to do (use $\alpha = .05$).

    b. If the true proportion of misshelved and lost books is actually .01, what is the probability that the inventory will be (unnecessarily) taken?

    c. If the true proportion is .05, what is the probability that the inventory will be postponed?

37. The article "Statistical Evidence of Discrimination" (*J. Amer. Stat. Assoc.*, 1982, pp. 773–783) discusses the court case *Swain v. Alabama* (1965), in which it was alleged that there was discrimination against blacks in grand jury selection. Census data suggested that 25% of those eligible for grand jury service were black, yet a random sample of 1050 called to appear for possible duty yielded only 177 blacks. Using a level .01 test, does this data argue strongly for a conclusion of discrimination?

38. A plan for an executive traveler's club has been developed by an airline on the premise that 5% of its current customers would qualify for membership. A random sample of 500 customers yielded 40 who would qualify.

    a. Using this data, test at level .01 the null hypothesis that the company's premise is correct against the alternative that it is not correct.

    b. What is the probability that when the test of part (a) is used, the company's premise will be judged correct when in fact 10% of all current customers qualify?

    c. What sample size would be sufficient to ensure that both α and the probability of part (b) are .01?

39. A manufacturer of plumbing fixtures has developed a new type of washerless faucet. Let $p = P$ (a randomly selected faucet of this type will develop a leak within 2 years under normal use). The manufacturer has decided to proceed with production unless it can be determined that p is too large; the borderline acceptable value of p is specified as .10. The manufacturer decides to subject n of these faucets to accelerated testing (approximating 2 years of normal use). With X = the number among the n faucets that leak before the test concludes, production will commence unless the observed X is too large. It is decided that if $p = .10$, the probability of not proceeding should be at most .10, while if $p = .30$ the probability of proceeding should be at most .10. Can $n = 10$ be used? $n = 20$? $n = 25$? What is the appropriate rejection region for the chosen n, and what are the actual error probabilities when this region is used?

40. (Based on an article in the May 1978 issue of *Consumer Reports*.) To decide whether tennis enthusiasts strongly prefer gut strings to nylon strings in their tennis rackets, each of 20 randomly selected enthusiasts is given two tennis rackets that are identical except that one is strung with nylon and the other with gut. After several weeks of playing alternately with each racket, each of the enthusiasts will be asked to state a preference for one of the two rackets. Let p = the true proportion of enthusiasts who prefer gut strings and X = the number of enthusiasts in the sample who prefer gut. Because gut is more expensive and breaks sooner than nylon, the null hypothesis is that at most 50% of all enthusiasts prefer gut; we simplify this to $H_0: p = .5$, planning to reject H_0 only if experimental evidence strongly favors gut strings.

    a. Which of the rejection regions {15, 16, 17, 18, 19, 20}, {0, 1, 2, 3, 4, 5}, or {0, 1, 2, 3, 17, 18, 19, 20} is most appropriate, and why are the other two not appropriate?

    b. What is the probability of a type I error for the chosen region of part (a)? Does the region specify a level .05 test? Is it the best level .05 test?

    c. If 60% of all enthusiasts prefer gut, calculate the probability of a type II error using the appropriate region from part (a). Repeat if 80% of all enthusiasts prefer gut.

    d. If 13 out of the 20 players prefer gut, should H_0 be rejected using a significance level of .10?

P-values

41. Scientists think that robots will play a crucial r [...] were 14 defectives. If human assemblers have a factories in the next 20 years. Suppose that [...] rate of .03 (3%), does this data support the experiment to determine whether the use of rob [...] esis that the proportion of defectives is lower for weave computer cables is feasible, a robot was u: [...] than humans? Use a .01 significance level. assemble 500 cables. The cables were examine [...]

8.4 P-Values

One way to report the [...] of a [...] hesis testing analysis is to simply say whether the null hypothe [...] was [...] a specified level of significance. Thus, an investigator might st [...] that H_0 was [...] ejected at level of significance .05 or that use of a level .01 t [...] resulted in [...] t rejecting H_0. This type of statement is somewhat inadequate b [...] because it says nothing about whether the computed value of the test statistic just b [...] into the rejection region or whether it exceeded the critical value by a large amount. A related difficulty is that such a report imposes the specified significance level on other decision makers. In many decision situations, individuals may have different views concerning the consequences of a type I or type II error. Each individual would then want to select his or her own significance level—some selecting $\alpha = .05$, others .01, and so on—and reach a conclusion accordingly. This could result in some individuals rejecting H_0 while others conclude that the data does not show a strong enough contradiction of H_0 to justify its rejection.

Example 8.14   The true average time to initial relief of pain for the current best-selling pain reliever is known to be 10 min. Let μ denote the true average time to relief for a company's newly developed reliever. The company wishes to produce and market this reliever only if it provides quicker relief than does the current best-seller, so wishes to test $H_0: \mu = 10$ versus $H_a: \mu < 10$. Only if experimental evidence leads to rejection of H_0 will the new reliever be introduced. After weighing the relative seriousness of each type of error, a single level of significance must be agreed on and a decision—to reject H_0 and introduce the reliever or not to do so—made at that level.

Suppose the new reliever has been introduced. The company supports its claim of quicker relief by stating that, based on an analysis of experimental data, H_0: $\mu = 10$ was rejected in favor of $H_a: \mu < 10$ using level of significance $\alpha = .10$. Any individuals contemplating a switch to this new reliever would naturally want to reach their own conclusions concerning the validity of the claim. Individuals who are satisfied with the current best-seller would view a type I error (concluding that the new product provides quicker relief when it actually does not) as serious, so might wish to use $\alpha = .05$, .01, or even smaller levels. Unfortunately, the nature of the company's statement prevents an individual decision maker from reaching a conclusion at such a level. The company has imposed its own choice of significance level on others. The report could have been done in a manner that allowed each individual flexibility in drawing a conclusion at a personally selected α. ∎

A P-value conveys much information about the strength of evidence against H_0 and allows an individual decision maker to draw a conclusion at any specified level α. Before we give a general definition, consider how the conclusion in a hypothesis testing problem depends on the selected level α.

Example 8.15

3 MAR 97

The nicotine content problem discussed in Example 8.5 involved testing H_0: $\mu = 1.5$ versus H_a: $\mu > 1.5$. Because of the inequality in H_a, the rejection region is upper-tailed, with H_0 rejected if $z \geq z_\alpha$. Suppose $z = 2.10$. The accompanying table displays the rejection region for each of four different α's along with the resulting conclusion.

Level of Significance α	Rejection Region	Conclusion
.05	$z \geq 1.645$	Reject H_0
.025	$z \geq 1.96$	Reject H_0
.01	$z \geq 2.33$	Do not reject H_0
.005	$z \geq 2.58$	Do not reject H_0

For α relatively large, the z critical value z_α is not very far out in the upper tail; 2.10 exceeds the critical value, and so H_0 is rejected. However, as α decreases, the critical value increases. For small α, the z critical value is large; 2.10 is less than z_α, and H_0 is not rejected.

Recall that for an upper-tailed z test, α is just the area under the z curve to the right of the critical value z_α. That is, once α is specified, the critical value is chosen to capture upper-tail area α. Appendix Table A.3 shows that the area to the right of 2.10 is .0179. Using an α larger than .0179 corresponds to $z_\alpha < 2.10$. An α less than .0179 necessitates using a z critical value that exceeds 2.10. The decision at a particular level α thus depends on how the selected α compares to the tail area captured by the computed z. This is illustrated in Figure 8.4. Notice in particular that .0179, the captured tail area, is the smallest level α at which H_0 would be rejected, because using any smaller α results in a z critical value that exceeds 2.10, so that 2.10 is not in the rejection region. ∎

In general, suppose the probability distribution of a test statistic when H_0 is true has been determined. Then, for specified α, the rejection region is determined by finding a critical value or values that capture tail area α (upper, lower, or two-tailed, whichever is appropriate) under the probability distribution curve. The smallest α for which H_0 would be rejected is the tail area captured by the computed value of the test statistic. This smallest α is the P-value.

Definition

> The **P-value** is the smallest level of significance at which H_0 would be rejected when a specified test procedure is used on a given data set. Once the P-value has been determined, the conclusion at any particular level α results from comparing the P-value to α:
>
> 1. P-value $\leq \alpha \Rightarrow$ reject H_0 at level α.
> 2. P-value $> \alpha \Rightarrow$ do not reject H_0 at level α.

Figure 8.4 Relationship between α and tail area captured by computed z; (a) tail area captured by computed z; (b) when $\alpha > .0179$, $z_\alpha < 2.10$ and H_0 is rejected; (c) when $\alpha < .0179$, $z_\alpha > 2.10$ and H_0 is not rejected

It is customary to call the data *significant* when H_0 is rejected and *not significant* otherwise. The *P*-value is then the smallest level at which the data is significant. An easy way to visualize the comparison of the *P*-value with the chosen α is to draw a picture like that of Figure 8.5. The calculation of the *P*-value depends on whether the test is upper-, lower-, or two-tailed. However, once it has been calculated, the comparison with α does not depend on which type of test was used.

Figure 8.5 Comparing α and the *P*-value; (a) reject H_0 when α lies here; (b) do not reject H_0 when α lies here

Example 8.16
(Example 8.14 continued)
3MAR97

Suppose that when data from an experiment involving the new pain reliever was analyzed, the *P*-value for testing H_0: $\mu = 10$ versus H_a: $\mu < 10$ was calculated as .0384. Since $\alpha = .05$ is larger than the *P*-value (.05 lies in the interval (a) of Figure 8.5), H_0 would be rejected by anyone carrying out the test at level .05. However, at level .01 H_0 would not be rejected because .01 is smaller than the smallest level (.0384) at which H_0 can be rejected. ∎

Most standard computer packages for doing statistical analyses automatically calculate and print out a *P*-value when a hypothesis-testing analysis is performed. A conclusion can then be drawn directly from the output, without reference to a table of critical values.

A useful alternative definition equivalent to the one just given is as follows: The *P*-value is the probability, calculated assuming H_0 is true, of getting a test statistic value at least as extreme as the one actually obtained. Thus, if $z = 2.10$ for an upper-tailed z test, *P*-value $= P(Z \geq 2.10$, when H_0 is true$) = 1 - \Phi(2.10) = .0179$ as before. The more extreme the test statistic value, the smaller is the *P*-value and the more the data argues against H_0.

The *P*-Value for a z Test

The *P*-value for a z test (one based on a test statistic whose distribution when H_0 is true is at least approximately standard normal) is easily determined from the information in Appendix Table A.3. Consider an upper-tailed test and let z denote the computed value of the test statistic Z. The null hypothesis is rejected if $z \geq z_\alpha$, and the *P*-value is the smallest α for which this is the case. Since z_α increases as α decreases, the *P*-value is the value of α for which $z = z_\alpha$. That is, the *P*-value is just the area captured by the computed value z in the upper tail of the standard normal curve. The corresponding cumulative area is $\Phi(z)$, so in this case *P*-value $= 1 - \Phi(z)$.

An analogous argument for a lower-tailed test shows that the *P*-value is the area captured by the computed value z in the lower tail of the standard normal curve. More care must be exercised in the case of a two-tailed test. Suppose first that z is positive. Then the *P*-value is the value of α satisfying $z = z_{\alpha/2}$ (i.e., computed z = upper-tail critical value). This says that the area captured in the upper tail is half the *P*-value, so that *P*-value $= 2[1 - \Phi(z)]$. If z is negative, the *P*-value is the α for which $z = -z_{\alpha/2}$, or, equivalently, $-z = z_{\alpha/2}$, so *P*-value $= 2[1 - \Phi(-z)]$. Since $-z = |z|$ when z is negative, *P*-value $= 2[1 - \Phi(|z|)]$ for either positive or negative z.

$$
\text{P-value} = P = \begin{cases} 1 - \Phi(z) & \text{for an upper-tailed test} \\ \Phi(z) & \text{for a lower-tailed test} \\ 2[1 - \Phi(|z|)] & \text{for a two-tailed test} \end{cases}
$$

Each of these is the probability of getting a value at least as extreme as what was obtained (assuming H_0 true).

Example 8.17

3/MAR97

The manufacturer's recommended air pressure for a certain type of tire is 32 psi. A random sample of $n = 35$ such tires from different cars operated by a large rental company yields a sample average pressure and sample standard deviation of 31.4 and 1.2, respectively. Does this data suggest that the average pressure for all such tires on the company's cars differs from the recommended value?

Let μ denote the population average pressure. The hypotheses of interest are H_0: $\mu = 32$ versus H_a: $\mu \neq 32$. Because n is large, a large-sample z test can be used. The

P-value for z test

computed value of the test statistic is

$$z = \frac{\bar{x} - \mu_0}{s/\sqrt{n}} = \frac{31.4 - 32}{1.2/\sqrt{35}} = -2.96$$

The *P*-value for a two-tailed test is then

$$P\text{-value} = 2[1 - \Phi(|-2.96|)] = 2(1 - .9985) = .003$$

At significance level .01, H_0 is rejected (because the *P*-value is smaller than this level, as in (a) of Figure 8.5) in favor of the conclusion that the population average pressure differs from the recommended value. If, however, a test with level .001 is employed, H_0 cannot be rejected.

The P-Value for a t Test

Whereas the standard normal table gives critical values corresponding to a great many different tail areas, Appendix Table A.5 contains only seven critical values for each different *t* distribution. This implies that the exact *P*-value—a captured tail area for a one-tailed test or twice the captured area for a two-tailed test—cannot usually be obtained. In practice it is usually sufficient to say where the *P*-value lies with respect to the most frequently used significance levels.

Consider a *t* test based on 10 df, so that the relevant *t* critical values from Appendix Table A.5 are as follows:

Tail area	.10	.05	.025	.01	.005	.001	.0005
Critical value	1.372	1.812	2.228	2.764	3.169	4.144	4.587

Suppose first that the test is upper-tailed (rejection region $t \geq t_{\alpha,10}$) and that $t = 2.53$. This computed value is between the two critical values 2.228 and 2.764 corresponding to $\alpha = .025$ and .01, respectively. Therefore, *P*-value < .025 (H_0 is rejected at level .025, so the smallest α at which rejection is possible is less than .025) and .01 < *P*-value (H_0 cannot be rejected at level .01). This gives the bounds .01 < *P*-value < .025. If $t = .93$, then we can say only that H_0 cannot be rejected at level .10, so .10 < *P*-value (a lower bound). Similarly, if $t = 4.78$, H_0 can be rejected at level .0005 ($4.78 \geq t_{.0005,10}$), resulting in the upper bound *P*-value < .0005.

For a lower-tailed test, an upper and/or lower bound on the *P*-value results from changing the sign of *t* and proceeding as in the case of an upper-tailed test. Thus, if $t = -1.98$ for a 10 df test, .025 < *P*-value < .05 (H_0 can be rejected at level .05 but not at level .025). Finally, for a two-tailed test, remember that the level of significance associated with using a particular critical value is twice the corresponding upper-tail area. For example, $\alpha = .05$ for the rejection region $t \geq 2.228$ or $t \leq -2.228$. Suppose $t = 2.41$, which lies between the two critical values 2.228 (corresponding to $\alpha = .05$) and 2.764 (for $\alpha = .02$). Then .02 < *P*-value < .05. That is, for a two-tailed test, bounds result from doubling the tail areas at the top of the bounding columns. Similarly, if $t = -1.15$, H_0 cannot be rejected at level 2(.10) = .20, so .20 < *P*-value.

Several widely used statistical computer packages incorporate algorithms for computing captured *t*-curve areas. Resulting computer output then displays an exact *P*-value rather than just the bounds. These packages have similar capabilities for other test procedures to be encountered shortly.

Exercises   Section 8.4 (42–53)

42. For which of the given P-values would the null hypothesis be rejected when performing a level .05 test?

 a. .001 **b.** .021 **c.** .078 **d.** .047 **e.** .148

43. Pairs of P-values and significance levels, α, are given. For each pair, state whether the observed P-value would lead to rejection of H_0 at the given significance level.

 a. P-value = .084, $\alpha = .05$

 b. P-value = .003, $\alpha = .001$

 c. P-value = .498, $\alpha = .05$

 d. P-value = .084, $\alpha = .10$

 e. P-value = .039, $\alpha = .01$

 f. P-value = .218, $\alpha = .10$

44. Let μ denote the reaction time to a certain stimulus. For a large-sample z test of $H_0: \mu = 5$ versus $H_a: \mu > 5$, find the P-value associated with each of the given values of the test statistic.

 a. $z = 1.42$ **b.** .90 **c.** 1.96 **d.** 2.48 **e.** −.11

45. Newly purchased automobile tires of a certain type are supposed to be filled to a pressure of 30 lb/in.^{2}. Let μ denote the true average pressure. Find the P-value associated with each given z statistic value for testing $H_0: \mu = 30$ versus $H_a: \mu \neq 30$.

 a. 2.10 **b.** −1.75 **c.** −.55 **d.** 1.41 **e.** −5.3

46. A researcher collected data in order to test $H_0: \mu = 17$ versus $H_a: \mu > 17$. Place bounds on the P-value for each of the given t-test statistic values and associated degrees of freedom.

 a. $t = 1.84$, df = 14

 b. $t = 3.74$, df = 25

 c. $t = 2.42$, df = 13

 d. $t = 1.32$, df = 8

 e. $t = 2.67$, df = 45

47. Place bounds on the P-value for a two-tailed t test for each case.

 a. $t = 2.3$, df = 6

 b. $t = −3.0$, df = 14

 c. $t = 4.2$, df = 24

 d. $t = −1.3$, df = 17

48. Let μ denote true average serum receptor concentration for all pregnant women. The average for all women is known to be 5.63. The paper "Serum Transferrin Receptor for the Detection of Iron Deficiency in Pregnancy" (*Amer. J. Clinical Nutr.*, 1991, pp. 1077–1081) reports that P-value > .10 for a test of $H_0: \mu = 5.63$ versus $H_a: \mu \neq 5.63$ based on $n = 176$ pregnant women. Using a significant level of .01, what would you conclude?

49. An aspirin manufacturer fills bottles by weight rather than by count. Since each bottle should contain 100 tablets, the average weight per tablet should be 5 grains. Each of 100 tablets taken from a very large lot is weighed, resulting in a sample average weight per tablet of 4.87 grains and a sample standard deviation of .35 grain. Does this information provide strong evidence for concluding that the company is not filling its bottles as advertised? Test the appropriate hypotheses using $\alpha = .01$ by first computing the P-value and then comparing it to the specified significance level.

50. Because of variability in the manufacturing process, the actual yielding point of a sample of mild steel subjected to increasing stress will usually differ from the theoretical yielding point. Let p denote the true proportion of samples that yield before their theoretical yielding point. If on the basis of a sample it can be concluded that more than 20% of all specimens yield before the theoretical point, the production process will have to be modified.

 a. If 15 of 60 specimens yield before the theoretical point, what is the P-value when the appropriate test is used, and what would you advise the company to do?

 b. If the true percentage of "early yields" is actually 50% (so that the theoretical point is the median of the yield distribution) and a level .01 test is used, what is the probability that the company concludes a modification of the process is necessary?

51. The times of first sprinkler activation for a series of tests with fire prevention sprinkler systems using an aqueous film-forming foam were (in sec)

27, 41, 22, 27, 23, 35, 30, 33, 24, 27, 28, 22, 24

(see "Use of AFFF in Sprinkler Systems," *Fire Technology*, 1976, p. 5). The system has been designed so that true average activation time is at most 25 sec under such conditions. Does the data strongly contradict the validity of this design specification?

Test the relevant hypotheses using $\alpha = .05$ by first obtaining an upper and lower bound for the P-value of the data. What assumption are you making about the activation time distribution?

52. A spectrophotometer used for measuring CO concentration [ppm (parts per million) by volume] is checked for accuracy by taking readings on a manufactured gas (called span gas) in which the CO concentration is very precisely controlled at 70 ppm. If the readings suggest that the spectrophotometer is not working properly, it will have to be recalibrated. Assume that if properly calibrated, measured concentration for span gas samples is normally distributed. On the basis of the six readings, 85, 77, 82, 68, 72, and 69, is recalibration necessary? Test using $\alpha = .05$ by first obtaining as much information as you can about the P-value.

53. Federal officials are currently investigating the problems associated with disposal of hazardous wastes. One disposal site is the abandoned Stringfellow acid pits in Riverside County, California. The EPA had sampled water from 11 wells in nearby Glen Avon. Radiation levels from 38 to 67 pCi/L were observed (*Los Angeles Times*, May 31, 1984). The EPA standard for maximum allowable radiation level for drinking water is 15 pCi/L. Suppose the sample of 11 wells has resulted in a sample mean radiation level of 52.5 pCi/L and a sample standard deviation of 8. Use a level .01 test to determine whether the data strongly suggests that the mean radiation level exceeds the EPA standard by first placing a bound or bounds on the P-value.

8.5 Some Comments on Selecting a Test Procedure

Once the experimenter has decided on the question of interest and the method for gathering data (the design of the experiment), construction of an appropriate test procedure consists of three distinct steps:

1. Specify a test statistic (the function of the observed values that will serve as the decision maker).

2. Decide on the general form of the rejection region (typically reject H_0 for suitably large values of the test statistic, reject for suitably small values, or reject for either small or large values).

3. Select the specific numerical critical value or values that will separate the rejection region from the acceptance region (by obtaining the distribution of the test statistic when H_0 is true, and then selecting a level of significance).

In the examples thus far, both Steps 1 and 2 were carried out in an ad hoc manner through intuition. For example, when the underlying population was assumed normal with mean μ and known σ, we were led from $\bar{X}$ to the standardized test statistic

$$Z = \frac{\bar{X} - \mu_0}{\sigma/\sqrt{n}}$$

For testing $H_0: \mu = \mu_0$ versus $H_a: \mu > \mu_0$, intuition then suggested rejecting H_0 when z was large. Finally the critical value was determined by specifying the level of significance α and using the fact that Z has a standard normal distribution when H_0 is true. The reliability of the test in reaching a correct decision can be assessed by studying type II error probabilities.

Issues to be considered in carrying out Steps 1–3 encompass the following questions:

1. What are the practical implications and consequences of choosing a particular level of significance once the other aspects of a test procedure have been determined?

2. Does there exist a general principle, not dependent just on intuition, that can be used to obtain best or good test procedures?

3. When there exist two or more tests that are appropriate in a given situation, how can the tests be compared to decide which should be used?

4. If a test is derived under specific assumptions about the distribution or population being sampled, how well will the test procedure work when the assumptions are violated?

Statistical Versus Practical Significance

While the process of reaching a decision by using the methodology of classical hypothesis testing involves selecting a level of significance and then rejecting or not rejecting H_0 at that level α, simply reporting the α used and the decision reached conveys little of the information contained in the sample data. Especially when the results of an experiment are to be communicated to a large audience, rejection of H_0 at level .05 will be much more convincing if the observed value of the test statistic greatly exceeds the 5% critical value than if it barely exceeds that value. This is precisely what led to the notion of P-value as a way of reporting significance without imposing a particular α on others who might wish to draw their own conclusions.

Even if a P-value is included in a summary of results, however, there may be difficulty in interpreting this value and in making a decision. This is because a small P-value, which would ordinarily indicate **statistical significance** in that it would strongly suggest rejection of H_0 in favor of H_a, may be the result of a large sample size in combination with a departure from H_0 that has little **practical significance**. In many experimental situations, only departures from H_0 of large magnitude would be worthy of detection, whereas a small departure from H_0 would have little practical significance.

Consider as an example testing H_0: $\mu = 100$ versus H_a: $\mu > 100$ where μ is the mean of a normal population with $\sigma = 10$. Suppose a true value of $\mu = 101$ would not represent a serious departure from H_0 in the sense that not rejecting H_0 when $\mu = 101$ would be a relatively inexpensive error. For a reasonably large sample size n, this μ would lead to an $\bar{x}$ value near 101, so we would not want this sample evidence to argue strongly for rejection of H_0 when $\bar{x} = 101$ is observed. For various sample sizes, Table 8.1 records both the P-value when $\bar{x} = 101$ and also the probability of not rejecting H_0 at level .01 when $\mu = 101$.

The second column in Table 8.1 shows that even for moderately large-sample sizes, the P-value of $\bar{x} = 101$ argues very strongly for rejection of H_0, whereas the observed $\bar{x}$ itself suggests that in practical terms the true value of μ differs little from the null value $\mu_0 = 100$. The third column points out that even when there is little

Table 8.1 An illustration of the effect of sample size on P-values and β

n	P-Value When $\bar{x} = 101$	$\beta(101)$ for Level .01 Test
25	.3085	.9664
100	.1587	.9082
400	.0228	.6293
900	.0013	.2514
1600	.0000335	.0475
2500	.000000297	.0038
10,000	7.69×10^{-24}	.0000

practical difference between the true μ and the null value, for a fixed level of significance a large-sample size will almost always lead to rejection of the null hypothesis at that level. To summarize, *one must be especially careful in interpreting evidence when the sample size is large, since any small departure from H_0 will almost surely be detected by a test, yet such a departure may have little practical significance.*

The Likelihood Ratio Principle

Let $x_1, x_2, \ldots, x_n$ be the observations in a random sample from a probability distribution $f(x; \theta)$. The joint distribution evaluated at these sample values is the product $f(x_1; \theta) \cdot f(x_2; \theta) \cdots \cdots f(x_n; \theta)$. As in the discussion of maximum likelihood estimation, the *likelihood function* is this joint distribution regarded as a function of θ. Consider testing H_0: θ is in Ω_0 versus H_a: θ is in Ω_a, where Ω_0 and Ω_a are disjoint (e.g., H_0: $\theta \leq 100$ versus H_a: $\theta > 100$). The **likelihood ratio principle** for test construction proceeds as follows:

1. Find the largest value of the likelihood for any θ in Ω_0 (by finding the maximum likelihood estimate within Ω_0 and substituting back into the likelihood function).

2. Find the largest value of the likelihood for any θ in Ω_a.

3. Form the ratio

$$\lambda(x_1, \ldots, x_n) = \frac{\text{Maximum likelihood for } \theta \text{ in } \Omega_0}{\text{Maximum likelihood for } \theta \text{ in } \Omega_a}$$

The ratio $\lambda(x_1, \ldots, x_n)$ is called the *likelihood ratio statistic value*. The test procedure consists of rejecting H_0 when this ratio is small. That is, a constant k is chosen, and H_0 is rejected if $\lambda(x_1, \ldots, x_n) \leq k$. Thus, H_0 is rejected when the denominator of λ greatly exceeds the numerator, indicating that the data is much more consistent with H_a than with H_0.

The constant k is selected to yield the desired type I error probability. Often the inequality $\lambda \leq k$ can be manipulated to yield a simpler equivalent condition. For example, for testing H_0: $\mu \leq \mu_0$ versus H_a: $\mu > \mu_0$ in the case of normality, $\lambda \leq k$ is equivalent to $t \geq c$. Thus, with $c = t_{\alpha, n-1}$, the likelihood ratio test is the one-sample t test.

The likelihood ratio principle can also be applied when the X_i's have different distributions and even when they are dependent, though the likelihood function can be complicated in such cases. Many of the test procedures to be presented in subsequent chapters are obtained from the likelihood ratio principle. These tests often turn out to minimize β among all tests that have the desired α, so are truly best tests. For more details and some worked examples, refer to the book by DeGroot or the one by Hogg and Craig listed in the Chapter 6 bibliography.

A practical limitation on the use of the likelihood ratio principle is that in order to construct the likelihood ratio test statistic, the form of the probability distribution from which the sample comes must be specified. To derive the t test from the likelihood ratio principle, the investigator must assume a normal pdf. If an investigator is willing to assume that the distribution is symmetric but does not want to be specific about its exact form (such as normal, uniform, or Cauchy), then the principle fails because there is no way to write a joint pdf simultaneously valid for all symmetric distributions. In Chapter 15, we will present several **distribution-free** test procedures, so called because the probability of a type I error is controlled simultaneously for many different underlying distributions. These procedures are useful when the investigator has limited knowledge of the underlying distribution. We shall also say more about issues 3 and 4 listed at the outset of this section.

Exercises   Section 8.5 (54–55)

54. Reconsider the paint-drying problem discussed in Example 8.2. The hypotheses were H_0: $\mu = 75$ versus H_a: $\mu < 75$, with σ assumed to have value 9.0. Consider the alternative value $\mu = 74$, which in the context of the problem would presumably not be a practically significant departure from H_0.

   a. For a level .01 test, compute β at this alternative for sample sizes $n = 100$, 900, and 2500.

   b. If the observed value of $\bar{X}$ is $\bar{x} = 74$, what can you say about the resulting P-value when $n = 2500$? Is the data statistically significant at any of the standard values of α?

   c. Would you really want to use a sample size of 2500 along with a level .01 test (leaving aside the cost of such an experiment)? Explain.

55. Consider the large-sample level .01 test in Section 8.3 for testing H_0: $p = .2$ against H_a: $p > .2$.

   a. For the alternative value $p = .21$, compute $\beta(.21)$ for sample sizes $n = 100, 2500, 10,000, 40,000$, and 90,000.

   b. For $\hat{p} = x/n = .21$, compute the P-value when $n = 100, 2500, 10,000$, and 40,000.

   c. In most situations, would it be reasonable to use a level .01 test in conjunction with a sample size of 40,000? Why, or why not?

Supplementary Exercises Chapter 8 (56–77)

56. A sample of 50 lenses used in eyeglasses yields a sample mean thickness of 3.05 mm and a sample standard deviation of .34 mm. The desired true average thickness of such lenses is 3.20 mm. Does the data strongly suggest that the true average thickness of such lenses is something other than what is desired? Test using $\alpha = .05$.

57. In Exercise 56, suppose the experimenter had believed before collecting the data that the value of σ was approximately .30. If the experimenter wished the probability of a type II error to be .05 when $\mu = 3.00$, was a sample size 50 unnecessarily large?

58. One method for straightening wire prior to coiling it to make a spring is called "roller straightening." The paper "The Effect of Roller and Spinner Wire Straightening on Coiling Performance and Wire Properties" (*Springs*, 1987, pp. 27–28) reports on the tensile properties of wire. Suppose a sample of 16 wires is selected and each is tested to determine tensile strength (N/mm^2). The resulting sample mean and standard deviation are 2160 and 30, respectively.

   a. The mean tensile strength for springs made using spinner straightening is 2150 N/mm^2. What hypotheses should be tested to determine if the mean tensile strength for the roller method exceeds 2150?

   b. Assuming that the tensile strength distribution is approximately normal, what test statistic would you use to test the hypotheses in part (a)?

   c. What is the value of the test statistic for this data?

   d. Place bounds on the *P*-value associated with the value of the test statistic computed in part (c).

   e. For a level .05 test, what conclusion would you reach?

59. A new method for measuring phosphorus levels in soil is described in the paper "A Rapid Method to Determine Total Phosphorus in Soils" (*Soil Sci. Amer. J.*, 1988, pp. 1301–1304). Suppose a sample of 11 soil specimens, each with a true phosphorus content of 548 mg/kg, is analyzed using the new method. The resulting sample mean and standard deviation for phosphorus level are 587 and 10, respectively.

   a. Is there evidence that the mean phosphorus level reported by the new method differs significantly from the true value of 548 mg/kg? Use $\alpha = .05$.

   b. What assumptions must you make in order for the test in part (a) to be appropriate?

60. The paper "Orchard Floor Management Utilizing Soil-Applied Coal Dust for Frost Protection" (*Agri. and Forest Meteorology*, 1988, pp. 71–82) reports the following values for soil heat flux of eight plots covered with coal dust.

34.7 35.4 34.7 37.7 32.5 28.0 18.4 24.9

The mean soil heat flux for plots covered only with grass is 29.0. Assuming that the heat-flux distribution is approximately normal, does the data suggest that the coal dust is effective in increasing the mean heat flux over that for grass? Test the appropriate hypotheses using $\alpha = .05$.

61. The paper "Caffeine Knowledge, Attitudes, and Consumption in Adult Women" (*J. of Nutrition Educ.*, 1992, pp. 179–184) reports the following summary data on daily caffeine consumption for a sample of adult women: $n = 47$, $\bar{x} = 215$ mg, $s = 235$ mg, and range: 5 to 1176.

   a. Does it appear plausible that the population distribution of daily caffeine consumption is normal? Is it necessary to assume a normal population distribution to test hypotheses about the value of the population mean consumption? Explain your reasoning.

   b. Suppose it had previously been believed that mean consumption was at most 200 mg. Does the given data contradict this prior belief? Test the appropriate hypotheses at significance level .10 and include a *P*-value in your analysis.

62. The accompanying output resulted when MINITAB was used to test the appropriate hypotheses about true average activation time based on the data in Exercise 51. Use this information to reach a conclusion at significance level .05 and also at level .01.

```
TEST OF MU = 25.000 VS MU G.T. 25.000
        N   MEAN  STDEV  SE MEAN     T  P VALUE
time   13  27.923  5.619    1.559  1.88   0.043
```

63. The true average breaking strength of ceramic insulators of a certain type is supposed to be at least 10 psi. They will be used for a particular application unless sample data indicates conclusively that this specification has not been met. A test of hypotheses using $\alpha = .01$ is to be based on a random sample of ten insulators. Assume that the breaking-strength distribution is normal with unknown standard deviation.

   a. If the true standard deviation is .80, how likely is it that insulators will be judged satisfactory when true average breaking strength is actually only 9.5? Only 9.0?

   b. What sample size would be necessary to have a 75% chance of detecting that true average breaking strength is 9.5 when the true standard deviation is .80?

64. The accompanying observations on residual flame time (sec) for strips of treated children's nightwear were given in the paper "An Introduction to Some Precision and Accuracy of Measurement Problems" (*J. of Testing and Eval.*, 1982, pp. 132–140). Suppose a true average flame time of at most 9.75 had been mandated. Does the data suggest that this condition has not been met? Carry out an appropriate test after first investigating the plausibility of any assumptions that underlie your method of inference.

9.85   9.93   9.75   9.77   9.67   9.87   9.67
9.94   9.85   9.75   9.83   9.92   9.74   9.99
9.88   9.95   9.95   9.93   9.92   9.89

65. In an experiment to test the effects of hormones on growth of beef cattle, 200 mg of progesterone and 20 mg of estradiol benzoate are implanted in the outer ear of 16 randomly selected steers, each weighing approximately 500 lb. The sample average weight gain per day for a certain number of days is found to be 2.79 lb, and the sample standard deviation is .41 lb per day. Does this data strongly suggest that the true average daily weight gain for steers treated with the hormone implant exceeds 2.50? Making any necessary assumptions, test the relevant hypotheses using $\alpha = .001$.

66. The incidence of a certain type of chromosome defect in the U.S. adult male population is believed to be 1 in 80. A random sample of 600 individuals in U.S. penal institutions reveals 12 who have such defects. Can it be concluded that the incidence rate of this defect among prisoners differs from the presumed rate for the entire adult male population?

   a. State and test the relevant hypotheses using $\alpha = .05$. What type of error might you have made in reaching a conclusion?

   b. What *P*-value is associated with this test? Based on this *P*-value, could H_0 be rejected at significance level .20?

67. In an investigation of the toxin produced by a certain poisonous snake, a researcher prepared 26 different vials, each containing 1 g of the toxin, and then determined the amount of antitoxin needed to neutralize the toxin. The sample average amount of antitoxin necessary was found to be 1.89 mg, and the sample standard deviation was .42. Previous research had indicated that the true average neutralizing amount was 1.75 mg/g of toxin. Does the new data contradict the value suggested by prior research?

   a. Test the relevant hypotheses at level .05.

   b. What can you say about the *P*-value of the data?

68. The sample average unrestrained compressive strength for 45 specimens of a particular type of brick was computed to be 3107 psi, and the sample standard deviation was 188. The distribution of unrestrained compressive strength may be somewhat skewed. Does the data strongly indicate that the true average unrestrained compressive strength is less than the design value of 3200? Test using $\alpha = .001$.

69. To test the ability of auto mechanics to identify simple engine problems, an automobile with a single such problem was taken in turn to 72 different car repair facilities. Only 42 of the 72 mechanics who worked on the car correctly identified the problem. Does this strongly indicate that the true proportion of mechanics who could identify this problem is less than .75? Compute the *P*-value and reach a conclusion accordingly.

70. When $X_1, X_2, \ldots, X_n$ are independent Poisson variables, each with parameter λ, and n is large, the sample mean $\bar{X}$ has approximately a normal distribution with $\mu = E(\bar{X}) = \lambda$ and $\sigma^2 = V(\bar{X}) = \lambda/n$. This implies that

$$Z = \frac{\bar{X} - \lambda}{\sqrt{\lambda/n}}$$

has approximately a standard normal distribution. For testing $H_0: \lambda = \lambda_0$, we can replace λ by λ_0 in the equation for Z to obtain a test statistic. This statistic is actually preferred to the large-sample statistic with denominator $S/\sqrt{n}$ (when the X_i's are Poisson) because it is tailored explicitly to the Poisson assumption. If the number of requests for consulting received by a certain statistician during a 5-day workweek has a Poisson distribution and the total number of consulting requests during a 36-week period is 160, does this suggest that the true average number of weekly requests exceeds 4.0? Test using $\alpha = .02$.

71. A hot-tub manufacturer advertises that with its heating equipment, a temperature of 100°F can be achieved in at most 15 min. A random sample of 32 tubs is selected, and the time necessary to achieve a 100°F temperature is determined for each tub. The sample average time and sample standard deviation are 17.5 min and 2.2 min, respectively. Does this data cast doubt on the company's claim? Compute the *P*-value and use it to reach a conclusion at level .05.

72. Chapter 7 presented a CI for the variance σ^2 of a normal population distribution. The key result there was that the rv $\chi^2 = (n - 1)S^2/\sigma^2$ has a chi-squared distribution with $n - 1$ df. Consider the null hypothesis H_0: $\sigma^2 = \sigma_0^2$ (equivalently, $\sigma = \sigma_0$). Then when H_0 is true, the test statistic $\chi^2 = (n - 1)S^2/\sigma_0^2$ has a chi-squared distribution with $n - 1$ df. If the relevant alternative is H_a: $\sigma^2 > \sigma_0^2$, rejecting H_0 if $(n - 1)s^2/\sigma_0^2 \geq \chi_{\alpha,n-1}^2$ gives a test with significance level α. To ensure reasonably uniform characteristics for a particular application, it is desired that the true standard deviation of the softening point of a certain type of petroleum pitch be at most .50°C. The softening points of ten different specimens were determined, yielding a sample standard deviation of .58°C. Does this strongly contradict the uniformity specification? Test the appropriate hypotheses using $\alpha = .01$.

73. Referring to Exercise 72, suppose an investigator wishes to test H_0: $\sigma^2 = .04$ versus H_a: $\sigma^2 < .04$ based on a sample of 21 observations. The computed value of $20s^2/.04$ is 8.58. Place bounds on the P-value and then reach a conclusion at level .01.

74. When the population distribution is normal and n is large, the sample standard deviation S has approximately a normal distribution with $E(S) \approx \sigma$ and $V(S) \approx \sigma^2/(2n)$. We already know that in this case, for any n, $\overline{X}$ is normal with $E(\overline{X}) = \mu$ and $V(\overline{X}) = \sigma^2/n$.

a. Assuming that the underlying distribution is normal, what is an approximately unbiased estimator of the 99th percentile $\theta = \mu + 2.33\sigma$?

b. When the X_i's are normal, it can be shown that $\overline{X}$ and S are independent rv's (one measures location while the other measures spread). Use this to compute $V(\hat{\theta})$ and $\sigma_{\hat{\theta}}$ for the estimator $\hat{\theta}$ of part (a). What is the estimated standard error $\hat{\sigma}_{\hat{\theta}}$?

c. Write a test statistic for testing H_0: $\theta = \theta_0$ that has approximately a standard normal distribution when H_0 is true. If soil pH is normally distributed in a certain region and 64 soil samples yield $\bar{x} = 6.33$, $s = .16$, does this provide strong evidence for concluding that at most 99% of all possible samples would have a pH of less than 6.75? Test using $\alpha = .01$.

75. Let $X_1, X_2, \ldots, X_n$ be a random sample from an exponential distribution with parameter λ. Then it can be shown that $2\lambda\Sigma X_i$ has a chi-squared distribution with $\nu = 2n$ (by first showing that $2\lambda X_i$ has a chi-squared distribution with $\nu = 2$).

a. Use this fact to obtain a test statistic and rejection region that together specify a level α test for H_0: $\mu = \mu_0$ versus each of the three commonly encountered alternatives. [HINT: $E(X_i) = \mu = 1/\lambda$, so $\mu = \mu_0$ is equivalent to $\lambda = 1/\mu_0$.]

b. Suppose that ten identical components, each having exponentially distributed time until failure, are tested. The resulting failure times are

95 16 11 3 42 71 225 64 87 123

Use the test procedure of part (a) to decide if the data strongly suggests that the true average lifetime is less than the previously claimed value of 75.

76. Suppose the population distribution is normal with known σ. Let γ be such that $0 < \gamma < \alpha$. For testing H_0: $\mu = \mu_0$ versus H_a: $\mu \neq \mu_0$, consider the test that rejects H_0 if either $z \geq z_\gamma$ or $z \leq -z_{\alpha - \gamma}$, where the test statistic is $Z = (\overline{X} - \mu_0)/(\sigma/\sqrt{n})$.

a. Show that $P(\text{type I error}) = \alpha$.

b. Derive an expression for $\beta(\mu')$. (HINT: Express the test in the form "reject H_0 if either $\bar{x} \geq c_1$ or $\leq c_2$.")

c. Let $\Delta > 0$. For what values of γ (relative to α) will $\beta(\mu_0 + \Delta) < \beta(\mu_0 - \Delta)$?

77. After a period of apprenticeship, an organization gives an exam that must be passed to be eligible for membership. Let $p = P(\text{randomly chosen apprentice passes})$. The organization wishes an exam that most but not all should be able to pass, so decides that $p = .90$ is desirable. For a particular exam, the relevant hypotheses are H_0: $p = .90$ versus the alternative H_a: $p \neq .90$. Suppose ten people take the exam and let $X = $ the number who pass.

a. Does the lower-tailed region $\{0, 1, \ldots, 5\}$ specify a level .01 test?

b. Show that even though H_a is two-sided, no two-tailed test is a level .01 test.

c. Sketch a graph of $\beta(p')$ as a function of p' for this test. Is this desirable?

Bibliography

See the bibliography at the end of Chapter 7.

9

Inferences Based on Two Samples

$$\Sigma\Sigma\Sigma\Sigma\Sigma\Sigma\Sigma\Sigma\Sigma\Sigma\Sigma\Sigma\Sigma\Sigma\Sigma\Sigma\Sigma\Sigma\Sigma$$

Introduction

Chapters 7 and 8 presented confidence intervals (CIs) and hypothesis-testing procedures for a single population parameter (μ, p, and σ^2). Here we extend these methods to problem situations involving the means, proportions, and variances of two different population distributions.

9.1   z Tests and Confidence Intervals for a Difference Between Two Population Means

The inferences discussed in this section concern a difference $\mu_1 - \mu_2$ between the means of two different population distributions. An investigator might, for example, wish to test hypotheses about the difference between true average breaking strengths of two different types of corrugated fiberboard. One such hypothesis would state that $\mu_1 - \mu_2 = 0$; that is, that $\mu_1 = \mu_2$. Alternatively, it may be appropriate to estimate $\mu_1 - \mu_2$ by computing a 95% CI. Such inferences are based on a sample of strength observations for each type of fiberboard.

Basic Assumptions

1. $X_1, X_2, \ldots, X_m$ is a random sample from a population with mean μ_1 and variance σ_1^2.
2. $Y_1, Y_2, \ldots, Y_n$ is a random sample from a population with mean μ_2 and variance σ_2^2.
3. The X and Y samples are independent of one another.

The natural estimator of $\mu_1 - \mu_2$ is $\bar{X} - \bar{Y}$, the difference between the corresponding sample means. The test statistic results from standardizing this estimator, so we need expressions for the expected value and standard deviation of $\bar{X} - \bar{Y}$.

Proposition

> The expected value of $\bar{X} - \bar{Y}$ is $\mu_1 - \mu_2$, so $\bar{X} - \bar{Y}$ is an unbiased estimator of $\mu_1 - \mu_2$. The standard deviation of $\bar{X} - \bar{Y}$ is
>
> $$\sigma_{\bar{X} - \bar{Y}} = \sqrt{\frac{\sigma_1^2}{m} + \frac{\sigma_2^2}{n}}$$

Proof Both these results depend on the rules of expected value and variance presented in Chapter 5. Since the expected value of a difference is the difference of expected values,

$$E(\bar{X} - \bar{Y}) = E(\bar{X}) - E(\bar{Y}) = \mu_1 - \mu_2$$

Because the X and Y samples are independent, $\bar{X}$ and $\bar{Y}$ are independent quantities, so the variance of the difference is the *sum* of $V(\bar{X})$ and $V(\bar{Y})$:

$$V(\bar{X} - \bar{Y}) = V(\bar{X}) + V(\bar{Y}) = \frac{\sigma_1^2}{m} + \frac{\sigma_2^2}{n}$$

The standard deviation of $\bar{X} - \bar{Y}$ is the square root of this expression.   ∎

If we think of $\mu_1 - \mu_2$ as a parameter θ, then its estimator is $\hat{\theta} = \bar{X} - \bar{Y}$ with standard deviation $\sigma_{\hat{\theta}}$ given by the above proposition. When σ_1^2 and σ_2^2 both have known values, the test statistic will have the form $(\hat{\theta} - \text{null value})/\sigma_{\hat{\theta}}$; this form of a test statistic was used in several one-sample problems in the previous chapter. When σ_1^2 and σ_2^2 are unknown, the sample variances must be used to estimate $\sigma_{\hat{\theta}}$.

Test Procedures for Normal Populations with Known Variances

In Chapters 7 and 8, the first CI and test procedure for a population mean μ were based on the assumption that the population distribution was normal with the value of the population variance σ^2 known to the investigator. Similarly, we first assume here that *both* population distributions are normal and that the values of *both* σ_1^2 and σ_2^2 are known. Situations in which one or both of these assumptions can be dispensed with will be presented shortly.

Because the population distributions are normal, both $\bar{X}$ and $\bar{Y}$ have normal distributions. This implies that $\bar{X} - \bar{Y}$ is normally distributed, with expected value $\mu_1 - \mu_2$ and standard deviation $\sigma_{\bar{X} - \bar{Y}}$ given in the foregoing proposition. Standardizing $\bar{X} - \bar{Y}$ gives the standard normal variable

$$Z = \frac{\bar{X} - \bar{Y} - (\mu_1 - \mu_2)}{\sqrt{\frac{\sigma_1^2}{m} + \frac{\sigma_2^2}{n}}} \tag{9.1}$$

In a hypothesis-testing problem, the null hypothesis will state that $\mu_1 - \mu_2$ has a specified value. Denoting this null value by Δ_0, the null hypothesis becomes H_0: $\mu_1 - \mu_2 = \Delta_0$. Often $\Delta_0 = 0$, in which case H_0 says that $\mu_1 = \mu_2$. A test statistic results from replacing $\mu_1 - \mu_2$ in Expression (9.1) by the null value Δ_0. Because the test statistic Z is obtained by standardizing $\bar{X} - \bar{Y}$ under the assumption that H_0 is true, it has a standard normal distribution in this case. Consider the alternative hypothesis H_a: $\mu_1 - \mu_2 > \Delta_0$. A value $\bar{x} - \bar{y}$ that considerably exceeds Δ_0 (the expected value of $\bar{X} - \bar{Y}$ when H_0 is true) provides evidence against H_0 and for H_a. Such a value of $\bar{x} - \bar{y}$ corresponds to a positive and large value of z. Thus, H_0 should be rejected in favor of H_a if z is greater than or equal to an appropriately chosen critical value. Because the test statistic Z has a standard normal distribution when H_0 is true, the upper-tailed rejection region $z \geq z_\alpha$ gives a test with significance level (type I error probability) α. Rejection regions for H_a: $\mu_1 - \mu_2 < \Delta_0$ and H_a: $\mu_1 - \mu_2 \neq \Delta_0$ that yield tests with desired significance level α are lower-tailed and two-tailed, respectively.

Null hypothesis: H_0: $\mu_1 - \mu_2 = \Delta_0$

Test statistic value: $z = \dfrac{\bar{x} - \bar{y} - \Delta_0}{\sqrt{\dfrac{\sigma_1^2}{m} + \dfrac{\sigma_2^2}{n}}}$

Alternative hypothesis	*Rejection region for level α test*
H_a: $\mu_1 - \mu_2 > \Delta_0$	$z \geq z_\alpha$
H_a: $\mu_1 - \mu_2 < \Delta_0$	$z \leq -z_\alpha$
H_a: $\mu_1 - \mu_2 \neq \Delta_0$	either $z \geq z_{\alpha/2}$ or $z \leq -z_{\alpha/2}$

Because these are z tests, a P-value is computed as it was for the z tests in Chapter 8 (e.g., P-value $= 1 - \Phi(z)$ for an upper-tailed test).

Example 9.1

3/5

Analysis of a random sample consisting of $m = 20$ specimens of cold-rolled steel to determine yield strengths resulted in a sample average strength of $\bar{x} = 29.8$ ksi. A second random sample of $n = 25$ two-side galvanized steel specimens gave a sample average strength of $\bar{y} = 34.7$ ksi. Assuming that the two yield-strength distributions are normal with $\sigma_1 = 4.0$ and $\sigma_2 = 5.0$ (suggested by a graph in the paper "Zinc-Coated Sheet Steel: An Overview," *Automotive Engr.*, Dec. 1984, pp. 39–43), does the data indicate that the corresponding true average yield strengths μ_1 and μ_2 are different? Let's carry out a test at significance level $\alpha = .01$.

1. The parameter of interest is $\mu_1 - \mu_2$, the difference between the true average strengths for the two types of steel.

2. The null hypothesis is H_0: $\mu_1 - \mu_2 = 0$.

3. The alternative hypothesis is H_a: $\mu_1 - \mu_2 \neq 0$; if H_a is true, then μ_1 and μ_2 are different.

4. With $\Delta_0 = 0$, the test statistic value is

$$z = \frac{\bar{x} - \bar{y}}{\sqrt{\dfrac{\sigma_1^2}{m} + \dfrac{\sigma_2^2}{n}}}$$

5. The inequality in H_a implies that the test is two-tailed. For $\alpha = .01$, $\alpha/2 = .005$ and $z_{\alpha/2} = z_{.005} = 2.58$. H_0 will be rejected if $z \geq 2.58$ or if $z \leq -2.58$.

6. Substituting $m = 20$, $\bar{x} = 29.8$, $\sigma_1^2 = 16.0$, $n = 25$, $\bar{y} = 34.7$, and $\sigma_2^2 = 25.0$ into the formula for z yields

$$z = \frac{29.8 - 34.7}{\sqrt{\dfrac{16.0}{20} + \dfrac{25.0}{25}}} = \frac{-4.90}{1.34} = -3.66$$

That is, the observed value of $\bar{x} - \bar{y}$ is more than 3 SDs below what would be expected were H_0 true.

7. Since $-3.66 < -2.58$, the computed z does fall in the lower tail of the rejection region. H_0 is therefore rejected at level .01 in favor of the conclusion that $\mu_1 \neq \mu_2$. The sample data strongly suggests that the true average yield strength for cold-rolled steel differs from that for galvanized steel. ■

Example 9.2

3/5

A letter in the *Journal of the American Medical Association* (May 19, 1978) reports that of 215 male physicians who were Harvard graduates and died between November 1974 and October 1977, the 125 in full-time practice lived an average of 48.9 years beyond graduation, while the 90 with academic affiliations lived an average of 43.2 years beyond graduation. Does the data suggest that the mean lifetime after graduation for doctors in full-time practice exceeds the mean lifetime for those who obtain an academic affiliation (if so, those medical students who say that they are "dying to obtain an academic affiliation" may be closer to the truth than they realize; put another way, is "publish or perish" really "publish and perish")?

Let μ_1 denote the true average number of years lived beyond graduation for physicians in full-time practice, and let μ_2 denote the same quantity for physicians with academic affiliations. Assuming the 125 and 90 physicians to be random samples from populations 1 and 2, respectively (which may not be reasonable if there is reason to believe that Harvard graduates have special characteristics that differentiate them from all other physicians—in this case inferences would be restricted just to the "Harvard populations"), we have $\bar{x} = 48.9$ and $\bar{y} = 43.2$. To apply the above test procedure, we must know σ_1^2 and σ_2^2. The letter from which the data was taken gave no information about variances, so for illustration assume that $\sigma_1 = 14.6$ and $\sigma_2 = 14.4$. The hypotheses are H_0: $\mu_1 - \mu_2 = 0$ versus H_a: $\mu_1 - \mu_2 > 0$, so Δ_0 is zero. The computed value of the test statistic is

$$z = \frac{48.9 - 43.2}{\sqrt{\dfrac{(14.6)^2}{125} + \dfrac{(14.4)^2}{90}}} = \frac{5.70}{\sqrt{1.70 + 2.30}} = 2.85$$

The P-value for an upper-tailed test is $1 - \Phi(2.85) = .0022$. At significance level .01, H_0 is rejected (because $\alpha > P$-value) in favor of the conclusion that $\mu_1 - \mu_2 > 0$ ($\mu_1 > \mu_2$). This is consistent with the information reported in the letter.   ∎

Using a Comparison to Identify Causality

Investigators are often interested in comparing either the effects of two different treatments on a response or the response after treatment with the response after no treatment (treatment vs. control). If the individuals or objects to be used in the comparison are not assigned by the investigators to the two different conditions, the study is said to be **observational.** The data in Example 9.2 is the result of a **retrospective** observational study; the investigator did not start out by selecting a sample of doctors and assigning some to the "academic affiliation" treatment and the others to the "full-time practice" treatment, but instead identified members of the two groups by looking backward in time (through obituaries!) to past records.

The difficulty with drawing conclusions based on an observational study is that while statistical analysis may indicate a significant difference in response between the two groups, the difference may be due to some underlying factors that had not been controlled rather than to any difference in treatments. Was the observed difference in Example 9.2 really due to a difference in the type of medical practice after graduation, or is there some other underlying factor (e.g., age at graduation) that might also furnish a plausible explanation for the difference? Observational studies have been used to argue for a causal link between smoking and lung cancer. There are many studies that show that the incidence of lung cancer is significantly higher among smokers than among nonsmokers. However, individuals had decided whether to become smokers long before investigators arrived on the scene, and factors in making this decision may have played a causal role in the contraction of lung cancer.

A **randomized controlled experiment** results when investigators assign subjects to the two treatments in a random fashion. When statistical significance is observed in such an experiment, the investigator and other interested parties will have more confidence in the conclusion that the difference in response has been caused by a difference in treatments. A very famous example of this type of experiment and conclusion is the Salk polio vaccine experiment described in Section 9.4. These issues are discussed at greater length in the (nonmathematical) books by Moore and by Freedman, et al. listed in the Chapter 1 references.

β and the Choice of Sample Size

The probability of a type II error is easily calculated when both population distributions are normal with known values of σ_1 and σ_2. Consider the case in which the alternative hypothesis is H_a: $\mu_1 - \mu_2 > \Delta_0$ and let Δ' denote a value of $\mu_1 - \mu_2$ that exceeds Δ_0 (a value for which H_0 is false). The upper-tailed rejection region $z \geq z_\alpha$ can be reexpressed in the form $\bar{x} - \bar{y} \geq \Delta_0 + z_\alpha \sigma_{\bar{X}-\bar{Y}}$. Thus, the probability of a type II error when $\mu_1 - \mu_2 = \Delta'$ is

$$\beta(\Delta') = P(\text{not rejecting } H_0 \text{ when } \mu_1 - \mu_2 = \Delta')$$
$$= P(\bar{X} - \bar{Y} < \Delta_0 + z_\alpha \sigma_{\bar{X}-\bar{Y}} \text{ when } \mu_1 - \mu_2 = \Delta')$$

When $\mu_1 - \mu_2 = \Delta'$, $\bar{X} - \bar{Y}$ is normally distributed with mean value Δ' and standard deviation $\sigma_{\bar{X}-\bar{Y}}$ (the same standard deviation as when H_0 is true); using these values to standardize the inequality in parentheses gives β.

Alternative hypothesis	$\beta(\Delta') = P(type$ II $error$ $when$ $\mu_1 - \mu_2 = \Delta')$
$H_a: \mu_1 - \mu_2 > \Delta_0$	$\Phi\left(z_\alpha - \dfrac{\Delta' - \Delta_0}{\sigma}\right)$
$H_a: \mu_1 - \mu_2 < \Delta_0$	$1 - \Phi\left(-z_\alpha - \dfrac{\Delta' - \Delta_0}{\sigma}\right)$
$H_a: \mu_1 - \mu_2 \neq \Delta_0$	$\Phi\left(z_{\alpha/2} - \dfrac{\Delta' - \Delta_0}{\sigma}\right) - \Phi\left(-z_{\alpha/2} - \dfrac{\Delta' - \Delta_0}{\sigma}\right)$
where $\sigma = \sigma_{\bar{X}-\bar{Y}} = \sqrt{(\sigma_1^2/m) + (\sigma_2^2/n)}$	

Example 9.3
(Example 9.1
continued)

Suppose that when μ_1 and μ_2 (the true average yield strengths for the two types of steel) differ by as much as 5, the probability of detecting such a departure from H_0 should be .90. Does a level .01 test with sample sizes $m = 20$ and $n = 25$ satisfy this condition? The value of σ for these sample sizes (the denominator of z) was previously calculated as 1.34. The probability of a type II error for the two-tailed level .01 test when $\mu_1 - \mu_2 = \Delta' = 5$ is

$$\beta(5) = \Phi\left(2.58 - \frac{5 - 0}{1.34}\right) - \Phi\left(-2.58 - \frac{5 - 0}{1.34}\right)$$

$$= \Phi(-1.15) - \Phi(-6.31) = .1251$$

It is easy to verify that $\beta(-5) = .1251$ also (because the rejection region is symmetric). Thus, the probability of detecting such a departure is $1 - \beta(5) = .8749$. Because this is somewhat less than .9, slightly larger sample sizes should be used.   ∎

As in Chapter 8, sample sizes m and n can be determined that will satisfy both $P(type$ I $error) = $ a specified α and $P(type$ II $error$ when $\mu_1 - \mu_2 = \Delta') = $ a specified β. For an upper-tailed test, equating the above expression for $\beta(\Delta')$ to the specified value of β gives

$$\frac{\sigma_1^2}{m} + \frac{\sigma_2^2}{n} = \frac{(\Delta' - \Delta_0)^2}{(z_\alpha + z_\beta)^2}$$

When the two sample sizes are equal, this equation yields

$$m = n = \frac{(\sigma_1^2 + \sigma_2^2)(z_\alpha + z_\beta)^2}{(\Delta' - \Delta_0)^2}$$

These expressions are also correct for a lower-tailed test, whereas α is replaced by $\alpha/2$ for a two-tailed test.

Large-Sample Tests

The assumptions of normal population distributions and known values of σ_1 and σ_2 are unnecessary when both sample sizes are large. In this case, the Central Limit Theorem guarantees that $\bar{X} - \bar{Y}$ has approximately a normal distribution regardless of the underlying population distributions. Furthermore, using S_1^2 and S_2^2 in place of σ_1^2 and σ_2^2 in Expression (9.1) gives a variable whose distribution is approximately standard normal:

$$Z = \frac{\bar{X} - \bar{Y} - (\mu_1 - \mu_2)}{\sqrt{\dfrac{S_1^2}{m} + \dfrac{S_2^2}{n}}}$$

A large-sample test statistic results from replacing $\mu_1 - \mu_2$ by Δ_0, the expected value of $\bar{X} - \bar{Y}$ when H_0 is true. This statistic Z then has approximately a standard normal distribution when H_0 is true, so level α tests are obtained by using z critical values exactly as before.

Use of the test statistic value

$$z = \frac{\bar{x} - \bar{y} - \Delta_0}{\sqrt{\dfrac{s_1^2}{m} + \dfrac{s_2^2}{n}}}$$

along with the previously stated upper-, lower-, and two-tailed rejection regions based on z critical values gives large-sample tests whose significance levels are approximately α. These tests are usually appropriate if both $m > 30$ and $n > 30$. A P-value is computed exactly as it was for our earlier z tests.

Example 9.4   In selecting a sulphur concrete for roadway construction in regions that experience heavy frost, it is important that the chosen concrete have a low value of thermal conductivity in order to minimize subsequent damage due to changing temperatures. Suppose two types of concrete, a graded aggregate and a no-fines aggregate, are being considered for a certain road. Table 9.1 summarizes data from an experiment carried out to compare the two types of concrete. Does this information suggest that true

Table 9.1 Data for Example 9.4

Type	Sample Size	Sample Average Conductivity	Sample SD
Graded	35	.497	.187
No-fines	35	.359	.158

average conductivity for the graded concrete exceeds that for the no-fines concrete? Let's use a test with $\alpha = .01$.

Let μ_1 and μ_2 denote the true average thermal conductivity for the graded aggregate and no-fines aggregate concrete, respectively. The two hypotheses are H_0: $\mu_1 - \mu_2 = 0$ versus H_a: $\mu_1 - \mu_2 > 0$. H_0 will be rejected if $z \geq z_{.01} = 2.33$. We calculate

$$z = \frac{(.497 - .359)}{\sqrt{\dfrac{(.187)^2}{35} + \dfrac{(.158)^2}{35}}} = \frac{.138}{.041} = 3.37$$

Since $3.37 \geq 2.33$, H_0 is rejected at significance level .01. Alternatively, the P-value for an upper-tailed z test is

$$P\text{-value} = 1 - \Phi(z) = 1 - \Phi(3.37) = .0004$$

H_0 should be rejected not only for a test with $\alpha = .01$ but also for $\alpha = .001$ or any other α exceeding .0004. The data argues strongly for the conclusion that true average thermal conductivity for the graded concrete does exceed that for the no-fines concrete. ∎

Confidence Intervals for $\mu_1 - \mu_2$

When both population distributions are normal, standardizing $\bar{X} - \bar{Y}$ gives a random variable Z with a standard normal distribution. Since the area under the z curve between $-z_{\alpha/2}$ and $z_{\alpha/2}$ is $1 - \alpha$, it follows that

$$P\left(-z_{\alpha/2} < \frac{\bar{X} - \bar{Y} - (\mu_1 - \mu_2)}{\sqrt{\dfrac{\sigma_1^2}{m} + \dfrac{\sigma_2^2}{n}}} < z_{\alpha/2} \right) = 1 - \alpha$$

Manipulation of the inequalities inside the parentheses to isolate $\mu_1 - \mu_2$ yields the equivalent probability statement

$$P\left(\bar{X} - \bar{Y} - z_{\alpha/2}\sqrt{\frac{\sigma_1^2}{m} + \frac{\sigma_2^2}{n}} < \mu_1 - \mu_2 < \bar{X} - \bar{Y} + z_{\alpha/2}\sqrt{\frac{\sigma_1^2}{m} + \frac{\sigma_2^2}{n}} \right)$$
$$= 1 - \alpha$$

This implies that a $100(1 - \alpha)\%$ CI for $\mu_1 - \mu_2$ has lower limit $\bar{x} - \bar{y} - z_{\alpha/2} \cdot \sigma_{\bar{X}-\bar{Y}}$ and upper limit $\bar{x} - \bar{y} + z_{\alpha/2} \cdot \sigma_{\bar{X}-\bar{Y}}$, where $\sigma_{\bar{X}-\bar{Y}}$ is the square-root expression. This interval is a special case of the general formula $\hat{\theta} \pm z_{\alpha/2} \cdot \sigma_{\hat{\theta}}$.

If both m and n are large, the CLT implies that the above interval is valid even without the assumption of normal populations; in this case, the confidence level is *approximately* $100(1 - \alpha)\%$. Furthermore, use of the sample variances S_1^2 and S_2^2 in the standardized variable Z yields a valid interval in which s_1^2 and s_2^2 replace σ_1^2 and σ_2^2.

Provided that m and n are both large, a CI for $\mu_1 - \mu_2$ with a confidence level of approximately $100(1 - \alpha)\%$ is

$$\bar{x} - \bar{y} \pm z_{\alpha/2} \sqrt{\frac{s_1^2}{m} + \frac{s_2^2}{n}}$$

where $-$ gives the lower limit and $+$ the upper limit of the interval.

The standard rule of thumb for characterizing sample sizes as large is again $m > 30$ and $n > 30$.

Example 9.5

3/5

Tensile strength tests were carried out on two different grades of wire rod ("Fluidized Bed Patenting of Wire Rods," *Wire J.*, June 1977, pp. 56–61), resulting in the data in Table 9.2.

Table 9.2 Data for Example 9.5

Grade	Sample Size	Sample Mean (kg/mm^2)	Sample SD
AISI 1064	$m = 129$	$\bar{x} = 107.6$	$s_1 = 1.3$
AISI 1078	$n = 129$	$\bar{y} = 123.6$	$s_2 = 2.0$

Let μ_1 and μ_2 denote the true average tensile strengths for grades 1064 and 1078, respectively. Then a 95% CI for $\mu_1 - \mu_2$ is

$$107.6 - 123.6 \pm 1.96 \sqrt{\frac{(1.3)^2}{129} + \frac{(2.0)^2}{129}} = -16.0 \pm .41$$

$$= (-16.41, \quad -15.59)$$

That is, we can be highly confident that μ_2 exceeds μ_1 by between 15.59 and 16.41. This CI is relatively narrow, suggesting that $\mu_1 - \mu_2$ has been precisely estimated, because both sample sizes are quite large and both sample standard deviations relatively small. ∎

If the variances σ_1^2 and σ_2^2 are at least approximately known and the investigator uses equal sample sizes, then the sample size n for each sample that yields a $100(1 - \alpha)\%$ interval of length L is

$$n = \frac{4z_{\alpha/2}^2(\sigma_1^2 + \sigma_2^2)}{L^2}$$

which will generally have to be rounded up to an integer.

Exercises   Section 9.1   (1–16)

1. An article in the November 1983 *Consumer Reports* compared various types of batteries. The average lifetimes of Duracell Alkaline AA batteries and Eveready Energizer Alkaline AA batteries were given as 4.1 hours and 4.5 hours, respectively. Suppose these are the population average lifetimes.

   a. Let $\bar{X}$ be the sample average lifetime of 100 Duracell batteries and $\bar{Y}$ be the sample average lifetime of 100 Eveready batteries. What is the mean value of $\bar{X} - \bar{Y}$ (i.e., where is the distribution of $\bar{X} - \bar{Y}$ centered)? How does your answer depend on the specified sample sizes?

   b. Suppose the population standard deviations of lifetime are 1.8 hours for Duracell batteries and 2.0 hours for Eveready batteries. With the sample sizes given in part (a), what is the variance of the statistic $\bar{X} - \bar{Y}$, and what is its standard deviation?

   c. For the sample sizes given in part (a), draw a picture of the approximate distribution curve of $\bar{X} - \bar{Y}$ (include a measurement scale on the horizontal axis). Would the shape of the curve necessarily be the same for sample sizes of 10 batteries of each type? Explain.

2. Let μ_1 and μ_2 denote true average tread lives for two competing brands of size FR78-15 radial tires. Test H_0: $\mu_1 - \mu_2 = 0$ versus H_a: $\mu_1 - \mu_2 \neq 0$ at level .05 using the following data: $m = 40$, $\bar{x} = 36{,}500$, $s_1 = 2200$, $n = 40$, $\bar{y} = 33{,}400$, and $s_2 = 1900$.

3. Let μ_1 denote true average tread life for a premium brand of FR78-15 radial tire and let μ_2 denote the true average tread life for an economy brand of the same size. Test H_0: $\mu_1 - \mu_2 = 10{,}000$ versus H_a: $\mu_1 - \mu_2 > 10{,}000$ at level .01 using the following data: $m = 40$, $\bar{x} = 36{,}500$, $s_1 = 2200$, $n = 40$, $\bar{y} = 23{,}800$, and $s_2 = 1500$.

4. Use the data of Exercise 2 to compute a 95% CI for $\mu_1 - \mu_2$. Does the resulting interval suggest that $\mu_1 - \mu_2$ has been precisely estimated?

5. Persons having Reynaud's syndrome are apt to suffer a sudden impairment of blood circulation in fingers and toes. In an experiment to study the extent of this impairment, each subject immersed a forefinger in water and the resulting heat output (cal/cm²/min) was measured. For $m = 10$ subjects with the syndrome, the average heat output was $\bar{x} = .64$, and for $n = 10$ nonsufferers, the average output was 2.05. Let μ_1 and μ_2 denote the true average heat outputs for the two types of subjects. Assume that the two distributions of heat output are normal with $\sigma_1 = .2$ and $\sigma_2 = .4$.

   a. Test H_0: $\mu_1 - \mu_2 = -1.0$ versus H_a: $\mu_1 - \mu_2 < -1.0$ at level .01 (H_a says that the calorie output for sufferers is more than 1 cal/cm²/min below that for nonsufferers).

   b. Compute the P-value for the value of Z obtained in part (a).

   c. What is the probability of a type II error when the actual difference between μ_1 and μ_2 is $\mu_1 - \mu_2 = -1.2$?

   d. Assuming that $m = n$, what sample sizes are required to ensure that $\beta = .1$ when $\mu_1 - \mu_2 = -1.2$?

6. An experiment to compare the tension bond strength of polymer latex modified mortar (Portland cement mortar to which polymer latex emulsions have been added during mixing) to that of unmodified mortar resulted in $\bar{x} = .18.12$ kgf/cm² for the modified mortar ($m = 40$) and $\bar{y} = 16.87$ kgf/cm² for the unmodified mortar ($n = 32$). Let μ_1 and μ_2 be the true average tension bond strengths for the modified and unmodified mortars, respectively.

   a. Assuming that $\sigma_1 = 1.6$ and $\sigma_2 = 1.4$, test H_0: $\mu_1 - \mu_2 = 0$ versus H_a: $\mu_1 - \mu_2 > 0$ at level .01.

   b. Compute the probability of a type II error for the test of part (a) when $\mu_1 - \mu_2 = 1$.

   c. Suppose the investigator decided to use a level .05 test and wished $\beta = .10$ when $\mu_1 - \mu_2 = 1$. If $m = 40$, what value of n is necessary?

   d. How would the analysis and conclusion of part (a) change if σ_1 and σ_2 were unknown but $s_1 = 1.6$ and $s_2 = 1.4$?

7. Are male college students more easily bored than their female counterparts? This question was examined in the paper "Boredom in Young Adults—Gender and Cultural Comparisons" (*J. of Cross-Cultural Psych.*, 1991, pp. 209–223). The authors administered a scale called the Boredom Proneness Scale to 97 male and 148 female U.S. college students. Does the accompanying data support the research hypothesis that the mean Boredom Proneness Rating is higher for men than for women?

Test the appropriate hypotheses using a .05 significance level.

Gender	Sample Size	Sample Mean	Sample SD
Males	97	10.40	4.83
Females	148	9.26	4.68

8. The accompanying table summarizes some data from an experiment carried out to study various characteristics of anchor bolts ("Ultimate Load Capacities of Expansion Anchor Bolts," *J. Energy Engr.*, 1993, pp. 139–158):

Bolt Diameter (in.)	Sample Size	Sample Mean Shear Strength (kip)	Sample SD
3/8	78	4.25	1.30
1/2	88	7.14	1.68

   a. Carry out a test with $\alpha = .05$ to decide whether true average strength for ½-in. bolts exceeds that for ⅜-in. bolts by more than 2 kips. Include a *P*-value in your analysis.

   b. Calculate and interpret a 95% CI for the difference in true average shear strengths.

9. An experiment was performed to compare the fracture toughness of high-purity 18 Ni maraging steel with commercial-purity steel of the same type (*Corrosion Science*, 1971, pp. 723–736). For $m = 32$ specimens, the sample average toughness was $\bar{x} = 65.6$ for the high-purity steel, while $\bar{y} = 59.8$ for $n = 38$ specimens of commercial steel. Because the high-purity steel is more expensive, its use for a certain application can be justified only if its fracture toughness exceeds that of commercial-purity steel by more than 5.

   a. Assuming that $\sigma_1 = 1.2$ and $\sigma_2 = 1.1$, test the relevant hypotheses using $\alpha = .001$.

   b. Compute β for the test conducted in part (a) when $\mu_1 - \mu_2 = 6$.

   c. If $s_1 = 1.2$ and $s_2 = 1.1$ (rather than σ_1 and σ_2), how do the analysis and conclusion of part (a) change?

10. Blood lead level was determined for a sample of 152 male hazardous-waste workers aged 20–30 and also for a sample of 86 female workers, giving mean ± standard error = 5.5 ± 0.3 for the men and 3.8 ± 0.2 for the women ("Temporal Changes in Blood Lead Levels of Hazardous Waste Workers in New Jersey, 1984–1987," *Environ. Monitoring and Assessment*, 1993, pp. 99–

107). Calculate an estimate of the difference between true average blood lead levels for male and female workers and do so in a way that provides information about reliability and precision.

11. The accompanying table gives summary data on cube compressive strength (N/mm²) for concrete specimens made with a pulverized fuel-ash mix ("A Study of Twenty-Five-Year-Old Pulverized Fuel Ash Concrete Used in Foundation Structures," *Proc. Inst. Civ. Engrs.*, Mar. 1985, pp. 149–165):

Age (days)	Sample Size	Sample Mean	Sample SD
7	68	26.99	4.89
28	74	35.76	6.43

Calculate a 99% CI for the difference between true average 7-day strength and true average 28-day strength.

12. A mechanical engineer wishes to compare strength properties of steel beams with similar beams made with a particular alloy. The same number of beams, *n*, of each type will be tested. Each beam will be set in a horizontal position with a support on each end, a force of 2500 lb will be applied at the center, and the deflection will be measured. From past experience with such beams, the engineer is willing to assume that the true standard deviation of deflection for both types of beam is .05 in. Because the alloy is more expensive, the engineer wishes to test at level .01 whether it has smaller average deflection than the steel beam. What value of *n* is appropriate if the desired type II error probability is .05 when the difference in true average deflection favors the alloy by .04 in.?

13. The level of monoamine oxidase (MAO) activity in blood platelets (nm/mg protein/h) was determined for each individual in a sample of 33 chronic schizophrenics, resulting in $\bar{x} = 2.69$ and $s_1 = 2.30$, as well as for 35 normal subjects, resulting in $\bar{y} = 6.35$, $s_2 = 4.03$. Does this data strongly suggest that true average MAO activity for normal subjects is more than twice the activity level for schizophrenics? Derive a test procedure and carry out the test using $\alpha = .01$. [HINT: H_0 and H_a here have a different form from the three standard cases. Let μ_1 and μ_2 refer to true average MAO activity for schizophrenics and normals, respectively, and consider the parameter $\theta = 2\mu_1 - \mu_2$. Write H_0 and H_a in terms of θ, estimate θ, and derive $\hat{\sigma}_\theta$ ("Reduced Monoamine Oxidase Activity in Blood Platelets from Schizophrenic Patients," *Nature*, July 28, 1972, pp. 225–226).]

14. Show for the upper-tailed test with σ_1 and σ_2 known that as either m or n is increased, β decreases when $\mu_1 - \mu_2 > \Delta_0$.

15. For the case of equal sample sizes ($m = n$) and fixed α, what happens to the necessary sample size n as β is decreased, where β is the desired type II error probability at a fixed alternative?

16. To decide whether two different types of steel have the same true average fracture toughness values, n specimens of each type are tested, yielding the following results:

Type	Sample Average	Sample SD
1	60.1	1.0
2	59.9	1.0

Calculate the P-value for the appropriate two-sample z test assuming that the above data was based on $n = 100$. Then repeat the calculation for $n = 400$. Is the small P-value for $n = 400$ indicative of a difference that has practical significance? Would you have been satisfied with just a report of the P-value? Comment briefly.

9.2

The Two-Sample t Test and Confidence Interval

In real problems, it is virtually always the case that the values of the population variances are unknown. In the previous section, we illustrated for large sample sizes the use of a test procedure and CI in which the sample variances were used in place of the population variances. In fact, for large samples, the CLT allows us to use these methods even when the two populations of interest are not normal.

There are many problems, though, in which at least one sample size is small and the population variances have unknown values. In the absence of the CLT, we proceed by making two specific assumptions about the underlying population distributions.

The use of inferential procedures that follow from these assumptions is then restricted to situations in which the assumptions are at least approximately satisfied.

Assumptions

> **1.** Both populations are normal, so that $X_1, X_2, \ldots, X_m$ is a random sample from a normal distribution and so is $Y_1, \ldots, Y_n$ (with the X's and Y's independent of one another).
>
> **2.** The values of the two population variances σ_1^2 and σ_2^2 are equal, so that their common value can be denoted by σ^2 (which is unknown).

The first assumption is of course analogous to what we assumed in connection with the one-sample t test and t CI, but the second assumption is new. Remember that the mean μ of a normal population specifies the location or center of the bell-shaped curve, while the variance σ^2 controls its spread. Assumption 2 then says that we are not comparing the locations of just any two normal populations but only populations that are known (or believed) a priori to have the same spread. When Assumption 2 is not reasonable, there is some controversy among statisticians as to how data should be analyzed. One possible approach is briefly described at the end of this section. Some authors recommend the use of a formal preliminary test for $\sigma_1^2 = \sigma_2^2$, but there are technical difficulties associated with this. Our approach is simply to "eyeball" the two sample variances; if they are of roughly the same order of magnitude, then one can be comfortable in using this test or CI.

The Pooled Estimator of σ^2

The natural estimator of $\mu_1 - \mu_2$ is still $\bar{X} - \bar{Y}$, but now the variance of this estimator can be expressed as

$$V(\bar{X} - \bar{Y}) = \frac{\sigma^2}{m} + \frac{\sigma^2}{n} = \sigma^2\left(\frac{1}{m} + \frac{1}{n}\right) \tag{9.2}$$

Because both the populations are normal, $\bar{X} - \bar{Y}$ is too, so if we standardize using the square root of (9.2), the result is a standard normal variable. The major result underlying inferential procedures in this section is that if σ in the denominator of the standardized variable is replaced by an appropriate estimator, the resulting variable will have a t distribution.

Because σ^2 is the variance of both the X distribution and the Y distribution, the best estimator should depend on both the X_i's and the Y_j's. Furthermore, more weight should be given to the sample corresponding to the larger of the two sample sizes. Both S_1^2 and S_2^2, the two sample variances, are estimators of σ^2; a better estimator than either one individually is the following weighted average of the two.

Definition

𝔰⎰𝔵

> The **pooled estimator** of the common variance σ^2, denoted by S_p^2, is defined by
>
> $$S_p^2 = \frac{(m-1)}{m+n-2}S_1^2 + \frac{(n-1)}{m+n-2}S_2^2 = \frac{(m-1)S_1^2 + (n-1)S_2^2}{m+n-2} \tag{9.3}$$
>
> The pooled estimator of σ is S_p.

The middle expression in (9.3) shows that S_p^2 has the form $\lambda S_1^2 + (1 - \lambda)S_2^2$ where $0 < \lambda < 1$. If, for example, $m = 15$ and $n = 7$, then $\lambda = .7$ and $1 - \lambda = .3$. Only if $m = n$ will $\lambda = 1 - \lambda = .5$, in which case S_p^2 is the ordinary unweighted average of S_1^2 and S_2^2.

Since $(m-1)S_1^2 = \Sigma(X_i - \bar{X})^2$ and $(n-1)S_2^2 = \Sigma(Y_j - \bar{Y})^2$, the numerator of S_p^2 is $\Sigma(X_i - \bar{X})^2 + \Sigma(Y_j - \bar{Y})^2$ (this would not be the case if the weighting factors were m and n, respectively). The X sample contributes $m - 1$ degrees of freedom (df) to the estimator and the Y sample contributes $n - 1$ df, for a total of $m + n - 2$ df.

The t Statistic and Test

Because of Expression (9.2) the variable

$$Z = \frac{\bar{X} - \bar{Y} - (\mu_1 - \mu_2)}{\sigma\sqrt{\dfrac{1}{m} + \dfrac{1}{n}}}$$

has a standard normal distribution. Substitution of S_p in place of σ yields the following key result on which inferences are based.

Theorem

Under Assumptions 1 and 2 of this section,

$$T = \frac{\bar{X} - \bar{Y} - (\mu_1 - \mu_2)}{S_p \sqrt{\dfrac{1}{m} + \dfrac{1}{n}}} \tag{9.4}$$

has a t distribution with $m + n - 2$ df.

The two-sample t statistic for testing $H_0: \mu_1 - \mu_2 = \Delta_0$ results from replacing $\mu_1 - \mu_2$ in (9.4) by Δ_0. The rejection regions for the various alternatives are similar to those in the one-sample case; each region uses a t critical value based on $m + n - 2$ df.

The Pooled t Test

Null hypothesis: $H_0: \mu_1 - \mu_2 = \Delta_0$

Test statistic value: $t = \dfrac{\bar{x} - \bar{y} - \Delta_0}{s_p \sqrt{\dfrac{1}{m} + \dfrac{1}{n}}}$

Alternative hypothesis	Rejection region for level α test
$H_a: \mu_1 - \mu_2 > \Delta_0$	$t \geq t_{\alpha, m+n-2}$
$H_a: \mu_1 - \mu_2 < \Delta_0$	$t \leq -t_{\alpha, m+n-2}$
$H_a: \mu_1 - \mu_2 \neq \Delta_0$	either $t \geq t_{\alpha/2, m+n-2}$ or $t \leq -t_{\alpha/2, m+n-2}$

Example 9.6

The paper "Effect of Internal Gas Pressure on the Compression Strength of Beverage Cans and Plastic Bottles" (*J. Testing and Evaluation*, 1993, pp. 129–131) includes the accompanying data on compression strength (lb) for a sample of 12-oz aluminum cans filled with strawberry drink and another sample filled with cola. Does the data suggest that the extra carbonation of cola results in a higher average compression strength?

Beverage	Sample Size	Sample Mean	Sample SD
Strawberry drink	15	540	21
Cola	15	554	15

The parameter of interest here is $\mu_1 - \mu_2$, the difference between true average compression strength for strawberry drink-filled cans and for cola-filled cans. The relevant hypotheses are

$$H_0: \mu_1 - \mu_2 = 0 \text{ versus } H_a: \mu_1 - \mu_2 < 0$$

and the formula for the test statistic value is

$$t = \frac{\bar{x} - \bar{y}}{s_p \sqrt{\dfrac{1}{m} + \dfrac{1}{n}}}$$

The inequality in H_a implies a lower-tailed test: Reject H_0 if $t \le -t_{\alpha, m+n-2} = -t_{\alpha, 28}$. A significance level of .01 necessitates using $-t_{.01, 28} = -2.467$. The computations are as follows:

$$s_p^2 = \frac{(15 - 1)(21)^2 + (15 - 1)(15)^2}{15 + 15 - 2} = \frac{9324}{28} = 333, \qquad s_p = 18.25$$

$$t = \frac{540 - 554}{18.25 \sqrt{\dfrac{1}{15} + \dfrac{1}{15}}} = \frac{-14}{6.664} = -2.10$$

Since $-2.10 > -2.467$, t is not in the rejection region, so H_0 cannot be rejected. The increased carbonation level has not been shown to affect compressive strength. From the t table,

$$t_{.025, 28} = 2.048 < 2.10 < t_{.01, 28} = 2.467$$

implying that $.01 < P\text{-value} < .025$. Thus, H_0 would have been rejected had a significance level of .05 been used.

Notice also that if the 1 and 2 labels are reversed, so that μ_1 now refers to cola, the alternative hypothesis becomes $\mu_1 - \mu_2 > 0$, an upper-tailed test is appropriate, and $t = 2.10 < 2.467$ results in the same conclusion as before. ■

β for the Pooled t Test ~~omit~~

In Chapter 8, we showed how β, the probability of a type II error, could be determined for a one-sample t test using the curves in Appendix Table A.13. These same curves can be used to obtain β for the pooled t test when either $\alpha = .05$ or .01. Suppose, for example, that μ_1 is the true average fracture toughness of high-purity steel specimens and μ_2 is the true average fracture toughness of commercial-steel specimens. An investigator wishes to test $H_0: \mu_1 - \mu_2 = 0$ versus $H_a: \mu_1 - \mu_2 > 0$ using a level .05 test based on sample sizes $m = 15$ and $n = 15$ (so the probability of rejecting H_0 when H_0 is true is only .05). The value $\mu_1 - \mu_2 = 5$ might be viewed as a substantial departure from H_0, so the investigator would want the probability of not rejecting H_0 when $\mu_1 - \mu_2 = 5$ to be small. This probability is exactly β for the value $\mu_1 - \mu_2 = 5$. The value of β for the one-sample t test depended on σ, the population standard deviation. Similarly, β for the pooled t test depends on σ, the common value of the two population standard deviations. To determine β, first specify the level of significance α, the two sample sizes, and a plausible value of σ. Then select an alternative value Δ' for which β is desired and compute

$$d = \frac{|\Delta' - \Delta_0|}{\sigma} \cdot \sqrt{\frac{mn}{(m + n)(m + n - 1)}}$$

Now locate the appropriate set of curves in Appendix Table A.13 ($\alpha = .05$ or $.01$ for a one- or two-tailed test), move over to the value of d on the horizontal axis, move up to the $m + n - 2$ df curve, and finally look over to the vertical axis to read the value of β.

Example 9.7 It is important that fabric used in children's clothing be fire-resistant. Suppose an investigator wishes to compare burn times of two different fabrics. A number of fabric specimens will be used in the experiment, and the burn time for each one will be recorded. Assuming that burn times are normally distributed with $\sigma_1 = \sigma_2$, the pooled t test is appropriate for testing $H_0: \mu_1 - \mu_2 = 0$ versus $H_a: \mu_1 - \mu_2 \neq 0$. The chosen sample sizes are $m = 15$ and $n = 15$. If the true standard deviation of each fabric's burn time is $\sigma = 1.50$ and μ_1 actually exceeds μ_2 by 2 ($\Delta' = 2$), then

$$d = \frac{2}{1.5} \sqrt{\frac{(15)(15)}{(30)(29)}} = .68$$

The number of degrees of freedom for the test is $m + n - 2 = 28$. The curve for 28 df does not appear in Appendix Table A.13, but it is very close to the 29 df curve, which does appear. Using a level $\alpha = .05$ two-tailed test gives $\beta \approx .08$, whereas using $\alpha = .01$ gives $\beta \approx .2$. At level .05, if the difference between μ_1 and μ_2 is 2 rather than 0, two samples of 15 in size will result in H_0 being incorrectly not rejected only 8% of the time. The test has good ability to detect this sort of departure from H_0. Of course, if the actual value of σ is greater than 1.5, the value of d would be smaller, and β would be larger. The greater the variability in the populations, the more difficult it is to draw the right conclusion. ∎

The t Interval for $\mu_1 - \mu_2$

To obtain a confidence interval for the difference between means of two normal populations when at least one sample size is small, we assume that the two population variances are equal. Then the variable

$$T = \frac{\bar{X} - \bar{Y} - (\mu_1 - \mu_2)}{S_p \sqrt{1/m + 1/n}}$$

has a t distribution with $m + n - 2$ df (where S_p^2 is the pooled estimator of the common variance σ^2), so

$$P(-t_{\alpha/2, m+n-2} \leq T \leq t_{\alpha/2, m+n-2}) = 1 - \alpha \tag{9.5}$$

Manipulation of the inequalities inside the parentheses in (9.5) in the same manner as before yields the desired interval.

The Pooled t Confidence Interval

A $100(1 - \alpha)\%$ CI for $\mu_1 - \mu_2$ is

$$\bar{x} - \bar{y} \pm t_{\alpha/2,\ m+n-2} \cdot s_p \sqrt{\frac{1}{m} + \frac{1}{n}}$$

Example 9.8

A random sample of 15 alumina ceramic insulators doped in a certain manner yielded a sample average holdoff voltage (kV) of 110 and a sample standard deviation of 24, whereas a random sample of 76 plain alumina ceramic insulators resulted in a sample average voltage and sample standard deviation of 101 and 22, respectively (data from "The Effect of Doping on the Voltage Holdoff Performance of Alumina Insulators in Vacuum," *IEEE Transactions on Electrical Insulation*, 1985, pp. 505–509). Assuming that the conditions for use of the pooled t CI are met (because one sample size is small, a z CI is not appropriate), let's calculate a 95% CI for difference between the true average voltage for doped insulators and that for plain insulators. The pooled estimate of σ^2 is $s_p^2 = [14(24)^2 + 75(22)^2]/89 = 498.5$, so $s_p = 22.3$. A 95% confidence level requires $t_{.025,89} \approx 1.99$. Thus,

$$110 - 101 \pm (1.99)(22.3)\sqrt{\frac{1}{15} + \frac{1}{76}} = 9 \pm 12.5 = (-3.5, \quad 21.5)$$

At the 95% confidence level, it is plausible that μ_1 exceeds μ_2 by as much as 21.5 or that μ_2 exceeds μ_1 by as much as 3.5. Since the CI includes zero, it is plausible that $\mu_1 = \mu_2$. ∎

A Test Procedure When $\sigma_1^2 \neq \sigma_2^2$

When Assumptions 1 and 2 are satisfied, there is a general consensus among statisticians that the two-sample t test described above should be used. In fact, it has been shown that if the distributions being sampled are not too nonnormal and/or the two variances are not too different from one another, then the t test works reasonably well in the sense that the actual level of significance is approximately the specified α, and the test continues to have small β. Statisticians customarily summarize these properties by saying that the t test is *robust* in the presence of mild departures from assumptions.

If, however, σ_1^2 and σ_2^2 are very different from one another, then using the t test, which involves a pooled estimate of a single parameter σ^2, will tend to yield erroneous conclusions. This is especially true when sample sizes are different—the test is more robust to departures from assumptions when $m = n$ than when $m \neq n$. Unfortunately, even when the populations are still assumed normal, there is no test procedure known to have very good properties. The following procedure (called the Smith–Satterthwaite test) is known to be approximately a level α test, but its type II error probabilities have proved difficult to study, so it is not known whether the test is in any sense a best test. The test statistic has approximately a t distribution when H_0 is true, but the number of degrees of freedom v is estimated from the data (v will not usually be an integer, so must be rounded to obtain a critical value from the t table).

Test statistic value: $t' = \dfrac{\bar{x} - \bar{y} - \Delta_0}{\sqrt{\dfrac{s_1^2}{m} + \dfrac{s_2^2}{n}}}$

$$
\text{Degrees of freedom:} \quad \nu = \frac{\left(\dfrac{s_1^2}{m} + \dfrac{s_2^2}{n}\right)^2}{\dfrac{(s_1^2/m)^2}{m-1} + \dfrac{(s_2^2/n)^2}{n-1}}
$$

Alternative hypothesis	Rejection region for approximate level α test
$H_a: \mu_1 - \mu_2 > \Delta_0$	$t' \geq t_{\alpha,\nu}$
$H_a: \mu_1 - \mu_2 < \Delta_0$	$t' \leq -t_{\alpha,\nu}$
$H_a: \mu_1 - \mu_2 \neq \Delta_0$	either $t' \geq t_{\alpha/2,\nu}$ or $t' \leq -t_{\alpha/2,\nu}$

Example 9.9

A paper in the *Journal of Nervous and Mental Disorders* (1968, vol. 146, pp. 136–146) reported the following data on the amount of dextroamphetamine excreted by a sample of children having organically related disorders and a sample of children with nonorganic disorders (dextroamphetamine is a drug commonly used to treat hyperkinetic children).

Organic   17.53   20.60   17.62   28.93   27.10
Nonorganic   15.59   14.76   13.32   12.45   12.79

(Observations refer to percentage of recovery of the drug 7 hours after its administration.) The summary values are $\bar{x} = 22.36$, $\bar{y} = 13.78$, $s_1^2 = 28.63$, and $s_2^2 = 1.80$. The data suggests that there is much less variability in percentage of recovery for "nonorganic" children than for "organic" children. To use T' to test H_0: $\mu_1 - \mu_2 = 0$ versus $H_a: \mu_1 - \mu_2 \neq 0$, where μ_1 and μ_2 refer to the true average percentages of recovery for the two conditions, we need

$$
\nu = \frac{\left(\dfrac{28.63}{5} + \dfrac{1.80}{5}\right)^2}{\dfrac{(28.63/5)^2}{4} + \dfrac{(1.80/5)^2}{4}} = \frac{37.04}{8.20 + .03} = 4.50 \approx 5
$$

$$
t' = \frac{22.36 - 13.78}{\sqrt{\dfrac{28.63}{5} + \dfrac{1.80}{5}}} = \frac{8.58}{2.47} = 3.47
$$

Since $t_{.005,5} = 4.032$, the data is not significant at level .01. If the pooled two-sample t test had been used, the computed t value would be $t = 3.48$, which would exceed $t_{.005,8}$. ∎

When an investigator can reasonably assume that two populations have the same shape and spread (so $\sigma_1^2 = \sigma_2^2$) but may differ with respect to location and yet the investigator does not wish to impose the assumption of normality, the "distribution-free" procedures presented in Chapter 15 can be used.

Exercises   Section 9.2 (17–30)

17. Let μ_1 and μ_2 denote true average stopping distances for two different types of cars traveling at 50 mph. Assuming normality and $\sigma_1 = \sigma_2$, test $H_0: \mu_1 - \mu_2 = 0$ versus $H_a: \mu_1 - \mu_2 \neq 0$ at level .05 using the following data: $m = 6, \bar{x} = 122.7, s_1 = 5.59, n = 5, \bar{y} = 129.3,$ and $s_2 = 5.25$.

18. Suppose μ_1 and μ_2 are true mean stopping distances at 50 mph for cars of a certain type equipped with two different types of braking systems. Use the pooled t test at significance level .01 to test $H_0: \mu_1 - \mu_2 = -10$ versus $H_a: \mu_1 - \mu_2 < -10$ for the following data: $m = 6,$ $\bar{x} = 115.7, s_1 = 5.03, n = 6, \bar{y} = 129.3,$ and $s_2 = 5.38$.

19. Use the data of Exercise 18 to calculate a 95% CI for the difference between true average stopping distance for cars equipped with system 1 and cars equipped with system 2. Does the interval suggest that precise information about the value of this difference is available?

20. The paper "Production of Soluble Organic Nitrogen During Activated Sludge Treatment" (*J. Water Pollution Control Fed.*, 1981, pp. 99–112) reports that for a sample consisting of 14 observations from one type of activated sludge culture, the sample average soluble chemical oxygen demand (SCOD, in mg/L) was 18.1 and the sample standard deviation was 6.0, while for a sample of 16 observations from another type of culture, the sample average SCOD was 15.9 and the sample standard deviation was 5.0. Let μ_1 and μ_2 denote the true average SCOD values for these two types of cultures. Assuming that both the sampled distributions are normal with $\sigma_1 = \sigma_2$, use a level .05 test to decide whether sample data strongly suggests that μ_1 and μ_2 differ from one another.

21. Tennis elbow is thought to be aggravated by the impact experienced when hitting the ball. The paper "Forces on the Hand in the Tennis One-Handed Backhand" (*Int. J. of Sport Biomechanics*, 1991, pp. 282–292) reports the force (N) on the hand just after impact on a one-handed backhand drive for six advanced players and for eight intermediate players. Summary data appears in the accompanying table.

Type of Player	Sample Size	Sample Mean	Sample SD
Advanced	6	40.3	11.3
Intermediate	8	21.4	8.3

The authors of the paper assumed in their analysis of the data that both force distributions (advanced and intermediate) were normal with the same standard deviation σ.

a. Calculate a point estimate of σ.

b. Use the accompanying MINITAB output to decide whether average force after impact is greater for advanced players than it is for intermediate players.

```
TWOSAMPLE T FOR MU1 VS MU2
      N    MEAN   STDEV   SE MEAN
1     6    40.3   11.3      4.6
2     8    21.40   8.30     2.9
95 PCT CI FOR MU1 - MU2:
(7.5, 30.3)
TTEST MU1 = MU2 (VS GT):
T= 3.62 P=0.0018 DF= 12
POOLED STDEV = 9.66
```

c. MINITAB output shows that a 95% CI for $\mu_1 - \mu_2$ is (7.5, 30.3). Use only this information along with the fact that $t_{.005,12} = 3.055$ to calculate a 99% CI and interpret the resulting interval.

22. The paper "The Influence of Corrosion Inhibitor and Surface Abrasion on the Failure of Aluminum-Wired Twist-On Connections" (*IEEE Trans. Components, Hybrids, and Manuf. Tech.*, 1984, pp. 20–25) reported the following summary data on potential drop measurements for one sample of connectors wired with alloy aluminum and another sample wired with EC aluminum:

Type	Sample Size	Sample Mean	Sample SD
Alloy	20	17.5	.55
EC	20	16.9	.49

Does this data suggest that true average potential drop for alloy connections is higher than for EC connections (as stated in the paper)? Carry out the appropriate test using a significance level of .01. In reaching your conclusion, what type of error might you have committed?

23. The flatwise bonding strength (lb/in.) was determined for dowel joints of two different types used in wood furniture frame construction, with the second type having greater rail thickness (1.5 in.) than the first type (1.25 in.).

1.25 in. $m = 10$ $\bar{x} = 1215.6$ $s_1 = 137.4$
1.5 in. $n = 10$ $\bar{y} = 1376.4$ $s_2 = 155.0$

a. Does the larger rail thickness of the second type of dowel result in a greater true average bonding strength than for dowels of the first type? Test the relevant hypotheses at level .05.

b. Place upper and lower bounds on the P-value of the data.

c. Suppose that prior to the experiment, the investigator had believed $\sigma = 150$. What would β be for $\mu_1 - \mu_2 = 100$?

d. Assuming $m = n$, what value would be required to yield $\beta = .50$ when $\mu_1 - \mu_2 = 100$?

24. Two agents used in the nozzles of fire-fighting spray systems are compared with respect to the amount of foam expansion. A random sample of five observations with the agent AFFF (aqueous film-forming foam) yields a sample mean of 4.7 and a sample standard deviation of .6. A second random sample of five observations with ATC (alcohol-type concentrates) results in a mean and standard deviation of 6.9 and .8, respectively. Does this data suggest that true average foam expansion for ATC exceeds that for AFFF by more than 1? Stating any assumptions necessary to your analysis, give as much information as you can concerning the P-value. Then use this information to reach a conclusion at significance level .01. (The sample sizes and means are given in "Enhancement of Fire Protection with Directional Cooling Spray Nozzles by Use of AFFF," *Fire Tech.*, 1983, pp. 5–13.)

25. In an experiment to study the effects of liming and urea fertilizer application on dimethoate (a pesticide) retention by loamy soil, the following percentages of dimethoate recovery were observed:

Soil treated with lime
28.5 24.7 26.2 23.9 29.6

Soil treated with urea
38.7 41.6 35.9 41.8 43.2

Compute a 95% CI for the difference in true average percentage of dimethoate recovery between the two treatments.

26. The article cited in Example 9.6 also gave the following information on compression strength for $m = 15$ 2-L bottles filled with grape soda and $n = 15$ bottles filled with punch: $\bar{x} = 104$, $s_1 = 14$, $\bar{y} = 197$, and $s_2 = 18$. Does true average strength for bottles filled with punch

appear to exceed that for soda-filled bottles by more than 75 lb? State and test the appropriate hypotheses using $\alpha = .05$.

27. Nine observations of surface-soil pH were made at each of two different locations at the Central Soil Salinity Research Institute experimental farm, and the resulting data appeared in the article "Sodium–Calcium Exchange Equilibria in Soils as Affected by Calcium Carbonate and Organic Matter" (*Soil Sci.*, 1984, p. 109). Does the data suggest that the true mean soil pH values differ for the two locations? Test the appropriate hypotheses using a .05 significance level. Be sure to state any assumptions necessary for the validity of your test.

Location			pH		
A	8.53	8.52	8.01	7.99	7.93
	7.89	7.85	7.82	7.80	
B	7.85	7.73	7.58	7.40	7.35
	7.30	7.27	7.27	7.23	

28. The paper "Pine Needles as Sensors of Atmospheric Pollution" (*Environ. Monitoring*, 1982, pp. 273–286) reports on the use of neutron-activity analysis to determine pollutant concentrations in pine needles. According to the paper's authors, "These observations strongly indicated that for those elements which are determined well by the analytical procedures, the distribution of concentrations is lognormal. Accordingly, in tests of significance the logarithms of concentrations will be used." The given data refers to bromine concentration in needles taken from a site near an oil-fired steam plant and from a relatively clean site. The summary values are means and standard deviations of the log-transformed observations. Let μ_1 be the true average log concentration at the first site and define μ_2 analogously for the second site. Test the hypothesis that the two concentration distribution means are equal against the alternative that they are different by using the pooled t test at level .05 with the log-transformed data.

Site	Sample Size	Mean Log Concentration	Standard Deviation of Log Concentration
Steam plant	8	18	4.9
Clean	9	11	4.6

29. High-nitrate intake in food consumption has been shown to have a number of deleterious effects, including lower thyroxin production, increased incidence of cyanosis in newborns, and lower milk production in dairy cows. The following data is the result of an experiment to measure the percentage of weight gain for young laboratory mice given a standard diet and mice given 2000 ppm nitrate in their drinking water.

Nitrate   12.7 19.3 20.5 10.5 14.0
           10.8 16.6 14.0 17.2
Control   18.2 32.9 10.0 14.3 16.2
           27.6 15.7

Use the test procedure with statistic T' to decide whether the data indicates at level .01 that a heavy dose of nitrate retards true average percentage weight gain in mice.

30. An experiment is to be performed to assess the effect of companionship at mealtime on food consumption. The amount of food consumed by each of n mice eating alone will be determined. Then each of n similar mice will be allowed to eat with a companion mouse trained to eat continuously. The research hypothesis is that the presence of such a companion will increase true average food consumption, so the investigators plan to use a level .01 upper-tailed test. What value of n is appropriate if σ is taken to be 1.0 and it is desired that the type II error probability should be $\beta = .10$ when the increase in true average food consumption is as much as 1?

9.3   Analysis of Paired Data

In Sections 9.1 and 9.2, we considered testing for a difference between two means μ_1 and μ_2. This was done by utilizing the results of a random sample $X_1, X_2, \ldots, X_m$ from the distribution with mean μ_1 and a completely independent (of the X's) sample $Y_1, \ldots, Y_n$ from the distribution with mean μ_2. That is, either m individuals were selected from population 1 and n different individuals from population 2, or m individuals (or experimental objects) were given one treatment and another set of n individuals were given the other treatment. In contrast, there are a number of experimental situations in which there is only one set of n individuals or experimental objects, and two observations are made on each individual or object.

Example 9.10   Trace metals in drinking water affect the flavor, and unusually high concentrations can pose a health hazard. The paper "Trace Metals of South Indian River" (*Envir. Studies*, 1982, pp. 62–66) reports on a study in which six river locations were selected (six experimental objects) and the zinc concentration (mg/L) determined for both surface water and bottom water at each location. The six pairs of observations are displayed in the accompanying table. Does the data suggest that true average concentration in bottom water exceeds that of surface water?

	Location					
	1	*2*	*3*	*4*	*5*	*6*
Zinc concentration in bottom water (x)	.430	.266	.567	.531	.707	.716
Zinc concentration in surface water (y)	.415	.238	.390	.410	.605	.609
Difference	.015	.028	.177	.121	.102	.107

Figure 9.1a displays a plot of this data. At first glance, there appears to be little difference between the *x* and *y* samples. From location to location, there is a great deal of variability in each sample, and it looks as though any differences between the samples can be attributed to this variability. However, when the observations are identified by location, as in Figure 9.1b, a different view emerges. At each location, bottom concentration exceeds surface concentration. This is confirmed by the fact that all *x* − *y* differences (bottom water concentration − surface water concentration) displayed in the bottom row of the data table are positive. As we will see, a correct analysis of this data focuses on these differences.

Figure 9.1 Plot of paired data from Example 9.10 (a) observations not identified by location, (b) observations identified by location. ■

Basic Assumptions

> The data consists of *n* independently selected pairs (X_1, Y_1), (X_2, Y_2), . . . , (X_n, Y_n), with $E(X_i) = \mu_1$ and $E(Y_i) = \mu_2$. Let $D_1 = X_1 - Y_1$, $D_2 = X_2 - Y_2$, . . . , $D_n = X_n - Y_n$, so the D_i's are the differences within pairs. Then the D_i's are assumed to be normally distributed with variance σ_D^2 (this is usually a consequence of the X_i's and Y_i's themselves being normally distributed).

We are again interested in testing hypotheses about the difference $\mu_1 - \mu_2$. The denominator of the two-sample *t* test was obtained by first applying the rule $V(\bar{X} - \bar{Y}) = V(\bar{X}) + V(\bar{Y})$. However, with paired data the *X* and *Y* observations within each pair are often not independent, so that $\bar{X}$ and $\bar{Y}$ are not independent of one another, and the rule is not valid. We must therefore abandon the two-sample *t* test and look for an alternative method of analysis.

The Paired t Test

Because different pairs are independent, the D_i's are independent of one another. If we let $D = X - Y$, where *X* and *Y* are the first and second observation, respectively, within an arbitrary pair, then the expected difference is

$$\mu_D = E(X - Y) = E(X) - E(Y) = \mu_1 - \mu_2$$

(the rule of expected values used here is valid even when *X* and *Y* are dependent). Thus, any hypothesis about $\mu_1 - \mu_2$ can be phrased as a hypothesis about the mean difference μ_D. But since the D_i's constitute a normal random sample (of differences) with mean μ_D, hypotheses about μ_D can be tested using a one-sample t test. That is, *to test hypotheses about $\mu_1 - \mu_2$ when data is paired, form the differences D_1, D_2, . . . , D_n and carry out a one-sample t test (based on $n - 1$ df) on the differences.*

The Paired t Test

Null hypothesis: $H_0: \mu_D = \Delta_0$ (where $D = X - Y$ is the difference between the first and second observations within a pair ($\mu_D = \mu_1 - \mu_2$))

Test statistic value: $t = \dfrac{\bar{d} - \Delta_0}{s_D/\sqrt{n}}$ (where $\bar{d}$ and s_D are the sample mean and standard deviation, respectively, of the d_i's)

Alternative hypothesis	Rejection region for level α test
$H_a: \mu_D > \Delta_0$	$t \geq t_{\alpha, n-1}$
$H_a: \mu_D < \Delta_0$	$t \leq -t_{\alpha, n-1}$
$H_a: \mu_D \neq \Delta_0$	either $t \geq t_{\alpha/2, n-1}$ or $t \leq -t_{\alpha/2, n-1}$

This paired test is valid even if $\sigma_1^2 \neq \sigma_2^2$, since the differences are still normally distributed and s_D^2 estimates $\sigma_D^2 = V(X - Y)$. While a two-sample t test would be based on $2n - 2$ df, the paired t test uses only $n - 1$ df. To ensure a correct analysis, we give up $n - 1$ df. There are situations in which an experiment can be run in either a paired or an unpaired (independent samples) manner; we will shortly discuss the issues involved in choosing between the two.

Example 9.11
(Example 9.10 continued)

With μ_1 and μ_2 denoting true average zinc concentrations for bottom and surface water, respectively, $\mu_D = \mu_1 - \mu_2$ is the true average difference between bottom and surface concentrations. The hypotheses of interest are $H_0: \mu_D = 0$ versus $H_a: \mu_D > 0$. The test statistic value is $t = \bar{d}/(s_D/\sqrt{n})$, and H_0 will be rejected if $t \geq t_{\alpha, n-1}$. Since $n - 1 = 5$, a level .01 test rejects H_0 in favor of H_a if $t \geq t_{.01,5} = 3.365$. Using $\Sigma d_i = .550$ and $\Sigma d_i^2 = .068832$,

$$\bar{d} = \frac{\Sigma d_i}{n} = \frac{.550}{6} = .0917$$

$$s_D^2 = \frac{\Sigma d_i^2 - (\Sigma d_i)^2/n}{n - 1} = \frac{.068832 - (.550)^2/6}{5} = .003683$$

and $s_D = \sqrt{.003683} = .0607$. Thus,

$$t = .0917/(.0607/\sqrt{6}) = .0917/.0248 = 3.70$$

Since $3.70 \geq 3.365$, H_0 is rejected at level .01. The data strongly suggests that true average zinc concentration for bottom water exceeds that for top water. ■

Example 9.12

Scientists and engineers frequently wish to compare two different techniques for measuring or determining the value of a variable. In such situations, interest centers on testing whether the mean difference in measurements is zero. The paper "Evaluation of the Deuterium Dilution Technique Against the Test Weighing Procedure for the Determination of Breast Milk Intake" (*Amer. J. Clinical Nutr.*, 1983, pp. 996–1003) reports the accompanying data on amount of milk ingested by each of 14 randomly selected infants.

	Infant						
	1	*2*	*3*	*4*	*5*	*6*	*7*
Isotopic method	1509	1418	1561	1556	2169	1760	1098
Test-weighing method	1498	1254	1336	1565	2000	1318	1410
Difference	11	164	225	−9	169	442	−312

	Infant						
	8	*9*	*10*	*11*	*12*	*13*	*14*
Isotopic method	1198	1479	1281	1414	1954	2174	2058
Test-weighing method	1129	1342	1124	1468	1604	1722	1518
Difference	69	137	157	−54	350	452	540

A normal probability plot of the differences (Figure 9.2) is quite straight, indicating that the assumption of a normal difference distribution is plausible. Let's use the paired *t* test at level .05 to see whether the true average difference in measured intake for the two methods is zero.

Figure 9.2 Normal probability plot of differences from Example 9.12

With μ_D denoting the true average difference between intake value measured by the isotopic method and by test weighing, the relevant hypotheses are H_0: $\mu_D = 0$ versus H_a: $\mu_D \neq 0$. The test is therefore two-tailed. From the *t* table, $t_{.025,13} = 2.160$, so H_0 will be rejected if $t \geq 2.160$ or if $t \leq -2.160$. The sample average difference and sample standard deviation of the differences are $\bar{d} = 167.21$ and $s_D = 228.21$, so

$$t = (167.21 - 0)/(228.21/\sqrt{14}) = 167.21/60.99 = 2.74$$

Because $2.74 \geq 2.160$, H_0 is rejected in favor of the conclusion that the true average difference is something other than zero.   ∎

When the number of pairs is large, the assumption of a normal difference distribution is not necessary. The CLT validates the resulting z test.

A Confidence Interval for μ_D

In the same way that the t CI for a single population mean μ is based on the t variable $T = (\bar{X} - \mu)/(S/\sqrt{n})$, a t confidence interval for $\mu_D \ (= \mu_1 - \mu_2)$ is based on the fact that

$$T = \frac{\bar{D} - \mu_D}{S_D/\sqrt{n}}$$

has a t distribution with $n - 1$ df. Manipulation of this t variable, as in earlier derivations of CIs, yields the following $100(1 - \alpha)\%$ CI:

$$\bar{d} \pm t_{\alpha/2, n-1} \cdot s_D/\sqrt{n}$$

When n is small, the validity of this interval requires that the distribution of differences be at least approximately normal. For large n, the CLT ensures that the resulting z interval is valid without any restrictions on the distribution of differences.

Example 9.13   In an experiment reported in the article "Emotional Attributes of Color: A Comparison of Violet and Green" (*Perceptual and Motor Skills*, 1971, pp. 403–406), 14 subjects were exposed to 60-sec intervals of alternating violet and green light for a total of 6 minutes. Galvanic skin response during the first 12 sec of exposure (calibrations) was measured and averaged over exposure intervals. The results appear in Table 9.3.

Table 9.3 Data for Example 9.13

	Subject													
	1	*2*	*3*	*4*	*5*	*6*	*7*	*8*	*9*	*10*	*11*	*12*	*13*	*14*
Violet	3.0	3.7	4.0	3.2	3.6	3.5	4.2	3.8	3.7	3.4	3.6	3.8	3.4	3.4
Green	2.2	2.7	3.1	2.9	3.3	2.6	2.9	2.8	3.2	2.5	3.5	3.0	2.3	3.5
d_i	.8	1.0	.9	.3	.3	.9	1.3	1.0	.5	.9	.1	.8	1.1	−.1

The computed summary quantities are $\bar{d} = .70$ and $s_D = .41$. With $t_{.025,13} = 2.160$, a 95% CI for μ_D is

$$.70 \pm (2.160)(.41)/\sqrt{14} = .70 \pm .24 = (.46, \quad .94)$$

We can be highly confident that the true average difference between galvanic skin response to violet and response to green is between .46 and .94. ∎

Paired Data and the Two-Sample t Test

The numerators of the paired t and pooled t statistics are identical, since $\bar{d} = \Sigma d_i/n = [\Sigma(x_i - y_i)]/n = (\Sigma x_i)/n - (\Sigma y_i)/n = \bar{x} - \bar{y}$. The difference between the two test statistics is due entirely to the denominators. Each test statistic is obtained by standardizing $\bar{X} - \bar{Y}$ $(= \bar{D})$, but in the presence of dependence the pooled t standardization is incorrect. To see this, recall from Section 5.4 that

$$V(X \pm Y) = V(X) + V(Y) \pm 2 \, \text{Cov}(X, Y)$$

Since the correlation between X and Y is

$$\rho = \text{Corr}(X, Y) = \text{Cov}(X, Y)/[\sqrt{V(X)} \cdot \sqrt{V(Y)}]$$

it follows that

$$V(X - Y) = \sigma^2 + \sigma^2 - 2\rho\sigma^2 = 2\sigma^2(1 - \rho)$$

Applying this to $\bar{X} - \bar{Y}$ yields

$$V(\bar{X} - \bar{Y}) = V(\bar{D}) = V\left(\frac{1}{n}\Sigma D_i\right) = \frac{V(D_i)}{n} = \frac{2\sigma^2(1 - \rho)}{n}$$

The two-sample t test is based on the assumption of independence, in which case $\rho = 0$. But in many paired experiments, there will be a strong *positive* dependence between X and Y (large X associated with large Y), so that ρ will be positive and the variance of $\bar{X} - \bar{Y}$ will be smaller than $2\sigma^2/n$. Thus, *whenever there is positive dependence within pairs, the denominator for the paired t statistic should be smaller than for t of the independent-samples test.* Often pooled t will be much closer to zero than paired t, considerably understating the significance of the data.

Paired Versus Unpaired Experiments

In our examples paired data resulted from two observations on the same individual (Examples 9.12 and 9.13) or experimental object (location in Example 9.10). Even when this cannot be done, paired data with dependence within pairs can be obtained by matching individuals or objects on one or more characteristics thought to influence responses. For example, in a medical experiment to compare the efficacy of two drugs for lowering blood pressure, the experimenter's budget might allow for the treatment of 20 patients. If 10 patients are randomly selected for treatment with the first drug and another 10 independently selected for treatment with the second drug, an independent-samples experiment results.

However, the experimenter, knowing that blood pressure is influenced by age and weight, might decide to pair off patients so that within each of the resulting 10 pairs, age and weight were approximately equal (though there might be sizable differences between pairs). Then each drug would be given to a different patient within each pair for a total of 10 observations on each drug.

Without this matching (or "blocking"), one drug might appear to outperform the other just because patients in one sample were lighter and younger and thus more susceptible to a decrease in blood pressure than the heavier and older patients in the second sample. However, there is a price to be paid for pairing—a smaller number of degrees of freedom for the paired analysis—so we must ask when one type of experiment should be preferred to the other.

There is no straightforward and precise answer to this question, but there are some useful guidelines. The first point is that if we have a choice between two t tests that are both valid (and carried out at the same level of significance α), we should prefer the test that has the larger number of degrees of freedom. The reason for this is that a larger number of degrees of freedom means smaller β for any fixed alternative value of the parameter or parameters. That is, for a fixed type I error probability, the probability of a type II error is decreased by increasing degrees of freedom. This is readily verified by examining the β curves in Appendix Table A.13.

However, if the experimental units are quite heterogeneous in their responses (large σ^2), it will be difficult to detect small but significant differences between two treatments. This is essentially what happened in the data set in Example 9.10—for both "treatments" (bottom water and top water) there is great between-location variability, which tends to mask differences in treatments within locations. If there is a high positive correlation within experimental units or subjects, the variance of $\bar{D} = \bar{X} - \bar{Y}$ will be $2\sigma^2(1 - \rho)/n$, which will be much smaller than the unpaired variance $2\sigma^2/n$. Because of this reduced variance due to pairing, it will be easier to detect a difference than if independent samples are used. The pros and cons of pairing can now be summarized as follows.

1. If there is great heterogeneity between subjects (large σ^2) and a large correlation within subjects (large positive ρ), then the loss in degrees of freedom will be compensated for by the increased precision associated with pairing, so a paired experiment is preferable to an independent-samples experiment.
2. If the experimental units are relatively homogeneous (small σ^2) and the correlation within pairs is not large, the gain in precision due to pairing will be outweighed by the decrease in degrees of freedom, so an independent-samples experiment should be used.

Of course, the magnitude of σ^2 and ρ will not usually be known very precisely, so an investigator will be required to make a seat-of-the-pants judgment as to whether Situation 1 or 2 obtains. In general, if the number of observations that can be obtained is large, then a loss in degrees of freedom (e.g., from 40 to 20) will not be serious;

but if the number is small, then the loss (say, from 16 to 8) because of pairing may be serious if not compensated for by increased precision. In making such a decision, the β curves for the t tests should be consulted.

Exercises Section 9.3 (31–40)

31. Consider the accompanying data on breaking load (kg/25 mm width) for various fabrics in both an unabraded condition and an abraded condition ("The Effect of Wet Abrasive Wear on the Tensile Properties of Cotton and Polyester-Cotton Fabrics," *J. Testing and Evaluation*, 1993, pp. 84–93). Use the paired t test, as did the authors of the cited paper, to test H_0: $\mu_D = 0$ versus H_a: $\mu_D > 0$ at significance level .01.

Fabric

	1	2	3	4	5	6	7	8
U	36.4	55.0	51.5	38.7	43.2	48.8	25.6	49.8
A	28.5	20.0	46.0	34.5	36.5	52.5	26.5	46.5

32. Two types of fish attractors, one made from vitrified clay pipes and the other from cement blocks and brush, were used during 16 different time periods spanning 4 years at Lake Tohopekaliga, Florida ("Two Types of Fish Attractors Compared in Lake Tohopekaliga, Florida," *Trans. Amer. Fisheries Soc.*, 1978, pp. 689–695). The following observations are of fish caught per fishing day.

Period

	1	2	3	4	5	6	7	8
Pipe	6.64	7.89	1.83	.42	.85	.29	.57	.63
Brush	9.73	8.21	2.17	.75	1.61	.75	.83	.56

Period

	9	10	11	12	13	14	15	16
Pipe	.32	.37	.00	.11	4.86	1.80	.23	.58
Brush	.76	.32	.48	.52	5.38	2.33	.91	.79

Does one attractor appear to be more effective on average than the other?

a. Use the paired t test with $\alpha = .01$ to test H_0: $\mu_D = 0$ versus H_a: $\mu_D \neq 0$.

b. What happens if the two-sample t test is used ($s_1 = 2.48$ and $s_2 = 2.91$)?

33. The paper "Relative Controllability of Dissimilar Cars" (*Human Factors*, 1962, pp. 375–380) reports results of an experiment to compare handling ability for two cars having quite different lengths, wheelbases, and turning radii. The observations are time in seconds required for subjects to parallel park each car.

Subject

	1	2	3	4	5	6	7
Car A	37.0	25.8	16.2	24.2	22.0	33.4	23.8
Car B	17.8	20.2	16.8	41.4	21.4	38.4	16.8

Subject

	8	9	10	11	12	13	14
Car A	58.2	33.6	24.4	23.4	21.2	36.2	29.8
Car B	32.2	27.8	23.2	29.6	20.6	32.2	53.8

Does the data suggest that the average person will more easily handle one car than the other? Test the relevant hypotheses using $\alpha = .10$.

34. In an experiment designed to study the effects of illumination level on task performance ("Performance of Complex Tasks Under Different Levels of Illumination," *J. Illuminating Eng.*, 1976, pp. 235–242), subjects were required to insert a fine-tipped probe into the eyeholes of ten needles in rapid succession both for a low light level with a black background and a higher level with a white background. Each data value is the time (sec) required to complete the task.

Subject

	1	2	3	4	5
Black	25.85	28.84	32.05	25.74	20.89
White	18.23	20.84	22.96	19.68	19.50

Subject

	6	7	8	9
Black	41.05	25.01	24.96	27.47
White	24.98	16.61	16.07	24.59

Does the data indicate that the higher level of illumination yields a decrease of more than 5 in true average task completion time?

35. The paper "Selection of a Method to Determine Residual Chlorine in Sewage Effluents" (*Water and Sewage Works*, 1971, pp. 360–364) reports the results of an experiment in which two different methods for determining chlorine content were used on samples of Cl_2-demand-free water for various doses and contact times. Observations are in mg/L.

	Sample			
	1	*2*	*3*	*4*
MSI method	.39	.84	1.76	3.35
SIB method	.36	1.35	2.56	3.92

	Sample			
	5	*6*	*7*	*8*
MSI method	4.69	7.70	10.52	10.92
SIB method	5.35	8.33	10.70	10.91

Construct a 99% CI for the difference in true average residual chlorine readings between the two methods.

36. It has been estimated that between 1945 and 1971, as many as 2 million children were born to mothers treated with diethylstilbestrol (DES), a nonsteroidal estrogen recommended for pregnancy maintenance. The FDA banned this drug in 1971 because research indicated a link with the incidence of cervical cancer. The paper "Effects of Prenatal Exposure to Diethylstilbestrol (DES) on Hemispheric Laterality and Spatial Ability in Human Males" (*Hormones and Behavior*, 1992, pp. 62–75) discussed a study in which 10 males exposed to DES and their unexposed brothers underwent various tests. Here we give summary data on the results of a spatial ability test: $\bar{x} = 12.6$ (exposed), $\bar{y} = 13.7$, and standard error of mean difference = .5. Test at level .05 to see whether exposure is associated with reduced spatial ability by first obtaining as much information as you can about the *P*-value.

37. The article "Sex and Race Discrimination in the New-Car Showroom: A Fact or Myth" (*J. Consumer Affairs*, 1977, pp. 107–113) reports the results of an experiment in which individuals of different races and sexes visited car dealerships to request the best possible deal on a certain car. Consider the following representative (hypothetical) data; each value is the agreed price in dollars:

	Dealership				
	1	*2*	*3*	*4*	*5*
Black woman	4459	4320	4268	4585	4736
White man	4348	4385	4231	4516	4550

	Dealership			
	6	*7*	*8*	*9*
Black woman	4262	4440	4398	4823
White man	4203	4285	4408	4570

Compute a 95% CI for the true average difference between best price offered black females and white males.

38. The urinary fluoride concentration (ppm) was determined for 11 randomly chosen livestock both at the beginning of and in the middle of their grazing period in a region previously exposed to fluoride pollution ("Fluoride Pollution Caused by a Brickworks in the Flemish Countryside of Belgium," *Int. J. Environmental Studies*, 1978, pp. 245–252).

	Subject					
	1	*2*	*3*	*4*	*5*	*6*
Beginning	24.7	46.1	18.5	29.5	26.3	33.9
Middle	12.4	14.1	7.6	9.5	19.7	10.6

	Subject				
	7	*8*	*9*	*10*	*11*
Beginning	23.1	20.7	18.0	19.3	23.0
Middle	9.1	11.5	13.3	8.3	15.0

Does the data suggest that there has been a decrease in the true average urinary fluoride concentration during the period under consideration? (NOTE: $s_1 = 8.31$ and $s_2 = 3.53$, suggesting that perhaps $\sigma_1 \neq \sigma_2$, but the paired *t* test is still valid.)

39. Referring to Exercise 34, compute an interval estimate for the difference between true average task time under the high illumination level and true average time under the low level.

40. Construct a paired data set for which $t = \infty$, so that the data is highly significant when the correct analysis is used, yet *t* for the two-sample *t* test is quite near zero, so the incorrect analysis yields an insignificant result.

9.4 Inferences Concerning a Difference Between Population Proportions

Having presented methods for comparing the means of two different populations, we now turn to the comparison of two population proportions. The notation for this problem is an extension of the notation used in the corresponding one-population problem. We let p_1 and p_2 denote the proportions of individuals in populations 1 and 2, respectively, who possess a particular characteristic. Alternatively, if we use the label S for an individual who possesses the characteristic of interest (does favor a particular proposition, has read at least one book within the last month, etc.), then p_1 and p_2 represent the probabilities of seeing the label S on a randomly chosen individual from populations 1 and 2, respectively.

We will assume the availability of a sample of m individuals from the first population and n from the second. The variables X and Y will represent the number of individuals in each sample possessing the characteristic that defines p_1 and p_2. Provided the population sizes are much larger than the sample sizes, the distribution of X can be taken to be binomial with parameters m and p_1, and similarly, Y is taken to be a binomial variable with parameters n and p_2. Furthermore, the samples are assumed to be independent of one another, so that X and Y are independent rv's.

The obvious estimator for $p_1 - p_2$, the difference in population proportions, is the corresponding difference in sample proportions $X/m - Y/n$. With $\hat{p}_1 = X/m$ and $\hat{p}_2 = Y/n$, the estimator of $p_1 - p_2$ can be expressed as $\hat{p}_1 - \hat{p}_2$.

Proposition

Let $X \sim \text{Bin}(m, p_1)$ and $Y \sim \text{Bin}(n, p_2)$ with X and Y independent variables. Then,

$$E(\hat{p}_1 - \hat{p}_2) = p_1 - p_2$$

so $\hat{p}_1 - \hat{p}_2$ is an unbiased estimator of $p_1 - p_2$, and

$$V(\hat{p}_1 - \hat{p}_2) = \frac{p_1 q_1}{m} + \frac{p_2 q_2}{n} \qquad \text{(where } q_i = 1 - p_i) \qquad (9.6)$$

Proof Since $E(X) = mp_1$ and $E(Y) = np_2$,

$$E\left(\frac{X}{m} - \frac{Y}{n}\right) = \frac{1}{m} E(X) - \frac{1}{n} E(Y) = \frac{1}{m} mp_1 - \frac{1}{n} np_2 = p_1 - p_2$$

Since $V(X) = mp_1 q_1$, $V(Y) = np_2 q_2$, and X and Y are independent,

$$V\left(\frac{X}{m} - \frac{Y}{n}\right) = V\left(\frac{X}{m}\right) + V\left(\frac{Y}{n}\right) = \frac{1}{m^2} V(X) + \frac{1}{n^2} V(Y) = \frac{p_1 q_1}{m} + \frac{p_2 q_2}{n} \qquad \blacksquare$$

We will focus first on situations in which both m and n are large. Then because $\hat{p}_1$ and $\hat{p}_2$ individually have approximately normal distributions, the estimator $\hat{p}_1 - \hat{p}_2$

also has approximately a normal distribution. Using the mean value and variance from the above proposition to standardize $\hat{p}_1 - \hat{p}_2$ yields a variable Z whose distribution is approximately standard normal:

$$Z = \frac{\hat{p}_1 - \hat{p}_2 - (p_1 - p_2)}{\sqrt{\dfrac{p_1 q_1}{m} + \dfrac{p_2 q_2}{n}}}$$

A Large-Sample Test Procedure

Analogously to the hypotheses for $\mu_1 - \mu_2$, the most general null hypothesis an investigator might consider would be of the form $H_0\colon p_1 - p_2 = \Delta_0$, where Δ_0 is again a specified number. Although for population means the case $\Delta_0 \neq 0$ presented no difficulties, for population proportions the cases $\Delta_0 = 0$ and $\Delta_0 \neq 0$ must be considered separately. Since the vast majority of actual problems of this sort involve $\Delta_0 = 0$ (i.e., the null hypothesis $p_1 = p_2$), we will concentrate on this case. When $H_0\colon p_1 - p_2 = 0$ is true, let p denote the common value of p_1 and p_2 (and similarly for q). Then the standardized variable

$$Z = \frac{\hat{p}_1 - \hat{p}_2 - 0}{\sqrt{pq\left(\dfrac{1}{m} + \dfrac{1}{n}\right)}} \qquad (9.7)$$

has approximately a standard normal distribution when H_0 is true. However, this Z cannot serve as a test statistic because the value of p is unknown—H_0 asserts only that there is a common value of p, but does not say what that value is. To obtain a test statistic having approximately a standard normal distribution when H_0 is true (so that use of an appropriate z critical value specifies a level α test), p must be estimated from the sample data.

Assuming then that $p_1 = p_2 = p$, instead of separate samples of size m and n from two different populations (two different binomial distributions), we really have a single sample of size $m + n$ from one population with proportion p. Since the total number of individuals in this combined sample having the characteristic of interest is $X + Y$, the estimator of p is

$$\hat{p} = \frac{X + Y}{m + n} = \frac{m}{m + n}\hat{p}_1 + \frac{n}{m + n}\hat{p}_2 \qquad (9.8)$$

The second expression for $\hat{p}$ shows that it is actually a weighted average of estimators $\hat{p}_1$ and $\hat{p}_2$ obtained from the two samples. If we take (9.8) (with $\hat{q} = 1 - \hat{p}$) and substitute back into (9.7), the resulting statistic has approximately a standard normal distribution when H_0 is true.

Null hypothesis: $H_0\colon p_1 - p_2 = 0$

Test statistic value (large samples): $z = \dfrac{\hat{p}_1 - \hat{p}_2}{\sqrt{\hat{p}\hat{q}(1/m + 1/n)}}$

Alternative hypothesis   Rejection region for approximate level α *test*

$H_a: p_1 - p_2 > 0$	$z \geq z_\alpha$
$H_a: p_1 - p_2 < 0$	$z \leq -z_\alpha$
$H_a: p_1 - p_2 \neq 0$	either $z \geq z_{\alpha/2}$ or $z \leq -z_{\alpha/2}$

A *P*-value is calculated in the same way as for earlier z tests.

Example 9.14

2/10

Some defendants in criminal proceedings plead guilty and are sentenced without a trial, whereas others who plead innocent are subsequently found guilty and then are sentenced. In recent years, legal scholars have speculated as to whether sentences of those who plead guilty differ in severity from sentences for those who plead innocent and are subsequently judged guilty. Consider the accompanying data on defendants from San Francisco County accused of robbery, all of whom had previous prison records ("Does It Pay to Plead Guilty? Differential Sentencing and the Functioning of Criminal Courts," *Law and Society Rev.*, 1981–1982, pp. 45–69). Does this data suggest that the proportion of all defendants in these circumstances who plead guilty and are sent to prison differs from the proportion who are sent to prison after pleading innocent and being found guilty?

	Plea	
	Guilty	*Not Guilty*
Number judged guilty	$m = 191$	$n = 64$
Number sentenced to prison	$x = 101$	$y = 56$
Sample proportion	$\hat{p}_1 = .529$	$\hat{p}_2 = .875$

Let p_1 and p_2 denote the two population proportions. The hypotheses of interest are $H_0: p_1 - p_2 = 0$ versus $H_a: p_1 - p_2 \neq 0$. At level .01, H_0 should be rejected if either $z \geq z_{.005} = 2.58$ or if $z \leq -2.58$. The combined estimate of the common success proportion is $\hat{p} = (101 + 56)/(191 + 64) = .616$. The value of the test statistic is then

$$z = \frac{.529 - .875}{\sqrt{(.616)(.384)(1/191 + 1/64)}} = \frac{-.346}{.070} = -4.94$$

Since $-4.94 \leq -2.58$, H_0 must be rejected.

The *P*-value for a two-tailed z test is

$$P\text{-value} = 2[1 - \Phi(|z|)] = 2[1 - \Phi(4.94)] < 2[1 - \Phi(3.49)] = .0004$$

A more extensive standard normal table yields *P*-value $\approx .0000006$. This *P*-value is so minuscule that at any reasonable level α, H_0 should be rejected. The data very strongly suggests that $p_1 \neq p_2$ and, in particular, that initially pleading guilty may be a good strategy as far as avoiding prison is concerned.

The cited article also reported data on defendants in several other counties. The authors broke down the data by type of crime (burglary or robbery) and by nature of prior record (none, some but no prison, and prison). In every case, the conclusion was the same: Among defendants judged guilty, those who pleaded that way were less likely to receive prison sentences. ■

Type II Error Probabilities and Sample Sizes

Here the determination of β is a bit more cumbersome than it was for other large-sample tests. The reason is that the denominator of Z is an estimate of the standard deviation of $\hat{p}_1 - \hat{p}_2$ assuming that $p_1 = p_2 = p$. When H_0 is false, $\hat{p}_1 - \hat{p}_2$ must be restandardized using

$$\sigma_{\hat{p}_1 - \hat{p}_2} = \sqrt{\frac{p_1 q_1}{m} + \frac{p_2 q_2}{n}} \tag{9.9}$$

The form of σ implies that β is not a function of just $p_1 - p_2$, so we denote it by $\beta(p_1, p_2)$.

Alternative hypothesis	$\beta(p_1, p_2)$
$H_a: p_1 - p_2 > 0$	$\Phi\left[\dfrac{z_\alpha \sqrt{\bar{p}\bar{q}(1/m + 1/n)} - (p_1 - p_2)}{\sigma} \right]$
$H_a: p_1 - p_2 < 0$	$1 - \Phi\left[\dfrac{-z_\alpha \sqrt{\bar{p}\bar{q}(1/m + 1/n)} - (p_1 - p_2)}{\sigma} \right]$
$H_a: p_1 - p_2 \neq 0$	$\Phi\left[\dfrac{z_{\alpha/2} \sqrt{\bar{p}\bar{q}(1/m + 1/n)} - (p_1 - p_2)}{\sigma} \right]$ $- \Phi\left[\dfrac{-z_{\alpha/2} \sqrt{\bar{p}\bar{q}(1/m + 1/n)} - (p_1 - p_2)}{\sigma} \right]$

where $\bar{p} = (mp_1 + np_2)/(m + n)$, $\bar{q} = (mq_1 + nq_2)/(m + n)$, and σ is given by (9.9).

Proof For the upper-tailed test ($H_a: p_1 - p_2 > 0$),

$$\beta(p_1, p_2) = P[\hat{p}_1 - \hat{p}_2 < z_\alpha \sqrt{\hat{p}\hat{q}(1/m + 1/n)}]$$
$$= P\left[\frac{(\hat{p}_1 - \hat{p}_2) - (p_1 - p_2)}{\sigma} < \frac{z_\alpha \sqrt{\hat{p}\hat{q}(1/m + 1/n)} - (p_1 - p_2)}{\sigma} \right]$$

When m and n are both large,

$$\hat{p} = (m\hat{p}_1 + n\hat{p}_2)/(m + n) \approx (mp_1 + np_2)/(m + n) = \bar{p}$$

and $\hat{q} \approx \bar{q}$, which yields the above (approximate) expression for $\beta(p_1, p_2)$. ■

Alternatively, for specified p_1, p_2 with $p_1 - p_2 = d$, the sample sizes necessary to achieve $\beta(p_1, p_2) = \beta$ can be determined. For example, for the upper-tailed test, we equate $-z_\beta$ to the argument of $\Phi(\cdot)$ (i.e., what's inside the parentheses) in the foregoing box. If $m = n$, there is a simple expression for the common value.

For the case $m = n$, the level α test has type II error probability β at the alternative values p_1, p_2 with $p_1 - p_2 = d$ when

$$n = \frac{\left[z_\alpha \sqrt{(p_1 + p_2)(q_1 + q_2)/2} + z_\beta \sqrt{p_1 q_1 + p_2 q_2} \right]^2}{d^2} \tag{9.10}$$

for an upper- or lower-tailed test, with $\alpha/2$ replacing α for a two-tailed test.

Example 9.15

One of the truly impressive applications of statistics occurred in connection with the design of the 1954 Salk polio vaccine experiment and analysis of the resulting data. Part of the experiment focused on the efficacy of the vaccine in combating paralytic polio. Because it was thought that without a control group of children, there would be no sound basis for assessment of the vaccine, it was decided to administer the vaccine to one group and a placebo injection (visually indistinguishable from the vaccine but known to have no effect) to a control group. For ethical reasons and also because it was thought that the knowledge of vaccine administration might have an effect on treatment and diagnosis, the experiment was conducted in a **double-blind** manner. That is, neither the individuals receiving injections nor those administering them actually knew who was receiving vaccine and who was receiving the placebo (samples were numerically coded)—remember, at that point it was not at all clear whether the vaccine was beneficial.

Letting p_1 and p_2 be the probabilities of a child getting paralytic polio for the control and treatment condition, respectively, the objective was to test H_0: $p_1 - p_2 = 0$ versus H_a: $p_1 - p_2 > 0$ (the alternative states that a vaccinated child is less likely to contract polio than an unvaccinated child). Supposing the true value of p_1 is .0003 (an incidence rate of 30 per 100,000), the vaccine would be a significant improvement if the incidence rate was halved—that is, $p_2 = .00015$. Using a level $\alpha = .05$ test, it would then be reasonable to ask for sample sizes for which $\beta = .1$ when $p_1 = .0003$ and $p_2 = .00015$. Assuming equal sample sizes, the required n is obtained from (9.10) as

$$n = \frac{[1.645\sqrt{(.5)(.00045)(1.99955)} + 1.28\sqrt{(.00015)(.99985) + (.0003)(.9997)}]^2}{(.0003 - .00015)^2}$$

$$= [(.0349 + .0271)/.00015]^2 \approx 171,000$$

The actual data for this experiment appears below. Sample sizes of approximately 200,000 were used. The reader can easily verify that $z = 6.43$, a highly significant value. The vaccine was judged a resounding success! An excellent expository article describing the experiment and results appears in *Statistics: A Guide to the Unknown*, edited by Judith Tanur et al. (see Chapter 1 bibliography).

Placebo $m = 201{,}229$, $x =$ number of cases of paralytic polio $= 110$
Vaccine $n = 200{,}745$, $y = 33$ ■

A Large-Sample Confidence Interval for $p_1 - p_2$

As with means, many two-sample problems involve the objective of comparison through hypothesis testing, but sometimes an interval estimate for $p_1 - p_2$ is appropriate. Both $\hat{p}_1 = X/m$ and $\hat{p}_2 = Y/n$ have approximate normal distributions when m and n are both large. If we identify θ with $p_1 - p_2$, then $\hat{\theta} = \hat{p}_1 - \hat{p}_2$ satisfies the conditions necessary for obtaining a large-sample CI. In particular, the estimated standard deviation of $\hat{\theta}$ is $\sqrt{(\hat{p}_1\hat{q}_1/m) + (\hat{p}_2\hat{q}_2/n)}$. The $100(1 - \alpha)\%$ interval $\hat{\theta} \pm z_{\alpha/2} \cdot \hat{\sigma}_{\hat{\theta}}$ then becomes

$$\hat{p}_1 - \hat{p}_2 \pm z_{\alpha/2} \sqrt{\frac{\hat{p}_1\hat{q}_1}{m} + \frac{\hat{p}_2\hat{q}_2}{n}}$$

Notice that the estimated standard deviation of $\hat{p}_1 - \hat{p}_2$ (the square-root expression) is different here from what it was for hypothesis testing when $\Delta_0 = 0$.

Example 9.16   The paper "The Association of Marijuana Use with Outcome of Pregnancy" (*Amer. J. Public Health*, 1983, pp. 1161–1164) reports the accompanying data on incidence of major malfunctions among newborns both for mothers who were marijuana users and for mothers who did not use marijuana.

	User	Nonuser
Sample size	1246	11,178
Number of major malfunctions	42	294
$\hat{p}$	.0337	.0263

Let p_1 denote the proportion of births that involve a major malfunction among all mothers who are marijuana users, and define p_2 similarly for nonusers. A 99% CI for $p_1 - p_2$ requires

$$2.58\sqrt{\frac{\hat{p}_1\hat{q}_1}{m} + \frac{\hat{p}_2\hat{q}_2}{n}} = 2.58\sqrt{\frac{(.0337)(.9663)}{1246} + \frac{(.0263)(.9737)}{11{,}178}}$$
$$= (2.58)(.00533) = .0138$$

The confidence interval is then $.0337 - .0263 \pm .0138 = .0074 \pm .0138 = (-.0064, .0212)$. This interval is quite narrow, which suggests that $p_1 - p_2$ has been precisely estimated. Notice that the interval contains zero, so this is one plausible value for $p_1 - p_2$.   ■

Small-Sample Inferences

On occasion an inference concerning $p_1 - p_2$ may have to be based on samples for which at least one sample size is small. Appropriate methods for such situations are not as straightforward as those for large samples, and there is more controversy among statisticians as to recommended procedures. One frequently used test, called the Fisher–Irwin test, is based on the hypergeometric distribution. A statistician can be consulted for more information.

Exercises   Section 9.4 (41–50)

41. Is someone who switches brands because of a financial inducement less likely to remain loyal than someone who switches without inducement? Let p_1 and p_2 denote the true proportions of switchers to a certain brand with and without inducement, respectively, who subsequently make a repeat purchase. Test $H_0: p_1 - p_2 = 0$ versus $H_a: p_1 - p_2 < 0$ using $\alpha = .01$ and the following data:

$m = 200$, number of successes $= 30$
$n = 600$, number of successes $= 180$

(Similar data appears in "Impact of Deals and Deal Retraction on Brand Switching," *J. Marketing*, 1980, pp. 62–70.)

42. A sample of 300 urban adult residents of a particular state revealed 63 who favored increasing the highway speed limit from 55 to 65 mph, while a sample of 180 rural residents yielded 75 who favored the increase. Does this data indicate that the sentiment for increasing the speed limit is different for the two groups of residents?

   a. Test $H_0: p_1 = p_2$ versus $H_a: p_1 \neq p_2$ using $\alpha = .05$, where p_1 refers to the urban population.

   b. If the true proportions favoring the increase are actually $p_1 = .20$ (urban) and $p_2 = .40$ (rural), what is the probability that H_0 will be rejected using a level .05 test with $m = 300$, $n = 180$?

43. It is thought that the front cover and the nature of the first question on mail surveys influence the response rate. The paper "The Impact of Cover Design and First Questions on Response Rates for a Mail Survey of Skydivers" (*Leisure Sciences*, 1991, pp. 67–76) tested this theory by experimenting with different cover designs. One cover was plain; the other used a picture of a skydiver. The researchers speculated that the return rate would be lower for the plain cover.

Cover	Number Sent	Number Returned
Plain	207	104
Skydiver	213	109

Does this data support the researchers' hypothesis? Test the relevant hypotheses using $\alpha = .10$ by first calculating a *P*-value.

44. Do teachers find their work rewarding and satisfying? The paper "Work-Related Attitudes" (*Psychological Reports*, 1991, pp. 443–450) reports the results of a survey of 395 elementary school teachers and 266 high school teachers. Of the elementary school teachers, 224 said they were very satisfied with their jobs, whereas 126 of the high school teachers were very satisfied with their work. Estimate the difference between the proportion of all elementary school teachers who are satisfied and all high school teachers who are satisfied by calculating a CI.

45. A random sample of 5726 telephone numbers from a certain region taken in March 1992 yielded 1105 that were unlisted, and 1 year later a sample of 5384 yielded 980 unlisted numbers.

   a. Test at level .10 to see if there is a difference in true proportions of unlisted numbers between the two years.

   b. If $p_1 = .20$ and $p_2 = .18$, what sample sizes ($m = n$) would be necessary to detect such a difference with probability .90?

46. Ionizing radiation is being given increasing attention as a method for preserving horticultural products. The paper "The Influence of Gamma-Irradiation on the Storage Life of Red Variety Garlic" (*J. of Food Processing and Preservation*, 1983, 179–183) reports that 153 of 180 irradiated garlic bulbs were marketable (no external sprouting, rotting, or softening) 240 days after treatment, while only 119 of 180 untreated bulbs were marketable after this length of time. Does this data suggest that ionizing radiation is beneficial as far as marketability is concerned?

47. In medical investigations the ratio $\theta = p_1/p_2$ is often of more interest than the difference $p_1 - p_2$ (e.g., individuals given treatment 1 are how many times as likely to recover as those given treatment 2?). Let $\hat{\theta} = \hat{p}_1/\hat{p}_2$.

When both sample sizes are large, $\hat{\theta}$ has approximately a normal distribution with mean value θ. Application of the technique described in Exercise 91 of Chapter 5 gives

$$V(\hat{\theta}) \approx \frac{p_1(1 - p_1)}{n_1 p_2^2} + \frac{p_1^2(1 - p_2)}{n_2 p_2^3}$$

Return to the heart attack data of Example 1.3, and calculate an interval of plausible values for θ at the 95% confidence level. What does this interval suggest about the efficacy of the aspirin treatment?

48. Sometimes experiments involving success or failure responses are run in a paired or before/after manner. Suppose that before a major policy speech by a political candidate, n individuals are selected and asked whether (S) or not (F) they favor the candidate. Then after the speech the same n people are asked the same question. The responses can be entered in a table as follows:

		After	
		S	F
Before	S	X_1	X_2
	F	X_3	X_4

where $X_1 + X_2 + X_3 + X_4 = n$. Let p_1, p_2, p_3, and p_4 denote the four cell probabilities, so that $p_1 = P(S$ before and S after), and so on. We wish to test the hypothesis that the true proportion of supporters (S) after the speech has not increased against the alternative that it has increased.

a. State the two hypotheses of interest in terms of p_1, p_2, p_3, and p_4.

b. Construct an estimator for the after/before difference in success probabilities.

c. When n is large, it can be shown that the rv $(X_i - X_j)/n$ has approximately a normal distribution with variance given by $[p_i + p_j - (p_i - p_j)^2]/n$. Use this to construct a test statistic with approximately a standard normal distribution when H_0 is true (the result is called McNemar's test).

d. If $x_1 = 350, x_2 = 150, x_3 = 200$, and $x_4 = 300$, what do you conclude?

49. Two different types of alloy, A and B, have been used to manufacture experimental specimens of a small tension link to be used in a certain engineering application. The ultimate strength (ksi) of each specimen was determined, and the results are summarized in the accompanying frequency distribution.

	A	B
26–under 30	6	4
30–under 34	12	9
34–under 38	15	19
38–under 42	7	10
	$m = 40$	$n = 42$

Compute a 95% CI for the difference between the true proportions of all specimens of alloys A and B that have an ultimate strength of at least 34 ksi.

50. Suppose a 95% CI for $p_1 - p_2$ is to be constructed based on equal sample sizes from the two populations. For what value of n ($= m$) will the resulting interval have length at most .1 irrespective of the results of the sampling?

Inferences Concerning Two Population Variances

Methods for comparing two population variances (or standard deviations) are occasionally needed, though such problems arise much less frequently than those involving means or proportions. For the case in which the populations under investigation are normal, the procedures are based on a new family of probability distributions.

The F Distribution

3/10

The F probability distribution has two parameters, denoted by v_1 and v_2. The parameter v_1 is called the *number of numerator degrees of freedom*, and v_2 is the *number of denominator degrees of freedom;* here v_1 and v_2 are positive integers. A random variable that has an F distribution cannot assume a negative value. Since the density function is complicated and will not be used explicitly, we omit the formula. There is an important connection between an F variable and chi-squared variables. If X_1 and X_2 are independent chi-squared rv's with v_1 and v_2 df, respectively, then the rv

$$F = \frac{X_1/v_1}{X_2/v_2} \qquad\qquad (9.11)$$

the ratio of the two chi-squared variables divided by their respective degrees of freedom, can be shown to have an F distribution.

Figure 9.3 illustrates the graph of a typical F density function. Analogous to the notation $t_{\alpha,v}$ and $\chi^2_{\alpha,v}$, we use F_{α,v_1,v_2} for the point on the axis that captures α of the area under the F density curve with v_1 and v_2 df in the upper tail. The density curve is not symmetric, so it would seem that both upper- and lower-tail critical values must be tabulated. This is not necessary, though, because of the following property.

$$F_{1-\alpha,v_1,v_2} = 1/F_{\alpha,v_2,v_1} \qquad\qquad (9.12)$$

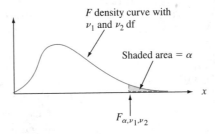

Figure 9.3 An F density curve

Appendix Table A.7 gives F_{α,v_1,v_2} for $\alpha = .05$ and $.01$, and various values of v_1 (in different columns of the table) and v_2 (in different rows of the table). For example, $F_{.05,6,10} = 3.22$ and $F_{.05,10,6} = 4.06$. To obtain $F_{.95,6,10}$, the number which captures $.95$ of the area to its right (and thus $.05$ to the left) under the F curve with $v_1 = 6$ and $v_2 = 10$, we use (9.12): $F_{.95,6,10} = 1/F_{.05,10,6} = 1/4.06 = .246$.

Inferential Methods

A test procedure for hypotheses concerning the ratio σ_1^2/σ_2^2 as well as a CI for this ratio are based on the following result.

Theorem

3/10

Let $X_1, \ldots, X_m$ be a random sample from a normal distribution with variance σ_1^2, let $Y_1, \ldots, Y_n$ be another random sample (independent of the X_i's) from a normal distribution with variance σ_2^2, and let S_1^2 and S_2^2 denote the two sample variances. Then the rv

$$F = \frac{S_1^2/\sigma_1^2}{S_2^2/\sigma_2^2} \tag{9.13}$$

has an F distribution with $v_1 = m - 1$ and v_2 and $n - 1$.

This theorem results from combining (9.11) with the fact that the variables $(m - 1)S_1^2/\sigma_1^2$ and $(n - 1)S_2^2/\sigma_2^2$ each have a chi-squared distribution with $m - 1$ and $n - 1$ df, respectively (see Section 7.4). Because F involves a ratio rather than a difference, the test statistic is the ratio of sample variances; the claim that $\sigma_1^2 = \sigma_2^2$ is then rejected if the ratio differs by too much from 1.

Null hypothesis: $H_0: \sigma_1^2 = \sigma_2^2$

Test statistic value: $f = s_1^2/s_2^2$

Alternative hypothesis	Rejection region for a level α test
$H_a: \sigma_1^2 > \sigma_2^2$	$f \geq F_{\alpha,m-1,n-1}$
$H_a: \sigma_1^2 < \sigma_2^2$	$f \leq F_{1-\alpha,m-1,n-1}$
$H_a: \sigma_1^2 \neq \sigma_2^2$	either $f \geq F_{\alpha/2,m-1,n-1}$ or $f \leq F_{1-\alpha/2,m-1,n-1}$

Since critical values are tabled only for $\alpha = .05$ and .01, the two-tailed test can be performed only at levels .10 or .02. More extensive F tables for testing at other levels are available elsewhere.

Example 9.17

3/10

On the basis of data reported in a 1979 article in the *Journal of Gerontology* ("Serum Ferritin in an Elderly Population," pp. 521–524), the authors concluded that the ferritin distribution in the elderly had a smaller variance than in the younger adults. (Serum ferritin is used in diagnosing iron deficiency.) For a sample of 28 elderly men, the sample standard deviation of serum ferritin (mg/L) was $s_1 = 52.6$; for 26 young men, the sample standard deviation was $s_2 = 84.2$. Does this data support the above conclusion as applied to men?

Let σ_1^2 and σ_2^2 denote the variance of the serum ferritin distributions for elderly men and young men, respectively. We wish to test $H_0: \sigma_1^2 = \sigma_2^2$ versus $H_a: \sigma_1^2 < \sigma_2^2$. At level .01, H_0 will be rejected if $f \leq F_{.99,27,25}$. To obtain the critical value, we need $F_{.01,25,27}$. From the F table, $F_{.01,25,27} \approx 2.54$ (since $F_{.01,24,27} = 2.55$), so $F_{.99,27,25} \approx 1/2.54 = .394$. The computed value of F is $(52.6)^2/(84.2)^2 = .390$. Since $.390 \leq .394$, H_0 is rejected at level .01 in favor of H_a, so variability does appear to be greater in young men than in elderly men. ∎

The validity of the two-sample t test rested on the assumption that $\sigma_1^2 = \sigma_2^2$. It is tempting to use the F test as a preliminary test to decide whether the two-sample

t test should be used (use it if H_0: $\sigma_1^2 = \sigma_2^2$ cannot be rejected in favor of H_a: $\sigma_1^2 \neq \sigma_2^2$), but we do not recommend this. The reason is that the F test may yield a significant value of F not only when H_0 is false but also when the underlying populations are not normal—it is sensitive to departures from normality as well as from H_0. Since the t test is a good test as long as the underlying populations are not too nonnormal, using a preliminary F test might result in not using the t test for the wrong reason. An additional complication is that the compound procedure (F at level α followed by t at level α) will not necessarily have level of significance α.

The CI for σ_1^2/σ_2^2 is based on replacing F in the probability statement

$$P(F_{1 - \alpha/2, v_1, v_2} < F < F_{\alpha/2, v_1, v_2}) = 1 - \alpha$$

by the F variable (9.13) and manipulating the inequalities to isolate σ_1^2/σ_2^2. An interval for σ_1/σ_2 results from taking the square root of each limit. The details are left for an exercise.

Exercises   Section 9.5 (51–56)

51. Obtain or compute the following quantities:

a. $F_{.05, 5, 8}$ **b.** $F_{.05, 8, 5}$ **c.** $F_{.95, 5, 8}$ **d.** $F_{.95, 8, 5}$

e. The 99th percentile of the F distribution with $v_1 = 10$, $v_2 = 12$

f. The 1st percentile of the F distribution with $v_1 = 10$, $v_2 = 12$

g. $P(F \leq 6.16)$ for $v_1 = 6$, $v_2 = 4$

h. $P(.177 \leq F \leq 4.74)$, for $v_1 = 10$, $v_2 = 5$

52. The sample standard deviation of sodium concentration in whole blood (mEq/L) for $m = 20$ marine eels was found to be $s_1 = 40.5$, while the sample standard deviation of concentration for $n = 20$ freshwater eels was $s_2 = 32.1$ ("Ionic Composition of the Plasma and Whole Blood of Marine and Freshwater Eels," *Comp. Biochemistry and Physiology*, 1974, pp. 541–544). Assuming normality of the two concentration distributions, test at level .10 to see if the data suggests any difference between concentration variances for the two types of eels.

53. A general contractor is considering purchasing lumber from one of two different suppliers. A sample of 12 2 in. × 4 in. boards of a certain length is obtained from each supplier and the length of each board is measured. The sample standard deviation of length for the first supplier's boards is found to be $s_1 = .13$ in., while $s_2 = .17$ in. for the second supplier. Does this data indicate that the lengths of one supplier's 2 in. × 4 in. boards are subject to more variability than those of the other supplier? Test using $\alpha = .02$, assuming normality.

54. Toxaphene is an insecticide that has been identified as a pollutant in the Great Lakes ecosystem. To investigate the effect of toxaphene exposure on animals, groups of rats were given toxaphene in their diet. The paper "Reproduction Study of Toxaphene in the Rat" (*J. of Environ. Sci. Health*, 1988, pp. 101–126) reports weight gains (in grams) for rats given a low dose (4 ppm) and for control rats whose diet did not include the insecticide. The sample standard deviation for 23 female control rats was 32 g and for 20 female low-dose rats was 54 g. Does this data suggest that there is more variability in low-dose weight gains than in control weight gains? Assuming normality, carry out a test of hypotheses at significance level .05.

55. In a study of copper deficiency in cattle, the copper values (μg Cu/100 mL blood) were determined both for cattle grazing in an area known to have well-defined molybdenum anomalies (metal values in excess of the normal range of regional variation) and for cattle grazing in a nonanomalous area ("An Investigation into Copper Deficiency in Cattle in the Southern Pennines," *J. Agricultural Soc. Cambridge*, 1972, pp. 157–163), resulting in $s_1 = 21.5$ ($m = 48$) for the anomalous condition and $s_2 = 19.45$ ($n = 45$) for the nonanomalous condition. Test for the equality versus inequality of population variances ($\alpha = .05$).

56. The article "Enhancement of Compressive Properties of Failed Concrete Cylinders with Polymer Impregnation" (*J. Testing and Evaluation*, 1977, pp. 333–337) reports the following data on impregnated compressive modulus (psi $\times 10^6$) when two different polymers were used to repair cracks in failed concrete.

| Epoxy | 1.75 | 2.12 | 2.05 | 1.97 |
| MMA prepolymer | 1.77 | 1.59 | 1.70 | 1.69 |

Obtain a 90% CI for the ratio of variances by first using the method suggested in the text to obtain a general confidence interval formula.

Supplementary Exercises Chapter 9 (57–79)

57. The accompanying summary data on compression strength (lb) for $12 \times 10 \times 8$ in. boxes appeared in the paper "Compression of Single-Wall Corrugated Shipping Containers Using Fixed and Floating Test Platens" (*J. Testing and Evaluation*, 1992, pp. 318–320). The authors stated that "the difference between the compression strength using fixed and floating platen methods was found to be small compared to normal variation in compression strength between identical boxes." Do you agree?

Method	Sample Size	Sample Mean	Sample SD
Fixed	10	807	27
Floating	10	757	41

58. The authors of the paper "Dynamics of Canopy Structure and Light Interception in *Pinus elliotti*, North Florida" (*Ecological Monographs*, 1991, pp. 33–51) planned an experiment to determine the effect of fertilizer on a measure of leaf area. A number of plots were available for the study, and half were selected at random to be fertilized. To ensure that the plots to receive the fertilizer and the control plots were similar, prior to beginning the experiment tree density (the number of trees per hectare) was recorded for eight plots to be fertilized and eight control plots, resulting in the given data. MINITAB output follows.

```
TWOSAMPLE T
N MEAN STDEV SE MEAN
8 1184   126      44
8 1196   118      42


95 PCT CI FOR MU1 - MU2: (-143,
  119)


TTEST MU1 = MU2 (VS NE): T =
-0.20 P=0.85 DF= 14


POOLED STDEV = 122
```

Fertilizer plots   1024 1216 1312 1280
                   1216 1312   992 1120
Control plots   1104 1072 1088 1328
                   1376 1280 1120 1200

Would you conclude that there is a significant difference in the mean tree density for fertilizer and control plots? Use $\alpha = .05$. Also interpret the CI given on the output.

59. The accompanying data was obtained in a study to evaluate the liquefaction potential at a proposed nuclear power station ("Cyclic Strengths Compared for Two Sampling Techniques," *J. Geotechnical Division, Am. Soc. Civil Engrs. Proceedings*, 1981, pp. 563–576). Prior to cyclic strength testing, soil samples were gathered using both a pitcher tube method and a block method, resulting in the following observed values of dry density (lb/ft^3).

Pitcher sampling   101.1 111.1 107.6   98.1
                    99.5   98.7 103.3 108.9
                   109.1 104.1 110.0   98.4
                   105.1 104.5 105.7 103.3
                   100.3 102.6 101.7 105.4
                    99.6 103.3 102.1 104.3

Block sampling   107.1 105.0   98.0   97.9
                   103.3 104.6 100.1   98.2
                    97.9 103.2   96.9

Calculate a 95% CI for the difference between true average dry densities using the two sampling methods.

60. The accompanying table summarizes data on body weight gain (g) for both a sample of control animals and a sample of animals given a 1 mg/pellet dose of a certain soft steroid ("The Soft Drug Approach," *CHEMTECH*, 1984, pp. 28–38).

Treatment	Sample Size	Sample Mean	Sample SD
Control	10	40.5	2.5
Soft steroid	8	32.8	2.6

Does the data indicate that true average weight gain for control animals differs from that for animals given the soft steroid? Carry out an appropriate test.

61. A sample of both surface soil and subsoil was taken from eight randomly selected agricultural locations in a particular county. The soil samples were analyzed to determine both surface pH and subsoil pH, with the following results:

	Location			
	1	*2*	*3*	*4*
Surface pH	6.55	5.98	5.59	6.17
Subsoil pH	6.78	6.14	5.80	5.91

	Location			
	5	*6*	*7*	*8*
Surface pH	5.92	6.18	6.43	5.68
Subsoil pH	6.10	6.01	6.18	5.88

Compute a 90% CI for the true average difference between surface and subsoil pH for agricultural land in this county. What assumption(s) have you made as a basis for your analysis?

62. The accompanying summary data on the ratio of strength to cross-sectional area for knee extensors is taken from the paper "Knee Extensor and Knee Flexor Strength: Cross-Sectional Area Ratios in Young and Elderly Men" (*J. of Gerontology*, 1992, pp. M204–M210).

Group	Sample Size	Sample Mean	Standard Error
Young men	13	7.47	.22
Elderly men	12	6.71	.28

Does this data suggest that the true average ratio for young men exceeds that for elderly men? Carry out a test of appropriate hypotheses using $\alpha = .05$. Be sure to state any assumptions necessary for your analysis.

63. The accompanying data on response time appeared in the paper "The Extinguishment of Fires Using Low-Flow Water Hose Streams—Part II" (*Fire Technology*, 1991, pp. 291–320).

Good visibility	.43	1.17	.37	.47	.68	.58	.50	2.75
Poor visibility	1.47	.80	1.58	1.53	4.33	4.23	3.25	3.22

The authors analyzed the data with the pooled *t* test. Does the use of this test appear justified? (HINT: Check

for normality. The normal scores for $n = 8$ are −1.53, −.89, −.49, −.15, .15, .49, .89, and 1.53.)

64. The paper "The Relationship Between Distress and Delight in Males' and Females' Reactions to Frightening Films" (*Human Communication Research*, 1991, pp. 625–637) reports that investigators measured emotional responses of 50 men and 60 women after viewing a segment from a horror film. The paper included the following statement: "Females were much more likely to express distress than were males. While males did express higher levels of delight than females, the difference was not statistically significant." The accompanying summary data was also contained in the paper.

	Distress Index		Delight Index	
Gender	Mean	SD	Mean	SD
Men	31.2	10.0	12.02	3.65
Women	40.4	9.1	9.09	5.55
	P-value < .001		Not significant (*P*-value > .05)	

Give a brief discussion of the relevant hypotheses and conclusions.

65. An experimenter wishes to obtain a CI for the difference between true average breaking strength for cables manufactured by company I and by company II. Suppose breaking strength is normally distributed for both types of cable with $\sigma_1 = 30$ psi and $\sigma_2 = 20$ psi.

a. If costs dictate that the sample size for the type I cable should be three times the sample size for the type II cable, how many observations are required if the 99% CI is to be no longer than 20 psi?

b. Suppose a total of 400 observations is to be made. How many of the observations should be made on type I cable samples if the length of the resulting interval is to be a minimum?

66. An experiment to determine the effects of temperature on the survival of insect eggs was described in the paper "Development Rates and a Temperature-Dependent Model of Pales Weevil" (*Environ. Entomology*, 1987, pp. 956–962). At 11°C, 73 of 91 eggs survived to the next stage of development. At 30°C, 102 of 110 eggs survived. Do the results of this experiment suggest that the survival rate (proportion surviving) differs for the two temperatures? Calculate the *P*-value and use it to test the appropriate hypotheses.

67. The insulin-binding capacity (pmol/mg protein) was measured for a sample of diabetic rats treated with a low dose of insulin and another sample of diabetic rats treated with a high dose, yielding the following data:

Low dose $m = 8$, $\bar{x} = 1.98$, $s_1 = .51$
High dose $n = 12$, $\bar{y} = 1.30$, $s_2 = .35$

(*J. Clinical Investigation*, 1978, pp. 552–560.)

a. Does the data indicate that there is any difference in true average insulin-binding capacity due to the dosage level? Use a test with $\alpha = .001$.

b. What can be said about the *P*-value?

68. In the article referenced in Exercise 67, there were actually four experimental treatments: control nondiabetic, untreated diabetic, low-dose diabetic, and high-dose diabetic. Denote the sample size for the *i*th treatment by n_i and the sample variance by $S_i^2 (i = 1, 2, 3, 4)$. Assuming that the true variance for each treatment is σ^2, construct a pooled estimator of σ^2 that is unbiased, and verify using rules of expected value that it is indeed unbiased. What is your estimate for the following actual data?

	Treatment			
	1	*2*	*3*	*4*
Sample size	16	18	8	12
Sample SD	.64	.81	.51	.35

69. Suppose a level .05 test of $H_0: \mu_1 - \mu_2 = 0$ versus $H_a: \mu_1 - \mu_2 > 0$ is to be performed, assuming $\sigma_1 = \sigma_2 = 10$ and normality of both distributions, using equal sample sizes ($m = n$). Evaluate the probability of a type II error when $\mu_1 - \mu_2 = 1$ and $n = 25, 100, 2500,$ and $10,000$. Can you think of real problems in which the difference $\mu_1 - \mu_2 = 1$ has little practical significance? Would sample sizes of $n = 10,000$ be desirable in such problems?

70. The following data refers to airborne bacteria count (number of colonies/ft³) both for $m = 8$ carpeted hospital rooms and for $n = 8$ uncarpeted rooms ("Microbial Air Sampling in a Carpeted Hospital," *J. Environmental Health*, 1968, p. 405). Does there appear to be a difference in true average bacteria count between carpeted and uncarpeted rooms?

Carpeted   11.8 8.2 7.1 13.0 10.8 10.1 14.6 14.0
Uncarpeted 12.1 8.3 3.8 7.2 12.0 11.1 10.1 13.7

Suppose you later learned that all carpeted rooms were in a veterans' hospital while all uncarpeted rooms were in a children's hospital. Would you be able to assess the effect of carpeting? Comment.

71. In a study of the relationship between nightmare frequency and sex, each of 160 men and 192 women was asked whether nightmares occurred often (at least one per month) or seldom, with the following results:

Men $m = 160$, $x = 55$ (often = success)
Women $n = 192$, $y = 60$

Does the data indicate at level .05 that there is any difference in the true proportions of men and women who often have nightmares? ("Personality Characteristics of Nightmare Sufferers," *J. Nervous and Mental Diseases*, 1971, pp. 29–31).

72. McNemar's test, developed in Exercise 48, can also be used when individuals are paired (matched) to yield n pairs and then one member of each pair is given treatment 1 and the other is given treatment 2. Then X_1 is the number of pairs in which both treatments were successful, and similarly for X_2, X_3, and X_4. The test statistic for testing equal efficacy of the two treatments is given by $(X_2 - X_3)/\sqrt{(X_2 + X_3)}$, which has approximately a standard normal distribution when H_0 is true. Use this to test whether the drug ergotamine is effective in the treatment of migraine headaches.

		Ergotamine	
		S	F
Placebo	S	44	34
	F	46	30

The data is fictitious, but the conclusion agrees with that in the paper "Controlled Clinical Trial of Ergotamine Tartrate" (*British Med. J.*, 1970, pp. 325–327).

73. Tardive dyskinesia denotes a syndrome comprising a variety of abnormal involuntary movements assumed to follow long-term use of antipsychotic drugs. In an experiment to see whether the drug deanol produced a greater improvement from baseline scores than a placebo treatment ("Double Blind Evaluation of Deanol in Tardive Dyskinesia," *J. Amer. Med. Assn.*, 1978, pp. 1997–1998), the two treatments were administered for 4 weeks each in random order to 14 patients, resulting in the following total severity index (TSI) scores:

	Patient						
	1	*2*	*3*	*4*	*5*	*6*	*7*
Deanol	12.4	6.8	12.6	13.2	12.4	7.6	12.1
Placebo	9.2	10.2	12.2	12.7	12.1	9.0	12.4

	Patient						
	8	9	10	11	12	13	14
Deanol	5.9	12.0	1.1	11.5	13.0	5.1	9.6
Placebo	5.9	8.5	4.8	7.8	9.1	3.5	6.4

(The investigators attributed placebo improvement to frequent attention and the nature of the experiment.) Does the data indicate that on average deanol yields a higher TSI than the placebo treatment? Test using $\alpha = .10$.

74. A random sample of 200 drivers from a particular county who drive a foreign car yielded 115 who use their seatbelts regularly, while another sample of 300 drivers who drive a domestic model yielded 154 who use their seatbelts regularly. Calculate a 99% CI for the difference in proportions of those using seatbelts regularly between drivers of foreign cars and drivers of domestic cars.

75. The paper "Evaluating Variability in Filling Operations" (*Food Tech.*, 1984, pp. 51–55) describes two different filling operations used in a ground-beef packing plant. Both filling operations were set to fill packages with 1400 g of ground beef. In a random sample of size 30 taken from each filling operation, the resulting means and standard deviations were 1402.24 g and 10.97 g for operation 1 and 1419.63 g and 9.96 g for operation 2.

a. Using a .05 significance level, is there sufficient evidence to indicate that the true mean weight of the packages differs for the two operations?

b. Does the data from operation 1 suggest that the true mean weight of packages produced by operation 1 is higher than 1400 g? Use a .05 significance level.

76. In Section 9.2 we described a procedure for testing H_0: $\mu_1 - \mu_2 = \Delta_0$ based on a variable T' having approximately a t distribution with degree of freedom v estimated from the data.

a. Use the variable T' to derive an approximate $100(1 - \alpha)\%$ CI for $\mu_1 - \mu_2$ when $\sigma_1 \neq \sigma_2$ (but the populations are still normal).

b. Compute the 95% interval of part (a) for the data in Exercise 54, in which μ_1 and μ_2 refer to the true

average compressive modulus for epoxy polymer and MMA prepolymer, respectively, used to repair failed concrete.

77. Let $X_1, \ldots, X_m$ be a random sample from a Poisson distribution with parameter λ_1 and let $Y_1, \ldots, Y_n$ be a random sample from another Poisson distribution with parameter λ_2. We wish to test H_0: $\lambda_1 - \lambda_2 = 0$ against one of the three standard alternatives. Since $\mu = \lambda$ for a Poisson distribution, when m and n are large the large-sample z test of Section 9.1 can be used. However, the fact that $V(\bar{X}) = \lambda/n$ suggests that a different denominator should be used in standardizing $\bar{X} - \bar{Y}$ Develop a large-sample test procedure appropriate to this problem and then apply it to the following data to test whether the plant densities for a particular species are equal in two different regions (where each observation is the number of plants found in a randomly located square sampling quadrat having area $1m^2$, so for region 1, there were 40 quadrats in which one plant was observed, etc.).

	Frequency								
	0	1	2	3	4	5	6	7	
Region 1	28	40	28	17	8	2	1	1	$m = 125$
Region 2	14	25	30	18	49	2	1	1	$n = 140$

78. Referring to Exercise 77, develop a large-sample confidence interval formula for $\lambda_1 - \lambda_2$ and calculate the interval for the data given there using a confidence level of 95%.

79. Borderline and mildly retarded children attending a hospital developmental evaluation clinic were divided into two groups on the basis of the presence or absence of a probable etiological factor causing the retardation. Blood-lead concentration was measured for each child, yielding the following data:

Factor absent   25.5 23.2 27.6 24.3 26.1 25.0

Factor present   21.2 19.8 20.3 21.0 19.6

Does the data indicate any difference in the extent of variability in blood-lead concentration for the two types of children?

Bibliography

See the bibliography at the end of Chapter 7.

10 The Analysis of Variance

ΣΣΣΣΣΣΣΣΣΣΣΣΣΣΣΣΣΣΣΣ

Introduction

In studying methods for the analysis of quantitative data, we first focused on problems involving a single sample of numbers and then turned to a comparative analysis of two different such samples. In one-sample problems, the data consisted of observations on or responses from individuals or experimental objects randomly selected from a single population. In two-sample problems, either the two samples were drawn from two different populations and the parameters of interest were the population means, or else two different treatments were applied to experimental units (individuals or objects) selected from a single population; in this latter case, the parameters of interest were referred to as true treatment means.

The **analysis of variance,** or more briefly **ANOVA,** refers broadly to a collection of experimental situations and statistical procedures for the analysis of quantitative responses from experimental units. The simplest ANOVA problem is referred to variously as a **single-factor, single-classification,** or **one-way ANOVA** and involves the analysis either of data sampled from more than two numerical populations (distributions) or data from experiments in which more than two treatments have been used. The characteristic that differentiates the treatments or populations from one another is called the **factor** under study, and the different treatments or populations are referred to as the **levels** of the factor. Examples of such situations include the following:

1. An experiment to study the effects of five different brands of gasoline on automobile engine operating efficiency (mpg)

2. An experiment to study the effects of the presence of four different sugar solutions (glucose, sucrose, fructose, and a mixture of the three) on bacterial growth

3. An experiment to assess the effects of different amounts of a particular psychedelic drug on manual dexterity

4. An experiment to decide whether varying the tracking weight of a stereo cartridge has any effect on record lifetime

In item 1 the factor of interest is gasoline brand, and there are five different levels of the factor. In item 2 the factor is sugar, with four levels (or five, if a control solution containing no sugar is used). In both items 1 and 2, the factor is qualitative in nature, and the levels correspond to possible categories of the factor. In items 3 and 4, the factors are concentration of drug and tracking weight, respectively, and both these factors are quantitative in nature, so that the levels identify different settings of the factor. When the factor of interest is quantitative, statistical techniques from regression analysis (discussed in Chapters 12 and 13) can also be used to analyze the data.

In this chapter, we focus on single-factor ANOVA. Section 10.1 presents the F test for testing the null hypothesis that the population or treatment means are identical. Section 10.2 considers further analysis of the data when H_0 has been rejected. Section 10.3 covers some other aspects of single-factor ANOVA. The next chapter introduces ANOVA experiments involving more than a single factor.

10.1   Single-Factor ANOVA

Single-factor ANOVA focuses on a comparison of more than two population or treatment means. Let

I = the number of populations or treatments being compared

μ_1 = the mean of population 1 or the true average response when treatment 1 is applied

$$\vdots$$

μ_I = the mean of population I or the true average response when treatment I is applied

Then the hypotheses of interest are

$$H_0: \mu_1 = \mu_2 = \cdots = \mu_I$$

versus

$$H_a: \text{at least two of the } \mu_i\text{'s are different}$$

If $I = 4$, H_0 is true only if all four μ_i's are identical. H_a would be true, for example, if $\mu_1 = \mu_2 \neq \mu_3 = \mu_4$, if $\mu_1 = \mu_3 = \mu_4 \neq \mu_2$, or if all four μ_i's differ from one another.

A test of these hypotheses requires that we have available a random sample from each population or treatment.

Example 10.1   The paper "Compression of Single-Wall Corrugated Shipping Containers Using Fixed and Floating Test Platens" (*J. Testing and Evaluation*, 1992, pp. 318–320) describes an experiment in which several different types of boxes were compared with respect to compression strength (lb). Table 10.1 presents the results of a single-factor ANOVA experiment involving $I = 4$ types of boxes (the sample means and standard deviations are in good agreement with values given in the paper).

Table 10.1 The data and summary quantities for Example 10.1

Type of Box	Compression Strength (lb)	Sample Mean	Sample SD
1	655.5 788.3 734.3 721.4 679.1 699.4	713.00	46.55
2	789.2 772.5 786.9 686.1 732.1 774.8	756.93	40.34
3	737.1 639.0 696.3 671.7 717.2 727.1	698.07	37.20
4	535.1 628.7 542.4 559.0 586.9 520.0	562.02	39.87
	"Grand" mean =	682.50	

With μ_i denoting the true average compression strength for boxes of type i ($i = 1$, 2, 3, 4), the null hypothesis is H_0: $\mu_1 = \mu_2 = \mu_3 = \mu_4$. Figure 10.1a shows a comparative boxplot for the four samples. There is a substantial amount of overlap among observations on the first three types of boxes, but compression strengths for the fourth type appear considerably smaller than for the other types. This suggests that H_0 is not true. The comparative boxplot in Figure 10.1b is based on adding 120 to each observation in the fourth sample (giving mean 682.02 and the same standard deviation) and leaving the other observations unaltered. It is no longer obvious whether H_0 is true or false. In situations such as this, we need a formal test procedure. ∎

Notation and Assumptions

In two-sample problems, we used the letters X and Y to differentiate the observations in one sample from those in the other. Because this is cumbersome for three or more samples, it is customary to use a single letter with two subscripts. The first subscript identifies the sample number, corresponding to the population or treatment being sampled, and the second subscript denotes the position of the observation within that sample. Let

$X_{i,j}$ = the random variable (rv) that denotes the jth measurement taken from the ith population, or the measurement taken on the jth experimental unit that receives the ith treatment,

$x_{i,j}$ = the observed value of $X_{i,j}$ when the experiment is performed.

The observed data is usually displayed in a rectangular table such as Table 10.1. There samples from the different populations appear in different rows of the table, and $x_{i,j}$ is the jth number in the ith row. For example, $x_{2,3} = 786.9$ (the third observation from the second population), and $x_{4,1} = 535.1$. When there is no ambiguity, we will write x_{ij} rather than $x_{i,j}$ (e.g., if there were 15 observations on each of 12 treatments, x_{112} could mean $x_{1,12}$ or $x_{11,2}$). It is assumed that the X_{ij}'s within any particular sample are independent—a random sample from the ith population or treatment distribution—and that different samples are independent of one another.

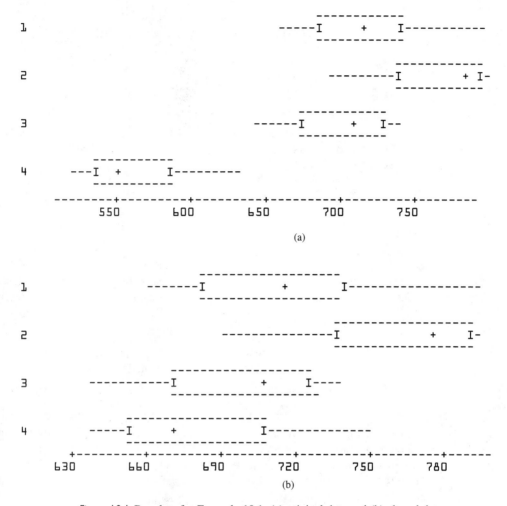

Figure 10.1 Boxplots for Example 10.1: (a) original data and (b) altered data

In some experiments, different samples contain different numbers of observations. However, the concepts and methods of single-factor ANOVA are most easily developed for the case of equal sample sizes. Restricting ourselves for the moment to this case, let J denote the number of observations in each sample ($J = 6$ in Example 10.1). The data set consists of IJ observations. The individual sample means will be denoted by $\bar{X}_{1.}, \bar{X}_{2.}, \ldots, \bar{X}_{I.}$. That is,

$$\bar{X}_{i.} = \frac{\sum_{j=1}^{J} X_{ij}}{J}, \qquad i = 1, 2, \ldots, I$$

The dot in place of the second subscript signifies that we have added over all values of that subscript while holding the other subscript value fixed, and the horizontal bar indicates division by J to obtain an average. Similarly, the average of all IJ observations, called the **grand mean,** is

$$\overline{X}_{..} = \frac{\sum\limits_{i=1}^{I} \sum\limits_{j=1}^{J} X_{ij}}{IJ}$$

For the strength data in Table 10.1, $\bar{x}_{1.} = 713.00$, $\bar{x}_{2.} = 756.93$, $\bar{x}_{3.} = 698.07$, $\bar{x}_{4.} = 562.02$, and $\bar{x}_{..} = 682.50$. Additionally, let $S_1^2, S_2^2, \ldots, S_I^2$ represent the sample variances:

$$S_i^2 = \frac{\sum\limits_{j=1}^{J} (X_{ij} - \overline{X}_{i.})^2}{J - 1}, \qquad i = 1, 2, \ldots, I$$

From Example 10.1, $s_1 = 46.55$, $s_1^2 = 2166.90$, and so on.

The assumptions that underlie the standard method of analysis in single-factor ANOVA are extensions of those on which the two-sample pooled t test was based.

Assumptions

> The I population or treatment distributions are all normal with the same variance σ^2. That is, each X_{ij} is normally distributed with
>
> $$E(X_{ij}) = \mu_i, \quad V(X_{ij}) = \sigma^2$$

The I sample standard deviations will generally differ somewhat even when the corresponding σ's are identical. In Example 10.1, the largest among s_1, s_2, s_3, and s_4 is about 1.25 times the smallest. A rough rule of thumb is that if the largest s is not much more than two times the smallest, it is reasonable to assume equal σ^2's.

In earlier chapters, a normal probability plot was suggested for checking normality. The individual sample sizes in ANOVA are typically too small for I separate plots to be informative. A single plot can be constructed by subtracting $\bar{x}_{1.}$ from each observation in the first sample, $\bar{x}_{2.}$ from each observation in the second, and so on, and then plotting these IJ deviations against the normal scores. Figure 10.2 gives such a plot for the data of Example 10.1. The straightness of the pattern gives strong support to the normality assumption.

The Test Statistic

If H_0 is true, the J observations in each sample come from a normal population distribution with the *same* mean value μ, in which case the sample means $\bar{x}_{1.}, \ldots, \bar{x}_{I.}$ should be reasonably close. The test procedure is based on comparing a measure

Figure 10.2 A normal probability plot based on the data of Example 10.1

of differences among the $\bar{x}_{i\cdot}$'s ("between-samples" variation) to a measure of variation calculated from *within* each of the samples.

Definition

Mean square for treatments is given by

$$MSTr = \frac{J}{I-1}[(\bar{X}_{1\cdot} - \bar{X}_{\cdot\cdot})^2 + (\bar{X}_{2\cdot} - \bar{X}_{\cdot\cdot})^2 + \cdots + (\bar{X}_{I\cdot} - \bar{X}_{\cdot\cdot})^2]$$

$$= \frac{J}{I-1} \sum_i (\bar{X}_{i\cdot} - \bar{X}_{\cdot\cdot})^2$$

and **mean square for error** is

$$MSE = \frac{S_1^2 + S_2^2 + \cdots + S_I^2}{I}$$

The test statistic for single-factor ANOVA is $F = MSTr/MSE$.

The terminology "mean square" will be explained shortly. Notice that uppercase $\bar{X}$'s and S^2's are used, so *MSTr* and *MSE* are defined as statistics. We will follow tradition and also use *MSTr* and *MSE* (rather than *mstr* and *mse*) to denote the calculated values of these statistics. Each S_i^2 assesses variation within a particular sample, so *MSE* is a measure of within-samples variation.

What kind of value of F provides evidence for or against H_0? The following box presents an important property of $E(MSTr)$ and $E(MSE)$, the expected values of these two statistics.

Proposition

> When H_0 is true,
>
> $E(MSTr) = E(MSE) = \sigma^2$
>
> whereas when H_0 is false,
>
> $E(MSTr) > E(MSE) = \sigma^2$
>
> That is, both statistics are unbiased for estimating the common population variance σ^2 when H_0 is true, but $MSTr$ tends to overestimate σ^2 when H_0 is false.

The unbiasedness of MSE is a consequence of $E(S_i^2) = \sigma^2$ whether H_0 is true or false. When H_0 is true, each $\bar{X}_{i\cdot}$ has the same mean value μ and variance σ^2/J, so $\Sigma(\bar{X}_{i\cdot} - \bar{X}_{\cdot\cdot})^2/(I - 1)$, the "sample variance" of the $\bar{X}_{i\cdot}$'s, estimates σ^2/J unbiasedly; multiplying this by J gives $MSTr$ as an unbiased estimator of σ^2 itself. The $\bar{X}_{i\cdot}$'s tend to spread out more when H_0 is false than when it is true, tending to inflate the value of $MSTr$ in this case. Thus, a value of F that greatly exceeds 1, corresponding to $MSTr$ much larger than MSE, casts considerable doubt on H_0. The form of the rejection region is $f \geq c$. The cutoff c should be chosen to give $P(F \geq c$ when H_0 is true$) = \alpha$, the desired significance level. We therefore need the distribution of F when H_0 is true.

F Distributions and the F Test

In Chapter 9, we introduced a family of probability distributions called F distributions. An F distribution arises in connection with a ratio in which there is one number of degrees of freedom (df) associated with the numerator and another number of degrees of freedom associated with the denominator. Let ν_1 and ν_2 denote the number of numerator and denominator degrees of freedom, respectively, for a variable with an F distribution. Both ν_1 and ν_2 are positive integers. Figure 10.3 pictures an F density curve and the corresponding upper-tail critical value F_{α,ν_1,ν_2}. Appendix Table A.7 gives these critical values for $\alpha = .05$ and for $\alpha = .01$. Values of ν_1 are identified with different columns of the table, and the rows are labeled with various values of ν_2. For example, the F critical value that captures upper-tail area .05 under the F curve with $\nu_1 = 4$ and $\nu_2 = 6$ is $F_{.05,4,6} = 4.53$, whereas $F_{.05,6,4} = 6.16$. When a test of hypotheses based on an F distribution is carried out using one of the better statistical computer packages, a P-value will be computed. This allows the use of an α other than .05 or .01.

The key theoretical result is that the test statistic F has an F distribution when H_0 is true.

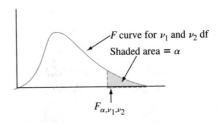

Figure 10.3 An F curve and critical value F_{α,ν_1,ν_2}

Theorem

Let $F = MSTr/MSE$ be the test statistic in a single-factor ANOVA problem involving I populations or treatments with a random sample of J observations from each one. When H_0 is true and the basic assumptions of this section are satisfied, F has an F distribution with $\nu_1 = I - 1$ and $\nu_2 = I(J - 1)$. With f denoting the computed value of F, the rejection region $f \geq F_{\alpha,I-1,I(J-1)}$ then specifies a test with significance level α.

The rationale for $\nu_1 = I - 1$ is that while $MSTr$ is based on the I deviations $\bar{X}_1. - \bar{X}.., \ldots, \bar{X}_I. - \bar{X}.., \Sigma(\bar{X}_i. - \bar{X}..) = 0$ so only $I - 1$ of these are freely determined. Because each sample contributes $J - 1$ df to MSE and these samples are independent, $\nu_2 = (J - 1) + \cdots + (J - 1) = I(J - 1)$.

Example 10.2
(Example 10.1 continued)

The values of I and J for the strength data are 4 and 6, respectively, so numerator df $= I - 1 = 3$ and denominator df $= I(J - 1) = 20$. At significance level .05, H_0: $\mu_1 = \mu_2 = \mu_3 = \mu_4$ will be rejected in favor of the conclusion that at least two μ_i's are different if $f \geq F_{.05,3,20} = 3.10$. The grand mean is $\bar{x}.. = \Sigma\Sigma x_{ij}/(IJ) = 682.50$

$$MSTr = \frac{6}{4-1}\,[(713.00 - 682.50)^2 + (756.93 - 682.50)^2$$
$$+ (698.07 - 682.50)^2 + (562.02 - 682.50)^2] = 42{,}455.86$$

$$MSE = \frac{1}{4}\,[(46.55)^2 + (40.34)^2 + (37.20)^2 + (39.87)^2] = 1691.92$$

$$f = MSTr/MSE = 42{,}455.86/1691.92 = 25.09$$

Since $25.09 \geq 3.10$, H_0 is resoundingly rejected at significance level .05. True average compression strength does appear to depend on box type. In fact, P-value $=$ area under F curve to right of $25.09 = .000$, so H_0 would be rejected at any reasonable significance level. ∎

Computational Formulas for ANOVA

The calculations leading to f can be done efficiently by using formulas similar to the computing formula for a single sample variance s^2. Let $x_i.$ represent the *sum* (not average, since there is no bar) of the x_{ij}'s for i fixed (sum of the numbers in the ith row of the table) and $x_{..}$ denote the sum of *all* the x_{ij}'s (the **grand total**).

<table>
<tr><td>Definition</td><td>

The **total sum of squares (SST), treatment sum of squares (SSTr),** and **error sum of squares (SSE)** are given by

$$SST = \sum_{i=1}^{I} \sum_{j=1}^{J} (x_{ij} - \bar{x}_{..})^2 = \sum_{i=1}^{I} \sum_{j=1}^{J} x_{ij}^2 - \frac{1}{IJ} x_{..}^2$$

$$SSTr = \sum_{i=1}^{I} \sum_{j=1}^{J} (\bar{x}_{i.} - \bar{x}_{..})^2 = \frac{1}{J} \sum_{i=1}^{I} x_{i.}^2 - \frac{1}{IJ} x_{..}^2$$

$$SSE = \sum_{i=1}^{I} \sum_{j=1}^{J} (x_{ij} - \bar{x}_{i.})^2, \qquad \text{where } x_{i.} = \sum_{j=1}^{J} x_{ij}, \quad x_{..} = \sum_{i=1}^{I} \sum_{j=1}^{J} x_{ij}$$

</td></tr>
</table>

The sum of squares $SSTr$ appears in the numerator of F, and SSE appears in the denominator of F; the reason for defining SST will be apparent shortly.

The expressions on the far right-hand side of SST and $SSTr$ are the computing formulas for these sums of squares. Both SST and $SSTr$ involve $x_{..}^2/(IJ)$ (the square of the grand total divided by IJ), which is usually called the **correction factor for the mean.** After the correction factor is computed, SST is obtained by squaring each number in the data table, adding these squares together, and subtracting the correction factor. $SSTr$ results from squaring each row total, summing them, dividing by J, and subtracting the correction factor.

A computing formula for SSE is a consequence of a simple relationship among the three sums of squares.

<table>
<tr><td>Fundamental Identity</td><td>

$$SST = SSTr + SSE \qquad\qquad (10.1)$$

</td></tr>
</table>

Thus, if any two of the sums of squares are computed, the third can be obtained through (10.1); SST and $SSTr$ are easiest to compute, and then $SSE = SST - SSTr$. The proof follows from squaring both sides of the relationship

$$x_{ij} - \bar{x}_{..} = (x_{ij} - \bar{x}_{i.}) + (\bar{x}_{i.} - \bar{x}_{..}) \qquad\qquad (10.2)$$

and summing over all i and j. This gives SST on the left, and $SSTr$ and SSE as the two extreme terms on the right. The cross-product term is easily seen to be zero.

The interpretation of the fundamental identity is an important aid to an understanding of ANOVA. SST is a measure of the total variation in the data—the sum of all squared deviations about the grand mean. The identity says that this total variation can be partitioned into two pieces. SSE measures variation that would be present (within rows) even if H_0 is true and is thus the part of total variation that is *unexplained* by the truth or falsity of H_0. $SSTr$ is the amount of variation (between rows) that *can be explained* by possible differences in the μ_i's. If explained variation is large relative to unexplained variation, then H_0 is rejected in favor of H_a.

Once *SSTr* and *SSE* are computed, each is divided by its associated df to obtain a mean square (mean in the sense of average). Then *F* is the ratio of the two mean squares.

$$MSTr = \frac{SSTr}{I-1}, \quad MSE = \frac{SSE}{I(J-1)}, \quad F = \frac{MSTr}{MSE} \tag{10.3}$$

The computations are often summarized in a tabular format, called an **ANOVA table,** as displayed in Table 10.2.

Table 10.2 An ANOVA table

Source of Variation	df	Sum of Squares	Mean Square	f
Treatments	$I-1$	SSTr	$MSTr = SSTr/(I-1)$	$\frac{MSTr}{MSE}$
Error	$I(J-1)$	SSE	$MSE = SSE/[I(J-1)]$	
Total	$IJ-1$	SST		

Example 10.3 The accompanying data resulted from an experiment comparing the degree of soiling for fabric copolymerized with three different mixtures of methacrylic acid (similar data appeared in the paper "Chemical Factors Affecting Soiling and Soil Release from Cotton DP Fabric," *American Dyestuff Reporter,* 1983, pp. 25–30).

						$x_{i\cdot}$	$\bar{x}_{i\cdot}$
Mixture 1	.56	1.12	.90	1.07	.94	4.59	.918
Mixture 2	.72	.69	.87	.78	.91	3.97	.794
Mixture 3	.62	1.08	1.07	.99	.93	4.69	.938

$$x_{\cdot\cdot} = 13.25$$

Let μ_i denote the true average degree of soiling when mixture *i* is used ($i = 1, 2, 3$). The null hypothesis $H_0: \mu_1 = \mu_2 = \mu_3$ states that the true average degree of soiling is identical for the three mixtures. We will carry out a test at significance level .01 to see whether H_0 should be rejected in favor of the assertion that true average degree of soiling is not the same for all mixtures. Since $I - 1 = 2$ and $I(J - 1) = 12$, the *F* critical value for the rejection region is $F_{.01,2,12} = 6.93$. Squaring each of the 15 observations and summing gives $\Sigma\Sigma x_{ij}^2 = (.56)^2 + (1.12)^2 + \cdots + (.93)^2 = 12.1351$. The values of the three sums of squares are

$$SST = 12.1351 - (13.25)^2/15 = 12.1351 - 11.7042 = .4309$$

$$SSTr = \frac{1}{5}[(4.59)^2 + (3.97)^2 + (4.69)^2] - 11.7042$$

$$= 11.7650 - 11.7042 = .0608$$

$$SSE = .4309 - .0608 = .3701$$

The remaining computations are summarized in the accompanying ANOVA table. Because $f = .99$ is not at least $F_{.01,2,12} = 6.93$, H_0 is not rejected at significance level .01. The mixtures appear to be indistinguishable with respect to degree of soiling.

Source of Variation	df	Sum of Squares	Mean Square	f
Treatments	2	.0608	.0304	.99
Error	12	.3701	.0308	
Total	14	.4309		

When the F test causes H_0 to be rejected, the experimenter will often be interested in further analysis to decide which μ_i's differ from which others. Procedures for doing this are called multiple comparison procedures, and several are described in the next two sections.

Exercises   Section 10.1 (1–10)

1. In an experiment to compare the tensile strengths of $I = 5$ different types of copper wire, $J = 4$ samples of each type were used. The between-samples and within-samples estimates of σ^2 were computed as $MSTr = 2573.3$ and $MSE = 1394.2$, respectively. Use the F test at level .05 to test H_0: $\mu_1 = \mu_2 = \mu_3 = \mu_4 = \mu_5$ versus H_a: at least two μ_i's are unequal.

2. Suppose that the compression strength observations on the fourth type of box in Example 10.1 had been 655.1, 748.7, 662.4, 679.0, 706.9, and 640.0 (obtained by adding 120 to each previous x_{4j}). Assuming no change in the remaining observations, carry out an F test with $\alpha = .05$.

3. The lumen output was determined for each of $I = 3$ different brands of 60-watt soft-white lightbulbs, with $J = 8$ bulbs of each brand tested. The sums of squares were computed as $SSE = 4773.3$ and $SSTr = 591.2$. State the hypotheses of interest (including word definitions of parameters), and use the F test of ANOVA ($\alpha = .05$) to decide whether there are any differences in true average lumen outputs among the three brands for this type of bulb.

4. In a study to assess the effects of malaria infection on mosquito hosts ("Plasmodium Cynomolgi: Effects of Malaria Infection on Laboratory Flight Performance of Anopheles Stephensi Mosquitos," *Experimental Parasitology*, 1977, pp. 397–404), mosquitos were fed on either infective or noninfective rhesus monkeys. Subsequently the distance they flew during a 24-hour period was measured using a flight mill. The mosquitos were divided into four groups of eight mosquitos each: infective rhesus and sporozites present (IRS), infective rhesus and oocysts present (IRD), infective rhesus and no infection developed (IRN), and noninfective (C). The summary data values are $\bar{x}_{1.} = 4.39$ (IRS), $\bar{x}_{2.} = 4.52$ (IRD), $\bar{x}_{3.} = 5.49$ (IRN), $\bar{x}_{4.} = 6.36$ (C), $\bar{x}_{..} = 5.19$, and $\Sigma\Sigma x_{ij}^2 = 911.91$. Use the ANOVA F test at level .05 to decide whether there are any differences between true average flight times for the four treatments.

5. Consider the following summary data on the modulus of elasticity ($\times 10^6$ psi) for lumber of three different grades (in close agreement with values in the paper "Bending Strength and Stiffness of Second-Growth Douglas-Fir Dimension Lumber" (*Forest Products J.*, 1991, pp. 35–43), except that the sample sizes there were larger).

Grade	J	$\bar{x}_{i.}$	s_i
1	10	1.63	.27
2	10	1.56	.24
3	10	1.42	.26

Use this data and a significance level of .01 to test the null hypothesis of no difference in mean modulus of elasticity for the three grades.

6. The article "Origin of Precambrian Iron Formations" (*Econ. Geology*, 1964, pp. 1025–1057) reports the following data on total Fe for four types of iron formation (1 = carbonate, 2 = silicate, 3 = magnetite, 4 = hematite).

1 20.5 28.1 27.8 27.0 28.0
   25.2 25.3 27.1 20.5 31.3
2 26.3 24.0 26.2 20.2 23.7
   34.0 17.1 26.8 23.7 24.9
3 29.5 34.0 27.5 29.4 27.9
   26.2 29.9 29.5 30.0 35.6
4 36.5 44.2 34.1 30.3 31.4
   33.1 34.1 32.9 36.3 25.5

Carry out an analysis of variance F test at significance level .01, and summarize the results in an ANOVA table.

7. In an experiment to investigate the performance of four different brands of spark plugs intended for use on a 125-cc two-stroke motorcycle, five plugs of each brand were tested and the number of miles (at a constant speed) until failure was observed. The partial ANOVA table for the data appears below. Fill in the missing entries, state the relevant hypotheses, and carry out a test.

Source	df	Sum of Squares	Mean Square	f
Brand				
Error			14,713.69	
Total		310,500.76		

8. A study of the properties of metal plate-connected trusses used for roof support ("Modeling Joints Made with Light-Gauge Metal Connector Plates," *Forest Products J.*, 1979, pp. 39–44) yielded the following observations on axial stiffness index (kips/in.) for plate lengths 4, 6, 8, 10, and 12 in.

4 309.2 409.5 311.0 326.5 316.8
   349.8 309.7
6 402.1 347.2 361.0 404.5 331.0
   348.9 381.7

8 392.4 366.2 351.0 357.1 409.9
   367.3 382.0
10 346.7 452.9 461.4 433.1 410.6
   384.2 362.6
12 407.4 441.8 419.9 410.7 473.4
   441.2 465.8

Does variation in plate length have any effect on true average axial stiffness? State and test the relevant hypotheses using analysis of variance with $\alpha = .01$. Display your results in an ANOVA table. (HINT: $\Sigma\Sigma x_{ij}^2 = 5{,}241{,}420.79$.)

9. Six samples of each of four types of cereal grain grown in a certain region were analyzed to determine thiamin content, resulting in the following data ($\mu g/g$):

Wheat   5.2 4.5 6.0 6.1 6.7 5.8
Barley 6.5 8.0 6.1 7.5 5.9 5.6
Maize   5.8 4.7 6.4 4.9 6.0 5.2
Oats   8.3 6.1 7.8 7.0 5.5 7.2

Does this data suggest that at least two of the grains differ with respect to true average thiamin content? Use a level $\alpha = .05$ test.

10. In single-factor ANOVA with I treatments and J observations per treatment, let $\mu = (1/I)\Sigma\mu_i$.

a. Express $E(\overline{X}_{..})$ in terms of μ. [HINT: $\overline{X}_{..} = (1/I)\Sigma\overline{X}_{i.}$]

b. Compute $E(\overline{X}_{i.}^2)$. [HINT: For any rv Y, $E(Y^2) = V(Y) + [E(Y)]^2$.]

c. Compute $E(\overline{X}_{..}^2)$.

d. Compute $E(SSTr)$ and then show that
$$E(MSTr) = \sigma^2 + \frac{J}{I-1}\Sigma(\mu_i - \mu)^2$$

e. Using the result of part (d), what is $E(MSTr)$ when H_0 is true? When H_0 is false, how does $E(MSTr)$ compare to σ^2?

Multiple Comparisons in ANOVA

10.2

When the computed value of the F statistic in single-factor ANOVA is not significant, the analysis is terminated because no differences among the μ_i's have been identified. But when H_0 is rejected, the investigator will usually want to know which of the μ_i's are different from one another. A method for carrying out this further analysis is called a **multiple comparisons procedure.** There are a number of such procedures in the statistics literature. Here we present one that many statisticians recommend for deciding for each i and j whether $\mu_i = \mu_j$.

Tukey's Procedure (the *T* Method)

Tukey's procedure involves the use of another probability distribution called the **Studentized range distribution.** The distribution depends on two parameters: a numerator df m and a denominator df ν. Let $Q_{\alpha,m,\nu}$ denote the upper-tail α critical value of the Studentized range distribution with m numerator df and ν denominator df (analogous to F_{α,ν_1,ν_2}). Values of $Q_{\alpha,m,\nu}$ are given in Appendix Table A.8. Then $Q_{\alpha,I,I(J-1)}$ can be used to obtain simultaneous confidence intervals for all pairwise differences $\mu_i - \mu_j$. Notice that numerator df for the appropriate Q critical value is I, the number of treatments or populations, and not $I - 1$, as it was for F.

Proposition

With probability $1 - \alpha$,

$$\bar{X}_{i\cdot} - \bar{X}_{j\cdot} - Q_{\alpha,I,I(J-1)} \sqrt{MSE/J} \leq \mu_i - \mu_j$$
$$\leq \bar{X}_{i\cdot} - \bar{X}_{j\cdot} + Q_{\alpha,I,I(J-1)} \sqrt{MSE/J} \qquad (10.4)$$

for *every* i and j ($i = 1, \ldots, I$ and $j = 1, \ldots, I$) with $i \neq j$.

When the computed $\bar{x}_{i\cdot}$, $\bar{x}_{j\cdot}$, and MSE are substituted into (10.4), the result is a collection of **simultaneous confidence statements** about the true values of all differences $\mu_i - \mu_j$ between true treatment means. *Each interval from (10.4) that does not include zero yields the conclusion that μ_i and μ_j differ significantly at level α.*

The computation of all intervals in (10.4) looks complicated, but a straightforward sequence of steps culminates in a pictorial summary of the conclusions:

The *T* Method for Identifying Significantly Different μ_i's

1. Select α and find $Q_{\alpha,I,I(J-1)}$ from Appendix Table A.8.
2. Determine $w = Q_{\alpha,I,I(J-1)} \cdot \sqrt{MSE/J}$.
3. List the sample means in increasing order and underline those pairs that differ by less than w. Any pair of sample means not underscored by the same line corresponds to a pair of true treatment means that are judged significantly different.

It is easily verified that a particular interval of the form (10.4) will contain zero if the corresponding pair of sample means is underscored by the same line.

Example 10.4

It is widely believed that although "house" brands (those sold only through a particular chain of stores) and regionally distributed brands of canned goods tend to be less expensive than their nationally distributed counterparts, they also tend to be less desirable in terms of quantity and quality of contents. To see whether such beliefs have any basis in fact, five different brands of fruit cocktail were chosen: two national brands (1 and 2), a regional brand (3), and a house brand from each of two large supermarket chains (4 and 5). Nine 16-oz cans of each brand were purchased, opened and drained of liquid, and their net contents weighed. The sample average weights for the five brands were $\bar{x}_{1.} = 14.5$, $\bar{x}_{2.} = 13.8$, $\bar{x}_{3.} = 13.3$, $\bar{x}_{4.} = 14.3$, and $\bar{x}_{5.} = 13.1$. The ANOVA table summarizing the first part of the analysis appears as Table 10.3.

Table 10.3 ANOVA table for Example 10.4

Source of Variation	df	Sum of Squares	Mean Square	f
Treatments (brands)	4	13.32	3.33	37.84
Error	40	3.53	.088	
Total	44	16.85		

Since $F_{.05,4,40} = 2.61$, H_0 is rejected (decisively) at level .05. We now use Tukey's procedure to look for significant differences among the μ_i's. From Appendix Table A.8, $Q_{.05,5,40} = 4.04$ (the second subscript on Q is I and not $I - 1$ as in F), so $w = 4.04\sqrt{.088/9} = .4$. The five sample means are arranged in increasing order, and every pair differing by less than .4 is underscored:

$\bar{x}_{5.}$	$\bar{x}_{3.}$	$\bar{x}_{2.}$	$\bar{x}_{4.}$	$\bar{x}_{1.}$
13.1	13.3	13.8	14.3	14.5

Thus, brands 1 and 4 are not significantly different from one another, but are significantly higher than the other three brands in their true average contents. Brand 2 is significantly better than 3 and 5 but worse than 1 and 4, and brands 3 and 5 do not differ significantly. Note that one house brand (4) is in the top group, while a national brand is in the lowest group. ∎

In Example 10.4 if $\bar{x}_{2.} = 14.15$ rather than 13.8 with the same computed w, then the configuration of underscored means would be

$\bar{x}_{5.}$	$\bar{x}_{3.}$	$\bar{x}_{2.}$	$\bar{x}_{4.}$	$\bar{x}_{1.}$
13.1	13.3	14.15	14.3	14.5

When there are no significant differences among a group of more than two means, it is customary to draw a continuous line underneath the entire group.

Example 10.5   A biologist wished to study the effects of ethanol on sleep time. A sample of 20 rats, matched for age and other characteristics, was selected, and each rat was given an oral injection having a particular concentration of ethanol per body weight. The rapid eye movement (REM) sleep time for each rat was then recorded for a 24-hour period, with the following results:

Treatment (concentration of ethanol)					$x_{i.}$	$\bar{x}_{i.}$
0 (control)	88.6	73.2	91.4	68.0	75.2	396.4	79.28
1 g/kg	63.0	53.9	69.2	50.1	71.5	307.7	61.54
2 g/kg	44.9	59.5	40.2	56.3	38.7	239.6	47.92
4 g/kg	31.0	39.6	45.3	25.2	22.7	163.8	32.76

$x_{..} = 1107.5$, $\bar{x}_{..} = 55.375$

Does the data indicate that the true average REM sleep time depends on the concentration of ethanol? (This example is based on an experiment reported in "Relationship of Ethanol Blood Level to REM and Non-REM Sleep Time and Distribution in the Rat," *Life Sciences,* 1978, pp. 839–846.)

The $\bar{x}_{i\cdot}$'s differ rather substantially from one another, but there is also a great deal of variability within each sample, so to answer the question precisely we must carry out the ANOVA. With $\Sigma\Sigma x_{ij}^2 = 68{,}697.6$ and correction factor $x_{\cdot\cdot}^2/(IJ) = (1107.5)^2/20 = 61{,}327.8$, the computing formulas yield

$$SST = 68{,}697.6 - 61{,}327.8 = 7369.8$$

$$SSTr = \frac{1}{5}\left[(396.40)^2 + (307.70)^2 + (239.60)^2 + (163.80)^2\right] - 61{,}327.8$$

$$= 67{,}210.2 - 61{,}327.8 = 5882.4$$

and

$$SSE = 7369.8 - 5882.4 = 1487.4$$

An SAS ANOVA table appears as Table 10.4. The last column gives the *P*-value, which is .0001. Using a significance level of .05, we reject the null hypothesis H_0: $\mu_1 = \mu_2 = \mu_3 = \mu_4$, since *P*-value = .0001 < .05 = α. True average REM sleep time does appear to depend on concentration level.

Table 10.4 SAS ANOVA table

	Analysis of Variance Procedure				
Dependent Variable: TIME					
Source	DF	Sum of Squares	Mean Square	F Value	Pr > F
Model	3	5882.35750	1960.78583	21.09	0.0001
Error	16	1487.40000	92.96250		
Corrected Total	19	7369.75750			

There are $I = 4$ treatments and 16 df for error, so $Q_{.05,4,16} = 4.05$ and $w = 4.05\sqrt{93.0/5} = 17.47$. Ordering the means and underscoring yields

$\bar{x}_{4\cdot}$	$\bar{x}_{3\cdot}$	$\bar{x}_{2\cdot}$	$\bar{x}_{1\cdot}$
32.76	47.92	61.54	79.28

The interpretation of this underscoring must be done with care, since we seem to have concluded that treatments 2 and 3 do not differ, 3 and 4 do not differ, yet 2 and 4 do differ. The suggested way of expressing this is to say that while evidence allows us to conclude that treatments 2 and 4 differ from one another, neither has been shown to be significantly different from 3. Treatment 1 has a significantly higher true average REM sleep time than any of the other treatments.

Figure 10.4 shows SAS output from the application of Tukey's procedure.

```
            Alpha= 0.05 df= 16   MSE= 92.9625
         Critical Value of Studentized Range= 4.046
          Minimum Significant Difference= 17.446
```

Means with the same letter are not significantly different.

Tukey Grouping			Mean	N	TREATMENT
		A	79.280	5	0(control)
		B	61.540	5	1 gm/kg
		B			
C		B	47.920	5	2 gm/kg
C					
C			32.760	5	4 gm/kg

Figure 10.4 Tukey's method using SAS   ■

The Interpretation of α in Multiple Comparison

Tukey's method involves the simultaneous construction of confidence intervals for all differences $\mu_i - \mu_j$ of pairs of treatment means. In the construction of confidence intervals (CIs) in the previous chapters, $\alpha = .05$ or 95% confidence referred to the error rate or confidence level for the individual statement about the parameter of interest. If an ANOVA experiment involves comparison of four treatments, then Tukey's procedure obtains simultaneously six different intervals (since there are six different pairs of treatment means). The error rate $\alpha = .05$ no longer refers to a particular interval, but instead refers to the experiment as a whole, so is called an **experimentwise error rate.** If Tukey's method were used on a great many different ANOVA data sets, then in approximately 95% of these experiments no erroneous claim would be made about any of the $\mu_i - \mu_j$'s, while in only 5% would at least one incorrect claim be made. That is, in 95% of all experiments, every CI constructed from (10.4) would include the true value of $\mu_i - \mu_j$, and only 5% of the time would at least one interval fail to cover the true value of a $\mu_i - \mu_j$. Because the confidence level for the entire set of comparisons of means is 95%, the confidence level for any particular comparison is larger than 95% and increases as the number of comparisons (i.e., treatment means) increases. This distinction between the experimentwise error rate or confidence level and a "per-comparison" error rate or confidence level is important and should be kept in mind when interpreting the results of any multiple comparisons analysis.

Confidence Intervals for Other Parametric Functions

In some situations, a CI is desired for a function of the μ_i's more complicated than a difference $\mu_i - \mu_j$. Let $\theta = \Sigma c_i \mu_i$, where the c_i's are constants. One such function is $\frac{1}{2}(\mu_1 + \mu_2) - \frac{1}{3}(\mu_3 + \mu_4 + \mu_5)$, which in the context of Example 10.4 measures the difference between fruit cocktail can contents for national and nonnational brands. Because the X_{ij}'s are normally distributed with $E(X_{ij}) = \mu_i$ and $V(X_{ij}) = \sigma^2$, $\hat{\theta} = \Sigma_i c_i \bar{X}_{i \cdot}$ is normally distributed, unbiased for θ, and

$$V(\hat{\theta}) = V\left(\sum_i c_i \bar{X}_{i.}\right) = \sum_i c_i^2 V(\bar{X}_{i.}) = \frac{\sigma^2}{J}\sum_i c_i^2$$

Estimating σ^2 by MSE and forming $\hat{\sigma}_{\hat{\theta}}$ results in a t variable $(\hat{\theta} - \theta)/\hat{\sigma}_{\hat{\theta}}$, which can be manipulated to obtain the following $100(1 - \alpha)\%$ confidence interval for $\Sigma c_i \mu_i$:

$$\sum c_i \bar{x}_{i.} \pm t_{\alpha/2, I(J-1)} \sqrt{\frac{MSE \; \Sigma c_i^2}{J}} \tag{10.5}$$

Example 10.6
(Example 10.4 continued)

The parametric function for comparing national and nonnational brands is $\theta = \frac{1}{2}(\mu_1 + \mu_2) - \frac{1}{3}(\mu_3 + \mu_4 + \mu_5)$, from which

$$\Sigma c_i^2 = \left(\frac{1}{2}\right)^2 + \left(\frac{1}{2}\right)^2 + \left(-\frac{1}{3}\right)^2 + \left(-\frac{1}{3}\right)^2 + \left(-\frac{1}{3}\right)^2 = \frac{5}{6}$$

With $\hat{\theta} = \frac{1}{2}(\bar{x}_{1.} + \bar{x}_{2.}) - \frac{1}{3}(\bar{x}_{3.} + \bar{x}_{4.} + \bar{x}_{5.}) = .583$ and $MSE = .088$, a 95% interval is

$$.583 \pm 2.021\sqrt{5(.088)/[(6)(9)]} = .583 \pm .182 = (.401, .765) \qquad \blacksquare$$

Exercises   Section 10.2 (11–21)

11. An experiment to compare the spreading rates of five different brands of yellow interior latex paint available in a particular area used four gallons ($J = 4$) of each paint. The sample average spreading rates (ft^2/gal) for the five brands were $\bar{x}_{1.} = 462.0$, $\bar{x}_{2.} = 512.8$, $\bar{x}_{3.} = 437.5$, $\bar{x}_{4.} = 469.3$, and $\bar{x}_{5.} = 532.1$. The computed value of F was found to be significant at level $\alpha = .05$. With $MSE = 272.8$, use Tukey's procedure to investigate significant differences in the true average spreading rates between brands.

12. In Exercise 11, suppose $\bar{x}_{3.} = 427.5$. Now which true average spreading rates differ significantly from one another? Be sure to use the method of underscoring to illustrate your conclusions and write a paragraph summarizing your results.

13. Repeat Exercise 12 supposing that $\bar{x}_{2.} = 502.8$ in addition to $\bar{x}_{3.} = 427.5$.

14. Use Tukey's procedure on the data in Exercise 4 to identify differences in true average flight times among the four types of mosquitos.

15. Use Tukey's procedure on the data of Exercise 6 to identify differences in true average total Fe among the four types of formations (use $MSE = 15.64$).

16. Use Tukey's procedure to investigate differences in true average axial stiffness for the data in Exercise 8.

17. Refer to Exercise 5 and compute a 95% t CI for $\theta = \frac{1}{2}(\mu_1 + \mu_2) - \mu_3$.

18. Consider the accompanying data on plant growth after the application of different types of growth hormone.

	1	13	17	7	14
	2	21	13	20	17
Hormone	3	18	15	20	17
	4	7	11	18	10
	5	6	11	15	8

  a. Perform an F test at level $\alpha = .05$.

  b. What happens when Tukey's procedure is applied?

19. Consider a single-factor ANOVA experiment in which $I = 3$, $J = 5$, $\bar{x}_{1.} = 10$, $\bar{x}_{2.} = 12$, and $\bar{x}_{3.} = 20$. Find a value of SSE for which $f > F_{.05,2,12}$, so that H_0: $\mu_1 = \mu_2 = \mu_3$ is rejected, yet when Tukey's procedure is applied none of the μ_i's can be said to differ significantly from one another.

20. Refer to Exercise 19 and suppose $\bar{x}_{1.} = 10$, $\bar{x}_{2.} = 15$, and $\bar{x}_{3.} = 20$. Can you now find a value of SSE that produces such a contradiction between the F test and Tukey's procedure?

21. The article "The Effect of Enzyme Inducing Agents on the Survival Times of Rats Exposed to Lethal Levels of Nitrogen Dioxide" (*Toxicology and Applied Pharma-*

cology, 1978, pp. 169–174) reports the following data on survival times for rats exposed to nitrogen dioxide (70 ppm) via different injection regimens. There were $J = 14$ rats in each group.

Regimen	$\bar{x}_i.(min)$	s_i
1. Control	166	32
2. 3-Methylcholanthrene	303	53
3. Allylisopropylacetamide	266	54
4. Phenobarbital	212	35
5. Chlorpromazine	202	34
6. *p*-Aminobenzoic Acid	184	31

a. Test the null hypothesis that true average survival time does not depend on injection regimen against the alternative that there is some dependence on injection regimen using $\alpha = .01$.

b. Suppose that $100(1 - \alpha)\%$ CIs for k different parametric functions are computed from the same ANOVA data set. Then it is easily verified that the simultaneous confidence level is at least $100(1 - k\alpha)\%$. Compute CIs with simultaneous confidence level at least 98% for $\mu_1 - \frac{1}{5}(\mu_2 + \mu_3 + \mu_4 + \mu_5 + \mu_6)$ and $\frac{1}{4}(\mu_2 + \mu_3 + \mu_4 + \mu_5) - \mu_6$.

10.3 More on Single-Factor ANOVA

In this section, we briefly consider some issues relating to single-factor ANOVA that were not dealt with in the first two sections. These include an alternative description of the model parameters, β for the F test, the relationship of the test to procedures previously considered, data transformation, a random effects model, and formulas for the case of unequal sample sizes.

An Alternative Description of the ANOVA Model

The assumptions of single-factor ANOVA can be described succinctly by means of the "model equation"

$$X_{ij} = \mu_i + \epsilon_{ij}$$

where ϵ_{ij} represents a random deviation from the population or true treatment mean μ_i. The ϵ_{ij}'s are assumed to be independent, normally distributed rv's (implying that the X_{ij}'s are also) with $E(\epsilon_{ij}) = 0$ (so that $E(X_{ij}) = \mu_i$) and $V(\epsilon_{ij}) = \sigma^2$ (from which $V(X_{ij}) = \sigma^2$ for every i and j). An alternative description of single-factor ANOVA will give added insight and suggest appropriate generalizations to models involving more than one factor. Define a parameter μ by

$$\mu = \frac{1}{I} \sum_{i=1}^{I} \mu_i$$

and the parameters $\alpha_1, \ldots, \alpha_I$ by

$$\alpha_i = \mu_i - \mu \qquad (i = 1, \ldots, I)$$

Then the treatment mean μ_i can be written as $\mu + \alpha_i$, where μ represents the true average overall response in the experiment, and α_i is the effect, measured as a departure from μ, due to the ith treatment. Whereas we initially had I parameters, we now have $I + 1$ ($\mu, \alpha_1, \ldots, \alpha_I$). However, because $\Sigma \alpha_i = 0$ (the average departure from the overall mean response is zero), only I of these new parameters are independently determined, so there are as many independent parameters as there were before. In terms of μ and the α_i's, the model becomes

$$X_{ij} = \mu + \alpha_i + \epsilon_{ij} \qquad (i = 1, \ldots, I \quad j = 1, \ldots, J)$$

In Chapter 11, we will develop analogous models for multifactor ANOVA. The claim that the μ_i's are identical is equivalent to the equality of the α_i's, and because $\Sigma\alpha_i = 0$, the null hypothesis becomes

$$H_0: \alpha_1 = \alpha_2 = \cdots = \alpha_I = 0$$

In Section 10.1, it was stated that $MSTr$ is an unbiased estimator of σ^2 when H_0 is true but otherwise tends to overestimate σ^2. More precisely,

$$E(MSTr) = \sigma^2 + \frac{J}{I-1} \Sigma\alpha_i^2$$

When H_0 is true, $\Sigma\alpha_i^2 = 0$ so $E(MSTr) = \sigma^2$ (MSE is unbiased whether or not H_0 is true). If $\Sigma\alpha_i^2$ is used as a measure of the extent to which H_0 is false, then a larger value of $\Sigma\alpha_i^2$ will result in a greater tendency for $MSTr$ to overestimate σ^2. In the next chapter, formulas for expected mean squares for multifactor models will be used to suggest how to form F ratios to test various hypotheses.

Proof of the formula for $E(MSTr)$ For any rv Y, $E(Y^2) = V(Y) + [E(Y)]^2$, so

$$E(SSTr) = E\left(\frac{1}{J}\sum_i X_{i\cdot}^2 - \frac{1}{IJ} X_{\cdot\cdot}^2\right) = \frac{1}{J}\sum_i E(X_{i\cdot}^2) - \frac{1}{IJ} E(X_{\cdot\cdot}^2)$$

$$= \frac{1}{J}\sum_i \{V(X_{i\cdot}) + [E(X_{i\cdot})]^2\} - \frac{1}{IJ} \{V(X_{\cdot\cdot}) + [E(X_{\cdot\cdot})]^2\}$$

$$= \frac{1}{J}\sum_i \{J\sigma^2 + [J(\mu + \alpha_i)]^2\} - \frac{1}{IJ}[IJ\sigma^2 + (IJ\mu)^2]$$

$$= I\sigma^2 + IJ\mu^2 + 2\mu J\sum_i \alpha_i + J\sum_i \alpha_i^2 - \sigma^2 - IJ\mu^2$$

$$= (I-1)\sigma^2 + J\sum_i \alpha_i^2, \qquad (\text{since } \Sigma\alpha_i = 0)$$

The result then follows from the relationship $MSTr = SSTr/(I - 1)$.   ∎

β for the F Test

Consider a set of parameter values $\alpha_1, \alpha_2, \ldots, \alpha_I$ for which H_0 is not true. The probability of a type II error, β, is the probability that H_0 is not rejected when that set is the set of true values. One might think that β would have to be determined separately for each different configuration of α_i's. Fortunately, since β for the F test depends on the α_i's and σ^2 only through $\Sigma\alpha_i^2/\sigma^2$, it can be simultaneously evaluated for many different alternatives. For example, $\Sigma\alpha_i^2 = 4$ for each of the following sets of α_i's for which H_0 is false, so β is identical for all three alternatives:

1. $\alpha_1 = -1$, $\alpha_2 = -1$, $\alpha_3 = 1$, $\alpha_4 = 1$
2. $\alpha_1 = -\sqrt{2}$, $\alpha_2 = \sqrt{2}$, $\alpha_3 = 0$, $\alpha_4 = 0$
3. $\alpha_1 = -\sqrt{3}$, $\alpha_2 = \sqrt{1/3}$, $\alpha_3 = \sqrt{1/3}$, $\alpha_4 = \sqrt{1/3}$

The quantity $J\Sigma\alpha_i^2/\sigma^2$ is called the **noncentrality parameter** for one-way ANOVA (because when H_0 is false the test statistic has a *noncentral F* distribution with this as one of its parameters), and β is a decreasing function of the value of this parameter. Thus, for fixed values of σ^2 and J, the null hypothesis is more likely to be rejected for alternatives far from H_0 (large $\Sigma\alpha_i^2$) than for alternatives close to H_0. For a fixed value of $\Sigma\alpha_i^2$, β decreases as the sample size J on each treatment increases and increases as the variance σ^2 increases (since greater underlying variability makes it more difficult to detect any given departure from H_0).

Because hand computation of β and sample size determination for the F test are quite difficult (as in the case of t tests), statisticians have constructed sets of curves from which β can be read. These are called *power curves*. Sets of curves for numerator df $\nu_1 = 3$ and $\nu_1 = 4$ are displayed in Figure 10.5* and Figure 10.6*, respectively. After the values of σ^2 and the α_i's for which β is desired are specified, these are used to compute the value of ϕ, where $\phi^2 = (J/I)\Sigma\alpha_i^2/\sigma^2$. We then enter the appropriate set of curves at the value of ϕ on the horizontal axis, move up to the curve associated with error df ν_2, and move over to the value of power on the vertical axis. Finally, $\beta = 1 - \text{power}$.

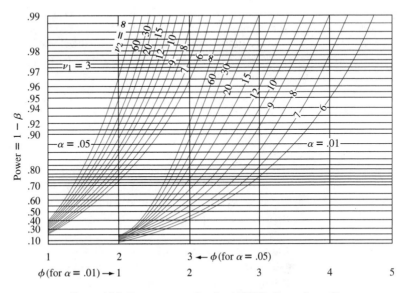

Figure 10.5 Power curves for the ANOVA F test ($\nu_1 = 3$)

*Reproduced with permission from E. S. Pearson and H. O. Hartley, "Charts of the Power Function for Analysis of Variance Tests, Derived from the Non-central F Distribution," *Biometrika*, vol. 38, 1951, p. 112.

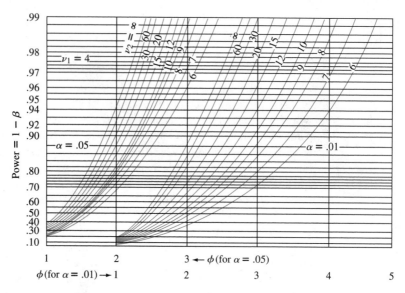

Figure 10.6 Power curves for the ANOVA F test ($\nu_1 = 4$)

Example 10.7   The effects of four different heat treatments on yield point (tons/in.^{2}) of steel ingots are to be investigated. A total of eight ingots will be cast using each treatment. Suppose the true standard deviation of yield point for any of the four treatments is $\sigma = 1$. How likely is it that H_0 will not be rejected at level .05 if three of the treatments have the same expected yield point and the other treatment has an expected yield point that is 1 ton/in.^{2} greater than the common value of the other three (i.e., the fourth yield is on average 1 SD above those for the first three treatments)?

Supposing that $\mu_1 = \mu_2 = \mu_3$ and $\mu_4 = \mu_1 + 1$, $\mu = (\Sigma\mu_i)/4 = \mu_1 + \frac{1}{4}$. Then $\alpha_1 = \mu_1 - \mu = -\frac{1}{4}$, $\alpha_2 = -\frac{1}{4}$, $\alpha_3 = -\frac{1}{4}$, $\alpha_4 = \frac{3}{4}$ so

$$\phi^2 = \frac{8}{4}[(-\tfrac{1}{4})^2 + (-\tfrac{1}{4})^2 + (-\tfrac{1}{4})^2 + (\tfrac{3}{4})^2] = \frac{3}{2}$$

and $\phi = 1.22$. The degrees of freedom are $\nu_1 = I - 1 = 3$ and $\nu_2 = I(J - 1) = 28$, so interpolating visually between $\nu_2 = 20$ and $\nu_2 = 30$ gives power $\approx .47$ and $\beta \approx .53$. This β is rather large, so we might decide to increase the value of J. How many ingots of each type would be required to yield $\beta \approx .05$ for the alternative under consideration? By trying different values of J we can verify that $J = 24$ will meet the requirement, but any smaller J will not.   ∎

As an alternative to the use of power curves, the SAS package has a function that calculates the cumulative area under a noncentral F curve (inputs F_α, numerator df, denominator df, and ϕ^2), and this area is β.

Relationship of the F Test to the t Test

When the number of treatments or populations is $I = 2$, all formulas and results connected with the F test still make sense, so ANOVA can be used to test $H_0: \mu_1 = \mu_2$ versus $H_a: \mu_1 \neq \mu_2$. In this case, a two-tailed two-sample t test can also be used.

Which test should be used? Somewhat surprisingly, the two are equivalent. It can be verified from the t and F tables that $t^2_{\alpha/2,\nu} = F_{\alpha,1,\nu}$, and a bit of algebra shows that the relationship between the two test statistics is also $T^2 = F$ in this case. Thus, the same conclusion (reject or not reject H_0) will be reached irrespective of which test is used. Of course, if the alternative is one-sided, then the one-tailed t test must be used, since the ANOVA alternative is always $H_a: \mu_1 \neq \mu_2$.

Single-Factor ANOVA When Sample Sizes Are Unequal

When the sample sizes from each population or treatment are not equal, let J_1, $J_2, \ldots, J_I$ denote the I sample sizes and let $n = \Sigma_i J_i$ denote the total number of observations. Define sums of squares (along with their computing formulas) by

$$SST = \sum_{i=1}^{I} \sum_{j=1}^{J_i} (X_{ij} - \bar{X}_{..})^2 = \sum_{i=1}^{I} \sum_{j=1}^{J_i} X_{ij}^2 - \frac{1}{n} X_{..}^2, \qquad df = n - 1$$

$$SSTr = \sum_{i=1}^{I} \sum_{j=1}^{J_i} (\bar{X}_{i.} - \bar{X}_{..})^2 = \sum_{i=1}^{I} \frac{1}{J_i} X_{i.}^2 - \frac{1}{n} X_{..}^2, \qquad df = I - 1$$

$$SSE = \sum_{i=1}^{I} \sum_{j=1}^{J_i} (X_{ij} - \bar{X}_{i.})^2 = SST - SSTr, \qquad df = \Sigma(J_i - 1) = n - I$$

Test statistic value:

$$f = \frac{MSTr}{MSE}, \qquad \text{where} \quad MSTr = \frac{SSTr}{I-1}, \quad MSE = \frac{SSE}{n-I}$$

Rejection region: $f \geq F_{\alpha, I-1, n-I}$

Example 10.8

The critical flicker frequency (cff) is the highest frequency (in cycles/sec) at which a person can detect the flicker in a flickering light source. At frequencies above the cff, the light source appears to be continuous even though it is actually flickering. An investigation carried out to see if true average cff depends on iris color yielded the following data (based on the article "The Effects of Iris Color on Critical Flicker Frequency," *J. General Psych.*, 1973, pp. 91–95).

Iris color		J_i	$x_{i.}$	$\bar{x}_{i.}$
1. Brown	26.8, 27.9, 23.7, 25.0, 26.3, 24.8, 25.7, 24.5	8	204.7	25.59
2. Green	26.4, 24.2, 28.0, 26.9, 29.1	5	134.6	26.92
3. Blue	25.7, 27.2, 29.9, 28.5, 29.4, 28.3	6	169.0	28.17
		$n = 19$	$x_{..} = 508.3$	

Let μ_1, μ_2, and μ_3 denote the true average cff's under the given experimental conditions for individuals with iris color brown, green, and blue, respectively. We will test $H_0: \mu_1 = \mu_2 = \mu_3$ against the alternative that at least two of these μ_i's differ. Numerator and denominator df are $I - 1 = 2$ and $n - I = 19 - 3 = 16$, respectively. $F_{.05,2,16} = 3.63$, so H_0 will be rejected at level .05 if $f \geq 3.63$. The sum of squares of all 19 observations is $\Sigma\Sigma x_{ij}^2 = 13{,}659.67$, and the correction factor is $(508.3)^2 / 19 = 13{,}598.36$. The three sums of squares are

$$SST = 13{,}659.67 - 13{,}598.36 = 61.31$$

$$SSTr = \frac{1}{8}(204.7)^2 + \frac{1}{5}(134.6)^2 + \frac{1}{6}(169.0)^2 - 13{,}598.36 = 23.00$$

$$SSE = 61.31 - 23.00 = 38.31$$

Further calculations are displayed in the accompanying ANOVA table. Because $4.81 \geq 3.63$, H_0 is rejected. True average cff does appear to depend on iris color—a conclusion consistent with that reached in the cited paper.

Source of Variation	df	Sum of Squares	Mean Square	f
Treatments (iris color)	2	23.00	11.50	4.81
Error	16	38.31	2.39	
Total	18	61.31		

    ■

Multiple Comparisons When Sample Sizes Are Unequal

There is more controversy among statisticians regarding which multiple comparisons procedure to use when sample sizes are unequal than there is in the case of equal sample sizes. The procedure that we present here is recommended in the excellent book *Beyond ANOVA: Basics of Applied Statistics* (see the chapter bibliography) for use when the I sample sizes $J_1, J_2, \ldots, J_I$ are reasonably close to one another ("mild imbalance"). It modifies Tukey's method by using averages of pairs of $1/J_i$'s in place of $1/J$.

Let

$$w_{ij} = Q_{\alpha, I, n-I} \cdot \sqrt{\frac{MSE}{2}\left(\frac{1}{J_i} + \frac{1}{J_j}\right)}$$

Then the probability is *approximately* $1 - \alpha$ that

$$\bar{X}_{i\cdot} - \bar{X}_{j\cdot} - w_{ij} \leq \mu_i - \mu_j \leq \bar{X}_{i\cdot} - \bar{X}_{j\cdot} + w_{ij}$$

for every i and j ($i = 1, \ldots, I$ and $j = 1, \ldots, I$) with $i \neq j$.

The simultaneous confidence level $100(1 - \alpha)\%$ is only approximate rather than exact as it is with equal sample sizes. The underscoring method can still be used, but now the w_{ij} factor used to decide if $\bar{x}_{i\cdot}$ and $\bar{x}_{j\cdot}$ can be connected will depend on J_i and J_j.

Example 10.9
(Example 10.8 continued)

The sample sizes for the critical flicker frequency data were $J_1 = 8$, $J_2 = 5$, $J_3 = 6$, and $I = 3$, $n - I = 16$, $MSE = 2.39$. A simultaneous confidence level of approximately 95% requires $Q_{.05,3,16} = 3.65$, from which

$$w_{12} = 3.65\sqrt{\frac{2.39}{2}\left(\frac{1}{8} + \frac{1}{5}\right)} = 2.27, \qquad w_{13} = 2.15, \quad w_{23} = 2.42$$

Since $\bar{x}_2 - \bar{x}_1 = 26.92 - 25.59 = 1.33 < w_{12}$, μ_1 and μ_2 are judged not significantly different. The accompanying underscoring scheme shows that the only significant difference is between the true average cff's for brown and blue irises.

1 (brown)	2 (green)	3 (blue)
25.59	26.92	28.17

∎

Data Transformation

The use of ANOVA methods can be invalidated by substantial differences in the variances $\sigma_1^2 \ldots , \sigma_I^2$ (which until now have been assumed equal with common value σ^2). It sometimes happens that $V(X_{ij}) = \sigma_i^2 = g(\mu_i)$, a known function of μ_i (so that when H_0 is false, the variances are not equal). For example, if X_{ij} has a Poisson distribution with parameter λ_i (approximately normal if $\lambda_i \geq 10$), then $\mu_i = \lambda_i$ and $\sigma_i^2 = \lambda_i$ so $g(\mu_i) = \mu_i$ is the known function. In such cases, one can often transform the X_{ij}'s to $h(X_{ij})$'s so that they will have approximately equal variances (while leaving the transformed variables approximately normal), and then the F test can be used on the transformed observations. The key idea in choosing a transformation $h(\cdot)$ is that often $V[h(X_{ij})] \approx V(X_{ij}) \cdot [h'(\mu_i)]^2 = g(\mu_i) \cdot [h'(\mu_i)]^2$. We wish the function $h(\cdot)$ for which $g(\mu_i) \cdot [h'(\mu_i)]^2 = c$ (a constant) for every i.

Proposition

> If $V(X_{ij}) = g(\mu_i)$, a known function of μ_i, then a transformation $h(X_{ij})$ that "stabilizes the variance" so that $V[h(X_{ij})]$ is approximately the same for each i is given by $h(x) \propto \int [g(x)]^{-1/2}dx$.

In the Poisson case, $g(x) = x$, so $h(x)$ should be proportional to $\int x^{-1/2}dx = 2x^{1/2}$. Thus, Poisson data should be transformed to $h(x_{ij}) = \sqrt{x_{ij}}$ before the analysis.

A Random Effects Model

The single-factor problems considered so far have all been assumed to be examples of a **fixed effects** ANOVA model. By this we mean that the chosen levels of the factor under study are the only ones considered relevant by the experimenter. The single-factor fixed effects model is

$$X_{ij} = \mu + \alpha_i + \epsilon_{ij}, \qquad \Sigma\alpha_i = 0 \tag{10.6}$$

where the ϵ_{ij}'s are random and both μ and the α_i's are fixed parameters whose values are unknown.

In some single-factor problems the particular levels studied by the experimenter are chosen, either by design or through sampling, from a large population of levels. For example, to study the effects on task performance time of using different operators on a particular machine, a sample of five operators might be chosen from a large pool of operators. Similarly, the effect of soil pH on the yield of maize plants might be studied by using soils with four specific pH values chosen from among the many possible pH levels. When the levels used are selected at random from a larger population of possible levels, the factor is said to be random rather than fixed, and the fixed effects model (10.6) is no longer appropriate. An analogous **random effects** model is obtained by replacing the fixed α_i's in (10.6) by random variables. The resulting model description is

$$X_{ij} = \mu + A_i + \epsilon_{ij} \quad \text{with } E(A_i) = E(\epsilon_{ij}) = 0$$
$$V(\epsilon_{ij}) = \sigma^2, \quad V(A_i) = \sigma_A^2 \tag{10.7}$$

all A_i's and ϵ_{ij}'s normally distributed and independent of one another.

The condition $E(A_i) = 0$ in (10.7) is similar to the condition $\Sigma \alpha_i = 0$ in (10.6); it states that the expected or average effect of the ith level measured as a departure from μ is zero.

For the random effects model (10.7), the hypothesis of no effects due to different levels is $H_0: \sigma_A^2 = 0$, which says that different levels of the factor contribute nothing to variability of the response. *Although the hypotheses in the single-factor fixed and random effects models are different, they are tested in exactly the same way,* by forming $F = MSTr/MSE$ and rejecting H_0 if $f \geq F_{\alpha, I-1, n-I}$. This can be justified intuitively by noting that $E(MSE) = \sigma^2$ (as for fixed effects), while

$$E(MSTr) = \sigma^2 + \frac{1}{I-1}\left(n - \frac{\Sigma J_i^2}{n}\right)\sigma_A^2 \tag{10.8}$$

where $J_1, J_2, \ldots, J_I$ are the sample sizes and $n = \Sigma J_i$. The factor in parentheses on the right side of (10.8) is nonnegative, so again $E(MSTr) = \sigma^2$ if H_0 is true and $E(MSTr) > \sigma^2$ if H_0 is false.

Example 10.10   The study of nondestructive forces and stresses in materials furnishes important information for efficient engineering design. The paper "Zero-Force Travel-Time Parameters for Ultrasonic Head-Waves in Railroad Rail" (*Materials Evaluation*, 1985, pp. 854–858) reports on a study of travel time for a certain type of wave that results from longitudinal stress of rails used for railroad track. Three measurements were made on each of six rails randomly selected from a population of rails. The investigators used random effects ANOVA to decide whether some variation in travel time could be attributed to "between-rail variability." The data appears in the accompanying table (each value, in nanoseconds, resulted from subtracting 36.1 μs from the original observation) along with the derived ANOVA table. The value of the

F ratio is highly significant, so H_0: $\sigma_A^2 = 0$ is rejected in favor of the conclusion that differences between rails is a source of travel-time variability.

				$x_i.$
1.	55	53	54	162
2.	26	37	32	95
3.	78	91	85	254
4.	92	100	96	288
5.	49	51	50	150
6.	80	85	83	248

$$x.. = 1197$$

Source of Variation	df	Sum of Squares	Mean Square	f
Treatments	5	9310.5	1862.1	115.2
Error	12	194.0	16.17	
Total	17	9504.5		

■

Exercises   Section 10.3 (22–34)

22. The following data refers to yield of tomatoes (kg/plot) for four different levels of salinity; salinity level here refers to electrical conductivity (EC), where the chosen levels were EC = 1.6, 3.8, 6.0, and 10.2 nmhos/cm:

1.6	59.5 53.3 56.8 63.1 58.7
3.8	55.2 59.1 52.8 54.5
6.0	51.7 48.8 53.9 49.0
10.2	44.6 48.5 41.0 47.3 46.1

Use the F test at level $\alpha = .05$ to test for any differences in true average yield due to the different salinity levels.

23. Apply the modified Tukey's method to the data in Exercise 22 to identify significant differences among the μ_i's.

24. The following partial ANOVA table is taken from the article "Perception of Spatial Incongruity" (*J. Nervous and Mental Disease,* 1961, p. 222) in which the abilities of three different groups to identify a perceptual incongruity were assessed and compared. All individuals in the experiment had been hospitalized to undergo psychiatric treatment. There were 21 individuals in the depressive group, 32 individuals in the functional "other" group, and 21 individuals in the brain-damaged group. Complete the ANOVA table and carry out the F test at level $\alpha = .01$.

Source	df	Sum of Squares	Mean Square	f
Groups			76.09	
Error				
Total		1123.14		

25. An article in the *Canadian Entomologist* ("Influence of Natural Diets and Larval Density on Gypsy Moth, Lymantria Dispor, Egg Mass Characteristics," 1977, pp. 1313–1318) reports the following data on egg mass diameters for moths reared on five different diets.

Diet	J_i	$\bar{x}_i.(mm)$	s_i
Red maple '74	13	1.134	.0252
Red oak/red maple	10	1.148	.0253
Red maple '75	20	1.159	.0179
Red oak	16	1.191	.0200
Red oak/white pine	16	1.217	.0160

a. Compute the grand mean $\bar{x}..$, $SSTr$, and $MSTr$. (HINT: $x.. = \Sigma J_i \bar{x}_i.$)

b. Compute SSE and MSE. [HINT: $SSE = \Sigma(J_i - 1)s_i^2$]

c. Does the data suggest that there are any differences between true average egg diameters for the different diets? Test using $\alpha = .05$.

d. Which means appear to differ significantly from one another?

26. Samples of six different brands of diet/imitation margarine were analyzed to determine the level of physiologically active polyunsaturated fatty acids (PAPFUA, in percentages), resulting in the following data:

Imperial	14.1 13.6 14.4 14.3	
Parkay	12.8 12.5 13.4 13.0 12.3	
Blue Bonnet	13.5 13.4 14.1 14.3	
Chiffon	13.2 12.7 12.6 13.9	
Mazola	16.8 17.2 16.4 17.3 18.0	
Fleischmann's	18.1 17.2 18.7 18.4	

(The preceding numbers are fictitious, but the sample means agree with data reported in the January 1975 issue of *Consumer Reports.*)

a. Use ANOVA to test for differences among the true average PAPFUA percentages for the different brands.

b. Compute CIs for all $(\mu_i - \mu_j)$'s.

c. Mazola and Fleischmann's are corn-based whereas the others are soybean-based. Compute a CI for

$$\frac{(\mu_1 + \mu_2 + \mu_3 + \mu_4)}{4} - \frac{(\mu_5 + \mu_6)}{2}$$

[HINT: Modify the expression for $V(\hat{\theta})$ that led to (10.5) in the previous section.]

27. Although tea is the world's most widely consumed beverage after water, little is known about its nutritional value. Folacin is the only B vitamin present in any significant amount in tea, and recent advances in assay methods have made accurate determination of folacin content feasible. Consider the accompanying data on folacin content for randomly selected specimens of the four leading brands of green tea.

Brand	Observations
1	7.9, 6.2, 6.6, 8.6, 8.9, 10.1, 9.6
2	5.7, 7.5, 9.8, 6.1, 8.4
3	6.8, 7.5, 5.0, 7.4, 5.3, 6.1
4	6.4, 7.1, 7.9, 4.5, 5.0, 4.0

(Based on "Folacin Content of Tea," *J. Amer. Dietetic Assoc.,* 1983, pp. 627–632.) Does this data suggest that true average folacin content is the same for all brands? Carry out a test using $\alpha = .05$.

28. For a single-factor ANOVA with sample sizes. J_i $(i = 1, 2, \ldots, I)$, show that $SSTr = \Sigma J_i(\bar{X}_{i\cdot} - \bar{X}_{\cdot\cdot})^2 = \Sigma_i J_i \bar{X}_{i\cdot}^2 - n\bar{X}_{\cdot\cdot}^2$, where $n = \Sigma J_i$.

29. When sample sizes are equal $(J_i = J)$, the parameters α_1, $\alpha_2, \ldots, \alpha_I$ of the alternative parameterization are restricted by $\Sigma \alpha_i = 0$. For unequal sample sizes, the most natural restriction is $\Sigma J_i \alpha_i = 0$. Use this to show that

$$E(MSTr) = \sigma^2 + \frac{1}{I-1}\Sigma J_i \alpha_i^2$$

What is $E(MSTr)$ when H_0 is true? [This expectation is correct if $\Sigma J_i \alpha_i = 0$ is replaced by the restriction $\Sigma \alpha_i = 0$ (or any other single linear restriction on the α_i's used to reduce the model to I independent parameters), but $\Sigma J_i \alpha_i = 0$ simplifies the algebra and yields natural estimates for the model parameters (in particular, $\hat{\alpha}_i = \bar{x}_{i\cdot} - \bar{x}_{\cdot\cdot}$).]

30. Reconsider Example 10.7 involving an investigation of the effects of different heat treatments on the yield point of steel ingots.

a. If $J = 8$ and $\sigma = 1$, what is β for a level .05 F test when $\mu_1 = \mu_2$, $\mu_3 = \mu_1 - 1$, $\mu_4 = \mu_1 + 1$?

b. For the alternative of part (a), what value of J is necessary to obtain $\beta = .05$?

c. If there are $I = 5$ heat treatments, $J = 10$, and $\sigma = 1$, what is β for the level .05 F test when four of the μ_i's are equal and the fifth differs by 1 from the other four?

31. When sample sizes are not equal, the noncentrality parameter is $\Sigma J_i \alpha_i^2/(2\sigma^2)$ and $\phi^2 = (1/I)\Sigma J_i \alpha_i^2/\sigma^2$. Referring to Exercise 22, what is the power of the test when $\mu_2 = \mu_3$, $\mu_1 = \mu_2 - \sigma$, $\mu_4 = \mu_2 + \sigma$?

32. In an experiment to compare the quality of four different brands of reel-to-reel recording tape, five 2400-ft reels of each brand (A–D) were selected and the number of flaws in each reel was determined.

A	10	5	12	14	8
B	14	12	17	9	8
C	13	18	10	15	18
D	17	16	12	22	14

It is believed that the number of flaws has approximately a Poisson distribution for each brand. Analyze the data at level .01 to see whether the expected number of flaws per reel is the same for each brand.

33. Suppose that X_{ij} is a binomial variable with parameters n and p_i (so approximately normal when $np_i \geq 5$ and $nq_i \geq 5$). Then since $\mu_i = np_i$, $V(X_{ij}) = \sigma_i^2 = np_i(1 - p_i) = \mu_i(1 - \mu_i/n)$. How should the X_{ij}'s be transformed so as to stabilize the variance? [HINT: $g(\mu_i) = \mu_i(1 - \mu_i/n)$.]

34. Simplify $E(MSTr)$ for the random effects model when $J_1 = J_2 = \cdots = J_I = J$.

Supplementary Exercises Chapter 10 (35–46)

35. An experiment was carried out to compare flow rates for four different types of nozzles.

   a. Sample sizes were 5, 6, 7, and 6, respectively, and calculations gave $f = 3.68$. State and test the relevant hypotheses using $\alpha = .01$

   b. Analysis of the data using a statistical computer package yielded P-value = .029. At level .01, what would you conclude, and why?

36. The paper "Computer-Assisted Instruction Augmented with Planned Teacher/Student Contacts" (*J. Exp. Educ.*, Winter, 1980–1981, pp. 120–126) compared five different methods for teaching descriptive statistics. The five methods were traditional lecture and discussion (*L/D*), programmed textbook instruction (*R*), programmed text with lectures (*R/L*), computer instruction (*C*), and computer instruction with lectures (*C/L*). Forty-five students were randomly assigned, 9 to each method. After completing the course, the students took a 1-hour exam. In addition, a 10-minute retention test was administered 6 weeks later. Summary quantities are given.

Method	Exam $\bar{x}_i$	s_i	Retention Test $\bar{x}_i$	s_i
L/D	29.3	4.99	30.20	3.82
R	28.0	5.33	28.80	5.26
R/L	30.2	3.33	26.20	4.66
C	32.4	2.94	31.10	4.91
C/L	34.2	2.74	30.20	3.53

The grand mean for the exam was 30.82, and the grand mean for the retention test was 29.30.

   a. Does the data suggest that there is a difference among the five teaching methods with respect to true mean exam score? Use $\alpha = .05$.

   b. Using a .05 significance level, test the null hypothesis of no difference among the true mean retention test scores for the five different teaching methods.

37. The article "Major Appliance Prices in the Chicago Area" (*J. Business*, 1977, pp. 231–235) reports the following data on prices for a certain size of refrigerator at a sample of stores in the Chicago area.

Brand	Sample Size	$\bar{x}_i$	s_i
Frigidaire	18	423.21	18.71
GE	28	418.13	21.15
Whirlpool	19	421.27	17.87

Does this data suggest any differences in the true average prices charged for the three brands?

38. An article in the British scientific journal *Nature* ("Sucrose Induction of Hepatic Hyperplasia in the Rat," August 25, 1972, p. 461) reports on an experiment in which each of five groups consisting of six rats was put on a diet with a different carbohydrate. At the conclusion of the experiment, the DNA content of the liver of each rat was determined (mg/g liver), with the following results:

Carbohydrate	$\bar{x}_i$
Starch	2.58
Sucrose	2.63
Fructose	2.13
Glucose	2.41
Maltose	2.49

Assuming also that $\Sigma\Sigma x_{ij}^2 = 183.4$, does the data indicate that true average DNA content is affected by the type of carbohydrate in the diet? Construct an ANOVA table and use a .05 level of significance.

39. Referring to Exercise 38, construct a t CI for

$$\theta = \mu_1 - (\mu_2 + \mu_3 + \mu_4 + \mu_5)/4$$

which measures the difference between the average DNA content for the starch diet and the combined average for the four other diets. Does the resulting interval include zero?

40. Referring to Exercise 38, what is β for the test when true average DNA content is identical for three of the starches and falls below this common value by 1 SD (σ) for the other two starches?

41. Four laboratories (1–4) are randomly selected from a large population, and each is asked to make three determinations of the percentage of methyl alcohol in specimens of a compound taken from a single batch. Based on the accompanying data, are differences among laboratories a source of variation in the percentage of methyl alcohol? State and test the relevant hypotheses using significance level .05.

1 85.06 85.25 84.87
2 84.99 84.28 84.88
3 84.48 84.72 85.10
4 84.10 84.55 84.05

42. Let $c_1, c_2, \ldots, c_I$ be numbers satisfying $\Sigma c_i = 0$. Then $\Sigma c_i \mu_i = c_1 \mu_1 + \cdots + c_I \mu_I$ is called a *contrast* in the μ_i's. Notice that with $c_1 = 1$, $c_2 = -1$, $c_3 = \cdots = c_I = 0$, $\Sigma c_i \mu_i = \mu_1 - \mu_2$, which implies that every pairwise difference between μ_i's is a contrast (so is, e.g., $\mu_1 - .5\mu_2 - .5\mu_3$). A method due to Scheffé gives simultaneous CIs with simultaneous confidence level $100(1 - \alpha)\%$ for *all* possible contrasts (an infinite number of them!). The interval for $\Sigma c_i \mu_i$ is

$$\Sigma c_i \bar{x}_{i\cdot} \pm (\Sigma c_i^2 / J_i)^{1/2} \cdot [(I - 1) \cdot MSE \cdot F_{\alpha, I-1, n-I}]^{1/2}$$

Using the critical flicker frequency data of Example 10.8, calculate the Scheffé intervals for the contrasts $\mu_1 - \mu_2$, $\mu_1 - \mu_3$, $\mu_2 - \mu_3$, and $.5\mu_1 + .5\mu_2 - \mu_3$ (this last contrast compares blue to the average of brown and

green). Which contrasts appear to differ significantly from 0, and why?

43. Four types of mortars—ordinary cement mortar (OCM), polymer impregnated mortar (PIM), resin mortar (RM), and polymer cement mortar (PCM)—were subjected to a compression test to measure strength (MPa). Three strength observations for each mortar type appeared in the paper "Polymer Mortar Composite Matrices for Maintenance-Free Highly Durable Ferrocement" (*J. Ferrocement*, 1984, pp. 337–345) and are reproduced below. Construct an ANOVA table. Using a .05 significance level, determine whether the data suggests that the true mean strength is not the same for all four mortar types. If you determine that the true mean strengths are not all equal, use Tukey's method to identify the significant differences.

OCM   32.15   35.53   34.20
PIM   126.32 126.80 134.79
RM   117.91 115.02 114.58
PCM   29.09   30.87   29.80

44. Suppose the x_{ij}'s are "coded" by $y_{ij} = cx_{ij} + d$. How does the value of the F statistic computed from the y_{ij}'s compare to the value computed from the x_{ij}'s? Justify your assertion.

45. In Example 10.10, subtract $\bar{x}_{i\cdot}$ from each observation in the ith sample ($i = 1, \ldots, 6$) to obtain a set of 18 residuals. Then construct a normal probability plot and comment on the plausibility of the normality assumption.

46. Consider a single-factor ANOVA in the case $I = 2$. Show that

a. $MSE = S_p^2$, the pooled estimator of σ^2 for the pooled t test.

b. $T^2 = F$, where T is the pooled t statistic for testing $H_0: \mu_1 - \mu_2 = 0$.

Bibliography

Miller, Rupert, *Beyond ANOVA: The Basics of Applied Statistics,* Wiley, New York, 1986. An excellent source of information about assumption checking and alternative methods of analysis.

Montgomery, Douglas, *Design and Analysis of Experiments*

(3rd ed.), Wiley, New York, 1991. A very up-to-date presentation of ANOVA models and methodology.

Neter, John, William Wasserman, and Michael Kutner, *Applied Linear Statistical Models* (3rd ed.), Irwin, Homewood, IL, 1990. The second half of this book contains a very well-presented survey of ANOVA; the level is comparable to that of the present text, but the discussion is more comprehensive, making the book an excellent reference.

Ott, Lyman, *An Introduction to Statistical Methods and Data Analysis* (4th ed.), Duxbury Press, Belmont, CA, 1993. Includes several chapters on ANOVA methodology that can profitably be read by students desiring a very nonmathematical exposition; there is a good chapter on various multiple comparison methods.

11 | Multifactor Analysis of Variance

Introduction

In the previous chapter, we used the analysis of variance (ANOVA) to test for equality of either I different population means or the true average responses associated with I different levels of a single factor (alternatively referred to as I different treatments). In many experimental situations, there are two or more factors that are of simultaneous interest. This chapter extends the methods of Chapter 10 to investigate such multifactor situations.

In the first two sections, we concentrate on the case of two factors of interest. We will use I to denote the number of levels of the first factor (A) and J to denote the number of levels of the second factor (B). Then there are IJ possible combinations consisting of one level of factor A and one of factor B; each such combination is called a treatment, so there are IJ different treatments. The number of observations made on treatment (i, j) will be denoted by K_{ij}. In Section 11.1 we present the model and analysis when $K_{ij} = 1$. An important special case of this type is a randomized block design, in which a single factor A is of primary interest but another factor, "blocks," is created in order to control for extraneous variability in experimental units or subjects. In Section 11.2, we focus on the case $K_{ij} = K > 1$, and mention briefly the difficulties associated with unequal K_{ij}'s.

Section 11.3 considers experiments involving more than two factors, including a Latin square design, which controls for the effects of two extraneous factors thought to influence the response variable. When the number of factors is large, an experiment consisting of at least one observation for each treatment would be expensive and time-consuming. An important special case, which we discuss in the last section, is that in which there are p factors, each of which has two levels. There are then 2^p different treatments, and we consider both the case in which observations are made on all these treatments (a complete design) and the case in which observations are made for only a selected subset of treatments (an incomplete design).

11.1 Two-Factor ANOVA with $K_{ij} = 1$

When factor A consists of I levels and factor B consists of J levels, there are IJ different combinations (pairs) of levels of the two factors, each called a treatment. With K_{ij} = the number of observations on the treatment consisting of factor A at level i and factor B at level j, we focus in this section on the case $K_{ij} = 1$, so that the data consists of IJ observations. We will first discuss the fixed effects model, in which the only levels of interest for the two factors are those actually represented in the experiment. The case in which one or both factors are random is discussed briefly at the end of the section.

Example 11.1

3|19

In a study on automobile traffic and air pollution reported in the *International Journal of Environmental Studies* ("Automobile Traffic and Air Pollution in a Developing Country," 1977, pp. 197–203), air samples taken at four different times and at five different locations were analyzed to obtain the amount of particulate matter present in the air (mg/m^3).

		Factor B: Location					Row
		1	2	3	4	5	Total
	Oct. 1975	76	67	81	56	51	331
Factor A:	Jan. 1976	82	69	96	59	70	376
Time	May 1976	68	59	67	54	42	290
	Sept. 1976	63	56	64	58	37	278
	Column total	289	251	308	227	200	1275

Is there any difference in true average amount of particulate matter present in the air due either to different sampling times or to different locations? ∎

As in single-factor ANOVA, double subscripts are used to identify random variables and observed values. Let

X_{ij} = the random variable (rv) denoting the measurement when factor A is held at level i and factor B is held at level j

x_{ij} = the observed value of X_{ij}

The x_{ij}'s are usually presented in a two-way table in which the ith row contains the observed values when factor A is held at level i and the jth column contains the observed values when factor B is held at level j. In the air-pollution experiment of Example 11.1, the number of levels of factor A is $I = 4$, the number of levels of factor B is $J = 5$, $x_{13} = 81$, $x_{42} = 56$, and so on.

Whereas in single-factor ANOVA we were interested only in row means and the grand mean, here we are interested also in column means. Let

$$\bar{X}_{i\cdot} = \begin{array}{c} \text{the average of measurements obtained} \\ \text{when factor } A \text{ is held at level } i \end{array} = \frac{\sum\limits_{j=1}^{J} X_{ij}}{J}$$

$$\bar{X}_{.j} = \begin{array}{c} \text{the average of measurements obtained} \\ \text{when factor } B \text{ is held at level } j \end{array} = \frac{\sum\limits_{i=1}^{I} X_{ij}}{I}$$

$$\bar{X}_{..} = \text{the grand mean} = \frac{\sum\limits_{i=1}^{I} \sum\limits_{j=1}^{J} X_{ij}}{IJ}$$

with observed values $\bar{x}_{i.}$, $\bar{x}_{.j}$, and $\bar{x}_{..}$. Totals rather than averages are denoted by omitting the horizontal bar (so $x_{.j} = \Sigma_i\, x_{ij}$, etc.). Intuitively, to see whether there is any effect due to the levels of factor A, we should compare the observed $\bar{x}_{i.}$'s with one another, while information about the different levels of factor B should come from the $\bar{x}_{.j}$'s.

The Model

Proceeding by analogy to single-factor ANOVA, one's first inclination in specifying a model is to let $\mu_{ij} = $ the true average response when factor A is at level i and factor B at level j, giving IJ mean parameters. Then let

$$X_{ij} = \mu_{ij} + \epsilon_{ij}$$

where ϵ_{ij} is the random amount by which the observed value differs from its expectation and the ϵ_{ij}'s are assumed normal and independent with common variance σ^2. Unfortunately, there is no valid test procedure for this choice of parameters. The reason is that under the alternative hypothesis of interest, the μ_{ij}'s are free to take on any values whatsoever, while σ^2 can be any value greater than zero, so that there are $IJ + 1$ freely varying parameters. But there are only IJ observations, so after using each x_{ij} as an estimate of μ_{ij}, there is no way to estimate σ^2.

To rectify this problem of a model having more parameters than observed values, we must specify a model that is realistic yet involves relatively few parameters.

Assume the existence of I parameters $\alpha_1, \alpha_2, \ldots, \alpha_I$ and J parameters $\beta_1, \beta_2, \ldots, \beta_J$ such that

$$X_{ij} = \alpha_i + \beta_j + \epsilon_{ij} \qquad (i = 1, \ldots, I, \quad j = 1, \ldots, J) \qquad (11.1)$$

so that

$$\mu_{ij} = \alpha_i + \beta_j \qquad (11.2)$$

Including σ^2, there are now $I + J + 1$ model parameters, so if $I \geq 3$ and $J \geq 3$, then there will be fewer parameters than observations (in fact, we will shortly modify (11.2) so that even $I = 2$ and/or $J = 2$ will be accommodated).

The model specified in (11.1) and (11.2) is called an **additive model** because each mean response μ_{ij} is the sum of an effect due to factor A at level i (α_i) and an effect due to factor B at level j (β_j). The difference between mean responses for factor A at level i and level i' when B is held at level j is $\mu_{ij} - \mu_{i'j}$. When the model is additive,

$$\mu_{ij} - \mu_{i'j} = (\alpha_i + \beta_j) - (\alpha_{i'} + \beta_j) = \alpha_i - \alpha_{i'}$$

which is independent of the level j of the second factor. A similar result holds for $\mu_{ij} - \mu_{ij'}$. Thus, additivity means that the difference in mean responses for two levels of one of the factors is the same for all levels of the other factor. Figure 11.1a shows a set of mean responses that satisfy the condition of additivity, and Figure 11.1b shows a nonadditive configuration of mean responses.

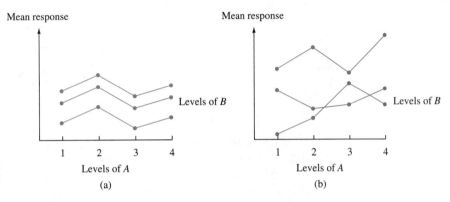

Figure 11.1 Mean responses for (a) an additive, and (b) a nonadditive model

Example 11.2
(Example 11.1 continued)

If we plot the observed x_{ij}'s in a manner analogous to that of Figure 11.1, the result is shown in Figure 11.2. While there is some "crossing over" in the observed x_{ij}'s, the configuration is reasonably representative of what would be expected under additivity with just one observation per treatment.

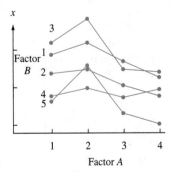

Figure 11.2 Plot of data from Example 11.2 ∎

Expression (11.2) is not quite the final model description because the α_i's and β_j's are not uniquely determined. Following are two different configurations of the α_i's and β_j's that yield the same additive μ_{ij}'s.

	$\beta_1 = 1$	$\beta_2 = 4$
$\alpha_1 = 1$	$\mu_{11} = 2$	$\mu_{12} = 5$
$\alpha_2 = 2$	$\mu_{21} = 3$	$\mu_{22} = 6$

	$\beta_1 = 2$	$\beta_2 = 5$
$\alpha_1 = 0$	$\mu_{11} = 2$	$\mu_{12} = 5$
$\alpha_2 = 1$	$\mu_{21} = 3$	$\mu_{22} = 6$

By subtracting any constant c from all α_i's and adding c to all β_j's, other configurations corresponding to the same additive model are obtained. This nonuniqueness is eliminated by use of the following model.

$$X_{ij} = \mu + \alpha_i + \beta_j + \epsilon_{ij}, \tag{11.3}$$

where $\displaystyle\sum_{i=1}^{I} \alpha_i = 0$, $\displaystyle\sum_{j=1}^{J} \beta_j = 0$, and the ϵ_{ij}'s are assumed independent, normally distributed, with mean 0 and common variance σ^2.

This is analogous to the alternative choice of parameters for single-factor ANOVA discussed in Section 10.3. It is not difficult to verify that (11.3) is an additive model in which the parameters are uniquely determined (for example, for the μ_{ij}'s mentioned above, $\mu = 4$, $\alpha_1 = -.5$, $\alpha_2 = .5$, $\beta_1 = -1.5$, and $\beta_2 = 1.5$). Notice that there are only $I - 1$ independently determined α_i's and $J - 1$ independently determined β_j's, so (including μ) (11.3) specifies $I + J - 1$ mean parameters.

The interpretation of the parameters of (11.3) is straightforward: μ is the true grand mean (mean response averaged over all levels of both factors), α_i is the effect of factor A at level i (measured as a deviation from μ), and β_j is the effect of factor B at level j. Unbiased (and maximum likelihood) estimators for these parameters are

$$\hat{\mu} = \bar{X}_{..}, \quad \hat{\alpha}_i = \bar{X}_{i.} - \bar{X}_{..}, \quad \text{and} \quad \hat{\beta}_j = \bar{X}_{.j} - \bar{X}_{..}$$

There are two different hypotheses of interest in a two-factor experiment with $K_{ij} = 1$. The first, denoted by H_{0A}, states that the different levels of factor A have no effect on true average response, while the second, denoted by H_{0B}, states that there is no factor B effect:

$$\begin{aligned} H_{0A}:\ &\alpha_1 = \alpha_2 = \cdots = \alpha_I = 0 \\ &\text{versus } H_{aA}: \text{ at least one } \alpha_i \neq 0 \\ H_{0B}:\ &\beta_1 = \beta_2 = \cdots = \beta_J = 0 \\ &\text{versus } H_{aB}: \text{ at least one } \beta_j \neq 0 \end{aligned} \tag{11.4}$$

(No factor A effect implies all α_i's are equal, so they must all be 0 since they sum to 0, and similarly for the β_j's.)

The Test Procedures

The description and analysis now follow closely that for single-factor ANOVA. The relevant sums of squares and their computing forms are given by

$3|1$

$$SST = \sum_{i=1}^{I} \sum_{j=1}^{J} (X_{ij} - \bar{X}..)^2 = \sum_{i=1}^{I} \sum_{j=1}^{J} X_{ij}^2 - \frac{1}{IJ} X_{..}^2, \qquad df = IJ - 1$$

$$SSA = \sum_{i=1}^{I} \sum_{j=1}^{J} (\bar{X}_{i\cdot} - \bar{X}..)^2 = \frac{1}{J} \sum_{i=1}^{I} X_{i\cdot}^2 - \frac{1}{IJ} X_{..}^2, \qquad df = I - 1$$

$$SSB = \sum_{i=1}^{I} \sum_{j=1}^{J} (\bar{X}_{\cdot j} - \bar{X}..)^2 = \frac{1}{I} \sum_{j=1}^{J} X_{\cdot j}^2 - \frac{1}{IJ} X_{..}^2, \qquad df = J - 1 \qquad (11.5)$$

$$SSE = \sum_{i=1}^{I} \sum_{j=1}^{J} (X_{ij} - \bar{X}_{i\cdot} - \bar{X}_{\cdot j} + \bar{X}..)^2, \qquad df = (I - 1)(J - 1)$$

and the fundamental identity

$$SST = SSA + SSB + SSE \qquad (11.6)$$

allows SSE to be determined by subtraction.

The expression for SSE results from replacing μ, α_i, and β_j in $\Sigma[X_{ij} - (\mu + \alpha_i + \beta_j)]^2$ by their estimators. Error df is $IJ -$ number of mean parameters estimated $= IJ - [1 + (I - 1) + (J - 1)] = (I - 1)(J - 1)$. As in single-factor ANOVA, total variation is split into a part (SSE) that is not explained by either the truth or the falsity of H_{0A} or H_{0B}, and two parts that can be explained by possible falsity of the two null hypotheses.

Statistical theory now says that if we form F ratios as in single-factor ANOVA, when H_{0A} (H_{0B}) is true, the corresponding F ratio has an F distribution with numerator df $= I - 1$ $(J - 1)$ and denominator df $= (I - 1)(J - 1)$.

Hypotheses: H_{0A} versus H_{aA}

Test statistic value: $f_A = \dfrac{MSA}{MSE}$

Rejection region: $f_A \geq F_{\alpha, I-1, (I-1)(J-1)}$

Hypotheses: H_{0B} versus H_{aB}

Test statistic value: $f_B = \dfrac{MSB}{MSE}$

Rejection region: $f_B \geq F_{\alpha, J-1, (I-1)(J-1)}$

Example 11.3
(Example 11.2
continued)

The $x_i.$'s (row totals) and $x_{.j}$'s (column totals) for the particulate matter data are displayed along the right and bottom margins of the data table given earlier. In addition, $\Sigma\Sigma x_{ij}^2 = 84{,}853$ and the correction factor is $x_{..}^2/(IJ) = (1275)^2/20 = 81{,}281.25$. The sums of squares are then

$$SST = 84{,}853 - 81{,}281.25 = 3571.75$$

$$SSA = \frac{1}{5}[(331)^2 + (376)^2 + (290)^2 + (278)^2] - 81{,}281.25 = 1182.95$$

$$SSB = \frac{1}{4}[(289)^2 + (251)^2 + (308)^2 + (227)^2 + (200)^2] - 81{,}281.25$$

$$= 1947.50$$

$$SSE = 3571.75 - (1182.95 + 1947.50) = 441.30$$

The accompanying ANOVA table (Table 11.1) summarizes further calculations.

Table 11.1 ANOVA table for Example 11.3

Source of Variation	df	Sum of Squares	Mean Square	f
Factor A (time)	$I - 1 = 3$	$SSA = 1182.95$	$MSA = 394.32$	$f_A = 10.72$
Factor B (location)	$J - 1 = 4$	$SSB = 1947.50$	$MSB = 486.88$	$f_B = 13.24$
Error	$(I - 1)(J - 1) = 12$	$SSE = 441.30$	$MSE = 36.78$	
Total	$IJ - 1 = 19$	$SST = 3571.75$		

The critical value for testing H_{0A} at level of significance .05 is $F_{.05,3,12} = 3.49$. Since $10.72 \geq 3.49$, H_{0A} is rejected in favor of the claim that the average amount of particulate matter does vary with time of sampling. Similarly, $F_{.05,4,12} = 3.26$ and $13.24 \geq 3.26$, so H_{0B} is rejected at significance level .05 in favor of the assertion that pollution varies with sampling location. ∎

Expected Mean Squares

The plausibility of using the F tests just described is demonstrated by computing the expected mean squares. After some tedious algebra,

$$E(MSE) = \sigma^2 \text{ (when the model is additive)}$$

$$E(MSA) = \sigma^2 + \frac{J}{I-1}\sum_{i=1}^{I}\alpha_i^2$$

$$E(MSB) = \sigma^2 + \frac{I}{J-1}\sum_{j=1}^{J}\beta_j^2$$

When H_{0A} is true, MSA is an unbiased estimator of σ^2, so F is a ratio of two unbiased estimators of σ^2. When H_{0A} is false, MSA tends to overestimate σ^2, so H_{0A} should be rejected when the ratio F_A is too large. Similar comments apply to MSB and H_{0B}.

Multiple Comparisons in Two-Factor ANOVA

When either H_{0A} or H_{0B} has been rejected, Tukey's procedure can be used to identify significant differences between the levels of the factor under investigation. The steps in the analysis are identical to those for a single-factor ANOVA:

1. For comparing levels of factor A, obtain $Q_{\alpha, I, (I-1)(J-1)}$.
    For comparing levels of factor B, obtain $Q_{\alpha, J, (I-1)(J-1)}$.

2. Compute

    $w = Q \cdot$ (estimated standard deviation of the sample
        means being compared)

    $$= Q_{\alpha, I, (I-1)(J-1)} \cdot \sqrt{MSE/J} \quad \text{for factor } A \text{ comparisons} \quad \text{or}$$
    $$Q_{\alpha, J, (I-1)(J-1)} \cdot \sqrt{MSE/I} \quad \text{for factor } B \text{ comparisons}$$

    (because, e.g., the standard deviation of $\bar{X}_{i\cdot}$ is $\sigma/\sqrt{J}$).

3. Arrange the sample means in increasing order, underscore those pairs differing by less than w, and identify pairs not underscored by the same line as corresponding to significantly different levels of the given factor.

Example 11.4
(Example 11.3
continued)

Both the hypothesis of no sampling time effects and that of no location effects were rejected at level .05. To identify significant differences among the four sampling times (factor A), we need $Q_{.05,4,12} = 4.20$ and $w = 4.20\sqrt{36.78/5} = 11.39$. The four-factor A sample means (row averages) are now listed in increasing order and any pair differing by less than 11.39 is underscored by a line segment:

$\bar{x}_{4\cdot}$	$\bar{x}_{3\cdot}$	$\bar{x}_{1\cdot}$	$\bar{x}_{2\cdot}$
55.60	58.00	66.20	75.20

We conclude that time 2 differs significantly from both time 3 and time 4, but no other significant differences are identified. For factor B (location), $w = 13.68$ and the underscoring is as follows:

$\bar{x}_{\cdot 5}$	$\bar{x}_{\cdot 4}$	$\bar{x}_{\cdot 2}$	$\bar{x}_{\cdot 1}$	$\bar{x}_{\cdot 3}$
50.00	56.75	62.75	72.25	77.00

The location pairs identified as significantly different from one another are 1 and 5, 3 and 5, 1 and 4, 3 and 4, and finally 2 and 3. Here the underscoring pattern makes interpretation difficult. ∎

Randomized Block Experiments

In using single-factor ANOVA to test for the presence of effects due to the I different treatments under study, once the IJ subjects or experimental units have been chosen, treatments should be allocated in a completely random fashion. That is, J subjects should be chosen at random for the first treatment, then another sample of J chosen at random from the remaining $IJ - J$ subjects for the second treatment, and so on.

It frequently happens, though, that subjects or experimental units exhibit heterogeneity with respect to other variables that may affect the observed responses.

When this is the case, the presence or absence of a significant F value may be due to this extraneous variation rather than to the presence or absence of factor effects. This was the reason for introducing a paired experiment in Chapter 9. The analogy to a paired experiment when $I > 2$ is called a **randomized block** experiment. An extraneous factor, "blocks," is constructed by dividing the IJ units into J groups with I units in each group. This grouping or blocking is done in such a way that within each block, the I units are homogeneous with respect to other factors thought to affect the responses. Then within each homogeneous block, the I treatments are randomly assigned to the I units or subjects in the block.

Example 11.5

A consumer product-testing organization wished to compare the annual power consumption for five different brands of dehumidifier. Because power consumption depends on the prevailing humidity level, it was decided to monitor each brand at four different levels ranging from moderate to heavy humidity (thus blocking on humidity level). Within each level, brands were randomly assigned to the five selected locations. The resulting amount of power consumption (annual kwh) appears in Table 11.2.

Table 11.2 Power consumption data for Example 11.5

Treatments (brands)	Blocks (humidity level) 1	2	3	4	$x_{i.}$	$\bar{x}_{i.}$
1	685	792	838	875	3190	797.50
2	722	806	893	953	3374	843.50
3	733	802	880	941	3356	839.00
4	811	888	952	1005	3656	914.00
5	828	920	978	1023	3749	937.25
$x_{.j}$	3779	4208	4541	4797	17,325	

Since $\Sigma\Sigma x_{ij}^2 = 15{,}178{,}901.00$ and $x_{..}^2/(IJ) = 15{,}007{,}781.25$,

$$SST = 15{,}178{,}901.00 - 15{,}007{,}781.25 = 171{,}119.75$$

$$SSA = \frac{1}{4}[60{,}244{,}049] - 15{,}007{,}781.25 = 53{,}231.00$$

$$SSB = \frac{1}{5}[75{,}619{,}995] - 15{,}007{,}781.25 = 116{,}217.75$$

and

$$SSE = 171{,}119.75 - 53{,}231.00 - 116{,}217.75 = 1671.00$$

The ANOVA calculations are summarized in Table 11.3.

Table 11.3 ANOVA table for Example 11.5

Source of Variation	df	Sum of Squares	Mean Square	f
Treatments (brands)	4	53,231.00	13,307.75	$f_A = 95.57$
Blocks	3	116,217.75	38,739.25	$f_B = 278.20$
Error	12	1671.00	139.25	
Total	19	171,119.75		

Since $F_{.05,4,12} = 3.26$ and $f_A = 95.57 \geq 3.26$, H_0 is rejected in favor of H_a, and we conclude that power consumption does depend on the brand of humidifier. To identify significantly different brands, we use Tukey's procedure. $Q_{.05,5,12} = 4.51$ and $w = 4.51\sqrt{139.25/4} = 26.6$.

$\bar{x}_{1\cdot}$	$\bar{x}_{3\cdot}$	$\bar{x}_{2\cdot}$	$\bar{x}_{4\cdot}$	$\bar{x}_{5\cdot}$
797.50	839.00	843.50	914.00	937.25

The underscoring indicates that the brands can be divided into three groups with respect to power consumption.

Because the block factor is of secondary interest, $F_{.05,3,12}$ is not needed, though the computed value of F_B is clearly highly significant. Figure 11.3 shows SAS output for this data. Notice that in the first part of the ANOVA table, the sums of squares

Analysis of Variance Procedure

Dependent Variable: POWERUSE

Source	DF	Sum of Squares	Mean Square	F Value	Pr > F
Model	7	169448.750	24206.964	173.84	0.0001
Error	12	1671.000	139.250		
Corrected Total	19	171119.750			

	R-Square	C.V.	Root MSE	POWERUSE Mean
	0.990235	1.362242	11.8004	866.25000

Source	DF	Anova SS	Mean Square	F Value	PR > F
BRAND	4	53231.000	13307.750	95.57	0.0001
HUMIDITY	3	116217.750	38739.250	278.20	0.0001

Alpha= 0.05 df= 12 MSE= 139.25
Critical Value of Studentized Range= 4.508
Minimum Significant Difference= 26.597

Means with the same letter are not significantly different.

Tukey Grouping		Mean	N	BRAND
A		937.250	4	5
A				
A		914.000	4	4
B		843.500	4	2
B				
B		839.000	4	3
C		797.500	4	1

Figure 11.3 SAS output for power consumption data

(SSs) for treatments (brands) and blocks (humidity levels) are combined into a single "model" SS.   ■

In many experimental situations in which treatments are to be applied to subjects, a single subject can receive all I of the treatments. Blocking is then often done on the subjects themselves to control for variability between subjects; each subject is then said to act as its own control. Social scientists sometimes refer to such experiments as repeated-measures designs. The "units" within a block are then the different "instances" of treatment application. Similarly, blocks are often taken as different time periods, locations, or observers.

Example 11.6   The data in Table 11.4 appeared in the paper "Compounding of Discriminative Stimuli from the Same and Different Sensory Modalities" (*J. Experimental Analysis Behavior*, 1971, pp. 337–342). Rat response was maintained by fixed interval schedules of reinforcement in the presence of a tone or two separate lights. The lights were either of moderate (L1) or low intensity (L2). Observations are given as the mean number of responses emitted by each subject during single and compound stimuli presentations over a 4-day period.

Table 11.4 Response data for Example 11.6

Stimulus	Subject 1	2	3	4	$x_{i\cdot}$	$\bar{x}_{i\cdot}$
L1	8.0	17.3	52.0	22.0	99.3	24.8
L2	6.9	19.3	63.7	21.6	111.5	27.9
Tone (T)	9.3	18.8	60.0	28.3	116.4	29.1
L1 + L2	9.2	24.9	82.4	44.9	161.4	40.3
L1 + T	12.0	31.7	83.8	37.4	164.9	41.2
L2 + T	9.4	33.6	96.6	40.6	180.2	45.1
$x_{\cdot j}$	54.8	145.6	438.5	194.8	833.7	

With $\Sigma\Sigma x_{ij}^2 = 44{,}614.21$, $SST = 15{,}653.56$, $SSA = 1428.28$, $SSB = 13{,}444.63$, and $SSE = 780.65$. Further calculations are summarized in Table 11.5.

Table 11.5 ANOVA table for Example 11.6

Source of Variation	df	Sum of Squares	Mean Square	f
Stimuli (A)	5	1428.28	285.66	$f_A = 5.49$
Subjects (B)	3	13,444.63	4481.54	$f_B = 86.12$
Error	15	780.65	52.04	
Total	23	15,653.56		

Since $F_{.05,5,15} = 2.90$ and $5.49 \geq 2.90$, we conclude that there are differences in the true average responses associated with the different stimuli. For Tukey's procedure, $w = 4.59\sqrt{52.04/4} = 16.56$.

$\bar{x}_{1.}$	$\bar{x}_{2.}$	$\bar{x}_{3.}$	$\bar{x}_{4.}$	$\bar{x}_{5.}$	$\bar{x}_{6.}$
24.8	27.9	29.1	40.3	41.2	45.1

Thus, both L1 and L2 are significantly different from L2 + T, and there are no other significant differences among the stimuli.   ■

In most randomized block experiments in which subjects serve as blocks, the subjects actually participating in the experiment are selected from a large population. The subjects then contribute random rather than fixed effects. This does not affect the procedure for comparing treatments when $K_{ij} = 1$ (one observation per "cell," as in this section), but the procedure is altered if $K_{ij} = K > 1$. We will shortly consider two-factor models in which effects are random.

More on Blocking

When $I = 2$, either the F test or the paired differences t test can be used to analyze the data. The resulting conclusion will not depend on which procedure is used, since $T^2 = F$ and $t^2_{\alpha/2,\nu} = F_{\alpha,1,\nu}$.

Just as with pairing, blocking entails both a potential gain and a potential loss in precision. If there is a great deal of heterogeneity in experimental units, the value of the variance parameter σ^2 in the one-way model will be large. The effect of blocking is to filter out the variation σ^2 in the two-way model appropriate for a randomized block experiment. Other things being equal, a smaller value of σ^2 results in a test that is more likely to detect departures from H_0 (i.e., a test with greater power).

However, other things are not equal here, since the single-factor F test is based on $I(J - 1)$ degrees of freedom (df) for error, while the two-factor F test is based on $(I - 1)(J - 1)$ df for error. Fewer degrees of freedom for error results in a decrease in power, essentially because the denominator estimator of σ^2 is not as precise. This loss in degrees of freedom can be especially serious if the experimenter can afford only a small number of observations. Nevertheless, if it appears that blocking will significantly reduce variability, it is probably worth the loss in degrees of freedom.

Models for Random Effects

In many experiments, the actual levels of a factor used in the experiment, rather than being the only ones of interest to the experimenter, have been selected from a much larger population of possible levels of the factor. In a two-factor situation, when this is the case for both factors, a **random effects model** is appropriate. The case in which the levels of one factor are the only ones of interest and the levels of the other factor are selected from a population of levels leads to a **mixed effects model**. The two-factor random effects model when $K_{ij} = 1$ is

$$X_{ij} = \mu + A_i + B_j + \epsilon_{ij} \qquad (i = 1, \ldots, I, \quad j = 1, \ldots, J)$$

where the A_i's, B_j's, and ϵ_{ij}'s are all independent, normally distributed rv's with mean 0 and variances σ^2_A, σ^2_B, and σ^2, respectively. The hypotheses of interest are then H_{0A}:

$\sigma_A^2 = 0$ (expected response does not depend on which level of factor A is selected) versus H_{aA}: $\sigma_A^2 > 0$ and H_{0B}: $\sigma_B^2 = 0$ versus H_{aB}: $\sigma_B^2 > 0$. Whereas $E(MSE) = \sigma^2$ as before, the expected mean squares for factors A and B are now

$$E(MSA) = \sigma^2 + J\sigma_A^2, \quad E(MSB) = \sigma^2 + I\sigma_B^2$$

Thus, when H_{0A} (H_{0B}) is true, F_A (F_B) is still a ratio of two unbiased estimators of σ^2. It can be shown that a level α test for H_{0A} versus H_{aA} still rejects H_{0A} if $f_A \geq F_{\alpha, I-1, (I-1)(J-1)}$, and similarly the same procedure as before is used to decide between H_{0B} and H_{aB}.

For the case in which factor A is fixed and factor B is random, the mixed model is

$$X_{ij} = \mu + \alpha_i + B_j + \epsilon_{ij} \qquad (i = 1, \ldots, I, \quad j = 1, \ldots, J)$$

where $\Sigma\alpha_i = 0$ and the B_j's and ϵ_{ij}'s are normally distributed with mean 0 and variances σ_B^2 and σ^2, respectively. Now the two null hypotheses are

$$H_{0A}: \alpha_1 = \cdots = \alpha_I = 0 \qquad \text{and} \qquad H_{0B}: \sigma_B^2 = 0$$

with expected mean squares

$$E(MSE) = \sigma^2, \quad E(MSA) = \sigma^2 + \frac{J}{I-1}\Sigma\alpha_i^2, \quad E(MSB) = \sigma^2 + I\sigma_B^2$$

The test procedures for H_{0A} versus H_{aA} and H_{0B} versus H_{aB} are exactly as before. For example, in the analysis of the air-pollution data in Example 11.1, if the five locations were randomly selected, then because $f_B = 13.24$ and $F_{.05,4,12} = 3.26$, H_{0B}: $\sigma_B^2 = 0$ is rejected in favor of H_{aB}: $\sigma_B^2 > 0$. An estimate of the "variance component" σ_B^2 is then given by $(MSB - MSE)/I = 112.53$.

Summarizing, when $K_{ij} = 1$, although the hypotheses and expected mean squares differ from the case of both effects fixed, the test procedures are identical.

Exercises   Section 11.1 (1–13)

1. The number of miles of useful tread wear (in 1000s) was determined for tires of each of five different makes of subcompact car (factor A, with $I = 5$) in combination with each of four different brands of radial tires (factor B, with $J = 4$), resulting in $IJ = 20$ observations. The values $SSA = 30.6$, $SSB = 44.1$, and $SSE = 59.2$ were then computed. Assume that an additive model is appropriate.

   a. Test H_0: $\alpha_1 = \alpha_2 = \alpha_3 = \alpha_4 = \alpha_5 = 0$ (no differences in true average tire lifetime due to makes of

   cars) versus H_a: at least one $\alpha_i \neq 0$ using a level .05 test.

   b. Test H_0: $\beta_1 = \beta_2 = \beta_3 = \beta_4 = 0$ (no differences in true average tire lifetime due to brands of tires) versus H_a: at least one $\beta_j \neq 0$ using a level .05 test.

2. Four different coatings are being considered for corrosion protection of metal pipe. The pipe will be buried in three different types of soil. To investigate whether the amount of corrosion depends either on the coating or on

the type of soil, 12 pieces of pipe are selected. Each piece is coated with one of the four coatings and buried in one of the three types of soil for a fixed time, after which the amount of corrosion (depth of maximum pits, in .0001 in.) is determined. The data appears in the accompanying table:

Soil Type (B)

		1	2	3
	1	64	49	50
Coating (A)	2	53	51	48
	3	47	45	50
	4	51	43	52

a. Assuming the validity of the additive model, carry out the ANOVA analysis using an ANOVA table to see whether the amount of corrosion depends on either the type of coating used or the type of soil. Use $\alpha = .05$.

b. Compute $\hat{\mu}$, $\hat{\alpha}_1$, $\hat{\alpha}_2$, $\hat{\alpha}_3$, $\hat{\alpha}_4$, $\hat{\beta}_1$, $\hat{\beta}_2$, and $\hat{\beta}_3$.

3. The paper "Adiabatic Humidification of Air with Water in a Packed Tower" (Chem. Eng. Prog., 1952, pp. 362–370) reports data on gas film heat transfer coefficient (Btu/hr ft^2 on °F) as a function of gas rate (factor A) and liquid rate (factor B).

		B			
		1(190)	2(250)	3(300)	4(400)
	1(200)	200	226	240	261
	2(400)	278	312	330	381
A	3(700)	369	416	462	517
	4(1100)	500	575	645	733

a. After constructing an ANOVA table, test at level .01 both the hypothesis of no gas-rate effect against the appropriate alternative and the hypothesis of no liquid-rate effect against the appropriate alternative.

b. Use Tukey's procedure to investigate differences in expected heat transfer coefficient due to different gas rates.

c. Repeat part (b) for liquid rates.

4. In an experiment to see whether the amount of coverage of light-blue interior latex paint depends either on the brand of paint or on the brand of roller used, 1 gallon of each of four brands of paint was applied using each of three brands of roller, resulting in the following data (number of square feet covered).

Roller Brand

		1	2	3
	1	454	446	451
Paint	2	446	444	447
Brand	3	439	442	444
	4	444	437	443

a. Construct the ANOVA table. [HINT: The computations can be expedited by subtracting 400 (or any other convenient number) from each observation. This does not affect the final results.]

b. State and test hypotheses appropriate for deciding whether paint brand has any effect on coverage. Use $\alpha = .05$.

c. Repeat part (b) for brand of roller.

d. Use Tukey's method to identify significant differences among brands. Is there one brand that seems clearly preferable to the others?

5. In an experiment to assess the effect of the angle of pull on the force required to cause separation in electrical connectors, four different angles (factor A) were used and each of a sample of five connectors (factor B) was pulled once at each angle ("A Mixed Model Factorial Experiment in Testing Electrical Connectors," Industrial Quality Control, 1960, pp. 12–16). The data appears in the accompanying table:

		B				
		1	2	3	4	5
	0°	45.3	42.2	39.6	36.8	45.8
	2°	44.1	44.1	38.4	38.0	47.2
A	4°	42.7	42.7	42.6	42.2	48.9
	6°	43.5	45.8	47.9	37.9	56.4

Does the data suggest that true average separation force is affected by the angle of pull? State and test the appropriate hypotheses at level .01 by first constructing an ANOVA table ($SST = 396.13$, $SSA = 58.16$, and $SSB = 246.97$).

6. A particular county employs three assessors who are responsible for determining the value of residential property in the county. To see whether these assessors differ systematically in their assessments, 5 houses are selected, and each assessor is asked to determine the market value of each house. With factor A denoting assessors ($I = 3$) and factor B denoting houses ($J = 5$), suppose $SSA = 11.7$, $SSB = 113.5$, and $SSE = 25.6$.

a. Test H_0: $\alpha_1 = \alpha_2 = \alpha_3 = 0$ at level .05 (H_0 states that there are no systematic differences among assessors).

b. Explain why a randomized block experiment with only 5 houses was used rather than a one-way ANOVA experiment involving a total of 15 different houses with each assessor asked to assess 5 different houses (a different group of 5 for each assessor).

7. The article "Rate of Stuttering Adaptation Under Two Electro-Shock Conditions" (*Behavior Research Therapy*, 1967, pp. 49–54) gives adaptation scores for three different treatments: (1) no shock, (2) shock following each stuttered word, and (3) shock during each moment of stuttering. These treatments were used on each of 18 stutterers.

a. Summary statistics include $x_{1.} = 905$, $x_{2.} = 913$, $x_{3.} = 936$, $x_{..} = 2754$, $\Sigma_j x_{.j}^2 = 430{,}295$, and $\Sigma\Sigma x_{ij}^2 = 143{,}930$. Construct the ANOVA table and test at level .05 to see whether true average adaptation score depends on the treatment given.

b. Judging from the F ratio for subjects (factor B), do you think that blocking on subjects was effective in this experiment? Explain.

8. The accompanying table gives plasma epinephrine concentration for ten experimental subjects during (1) isoflurane, (2) halothane, and (3) cyclopropane anesthesia ("Sympathoadrenal and Hemodynamic Effects of Isoflurane, Halothane, and Cyclopropane in Dogs," *Anesthesiology*, 1974, pp. 465–470).

a. Does the choice of anesthetic affect true average concentration? Test H_0: $\alpha_1 = \alpha_2 = \alpha_3 = 0$ at level .05 after constructing the ANOVA table.

Subject (B)

		1	2	3	4	5
Anesthetic (A)	1	.28	.51	1.00	.39	.29
	2	.30	.39	.63	.38	.21
	3	1.07	1.35	.69	.28	1.24

		6	7	8	9	10
Anesthetic (A)	1	.36	.32	.69	.17	.33
	2	.88	.39	.51	.32	.42
	3	1.53	.49	.56	1.02	.30

$$\Sigma\Sigma x_{ij}^2 = 13.7980$$

b. Use Tukey's procedure to investigate significant differences among the anesthetics.

9. The paper "The Effects of a Pneumatic Stool and a One-Legged Stool on Lower Limb Joint Load and Muscular Activity During Sitting and Rising" (*Ergonomics*, 1993, pp. 519–535) gives the accompanying data on the effort required of a subject to arise from four different types of stools (Borg scale). Perform an analysis of variance using $\alpha = .05$ and follow this with a multiple comparisons analysis if appropriate.

Subject

		1	2	3	4	5	6	7	8	9	$\bar{x}_{i.}$
Type of Stool	1	12	10	7	7	8	9	8	7	9	8.55
	2	15	14	14	11	11	11	12	11	13	12.44
	3	12	13	13	10	8	11	12	8	10	10.77
	4	10	12	9	9	7	10	11	7	8	9.22

10. Suppose that in the experiment described in Exercise 6 the five houses had actually been selected at random from among those of a certain age and size, so that factor B is random rather than fixed. Test H_0: $\sigma_B^2 = 0$ versus H_a: $\sigma_B^2 > 0$ using a level .01 test.

11. a. Show that a constant d can be added to (or subtracted from) each x_{ij} without affecting any of the ANOVA sums of squares.

b. Suppose that each x_{ij} is multiplied by a nonzero constant c. How does this affect the ANOVA sums of squares? How does this affect the values of the F statistics F_A and F_B? What effect does "coding" the data by $y_{ij} = cx_{ij} + d$ have on the conclusions resulting from the ANOVA procedures?

12. Use the fact that $E(X_{ij}) = \mu + \alpha_i + \beta_j$ with $\Sigma\alpha_i = \Sigma\beta_j = 0$ to show that $E(\bar{X}_{i.} - \bar{X}_{..}) = \alpha_i$, so that $\hat{\alpha}_i = \bar{X}_{i.} - \bar{X}_{..}$ is an unbiased estimator for α_i.

13. The power curves of Figures 10.5 and 10.6 can be used to obtain $\beta = P(\text{type II error})$ for the F test in two-factor ANOVA. For fixed values of $\alpha_1, \alpha_2, \ldots, \alpha_I$, the quantity $\phi^2 = (J/I)\Sigma\alpha_i^2/\sigma^2$ is computed. Then the figure corresponding to $\nu_1 = I - 1$ is entered on the horizontal axis at the value ϕ, the power is read on the vertical axis from the curve labeled $\nu_2 = (I - 1)(J - 1)$, and $\beta = 1 - \text{power}$.

a. For the corrosion experiment described in Exercise 2, find β when $\alpha_1 = 4$, $\alpha_2 = 0$, $\alpha_3 = \alpha_4 = -2$, and $\sigma = 4$. Repeat if $\alpha_1 = 6$, $\alpha_2 = 0$, $\alpha_3 = \alpha_4 = -3$, and $\sigma = 4$.

b. By symmetry, what is β for the test of H_{0B} versus H_{aB} in Example 11.1 when $\beta_1 = 4$, $\beta_2 = \beta_3 = \beta_4 = \beta_5 = -1$, and $\sigma = 4$?

11.2

3/21

In Section 11.1, we analyzed data from a two-factor experiment in which there was one observation for each of the IJ combinations of levels of the two factors. To obtain valid test procedures, the μ_{ij}'s were assumed to have an additive structure with $\mu_{ij} = \mu + \alpha_i + \beta_j$, $\Sigma\alpha_i = \Sigma\beta_j = 0$. Additivity means that the difference in true average responses for any two levels of the factors is the same for each level of the other factor. For example, $\mu_{ij} - \mu_{i'j} = (\mu + \alpha_i + \beta_j) - (\mu + \alpha_{i'} + \beta_j) = \alpha_i - \alpha_{i'}$ independent of the level j of the second factor. This is shown in Figure 11.1a, in which the lines connecting true average responses are parallel.

Figure 11.1b depicts a set of true average responses that does not have additive structure. The lines connecting these μ_{ij}'s are not parallel, which means that the difference in true average responses for different levels of one factor does depend on the level of the other factor. When additivity does not hold, we say that there is **interaction** between the different levels of the factors. The assumption of additivity allowed us in Section 11.1 to obtain an estimator of the random error variance σ^2 (*MSE*) that was unbiased whether or not either null hypothesis of interest was true. When $K_{ij} > 1$ for at least one (i, j) pair, a valid estimator of σ^2 can be obtained without assuming additivity. In specifying the appropriate model and deriving test procedures, we will focus on the case $K_{ij} = K > 1$, so the number of observations per "cell" (for each combination of levels) is constant.

Parameters for the Fixed Effects Model with Interaction

Rather than use the μ_{ij}'s themselves as model parameters, it is usual to use an equivalent set that reveals more clearly the role of interaction. Let

$$\mu = \frac{1}{IJ}\sum_i \sum_j \mu_{ij}, \quad \mu_{i\cdot} = \frac{1}{J}\sum_j \mu_{ij}, \quad \text{and} \quad \mu_{\cdot j} = \frac{1}{I}\sum_i \mu_{ij} \tag{11.7}$$

Thus μ is the expected response averaged over all levels of both factors (the true grand mean), $\mu_{i\cdot}$ is the expected response averaged over levels of the second factor when the first factor A is held at level i, and similarly for $\mu_{\cdot j}$. Now define

$$\alpha_i = \mu_{i\cdot} - \mu = \text{the effect of factor } A \text{ at level } i$$
$$\beta_j = \mu_{\cdot j} - \mu = \text{the effect of factor } B \text{ at level } j \tag{11.8}$$
$$\gamma_{ij} = \mu_{ij} - (\mu + \alpha_i + \beta_j)$$

from which

$$\mu_{ij} = \mu + \alpha_i + \beta_j + \gamma_{ij} \tag{11.9}$$

The model is additive if and only if all γ_{ij}'s = 0. The γ_{ij}'s are referred to as the **interaction parameters.** The α_i's are called the **main effects for factor A,** while the β_j's are the **main effects for factor B.** Although there are I α_i's, J β_j's, and IJ γ_{ij}'s in addition to μ, the conditions $\Sigma\alpha_i = 0$, $\Sigma\beta_j = 0$, $\Sigma_j \gamma_{ij} = 0$ for any i, and $\Sigma_i \gamma_{ij} = 0$ for any j [all by virtue of (11.7) and (11.8)] imply that only IJ of these new parameters are independently determined: μ, $I - 1$ of the α_i's, $J - 1$ of the β_j's, and $(I - 1)(J - 1)$ of the γ_{ij}'s.

There are now three sets of hypotheses that will be considered:

H_{0AB}: $\gamma_{ij} = 0$ for all i, j versus H_{aAB}: at least one $\gamma_{ij} \neq 0$

H_{0A}: $\alpha_1 = \cdots = \alpha_I = 0$ versus H_{aA}: at least one $\alpha_i \neq 0$

H_{0B}: $\beta_1 = \cdots = \beta_J = 0$ versus H_{aB}: at least one $\beta_j \neq 0$

The no-interaction hypothesis H_{0AB} is usually tested first. If H_{0AB} is not rejected, then the other two hypotheses can be tested to see whether the main effects are significant. If H_{0AB} is rejected and H_{0A} is then tested and not rejected, the resulting model $\mu_{ij} = \mu + \beta_j + \gamma_{ij}$ does not lend itself to straightforward interpretation. In such a case it is best to construct a picture similar to that of Figure 11.1b to try to visualize the way in which the factors interact.

Notation, Model, and Analysis

We now use triple subscripts for both random variables and observed values, with X_{ijk} and x_{ijk} referring to the kth observation (replication) when factor A is at level i and factor B is at level j. The model is then

$$X_{ijk} = \mu + \alpha_i + \beta_j + \gamma_{ij} + \epsilon_{ijk} \qquad (11.10)$$
$$i = 1, \ldots, I, \quad j = 1, \ldots, J, \quad k = 1, \ldots, K$$

where the ϵ_{ij}'s are independent and normally distributed, each with mean 0 and variance σ^2.

Again a dot in place of a subscript means that we have summed over all values of that subscript, while a horizontal bar denotes averaging. Thus, $X_{ij\cdot}$ is the total of all K observations made for factor A at level i and factor B at level j [all observations in the (i, j)th cell], and $\overline{X}_{ij\cdot}$ is the average of these K observations.

Example 11.7   Three different varieties of tomato (Harvester, Pusa Early Dwarf, and Ife No. 1) and four different plant densities (10, 20, 30, and 40 thousand plants per hectare) are being considered for planting in a particular region. To see whether either variety or plant density affects yield, each combination of variety and plant density is used in three different plots, resulting in the data on yields in Table 11.6 (based on the article "Effects of Plant Density on Tomato Yields in Western Nigeria," *Experimental Agriculture*, 1976, pp. 43–47):

Table 11.6 Yield data for Example 11.7

Variety	Planting Density				$x_{i..}$	$\bar{x}_{i..}$
	10,000	*20,000*	*30,000*	*40,000*		
H	10.5, 9.2, 7.9	12.8, 11.2, 13.3	12.1, 12.6, 14.0	10.8, 9.1, 12.5	136.0	11.33
Ife	8.1, 8.6, 10.1	12.7, 13.7, 11.5	14.4, 15.4, 13.7	11.3, 12.5, 14.5	146.5	12.21
P	16.1, 15.3, 17.5	16.6, 19.2, 18.5	20.8, 18.0, 21.0	18.4, 18.9, 17.2	217.5	18.13
$x_{.j.}$	103.3	129.5	142.0	125.2	500.00	
$\bar{x}_{.j.}$	11.48	14.39	15.78	13.91		13.89

Here $I = 3$, $J = 4$, and $K = 3$, for a total of $IJK = 36$ observations. ∎

To test the hypotheses of interest, we again define sums of squares and present computing formulas:

$3/\lambda$

$$SST = \sum_i \sum_j \sum_k (X_{ijk} - \bar{X}...)^2 = \sum_i \sum_j \sum_k X_{ijk}^2 - \frac{1}{IJK} X_{...}^2, \qquad df = IJK - 1$$

$$SSE = \sum_i \sum_j \sum_k (X_{ijk} - \bar{X}_{ij.})^2$$

$$= \sum_i \sum_j \sum_k X_{ijk}^2 - \frac{1}{K} \sum_i \sum_j X_{ij.}^2, \qquad df = IJ(K - 1)$$

$$SSA = \sum_i \sum_j \sum_k (\bar{X}_{i..} - \bar{X}...)^2 = \frac{1}{JK} \sum_i X_{i..}^2 - \frac{1}{IJK} X_{...}^2, \qquad df = I - 1$$

$$SSB = \sum_i \sum_j \sum_k (\bar{X}_{.j.} - \bar{X}...)^2 = \frac{1}{IK} \sum_j X_{.j.}^2 - \frac{1}{IJK} X_{...}^2, \qquad df = J - 1$$

$$SSAB = \sum_i \sum_j \sum_k (X_{ij.} - \bar{X}_{i..} - \bar{X}_{.j.} + \bar{X}...)^2, \qquad df = (I - 1)(J - 1)$$

and the fundamental identity

$$SST = SSA + SSB + SSAB + SSE$$

implies that **interaction sum of squares** *SSAB* can be obtained by subtraction.

Total variation is thus partitioned into four pieces: unexplained (*SSE*—which would be present whether or not any of the three null hypotheses was true) and three pieces that may be explained by the truth or falsity of the three H_0's. Each of four mean squares is defined by $MS = SS/df$. The expected mean squares suggest that each set of hypotheses should be tested using the appropriate ratio of mean squares with MSE in the denominator:

$$E(MSE) = \sigma^2$$

$$E(MSA) = \sigma^2 + \frac{JK}{I-1}\sum_{i=1}^{I}\alpha_i^2$$

$$E(MSB) = \sigma^2 + \frac{IK}{J-1}\sum_{j=1}^{J}\beta_j^2$$

$$E(MSAB) = \sigma^2 + \frac{K}{(I-1)(J-1)}\sum_{i=1}^{I}\sum_{j=1}^{J}\gamma_{ij}^2$$

Each of the three mean square ratios can be shown to have an F distribution when the associated H_0 is true, which yields the following level α test procedures:

Hypotheses	Test Statistic Value	Rejection Region
H_{0A} versus H_{aA}	$f_A = \dfrac{MSA}{MSE}$	$f_A \geq F_{\alpha,I-1,IJ(K-1)}$
H_{0B} versus H_{aB}	$f_B = \dfrac{MSB}{MSE}$	$f_B \geq F_{\alpha,J-1,IJ(K-1)}$
H_{0AB} versus H_{aAB}	$f_{AB} = \dfrac{MSAB}{MSE}$	$f_{AB} \geq F_{\alpha,(I-1)(J-1),IJ(K-1)}$

As before, the results of the analysis are summarized in an ANOVA table.

Example 11.8
(Example 11.7 continued)

From the given data, $x_{...}^2 = (500)^2 = 250{,}000$,

$$\sum_i\sum_j\sum_k x_{ijk}^2 = (10.5)^2 + (9.2)^2 + \cdots + (18.9)^2 + (17.2)^2 = 7404.80$$

$$\sum_i x_{i..}^2 = (136.0)^2 + (146.5)^2 + (217.5)^2 = 87{,}264.50$$

and

$$\sum_j x_{.j.}^2 = 63{,}280.18$$

The cell totals ($x_{ij.}$'s) are

	10,000	20,000	30,000	40,000
H	27.6	37.3	38.7	32.4
Ife	26.8	37.9	43.5	38.3
P	48.9	54.3	59.8	54.5

from which $\Sigma_i\Sigma_j x_{ij}^2 = (27.6)^2 + \cdots + (54.5)^2 = 22{,}100.28$. Then

$$SST = 7404.80 - \frac{1}{36}(250{,}000) = 7404.80 - 6944.44 = 460.36$$

$$SSA = \frac{1}{12}(87{,}264.50) - 6944.44 = 327.60$$

$$SSB = \frac{1}{9}(63{,}280.18) - 6944.44 = 86.69$$

$$SSE = 7404.80 - \frac{1}{3}(22{,}100.28) = 38.04$$

and

$$SSAB = 460.36 - 327.60 - 86.69 - 38.04 = 8.03$$

Table 11.7 now summarizes the computations:

Table 11.7 ANOVA table for Example 11.8

Source of Variation	df	Sum of Squares	Mean Square	f
Varieties	2	327.60	163.8	$f_A = 103.02$
Density	3	86.69	28.9	$f_B = 18.18$
Interaction	6	8.03	1.34	$f_{AB} = .84$
Error	24	38.04	1.59	
Total	35	460.36		

Since $F_{.01,6,24} = 3.67$ and $f_{AB} = .84$ is not ≥ 3.67, H_{0AB} cannot be rejected at level .01, so we conclude that the interaction effects are not significant. Now the presence or absence of main effects can be investigated. Since $F_{.01,2,24} = 5.61$ and $f_A = 103.02 \geq 5.61$, H_{0A} is rejected at level .01 in favor of the conclusion that different varieties do affect the true average yields. Similarly, $f_B = 18.18 \geq 4.72 = F_{.01,3,24}$, so we conclude that true average yield also depends on plant density. ∎

Multiple Comparisons

When the no-interaction hypothesis H_{0AB} is not rejected and at least one of the two main effect null hypotheses is rejected, Tukey's method can be used to identify significant differences in levels. For identifying differences among the α_i's when H_{0A} is rejected,

1. Obtain $Q_{\alpha,I,IJ(K-1)}$, where the second subscript I identifies the number of levels being compared and the third subscript refers to the number of degrees of freedom for error.

2. Compute $w = Q\sqrt{MSE/(JK)}$, where JK is the number of observations averaged to obtain each of the $\bar{x}_{i..}$'s compared in Step 3.

3. Order the $\bar{x}_{i\cdot\cdot}$'s from smallest to largest and, as before, underscore all pairs that differ by less than w. Pairs not underscored correspond to significantly different levels of factor A.

To identify different levels of factor B when H_{0B} is rejected, replace the second subscript in Q by J, replace JK by IK in w, and replace $\bar{x}_{i\cdot\cdot}$ by $\bar{x}_{\cdot j\cdot}$.

Example 11.9

(Example 11.8 continued)

For factor A (varieties), $I = 3$, so with $\alpha = .01$ and $IJ(K - 1) = 24$, $Q_{.01,3,24} = 4.55$. Then $w = 4.55\sqrt{1.59/12} = 1.66$, so ordering and underscoring gives

$\bar{x}_{1\cdot\cdot}$	$\bar{x}_{2\cdot\cdot}$	$\bar{x}_{3\cdot\cdot}$
11.33	12.21	18.13

The Harvester and Ife varieties do not appear to differ significantly from one another in effect on true average yield, but both differ from the Pusa variety.

For factor B (density), $J = 4$ so $Q_{.01,4,24} = 4.91$ and $w = 4.91\sqrt{1.59/9} = 2.06$.

$\bar{x}_{\cdot 1\cdot}$	$\bar{x}_{\cdot 4\cdot}$	$\bar{x}_{\cdot 2\cdot}$	$\bar{x}_{\cdot 3\cdot}$
11.48	13.91	14.39	15.78

Thus, with experimentwise error rate .01, which is quite conservative, only the lowest density appears to differ significantly from all others. Even with $\alpha = .05$ (so that $w = 1.64$), densities 2 and 3 cannot be judged significantly different from one another in their effect on yield.   ∎

Models with Mixed and Random Effects

In some problems, the levels of either factor may have been chosen from a large population of possible levels, so that the effects contributed by the factor are random rather than fixed. As in Section 11.1, if both factors contribute random effects, the model is referred to as a random effects model, while if one factor is fixed and the other is random, a mixed effects model results. We will consider here the analysis for a mixed effects model in which factor A (rows) is the fixed factor and factor B (columns) is the random factor. The case in which both factors are random is dealt with in the exercises.

The mixed effects model in this situation is

$$X_{ijk} = \mu + \alpha_i + B_j + G_{ij} + \epsilon_{ijk}$$
$$i = 1, \ldots, I, \quad j = 1, \ldots, J, \quad k = 1, \ldots, K$$

Here μ and α_i's are constants with $\Sigma \alpha_i = 0$, and the B_j's, G_{ij}'s, and ϵ_{ijk}'s are normally distributed random variables with expected value 0 and variances σ_B^2, σ_G^2, and σ^2, respectively. Because the I levels of the fixed factor A are the only ones under consideration, we also assume that $\Sigma_i G_{ij} = 0$ (this implies that the G_{ij}'s for fixed j are not independent of one another but are negatively correlated). The three hypotheses of interest are then

$$H_{0A}: \alpha_1 = \alpha_2 = \cdots = \alpha_I = 0 \quad \text{versus} \quad H_{aA}: \text{at least one } \alpha_i \neq 0$$

$$H_{0B}: \sigma_B^2 = 0 \qquad\qquad\qquad \text{versus} \quad H_{aB}: \sigma_B^2 > 0$$

$$H_{0G}: \sigma_G^2 = 0 \qquad\qquad\qquad \text{versus} \quad H_{aG}: \sigma_G^2 > 0$$

It is customary to test H_{0A} and H_{0B} only if the no-interaction hypothesis H_{0G} cannot be rejected.

The relevant sums of squares and mean squares needed for the test procedures are defined and computed exactly as in the fixed effects case. The expected mean squares for the mixed model are

$$E(MSE) = \sigma^2$$

$$E(MSA) = \sigma^2 + \frac{IK}{I-1}\sigma_G^2 + \frac{JK}{I-1}\Sigma\alpha_i^2$$

$$E(MSB) = \sigma^2 + IK\sigma_B^2$$

and

$$E(MSAB) = \sigma^2 + \frac{IK}{I-1}\sigma_G^2$$

Thus, to test the no-interaction hypothesis, the ratio $f_{AB} = MSAB/MSE$ is again appropriate, with H_{0G} rejected if $f_{AB} \geq F_{\alpha,(I-1)(J-1),IJ(K-1)}$. Similarly, the F ratio and test procedure for testing H_{0B} are exactly as they were in the fixed effects case. However, for testing H_{0A} versus H_{aA}, the expected mean squares suggest that while the numerator of the F ratio should still be MSA, the denominator should be $MSAB$ rather than MSE, and this is indeed the case:

For testing H_{0A} versus H_{aA} (factors A fixed, B random), the test statistic value is $f_A = MSA/MSAB$, and the rejection region is $f_A \geq F_{\alpha,I-1,(I-1)(J-1)}$.

Example 11.10

A study was carried out to compare the writing lifetimes of four premium brands of pens. It was thought that the writing surface might affect lifetime, so three different surfaces were randomly selected. A writing machine was used to ensure that conditions were otherwise homogeneous (e.g., constant pressure and a fixed angle). Table 11.8 shows the two lifetimes (min) obtained for each brand–surface combination.

Table 11.8 Lifetime data for Example 11.10

		Writing Surface			
		1	*2*	*3*	$x_{i..}$
	1	709, 659	713, 726	660, 645	4112
	2	668, 685	722, 740	692, 720	4227
Brand of Pen	*3*	659, 685	666, 684	678, 750	4122
	4	698, 650	704, 666	686, 733	4137
	$x_{\cdot j \cdot}$	5413	5621	5564	16,598

Additional summary quantities are $\Sigma_i\Sigma_j\Sigma_k\, x_{ijk}^2 = 11{,}499{,}492$ and $\Sigma_i\Sigma_j\, x_{ij.}^2 = 22{,}982{,}552$.

Factor A, brand of pen, is fixed here, since there are only four brands of interest to the experimenters and they are all represented in the experiment. Factor B, however, is random, since it is not the three specific surfaces actually chosen that are of interest; instead, the question is whether there is variability in lifetime associated with the population of all surfaces. The three null hypotheses are then H_{0A}: $\alpha_1 = \alpha_2 = \alpha_3 = \alpha_4 = 0$, H_{0B}: $\sigma_B^2 = 0$, and H_{0G}: $\sigma_G^2 = 0$. The ANOVA table is Table 11.9; remember that $f_A = MSA/MSAB$, not MSA/MSE as with fixed effects.

Table 11.9 ANOVA table for Example 11.10

Source of Variation	df	Sum of Squares	Mean Square	f
Brands (A)	3	1387.50	462.50	$f_A = .34$
Surfaces (B)	2	2888.08	1444.04	$f_B = 2.11$
Interaction	6	8100.25	1350.04	$f_{AB} = 1.97$
Error	12	8216.00	684.67	
Total	23	20,591.83		

None of the three computed F's exceeds $F_{.05}$ for appropriate numerator and denominator degrees of freedom, so none of the three null hypotheses can be rejected. In particular, the data suggests no differences in lifetime among the four brands of pen and no variability in lifetime due to different writing surfaces.   ∎

When at least two of the K_{ij}'s are unequal, the ANOVA computations are much more complex than for the case $K_{ij} = K$, and there are no nice formulas for the appropriate test statistics. One of the chapter references can be consulted for more information.

Exercises   Section 11.2 (14–24)

14. In an experiment to assess the effects of curing time (factor A) and type of mix (factor B) on the compressive strength of hardened cement cubes, three different curing times were used in combination with four different mixes, with three observations obtained for each of the 12 curing time/mix combinations. The resulting sums of squares were computed to be $SSA = 30{,}763.0$, $SSB = 34{,}185.6$, $SSE = 97{,}436.8$, and $SST = 205{,}966.6$.

a. Construct an ANOVA table.

b. Test at level .05 the null hypothesis H_{0AB}: all γ_{ij}'s $= 0$ (no interaction of factors) against H_{aAB}: at least one $\gamma_{ij} \neq 0$.

c. Test at level .05 the null hypothesis H_{0A}: $\alpha_1 = \alpha_2 = \alpha_3 = 0$ (factor A main effects are absent) against H_{aA}: at least $\alpha_i \neq 0$.

d. Test H_{0B}: $\beta_1 = \beta_2 = \beta_3 = \beta_4 = 0$ versus H_{aB}: at least one $\beta_j \neq 0$ using a level .05 test.

 The values of the $\bar{x}_{i..}$'s were $\bar{x}_{1..} = 4010.88$, $\bar{x}_{2..} = 4029.10$, and $\bar{x}_{3..} = 3960.02$. Use Tukey's procedure to investigate significant differences among the three curing times.

15. The paper "Towards Improving the Properties of Plaster Moulds and Castings" (*J. Engr. Manuf.*, 1991, pp. 265–269) describes several ANOVAs carried out to study how the amount of carbon fiber and sand additions affect various characteristics of the molding process.

Here we give data on casting hardness and on wet-mold strength.

Sand Addition (%)	Carbon Fiber Addition (%)	Casting Hardness	Wet-Mold Strength
0	0	61.0	34.0
0	0	63.0	16.0
15	0	67.0	36.0
15	0	69.0	19.0
30	0	65.0	28.0
30	0	74.0	17.0
0	.25	69.0	49.0
0	.25	69.0	48.0
15	.25	69.0	43.0
15	.25	74.0	29.0
30	.25	74.0	31.0
30	.25	72.0	24.0
0	.50	67.0	55.0
0	.50	69.0	60.0
15	.50	69.0	45.0
15	.50	74.0	43.0
30	.50	74.0	22.0
30	.50	74.0	48.0

a. An ANOVA for wet-mold strength gives SS-Sand = 705, SSFiber = 1278, SSE = 843, and SST = 3105. Test for the presence of any effects using $\alpha = .05$.

b. Carry out an ANOVA on the casting hardness observations using $\alpha = .05$.

c. Plot sample mean hardness against sand percentage for different levels of carbon fiber. Is the plot consistent with your analysis in part (b)?

16. The accompanying data resulted from an experiment carried out to investigate whether yield from a certain chemical process depended either on the formulation of a particular input or on mixer speed.

		Speed		
		60	70	80
		189.7	185.1	189.0
	1	188.6	179.4	193.0
		190.1	177.3	191.1
Formulation				
		165.1	161.7	163.3
	2	165.9	159.8	166.6
		167.6	161.6	170.3

A statistical computer package gave SSForm = 2253.44, SSSpeed = 230.81, SSForm*Speed = 18.58, and SSE = 71.87.

a. Does there appear to be interaction between the factors?

b. Does yield appear to depend either on formulation or speed?

c. Calculate estimates of the main effects.

d. The *fitted values* are $\hat{x}_{ijk} = \hat{\mu} + \hat{\alpha}_i + \hat{\beta}_j + \hat{\gamma}_{ij}$, and the *residuals* are $x_{ijk} - \hat{x}_{ijk}$. Verify that the residuals are .23, −.87, .63, 4.50, −1.20, −3.30, −2.03, 1.97, .07, −1.10, −.30, 1.40, .67, −1.23, .57, −3.43, −.13, and 3.57.

e. Construct a normal probability plot from the residuals given in part (d). Do the ϵ_{ijk}'s appear to be normally distributed?

17. The accompanying data table gives observations on total acidity of coal samples of three different types, with determinations made using three different concentrations of ethanolic NaOH ("Chemistry of Brown Coals," *Australian J. Applied Science,* 1958, pp. 375–379).

		Type of Coal		
		Morwell	Yallourn	Maddingley
NaOH Conc.	.404N	8.27, 8.17	8.66, 8.61	8.14, 7.96
	.626N	8.03, 8.21	8.42, 8.58	8.02, 7.89
	.786N	8.60, 8.20	8.61, 8.76	8.13, 8.07

Additionally, $\Sigma_i\Sigma_j\Sigma_k\, \bar{x}_{ijk}^2 = 1240.1525$ and $\Sigma_i\Sigma_j x_{ij\cdot}^2 = 2479.9991$.

a. Assuming both effects to be fixed, construct an ANOVA table, test for the presence of interaction, and then test for the presence of main effects for each factor (all using level .01).

b. Use Tukey's procedure to identify significant differences among the types of coal.

18. The current (in μA) necessary to produce a certain level of brightness of a television tube was measured for two different types of glass and three different types of phosphor, resulting in the accompanying data ("Fundamentals of Analysis of Variance," *Industrial Quality Control,* 1956, pp. 5–8):

		Phosphor Type		
		1	2	3
Glass Type	1	280, 290, 285	300, 310, 295	270, 285, 290
	2	230, 235, 240	260, 240, 235	220, 225, 230

Assuming that both factors are fixed, test H_{0AB} versus H_{aAB} at level .01. Then if H_{0AB} cannot be rejected, test the two sets of main effect hypotheses.

19. In an experiment to investigate the effect of "cement factor" (number of sacks of cement per cubic yard) on flexural strength of the resulting concrete ("Studies of Flexural Strength of Concrete. Part 3: Effects of Variation in Testing Procedure," *Proceedings ASTM*, 1957, pp. 1127–1139), $I = 3$ different factor values were used, $J = 5$ different batches of cement were selected, and $K = 2$ beams were cast from each cement factor/batch combination. Summary values include $\Sigma\Sigma\Sigma x_{ijk}^2 = 12,280,103$, $\Sigma\Sigma x_{ij\cdot}^2 = 24,529,699$, $\Sigma x_{i\cdot\cdot}^2 = 122,380,901$, $\Sigma x_{\cdot j\cdot}^2 = 73,427,483$, and $x_{\cdots} = 19,143$.

a. Construct the ANOVA table.

b. Assuming a mixed model with cement factor (A) fixed and batches (B) random, test the three pairs of hypotheses of interest at level .05.

20. The accompanying data was obtained in an experiment to investigate whether compressive strength of concrete cylinders depends on the type of capping material used or variability in different batches ("The Effect of Type of Capping Material on the Compressive Strength of Concrete Cylinders," *Proceedings ASTM*, 1958, pp. 1166–1186). Each number is a cell total $(x_{ij\cdot})$ based on $K = 3$ observations.

Batch

		1	2	3	4	5
	1	1847	1942	1935	1891	1795
Capping Material	2	1779	1850	1795	1785	1626
	3	1806	1892	1889	1891	1756

In addition, $\Sigma\Sigma\Sigma x_{ijk}^2 = 16,815,853$ and $\Sigma\Sigma x_{ij\cdot}^2 = 50,433,409$. Obtain the ANOVA table and then test at level .01 the hypotheses H_{0G} versus H_{aG}, H_{0A} versus H_{aA}, and H_{0B} versus H_{aB}, assuming that capping is a fixed effect and batches is a random effect.

21. Show that $E(\bar{X}_{i\cdot\cdot} - \bar{X}_{\cdots}) = \alpha_i$, so that $\bar{X}_{i\cdot\cdot} - \bar{X}_{\cdots}$ is an unbiased estimator for α_i (in the fixed effects model).

22. With $\hat{\gamma}_{ij} = \bar{X}_{ij\cdot} - \bar{X}_{i\cdot\cdot} - \bar{X}_{\cdot j\cdot} + \bar{X}_{\cdots}$, show that $\hat{\gamma}_{ij}$ is an unbiased estimator for γ_{ij} (in the fixed effects model).

23. Show how a $100(1 - \alpha)\%$ t CI for $\alpha_i - \alpha_{i'}$ can be obtained. Then compute a 95% interval for $\alpha_2 - \alpha_3$ using the data from Exercise 17. [HINT: With $\theta = \alpha_2 - \alpha_3$, the result of Exercise 21 indicates how to obtain $\hat{\theta}$. Then compute $V(\hat{\theta})$ and $\sigma_{\hat{\theta}}$, and obtain an estimate of $\sigma_{\hat{\theta}}$ by using $\sqrt{MSE}$ to estimate σ (which identifies the appropriate number of df).]

24. When both factors are random in a two-way ANOVA experiment with K replications per combination of factor levels, the expected mean squares are $E(MSE) = \sigma^2$, $E(MSA) = \sigma^2 + K\sigma_G^2 + JK\sigma_A^2$, $E(MSB) = \sigma^2 + K\sigma_G^2 + IK\sigma_B^2$, and $E(MSAB) = \sigma^2 + K\sigma_G^2$.

a. What F ratio is appropriate for testing $H_{0G}: \sigma_G^2 = 0$ versus $H_{aG}: \sigma_G^2 > 0$?

b. Answer (a) for testing $H_{0A}: \sigma_A^2 = 0$ versus $H_{aA}: \sigma_A^2 > 0$ and $H_{0B}: \sigma_B^2 = 0$ versus $H_{aB}: \sigma_B^2 > 0$.

11.3   Three-Factor ANOVA

To indicate the nature of models and analyses when ANOVA experiments involve more than two factors, we will focus here on the case of three fixed factors—*A*, *B*, and *C*. The numbers of levels of the three factors will be denoted by *I*, *J*, and *K*, respectively, and L_{ijk} = the number of observations made with factor *A* at level *i*, factor *B* at level *j*, and factor *C* at level *k*. As with two-factor ANOVA, the analysis is quite complicated when the L_{ijk}'s are not all equal, so we further specialize to $L_{ijk} = L$. Then X_{ijkl} and x_{ijkl} denote the observed value, before and after the experiment is performed, of the *l*th replication ($l = 1, 2, \ldots, L$) when the three factors are fixed at levels *i*, *j*, and *k*.

To understand the parameters that will appear in the three-factor ANOVA model, first recall that in two-factor ANOVA with replications, $E(X_{ijk}) = \mu_{ij} = \mu + \alpha_i + \beta_j + \gamma_{ij}$, where the restrictions $\Sigma_i \alpha_i = \Sigma_j \beta_j = 0$, $\Sigma_i \gamma_{ij} = 0$ for every j, and $\Sigma_j \gamma_{ij} = 0$ for every i were necessary to obtain a unique set of parameters. If we use dot subscripts on the μ_{ij}'s to denote averaging (rather than summation), then

$$\mu_{i\cdot} - \mu_{\cdot\cdot} = \frac{1}{J}\sum_j \mu_{ij} - \frac{1}{IJ}\sum_i \sum_j \mu_{ij} = \alpha_i$$

is the effect of factor A at level i averaged over levels of factor B, while

$$\mu_{ij} - \mu_{\cdot j} = \mu_{ij} - \frac{1}{I}\sum_i \mu_{ij} = \alpha_i + \gamma_{ij}$$

is the effect of factor A at level i specific to factor B at level j. If the effect of A at level i depends on the level of B, then there is interaction between the factors, and the γ_{ij}'s are not all zero. In particular,

$$\mu_{ij} - \mu_{\cdot j} - \mu_{i\cdot} + \mu_{\cdot\cdot} = \gamma_{ij} \tag{11.11}$$

The Three-Factor Fixed Effects Model

The model for three-factor ANOVA with $L_{ijk} = L$ is

$$X_{ijkl} = \mu_{ijk} + \epsilon_{ijkl} \qquad i = 1, \ldots, I, \quad j = 1, \ldots, J, \tag{11.12}$$
$$k = 1, \ldots, K, \quad l = 1, \ldots, L$$

where the ϵ_{ijkl}'s are normally distributed with mean 0 and variance σ^2, and

$$\mu_{ijk} = \mu + \alpha_i + \beta_j + \delta_k + \gamma_{ij}^{AB} + \gamma_{ik}^{AC} + \gamma_{jk}^{BC} + \gamma_{ijk} \tag{11.13}$$

The restrictions necessary to obtain uniquely defined parameters are that the sum over any subscript of any parameter on the right-hand side of (11.13) equal 0.

The parameters γ_{ij}^{AB}, γ_{ik}^{AC}, and γ_{jk}^{BC} are called two-factor interactions, while γ_{ijk} is called a three-factor interaction; the α_i's, β_j's, and δ_k's are the main effects parameters. For any fixed level k of the third factor, analogous to (11.11),

$$\mu_{ijk} - \mu_{i\cdot k} - \mu_{\cdot jk} + \mu_{\cdot\cdot k} = \gamma_{ij}^{AB} + \gamma_{ijk}$$

is the interaction of the ith level of A with the jth level of B specific to the kth level of C, while

$$\mu_{ij\cdot} - \mu_{i\cdot\cdot} - \mu_{\cdot j\cdot} + \mu_{\cdot\cdot\cdot} = \gamma_{ij}^{AB}$$

is the interaction between A at level i and B at level j averaged over levels of C. If the interaction of A at level i and B at level j does not depend on k, then all γ_{ijk}'s equal $\cap$ Thus nonzero γ_{ijk}'s represent nonadditivity of the two-factor γ_{ij}^{AB}'s over the various levels of the third factor C. If the experiment included more than three factors, there would be corresponding higher-order interaction terms with analogous interpretations. Note that in the above argument, if we had considered fixing the level of either A or B (rather than C, as was done) and examining the γ_{ijk}'s, their interpretation would be the same—if any of the interactions of two factors depend on the level of the third factor, then there are nonzero γ_{ijk}'s.

The Analysis of a Three-Factor Experiment

When $L > 1$, there is a sum of squares for each main effect, each two-factor interaction, and the three-factor interaction. To write these in a way that indicates how sums of squares are defined when there are more than three factors, note that any of the model parameters in (11.13) can be estimated unbiasedly by averaging X_{ijkl} over appropriate subscripts and taking differences. Thus,

$$\hat{\mu} = \bar{X}...., \quad \hat{\alpha}_i = \bar{X}_{i...} - \bar{X}...., \quad \hat{\gamma}_{ij}^{AB} = \bar{X}_{ij..} - \bar{X}_{i...} - \bar{X}_{.j..} + \bar{X}....,$$

$$\hat{\gamma}_{ijk} = \bar{X}_{ijk.} - \bar{X}_{ij..} - \bar{X}_{i.k.} - \bar{X}_{.jk.} + \bar{X}_{i...} + \bar{X}_{.j..} + \bar{X}_{..k.} - \bar{X}....$$

with other main effects and interaction estimators obtained by symmetry. Then sums of squares are

$$SST = \sum_i \sum_j \sum_k \sum_l (X_{ijkl} - \bar{X}....)^2, \qquad df = IJKL - 1$$

$$SSA = \sum_i \sum_j \sum_k \sum_l \hat{\alpha}_i^2 = JKL \sum_i (\bar{X}_{i...} - \bar{X}....)^2, \qquad df = I - 1$$

$$SSAB = \sum_i \sum_j \sum_k \sum_l (\hat{\gamma}_{ij}^{AB})^2 = KL \sum_i \sum_j (\bar{X}_{ij..} - \bar{X}_{i...} - \bar{X}_{.j..} + \bar{X}....)^2,$$

$$df = (I - 1)(J - 1)$$

$$SSABC = \sum_i \sum_j \sum_k \sum_l \hat{\gamma}_{ijk}^2 = L \sum_i \sum_j \sum_k \hat{\gamma}_{ijk}^2, \qquad df = (I - 1)(J - 1)(K - 1)$$

$$SSE = \sum_i \sum_j \sum_k \sum_l (X_{ijkl} - \bar{X}_{ijk.})^2, \qquad df = IJK(L - 1)$$

with the other main effect and two-factor interaction sums of squares obtained by symmetry. SST is the sum of the other eight SSs.

Even the computational formulas for these SSs are quite tedious to use, so we eschew them in favor of output from a statistical computer package. The current version of MINITAB, for example, will fit a three-factor model with fixed, mixed, or random effects.

Each sum of squares (excepting SST) when divided by its df gives a mean square, with

$$E(MSE) = \sigma^2$$

$$E(MSA) = \sigma^2 + \frac{JKL}{I - 1} \sum_i \alpha_i^2$$

$$E(MSAB) = \sigma^2 + \frac{KL}{(I - 1)(J - 1)} \sum_i \sum_j (\gamma_{ij}^{AB})^2$$

$$E(MSABC) = \sigma^2 + \frac{L}{(I-1)(J-1)(K-1)} \sum_i \sum_j \sum_k (\gamma_{ijk})^2$$

and similar expressions for the other expected mean squares. Main effect and interaction hypotheses are tested by forming F ratios with MSE in each denominator:

Null Hypothesis	Test Statistic Value	Rejection Region
H_{0A}: all α_i's = 0	$f_A = \dfrac{MSA}{MSE}$	$f_A \geq F_{\alpha, I-1, IJK(L-1)}$
H_{0AB}: all γ_{ij}^{AB}'s = 0	$f_{AB} = \dfrac{MSAB}{MSE}$	$f_{AB} \geq F_{\alpha,(I-1)(J-1),IJK(L-1)}$
H_{0ABC}: all γ_{ijk}'s = 0	$f_{ABC} = \dfrac{MSABC}{MSE}$	$f_{ABC} \geq F_{\alpha,(I-1)(J-1)(K-1),IJK(L-1)}$

Usually the main effect hypotheses are tested only if all interactions are judged not significant.

The above analysis assumes that $L_{ijk} = L > 1$. If $L = 1$, then as in the two-factor case, the highest-order interactions must be assumed to equal 0 in order to obtain an MSE that estimates σ^2. Setting $L = 1$ and disregarding the fourth subscript summation over l, the above formulas for sums of squares are still valid, and $SSE = \Sigma_i \Sigma_j \Sigma_k \hat{\gamma}_{ijk}^2$ with $\bar{X}_{ijk\cdot} = X_{ijk}$ in the expression for $\hat{\gamma}_{ijk}$.

Example 11.11   The following observations (body temperature – 100°F) were reported in an experiment to study heat tolerance of cattle ("The Significance of the Coat in Heat Tolerance of Cattle," *Australian J. Agriculture Research*, 1959, pp. 744–748). Measurements were made at four different periods (factor A, with $I = 4$) on two different strains of cattle (factor B, with $J = 2$) having four different types of coat (factor C, with $K = 4$); $L = 3$ observations were made for each of the $4 \times 2 \times 4 = 32$ combinations of levels of the three factors.

	B_1				B_2			
	C_1	C_2	C_3	C_4	C_1	C_2	C_3	C_4
A_1	3.6	3.4	2.9	2.5	4.2	4.4	3.6	3.0
	3.8	3.7	2.8	2.4	4.0	3.9	3.7	2.8
	3.9	3.9	2.7	2.2	3.9	4.2	3.4	2.9
A_2	3.8	3.8	2.9	2.4	4.4	4.2	3.8	2.0
	3.6	3.9	2.9	2.2	4.4	4.3	3.7	2.9
	4.0	3.9	2.8	2.2	4.6	4.7	3.4	2.8
A_3	3.7	3.8	2.9	2.1	4.2	4.0	4.0	2.0
	3.9	4.0	2.7	2.0	4.4	4.6	3.8	2.4
	4.2	3.9	2.8	1.8	4.5	4.5	3.3	2.0
A_4	3.6	3.6	2.6	2.0	4.0	4.0	3.8	2.0
	3.5	3.7	2.9	2.0	4.1	4.4	3.7	2.2
	3.8	3.9	2.9	1.9	4.2	4.2	3.5	2.3

The table of cell totals (x_{ijk}'s) for all combinations of the three factors is

x_{ijk}	B_1				B_2			
	C_1	C_2	C_3	C_4	C_1	C_2	C_3	C_4
A_1	11.3	11.0	8.4	7.1	12.1	12.5	10.7	8.7
A_2	11.4	11.6	8.6	6.8	13.4	13.2	10.9	7.7
A_3	11.8	11.7	8.4	5.9	13.1	13.1	11.1	6.4
A_4	10.9	11.2	8.4	5.9	12.3	12.6	11.0	6.5

Figure 11.4 displays plots of the corresponding cell means $\bar{x}_{ijk} = x_{ijk}/3$. We will return to these plots after considering tests of various hypotheses. The basis for these tests is the ANOVA table given in Table 11.10.

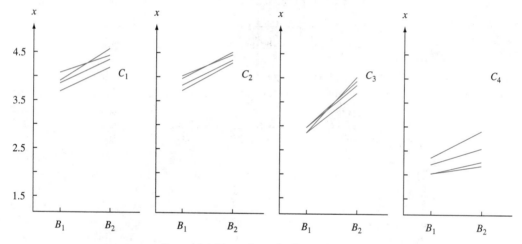

Figure 11.4 Plots of x_{ijk} for Example 11.11

Table 11.10 ANOVA table for Example 11.11

Source	df	Sum of Squares	Mean Square	f
A	$I - 1 = 3$	.49	.163	4.13
B	$J - 1 = 1$	6.45	6.45	163.29
C	$K - 1 = 3$	48.93	16.31	412.91
AB	$(I - 1)(J - 1) = 3$	.02	.0067	.170
AC	$(I - 1)(K - 1) = 9$	1.61	.179	4.53
BC	$(J - 1)(K - 1) = 3$	.88	.293	7.42
ABC	$(I - 1)(J - 1)(K - 1) = 9$	.25	.0278	.704
Error	$IJK(L - 1) = 64$	2.53	.0395	
Total	$IJKL - 1 = 95$	61.16		

Since $F_{.01,9,64} \approx 2.70$ and $f_{ABC} = MSABC/MSE = .704$ does not exceed 2.70, we conclude that three-factor interactions are not significant. However, although the AB interactions are also not significant, both AC and BC interactions as well as all main effects seem to be necessary in the model. When there are no ABC or AB interactions, a plot of the $\bar{x}_{ijk\cdot}$'s (= $\hat{\mu}_{ijk}$) separately for each level of C should reveal no substantial interactions (if only the ABC interactions are zero, plots are more difficult to interpret; see the article "Two-Dimensional Plots for Interpreting Interactions in the Three-Factor Analysis of Variance Model," *Amer. Statistician,* May 1979, pp. 63–69). ∎

Tukey's procedure can be used in three-factor (or more) ANOVA. The second subscript on Q is the number of sample means being compared, while the third is degrees of freedom for error.

Models with random and mixed effects can also be analyzed. Sums of squares and degrees of freedom are identical to the fixed effects case, but expected mean squares are of course different for the random main effects or interactions. A good reference is the book by Douglas Montgomery listed in the chapter references.

Latin Square Designs

When several factors are to be studied simultaneously, an experiment in which there is at least one observation for every possible combination of levels is referred to as a **complete layout.** If the factors are A, B, and C with I, J, and K levels, respectively, a complete layout requires at least IJK observations. Frequently an experiment of this size is either impracticable because of cost, time, or space constraints, or literally impossible. For example, if the response variable is sales of a certain product and the factors are different display configurations, different stores, and different time periods, then only one display configuration can realistically be used in a given store during a given time period.

A three-factor experiment in which fewer than IJK observations are made is called an **incomplete layout.** There are some incomplete layouts in which the pattern of combinations of factors is such that the analysis is straightforward. One such three-factor design is called a **Latin square.** It is appropriate when $I = J = K$ (e.g., four display configurations, four stores, and four time periods) and all two- and three-factor interaction effects are assumed absent. If the levels of factor A are identified with the rows of a two-way table and the levels of B with the columns of the table, then the defining characteristic of a Latin square design is that *every level of factor C appears exactly once in each row and exactly once in each column.* Pictured in Figure 11.5 are examples of 3×3, 4×4, and 5×5 Latin squares. There

Figure 11.5 Examples of Latin squares

are 12 different 3×3 Latin squares, and the number of different $N \times N$ Latin squares increases rapidly with N (e.g., every permutation of rows of a given Latin square yields a Latin square, and similarly for column permutations). It is recommended that the square actually used in a particular experiment be chosen at random from the set of all possible squares of the desired dimension; for further details, one of the chapter references can be consulted.

The Model and Analysis for Latin Squares

The letter N will be used to denote the common value of I, J, and K. Then a complete layout with one observation per combination would require N^3 observations, whereas a Latin square requires only N^2 observations. Once a particular square has been chosen, the value of k (the level of factor C) is completely determined by the values of i and j. To emphasize this, we use $x_{ij(k)}$ to denote the observed value when the three factors are at levels i, j, and k, respectively, with k taking on only one value for each i, j pair. The model is then

$$X_{ij(k)} = \mu + \alpha_i + \beta_j + \delta_k + \epsilon_{ij(k)}, \qquad i, j, k = 1, \ldots, N$$

where $\Sigma\alpha_i = \Sigma\beta_j = \Sigma\delta_k = 0$ and

the $\epsilon_{ij(k)}$'s are independent and normally distributed with mean 0 and variance σ^2.

We employ the following notation for totals and averages:

$$X_{i..} = \sum_j X_{ij(k)}, \quad X_{.j.} = \sum_i X_{ij(k)}, \quad X_{..k} = \sum_{i,j} X_{ij(k)}, \quad X_{...} = \sum_i \sum_j X_{ij(k)},$$

$$\bar{X}_{i..} = \frac{X_{i..}}{N}, \quad \bar{X}_{.j.} = \frac{X_{.j.}}{N}, \quad \bar{X}_{..k} = \frac{X_{..k}}{N}, \quad \bar{X}_{...} = \frac{X_{...}}{N^2}$$

Note that although $X_{i..}$ previously suggested a double summation, now it corresponds to a single sum over all j (and the associated values of k). The sums of squares used in the analysis are then

$$SST = \sum_i \sum_j (X_{ij(k)} - \bar{X}_{...})^2 = \sum_i \sum_j X_{ij(k)}^2 - \frac{X_{...}^2}{N^2}, \qquad df = N^2 - 1$$

$$SSA = \sum_i \sum_j (\bar{X}_{i..} - \bar{X}_{...})^2 = \frac{1}{N}\sum_i X_{i..}^2 - \frac{X_{...}^2}{N^2}, \qquad df = N - 1$$

$$SSB = \sum_i \sum_j (\bar{X}_{.j.} - \bar{X}_{...})^2 = \frac{1}{N}\sum_j X_{.j.}^2 - \frac{X_{...}^2}{N^2}, \qquad df = N - 1$$

$$SSC = \sum_i \sum_j (\bar{X}_{..k} - \bar{X}_{...})^2 = \frac{1}{N}\sum_k X_{..k}^2 - \frac{X_{...}^2}{N^2}, \qquad df = N - 1$$

$$SSE = \sum_i \sum_j [X_{ij(k)} - (\hat{\mu} + \hat{\alpha}_i + \hat{\beta}_j + \hat{\delta}_k)]^2$$
$$= \sum_i \sum_j (X_{ij(k)} - \bar{X}_{i..} - \bar{X}_{.j.} - \bar{X}_{..k} + 2\bar{X}...)^2, \qquad df = (N-1)(N-2)$$

$$SST = SSA + SSB + SSC + SSE$$

Each mean square is, of course, the ratio SS/df. For testing H_{0C}: $\delta_1 = \delta_2 = \cdots = \delta_N = 0$, the test statistic value is $f_C = MSC/MSE$, with H_{0C} rejected if $f_C \geq F_{\alpha,N-1,(N-1)(N-2)}$. The other two main effect null hypotheses are also rejected if the corresponding F ratio exceeds $F_{\alpha,N-1,(N-1)(N-2)}$.

If any of the null hypotheses is rejected, significant differences can be identified by using Tukey's procedure. After computing $w = Q_{\alpha,N,(N-1)(N-2)} \cdot \sqrt{MSE/N}$, pairs of sample means (the $\bar{x}_{i..}$'s, $\bar{x}_{.j.}$'s, or $\bar{x}_{..k}$'s) differing by more than w correspond to significant differences between associated factor effects (the α_i's, β_j's, or δ_k's).

The hypothesis H_{0C} is frequently the one of central interest, and a Latin square design is used to control for extraneous variation in the A and B factors, as was done by a randomized block design for the case of a single extraneous factor. Thus, in the product sales example mentioned earlier, variation due to both stores and time periods is controlled by a Latin square design, enabling an investigator to test for the presence of effects due to different product display configurations.

Example 11.12 In an experiment to investigate the effect of relative humidity on abrasion resistance of leather cut from a rectangular pattern ("The Abrasion of Leather," *J. Inter. Soc. Leather Trades' Chemists*, 1946, p. 287), a 6 × 6 Latin square was used to control for possible variability due to row and column position in the pattern. The six levels of relative humidity studied were 1 = 25%, 2 = 37%, 3 = 50%, 4 = 62%, 5 = 75%, and 6 = 87%, with the following results:

		\multicolumn{6}{c}{B (columns)}					
		1	2	3	4	5	6	$x_{i..}$
	1	[3]7.38	[4]5.39	[6]5.03	[2]5.50	[5]5.01	[1]6.79	35.10
	2	[2]7.15	[1]8.16	[5]4.96	[4]5.78	[3]6.24	[6]5.06	37.35
	3	[4]6.75	[6]5.64	[3]6.34	[5]5.31	[1]7.81	[2]8.05	39.90
A (rows)	4	[1]8.05	[3]6.45	[2]6.31	[6]5.46	[4]6.05	[5]5.51	37.83
	5	[6]5.65	[5]5.44	[1]7.27	[3]6.54	[2]7.03	[4]5.96	37.89
	6	[5]6.00	[2]6.55	[4]5.93	[1]8.02	[6]5.80	[3]6.61	38.91
	$x_{.j.}$	40.98	37.63	35.84	36.61	37.94	37.98	

Also, $x_{..1} = 46.10$, $x_{..2} = 40.59$, $x_{..3} = 39.56$, $x_{..4} = 35.86$, $x_{..5} = 32.23$, $x_{..6} = 32.64$, $x_{...} = 226.98$, and $\Sigma_i \Sigma_j x_{ij(k)}^2 = 1462.89$. Further computations are summarized in Table 11.11.

Table 11.11 ANOVA table for Example 11.12

Source of Variation	df	Sum of Squares	Mean Square	f
A (rows)	5	2.19	.438	2.50
B (columns)	5	2.57	.514	2.94
C (treatments)	5	23.53	4.706	26.89
Error	20	3.49	.175	
Total	35	31.78		

Since $F_{.05,5,20} = 2.71$ and $26.89 \geq 2.71$, H_{0C} is rejected in favor of the hypothesis that relative humidity does on average affect abrasion resistance.

To apply Tukey's procedure, $w = Q_{.05,6,20} \cdot \sqrt{MSE/6} = 4.45\sqrt{.175/6} = .76$. Ordering the $\bar{x}_{..k}$'s and underscoring yields

75%	87%	62%	50%	37%	25%
5.37	5.44	5.98	6.59	6.77	7.68

In particular, the lowest relative humidity appears to result in a true average abrasion resistance significantly higher than for any other relative humidity studied. ■

Exercises   Section 11.3 (25–35)

25. The output of a continuous extruding machine that coats steel pipe with plastic was studied as a function of the thermostat temperature profile (A, at three levels), type of plastic (B, at three levels), and the speed of the rotating screw that forces the plastic through a tube-forming die (C, at three levels). There were two replications ($L = 2$) at each combination of levels of the factors, yielding a total of 54 observations on output. The sums of squares were $SSA = 14,144.44$, $SSB = 5511.27$, $SSC = 244,696.39$, $SSAB = 1069.62$, $SSAC = 62.67$, $SSBC = 331.67$, $SSE = 3127.50$, and $SST = 270,024.33$.

a. Construct the ANOVA table.

b. Use appropriate F tests to show that none of the F ratios for two- or three-factor interactions is significant at level .05.

c. Which main effects appear significant?

d. With $x_{..1.} = 8242$, $x_{..2.} = 9732$, and $x_{..3.} = 11,210$, use Tukey's procedure to identify significant differences among the levels of factor C.

26. To see whether thrust force in drilling is affected by drilling speed (A), feed rate (B), or material used (C), an experiment using four speeds, three rates, and two materials was performed, with two samples ($L = 2$) drilled at each combination of levels of the three factors. Sums of squares were calculated as follows: $SSA = 19,149.73$, $SSB = 2,589,047.62$, $SSC = 157,437.52$, $SSAB = 53,238.21$, $SSAC = 9033.73$, $SSBC = 91,880.04$, $SSE = 56,819.50$, and $SST = 2,983,164.81$. Construct the ANOVA table and identify significant interactions using $\alpha = .01$. Is there any single factor that appears to have no effect on thrust force (does any factor appear nonsignificant in every effect in which it appears)?

27. The paper "An Analysis of Variance Applied to Screw Machines" (*Industrial Quality Control,* 1956, pp. 8–9) describes an experiment to investigate how the length of steel bars was affected by time of day (*A*), heat treatment applied (*B*), and screw machine used (*C*). The three times were 8:00 A.M., 11:00 A.M., and 3:00 P.M., and there were two treatments and four machines (a $3 \times 2 \times 4$ factorial experiment), resulting in the accompanying data [coded as 1000(length − 4.380), which does not affect the analysis].

B_1

	C_1	C_2	C_3	C_4
A_1	6, 9, 1, 3	7, 9, 5, 5	1, 2, 0, 4	6, 6, 7, 3
A_2	6, 3, 1, −1	8, 7, 4, 8	3, 2, 1, 0	7, 9, 11, 6
A_3	5, 4, 9, 6	10, 11, 6, 4	−1, 2, 6, 1	10, 5, 4, 8

B_1

	C_1	C_2	C_3	C_4
A_1	4, 6, 0, 1	6, 5, 3, 4	−1, 0, 0, 1	4, 5, 5, 4
A_2	3, 1, 1, −2	6, 4, 1, 3	2, 0, −1, 1	9, 4, 6, 3
A_3	6, 0, 3, 7	8, 7, 10, 0	0, −2, 4, −4	4, 3, 7, 0

Sums of squares include *SSAB* = 1.646, *SSAC* = 71.021, *SSBC* = 1.542, *SSE* = 447.500, and *SST* = 1037.833.

a. Construct the ANOVA table for this data.

b. Test to see whether any of the interaction effects are significant at level .05.

c. Test to see whether any of the main effects are significant at level .05 (i.e., H_{0A} versus H_{aA}, etc.).

d. Use Tukey's procedure to investigate significant differences among the four machines.

28. The following summary quantities were computed from an experiment involving four levels of nitrogen (*A*), two times of planting (*B*), and two levels of potassium (*C*) ("Use and Misuse of Multiple Comparison Procedures," *Agronomy J.,* 1977, pp.

205–208). Only one observation (N content, in percentage, of corn grain) was made for each of the 16 combinations of levels.

SSA = .22625, *SSB* = .000025, *SSC* = .0036, *SSAB* = .004325, *SSAC* = .00065, *SSBC* = .000625, *SST* = .2384.

a. Construct the ANOVA table.

b. Assume that there are no three-way interaction effects, so that *MSABC* is a valid estimate of σ^2, and test at level .05 for interaction and main effects.

c. The nitrogen averages are $\bar{x}_{1..} = 1.1200$, $\bar{x}_{2..} = 1.3025$, $\bar{x}_{3..} = 1.3875$, and $\bar{x}_{4..} = 1.4300$. Use Tukey's method to examine differences in percentage N among the nitrogen levels ($Q_{.05,4,3} = 6.82$).

29. The paper "Kolbe–Schmitt Carbonation of 2-Naphthol" (*Industrial and Eng. Chemistry: Process and Design Development,* 1969, pp. 165–173) presented the accompanying data on percentage yield of BON acid as a function of reaction time (1, 2, and 3 hours), temperature (30, 70, and 100°C), and pressure (30, 70, and 100 psi). Assuming that there is no three-factor interaction, so that *SSE* = *SSABC* provides an estimate of σ^2, MINITAB gave the accompanying ANOVA table. Carry out all appropriate tests.

B_1

	C_1	C_2	C_3
A_1	68.5	73.0	68.7
A_2	74.5	75.0	74.6
A_3	70.5	72.5	74.7

B_2

	C_1	C_2	C_3
A_1	72.8	80.1	72.0
A_2	72.0	81.5	76.0
A_3	69.5	84.5	76.0

B_3

	C_1	C_2	C_3
A_1	72.5	72.5	73.1
A_2	75.5	70.0	76.0
A_3	65.0	66.5	70.5

```
Analysis of Variance for Yield
Source         DF        SS        MS        F         P
time            2      42.112    21.056     8.76     0.010
temp            2     110.732    55.366    23.04     0.000
press           2      68.136    34.068    14.18     0.002
time*temp       4      67.761    16.940     7.05     0.010
time*press      4      35.184     8.796     3.66     0.056
temp*press      4     136.437    34.109    14.20     0.001
Error           8      19.223     2.403
Total          26     479.585
```

30. When factors A and B are fixed but factor C is random, the expected mean squares are

$$E(MSE) = \sigma^2$$

$$E(MSA) = \sigma^2 + JL\sigma_{AC}^2 + \frac{JKL}{I-1}\Sigma\alpha_i^2$$

$$E(MSB) = \sigma^2 + IL\sigma_{BC}^2 + \frac{IKL}{J-1}\Sigma\beta_j^2$$

$$E(MSC) = \sigma^2 + IJL\sigma_C^2$$

$$E(MSAB) = \sigma^2 + L\sigma_{ABC}^2 + \frac{KL}{(I-1)(J-1)}\Sigma_i\Sigma_j(\gamma_{ij}^{AB})^2$$

$$E(MSAC) = \sigma^2 + JL\sigma_{AC}^2$$

$$E(MSBC) = \sigma^2 + IL\sigma_{BC}^2$$

$$E(MSABC) = \sigma^2 + L\sigma_{ABC}^2$$

a. Based on these expected mean squares, what F ratios would you use to test H_0: $\sigma_{ABC}^2 = 0$; H_0: $\sigma_C^2 = 0$; H_0: $\gamma_{ij}^{AB} = 0$ for all i, j; and H_0: $\alpha_1 = \cdots = \alpha_I = 0$?

b. In an experiment to assess the effects of age, type of soil, and day of production on compressive strength of cement/soil mixtures, two ages (A), four types of soil (B), and 3 days $(C,$ assumed random) were used, with $L = 2$ observations made for each combination of factor levels. The resulting sums of squares were $SSA = 14,318.24$, $SSB = 9656.40$, $SSC = 2270.22$, $SSAB = 3408.93$, $SSAC = 1442.58$, $SSBC = 3096.21$, $SSABC = 2832.72$, and $SSE = 8655.60$. Obtain the ANOVA table and carry out all tests using level .01.

31. Because of potential variability in aging due to different castings and segments on the castings, a Latin square design with $N = 7$ was used to investigate the effect of heat treatment on aging. With A = castings, B = segments, C = heat treatments, summary statistics include $x_{...} = 3815.8$, $\Sigma x_{i..}^2 = 297,216.90$,

$\Sigma x_{.j.}^2 = 297,200.64$, $\Sigma x_{..k}^2 = 297,155.01$, and $\Sigma\Sigma x_{ij(k)}^2 = 297,317.65$. Compute the ANOVA table and test at level .05 the hypothesis that heat treatment has no effect on aging.

32. The article "The Responsiveness of Food Sales to Shelf Space Requirements" (*J. Marketing Research,* 1964, pp. 63–67) reports the use of a Latin square design to investigate the effect of shelf space on food sales. The experiment was carried out over a 6-week period using six different stores, resulting in the following data on sales of powdered coffee cream (with shelf space index in parentheses).

		Weeks		
		1	*2*	*3*
Stores	1	27 (5)	14 (4)	18 (3)
	2	34 (6)	31 (5)	34 (4)
	3	39 (2)	67 (6)	31 (5)
	4	40 (3)	57 (1)	39 (2)
	5	15 (4)	15 (3)	11 (1)
	6	16 (1)	15 (2)	14 (6)

		Weeks		
		4	*5*	*6*
Stores	1	35 (1)	28 (6)	22 (2)
	2	46 (3)	37 (2)	23 (1)
	3	49 (4)	38 (1)	48 (3)
	4	70 (6)	37 (4)	50 (5)
	5	9 (2)	18 (5)	17 (6)
	6	12 (5)	19 (3)	22 (4)

Construct the ANOVA table, and state and test at level .01 the hypothesis that shelf space does not affect sales against the appropriate alternative.

33. The article "Variation in Moisture and Ascorbic Acid Content from Leaf to Leaf and Plant to Plant in Turnip Greens" (*Southern Cooperative Services Bull.,* 1951, pp. 13–17) uses a Latin square design in which factor A is plant, factor B is leaf size (smallest to largest), factor C (in parentheses) is time of weighing, and the response variable is moisture content.

Leaf Size (B)

		1	2	3
	1	6.67 (5)	7.15 (4)	8.29 (1)
	2	5.40 (2)	4.77 (5)	5.40 (4)
Plant (A)	3	7.32 (3)	8.53 (2)	8.50 (5)
	4	4.92 (1)	5.00 (3)	7.29 (2)
	5	4.88 (4)	6.16 (1)	7.83 (3)

Leaf Size (B)

		4	5
	1	8.95 (3)	9.62 (2)
	2	7.54 (1)	6.93 (3)
Plant (A)	3	9.99 (4)	9.68 (1)
	4	7.85 (5)	7.08 (4)
	5	5.83 (2)	8.51 (5)

When all three factors are random, the expected mean squares are $E(MSA) = \sigma^2 + N\sigma_A^2$, $E(MSB) = \sigma^2 + N\sigma_B^2$, $E(MSC) = \sigma^2 + N\sigma_C^2$, and $E(MSE) = \sigma^2$. This implies that the F ratios for testing H_{0A}: $\sigma_A^2 = 0$, H_{0B}: $\sigma_B^2 = 0$, and H_{0C}: $\sigma_C^2 = 0$ are identical to those for fixed effects. Obtain the ANOVA table and test at level .05 to see if there is any variation in moisture content due to the factors.

34. The paper "An Assessment of the Effects of Treatment, Time, and Heat on the Removal of Erasable Pen Marks from Cotton and Cotton/Polyester Blend Fabrics (*J. Testing and Eval.*, 1991, pp. 394–397) reports the following sums of squares for the response variable *degree of removal of marks:* $SSA = 39.171$, $SSB = .665$, $SSC = 21.508$, $SSAB = 1.432$, $SSAC = 15.953$, $SSBC = 1.382$, $SSABC = 9.016$, and $SSE = 115.820$. Four different laundry treatments, three different types of pen, and six different fabrics were used in the experiment, and there were three observations for each treatment–pen–fabric combination. Perform an analysis of variance using $\alpha = .01$ for each test and state your conclusions (assume fixed effects for all three factors).

35. A four-factor ANOVA experiment was carried out to investigate the effects of fabric (A), type of exposure (B), level of exposure (C), and fabric direction (D) on extent of color change in exposed fabric as measured by a spectrocolorimeter. Two observations were made for each of the three fabrics, two types, three levels, and two directions, resulting in $MSA = 2207.329$, $MSB = 47.255$, $MSC = 491.783$, $MSD = .044$, $MSAB = 15.303$, $MSAC = 275.446$, $MSAD = .470$, $MSBC = 2.141$, $MSBD = .273$, $MSCD = .247$, $MSABC = 3.714$, $MSABD = 4.072$, $MSACD = .767$, $MSBCD = .280$, $MSE = .977$, and $MST = 93.621$ ("Accelerated Weathering of Marine Fabrics," *J. Testing and Eval.*, 1992, pp. 139–143). Assuming fixed effects for all factors, carry out an analysis of variance using $\alpha = .01$ for all tests and summarize your conclusions.

11.4   # 2^p Factorial Experiments

If an experimenter wishes to study simultaneously the effect of p different factors on a response variable and the factors have $I_1, I_2, \ldots, I_p$ levels, respectively, then a complete experiment requires at least $I_1 \cdot I_2 \cdots \cdot I_p$ observations. In such situations, the experimenter can often perform a "screening experiment" with each factor at only two levels to obtain preliminary information about factor effects. An experiment in which there are p factors, each at two levels, is referred to as a **2^p factorial experiment.** The analysis of data from such an experiment is computationally simpler than for more general factorial experiments. In addition, a 2^p experiment provides a simple setting for introducing the important concepts of confounding and fractional replications.

2^3 Experiments

As in Section 11.3, we let X_{ijkl} and x_{ijkl} refer to the observation from the lth replication with factors A, B, and C at levels i, j, and k, respectively. The model for this situation is

$$X_{ijkl} = \mu + \alpha_i + \beta_j + \delta_k + \gamma_{ij}^{AB} + \gamma_{ik}^{AC} + \gamma_{jk}^{BC} + \gamma_{ijk} + \epsilon_{ijkl} \qquad (11.14)$$

for $i = 1, 2$; $j = 1, 2$; $k = 1, 2$; $l = 1, \dots, n$. The ϵ_{ijkl}'s are assumed independent, normally distributed, with mean 0 and variance σ^2. Because there are only two levels of each factor, the side conditions on the parameters of (11.14) that uniquely specify the model are simply stated: $\alpha_1 + \alpha_2 = 0, \dots, \gamma_{11}^{AB} + \gamma_{21}^{AB} = 0$, $\gamma_{12}^{AB} + \gamma_{22}^{AB} = 0$, $\gamma_{11}^{AB} + \gamma_{12}^{AB} = 0$, $\gamma_{21}^{AB} + \gamma_{22}^{AB} = 0$, and the like. These conditions imply that there is only one functionally independent parameter of each type (for each main effect and interaction). For example, $\alpha_2 = -\alpha_1$, while $\gamma_{21}^{AB} = -\gamma_{11}^{AB}$, $\gamma_{12}^{AB} = -\gamma_{11}^{AB}$, and $\gamma_{22}^{AB} = \gamma_{11}^{AB}$. Because of this, each sum of squares in the analysis will have 1 df.

The parameters of the model can be estimated by taking averages over various subscripts of the X_{ijkl}'s and then forming appropriate linear combinations of the averages. For example,

$$\hat{\alpha}_1 = \bar{X}_{1\dots} - \bar{X}_{\dots}$$
$$= \frac{(X_{111\cdot} + X_{121\cdot} + X_{112\cdot} + X_{122\cdot} - X_{211\cdot} - X_{212\cdot} - X_{221\cdot} - X_{222\cdot})}{8n}$$

and

$$\hat{\gamma}_{11}^{AB} = \bar{X}_{11\cdot\cdot} - \bar{X}_{1\dots} - \bar{X}_{\cdot1\cdot\cdot} + \bar{X}_{\dots}$$
$$= \frac{(X_{111\cdot} - X_{121\cdot} - X_{211\cdot} + X_{221\cdot} + X_{112\cdot} - X_{122\cdot} - X_{212\cdot} + X_{222\cdot})}{8n}$$

Each estimator is, except for the factor $1/(8n)$, a linear function of the cell totals ($X_{ijk\cdot}$'s) in which each coefficient is $+1$ or -1, with an equal number of each; such functions are called **contrasts** in the $X_{ijk\cdot}$'s. Furthermore, the estimators satisfy the same side conditions satisfied by the parameters themselves. For example,

$$\hat{\alpha}_1 + \hat{\alpha}_2 = \bar{X}_{1\dots} - \bar{X}_{\dots} + \bar{X}_{2\dots} - \bar{X}_{\dots} = \bar{X}_{1\dots} + \bar{X}_{2\dots} - 2\bar{X}_{\dots}$$
$$= \frac{1}{4n}X_{1\dots} + \frac{1}{4n}X_{2\dots} - \frac{2}{8n}X_{\dots} = \frac{1}{4n}X_{\dots} - \frac{1}{4n}X_{\dots} = 0$$

Example 11.13 In an experiment to investigate the compressive strength properties of cement–soil mixtures, two different aging periods were used in combination with two different aging temperatures and two different soils. Two replications were made for each combination of levels of the three factors, resulting in the following data:

Age	Temperature	Soil 1	Soil 2
1	1	471, 413	385, 434
	2	485, 552	530, 593
2	1	712, 637	770, 705
	2	712, 789	741, 806

The computed cell totals are $x_{111.} = 884$, $x_{211.} = 1349$, $x_{121.} = 1037$, $x_{221.} = 1501$, $x_{112.} = 819$, $x_{212.} = 1475$, $x_{122.} = 1123$, and $x_{222.} = 1547$, so $x_{....} = 9735$. Then

$$\hat{\alpha}_1 = (884 - 1349 + 1037 - 1501 + 819 - 1475 + 1123 - 1547)/16$$

$$= -125.5625 = -\hat{\alpha}_2$$

$$\hat{\gamma}_{11}^{AB} = (884 - 1349 - 1037 + 1501 + 819 - 1475 - 1123 + 1547)/16$$

$$= -14.5625 = -\hat{\gamma}_{12}^{AB} = -\hat{\gamma}_{21}^{AB} = \hat{\gamma}_{22}^{AB}$$

The other parameter estimates can be computed in the same manner. ∎

Sums of Squares and Analysis for a 2^3 Experiment

The reason for computing parameter estimates is that sums of squares for the various effects are easily obtained from the estimates. For example,

$$SSA = \sum_i \sum_j \sum_k \sum_l \hat{\alpha}_i^2 = 4n \sum_{i=1}^{2} \hat{\alpha}_i^2 = 4n[\hat{\alpha}_1^2 + (-\hat{\alpha}_1)^2] = 8n\hat{\alpha}_1^2$$

and

$$SSAB = \sum_i \sum_j \sum_k \sum_l (\hat{\gamma}_{ij}^{AB})^2$$

$$= 2n \sum_{i-1}^{2} \sum_{j=1}^{2} (\hat{\gamma}_{ij}^{AB})^2 = 2n[(\hat{\gamma}_{11}^{AB})^2 + (-\hat{\gamma}_{11}^{AB})^2 + (-\hat{\gamma}_{11}^{AB})^2 + (\hat{\gamma}_{11}^{AB})^2]$$

$$= 8n(\hat{\gamma}_{11}^{AB})^2$$

Since each estimate is a contrast in the cell totals multiplied by $1/(8n)$, each sum of squares has the form (contrast)$^2/(8n)$. Thus, to compute the various sums of squares, we need to know the coefficients ($+1$ or -1) of the appropriate contrasts. The signs ($+$ or $-$) on each $x_{ijk.}$ in each effect contrast are most conveniently displayed in a table. We will use the notation (1) for the experimental condition $i = 1, j = 1$, $k = 1$, a for $i = 2, j = 1, k = 1$, ab for $i = 2, j = 2, k = 1$, and so on. If level 1 is thought of as "low" and level 2 as "high," any letter that appears denotes a high level of the associated factor. In Table 11.12, each column gives the signs for a particular effect contrast in the $x_{ijk.}$'s associated with the different experimental conditions.

In each of the first three columns, the sign is $+$ if the corresponding factor is at the high level and $-$ if it is at the low level. Every sign in the AB column is then the "product" of the signs in the A and B columns, with $(+)(+) = (-)(-) = +$ and $(+)(-) = (-)(+) = -$, and similarly for the AC and BC columns. Finally, the signs in

Table 11.12 Signs for computing effect contrasts

Experimental Condition	Cell Total	Factorial Effect						
		A	B	C	AB	AC	BC	ABC
(1)	$x_{111\cdot}$	−	−	−	+	+	+	−
a	$x_{211\cdot}$	+	−	−	−	−	+	+
b	$x_{121\cdot}$	−	+	−	−	+	−	+
ab	$x_{221\cdot}$	+	+	−	+	−	−	−
c	$x_{112\cdot}$	−	−	+	+	−	−	+
ac	$x_{212\cdot}$	+	−	+	−	+	−	−
bc	$x_{122\cdot}$	−	+	+	−	−	+	−
abc	$x_{222\cdot}$	+	+	+	+	+	+	+

the ABC column are the products of AB with C (or B with AC or A with BC). Thus, for example,

$$AC \text{ contrast} = +\,x_{111\cdot} - x_{211\cdot} + x_{121\cdot} - x_{221\cdot} - x_{112\cdot} + x_{212\cdot} - x_{122\cdot} + x_{222\cdot}.$$

Once the seven effect contrasts are computed,

$$SS(\text{effect}) = \frac{(\text{effect contrast})^2}{8n}$$

Even with a table of signs, calculation of the contrasts is tedious. An efficient computational technique, due to Yates, is as follows. Write in a column the eight cell totals in the **standard order** as given in the table of signs and establish three further columns. In each of these three columns, the first four entries are the sums of entries 1 and 2, 3 and 4, 5 and 6, 7 and 8 of the previous columns. The last four entries are the differences between entries 2 and 1, 4 and 3, 6 and 5, and 8 and 7 of the previous column. The last column then contains $x_{\cdots}$ and the seven effect contrasts in standard order. Squaring each contrast and dividing by $8n$ then gives the seven sums of squares.

Example 11.14
(Example 11.13 continued)

Since $n = 2$, $8n = 16$. Yates's method is illustrated in Table 11.13.

Table 11.13 Yates's method of computation

Treatment Condition	$x_{ijk\cdot}$	1	2	Effect Contrast	$SS = (\text{contrast})^2/16$
(1) $= x_{111\cdot}$	884	2233	4771	9735	
$a = x_{211\cdot}$	1349	2538	4964	2009	252,255.06
$b = x_{121\cdot}$	1037	2294	929	681	28,985.06
$ab = x_{221\cdot}$	1501	2670	1080	−233	3,393.06
$c = x_{112\cdot}$	819	465	305	193	2,328.06
$ac = x_{212\cdot}$	1475	464	376	151	1,425.06
$bc = x_{122\cdot}$	1123	656	−1	71	315.06
$abc = x_{222\cdot}$	1547	424	−232	−231	3,335.06
					292,036.42

From the original data, $\Sigma_i\Sigma_j\Sigma_k\Sigma_l x_{ijkl}^2 = 6{,}232{,}289$, and

$$\frac{x_{\cdots}^2}{16} = 5{,}923{,}139.06$$

so

$$SST = 6{,}232{,}289 - 5{,}923{,}139.06 = 309{,}149.94$$
$$SSE = SST - [SSA + \cdots + SSABC] = 309{,}149.94 - 292{,}036.42$$
$$= 17{,}113.52$$

The ANOVA calculations are summarized in Table 11.14.

Table 11.14 ANOVA table for Example 11.14

Source of Variation	df	Sum of Squares	Mean Square	f
A	1	252,255.06	252,255.06	117.92
B	1	28,985.06	28,985.06	13.55
C	1	2,328.06	2,328.06	1.09
AB	1	3,393.06	3,393.06	1.59
AC	1	1,425.06	1,425.06	.67
BC	1	315.06	315.06	.15
ABC	1	3,335.06	3,335.06	1.56
Error	8	17,113.52	2,139.19	
Total	15	309,149.94		

Figure 11.6 shows SAS output for this example. Only the P-values for age (A) and temperature (B) are less than .01, so only these effects are judged significant.   ∎

2^p Experiments for $p > 3$

Although the computations when $p > 3$ are quite tedious, the analysis parallels that of the three-factor case. For example, if there are four factors A, B, C, and D, there are 16 different experimental conditions. The first 8 in standard order are exactly those already listed for a three-factor experiment, while the second 8 are obtained by placing the letter d beside each condition in the first group. Yates's method is then initiated by computing totals across replications, listing these totals in standard order, and proceeding as before; with p factors, the pth column to the right of the treatment totals will give the effect contrasts.

For $p > 3$, there will often be no replications of the experiment (so only one complete replicate is available). One possible way to test hypotheses is to assume that certain higher-order effects are absent and then add the corresponding sums of squares to obtain an *SSE*. Such an assumption can, however, be misleading in the absence of prior knowledge (see p. 274 of the book by Montgomery listed

Analysis of Variance Procedure

Dependent Variable: STRENGTH

Source	DF	Sum of Squares	Mean Square	F Value	Pr > F
Model	7	292036.4375	41719.4911	19.50	0.0002
Error	8	17113.5000	2139.1875		
Corrected Total	15	309149.9375			

R-Square	C.V.	Root MSE	POWERUSE Mean
0.944643	7.601660	46.25135	608.437500

Source	DF	Anova SS	Mean Square	F Value	Pr > F
AGE	1	252255.0625	252255.0625	117.92	0.0001
TEMP	1	28985.0625	28985.0625	13.55	0.0062
AGE*TEMP	1	3393.0625	3393.0625	1.59	0.2434
SOIL	1	2328.0625	2328.0625	1.09	0.3273
AGE*SOIL	1	1425.0625	1425.0625	0.67	0.4380
TEMP*SOIL	1	315.0625	315.0625	0.15	0.7111
AGE*TEMP*SOIL	1	3335.0625	3335.0625	1.56	0.2471

Figure 11.6 SAS output for strength data

in the chapter bibliography). An alternative approach involves working directly with the effect contrasts. Each contrast has a normal distribution with the same variance. When a particular effect is absent, the expected value of the corresponding contrast is 0, but this is not so when the effect is present. The suggested method of analysis is to construct a normal probability plot of the effect contrasts (or, equivalently, the effect parameter estimates, since estimate = contrast/2^p when $n = 1$). Points corresponding to absent effects will tend to fall close to a straight line, whereas points associated with substantial effects will typically be far from this line.

Example 11.15 The accompanying data appeared in the paper "Quick and Easy Analysis of Unreplicated Factorials" (*Technometrics,* 1989, pp. 469–473). The four factors are A = acid strength, B = time, C = amount of acid, and D = temperature, and the response variable is the yield of isatin. The observations, in standard order, are .08, .04, .53, .43, .31, .09, .12, .36, .79, .68, .73, .08, .77, .38, .49, and .23. Table 11.15 displays the effect estimates as given in the paper (which used contrast/8 rather than contrast/16).

Table 11.15 Effect estimates for Example 11.15

Effect estimate	A	B	AB	C	AC	BC	ABC	D
	−.191	−.021	−.001	−.076	.034	−.066	.149	.274

Effect estimate	AD	BD	ABD	CD	ACD	BCD	$ABCD$
	−.161	−.251	−.101	−.026	−.006	.124	.019

A normal probability plot of the effect estimates, obtained from MINITAB, appears as Figure 11.7. All points in the plot fall close to the same straight line, suggesting complete absence of any effects (we will shortly give an example in which this is not the case).

Figure 11.7 A normal probability plot of effect estimates from Example 11.15   ■

Visual judgments of deviation from straightness in a normal probability plot are rather subjective. The paper cited in Example 11.15 describes a more objective technique for identifying significant effects in an unreplicated experiment.

Confounding

It is often not possible to carry out all 2^P experimental conditions of a 2^P factorial experiment in a homogeneous experimental environment. In such situations, it may be possible to separate the experimental conditions into 2^r homogeneous blocks ($r < p$), so that there are 2^{p-r} experimental conditions in each block. The blocks may, for example, correspond to different laboratories, different time periods, or different operators or work crews. In the simplest case, $p = 3$ and $r = 1$, so that there are two blocks with each block consisting of four of the eight experimental conditions.

As always, blocking is effective in reducing variation associated with extraneous sources. However, when the 2^P experimental conditions are placed in 2^r blocks, the price paid for this blocking is that $2^r - 1$ of the factor effects cannot be estimated. This is because $2^r - 1$ factor effects (main effects and/or interactions) are mixed up or **confounded** with the block effects. The allocation of experimental conditions to blocks is then usually done so that only higher-level interactions are confounded, while main effects and low-order interactions remain estimable and can be tested for.

To see how allocation to blocks is accomplished, consider first a 2^3 experiment with two blocks ($r = 1$) and four treatments per block. Suppose we select ABC as the effect to be confounded with blocks. Then any experimental condition having an odd number of letters in common with ABC, such as b (one letter) or abc (three letters), is placed in one block, while any condition having an even number of letters in common with ABC (where 0 is even) goes in the other block. Figure 11.8 shows this allocation of treatments to the two blocks.

Figure 11.8 Confounding ABC in a 2^3 experiment

In the absence of replications, the data from such an experiment would usually be analyzed by assuming that there were no two-factor interactions (additivity) and using $SSE = SSAB + SSAC + SSBC$ with 3 df to test for the presence of main effects. Alternatively, a normal probability plot of effect contrasts or effect parameter estimates could be examined. Most frequently, though, there are replications when just three factors are being studied. Suppose there are u replicates, resulting in a total of $2^r \cdot u$ blocks in the experiment. Then after subtracting from SST all sums of squares associated with effects not confounded with blocks (computed using Yates's method), the block sum of squares is computed using the $2^r \cdot u$ block totals and then subtracted to yield SSE (so there are $2^r \cdot u - 1$ df for blocks).

Example 11.16   The article "Factorial Experiments in Pilot Plant Studies" (*Industrial and Eng. Chemistry*, 1951, pp. 1300–1306) reports the results of an experiment to assess the effects of reactor temperature (A), gas throughput (B), and concentration of active constituent (C) on strength of the product solution (measured in arbitrary units) in a recirculation unit. Two blocks were used, with the ABC effect confounded with blocks, and there were two replications, resulting in the data in Figure 11.9. The four block × replication totals are 288, 212, 88, and 220, with a grand total of 808, so

$$SSBl = \frac{(288)^2 + (212)^2 + (88)^2 + (220)^2}{4} - \frac{(808)^2}{16} = 5204.00$$

	Replication 1				Replication 2		
Block 1		Block 2		Block 1		Block 2	
(1)	99	a	18	(1)	46	a	18
ab	52	b	51	ab	−47	b	62
ac	42	c	108	ac	22	c	104
bc	95	abc	35	bc	67	abc	36

Figure 11.9 Data for Example 11.16

The other sums of squares are computed by Yates's method using the eight experimental condition totals, resulting in the ANOVA table given as Table 11.16. By comparison with $F_{.05,1,6} = 5.99$, we conclude that only the main effects for A and C differ significantly from zero.

Table 11.16 ANOVA table for Example 11.16

Source of Variation	df	Sum of Squares	Mean Square	f
A	1	12,996	12,996	39.82
B	1	702.25	702.25	2.15
C	1	2,756.25	2,756.25	8.45
AB	1	210.25	210.25	.64
AC	1	30.25	30.25	.093
BC	1	25	25	.077
Blocks	3	5,204	1,734.67	5.32
Error	6	1,958	326.33	
Total	15	23,882		

■

Confounding Using More Than Two Blocks

In the case $r = 2$ (four blocks), three effects are confounded with blocks. The experimenter first chooses two defining effects to be confounded. For example, in a five-factor experiment (A, B, C, D, and E), the two three-factor interactions BCD and CDE might be chosen for confounding. The third effect confounded is then the **generalized interaction** of the two, obtained by writing the two chosen effects side by side and then cancelling any letters common to both: $(BCD)(CDE) = BE$. Notice that if ABC and CDE are chosen for confounding, their generalized interaction is $(ABC)(CDE) = ABDE$, so that no main effects or two-factor interactions are confounded.

Once the two defining effects have been selected for confounding, one block consists of all treatment conditions having an even number of letters in common with both defining effects, the second block consists of all conditions having an even number of letters in common with the first defining contrast and an odd number of letters in common with the second contrast, while the third and fourth blocks consist of the "odd/even" and "odd/odd" contrasts. In a five-factor experiment with defining effects ABC and CDE, this results in the allocation to blocks as shown in Figure 11.10 (with the number of letters in common with each defining contrast appearing beside each experimental condition).

The block containing (1) is called the **principal block.** Once it has been constructed, a second block can be obtained by selecting any experimental condition not in the principal block and obtaining its generalized interaction with every condition in the principal block. The other blocks are then constructed in the same way by first selecting a condition not in a block already constructed and finding generalized interactions with the principal block.

	Block 1			Block 2			Block 3			Block 4	
(1)	(0, 0)		d	(0, 1)		a	(1, 0)		c	(1, 1)	
ab	(2, 0)		e	(0, 1)		b	(1, 0)		ad	(1, 1)	
de	(0, 2)		ac	(2, 1)		cd	(1, 2)		ae	(1, 1)	
acd	(2, 2)		bc	(2, 1)		ce	(1, 2)		bd	(1, 1)	
ace	(2, 2)		abd	(2, 1)		ade	(1, 2)		be	(1, 1)	
bcd	(2, 2)		abe	(2, 1)		bde	(1, 2)		abc	(3, 1)	
bce	(2, 2)		acde	(2, 3)		abcd	(3, 2)		cde	(1, 3)	
abde	(2, 2)		bcde	(2, 3)		abce	(3, 2)		abcde	(3, 3)	

Figure 11.10 Four blocks in a 2^5 factorial experiment with defining effects ABC and CDE

For experimental situations with $p > 3$, there is often no replication, so sums of squares associated with nonconfounded higher-order interactions are usually pooled to obtain an error sum of squares that can be used in the denominators of the various F statistics. All computations can again be carried out using Yates's technique, with $SSBl$ being the sum of sums of squares associated with confounded effects.

When $r > 2$, one first selects r defining effects to be confounded with blocks, making sure that no one of the effects chosen is the generalized interaction of any other two selected. The additional $2^r - r - 1$ effects confounded with the blocks are then the generalized interactions of all effects in the defining set (including not only generalized interactions of pairs of effects, but also of sets of three, four, and so on). For more details the book by Montgomery can be consulted.

Fractional Replication

When the number p of factors is large, even a single replicate of a 2^p experiment can be expensive and time-consuming. For example, one replicate of a 2^6 factorial experiment involves an observation for each of the 64 different experimental conditions. An appealing strategy in such situations is to make observations for only a fraction of the 2^p conditions. Provided that care is exercised in the choice of conditions to be observed, much information about factor effects can still be obtained.

Suppose we decide to include only 2^{p-1} (half) of the 2^p possible conditions in our experiment; this is usually called a **half-replicate.** The price paid for this economy is twofold. First, information about a single effect (determined by the 2^{p-1} conditions selected for observation) is completely lost to the experimenter in the sense that no reasonable estimate of the effect is possible. Second, the remaining $2^p - 2$ main effects and interactions are paired up so that any one effect in a particular pair is confounded with the other effect in the same pair. For example, one such pair may be $\{A, BCD\}$, so that separate estimates of the A main effect and BCD interaction are not possible. It is desirable, then, to select a half-replicate for which main effects and low-order interactions are paired off (confounded) only with higher-order interactions rather than with one another.

The first step in selecting a half-replicate is to select a defining effect as the nonestimable effect. Suppose that in a five-factor experiment, $ABCDE$ is chosen as the defining effect. Now the $2^5 = 32$ possible treatment conditions are divided into two groups with 16 conditions each, one group consisting of all conditions having an odd number of letters in common with $ABCDE$ and the other containing an even number of letters in common with the defining contrast. Then either group of 16 conditions is used as the half-replicate. The "odd" group is

a, b, c, d, e, abc, abd, abe, acd, ace, ade, bcd, bce, bde, cde, abcde

Each main effect and interaction other than $ABCDE$ is then confounded with (**aliased** with) its generalized interaction with $ABCDE$. Thus $(AB)(ABCDE) = CDE$, so the AB interaction and CDE interaction are confounded with each other. The resulting **alias pairs** are

$\{A, BCDE\}, \quad \{B, ACDE\}, \quad \{C, ABDE\}, \quad \{D, ABCE\}, \quad \{E, ABCD\},$
$\{AB, CDE\}, \quad \{AC, BDE\}, \quad \{AD, BCE\}, \quad \{AE, BCD\}, \quad \{BC, ADE\},$
$\{BD, ACE\}, \quad \{BE, ACD\}, \quad \{CD, ABE\}, \quad \{CE, ABD\}, \quad \{DE, ABC\}$

Note in particular that every main effect is aliased with a four-factor interaction. Assuming these interactions to be negligible allows for testing for the presence of main effects.

To select a quarter-replicate of a 2^p factorial experiment (2^{p-2} of the 2^p possible treatment conditions), two defining effects must be selected. These two and their generalized interaction become the nonestimable effects. Instead of alias pairs as in the half-replicate, each remaining effect is now confounded with three other effects, each being its generalized interaction with one of the three nonestimable effects.

Example 11.17   The article "More on Planning Experiments to Increase Research Efficiency" (*Industrial and Eng. Chemistry,* 1970, pp. 60–65) reports on the results of a quarter-replicate of a 2^5 experiment in which the five factors were A = condensation temperature, B = amount of material B, C = solvent volume, D = condensation time, and E = amount of material E. The response variable was the yield of the chemical process. The chosen defining contrasts were ACE and BDE, with generalized interaction $(ACE)(BDE) = ABCD$. The remaining 28 main effects and interactions can now be partitioned into seven groups of four effects each such that the effects within a group cannot be assessed separately. For example, the generalized interactions of A with the nonestimable effects are $(A)(ACE) = CE$, $(A)(BDE) = ABDE$, and $(A)(ABCD) = BCD$, so one alias group is $\{A, CE, ABDE, BCD\}$. The complete set of alias groups is

$\{A, CE, ABDE, BCD\}, \quad \{B, ABCE, DE, ACD\}, \quad \{C, AE, BCDE, ABD\},$
$\{D, ACDE, BE, ABC\}, \quad \{E, AC, BD, ABCDE\}, \quad \{AB, BCE, ADE, CD\},$
$\{AD, CDE, ABE, BC\}$ ∎

Analysis of a Fractional Replicate

Once the defining contrasts have been chosen for a quarter-replicate, they are used as in the discussion of confounding to divide the 2^p treatment conditions into four groups of 2^{p-2} conditions each. Then any one of the four groups is selected as the set of conditions for which data will be collected. Similar comments apply to a $1/2^r$ replicate of a 2^p factorial experiment.

Having made observations for the selected treatment combinations, a table of signs similar to Table 11.12 is constructed. The table contains a row only for each of the treatment combinations actually observed rather than the full 2^p rows, and there is a single column for each alias group (since each effect in the group would have the same set of signs for the treatment conditions selected for observation). The signs in each column indicate as usual how contrasts for the various sums of squares are computed. Yates's method can also be used, but the rule for arranging observed conditions in standard order must be modified.

The difficult part of a fractional replication analysis typically involves deciding what to use for error sum of squares. Since there will usually be no replication (though one could observe, e.g., two replicates of a quarter-replicate), some effect sums of squares must be pooled to obtain an error sum of squares. In a half-replicate of a 2^8 experiment, for example, an alias structure can be chosen so that the eight main effects and 28 two-factor interactions are each confounded only with higher-order interactions and that there are an additional 27 alias groups involving only higher-order interactions. Assuming the absence of higher-order interaction effects, the resulting 27 sums of squares can then be added to yield an error sum of squares, allowing 1 df tests for all main effects and two-factor interactions. However, in many cases tests for main effects can be obtained only by pooling some or all of the sums of squares associated with alias groups involving two-factor interactions, while the corresponding two-factor interactions cannot be investigated.

Example 11.18
(Example 11.17 continued)

The set of treatment conditions chosen and resulting yields for the quarter-replicate of the 2^5 experiment were

e	*ab*	*ad*	*bc*	*cd*	*ace*	*bde*	*abcde*
23.2	15.5	16.9	16.2	23.8	23.4	16.8	18.1

The abbreviated table of signs is displayed in Table 11.17.

Table 11.17 Table of signs for Example 11.18

	A	B	C	D	E	AB	AD
e	−	−	−	−	+	+	+
ab	+	+	−	−	−	+	−
ad	+	−	−	+	−	−	+
bc	−	+	+	−	−	−	+
cd	−	−	+	+	−	+	−
ace	+	−	+	−	+	−	−
bde	−	+	−	+	+	−	−
abcde	+	+	+	+	+	+	+

With *SSA* denoting the sum of squares for effects in the alias group {*A, CE, ABDE, BCD*},

$$SSA = \frac{(-23.2 + 15.5 + 16.9 - 16.2 - 23.8 + 23.4 - 16.8 + 18.1)^2}{8} = 4.65$$

Similarly, $SSB = 53.56$, $SSC = 10.35$, $SSD = .91$, $SSE' = 10.35$ (the $'$ differentiates this quantity from error sum of squares SSE), $SSAB = 6.66$, and $SSAD = 3.25$, giving $SST = 4.65 + 53.56 + \cdots + 3.25 = 89.73$. To test for main effects, we use $SSE = SSAB + SSAD = 9.91$ with 2 df. The ANOVA table appears as Table 11.18.

Table 11.18 ANOVA table for Example 11.18

Source	df	Sum of Squares	Mean Square	f
A	1	4.65	4.65	.94
B	1	53.56	53.56	10.80
C	1	10.35	10.35	2.09
D	1	.91	.91	.18
E	1	10.35	10.35	2.09
Error	2	9.91	4.96	
Total	7	89.73		

Since $F_{.05,1,2} = 18.51$, none of the five main effects can be judged significant. Of course, with only 2 df for error, the test is not very powerful (i.e., it is quite likely to fail to detect the presence of effects). The article from *Industrial and Engineering Chemistry* from which the data came actually had an independent estimate of the standard error of the treatment effects based on prior experience, so used a somewhat different analysis. Our analysis was done here only for illustrative purposes, since one would ordinarily want many more than 2 df for error. ∎

As an alternative to *F* tests based on pooling sums of squares to obtain *SSE*, a normal probability plot of effect contrasts can be examined.

Example 11.19   An experiment was carried out to investigate shrinkage in the plastic casing material used for speedometer cables ("An Explanation and Critique of Taguchi's Contribution to Quality Engineering," *Quality and Reliability Engr. Intl.*, 1988, pp. 123–131). The engineers started with 15 factors: liner outside diameter, liner die, liner material, liner line speed, wire braid type, braiding tension, wire diameter, liner tension, liner temperature, coating material, coating die type, melt temperature, screen pack, cooling method, and line speed. It was suspected that only a few of these factors were important, so a screening experiment in the form of a 2^{15-11} factorial (a $1/2^{11}$ fraction of a 2^{15} factorial experiment) was carried out. The resulting alias structure is quite complicated; in particular, every main effect is confounded with two-factor interactions. The response variable was the percentage shrinkage for a cable specimen produced at designated levels of the factors.

Figure 11.11 displays a normal probability plot of the effect contrasts. All but two of the points fall quite close to a straight line. The discrepant points correspond to effects E = wire braid type and G = wire diameter, suggesting that these two factors are the only ones that affect the amount of shrinkage.

Figure 11.11 Normal probability plot of contrasts from Example 11.19   ∎

The subjects of factorial experimentation, confounding, and fractional replication encompass many models and techniques we have not discussed. For more information, the chapter references should be consulted.

Exercises   Section 11.4 (36–47)

36. The accompanying data resulted from an experiment to study the nature of dependence of welding current on the three factors: welding voltage, wire feed speed, and tip-to-workpiece distance. There were two levels of each factor (a 2^3 experiment) with two replications per combination of levels (the averages across replications agree with values given in the paper "A Study on Prediction of Welding Current in Gas Metal Arc

Welding," *J. Engr. Manuf.*, 1991, pp. 64–69). The first two given numbers are for the treatment (1), the next two for a, and so on in standard order: 200.0, 204.2, 215.5, 219.5, 272.7, 276.9, 299.5, 302.7, 166.6, 172.6, 186.4, 192.0, 232.6, 240.8, 253.4, 261.6.

a. Verify that the sums of squares are as given in the accompanying ANOVA table from MINITAB.

```
Analysis of Variance for current
Source            DF        SS        MS        F       P
Volt               1    1685.1    1685.1    102.38   0.000
Speed              1   21272.2   21272.2   1292.37   0.000
Dist               1    5076.6    5076.6    308.42   0.000
Volt*speed         1      36.6      36.6      2.22   0.174
Volt*dist          1       0.4       0.4      0.03   0.877
Speed*dist         1     109.2     109.2      6.63   0.033
Volt*speed*dist    1      23.5      23.5      1.43   0.266
Error              8     131.7      16.5
Total             15   28335.3
```

b. Which effects appear to be important, and why?

37. The accompanying data resulted from a 2^3 experiment with three replications per combination of treatments designed to study the effects of concentration of detergent (A), concentration of sodium carbonate (B), and concentration of sodium carboxymethyl cellulose (C) on cleaning ability of a solution in washing tests (a larger number indicates better cleaning ability than a smaller number).

Factor Levels

A	B	C	Condition	Observations
1	1	1	(1)	106, 93, 116
2	1	1	a	198, 200, 214
1	2	1	b	197, 202, 185
2	2	1	ab	329, 331, 307
1	1	2	c	149, 169, 135
2	1	2	ac	243, 247, 220
1	2	2	bc	255, 230, 252
2	2	2	abc	383, 360, 364

a. After obtaining cell totals x_{ijk}, compute estimates of β_1, γ_{11}^{AC}, and γ_{21}^{AC}.

b. Use the cell totals along with Yates's method to compute the effect contrasts and sums of squares. Then construct an ANOVA table and test all appropriate hypotheses using $\alpha = .05$.

38. In a study of processes used to remove impurities from cellulose goods ("Optimization of Rope-Range Bleaching of Cellulosic Fabrics," *Textile Research J.,* 1976, pp. 493–496), the following data resulted from a 2^4 experiment involving the desizing process. The four factors were enzyme concentration (A), pH (B), temperature (C), and time (D).

a. Use Yates's algorithm to obtain sums of squares and the ANOVA table.

b. Do there appear to be any second-, third-, or fourth-order interaction effects present? Explain your reasoning. Which main effects appear to be significant?

Treat-ment	En-zyme (g/l)	pH	Temp. (°C)	Time (h)	Starch % by Weight 1st Repl.	Starch % by Weight 2nd Repl.
(1)	.50	6.0	60.0	6	9.72	13.50
a	.75	6.0	60.0	6	9.80	14.04
b	.50	7.0	60.0	6	10.13	11.27
ab	.75	7.0	60.0	6	11.80	11.30
c	.50	6.0	70.0	6	12.70	11.37
ac	.75	6.0	70.0	6	11.96	12.05
bc	.50	7.0	70.0	6	11.38	9.92
abc	.75	7.0	70.0	6	11.80	11.10
d	.50	6.0	60.0	8	13.15	13.00
ad	.75	6.0	60.0	8	10.60	12.37
bd	.50	7.0	60.0	8	10.37	12.00
abd	.75	7.0	60.0	8	11.30	11.64
cd	.50	6.0	70.0	8	13.05	14.55
acd	.75	6.0	70.0	8	11.15	15.00
bcd	.50	7.0	70.0	8	12.70	14.10
abcd	.75	7.0	70.0	8	13.20	16.12

39. In Exercise 37, suppose a low water temperature has been used to obtain the data. The entire experiment is then repeated with a higher water temperature to obtain the following data. Use Yates's algorithm on the entire set of 48 observations to obtain the sums of squares and ANOVA table, and then test appropriate hypotheses at level .05.

Condition	Observations
d	144, 154, 158
ad	239, 227, 244
bd	232, 242, 246
abd	364, 362, 346
cd	194, 162, 203
acd	284, 295, 291
bcd	291, 287, 297
abcd	411, 406, 395

40. The following data on power consumption in electric furnace heats (kW consumed per ton of melted product) resulted from a 2^4 factorial experiment with three replicates ("Studies on a 10-cwt Arc Furnace," *J. Iron and Steel Institute,* 1956, p. 22). The factors were nature of roof A (low, high), power setting B (low, high), scrap used C (tube, plate), and charge D (700 lb, 1000 lb).

Treat-ment	x_{ijklm}	Treat-ment	x_{ijklm}
(1)	866, 862, 800	d	988, 808, 650
a	946, 800, 840	ad	966, 976, 876
b	774, 834, 746	bd	702, 658, 650
ab	709, 789, 646	abd	784, 700, 596
c	1017, 990, 954	cd	922, 808, 868
ac	1028, 906, 977	acd	1056, 870, 908
bc	817, 783, 771	bcd	798, 726, 700
abc	829, 806, 691	abcd	752, 714, 714

Construct the ANOVA table and test all hypotheses of interest using $\alpha = .01$.

41. The article "Statistical Design and Analysis of Quali-
fication Test Program for a Small Rocket Engine"
(*Industrial Quality Control*, 1964, pp. 14–18) presents
data from an experiment to assess the effects of vibration
(A), temperature cycling (B), altitude cycling (C), and
temperature for altitude cycling and firing (D) on thrust
duration. A subset of the data appears here (in the paper
there were four levels of D rather than just two). Use the
Yates method to obtain sums of squares and the ANOVA
table. Then assume that three- and four-factor interactions are absent, pool the corresponding sums of squares
to obtain an estimate of σ^2, and test all appropriate
hypotheses at level .05.

		D_1		D_2	
		C_1	C_2	C_1	C_2
A_1	B_1	21.60	21.60	11.54	11.50
	B_2	21.09	22.17	11.14	11.32
A_2	B_1	21.60	21.86	11.75	9.82
	B_2	19.57	21.85	11.69	11.18

42. a. In a 2^4 experiment, suppose two blocks are to be
used, and it is decided to confound the $ABCD$
interaction with the block effect. Which treatments
should be carried out in the first block [the one
containing the treatment (1)], and which treatments
are allocated to the second block?

b. In an experiment to investigate niacin retention in
vegetables as a function of cooking temperature (A),
sieve size (B), type of processing (C), and cooking
time (D), each factor was held at two levels. Two
blocks were used, with the allocation of blocks as
given in part (a) in order to confound only the
$ABCD$ interaction with blocks. Use Yates's procedure to obtain the ANOVA table for the accompanying data.

Treatment	x_{ijkl}	Treatment	x_{ijkl}
(1)	91	d	72
a	85	ad	78
b	92	bd	68
ab	94	abd	79
c	86	cd	69
ac	83	acd	75
bc	85	bcd	72
abc	90	$abcd$	71

c. Assume that all three-way interaction effects are
absent, so that the associated sums of squares can
be combined to yield an estimate of σ^2, and carry
out all appropriate tests at level .05.

43. a. An experiment was carried out to investigate the
effects on audio sensitivity of varying resistance
(A), two capacitances (B, C), and inductance
of a coil (D) in part of a television circuit. If
four blocks were used with four treatments per
block, and the defining effects for confounding
were AB and CD, which treatments appeared in
each block?

b. Suppose two replications of the experiment described in part (a) were performed, resulting in the
accompanying data. Obtain the ANOVA table and
test all relevant hypotheses at level .01.

Treat-ment	x_{ijkl1}	x_{ijkl2}	Treat-ment	x_{ijkl1}	x_{ijkl2}
(1)	618	598	d	598	585
a	583	560	ad	587	541
b	477	525	bd	480	508
ab	421	462	abd	462	449
c	601	595	cd	603	577
ac	550	589	acd	571	552
bc	505	484	bcd	502	508
abc	452	451	$abcd$	449	455

44. In an experiment involving four factors (A, B, C, and D)
and four blocks, show that at least one main effect or
two-factor interaction effect must be confounded with
the block effect.

45. a. In a seven-factor experiment ($A, \ldots, G$), suppose
a quarter-replicate is actually carried out. If the
defining effects are $ABCDE$ and $CDEFG$, what is
the third nonestimable effect and what treatments
are in the group containing (1)? What are the alias
groups of the seven main effects?

b. If the quarter-replicate is to be carried out using four
blocks (with eight treatments per block), what are
the blocks if the chosen confounding effects are
ACF and BDG?

46. Suppose that in the rocket thrust problem of Exercise 41,
enough resources had been available for only a half-
replicate of the 2^4 experiment.

a. If the effect $ABCD$ is chosen as the defining effect
for the replicate and the group of eight treatments
for which data is obtained includes treatment (1),
what other treatments are in the observed group and
what are the alias pairs?

b. Suppose the results of carrying out the experiment
as described in part (a) are as recorded below (given
in standard order after deleting the half not

observed). Assuming that two- and three-factor interactions are negligible, test at level .05 for the presence of main effects. Also construct a normal probability plot:

19.09 20.11 21.66 20.44
13.72 11.26 11.72 12.29

47. A half-replicate of a 2^5 experiment to investigate the effects of heating time (A), quenching time (B), drawing time (C), position of heating coils (D), and measurement position (E) on hardness of steel castings resulted in the accompanying data. Construct the ANOVA table and (assuming second- and higher-order interactions to be negligible) test at level .01 for the

presence of main effects. Also construct a normal probability plot.

Treat-ment	Observation	Treat-ment	Observation
a	70.4	acd	66.6
b	72.1	ace	67.5
c	70.4	ade	64.0
d	67.4	bcd	66.8
e	68.0	bce	70.3
abc	73.8	bde	67.9
abd	67.0	cde	65.9
abe	67.8	abcde	68.0

Supplementary Exercises Chapter 11 (48–55)

48. The results of a study on the effectiveness of line drying on the smoothness of fabric were summarized in the paper "Line-Dried vs. Machine-Dried Fabrics: Comparison of Appearance, Hand, and Consumer Acceptance" (*Home Econ. Research J.*, 1984, pp. 27–35). Smoothness scores were given for nine different types of fabric and five different drying methods: (1) machine dry, (2) line dry, (3) line dry followed by 15-min tumble, (4) line dry with softener, and (5) line dry with air movement. Regarding the different types of fabric as blocks, construct an ANOVA table. Using a .05 significance level, test to see if there is a difference in the true mean smoothness score for the drying methods.

Drying Method

		1	2	3	4	5
	Crepe	3.3	2.5	2.8	2.5	1.9
	Double knit	3.6	2.0	3.6	2.4	2.3
	Twill	4.2	3.4	3.8	3.1	3.1
	Twill mix	3.4	2.4	2.9	1.6	1.7
Fabric	Terry	3.8	1.3	2.8	2.0	1.6
	Broadcloth	2.2	1.5	2.7	1.5	1.9
	Sheeting	3.5	2.1	2.8	2.1	2.2
	Corduroy	3.6	1.3	2.8	1.7	1.8
	Denim	2.6	1.4	2.4	1.3	1.6

49. The water absorption of two types of mortar used to repair damaged cement was discussed in the paper "Polymer Mortar Composite Matrices for Maintenance-Free, Highly Durable Ferrocement" (*J. Ferrocement*, 1984, pp. 337–345). Specimens of ordinary cement mortar (OCM) and polymer cement mortar (PCM) were

submerged for varying lengths of time (5, 9, 24, or 48 hours) and water absorption (% by weight) was recorded. With mortar type as factor A (with two levels) and submersion period as factor B (with four levels), three observations were made for each factor-level combination. Data included in the paper was used to compute the sums of squares, which were $SSA = 322.667$, $SSB = 35.623$, $SSAB = 8.557$, and $SST = 372.113$. Use this information to construct an ANOVA table, and then use a .05 significance level to test the appropriate hypotheses.

50. Four plots were available for an experiment to compare clover accumulation for four different sowing rates ("Performance of Overdrilled Red Clover with Different Sowing Rates and Initial Grazing Managements," *N. Zeal. J. Exp. Ag.*, 1984, pp. 71–81). Since the four plots had been grazed differently prior to the experiment and it was thought that this might affect clover accumulation, a randomized block experiment was used with all four sowing rates tried on a section of each plot. Use the given data to test the null hypothesis of no difference in true mean clover accumulation (kg DM/ha) for the different sowing rates.

Sowing Rate (kg/ha)

		3.6	6.6	10.2	13.5
	1	1155	2255	3505	4632
	2	123	406	564	416
Plot	3	68	416	662	379
	4	62	75	362	564

51. A chemical engineer has carried out an experiment to study the effects of the fixed factors vat pressure (A), cooking time of pulp (B), and hardwood concentration (C) on the strength of paper. The experiment involved two pressures, four cooking times, three concentrations, and two observations at each combination of these levels. Calculated sums of squares are $SSA = 6.94$, $SSB = 5.61$, $SSC = 12.33$, $SSAB = 4.05$, $SSAC = 7.32$, $SSBC = 15.80$, $SSE = 14.40$, and $SST = 70.82$. Construct the ANOVA table and carry out appropriate tests at significance level .05.

52. The bond strength when mounting an integrated circuit on a metalized glass substrate was studied as a function of factor A = adhesive type, factor B = cure time, and factor C = conductor material (copper and nickel). The data follows, along with an ANOVA table from MINITAB. What conclusions can you draw from the data?

Copper			*Cure Time*		
		1	*2*	*3*
Adhesive	1	72.7	74.6	80.0
		80.0	77.5	82.7
	2	77.8	78.5	84.6
		75.3	81.1	78.3
	3	77.3	80.9	83.9
		76.5	82.6	85.0

Nickel		*1*	*2*	*3*
Adhesive	1	74.7	75.7	77.2
		77.4	78.2	74.6
	2	79.3	78.8	83.0
		77.8	75.4	83.9
	3	77.2	84.5	89.4
		78.4	77.5	81.2

53. The paper "Food Consumption and Energy Reqirements of Captive Bald Eagles" (*J. Wildlife Mgmt.*, 1982, pp. 646–654) investigated mean gross daily energy intake (the response variable) for different diet types (factor A, with three levels) and temperature (factor B, with three levels). Summary quantities given in the paper were used to generate data, resulting in $SSA = 18{,}138$, $SSB = 5182$, $SSAB = 1737$, $SST = 36{,}348$, and error df = 36. Construct an ANOVA table and test the relevant hypotheses.

54. Analogous to a Latin square, a Greco–Latin square design can be used when it is suspected that three extraneous factors may affect the response variable and all four factors (the three extraneous ones and the one of interest) have the same number of levels. In a Latin square, each level of the factor of interest (C) appears once in each row (with each level of A) and once in each column (with each level of B). In a Greco–Latin square, each level of factor D appears once in each row, in each column, and also with each level of the third extraneous factor C. Alternatively, the design can be used when the four factors are all of equal interest, the number of levels of each is N, and resources are available for only N^2 observations. A 5×5 square is pictured in (a), with (k, l) in each cell denoting the kth level of C and lth level of D. In (b) we present data on weight loss in silicon bars used for semiconductor material as a function of volume of etch (A), color of nitric acid in the etch solution (B), size of bars (C), and time in the etch solution (D) (from "Applications of Analytic Techniques to the Semiconductor Industry," Fourteenth Midwest Quality Control Conference, 1959).

Let $X_{ij(kl)}$ denote the observed weight loss when factor A is at level i, B is at level j, C is at level k, and D is at level l. Assuming no interaction between factors, total sum of squares SST (with $N^2 - 1$ df) can be partitioned into SSA, SSB, SSC, SSD, and SSE. Give expressions for

```
Analysis of Variance for strength
Source                          DF      SS       MS      F      P
Adhesive                         2   101.317  50.659   6.54  0.007
Curetime                         2   151.317  75.659   9.76  0.001
Conmater                         1     0.722   0.722   0.09  0.764
Adhesive*curetime                4    30.526   7.632   0.98  0.441
Adhesive*conmater                2     8.015   4.008   0.52  0.605
Curetime*conmater                2     5.952   2.976   0.38  0.687
Adhesive*curetime*conmater       4    33.298   8.325   1.07  0.398
Error                           18   139.515   7.751
Total                           35   470.663
```

these sums of squares, including computing formulas, obtain the ANOVA table for the given data, and test each of the four main effect hypotheses using $\alpha = .05$.

		B			
(C, D)	1	2	3	4	5

A	1	(1, 1)	(2, 3)	(3, 5)	(4, 2)	(5, 4)
	2	(2, 2)	(3, 4)	(4, 1)	(5, 3)	(1, 5)
	3	(3, 3)	(4, 5)	(5, 2)	(1, 4)	(2, 1)
	4	(4, 4)	(5, 1)	(1, 3)	(2, 5)	(3, 2)
	5	(5, 5)	(1, 2)	(2, 4)	(3, 1)	(4, 3)

(a)

65	82	108	101	126
84	109	73	97	83
105	129	89	89	52
119	72	76	117	84
97	59	94	78	106

(b)

55. In Exercise 3, after calculating $\hat{\mu}$, $\hat{\alpha}_i$, and $\hat{\beta}_j$, compute the residuals $x_{ij} - (\hat{\mu} + \hat{\alpha}_i + \hat{\beta}_j)$. Then construct a normal probability plot and comment.

Bibliography

Box, George, William Hunter, and Stuart Hunter, *Statistics for Experimenters,* Wiley, New York, 1978. Contains a wealth of suggestions and insights on data analysis based on the authors' extensive consulting experience.

DeVor, R., T. Chang, and J. W. Sutherland, *Statistical Quality Design and Control,* Macmillan, New York, 1992. Includes a modern survey of factorial and fractional factorial experimentation with a minimum of mathematics.

Hocking, Ronald, *The Analysis of Linear Models,* Brooks/Cole, Pacific Grove, CA, 1985. A very general treatment of analysis of variance written by one of the foremost authorities in this field.

Kleinbaum, David, Lawrence Kupper, and Keith Muller, *Applied Regression Analysis and Other Multivariable Methods* (2nd ed.), Duxbury Press, Boston, 1988. Contains an especially good discussion of problems associated with analysis of "unbalanced data"—that is, unequal K_{ij}'s.

Kuehl, Robert O., *Statistical Principles of Research Design and Analysis,* Wadsworth, Belmont, CA, 1994. An up-to-date and comprehensive treatment of designed experiments and analysis of the resulting data.

Montgomery, Douglas, *Design and Analysis of Experiments* (3rd ed.), Wiley, New York, 1991. See the Chapter 10 bibliography.

Neter, John, William Wasserman, and Michael Kutner, *Applied Linear Statistical Models* (3rd ed.), Irwin, Homewood, IL, 1990. See the Chapter 10 bibliography.

Vardeman, Stephen, *Statistics for Engineering Problem Solving,* PWS, Boston, 1994. A general introduction for engineers, with much descriptive and inferential methodology for data from designed experiments.

12 Simple Linear Regression and Correlation

$$\Sigma\Sigma\Sigma\Sigma\Sigma\Sigma\Sigma\Sigma\Sigma\Sigma\Sigma\Sigma\Sigma\Sigma\Sigma\Sigma\Sigma$$

Introduction

In the two-sample problems discussed in Chapter 9, we were interested in comparing values of parameters for the x distribution and the y distribution. Even when observations were paired, we did not try to use information about one of the variables in studying the other variable. This is precisely the objective of regression analysis: to exploit the relationship between two (or more) variables so that we can gain information about one of them through knowing values of the other(s).

Much of mathematics is devoted to studying variables that are *deterministically* related. Saying that x and y are related in this manner means that once we are told the value of x, the value of y is completely specified. For example, suppose we decide to rent a car for a weekend and that the rental cost is \$15.00 plus \$.10 per mile driven. If we let $x =$ the number of miles driven and $y =$ the amount we will pay the rental agency, then $y = 15 + .1x$. If we drive the car 100 miles ($x = 100$), then $y = 15 + .1(100) = 25$. As another example, if the initial velocity of a particle is v_0 and it undergoes constant acceleration a, then distance traveled $= y = v_0 x + \frac{1}{2}ax^2$ where $x =$ time.

There are many variables x and y that would appear to be related to one another, but not in a deterministic fashion. A familiar example to many students is given by variables $x =$ high school grade point average (GPA) and $y =$ college GPA. The value of y cannot be determined just from knowledge of x, and two different students could have the same x value but have very different y values. Yet there is a tendency for those students who have high (low) high school GPAs also to have high (low) college GPAs. Knowledge of a student's high school GPA should be quite helpful in enabling us to predict how that person will do in college.

Other examples of variables related in a nondeterministic fashion include $x =$ age of a child and $y =$ size of that child's vocabulary, $x =$ size of an engine in cubic centimeters and $y =$ fuel efficiency for an automobile equipped with that engine, and $x =$ applied tensile force and $y =$ amount of elongation in a metal strip.

Regression analysis is the part of statistics that deals with investigation of the relationship between two or more variables related in a nondeterministic fashion. In this chapter, we generalize the linear relation $y = \beta_0 + \beta_1 x$ to a linear probabilistic relationship, develop procedures for making inferences about the parameters of the model, and obtain a quantitative measure (the correlation coefficient) of the extent to which the two variables are related. In the next chapter, we will consider techniques for validating a particular model and investigate nonlinear relationships and relationships involving more than two variables.

12.1   The Simple Linear Regression Model

The simplest deterministic mathematical relationship between two variables x and y is a linear relationship $y = \beta_0 + \beta_1 x$. The set of pairs (x, y) for which $y = \beta_0 + \beta_1 x$ determines a straight line with slope β_1 and y-intercept β_0.* The objective of this section is to develop a linear probabilistic model.

If the two variables are not deterministically related, then for a fixed value of x, the value of the second variable is random. For example, if we are investigating the relationship between age of child and size of vocabulary and decide to select a child of age $x = 5.0$ years, then before the selection is made, vocabulary size is a random variable Y. After a particular 5-year-old child has been selected and tested, a vocabulary of 2000 words may result. We would then say that the observed value of Y associated with fixing $x = 5.0$ was $y = 2000$.

More generally, the variable whose value is fixed by the experimenter will be denoted by x and will be called the **independent variable**. For fixed x, the second variable will be random; we denote this random variable and its observed value by Y and y, respectively, and refer to it as the **dependent variable**.

Usually observations will be made for a number of settings of the independent variable. Let $x_1, x_2, \ldots, x_n$ denote values of the independent variable for which observations are made and let Y_i and y_i respectively denote the random variable and observed value associated with x_i. The available data then consists of the n pairs $(x_1, y_1), (x_2, y_2), \ldots, (x_n, y_n)$.

Example 12.1

The paper "A Study of Stainless Steel Stress-Corrosion Cracking by Potential Measurements" (*Corrosion*, 1962, pp. 425–432) reports on the relationship between applied stress (the independent variable x, in kg/mm^2) and time to fracture (the dependent variable y, in hours) for 18-8 stainless steel under uniaxial tensile stress in a 40% CaCl$_2$ solution at 100°C. Ten different settings of applied stress were used, and the resulting data values (as read from a graph which appeared in the paper) were

i	1	2	3	4	5	6	7	8	9	10
x_i	2.5	5	10	15	17.5	20	25	30	35	40
y_i	63	58	55	61	62	37	38	45	46	19

■

---

*The slope of a line is the change in y for a 1-unit increase in x. For example, if $y = -3x + 10$, then y decreases by 3 when x increases by 1, so the slope is -3. The y-intercept is the height at which the line crosses the vertical axis and is obtained by setting $x = 0$ in the equation.

In Example 12.1, the value of x_i increased as i increased, but this need not be the case. Furthermore, the values of the x_i's need not all be distinct. The investigators might have made several independent determinations of fracture time for a stress of 15 kg/mm^2.

A first step in regression analysis involving two variables is to construct a **scatter plot** of the observed data. In such a plot, each (x_i, y_i) is represented as a point plotted on a two-dimensional coordinate system. Figure 12.1 pictures a scatter plot of the data from Example 12.1. It is clear that no simple curve will pass through all the points in the plot. There is, however, a strong tendency for a large stress value to be associated with a small value of time-to-fracture, which suggests a relationship of some sort.

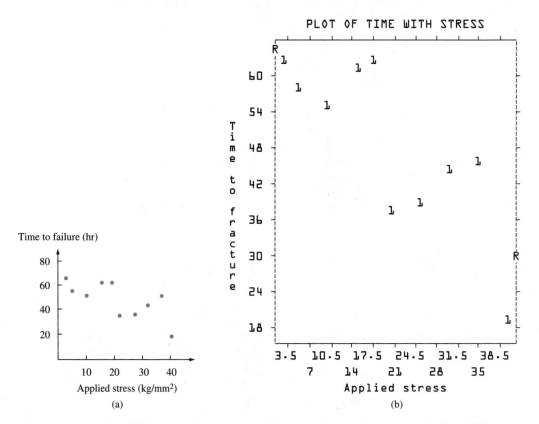

Figure 12.1 Scatter plots for data of Example 12.1: (a) hand drawn; (b) from SPSS

A Linear Probabilistic Model

For the deterministic model $y = \beta_0 + \beta_1 x$, the actual observed value of y is a linear function of x. The appropriate generalization of this to a probabilistic model assumes that *the expected value of Y is a linear function of x*, but that for fixed x, the variable Y differs from its expected value by a random amount.

> ### The Simple Linear Regression Model
>
> There exist parameters β_0, β_1, and σ^2 such that for any fixed value of the independent variable x, the dependent variable is related to x through the model equation
>
> $$Y = \beta_0 + \beta_1 x + \epsilon \qquad (12.1)$$
>
> The quantity ϵ in the model equation is a random variable, assumed to be normally distributed with $E(\epsilon) = 0$ and $V(\epsilon) = \sigma^2$.

The variable ϵ is usually referred to as the **random deviation** or **random error term** in the model. Without ϵ, any observed pair (x, y) would correspond to a point falling exactly on the line $y = \beta_0 + \beta_1 x$, called the **true regression line**. The inclusion of the random error term allows (x, y) to fall either above the true regression line (when $\epsilon > 0$) or below the line (when $\epsilon < 0$). The points $(x_1, y_1), \ldots, (x_n, y_n)$ resulting from n independent observations will then be scattered about the true regression line as illustrated in Figure 12.2. On occasion, the appropriateness of the simple linear regression model may be suggested by theoretical considerations (e.g., there is an exact linear relationship between the two variables, with ϵ representing measurement error). Much more frequently, though, the reasonableness of the model is indicated by a scatter plot exhibiting a substantial linear pattern (as in Figure 12.1).

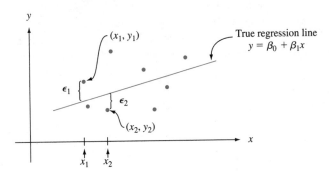

Figure 12.2 Points corresponding to observations from the simple linear regression model

Implications of the model equation (12.1) can best be understood with the aid of the following notation. Let x^* denote a particular value of the independent variable x and

$\mu_{Y \cdot x^*} =$ the expected (or mean) value of Y when $x = x^*$

$\sigma^2_{Y \cdot x^*} =$ the variance of Y when $x = x^*$

Alternative notation is $E(Y|x^*)$ and $V(Y|x^*)$. For example, in the context of Example 12.1, $\mu_{Y \cdot 20}$ would denote the expected value of time-to-fracture when applied stress is 20 kg/mm^2. If we think of an entire population of (x, y) pairs, then $\mu_{Y \cdot x^*}$ is the mean of all y values for which $x = x^*$, and $\sigma^2_{Y \cdot x^*}$ is a measure of how much these values of y spread out about the mean value. If, for example, $x =$ age of a child and $y =$ vocabulary size, then $\mu_{Y \cdot 5}$ is the average vocabulary size for all 5-year-old children in the population, and $\sigma_{Y \cdot 5}^2$ describes the amount of variability in vocabulary size for this part of the population. Once x is fixed, the only randomness on the right-hand side of the model equation (12.1) is in the random error ϵ, and its mean value and variance are 0 and σ^2, respectively, whatever the value of x. This implies that

$$\mu_{Y \cdot x^*} = E(\beta_0 + \beta_1 x^* + \epsilon) = \beta_0 + \beta_1 x^* + E(\epsilon) = \beta_0 + \beta_1 x^*$$
$$\sigma_{Y \cdot x^*}^2 = V(\beta_0 + \beta_1 x^* + \epsilon) = V(\beta_0 + \beta_1 x^*) + V(\epsilon) = 0 + \sigma^2 = \sigma^2$$

Replacing x^* in $\mu_{Y \cdot x^*}$ by x gives the relation $\mu_{Y \cdot x} = \beta_0 + \beta_1 x$, which says that the *mean value* of Y, rather than Y itself, is a linear function of x. The true regression line $y = \beta_0 + \beta_1 x$ is thus the *line of mean values*; its height above any particular x value is the expected value of Y for that value of x. The slope β_1 of the true regression line is interpreted as the *expected* change in Y associated with a 1-unit increase in the value of x. The second relation states that the amount of variability in the distribution of Y values is the same at each different value of x (homogeneity of variance). In the example involving age of a child and vocabulary size, the model implies that average vocabulary size changes linearly with age (hopefully β_1 is positive) and that the amount of variability in vocabulary size at any particular age is the same as at any other age. Finally, for fixed x, Y is the sum of a constant $\beta_0 + \beta_1 x$ and a normally distributed rv ϵ, so itself has a normal distribution. These properties are illustrated in Figure 12.3. The variance parameter σ^2 determines the extent to which each normal curve spreads out about its mean value (the height of the line). When σ^2 is small, an observed point (x, y) will almost always fall quite close to the true regression line, whereas observations may deviate considerably from their expected values (corresponding to points far from the line) when σ^2 is large.

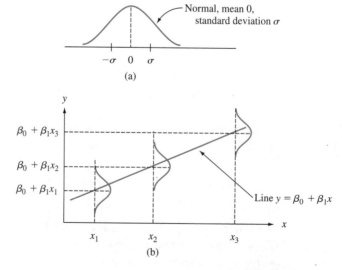

Figure 12.3 (a) Distribution of ϵ; (b) distribution of Y for different values of x

Example 12.2
(Example 12.1 continued) 3|31|97

Suppose the relationship between applied stress x and time-to-failure y is described by the simple linear regression model with true regression line $y = 65 - 1.2x$ and $\sigma = 8$. Then for any fixed value x^* of stress, time-to-failure has a normal distribution with mean value $65 - 1.2x^*$ and standard deviation 8. Roughly speaking, in the population consisting of all (x, y) points, the magnitude of a typical deviation from the true regression line is about 8. For $x = 20$, Y has mean value $\mu_{Y\cdot 20} = 65 - 1.2(20) = 41$, so

$$P(Y > 50 \text{ when } x = 20) = P\left(Z > \frac{50 - 41}{8}\right) = 1 - \Phi(1.13) = .1292$$

The probability that time-to-failure exceeds 50 when applied stress is 25 is, because $\mu_{Y\cdot 25} = 35$,

$$P(Y > 50 \text{ when } x = 25) = P\left(Z > \frac{50 - 35}{8}\right) = 1 - \Phi(1.88) = .0301$$

These probabilities are illustrated as the shaded areas in Figure 12.4.

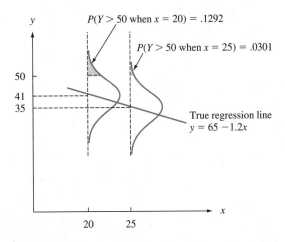

Figure 12.4 Probabilities based on the simple linear regression model

Suppose that Y_1 denotes an observation on time-to-failure made with $x = 25$ and Y_2 denotes an independent observation made with $x = 24$. Then $Y_1 - Y_2$ is normally distributed with mean value $E(Y_1 - Y_2) = \beta_1 = -1.2$, variance $V(Y_1 - Y_2) = \sigma^2 + \sigma^2 = 128$, and standard deviation $\sqrt{128} = 11.314$. The probability that Y_1 exceeds Y_2 is

$$P(Y_1 - Y_2 > 0) = P\left(Z > \frac{0 - (-1.2)}{11.314}\right) = P(Z > .11) = .4562$$

That is, even though we expected Y to decrease when x increases by 1 unit, it is not unlikely that the observed Y at $x + 1$ will be larger than the observed Y at x. ∎

Exercises   Section 12.1 (1–5)

1. The article "Some Field Experience in the Use of an Accelerated Method in Estimating 28-Day Strength of Concrete" (*J. Amer. Concrete Institute*, 1969, p. 895) considered regressing y = 28-day standard-cured strength (psi) against x = accelerated strength (psi). Suppose the equation of the true regression line is $y = 1800 + 1.3x$.

a. What is the expected value of 28-day strength when accelerated strength = 2500?

b. By how much can we expect 28-day strength to change when accelerated strength increases by 1 psi?

c. Answer part (b) for an increase of 100 psi.

d. Answer part (b) for a decrease of 100 psi.

2. Referring to Exercise 1, suppose that the standard deviation of the random deviation ϵ is 350 psi.

a. What is the probability that the observed value of 28-day strength will exceed 5000 psi when the value of accelerated strength is 2000?

b. Repeat part (a) with 2500 in place of 2000.

c. Consider making two independent observations on 28-day strength, the first for an accelerated strength of 2000 and the second for $x = 2500$. What is the probability that the second observation will exceed the first by more than 1000 psi?

d. Let Y_1 and Y_2 denote observations on 28-day strength when $x = x_1$ and $x = x_2$, respectively. By how much would x_2 have to exceed x_1 in order that $P(Y_2 > Y_1) = .95$?

3. The flow rate y (m^3/min) in a device used for air-quality measurement depends on the pressure drop x (in. of water) across the device's filter. Suppose that for x values between 5 and 20, the two variables are related according to the simple linear regression model with true regression line $y = -.12 + .095x$.

a. What is the expected change in flow rate associated with a 1-in. increase in pressure drop? Explain.

b. What change in flow rate can be expected when pressure drop decreases by 5 in.?

c. What is the expected flow rate for a pressure drop of 10 in.? A drop of 15 in.?

d. Suppose $\sigma = .025$ and consider a pressure drop of 10 in. What is the probability that the observed value of flow rate will exceed .835? That observed flow rate will exceed .840?

e. What is the probability that an observation on flow rate when pressure drop is 10 in. will exceed an observation on flow rate made when pressure drop is 11 in.?

4. Suppose the expected cost of a production run is related to the size of the run by the equation $y = 4000 + 10x$. Let Y denote an observation on the cost of a run. If the variables *size* and *cost* are related according to the simple linear regression model, could it be the case that $P(Y > 5500$ when $x = 100) = .05$ and $P(Y > 6500$ when $x = 200) = .10$? Explain.

5. Suppose that in a certain chemical process the reaction time y (hour) is related to the temperature (°F) in the chamber in which the reaction takes place according to the simple linear regression model with equation $y = 5.00 - .01x$ and $\sigma = .075$.

a. What is the expected change in reaction time for a 1°F increase in temperature? For a 10°F increase in temperature?

b. What is the expected reaction time when temperature is 200°F? When temperature is 250°F?

c. Suppose five observations are made independently on reaction time, each one for a temperature of 250°F. What is the probability that all five times are between 2.4 and 2.6 hours?

d. What is the probability that two independently observed reaction times for temperatures 1° apart are such that the time at the higher temperature exceeds the time at the lower temperature?

12.2   Estimating Model Parameters

We will assume in this and the next several sections that the variables x and y are related according to the simple linear regression model. The values of β_0, β_1, and

σ^2 will almost never be known to an investigator. Instead, sample data consisting of n observed pairs $(x_1, y_1), \ldots, (x_n, y_n)$ will be available, from which the model parameters and the true regression line itself can be estimated. These observations are assumed to have been obtained independently of one another. That is, y_i is the observed value of an rv Y_i, where $Y_i = \beta_0 + \beta_1 x_i + \epsilon_i$ and the n deviations $\epsilon_1, \epsilon_2, \ldots, \epsilon_n$ are independent rv's. Independence of $Y_1, Y_2, \ldots, Y_n$ follows from the independence of the ϵ_i's.

According to the model, the observed points will be distributed about the true regression line in a random manner. Figure 12.5 shows a typical plot of observed pairs along with two candidates for the estimated regression line, $y = a_0 + a_1 x$ and $y = b_0 + b_1 x$. Intuitively, the line $y = a_0 + a_1 x$ is not a reasonable estimate of the true line $y = \beta_0 + \beta_1 x$ because, if $y = a_0 + a_1 x$ were the true line, the observed points would almost surely have been closer to this line. The line $y = b_0 + b_1 x$ is a more plausible estimate because the observed points are scattered rather closely about this line.

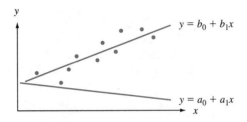

Figure 12.5 Two different estimates of the true regression line

Figure 12.5 and the foregoing discussion suggest that our estimate of $y = \beta_0 + \beta_1 x$ should be a line that provides in some sense a best fit to the observed data points. This is what motivates the principle of least squares, which can be traced back to the German mathematician Gauss (1777–1855). According to this principle, a line provides a good fit to the data if the vertical distances (deviations) from the observed points to the line are small (see Figure 12.6). The measure of the goodness of fit is the sum of the squares of these deviations. The best-fit line is then the one having the smallest possible sum of squared deviations.

Figure 12.6 Deviations of Example 12.1 data from line $y = b_0 + b_1 x$

3/31/97

Principle of Least Squares

The vertical deviation of the point (x_i, y_i) from the line $y = b_0 + b_1 x$ is

$$\text{height of point} - \text{height of line} = y_i - (b_0 + b_1 x_i)$$

The sum of squared vertical deviations from the points $(x_1, y_1), \ldots, (x_n, y_n)$ to the line is then

$$f(b_0, b_1) = \sum_{i=1}^{n} [y_i - (b_0 + b_1 x_i)]^2$$

The point estimates of β_0 and β_1, denoted by $\hat{\beta}_0$ and $\hat{\beta}_1$ and called the **least-squares estimates**, are those values that minimize $f(b_0, b_1)$. That is, $\hat{\beta}_0$ and $\hat{\beta}_1$ are such that $f(\hat{\beta}_0, \hat{\beta}_1) \leq f(b_0, b_1)$ for any b_0 and b_1. The **estimated regression line** or **least-squares line** is then the line whose equation is $y = \hat{\beta}_0 + \hat{\beta}_1 x$.

The minimizing values of b_0 and b_1 are found by taking partial derivatives of $f(b_0, b_1)$ with respect to both b_0 and b_1, equating them both to zero [analogously to $f'(b) = 0$ in univariate calculus], and solving the equations:

$$\frac{\partial f(b_0, b_1)}{\partial b_0} = \Sigma 2(y_i - b_0 - b_1 x_i)(-1) = 0$$

$$\frac{\partial f(b_0, b_1)}{\partial b_1} = \Sigma 2(y_i - b_0 - b_1 x_i)(-x_i) = 0$$

Cancellation of the -2 factor and rearrangement gives the following system of equations, called the **normal equations:**

$$n b_0 + (\Sigma x_i) b_1 = \Sigma y_i$$
$$(\Sigma x_i) b_0 + (\Sigma x_i^2) b_1 = \Sigma x_i y_i$$

The normal equations are linear in the two unknowns b_0 and b_1. Provided that at least two of the x_i's are different, the least-squares estimates are the unique solution to this system.

The least-squares estimates of the coefficients β_0 and β_1 of the true regression line are

$$b_1 = \hat{\beta}_1 = \frac{\Sigma(x_i - \bar{x})(y_i - \bar{y})}{\Sigma(x_i - \bar{x})^2} = \frac{n\,\Sigma x_i y_i - (\Sigma x_i)(\Sigma y_i)}{n\,\Sigma x_i^2 - (\Sigma x_i)^2} \tag{12.2}$$

$$b_0 = \hat{\beta}_0 = \frac{\Sigma y_i - \hat{\beta}_1 \Sigma x_i}{n} = \bar{y} - \hat{\beta}_1 \bar{x}$$

The expression for $\hat{\beta}_1$ on the right in (12.2) should be used for computation; it requires only the summary statistics Σx_i, Σy_i, Σx_i^2, $\Sigma x_i y_i$ (Σy_i^2 will be needed shortly) and minimizes the effects of rounding. In computing $\hat{\beta}_0$, use extra digits in $\hat{\beta}_1$ because, if Σx_i is large in magnitude, rounding will affect the final answer. We emphasize that *before* (12.2) *is used, the scatter plot should be examined to see if a linear probabilistic model is plausible.* If the points do not tend to cluster about a straight line with roughly the same degree of spread for all x, other models should be investigated. In practice, plots and regression calculations are usually done by using a statistical computer package.

Example 12.3

Plasma etching is essential to the fine-line pattern transfer in current semiconductor processes. The paper "Ion Beam-Assisted Etching of Aluminum with Chlorine" (*J. Electrochem. Soc.*, 1985, pp. 2010–2012) gives the accompanying data (read from a graph) on chlorine flow (x, in SCCM) through a nozzle used in the etching mechanism and etch rate (y, in 100A/min). The scatter plot (Figure 12.7) shows a substantial linear pattern, suggesting that the simple linear regression model is a reasonable way to relate x and y. The necessary summary quantities for the analysis are most conveniently obtained via a tabular format with columns for x, y, x^2, xy, and y^2:

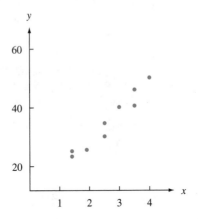

Figure 12.7 Scatter plot of the data from Example 12.3

x	1.5	1.5	2.0	2.5	2.5	3.0	3.5	3.5	4.0
y	23.0	24.5	25.0	30.0	33.5	40.0	40.5	47.0	49.0

Observation	x	y	x^2	xy	y^2
1	1.5	23.0	2.25	34.50	529.00
2	1.5	24.5	2.25	36.75	600.25
3	2.0	25.0	4.00	50.00	625.00
4	2.5	30.0	6.25	75.00	900.00
5	2.5	33.5	6.25	83.75	1122.25

Observation	x	y	x^2	xy	y^2
6	3.0	40.0	9.00	120.00	1600.00
7	3.5	40.5	12.25	141.75	1640.25
8	3.5	47.0	12.25	164.50	2209.00
9	4.0	49.0	16.00	196.00	2401.00
	24.0	312.5	70.50	902.25	11626.75
	↑	↑	↑	↑	↑
	Σx_i	Σy_i	Σx_i^2	$\Sigma x_i y_i$	Σy_i^2

The least-squares estimates are

$$\hat{\beta}_1 = \frac{9(902.25) - (24.0)(312.5)}{9(70.50) - (24.0)^2} = \frac{620.25}{58.50} = 10.602564$$

$$\hat{\beta}_0 = \frac{312.5 - (10.602564)(24.0)}{9} = \frac{58.038464}{9} = 6.448718$$

The equation of the estimated regression line is $y = 6.448718 + 10.602564x \approx 6.45 + 10.60x$. ■

The estimated regression line can immediately be used for two different purposes. For a fixed x value x^*, $\hat{\beta}_0 + \hat{\beta}_1 x^*$ (the height of the line above x^*) gives either (1) a point estimate of the expected value of Y when $x = x^*$ or (2) a prediction of the Y value that will result from a single new observation made at $x = x^*$.

Example 12.4
(Example 12.1 continued)

From the data on applied stress (x) and time-to-failure (y) given in the previous section, it is easily verified that $n = 10$, $\Sigma x_i = 200$, $\Sigma x_i^2 = 5412.5$, $\Sigma y_i = 484$, and $\Sigma x_i y_i = 8407.5$, so

$$\hat{\beta}_1 = \frac{(10)(8407.5) - (200)(484)}{(10)(5412.5) - (200)^2} = \frac{-12725}{14125} = -.900885$$

and

$$\hat{\beta}_0 = \frac{484 - (-.900885)(200)}{10} = 66.417699$$

The equation of the estimated regression line is thus $y = 66.42 - .901x$. A point estimate for expected time-to-failure when applied stress is 22.5 results from substituting this x value into the estimated equation:

$$\hat{\mu}_{Y \cdot 22.5} = \hat{\beta}_0 + \hat{\beta}_1(22.5) = 66.42 - .901(22.5) = 46.1$$

If the investigator were to make one more observation on time-to-failure when applied stress is 30, a prediction of the resulting failure time is

$$\hat{y} = (\text{predicted value of } y \text{ when } x = 30) = 66.42 - .901(30) = 39.4$$

The data set contained an observation made for $x = 30$; the corresponding y value was 45, which is larger than $\hat{y}$ because the point (30, 45) lies above the estimated regression line. ■

Estimating σ^2

The parameter σ^2 determines the amount of variability inherent in the regression model. A large value of σ^2 will lead to observed (x_i, y_i)'s that are quite spread out about the true regression line, whereas when σ^2 is small the observed points will tend to fall very close to the true line (see Figure 12.8). An estimate of σ^2 will be used in confidence interval (CI) formulas and hypothesis-testing procedures presented in the next two sections. Because the equation of the true line is unknown, the estimate is based on the extent to which the sample observations deviate from the estimated line. Many large deviations (residuals) suggest a large value of σ^2, whereas deviations all of which are small in magnitude suggest that σ^2 is small.

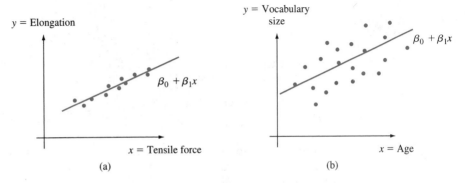

Figure 12.8 Typical sample for (a) small σ^2, (b) large σ^2

Definition

4/2/97

The **fitted** (or **predicted**) **values** $\hat{y}_1, \hat{y}_2, \ldots, \hat{y}_n$ are obtained by successively substituting $x_1, \ldots, x_n$ into the equation of the estimated regression line: $\hat{y}_1 = \hat{\beta}_0 + \hat{\beta}_1 x_1, \hat{y}_2 = \hat{\beta}_0 + \hat{\beta}_1 x_2, \ldots, \hat{y}_n = \hat{\beta}_0 + \hat{\beta}_1 x_n$. The **residuals** are the vertical deviations $y_1 - \hat{y}_1, y_2 - \hat{y}_2, \ldots, y_n - \hat{y}_n$ from the estimated line.

In words, the predicted value $\hat{y}_i$ is the value of y that we would predict or expect when using the estimated regression line with $x = x_i$; $\hat{y}_i$ is the height of the estimated regression line above the value x_i for which the ith observation was made. The residual $y_i - \hat{y}_i$ is the difference between the observed y_i and the predicted $\hat{y}_i$. If the residuals are all small in magnitude, then much of the variability in observed y values appears to be due to the linear relationship between x and y, while many large residuals suggest quite a bit of inherent variability in y relative to the amount due to the linear relation. Assuming that the line in Figure 12.6 is the least-squares line, the residuals are identified by the vertical line segments from the observed points to the line.

Example 12.5   An investigation of the relationship between traffic flow x (1000s of cars per 24 hours) and lead content y of bark on trees near the highway ($\mu g/g$ dry wt) yielded the data in the x_i and y_i columns below:

i	x_i	y_i	$\hat{y}_i$	$y_i - \hat{y}_i$
1	8.3	227	287.48	−60.48
2	8.3	312	287.48	24.52
3	12.1	362	424.98	−62.98
4	12.1	521	424.98	96.02
5	17.0	640	602.28	37.72
6	17.0	539	602.28	−63.28
7	17.0	728	602.28	125.72
8	24.3	945	866.43	78.57
9	24.3	738	866.43	−128.43
10	24.3	759	866.43	−107.43
11	33.6	1263	1202.94	60.06

The summary statistics are $\Sigma x_i = 198.3$, $\Sigma x_i^2 = 4198.03$, $\Sigma y_i = 7034$, $\Sigma y_i^2 = 5{,}390{,}382$, and $\Sigma x_i y_i = 149{,}354.4$, so

$$\hat{\beta}_1 = \frac{11(149{,}354.4) - (198.3)(7034)}{11(4198.03) - (198.3)^2} = 36.18385, \quad \hat{\beta}_0 = -12.84159$$

The estimated regression line is $y = -12.84 + 36.18x$. For numerical accuracy, the fitted values $\hat{y}_i$ are calculated from $\hat{y}_i = -12.84159 + 36.18385x_i$. The residuals should sum to 0 (a consequence of the first normal equation), but the sum here is .01 because of rounding. A positive residual results from a point lying above the estimated regression line ($y_i > \hat{y}_i$) and a negative residual from a point lying below the line. ∎

In much the same way that the deviations from the mean in a one-sample situation were combined to obtain the estimate $s^2 = \Sigma(x_i - \bar{x})^2/(n - 1)$, the estimate of σ^2 in regression analysis is based on squaring and summing the residuals. We will continue to use the symbol s^2 for this estimated variance, so don't confuse it with our earlier s^2.

Definition

4/24/97

> The **error sum of squares**, denoted by SSE, is
>
> $$SSE = \Sigma(y_i - \hat{y}_i)^2 = \Sigma[y_i - (\hat{\beta}_0 + \hat{\beta}_1 x_i)]^2$$
>
> and the estimate of σ^2 is
>
> $$\hat{\sigma}^2 = s^2 = \frac{SSE}{n - 2} = \frac{\Sigma(y_i - \hat{y}_i)^2}{n - 2}$$

The divisor $n - 2$ in s^2 is the number of degrees of freedom (df) associated with the estimate (or, equivalently, with error sum of squares). This is because to obtain s^2, the two parameters β_0 and β_1 must first be estimated, which results in a loss of 2 df (just as μ had to be estimated in one-sample problems, resulting in an estimated variance based on $n - 1$ df). Replacing each y_i in the formula for s^2 by the rv Y_i gives the estimator S^2. It can be shown that S^2 is an unbiased estimator for σ^2 (though the estimator S is not unbiased for σ).

Example 12.6
(Example 12.5 continued)

The residuals for the traffic flow/lead-content data were calculated earlier. The corresponding error sum of squares is

$$SSE = (-60.48)^2 + (24.52)^2 + \cdots + (60.06)^2 = 76{,}493.98$$

The estimate of σ^2 is then $\hat{\sigma}^2 = s^2 = 76{,}493.98/(11 - 2) = 8499.33$, and the estimated standard deviation is $\hat{\sigma} = s = \sqrt{8499.33} = 92.19$. Roughly speaking, 92.19 is the magnitude of a typical deviation from the estimated regression line. ∎

Computation of *SSE* from the defining formula involves much tedious arithmetic because both the predicted values and residuals must first be calculated. Use of the following computational formula does not require these quantities.

4/21/97

$$SSE = \Sigma y_i^2 - \hat{\beta}_0 \Sigma y_i - \hat{\beta}_1 \Sigma x_i y_i$$

This expression results from substituting $\hat{y}_i = \hat{\beta}_0 + \hat{\beta}_1 x_i$ into $\Sigma (y_i - \hat{y}_i)^2$, squaring the summand, carrying through the sum to the resulting three terms, and simplifying. This computational formula is especially sensitive to the effects of rounding in $\hat{\beta}_0$ and $\hat{\beta}_1$, so carrying as many digits as possible in intermediate computations will protect against round-off error.

Example 12.7

The paper "Promising Quantitative Nondestructive Evaluation Techniques for Composite Materials" (*Materials Evaluation*, 1985, pp. 561–565) reports on a study to investigate how the propagation of an ultrasonic stress wave through a substance depends on the properties of the substance. The accompanying data on fracture strength (*x*, as a percentage of ultimate tensile strength) and attenuation (*y*, in neper/cm, the decrease in amplitude of the stress wave) in fiberglass-reinforced polyester composites was read from a graph that appeared in the paper. The simple linear regression model is suggested by the substantial linear pattern in the scatter plot.

x	12	30	36	40	45	57	62	67	71	78	93	94	100	105
y	3.3	3.2	3.4	3.0	2.8	2.9	2.7	2.6	2.5	2.6	2.2	2.0	2.3	2.1

The necessary summary quantities are $n = 14$, $\Sigma x_i = 890$, $\Sigma x_i^2 = 67{,}182$, $\Sigma y_i = 37.6$, $\Sigma y_i^2 = 103.54$, and $\Sigma x_i y_i = 2234.30$, from which

$$\hat{\beta}_1 = \frac{14(2234.30) - (890)(37.6)}{14(67{,}182) - (890)^2} = \frac{-2183.80}{148{,}448} = -.0147109$$

and $\hat{\beta}_0 = 3.6209072$. The computational formula for *SSE* gives

$$SSE = 103.54 - (3.6209072)(37.6) - (-.0147109)(2234.30)$$
$$= .2624532$$

so $s^2 = .2624532/12 = .0218711$ and $s = .1479$. When $\hat{\beta}_0$ and $\hat{\beta}_1$ are rounded to three decimal places in the computational formula for *SSE*, the result is

$$SSE = 103.54 - (3.621)(37.6) - (-.015)(2234.30) = .905$$

which is more than three times the correct value.   ∎

The Coefficient of Determination

Figure 12.9 shows three different scatter plots of bivariate data. In all three plots, the heights of the different points vary substantially, indicating that there is much variability in observed y values. The points in the first plot all fall exactly on a straight line. In this case, all (100%) of the sample variation in y can be attributed to the fact that x and y are linearly related in combination with variation in x. The points in Figure 12.9b do not fall exactly on a line, but compared to overall y variability, the deviations from the least-squares line are small. It is reasonable to conclude in this case that much of the observed y variation can be attributed to the approximate linear relationship between the variables postulated by the simple linear regression model. When the scatter plot looks like that of Figure 12.9c, there is substantial variation about the least-squares line relative to overall y variation, so the simple linear regression model fails to explain variation in y by relating y to x.

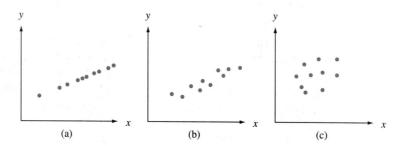

Figure 12.9 Using the model to explain y variation: (a) data for which all variation is explained; (b) data for which most variation is explained; (c) data for which little variation is explained

The error sum of squares *SSE* can be interpreted as a measure of how much variation in y is left unexplained by the model—that is, how much cannot be attributed to a linear relationship. In Figure 12.9a, $SSE = 0$, and there is no unexplained variation, whereas unexplained variation is small for the data of Figure 12.9b and much larger in Figure 12.9c. A quantitative measure of the total amount of variation in observed y values is given by the **total sum of squares**

$$SST = \Sigma(y_i - \bar{y})^2 = \Sigma y_i^2 - (\Sigma y_i)^2/n$$

Total sum of squares is the sum of squared deviations about the sample mean of the observed y values. Thus, the same number $\bar{y}$ is subtracted from each y_i in *SST*, whereas *SSE* involves subtracting each different predicted value $\hat{y}_i$ from the corresponding observed y_i. Just as *SSE* is the sum of squared deviations about the least-squares line $y = \hat{\beta}_0 + \hat{\beta}_1 x$, *SST* is the sum of squared deviations about the horizontal line at height $\bar{y}$ (since then vertical deviations are $y_i - \bar{y}$) as pictured in Figure 12.10. Furthermore, because the sum of squared deviations about the least-squares line is smaller than the sum of squared deviations about *any* other line, *SSE* < SST unless the horizontal line itself is the least-squares line. The ratio *SSE/SST* is the proportion of total variation that cannot be "explained by" the simple linear regression model, and $1 - SSE/SST$ (a number between 0 and 1) is the proportion of observed y variation explained by the model.

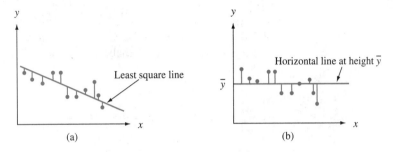

Figure 12.10 Sums of squares illustrated; (a) SSE = sum of squared deviations about the least-squares line; (b) SST = sum of squared deviations about the horizontal line

Definition

> The **coefficient of determination**, denoted by r^2, is given by
>
> $$r^2 = 1 - \frac{SSE}{SST}$$
>
> It is interpreted as the proportion of observed y variation that can be explained by the simple linear regression model (attributed to an approximate linear relationship between y and x).

The higher the value of r^2, the more successful is the simple linear regression model in explaining y variation. When regression analysis is done by a statistical computer package, either r^2 or $100r^2$ (the percentage of variation explained by the model) is a prominent part of the output. If r^2 is small, an analyst will usually want to search for an alternative model (either a nonlinear model or a multiple regression model that involves more than a single independent variable) that can more effectively explain y variation.

Example 12.8
(Example 12.7 continued)

The value of *SSE* for the fracture strength/attenuation data was previously calculated as .2625. The value of total sum of squares is

$$SST = \Sigma y_i^2 - (\Sigma y_i)^2/n = 103.54 - (37.6)^2/14 = 2.5571$$

The value of r^2 is thus $1 - .2625/2.5571 = 1 - .103 = .897$, so 89.7% of the observed variation in attenuation can be attributed to the approximate linear relationship between attenuation and fracture strength. This very high r^2 value confirms the initial impression gleaned from the scatter plot that the simple linear regression model will give a good fit to the data. ∎

The coefficient of determination can be written in a slightly different way by introducing a third sum of squares—**regression sum of squares, SSR**—given by $SSR = SST - SSE$. Regression sum of squares is interpreted as the amount of total variation that *is* explained by the model. Then we have

$$r^2 = 1 - SSE/SST = (SST - SSE)/SST = SSR/SST$$

the ratio of explained variation to total variation.

Figure 12.11 shows regression output from MINITAB for the strength/attenuation data of Examples 12.7 and 12.8. The values of the three sums of squares appear in the SS column of the "Analysis of Variance" table. The value of r^2 is .897 as calculated earlier.

Figure 12.11 MINITAB output for Examples 12.7 and 12.8

Example 12.9
(Example 12.4 continued)

The sums of squares for the applied stress and time-to-failure data given earlier are

$$SST = \Sigma y_i^2 - (\Sigma y_i)^2/n = 25{,}238 - (484)^2/10 = 1812.40$$

$$SSE = \Sigma y_i^2 - \hat{\beta}_0 \Sigma y_i - \hat{\beta}_1 \Sigma x_i y_i$$

$$= 25{,}238 - (66.417699)(484) - (-.900885)(8407.5) = 666.0243$$

and $SSR = 1812.40 - 666.02 = 1146.38$. Thus, $r^2 = 1146.38/1812.40 = .633$. In many social science studies, a model that explains 63.3% of observed y variation would be considered highly satisfactory, but in the physical sciences and engineering such is usually not the case. Because the scatter plot reveals no obvious curved pattern, the investigator would probably want to relate time-to-failure to attributes other than applied stress via a multiple regression model. ∎

Further Comments: Terminology and Scope of Regression Analysis

The term *regression analysis* was first used by Francis Galton in the late nineteenth century in connection with his work on the relationship between father's height x and son's height y. After collecting a number of pairs (x_i, y_i), Galton used the principle of least squares to obtain the equation of the estimated regression line with the objective of using it to predict son's height from father's height. In using the derived line, Galton found that if a father was above average in height, the son would also be expected to be above average in height, *but not by as much as the father was*. Similarly, the son of a shorter-than-average father would also be expected to be shorter than average, but not by as much as the father. Thus, the predicted height of a son was "pulled back in" toward the mean; because *regression* means a coming or going back, Galton adopted the terminology *regression line*. This phenomenon of being pulled back in toward the mean has been observed in many other situations (e.g., batting averages from year to year in baseball) and is called the **regression effect**.

Our discussion thus far has presumed that the independent variable is under the control of the investigator, so that only the dependent variable Y is random. This was not, however, the case with Galton's experiment; fathers' heights were not preselected, but instead both X and Y were random. Methods and conclusions of regression analysis can be applied both when the values of the independent variable are fixed in advance and when they are random, but because the derivations and interpretations are more straightforward in the former case, we will continue to work explicitly with it. For more commentary, see the excellent book by John Neter et al. listed in the chapter bibliography.

Finally, when a scatter plot suggests the appropriateness of the simple linear regression model, the model should not be assumed valid much outside the range of x observations in the sample. In Example 12.4, use of the estimated regression line to predict failure time when stress is 75 yields $\hat{y} = -1.16$, a ridiculous prediction. This is an extreme example of the danger of **extrapolation**, using the estimated relationship for x values well outside the range of observation. Even when the prediction is not ridiculous, it may be very much in error because the form of the model changes considerably for x values much above the largest value in the sample or much below the smallest value.

Exercises   Section 12.2 (6–21)

6. A study reported in the article "The Effects of Water Vapor Concentration on the Rate of Combustion of an Artificial Graphite in Humid Air Flow" (*Combustion and Flame*, 1983, pp. 107–118) gives data on $x =$ temperature of a nitrogen–oxygen mixture ($1000°F$) under specified conditions and $y =$ oxygen diffusivity. Summary quantities are $n = 9$, $\Sigma x_i = 12.6$, $\Sigma y_i = 27.68$, $\Sigma x_i^2 = 18.24$, $\Sigma x_i y_i = 40.968$, and $\Sigma y_i^2 = 93.3448$. As-

sume that the two variables are related according to the simple linear regression model.

a. Calculate the least-squares estimates of the slope and y-intercept of the population regression line.

b. Use the equation of the estimated regression line to predict what value of oxygen diffusivity would be observed when $x = 1.5$.

c. Give a point estimate of the true average diffusivity for a temperature of 1.75.

d. The observed value of diffusivity for a temperature of 1.5 was 3.39. Calculate the value of the corresponding residual.

7. *Lancet* (Mar. 29, 1975) reports on the relationship between ages of a number of children who had high levels of lead absorption and a measure of wrist flexor and extensor muscle function for these children ("Neuropsychological Dysfunction in Children with Chronic Low-Level Lead Absorption"). The measure involved the number of taps with a stylus on a single metal plate during a 10-sec period. Representative data follows, with x = age in months and y = taps/10 sec:

x	73	84	98	112	116	132	150	160	164	180
y	35	40	50	42	46	41	52	52	51	66

a. Write the system of normal equations for this data.

b. Compute $\hat{\beta}_0$ and $\hat{\beta}_1$ and verify that they solve the normal equations of part (a).

c. What would be your estimate of expected change in the measure of muscle function for a 1-month increase in age?

d. Answer part (c) for a 1-year increase in age.

e. Estimate the true average measure of muscle function for 10-year-old children in the population sampled.

8. The following summary statistics were obtained from a study that used regression analysis to investigate the relationship between pavement deflection and surface temperature of the pavement at various locations on a state highway. Here x = temperature (°F) and y = deflection adjustment factor ($y \geq 0$):

$n = 15, \quad \Sigma x_i = 1425, \quad \Sigma y_i = 10.68,$
$\Sigma x_i^2 = 139{,}037.25, \quad \Sigma x_i y_i = 987.645,$
$\Sigma y_i^2 = 7.8518$

(Many more than 15 observations were made in the study; the reference is "Flexible Pavement Evaluation and Rehabilitation," *Transportation Eng. J.*, 1977, pp. 75–85.)

a. Compute $\hat{\beta}_1, \hat{\beta}_0$, and the equation of the estimated regression line. Graph the estimated line.

b. What is the estimate of expected change in the deflection adjustment factor when temperature is increased by 1°F?

c. Suppose temperature was measured in °C rather than °F. What would be the estimated regression line? Answer part (b) for an increase of 1°C. (HINT: $°F = \frac{9}{5}°C + 32$; now substitute for the "old x" in terms of the "new x.")

d. Supposing a 200°F surface temperature was within the realm of possibility, would you use the estimated line of part (a) to predict deflection factor for this temperature? Why, or why not?

9. The following data is representative of that reported in the article "An Experimental Correlation of Oxides of Nitrogen Emissions from Power Boilers Based on Field Data" (*J. Eng. for Power*, July 1973, pp. 165–170), with x = burner area liberation rate (MBtu/hr-ft^2) and y = NO$_x$ emission rate (ppm):

x	100	125	125	150	150	200	200
y	150	140	180	210	190	320	280
x	250	250	300	300	350	400	400
y	400	430	440	390	600	610	670

a. Assuming that the simple linear regression model is valid, obtain the least-squares estimate of the true regression line.

b. What is the estimate of expected NO$_x$ emission rate when burner area liberation rate equals 225?

c. Estimate the amount by which you expect NO$_x$ emission rate to change when burner area liberation rate is decreased by 50.

d. Would you use the estimated regression line to predict emission rate for a liberation rate of 500? Why, or why not?

10. A number of studies have shown lichens (certain plants composed of an alga and a fungus) to be excellent bioindicators of air pollution. The paper "The Epiphytic Lichen Hypogymnia Physodes as a Biomonitor of Atmospheric Nitrogen and Sulphur Deposition in Norway" (*Envir. Monitoring and Assessment*, 1993, pp. 27–47) gives the following data (read from a graph) on x = NO$_3^-$ wet deposition (g N/m^2) and y = Lichen N (% dry weight):

x	.05	.10	.11	.12	.31	.37	.42
y	.48	.55	.48	.50	.58	.52	1.02
x	.58	.68	.68	.73	.85	.92	
y	.86	.86	1.00	.88	1.04	1.70	

```
The regression equation is
lichen N = 0.365 + 0.967 no3 depo

Predictor        Coef      Stdev    t-ratio        p
Constant      0.36510    0.09904       3.69    0.004
no3 depo      0.9668     0.1829        5.29    0.000

s = 0.1932      R-sq = 71.7%     R-sq (adj) = 69.2%

Analysis of Variance

SOURCE          DF         SS        MS        F        P
Regression       1     1.0427    1.0427    27.94    0.000
Error           11     0.4106    0.0373
Total           12     1.4533
```

The author used simple linear regression to analyze the data. Use the accompanying MINITAB output to answer the following questions:

a. What are the least-squares estimates of β_0 and β_1?

b. Predict Lichen N for an NO_3^- deposition value of .5.

c. What is the estimate of σ?

d. What is the value of total variation, and how much of it can be explained by the model relationship?

11. The paper "Effects of Bike Lanes on Driver and Bicyclist Behavior" (ASCE Transportation Eng. J., 1977, pp. 243–256) reports the results of a regression analysis with x = available travel space in feet (a convenient measure of roadway width, defined as the distance between a cyclist and the roadway center line) and separation distance y between a bike and a passing car (determined by photography). The data, for ten streets with bike lanes, appears below:

x	12.8	12.9	12.9	13.6	14.5
y	5.5	6.2	6.3	7.0	7.8
x	14.6	15.1	17.5	19.5	20.8
y	8.3	7.1	10.0	10.8	11.0

a. Verify that $\Sigma x_i = 154.20$, $\Sigma y_i = 80$, $\Sigma x_i^2 = 2452.18$, $\Sigma x_i y_i = 1282.74$, and $\Sigma y_i^2 = 675.16$.

b. Derive the equation of the estimated regression line.

c. What separation distance would you predict for another street that has 15.0 as its available travel space value?

d. What would be the estimate of expected separation distance for all streets having available travel space value 15.0?

12. Refer to the summary quantities given in Exercise 6 for the regression of y = oxygen diffusivity on x = temperature.

a. Calculate the value of SSE.

b. Calculate estimates of σ^2 and σ.

c. Calculate and interpret the value of r^2.

13. a. Use the summary quantities from Exercise 8 to estimate the standard deviation of the random deviation ϵ in the simple linear regression model.

b. Based on the summary quantities from Exercise 8, what proportion of variation in deflection adjustment factor can be explained by the simple linear regression relationship between adjustment factor and temperature?

14. a. Obtain SSE for the data in Exercise 9 from the defining formula ($SSE = \Sigma(y_i - \hat{y}_i)^2$) and compare to the value calculated from the computational formula.

b. Calculate the value of total sum of squares. Does the simple linear regression model appear to do an effective job of explaining variation in emission rate? Justify your assertion.

15. The accompanying data resulted from a study carried out to examine the relationship between a measure of the corrosion of iron (y) and the concentration of $NaPO_4$ (x, in ppm) ("Mechanism of Action of Acidified Sodium Phosphate Solution as a Corrosion Inhibitor of Iron in Tap Water," Brit. Corrosion J., 1979, pp. 176–178):

x	2.50	5.03	7.60	11.60	13.00	19.60
y	7.68	6.95	6.30	5.75	5.01	1.43
x	26.20	33.00	40.00	50.00	55.00	
y	.93	.72	.68	.65	.56	

a. Construct a scatter plot of the data. Does the simple linear regression model appear to be plausible?

b. Calculate the equation of the estimated regression line, use it to predict the value of the corrosion rate that would be observed for a concentration of 33 ppm, and calculate the corresponding residual.

c. Estimate the standard deviation of observations about the true regression line.

d. What percentage of sample variation in corrosion can be attributed to the model relationship?

16. The accompanying data was read from a graph that appeared in the paper "Reactions on Painted Steel Under the Influence of Sodium Chloride, and Combinations Thereof" (*Ind. Engr. Chem. Prod. Res. Dev.*, 1985, pp. 375–378). The independent variable is SO_2 deposition rate (mg/m^2/day) and the dependent variable is steel weight loss (g/m^2).

x	14	18	40	43	45	112
y	280	350	470	500	560	1200

a. Construct a scatter plot. Does the simple linear regression model appear to be reasonable in this situation?

b. Calculate the equation of the estimated regression line.

c. What percentage of observed variation in steel weight loss can be attributed to the model relationship in combination with variation in deposition rate?

d. Because the largest x value in the sample greatly

exceeds the others, this observation may have been very influential in determining the equation of the estimated line. Delete this observation and recalculate the equation. Does the new equation appear to differ substantially from the original one (you might consider predicted values)?

17. Show that b_1 and b_0 of expression (12.2) satisfy the normal equations.

18. Show that the "point of averages" $(\bar{x}, \bar{y})$ lies on the estimated regression line.

19. Suppose an investigator has data on the amount of shelf space x devoted to display of a particular product and sales revenue y for that product. Then the investigator may wish to fit a model for which the true regression line passes through $(0, 0)$. The appropriate model is $Y = \beta_1 x + \epsilon$. Assume that $(x_1, y_1), \ldots, (x_n, y_n)$ are observed pairs generated from this model and derive the least-squares estimator of β_1. (HINT: Write the sum of squared deviations as a function of b_1, a trial value, and use calculus to find the minimizing value of b_1.)

20. a. Consider the data in Exercise 7. Suppose that instead of the least-squares line passing through the points $(x_1, y_1), \ldots, (x_n, y_n)$ we wish the least-squares line passing through $(x_1 - \bar{x}, y_1), \ldots, (x_n - \bar{x}, y_n)$. Construct a scatter plot of the (x_i, y_i) points and then of the $(x_i - \bar{x}, y_i)$ points. Use the plots to explain intuitively how the two least-squares lines are related to one another.

b. Suppose that instead of the model $Y_i = \beta_0 + \beta_1 x_i + \epsilon_i (i = 1, \ldots, n)$, we wish to fit a model of the form $Y_i = \beta_0^* + \beta_1^*(x_i - \bar{x}) + \epsilon_i (i = 1, \ldots, n)$. What are the least-squares estimators of β_0^* and β_1^*, and how to they relate to $\hat{\beta}_0$ and $\hat{\beta}_1$?

21. Consider the following three data sets, in which the variables of interest are $x =$ commuting distance and $y =$ commuting time. Based on a scatter plot and the values of s and r^2, in which situation would simple linear regression be most (least) effective, and why?

Data Set	1		2		3	
	x	y	x	y	x	y
	15	42	5	16	5	8
	16	35	10	32	10	16
	17	45	15	44	15	22

(continued on next page)

Data Set		1		2		3	
		x	y	x	y	x	y
		18	42	20	45	20	23
		19	49	25	63	25	31
		20	46	50	115	50	60
$\Sigma(x_i - \bar{x})^2$		17.50		1270.8333		1270.8333	
$\Sigma(x_i - \bar{x})(y_i - \bar{y})$		29.50		2722.5		1431.6667	
$\hat{\beta}_1$		1.685714		2.142295		1.126557	
$\hat{\beta}_0$		13.666672		7.868852		3.196729	
SST		114.83		5897.5		1627.33	
SSE		65.10		65.10		14.48	

12.3 Inferences About the Slope Parameter β_1

In Section 12.2 we used the least-squares criterion to obtain point estimates for the two parameters β_0 and β_1 of the true regression line. The values of the x_i's are assumed to be chosen before the experiment is performed, so only the Y_i's are random. The estimators (statistics, and thus random variables) for β_0 and β_1 are obtained by replacing y_i by Y_i in (12.2):

$$\hat{\beta}_1 = \frac{\Sigma(x_i - \bar{x})(Y_i - \bar{Y})}{\Sigma(x_i - \bar{x})^2}$$

$$\hat{\beta}_0 = \frac{\Sigma Y_i - \hat{\beta}_1 \Sigma x_i}{n}$$

Similarly, the estimator for σ^2 results from replacing each y_i in the formula for s^2 by the rv Y_i:

$$\hat{\sigma}^2 = S^2 = \frac{\Sigma Y_i^2 - \hat{\beta}_0 \Sigma Y_i - \hat{\beta}_1 \Sigma x_i Y_i}{n - 2}$$

In earlier chapters, confidence intervals and hypothesis-testing procedures for any particular parameter were based on properties of an estimator for the parameter. The development of inferential procedures for β_1, the expected change in Y associated with a 1-unit increase in x, rests on certain facts about the distribution of the estimator $\hat{\beta}_1$ (the intercept parameter β_0 is usually of less interest than is β_1, so we concentrate on the latter parameter). The denominator of $\hat{\beta}_1$, $\Sigma(x_i - \bar{x})^2$, depends only on the x_i's and not on the Y_i's, so it is a constant c. Then because $\Sigma(x_i - \bar{x})\bar{Y} = \bar{Y}\Sigma(x_i - \bar{x}) = \bar{Y} \cdot 0 = 0$, the slope estimator can be written as

$$\hat{\beta}_1 = \frac{\Sigma(x_i - \bar{x})Y_i}{c} = \Sigma c_i Y_i, \qquad \text{where } c_i = (x_i - \bar{x})/c$$

That is, $\hat{\beta}_1$ is a linear function of the independent rv's $Y_1, Y_2, \ldots, Y_n$, each of which is normally distributed. Invoking properties of a linear function of random variables discussed in Chapter 5 leads to the following results.

1. The mean value of $\hat{\beta}_1$ is $E(\hat{\beta}_1) = \mu_{\hat{\beta}_1} = \beta_1$, so $\hat{\beta}_1$ is an unbiased estimator of β_1 (the distribution of $\hat{\beta}_1$ is always centered at the value of β_1).

2. The variance of $\hat{\beta}_1$ is

$$V(\hat{\beta}_1) = \sigma^2_{\hat{\beta}_1} = \frac{\sigma^2}{\Sigma(x_i - \bar{x})^2} = \frac{\sigma^2}{\Sigma x_i^2 - (\Sigma x_i)^2/n} \tag{12.3}$$

Replacing σ^2 by its estimate s^2 gives an estimate for $\sigma^2_{\hat{\beta}_1}$ (the estimated variance of β_1):

$$s^2_{\hat{\beta}_1} = \frac{s^2}{\Sigma(x_i - \bar{x})^2}$$

(this estimated variance can also be denoted by $\hat{\sigma}^2_{\hat{\beta}_1}$). The estimated standard deviation (standard error) of $\hat{\beta}_1$ is $s_{\hat{\beta}_1} = \sqrt{s^2_{\hat{\beta}_1}}$.

3. The estimator $\hat{\beta}_1$ has a normal distribution (because it is a linear function of independent normal rv's).

According to (12.3), the variance of $\hat{\beta}_1$ equals the variance σ^2 of the random error term—or, equivalently, of any Y_i—divided by $\Sigma(x_i - \bar{x})^2$. Because $\Sigma(x_i - \bar{x})^2$ is a measure of how spread out the x_i's are about $\bar{x}$, we conclude that making observations at x_i values that are quite spread out results in a more precise estimator of the slope parameter (smaller variance of $\hat{\beta}_1$) whereas values of x_i all close to one another imply a highly variable estimator. Of course, if the x_i's are spread out too far, a linear model may not be appropriate throughout the range of observation.

Many inferential procedures discussed earlier were based on standardizing an estimator by first subtracting its mean value and then dividing by its estimated standard deviation. In particular, test procedures and a CI for the mean μ of a normal population utilized the fact that the standardized variable $(\bar{X} - \mu)/(S/\sqrt{n})$—that is, $(\bar{X} - \mu)/S_{\bar{\mu}}$—had a t distribution with $n - 1$ df. A similar result here provides the key to further inferences concentrating β_1.

Theorem

The assumptions of the simple linear regression model imply that the standardized variable

$$T = \frac{\hat{\beta}_1 - \beta_1}{S/\sqrt{\Sigma x_i^2 - (\Sigma x_i)^2/n}} = \frac{\hat{\beta}_1 - \beta_1}{S_{\hat{\beta}_1}}$$

has a t distribution with $n - 2$ df.

A Confidence Interval for β_1

As in the derivation of earlier CIs, we begin with a probability statement:

$$P\left(-t_{\alpha/2,n-2} < \frac{\hat{\beta}_1 - \beta_1}{s_{\hat{\beta}_1}} < t_{\alpha/2,n-2}\right) = 1 - \alpha$$

Manipulation of the inequalities inside the parentheses to isolate β_1 and substitution of estimates in place of the estimators gives the CI formula.

> A $100(1 - \alpha)\%$ CI for the slope β_1 of the true regression line is
>
> $$\hat{\beta}_1 \pm t_{\alpha/2,n-2} \cdot s_{\hat{\beta}_1}$$

This interval has the same general form as did many of our previous intervals. It is centered at the point estimate of the parameter, and the amount it extends out to each side of the estimate depends on the desired confidence level (through the t critical value) and on the amount of variability in the estimator $\hat{\beta}_1$ (through $s_{\hat{\beta}_1}$, which will tend to be small when there is little variability in the distribution of $\hat{\beta}_1$ and large otherwise).

Example 12.10

Silane coupling agents have been used in the rubber industry to improve the performance of fillers in rubber compounds. The accompanying data on y = tensile modulus (in MPa, a measure of silane coupling effectiveness) and x = bound-rubber content (%) appears in the paper "The Effect of the Structure of Sulfur-Containing Silane Coupling Agents on Their Activity in Silica-Filled SBR" (*Rubber Chem. and Tech.*, 1984, pp. 675–685):

x	16.1	31.5	21.5	22.4	20.5	28.4	30.3	25.6	32.7	29.2	34.7
y	4.41	6.81	5.26	5.99	5.92	6.14	6.84	5.87	7.03	6.89	7.87

A scatter plot of the data exhibits a strong linear pattern, so it is reasonable to assume that bound-rubber content and tensile modulus are related according to the simple linear regression model. In this context, β_1 is the expected change in tensile modulus associated with a 1% increase in bound-rubber content. The necessary summary quantities are $\Sigma x_i = 292.90$, $\Sigma y_i = 69.03$, $\Sigma x_i^2 = 8141.75$, $\Sigma x_i y_i = 1890.200$, and $\Sigma y_i^2 = 442.1903$, from which

$$\hat{\beta}_1 = \frac{11(1890.200) - (292.90)(69.03)}{11(8141.75) - (292.90)^2} = \frac{573.313}{3768.84} = .152119$$

$$\hat{\beta}_0 = \frac{69.03 - (.152119)(292.90)}{11} = 2.224940$$

$$SSE = 442.1903 - (2.224940)(69.03) - (.152119)(1890.20)$$
$$= 1.067358$$

$$SST = 442.1903 - (69.03)^2/11 = 8.995673$$

$$r^2 = 1 - SSE/SST = 1 - 1.067/8.996 = .881$$

This high r^2 value provides further support for our choice of the simple linear regression model. A CI for β_1 requires $s^2 = SSE/(n - 2) = 1.067358/9 = .118595$, $s = \sqrt{.118595} = .344376 \approx .3444$, and

$$s_{\beta_1} = \frac{s}{\sqrt{\Sigma x_i^2 - (\Sigma x_i)^2/n}} = \frac{.3444}{\sqrt{8141.75 - (292.90)^2/11}} = \frac{.3444}{18.5100}$$
$$= .0186$$

The t critical value for a confidence level of 95% is $t_{.025,9} = 2.262$, and the interval is

$$.152 \pm (2.262)(.0186) = .152 \pm .042 = (.110, .194)$$

That is, we can be highly confident that when bound-rubber content is increased by 1%, the associated expected change in tensile modulus is between .110 and .194. Even with a high confidence level and small sample size, the narrowness of this interval indicates that β_1 has been rather precisely estimated.   ■

Hypothesis-Testing Procedures

As before, the null hypothesis in a test about β_1 will be an equality statement. The null value (value of β_1 claimed true by the null hypothesis) will be denoted by β_{10} (read "beta one nought," *not* "beta ten"). The test statistic results from replacing β_1 in the standardized variable T by the null value β_{10}—that is, from standardizing the estimator of β_1 under the assumption that H_0 is true. The test statistic thus has a t distribution with $n - 2$ df when H_0 is true, so the type I error probability is controlled at the desired level α by using an appropriate t critical value.

The most commonly encountered pair of hypotheses about β_1 is H_0: $\beta_1 = 0$ versus H_a: $\beta_1 \neq 0$. When this null hypothesis is true, $\mu_{Y \cdot x} = \beta_0$ independent of x, so knowledge of x gives no information about the value of the dependent variable. A test of these two hypotheses is often referred to as the *model utility test* in simple linear regression. Unless n is quite small, H_0 will be rejected, and the utility of the model confirmed precisely when r^2 is large. The simple linear regression model should not be used for further inferences (estimates of mean value or predictions of future values) unless the model utility test results in rejection of H_0 for a suitably small α.

Null hypothesis: H_0: $\beta_1 = \beta_{10}$

Test statistic value: $t = \dfrac{\hat{\beta}_1 - \beta_{10}}{s_{\beta_1}}$

Alternative Hypothesis	Rejection Region for Level α Test
H_a: $\beta_1 > \beta_{10}$	$t \geq t_{\alpha,n-2}$
H_a: $\beta_1 < \beta_{10}$	$t \leq -t_{\alpha,n-2}$
H_a: $\beta_1 \neq \beta_{10}$	Either $t \geq t_{\alpha/2,n-2}$ or $t \leq -t_{\alpha/2,n-2}$

The **model utility test** is the test of H_0: $\beta_1 = 0$ versus H_a: $\beta_1 \neq 0$, in which case the test statistic value is the t ratio $t = \hat{\beta}_1/s_{\beta_1}$.

Example 12.11

4/4/9%

The cleanliness of molten aluminum metal or alloy prior to casting is determined mainly by the hydrogen and inclusion content of the melt. The paper "Effect of Melt Cleanliness on the Properties of an Al-10 Wt Pct Si-10 Vol Pct SiC(p) Composite" (*Metallurgical Trans.*, 1993, pp. 1631–1645) reports on a study in which various tensile properties were related to x = volume fraction of oxides/inclusions (%). Here we present data (read from a graph) on y = elongation (%) of test bars. The authors state that the scatter plot shows a linear relationship and give the equation of the least-squares line. Let's use the MINITAB output of Figure 12.12 to carry out the model utility test at significance level $\alpha = .01$.

x	.10	.16	.31	.37	.37	.46	.50	.50	.60	.70
y	.96	1.10	.80	.84	.77	.87	.60	.87	.60	.61

x	.75	.80	.90	1.00	1.07	1.08	1.11	1.30	1.37	1.54
y	.70	.41	.40	.41	.45	.59	.25	.25	.08	.10

```
The regression equation is
elon = 1.07 - 0.649 volfrac
                                  s β̂₁        t = β̂₁/s β̂₁

Predictor     Coef      Stdev     t-ratio         P      P-value
Constant    1.06930    0.04966     21.53       0.000     for model
volfrac    -0.64884    0.05840    -11.11       0.000  ← utility test

s = 0.1049      R-sq = 87.3%    R-sq(adj) = 86.6%

Analysis of Variance

SOURCE       DF         SS         MS        F        P
Regression    1       1.3583    1.3583   123.42    0.000
Error        18       0.1981    0.0110
Total        19       1.5564
```

Figure 12.12 MINITAB output for Example 12.11

The parameter of interest is β_1, the expected change in percentage elongation associated with a 1% increase in the volume fraction of oxides/inclusions. $H_0: \beta_1 = 0$ will be rejected in favor of $H_a: \beta_1 \neq 0$ if the t ratio $t = \hat{\beta}_1/s_{\hat{\beta}_1}$ satisfies either $t \geq t_{\alpha/2, n-2} = t_{.005, 18} = 2.878$ or $t \leq -2.878$. From Figure 12.12, $\hat{\beta}_1 = -.64884$, $s_{\hat{\beta}_1} = .05840$, and

$$t = \frac{-.64884}{.05840} = -11.11 \qquad \text{(also on output)}$$

Clearly, $-11.11 \leq -2.878$, so H_0 is resoundingly rejected. Alternatively, the P-value is twice the area captured under the 18 df t curve to the left of -11.11. MINITAB gives P-value = .000, so H_0 should be rejected at any reasonable α. This confirmation of the simple linear regression model's utility gives us license to calculate various estimates and predictions as described in Section 12.4. ■

Regression and ANOVA

The splitting of the total sum of squares $\Sigma(y_i - \bar{y})^2$ into a part SSE, which measures unexplained variation, and a part SSR, which measures variation explained by the linear relationship, is strongly reminiscent of one-way ANOVA. In fact, the null hypothesis $H_0: \beta_1 = 0$ can be tested against $H_a: \beta_1 \neq 0$ by constructing an ANOVA table (Table 12.1) and rejecting H_0 if $f \geq F_{\alpha,1,n-2}$.

Table 12.1 ANOVA table for simple linear regression

Source of Variation	df	Sum of Squares	Mean Square	f
Regression	1	SSR	SSR	$\dfrac{SSR}{SSE/(n-2)}$
Error	$n-2$	SSE	$s^2 = \dfrac{SSE}{(n-2)}$	
Total	$n-1$	SST		

The F test gives exactly the same result as the model utility t test because $t^2 = f$ and $t^2_{\alpha/2,n-2} = F_{\alpha,1,n-2}$. Virtually all computer packages that have regression options include such an ANOVA table in the printout.

Example 12.12
(Example 12.10 continued)

The ANOVA table for the regression of tensile modulus on bound-rubber content appears in Table 12.2.

Table 12.2 ANOVA table for Example 12.12

Source of Variation	df	Sum of Squares	Mean Square	f
Regression	1	7.929	7.929	66.63
Error	9	1.067	.119	
Total	10	8.996		

Since $F_{.05,1,9} = 5.12$ and $66.63 \geq 5.12$, $H_0: \beta_1 = 0$ is rejected in favor of the alternative $H_a: \beta_1 \neq 0$. ∎

Exercises   Section 12.3 (22–35)

22. Reconsider the situation described in Exercise 1, in which x = accelerated strength of concrete and y = 28-day cured strength. Suppose the simple linear regression model is valid for x between 1000 and 4000 and that $\beta_1 = 1.25$ and $\sigma = 350$. Consider an experiment in which $n = 7$, and the x values at which observations are made are $x_1 = 1000$, $x_2 = 1500$, $x_3 = 2000$, $x_4 = 2500$, $x_5 = 3000$, $x_6 = 3500$, and $x_7 = 4000$.

a. Calculate $\sigma_{\hat{\beta}_1}$, the standard deviation of $\hat{\beta}_1$.

b. What is the probability that the estimated slope based on such observations will be between 1.00 and 1.50?

c. Suppose it is also possible to make a single observation at each of the $n = 11$ values $x_1 = 2000$, $x_2 = 2100$, ..., $x_{11} = 3000$. If a major objective is to estimate β_1 as accurately as possible, would the experiment with $n = 11$ be preferable to the one with $n = 7$?

23. Reconsider the summary quantities given in Exercise 8 for the regression of y = deflection factor on x = temperature.

   a. Compute the estimated standard deviation $s_{\hat{\beta}_1}$.

   b. Calculate a 95% CI for β_1, the expected change in deflection factor associated with a 1°F increase in temperature.

24. In a study of a reactive sputtering technique for the deposit of silicon nitride films on substrate material, the following measurements were obtained on P-etch rate (y) as a function of sputtering voltage (x) ("Preparation and Properties of Reactively Sputtered Silicon Nitride," *J. Vacuum Science and Technology*, 1967, pp. 37–40):

x	400	600	800	800	1000
y	44.0	39.9	35.0	33.8	29.1

Assuming that the variables are related according to the simple linear regression model, obtain a 99% CI for the expected change in P-etch rate when sputtering voltage is increased by 1 unit. Does the interval suggest that precise information about the value of this expected change is available?

25. Refer to the MINITAB output of Exercise 10, in which x = NO_3^- wet deposition and y = lichen N (%).

   a. Carry out the model utility test at level .01, using the rejection region approach.

   b. Repeat part (a) using the *P*-value approach.

   c. Suppose it had previously been believed that when NO_3^- wet deposition increases by .1 g N/m^2, the associated change in expected lichen N is at least .15%. Carry out a test of hypotheses at level .01 to decide whether the data contradicts this prior belief.

26. The paper "Root Dentine Transparency: Age Determination of Human Teeth Using Computerized Densitometric Analysis" (*Amer. J. Phys. Anthro.*, 1991, pp. 25–30) reports on an investigation of methods for age determination based on tooth characteristics. With x = percentage of root with transparent dentine and y = age (years), consider the following representative data for anterior teeth:

x	15	19	31	39	41	44	47	48	55	64
y	23	52	65	55	32	60	78	59	61	60

   a. Use the accompanying MINITAB output to calculate a 95% CI for the expected change in age associated with a 1% increase in transparent dentine content. What does the interval suggest about usefulness of the model?

   b. Carry out a test of model utility. Would you use the least-squares line to predict age from transparent dentine content? Explain.

27. An article in the *Journal of Public Health Engineering* reports the results of a regression analysis based on $n = 15$ observations in which x = filter application temperature (°C) and y = % efficiency of BOD removal. Calculated quantities include $\Sigma x_i = 402$, $\Sigma x_i^2 = 11,098$, $s = 3.725$, and $\hat{\beta}_1 = 1.7035$.

   a. Test at level .01 H_0: $\beta_1 = 1$, which states that the expected increase in % BOD removal is 1 when

```
The regression equation is
age = 32.1 + 0.555%

Predictor        Coef       Stdev     t-ratio         P
Constant        32.08       13.32        2.41     0.043
Percent        0.5549      0.3101        1.79     0.111

s = 14.30     R-sq = 28.6%     R-sq(adj) = 19.7%

Analysis of Variance

SOURCE          DF          SS          S         F         P
Regression       1       654.8      654.8      3.20     0.111
Error            8      1635.7      204.5
Total            9      2290.5
```

filter application temperature increases by 1°C, against the alternative H_a: $\beta_1 > 1$.

b. Compute a 99% CI for β_1, the expected increase in % BOD removal for a 1°C increase in filter application temperature.

28. The article "Hydrogen, Oxygen, and Nitrogen in Cobalt Metal" (*Metallurgia*, 1969, pp. 121–127) contains a plot of the following data pairs, where x = pressure of extracted gas (microns) and y = extraction time (min):

x	40	130	155	160	260	275	325	370	420	480
y	2.5	3.0	3.1	3.3	3.7	4.1	4.3	4.8	5.0	5.4

a. Estimate σ and the standard deviation of $\hat{\beta}_1$.

b. Suppose the investigators had believed prior to the experiment that $\beta_1 = .0060$. Does the data contradict this prior belief? Test using $\alpha = .10$.

29. Using the data from Exercise 15 on x = NaPO$_4$ concentration and y = corrosion rate, can it be concluded that the simple linear regression model specifies a useful relationship between the two variables? State and test the appropriate hypotheses at significance level .05.

30. Refer to the data on x = liberation rate and y = NO$_x$ emission rate given in Exercise 9.

a. Does the simple linear regression model specify a useful relationship between the two rates? Use the appropriate test procedure to obtain bounds on the P-value and then reach a conclusion at significance level .01.

b. Compute a 95% CI for the expected change in emission rate associated with a 10 MBtu/hr-ft^2 increase in liberation rate.

31. Carry out the model utility test using the ANOVA approach for the traffic flow/lead-content data of Example 12.5. Verify that it gives a result equivalent to that of the t test.

32. Use the rules of expected value to show that $\hat{\beta}_0$ is an unbiased estimator for β_0 (assuming that $\hat{\beta}_1$ is unbiased for β_1).

33. a. Verify that $E(\hat{\beta}_1) = \beta_1$ by using the rules of expected value from Chapter 5.

b. Use the rules of variance from Chapter 5 to verify the expression for $V(\hat{\beta}_1)$ given in this section.

34. Verify that if each x_i is multiplied by a positive constant c and each y_i is multiplied by another positive constant d, the t statistic for testing H_0: $\beta_1 = 0$ versus H_a: $\beta_1 \neq 0$ is unchanged in value (the value of $\hat{\beta}_1$ will change, which shows that the magnitude of $\hat{\beta}_1$ is not by itself indicative of model utility).

35. The probability of a type II error for the t test for H_0: $\beta_1 = \beta_{10}$ can be computed in the same manner as it was computed for the t tests of Chapter 8. If the alternative value of β_1 is denoted by β_1', the value of

$$d = \frac{\left| \beta_{10} - \beta_1' \right|}{\sigma \sqrt{\dfrac{n-1}{\Sigma x_i^2 - (\Sigma x_i)^2/n}}}$$

is first calculated, then the appropriate set of curves in Table A.13 is entered on the horizontal axis at the value of d, and β is read from the curve for $n - 2$ df. Use this to compute P(type II error) for the test of Exercise 27 when $\beta_1' = 2$ and $\sigma = 4$.

12.4   Inferences Concerning $\mu_{Y \cdot x}$ and the Prediction of Future Y Values

Let x^* denote a specified value of the independent variable x. Then once the estimates $\hat{\beta}_0$ and $\hat{\beta}_1$ have been calculated, $\hat{\beta}_0 + \hat{\beta}_1 x^*$ can be regarded either as a point estimate of $\mu_{Y \cdot x^*}$ (the expected or true average value of Y when $x = x^*$) or as a prediction of the Y value that will result from a single observation made when $x = x^*$. The point estimate or prediction by itself gives no information concerning how precisely $\mu_{Y \cdot x^*}$ has been estimated or Y has been predicted. This can be remedied by developing a CI for $\mu_{Y \cdot x^*}$ and a prediction interval (PI) for a single Y value.

If we think of $\hat{\beta}_0$ and $\hat{\beta}_1$ as random variables (before observations on the dependent variable have been made), then $\hat{\beta}_0 + \hat{\beta}_1 x^*$ is also a random variable. The CI and PI, as well as test procedures for hypotheses about $\mu_{Y \cdot x^*}$, are obtained by exploiting properties of the distribution of $\hat{\beta}_0 + \hat{\beta}_1 x^*$. Substitution of the expressions for $\hat{\beta}_0$ and $\hat{\beta}_1$ into $\hat{\beta}_0 + \hat{\beta}_1 x^*$ followed by some algebraic manipulation leads to the representation of $\hat{\beta}_0 + \hat{\beta}_1 x^*$ as a linear function of the Y_i's:

$$\hat{\beta}_0 + \hat{\beta}_1 x^* = \sum_{i=1}^{n} \left[\frac{1}{n} + \frac{(x^* - \bar{x})(x_i - \bar{x})}{\Sigma(x_i - \bar{x})^2} \right] Y_i = \sum_{i=1}^{n} d_i Y_i$$

The coefficients $d_1, d_2, \ldots, d_n$ in this linear function involve the x_i's and x^*, all of which are fixed. Application of the rules of Chapter 5 to this linear function gives the following properties.

---

1. The mean value of $\hat{\beta}_0 + \hat{\beta}_1 x^*$ is

   $$E(\hat{\beta}_0 + \hat{\beta}_1 x^*) = \mu_{\hat{\beta}_0 + \hat{\beta}_1 x^*} = \beta_0 + \beta_1 x^*$$

   Thus, $\hat{\beta}_0 + \hat{\beta}_1 x^*$ is an unbiased estimator for $\beta_0 + \beta_1 x^*$ (i.e., for $\mu_{Y \cdot x^*}$).

2. The variance of $\hat{\beta}_0 + \hat{\beta}_1 x^*$ is

   $$V(\hat{\beta}_0 + \hat{\beta}_1 x^*) = \sigma^2_{\hat{\beta}_0 + \hat{\beta}_1 x^*} = \sigma^2 \left[\frac{1}{n} + \frac{n(x^* - \bar{x})^2}{n \Sigma x_i^2 - (\Sigma x_i)^2} \right]$$

   and the standard deviation $\sigma_{\hat{\beta}_0 + \hat{\beta}_1 x^*}$ is the square root of this expression. The estimated standard deviation of $\hat{\beta}_0 + \hat{\beta}_1 x^*$, denoted by $s_{\hat{\beta}_0 + \hat{\beta}_1 x^*}$, results from replacing σ by its estimate s:

   $$s_{\hat{\beta}_0 + \hat{\beta}_1 x^*} = s \sqrt{\left[\frac{1}{n} + \frac{n(x^* - \bar{x})^2}{n \Sigma x_i^2 - (\Sigma x_i)^2} \right]}$$

3. $\hat{\beta}_0 + \hat{\beta}_1 x^*$ has a normal distribution.

The variance of $\hat{\beta}_0 + \hat{\beta}_1 x^*$ is smallest when $x^* = \bar{x}$ and increases as x^* moves away from $\bar{x}$ in either direction. Thus, the estimator of $\mu_{Y \cdot x^*}$ is more precise when x^* is near the center of the x_i's than when it is far from the x values at which observations have been made. This will imply that both the CI and PI are narrower for an x^* near $\bar{x}$ than for an x^* far from $\bar{x}$. Most statistical computer packages will provide both $\hat{\beta}_0 + \hat{\beta}_1 x^*$ and $s_{\hat{\beta}_0 + \hat{\beta}_1 x^*}$ for any specified x^*.

Inferences Concerning $\mu_{Y \cdot x^*}$

Just as inferential procedures for β_1 were based on the t variable obtained by standardizing β_1, a t variable obtained by standardizing $\hat{\beta}_0 + \hat{\beta}_1 x^*$ leads to a CI and test procedures here.

Theorem

The variable

$$T = \frac{\hat{\beta}_0 + \hat{\beta}_1 x^* - (\beta_0 + \beta_1 x^*)}{S_{\hat{\beta}_0 + \hat{\beta}_1 x^*}} \qquad (12.4)$$

has a t distribution with $n - 2$ df.

As for β_1 in the previous section, a probability statement involving this standardized variable can be manipulated to yield a confidence interval for $\mu_{Y \cdot x^*}$.

A $100(1 - \alpha)\%$ CI for $\mu_{Y \cdot x^*}$, the expected value of Y when $x = x^*$, is

$$\hat{\beta}_0 + \hat{\beta}_1 x^* \pm t_{\alpha/2, n-2} \cdot s_{\hat{\beta}_0 + \hat{\beta}_1 x^*} \qquad (12.5)$$

This CI is centered at the point estimate for $\mu_{Y \cdot x^*}$ and extends out to each side by an amount that depends on the confidence level and on the extent of variability in the estimator on which the point estimate is based.

Example 12.13   For a high-performance tissue machine used in processing of paper by paper mills, the following data was collected on machine speed x (m/min) and temperature in the drying hood y (°C) ("Gas Turbines for Process Improvement of Industrial Thermal Power Plants," *Combustion*, April 1976, pp. 35–41):

x	1000	1100	1200	1250	1300	1400	1450
y	220	280	350	375	450	470	500

To obtain a 99% CI for true average hood temperature when machine speed equals 1200, we first calculate $\Sigma x_i = 8700$, $\Sigma x_i^2 = 10,965,000$, $\Sigma y_i = 2645$, $\Sigma y_i^2 = 1,063,325$, $\Sigma x_i y_i = 3,384,750$, $\hat{\beta}_1 = .640$, $\hat{\beta}_0 = -417.70$, $s = 17.62$, and $\bar{x} = 1242.86$. The estimated standard deviation of $\hat{\beta}_0 + \hat{\beta}_1(1200)$ is

$$S_{\hat{\beta}_0 + \hat{\beta}_1(1200)} = 17.62 \sqrt{\frac{1}{7} + \frac{7(1200 - 1242.86)^2}{7(10,965,000) - (8700)^2}}$$

$$= 17.62\sqrt{.1429 + .0121} = 6.937$$

Using the t critical value $t_{.005,5} = 4.032$ gives the interval

$$-417.70 + (.640)(1200) \pm (4.032)(6.937) = 350.30 \pm 27.97$$

$$= (322.33, 378.27)$$

We can be highly confident that the value of $\mu_{Y \cdot 1200}$ is between 322.33 and 378.27. The interval is rather wide because the sample size is small, the confidence level is quite high, and the amount of variability in $\hat{\beta}_0 + \hat{\beta}_1(1200)$ is not insubstantial. A 99%

CI for $\mu_{Y \cdot 1300}$ would be somewhat wider than the above interval, since 1300 is farther from $\bar{x}$ than is 1200.   ■

In some situations, a CI is desired not just for a single x value but for two or more x values. Suppose an investigator wishes a CI both for $\mu_{Y \cdot v}$ and for $\mu_{Y \cdot w}$, where v and w are two different values of the independent variable. It is tempting to compute the interval (12.5) first for $x = v$ and then for $x = w$. Suppose we use $\alpha = .05$ in each computation to get two 95% intervals. Then if the variables involved in computing the two intervals were independent of one another, the joint confidence coefficient would be $(.95) \cdot (.95) \approx .90$.

However, the intervals are not independent because the same $\hat{\beta}_0$, $\hat{\beta}_1$, and S are used in each. We therefore cannot assert that the joint confidence coefficient for the two intervals is exactly 90%. It can be shown, though, that if the $100(1 - \alpha)\%$ CI (12.5) is computed both for $x = v$ and for $x = w$ to obtain joint CIs for $\mu_{Y \cdot v}$ and $\mu_{Y \cdot w}$, then *the joint confidence coefficient on the resulting pair of intervals is at least* $100(1 - 2\alpha)\%$. In particular, using $\alpha = .05$ results in a joint confidence coefficient of *at least* 90%, while using $\alpha = .01$ results in at least 98% confidence.

Example 12.14
(Example 12.13 continued)

The 99% CI for true average machine speed when hood temperature is 1200 was computed as (322.33, 378.27). The 99% CI for a hood temperature of 1350 is (413.11, 479.49). Thus, we can be at least 98% confident that both $\mu_{Y \cdot 1200}$ lies in (322.33, 378.27) and $\mu_{Y \cdot 1350}$ lies in (413.11, 479.49).   ■

The validity of these joint or simultaneous CIs rests on a probability result called the **Bonferroni inequality**, so the joint CIs are referred to as **Bonferroni intervals**. The method is easily generalized to yield joint intervals for k different $\mu_{Y \cdot x}$'s. *Using the interval (12.5) separately first for $x = x_1^*$, then for $x = x_2^*$, . . . , and finally for $x = x_k^*$ yields a set of k CIs for which the joint or simultaneous confidence level is guaranteed to be at least $100(1 - k\alpha)\%$.*

Tests of hypotheses about $\beta_0 + \beta_1 x^*$ are based on the test statistic T obtained by replacing $\beta_0 + \beta_1 x^*$ in the numerator of (12.4) by the null value μ_0. For example, $H_0: \beta_0 + \beta_1(1275) = 400$ in Example 12.13 says that when machine speed is 1275, expected (i.e., true average) temperature is 400. The test statistic value is $t = [\hat{\beta}_0 + \hat{\beta}_1(1275) - 400]/s_{\hat{\beta}_0 + \hat{\beta}_1(1275)}$, and the test is upper-, lower-, or two-tailed according to the inequality in H_a.

A Prediction Interval for a Future Value of Y

Analogous to the CI (12.5) for $\mu_{Y \cdot x^*}$, one frequently wishes to obtain an interval of plausible values for the value of Y associated with some future experiment when the independent variable has value x^*. For instance, in the example in which vocabulary size y is related to the age x of a child, for $x = 6$ years (12.5) would provide a CI for the true average vocabulary size of all 6-year-old children. Alternatively, we might wish an interval of plausible values for the vocabulary size of a particular 6-year-old child.

A CI refers to a parameter, or population characteristic, whose value is fixed but unknown to us. In contrast, a future value of Y is not a parameter but instead a random variable; for this reason we refer to an interval of plausible values for a future Y as a **prediction interval** rather than a confidence interval. The error of estimation is $\beta_0 + \beta_1 x^* - (\hat{\beta}_0 + \hat{\beta}_1 x^*)$, a difference between a fixed (but unknown) quantity and a random variable. The error of prediction is $Y - (\hat{\beta}_0 + \hat{\beta}_1 x^*)$, a difference between two random variables. There is thus more uncertainty in prediction than in estimation, so a PI will be wider than a CI. Because the future value Y is independent of the observed Y_i's,

$$
\begin{aligned}
V[Y - (\hat{\beta}_0 + \hat{\beta}_1 x^*)] &= \text{variance of prediction error} \\
&= V(Y) + V(\hat{\beta}_0 + \hat{\beta}_1 x^*) \\
&= \sigma^2 + \sigma^2\left[\frac{1}{n} + \frac{n(x^* - \bar{x})^2}{n\,\Sigma x_i^2 - (\Sigma x_i)^2}\right] \\
&= \sigma^2\left[1 + \frac{1}{n} + \frac{n(x^* - \bar{x})^2}{n\,\Sigma x_i^2 - (\Sigma x_i)^2}\right]
\end{aligned}
$$

Furthermore, because $E(Y) = \beta_0 + \beta_1 x^*$ and $E(\hat{\beta}_0 + \hat{\beta}_1 x^*) = \beta_0 + \beta_1 x^*$, the expected value of the prediction error is $E(Y - (\hat{\beta}_0 + \hat{\beta}_1 x^*)) = 0$. It can then be shown that the standardized variable

$$
T = \frac{Y - (\hat{\beta}_0 + \hat{\beta}_1 x^*)}{S\sqrt{1 + \dfrac{1}{n} + \dfrac{n(x^* - \bar{x})^2}{n\,\Sigma x_i^2 - (\Sigma x_i)^2}}}
$$

has a t distribution with $n - 2$ df. Substituting this T into the probability statement $P(-t_{\alpha/2, n-2} < T < t_{\alpha/2, n-2}) = 1 - \alpha$ and manipulating to isolate Y between the two inequalities yields the following interval.

A $100(1 - \alpha)\%$ PI for a future Y observation made when $x = x^*$ is

$$
\hat{\beta}_0 + \hat{\beta}_1 x^* \pm t_{\alpha/2, n-2} \cdot s\sqrt{1 + \frac{1}{n} + \frac{n(x^* - \bar{x})^2}{n\,\Sigma x_i^2 - (\Sigma x_i)^2}} \tag{12.6}
$$

$$
= \hat{\beta}_0 + \hat{\beta}_1 x^* \pm t_{\alpha/2, n-2} \cdot \sqrt{s^2 + s^2_{\hat{\beta}_0 + \hat{\beta}_1 x^*}}
$$

The interpretation of the prediction level $100(1 - \alpha)\%$ is identical to that of earlier confidence levels—if (12.6) is used repeatedly, in the long run the resulting intervals will actually contain the observed y values $100(1 - \alpha)\%$ of the time. Notice that the 1 underneath the initial square root symbol makes the PI (12.6) longer than the CI (12.5), though the intervals are both centered at $\hat{\beta}_0 + \hat{\beta}_1 x^*$. Also, as $n \to \infty$, the length of the CI approaches 0, while the length of the PI does not (because even with perfect knowledge of β_0 and β_1, there will still be uncertainty in prediction).

Example 12.15   The article "The Incorporation of Uranium and Silver by Hydrothermally Synthesized Galena" (*Econ. Geology*, 1964, pp. 1003–1024) reports on the determination of silver content of galena crystals grown in a closed hydrothermal system over a range of temperature. With x = crystallization temperature in °C and y = Ag_2S in mol %, the data follows:

x	398	292	352	575	568	450	550	408	484	350	503	600	600
y	.15	.05	.23	.43	.23	.40	.44	.44	.45	.09	.59	.63	.60

From the summary values $\Sigma x_i = 6130$, $\Sigma x_i^2 = 3{,}022.050$, $\Sigma y_i = 4.73$, $\Sigma y_i^2 = 2.1785$, and $\Sigma x_i y_i = 2418.74$, the regression estimates are computed to be $\hat{\beta}_1 = .00143$, $\hat{\beta}_0 = -.311$, and $s = .131$. If the next crystallization temperature chosen is 500° C, a 95% PI for the resulting silver content is

$$-.311 + (.00143)(500) \pm (2.201)(.131) \sqrt{1 + \frac{1}{13} + \frac{13(500 - 471.54)^2}{1{,}709{,}750}}$$

$$= .40 \pm .30 = (.10, .70)$$

which is a very wide interval. Figure 12.13 displays MINITAB output from a request for a simple linear regression analysis. Notice that both a CI for a mean y and a PI for a single y are given; in addition, Fit = $\hat{\beta}_0 + \hat{\beta}_1 x^*$ and Stdev.Fit = $s_{\hat{\beta}_0 + \hat{\beta}_1 x^*}$.

```
The regression equation is
ag2s = -0.311 + 0.00143 temp

Predictor          Coef           Stdev        t-ratio
Constant        -0.3115          0.1737         -1.79
temp          0.0014322       0.0003602          3.98

s = 0.1306          R-sq = 59.0%      R-sq(adj) = 55.2%

Analysis of Variance

SOURCE          DF              SS              MS
Regression       1          0.26978         0.26978
Error           11          0.18773         0.01707
Total           12          0.45751

 Fit    Stdev.Fit          95% C.I.                95% P.I.
.4046      0.0377   ( 0.3217, 0.4875)      ( 0.1053, 0.7039)
```

Figure 12.13 MINITAB output for Example 12.15   ∎

Exercises   Section 12.4 (36–48)

36. Reconsider the filter application temperature/% BOD removal experiment described in Exercise 27. In addition to information given there, $\hat{\beta}_0 = 8.2141$.

a. Compute a 90% CI for $\beta_0 + 25\beta_1$, the expected % BOD removal when filter application temperature is 25°C.

b. Test at level .01 the hypotheses $H_0: \beta_0 + 25\beta_1 = 50$ versus $H_a: \beta_0 + 25\beta_1 > 50$ (the alternative hypothesis states that expected % BOD removal exceeds 50 when filter application temperature is 25°C).

37. Reconsider the data of Example 12.3, in which x = chlorine flow and y = etch rate.

a. Calculate a 95% CI for $\mu_{Y \cdot 3.0}$, the true average etch rate when flow = 3.0. Has this average been precisely estimated?

b. Calculate a 95% PI for a single future observation on etch rate to be made when flow = 3.0. Is the prediction likely to be accurate?

c. Would the 95% CI and PI when flow = 2.5 be wider or narrower than the corresponding intervals of parts (a) and (b)? Answer without actually computing the intervals.

d. Would you recommend calculating a 95% PI for a flow of 6.0? Explain.

38. Does the data of Example 12.3 suggest that $\mu_{Y \cdot 2.5}$, the expected etch rate when flow = 2.5, is something other than 30? Test H_0: $\mu_{Y \cdot 2.5} = 30$ against the appropriate alternative, using a significance level of .01.

39. Refer to Example 12.11 in which x = volume fraction of oxides/inclusions and y = % elongation.

a. MINITAB gave $s_{\beta_0 + \beta_1(.40)} = .0311$ and $s_{\beta_0 + \beta_1(1.20)} = .0352$. Why is the former estimated standard deviation smaller than the latter one?

b. Use the MINITAB output from the example to calculate a 95% CI for expected % elongation when volume fraction = .40.

c. Use the MINITAB output to calculate a 95% PI for a single value of % elongation to be observed when volume fraction = 1.20.

40. You are told that a 95% CI for expected lead content when traffic flow is 15, based on a sample of $n = 10$ observations, is (462.1, 597.7). Calculate a CI with confidence level 99% for true expected lead content when traffic flow is 15.

41. An experiment to measure the macroscopic magnetic relaxation time in crystals (μsec) as a function of the strength of the external biasing magnetic field (KG) yielded the following data ("An Optical Faraday Rotation Technique for the Determination of Magnetic Relaxation Times," *IEEE Trans. Magnetics*, June 1968, pp. 175–178, with data read from a graph that appeared in the paper).

x	11.0	12.5	15.2	17.2	19.0	20.8
y	187	225	305	318	367	365
x	22.0	24.2	25.3	27.0	29.0	
y	400	435	450	506	558	

The summary statistics are $\Sigma x_i = 223.2$, $\Sigma y_i = 4116$, $\Sigma x_i^2 = 4877.50$, $\Sigma x_i y_i = 90,096.1$, and $\Sigma y_i^2 = 1,666,782$. Compute the following:

a. A 95% CI for expected relaxation time when field strength equals 18.

b. A 95% PI for future relaxation time when field strength equals 18.

c. Simultaneous CIs for expected relaxation time when field strength equals both 15 and 20; your joint confidence coefficient should be at least 90%.

42. Consider the following four intervals based on the data of Exercise 26:

a. A CI for mean age when $x = 35$

b. A PI for age when $x = 35$

c. A CI for mean age when $x = 42$

d. A PI for age when $x = 42$

Without computing any of these intervals, what can be said about their widths relative to one another?

43. Using the data in Example 12.13, obtain simultaneous CIs for true average hood temperature when machine speed equals 1200, 1250, and 1300; your joint confidence coefficient should be at least 90%.

44. Exercise 28 presented data on x = pressure of extracted gas and y = extraction time.

a. Does the data strongly contradict prior belief that when extraction pressure is 300, expected extraction time is at most 4 min? Use a level .01 test. What can you say about the P-value (upper and/or lower bound)?

b. Extraction time is to be observed once for a pressure of 200 and once for a pressure of 300. Obtain a pair of prediction intervals for the future observed values such that the joint prediction coefficient is at least 90%.

45. Use the data given in Exercise 15 to predict the amount of corrosion resulting from a single experimental run in which $NaPO_4$ concentration is 15.0. Your calculations should give information about the accuracy of such a prediction by indicating a range of plausible values.

46. The decline of water supplies in certain areas of the United States has created the need for increased understanding of relationships between economic factors such as crop yield and hydrologic and soil factors. The paper "Variability of Soil Water Properties and Crop Yield in a Sloped Watershed" (*Water Resources*

Bull., 1988, pp. 281–288) gives data on grain sorghum yield (y, in g/m-row) and distance upslope (x, in m) on a sloping watershed. Selected observations appear in the accompanying table:

x	0	10	20	30	45	50	70
y	500	590	410	470	450	480	510
x	80	100	120	140	160	170	190
y	450	360	400	300	410	280	350

a. Construct a scatter plot. Does the simple linear regression model appear to be plausible?

b. Carry out a test of model utility.

c. Estimate true average yield when distance upslope is 75 by giving an interval of plausible values.

47. Infestation of crops by insects has long been of great concern to farmers and agricultural scientists. The paper "Cotton Square Damage by the Plant Bug, *Lygus hesperus*, and Abscission Rates" (*J. of Econ. Entom.*, 1988, pp. 1328–1337) reports data on x = age of a cotton plant (days) and y = % damaged squares. Consider the accompanying n = 12 observations (read from a scatter plot in the paper):

x	9	12	12	15	18	18
y	11	12	23	30	29	52
x	21	21	27	30	30	33
y	41	65	60	72	84	93

a. Why is the relationship between x and y not deterministic?

b. Does a scatter plot suggest that the simple linear regression model will describe the relationship between the two variables?

c. The summary statistics are $\Sigma x_i = 246$, $\Sigma x_i^2 = 5742$, $\Sigma y_i = 572$, $\Sigma y_i^2 = 35{,}634$, and $\Sigma x_i y_i = 14{,}022$. Determine the equation of the least-squares line.

d. Predict the percentage of damaged squares when the age is 20 days by giving an interval of plausible values.

48. Verify that $V(\hat{\beta}_0 + \hat{\beta}_1 x)$ is indeed given by the expression in the text. [HINT: $V(\Sigma d_i Y_i) = \Sigma d_i^2 \cdot V(Y_i)$.]

12.5 Correlation

There are many situations in which the objective in studying the joint behavior of two variables is to see whether they are related, rather than to use one to predict the value of the other. In this section, we first develop the sample correlation coefficient r as a measure of how strongly related two variables x and y are in a sample and then relate r to the correlation coefficient ρ defined in Chapter 5.

The Sample Correlation Coefficient r

Given n pairs of observations $(x_1, y_1), (x_2, y_2), \ldots, (x_n, y_n)$, it is natural to speak of x and y having a positive relationship if large x's are paired with large y's and small x's with small y's. Similarly, if large x's are paired with small y's and small x's with large y's, then a negative relationship between the variables is implied. Consider the quantity

$$S_{xy} = \sum_{i=1}^{n} (x_i - \bar{x})(y_i - \bar{y})$$

Then if the relationship is strongly positive, an x_i above the mean $\bar{x}$ will tend to be paired with a y_i above the mean $\bar{y}$, so that $(x_i - \bar{x})(y_i - \bar{y}) > 0$, and this product will also be positive whenever both x_i and y_i are below their respective means. Thus, a positive relationship implies that s_{xy} will be positive. An analogous argument shows that when the relationship is negative, s_{xy} will be negative, since most of the products $(x_i - \bar{x})(y_i - \bar{y})$ will be negative. This is illustrated in Figure 12.14.

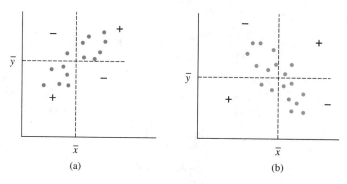

Figure 12.14 (a) Scatter plot with s_{xy} positive; (b) scatter plot with s_{xy} negative [+ means $(x_i - \bar{x})(y_i - \bar{y}) > 0$, and − means $(x_i - \bar{x})(y_i - \bar{y}) < 0$]

While s_{xy} seems a plausible measure of the strength of a relationship, we do not yet have any idea of how positive or negative it can be. Unfortunately, s_{xy} has a serious defect: By changing the units of measurement of either x or y, s_{xy} can be made either arbitrarily large in magnitude or arbitrarily close to zero. For example, if $s_{xy} = 25$ when x is measured in meters, then $s_{xy} = 25,000$ when x is measured in millimeters and .025 when x is expressed in kilometers. A reasonable condition to impose on any measure of how strongly x and y are related is that the calculated measure should not depend on the particular units used to measure them. This condition is achieved by modifying s_{xy} to obtain the sample correlation coefficient.

Definition

4/7/97

The **sample correlation coefficient** for the n pairs $(x_1, y_1), \ldots, (x_n, y_n)$ is

$$r = \frac{s_{xy}}{\sqrt{\Sigma(x_i - \bar{x})^2} \ \sqrt{\Sigma(y_i - \bar{y})^2}}$$

$$= \frac{n \, \Sigma x_i y_i - (\Sigma x_i)(\Sigma y_i)}{\sqrt{n \, \Sigma x_i^2 - (\Sigma x_i)^2} \ \sqrt{n \, \Sigma y_i^2 - (\Sigma y_i)^2}}$$

(12.7)

The second expression in (12.7) is more convenient for hand computation.

Example 12.16   The medical journal *Lancet* (April 2, 1977) reports results on water/lead concentration in the maternal home during pregnancy and blood/lead concentration (both in μmol/L) for a group of mentally retarded children. Representative data appears below:

x (blood)	1.98	1.44	2.02	1.20	1.57	1.82	1.45	1.80
y (water)	5.6	7.7	8.8	5.1	6.8	3.9	4.5	5.8

According to (12.7), r can be calculated from the five quantities Σx_i, Σx_i^2, Σy_i, Σy_i^2, and $\Sigma x_i y_i$; these are 13.28, 22.63, 48.2, 309.44, and 80.81, respectively. Thus,

$$r = \frac{8(80.81) - (13.28)(48.2)}{\sqrt{8(22.63) - (13.28)^2} \sqrt{8(309.44) - (48.2)^2}}$$

$$= \frac{6.38}{(2.16)(12.34)} = .239$$ ∎

Properties of r

The most important properties of r are as follows:

1. The value of r does not depend on which of the two variables under study is labeled x and which is labeled y.
2. The value of r is independent of the units in which x and y are measured.
3. $-1 \leq r \leq 1$
4. $r = 1$ if and only if (iff) all (x_i, y_i) pairs lie on a straight line with positive slope, and $r = -1$ iff all (x_i, y_i) pairs lie on a straight line with negative slope.
5. The square of the sample correlation coefficient gives the value of the coefficient of determination that would result from fitting the simple linear regression model—in symbols, $(r)^2 = r^2$.

Property 1 stands in marked contrast to what happens in regression analysis, where virtually all quantities of interest (the estimated slope, estimated y-intercept, s^2, etc.) depend on which of the two variables is treated as the dependent variable. However, Property 5 shows that the proportion of variation in the dependent variable explained by fitting the simple linear regression model does not depend on which variable plays this role.

Property 2 is equivalent to saying that r is unchanged if each x_i is replaced by cx_i and if each y_i is replaced by dy_i (a change in the scale of measurement), as well as if each x_i is replaced by $x_i - a$ and y_i by $y_i - b$ (which changes the location of zero on the measurement axis). This implies, for example, that r is the same whether temperature is measured in °F or °C.

Property 3 tells us that the maximum value of r, corresponding to the largest possible degree of positive relationship, is $r = 1$, while the most negative relationship is identified with $r = -1$. According to Property 4, the largest positive and largest negative correlations are achieved only when all points lie along a straight line. Any other configuration of points, even if the configuration suggests a deterministic relationship between variables, will yield an r value less than 1 in absolute magnitude. Thus, *r measures the degree of linear relationship* among variables. A value of *r* near 0 is not evidence of the lack of a strong relationship, but only the absence of a linear relation, so that such a value of *r* must be interpreted with caution. Figure 12.15 illustrates several configurations of points associated with different values of *r*.

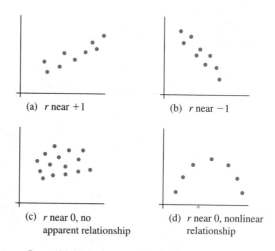

$$\text{(a)} \quad r \text{ near } +1 \qquad\qquad \text{(b)} \quad r \text{ near } -1$$

(c) *r* near 0, no apparent relationship   (d) *r* near 0, nonlinear relationship

Figure 12.15 Data plots for different values of *r*

A frequently asked question is, "When can it be said that there is a strong correlation between the variables, and when is the correlation weak?" A reasonable rule of thumb is to say that the correlation is weak if $0 \leq |r| \leq .5$, strong if $.8 \leq |r| \leq 1$, and moderate otherwise. It may surprise you that $r = .5$ is considered weak, but $r^2 = .25$ implies that in a regression of *y* on *x*, only 25% of observed *y* variation would be explained by the model. In Example 12.16, the correlation between blood/lead concentration and water/lead concentration would be described as weak.

The Population Correlation Coefficient ρ and Inferences About Correlation

The correlation coefficient *r* is a measure of how strongly related *x* and *y* are in the observed sample. We can think of the pairs (x_i, y_i) as having been drawn from a bivariate population of pairs, with (X_i, Y_i) having joint probability distribution $f(x, y)$. In Chapter 5, we defined the correlation coefficient $\rho(X, Y)$ by

$$\rho = \rho(X, Y) = \frac{\text{Cov}(X, Y)}{\sigma_X \cdot \sigma_Y}$$

where

$$\text{Cov}(X, Y) = \begin{cases} \sum_x \sum_y (x - \mu_X)(y - \mu_Y)f(x, y) & (X, Y) \text{ discrete} \\[2mm] \int_{-\infty}^{\infty}\int_{-\infty}^{\infty} (x - \mu_X)(y - \mu_Y)f(x, y) \, dx \, dy & (X, Y) \text{ continuous} \end{cases}$$

If we think of $f(x, y)$ as describing the distribution of pairs of values within the entire population, ρ becomes a measure of how strongly related x and y are in that population. Properties of ρ analogous to those for r were given in Chapter 5.

The population correlation coefficient ρ is a parameter or population characteristic, just as μ_X, μ_Y, σ_X, and σ_Y are, so we can use the sample correlation coefficient to make various inferences about ρ. In particular, r is a point estimate for ρ, and the corresponding estimator is

$$\hat{\rho} = R = \frac{\Sigma(X_i - \bar{X})(Y_i - \bar{Y})}{\sqrt{\Sigma(X_i - \bar{X})^2} \sqrt{\Sigma(Y_i - \bar{Y})^2}}$$

Example 12.17

In some locations, there is a strong association between concentrations of two different pollutants. The paper "The Carbon Component of the Los Angeles Aerosol: Source Apportionment and Contributions to the Visibility Budget" (*J. Air Pollution Control Fed.*, 1984, pp. 643–650) reports the accompanying data on ozone concentration x (ppm) and secondary carbon concentration y ($\mu g/m^3$):

x	.066	.088	.120	.050	.162	.186	.057	.100
y	4.6	11.6	9.5	6.3	13.8	15.4	2.5	11.8

x	.112	.055	.154	.074	.111	.140	.071	.110
y	8.0	7.0	20.6	16.6	9.2	17.9	2.8	13.0

The summary quantities are $n = 16$, $\Sigma x_i = 1.656$, $\Sigma y_i = 170.6$, $\Sigma x_i^2 = .196912$, $\Sigma x_i y_i = 20.0397$, and $\Sigma y_i^2 = 2253.56$, from which

$$r = \frac{16(20.0397) - (1.656)(170.6)}{\sqrt{16(.196912) - (1.656)^2} \sqrt{16(2253.56) - (170.6)^2}}$$

$$= \frac{38.1216}{(.6389)(83.3823)} = .716$$

The point estimate of the population correlation coefficient ρ between ozone concentration and secondary carbon concentration is $\hat{\rho} = r = .716$. ∎

In our previous work concerning inference, we typically did not have to make strong assumptions about the population distribution in order to obtain point estimators and estimates of parameters. However, to construct CIs and test procedures, we have usually assumed that the sample comes from a normal population. To test hypotheses about ρ, we must make an analogous assumption about the distribution of pairs of (x, y) values in the population. We are now assuming that *both X and Y* are random, whereas much of our regression work focused on x fixed by the experimenter.

Assumption

The joint probability distribution of (X, Y) is specified by

$$f(x, y) =$$

$$\frac{1}{2\pi \cdot \sigma_1 \sigma_2 \sqrt{1 - \rho^2}} \, e^{-[((x - \mu_1)/\sigma_1)^2 - 2\rho(x - \mu_1)(y - \mu_2)/\sigma_1\sigma_2 + ((y - \mu_2)/\sigma_2)^2]/[2(1 - \rho^2)]}$$

$$-\infty < x < \infty$$

$$-\infty < y < \infty \qquad (12.8)$$

where μ_1 and σ_1 are the mean and standard deviation of X, and μ_2 and σ_2 are the mean and standard deviation of Y; $f(x,y)$ is called the **bivariate normal probability distribution**.

The bivariate normal distribution is obviously rather complicated, but for our purposes we need only a passing acquaintance with several of its properties. The surface determined by $f(x, y)$ lies entirely above the x–y plane $[f(x, y) \geq 0]$ and has a three-dimensional bell- or mound-shaped appearance, as illustrated in Figure 12.16. If we slice through the surface with any plane perpendicular to the x–y plane and look at the shape of the curve sketched out on the "slicing plane," the result is a normal curve. More precisely, if $X = x$, it can be shown that the (conditional) distribution of Y is normal with mean $\mu_{Y \cdot x} = \mu_2 - \rho\mu_1\sigma_2/\sigma_1 + \rho\sigma_2 x/\sigma_1$ and variance $(1 - \rho^2)\sigma_2^2$. This is exactly the model used in simple linear regression with $\beta_0 = \mu_2 - \rho\mu_1\sigma_2/\sigma_1$, $\beta_1 = \rho\sigma_2/\sigma_1$, and $\sigma^2 = (1 - \rho^2)\sigma_2^2$ independent of x. The implication is that *if the observed pairs (x_i, y_i) are actually drawn from a bivariate normal distribution, then the simple linear regression model is an appropriate way of studying the behavior of Y for fixed x.* If $\rho = 0$, then $\mu_{Y \cdot x} = \mu_2$ independent of x; in fact, when $\rho = 0$ the joint probability density function $f(x, y)$ of (12.8) can be factored into a part involving x only and a part involving y only, which implies that X and Y are independent variables.

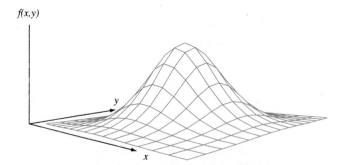

Figure 12.16 A graph of the bivariate normal pdf

Assuming that the pairs are drawn from a bivariate normal distribution allows us to test hypotheses about ρ and construct a CI. There is no completely satisfactory way to check the plausibility of the bivariate normality assumption. A partial check involves constructing two separate normal probability plots, one for the sample x_i's and another for the sample y_i's, since bivariate normality implies that the marginal distributions of both X and Y are normal. If either plot deviates substantially from a straight-line pattern, the following inferential procedures should not be used when the sample size n is small.

Procedure for Testing H_0: $\rho = 0$

Test statistic: $\quad T = \dfrac{R\sqrt{n-2}}{\sqrt{1-R^2}}$

When H_0 is true, T has a t distribution with $n - 2$ df.

Alternative Hypothesis	Rejection Region for Level α Test
H_a: $\rho > 0$	$t \geq t_{\alpha, n-2}$
H_a: $\rho < 0$	$t \leq -t_{\alpha, n-2}$
H_a: $\rho \neq 0$	Either $t \geq t_{\alpha/2, n-2}$ or $t \leq -t_{\alpha/2, n-2}$

Example 12.18
(Example 12.17 continued)

Separate normal probability plots of the 16 ozone-concentration measurements and 16 secondary carbon–concentration measurements are quite straight, so we assume that the 16 pairs were drawn from a bivariate normal population distribution and use the above t test to decide whether the data suggests a positive relationship between the two concentrations in the population consisting of all concentration pairs in the sampled locale. The hypotheses of interest are H_0: $\rho = 0$ versus H_a: $\rho > 0$. The necessary t critical value for a level .01 test is $t_{.01,14} = 2.624$. We previously computed $r = .716$, so

$$t = \frac{.716\sqrt{14}}{\sqrt{1 - (.716)^2}} = 3.84$$

Since $3.84 \geq 2.624$, we reject H_0 and conclude that the two concentrations are positively related in the population.

Figure 12.17 gives SPSS output for a correlation analysis. The P-value for a

```
Correlations: CARBON

    OZONE            .7155
                   (   .16)
                   P=  .002

(Coefficient / (Cases) / 2-tailed Significance)
```

Figure 12.17 SPSS output for Example 12.18

two-tailed test is given as .002, so the *P*-value for our test is .001. This agrees with $t_{.001,14} = 3.787$. ∎

Because ρ measures the extent to which there is a linear relationship between the two variables in the population, the null hypothesis $H_0: \rho = 0$ states that there is no such population relationship. In Section 12.3, we used the *t* ratio $\hat{\beta}_1/s_{\hat{\beta}_1}$ to test for a linear relationship between the two variables in the context of regression analysis. It turns out that the two test procedures are completely equivalent because $r\sqrt{n-2}/\sqrt{1-r^2} = \hat{\beta}_1/s_{\hat{\beta}_1}$. When interest lies only in assessing the strength of any linear relationship rather than in fitting a model and using it to estimate or predict, the test statistic formula just presented requires fewer computations than does the *t* ratio.

Other Inferences Concerning ρ

The procedure for testing $H_0: \rho = \rho_0$ when $\rho_0 \neq 0$ is not equivalent to any procedure from regression analysis. The test statistic is based on a transformation of *R* called the Fisher transformation.

Proposition

When $(X_1, Y_1), \ldots, (X_n, Y_n)$ is a sample from a bivariate normal distribution, the rv

$$V = \frac{1}{2} \ln\left(\frac{1+R}{1-R}\right) \tag{12.9}$$

has approximately a normal distribution with mean and variance

$$\mu_V = \frac{1}{2} \ln\left(\frac{1+\rho}{1-\rho}\right), \quad \sigma_V^2 = \frac{1}{n-3}$$

The rationale for the transformation is to obtain a function of *R* that has a variance independent of ρ; this would not be the case with *R* itself. Also, the transformation should not be used if *n* is quite small, since the approximation will not be valid.

Test statistic: $Z = \dfrac{V - \frac{1}{2} \ln[(1+\rho_0)/(1-\rho_0)]}{1/\sqrt{n-3}}$

Alternative Hypothesis	Rejection Region for Level α Test
$H_a: \rho > \rho_0$	$z \geq z_\alpha$
$H_a: \rho < \rho_0$	$z \leq -z_\alpha$
$H_a: \rho \neq \rho_0$	Either $z \geq z_{\alpha/2}$ or $z \leq -z_{\alpha/2}$

Example 12.19   The paper "A Study of a Partial Nutrient Removal System for Wastewater Treatment Plants" (*Water Research*, 1972, pp. 1389–1397) reports on a method of nitrogen

removal that involves the treatment of the supernatant from an aerobic digester. Both the influent total nitrogen x (mg/L) and the percentage y of nitrogen removed were recorded for 20 days, with resulting summary statistics $\Sigma x_i = 285.90$, $\Sigma x_i^2 = 4409.55$, $\Sigma y_i = 690.30$, $\Sigma y_i^2 = 29,040.29$, and $\Sigma x_i y_i = 10,818.56$. Does the data indicate that influent total nitrogen and percentage of nitrogen removed are at least moderately positively correlated?

Our earlier interpretation of moderate positive correlation was $.5 < \rho < .8$, so we wish to test $H_0: \rho = .5$ versus $H_a: \rho > .5$. The computed value of r is .733, so

$$\frac{1}{2} \ln\left(\frac{1 + .733}{1 - .733}\right) = .935, \quad \frac{1}{2} \ln\left(\frac{1 + .5}{1 - .5}\right) = .549$$

This gives $z = (.935 - .549)\sqrt{17} = 1.59$. Since $1.59 < 1.645$, at level .05 we cannot conclude that $\rho > .5$, so the relationship has not been proved to be even moderately strong (a somewhat surprising conclusion since $r = .73$, but when n is small a large r may result even when ρ is small). ∎

To obtain a CI for ρ, we first derive an interval for $\mu_V = \frac{1}{2} \ln[(1 + \rho)/(1 - \rho)]$. Standardizing V, writing a probability statement, and manipulating the resulting inequalities yields

$$\left(v - \frac{z_{\alpha/2}}{\sqrt{n - 3}}, \quad v + \frac{z_{\alpha/2}}{\sqrt{n - 3}}\right) \tag{12.10}$$

as a $100(1 - \alpha)\%$ interval for μ_V, where $v = \frac{1}{2} \ln[(1 + r)/(1 - r)]$. This interval can then be manipulated to yield a CI for ρ.

> The interval
> $$\left(\frac{e^{2c_1} - 1}{e^{2c_1} + 1}, \quad \frac{e^{2c_2} - 1}{e^{2c_2} + 1}\right)$$
> is a $100(1 - \alpha)\%$ CI for ρ, where c_1 and c_2 are the left and right endpoints, respectively, of the interval (12.10).

Example 12.20
(Example 12.19 continued)

The sample correlation coefficient between influent nitrogen and percentage nitrogen removed was $r = .733$, giving $v = .935$. With $n = 20$, a 95% interval for μ_V is $(.935 - 1.96/\sqrt{17}, .935 + 1.96/\sqrt{17}) = (.460, 1.410) = (c_1, c_2)$. The 95% interval for ρ is

$$\left[\frac{e^{2(.46)} - 1}{e^{2(.46)} + 1}, \quad \frac{e^{2(1.41)} - 1}{e^{2(1.41)} + 1}\right] = (.43, .89)$$

∎

In Chapter 5, we cautioned that a large value of the correlation coefficient (near 1 or -1) implies only association and not causation. This applies both to ρ and to r.

Exercises   Section 12.5 (49–57)

49. Observations were made on x = June–August precipitation (cm) and y = perennial grass production (kg/ha) for certain pasture areas during each year of a 10-year period, with the following results:

x	22.05	35.74	30.48	11.89	27.28
y	291	629	823	307	660
x	9.63	17.63	22.20	17.27	19.63
y	263	375	366	563	558

("Influence of Precipitation on Perennial Grass Production in the Semidesert Southwest," *Ecology*, 1975, pp. 981–986).

   a. Compute the sample correlation coefficient for this data.

   b. If you were to do a regression analysis, what would the value of the coefficient of determination be?

50. The paper "A Dual-Buffer Titration Method for Lime Requirement of Acid Mine-soils" (*J. of Environ. Qual.*, 1988, pp. 452–456) reports on the results of a study relating to revegetation of soil at mine reclamation sites. With x = KCl extractable aluminum and y = amount of lime required to bring soil pH to 7.0, data in the paper resulted in the following summary statistics: $n = 24$, $\Sigma x = 48.15$, $\Sigma x^2 = 155.4685$, $\Sigma y = 263.5$, $\Sigma y^2 = 3750.53$, and $\Sigma xy = 658.455$. Carry out a test at significance level .01 to see whether the population correlation coefficient is something other than 0.

51. The following summary quantities for x = particulate pollution ($\mu g/m^3$) and y = luminance (.01 cd/m^2) were calculated from a representative sample of data that appeared in the paper "Luminance and Polarization of the Sky Light at Seville (Spain) Measured in White Light" (*Atmos. Environ.*, 1988, pp. 595–599): $n = 15$, $\Sigma x = 860$, $\Sigma y = 348$, $\Sigma x^2 = 56,700$, $\Sigma y^2 = 8954$, and $\Sigma xy = 22,265$.

   a. Test to see whether there is a positive correlation between particulate pollution and luminance in the population from which the data was selected.

   b. What proportion of observed variation in luminance can be attributed to the approximate linear relationship between luminance and particulate pollution?

52. Hydrogen content is conjectured to be an important factor in porosity of aluminum alloy castings. The paper "The Reduced Pressure Test as a Measuring Tool in the

Evaluation of Porosity/Hydrogen Content in A1-7 Wt Pct Si-10 Vol Pct SiC(p) Metal Matrix Composite" (*Metallurgical Trans.*, 1993, pp. 1857–1868) gives the accompanying data on x = content and y = gas porosity for one particular measurement technique:

x	.18	.20	.21	.21	.21	.22	.23
y	.46	.70	.41	.45	.55	.44	.24
x	.23	.24	.24	.25	.28	.30	.37
y	.47	.22	.80	.88	.70	.72	.75

MINITAB gives the following output in response to a CORRELATION command:

```
Correlation    of    Hydrcon    and
Porosity = 0.449
```

   a. Test at level .05 to see whether the population correlation coefficient differs from 0.

   b. If a simple linear regression analysis had been carried out, what percentage of observed variation in porosity could be attributed to the model relationship?

53. Physical properties of six flame-retardant fabric samples were investigated in the paper "Sensory and Physical Properties of Inherently Flame-Retardant Fabrics" (*Textile Research*, 1984, pp. 61–68). Use the accompanying data and a .05 significance level to determine if a linear relationship exists between stiffness x (mg-cm) and thickness y (mm). Is the result of the test surprising in light of the value of r?

x	7.98	24.52	12.47	6.92	24.11	35.71
y	.28	.65	.32	.27	.81	.57

54. The article "Increases in Steroid Binding Globulins Induced by Tamoxifen in Patients with Carcinoma of the Breast" (*J. Endocrinology*, 1978, pp. 219–226) reports data on the effects of the drug tamoxifen on change in the level of cortisol-binding globulin of patients during treatment. With age = x and ΔCBG = y, summary values are $n = 26$, $\Sigma x_i = 1613$, $\Sigma(x_i - \bar{x})^2 = 3756.96$, $\Sigma y_i = 281.9$, $\Sigma(y_i - \bar{y})^2 = 465.34$, and $\Sigma x_i y_i = 16,731$.

   a. Compute a 90% CI for the true correlation coefficient ρ.

   b. Test H_0: $\rho = -.5$ versus H_a: $\rho < -.5$ at level .05.

   c. In a regression analysis of y on x, what proportion of variation in change of cortisol-binding globulin level could be explained by variation in patient age within the sample?

d. If you decide to perform a regression analysis with age as the dependent variable, what proportion of variation in age is explainable by variation in ΔCBG?

55. The paper "Chronological Trend in Blood Lead Levels" (*N. Engl. J. Med.*, 1983, pp. 1373–1377) gives the following data on y = average blood lead level of white children age 6 months to 5 years and x = amount of lead used in gasoline production (in 1000 tons) for ten 6-month periods:

x	48	59	79	80	95
y	9.3	11.0	12.8	14.1	13.6
x	95	97	102	102	107
y	13.8	14.6	14.6	16.0	18.2

a. Construct separate normal probability plots for x

and y. Do you think it is reasonable to assume that the (x, y) pairs are from a bivariate normal population?

b. Does the data provide sufficient evidence to conclude that there is a linear relationship between blood lead level and the amount of lead used in gasoline production? Use $\alpha = .01$.

56. A sample of $n = 500$ (x, y) pairs was collected and a test of $H_0: \rho = 0$ versus $H_a: \rho \neq 0$ was carried out. The resulting P-value was computed to be .00032.

a. What conclusion would be appropriate at level of significance .001?

b. Does this small P-value indicate that there is a very strong linear relationship between x and y (a value of ρ that differs considerably from 0)? Explain.

57. A sample of $n = 10,000$ (x, y) pairs resulted in $r = .022$. Test $H_0: \rho = 0$ versus $H_a: \rho \neq 0$ at level .05. Is the result statistically significant? Comment on the practical significance of your analysis.

Supplementary Exercises /Chapter 12 (58–71)

58. The paper "Refuse-Derived Fuel Evaluation in an Industrial Spreader-Stoker Boiler" (*J. Engr. for Gas Turbines and Power*, 1984, pp. 782–788) reports the accompanying data on x = % refuse-derived fuel (RDF) heat input and y = % efficiency for a certain boiler:

x	37	30	48	29	27	16	0	20
y	78.0	77.2	74.4	77.7	76.9	79.0	82.1	76.5

a. Obtain the equation of the estimated regression line.

b. Does the simple linear regression model specify a useful relationship between % RDF heat input and % efficiency? State and test the appropriate hypotheses.

c. To obtain an accurate estimate of β_1, would it have been preferable to make four observations at $x = 0$ and four observations at $x = 50$ (assuming that the model is valid for x between 0 and 50)? What about three observations at $x = 0$ and three at $x = 50$? Explain.

d. Estimate true average % efficiency when % RDF heat input is 25 using a 95% CI. Does it appear that true average % efficiency has been precisely estimated? Explain.

59. The accompanying data on x = diesel oil consumption rate measured by the drain–weigh method and y = rate measured by the Cl-trace method, both in g/hr, was read from a graph in the paper "A New Measurement Method of Diesel Engine Oil Consumption Rate" (*J. Society Auto Engr.*, 1985, pp. 28–33):

x	4	5	8	11	12	16	17	20	22	28	30	31	39
y	5	7	10	10	14	15	13	25	20	24	31	28	39

a. Assuming that x and y are related by the simple linear regression model, test $H_0: \beta_1 = 1$ versus $H_a: \beta_1 \neq 1$ using a significance level of .05.

b. Calculate the value of the sample correlation coefficient for this data.

60. The accompanying SAS output is based on data from the article "Evidence for and the Rate of Denitrification in the Arabian Sea" (*Deep Sea Research*, 1978, pp. 431–435). The variables under study are x = salinity level (%) and y = nitrate level (μM/L).

a. What is the sample size n? [HINT: Look for degrees of freedom for *SSE*.]

b. Calculate a point estimate of expected nitrate level when salinity level is 35.5.

```
Least-Squares Regression of Nitrate Level from Salinity

Model: MODEL1
Dependent Variable: NITRLVL

                    Analysis of Variance

                    Sum of        Mean          SS Resid
Source      DF      Squares       Square      F Value    Prob>F

Model       1       64.49622      64.49622     63.309    0.0002
Error       6       6.11253       1.01875
C Total     7       70.60875


          Root MSE    1.00933  R-square  0.9134
          Dep Mean   26.91250  Adj R-sq  0.8990
          C.V.        3.75043

                    Parameter Estimates

            DF    Parameter     Standard     T for H0:
Variable          Estimate        Error    Parameter=0   Prob>:T:

INTERCEP    1    326.976038    37.71380243     8.670      0.0001
SALINITY    1     -8.403964     1.05621381    -7.957      0.0002
```

c. Does there appear to be a useful linear relationship between the two variables?

d. What is the value of the sample correlation coefficient?

e. Would you use the simple linear regression model to draw conclusions when the salinity level is 40?

61. The paper "Root Regeneration and Early Growth of Red Oak Seedlings: Influence of Soil Temperature" (*Forest Science*, 1970, pp. 442–446) reports the results of a regression analysis in which the independent variable x was daily degree hours of soil heat and the dependent variable y was shoot elongation per seedling (cm). The accompanying data values were read from a graph that appeared in the paper:

x	300	350	400	400
y	5.8	4.5	5.9	6.2
x	450	450	480	480
y	6.0	7.5	6.1	8.6
x	530	530	580	580
y	8.9	8.2	14.2	11.9

x	620	620	670	700
y	11.1	11.5	14.5	14.8

$\Sigma x_i = 8140$, $\Sigma x_i^2 = 4{,}340{,}600$, $\Sigma y_i = 145.7$, $\Sigma y_i^2 = 1505.01$, $\Sigma x_i y_i = 79{,}574$

a. What proportion of variation in shoot elongation is explained by variation in soil heat?

b. Does there appear to be a useful linear relationship between the two variables? Test the appropriate hypotheses.

c. Can it be concluded that expected amount of shoot elongation differs significantly from 10 when daily degree hours of soil heat equals 500? Use a level .05 test.

62. Eye weight (g) and cornea thickness (μm) were recorded for nine randomly selected calves, and the resulting data from "The Collagens of the Developing Bovine Cornea" (*Exper. Eye Research*, 1984, pp. 639–652) is given. Use this data and a .05 significance level to test the null hypothesis of no correlation between eye weight and cornea thickness versus the alternative hypothesis of a positive correlation between eye weight and cornea thickness.

Eye Weight	.2	1.4	2.2	2.7	4.9
Thickness	416	673	733	801	957
Eye Weight	5.3	8.0	8.8	9.6	
Thickness	1035	883	736	567	

63. The accompanying data is a subset of the data that appeared in the paper "Radial Tension Strength of Pipe and Other Curved Flexural Members" (*J. Amer. Concrete Inst.*, 1980, pp. 33–39). The variables are age of a pipe specimen (x, in days) and load necessary to obtain a first crack (y, in 1000 lb/ft).

x	20	20	20	25	25
y	11.45	10.42	11.14	10.84	11.17
x	25	31	31	31	
y	10.54	9.47	9.19	9.54	

a. Calculate the equation of the estimated regression line.

b. Suppose a theoretical model suggests that the expected decrease in load associated with a 1-day increase in age is at most .10. Does the data contradict this assertion? State and test the appropriate hypotheses at significance level .05.

c. For purposes of estimating the slope of the true regression line as accurately as possible, would it have been preferable to make a single observation at each of the ages 20, 21, 22, . . . , 30, and 31? Explain.

d. Calculate an estimate of true average load to first crack when age is 28 days. Your estimate should convey information regarding precision of estimation.

64. An investigation was carried out to study the relationship between speed (ft/sec) and stride rate (number of steps taken/sec) among female marathon runners. Resulting summary quantities included $n = 11$, $\Sigma(\text{speed}) = 205.4$, $\Sigma(\text{speed})^2 = 3880.08$, $\Sigma(\text{rate}) = 35.16$, $\Sigma(\text{rate})^2 = 112.681$, and $\Sigma(\text{speed})(\text{rate}) = 660.130$.

a. Calculate the equation of the least-squares line that you would use to predict stride rate from speed.

b. Calculate the equation of the least-squares line that you would use to predict speed from stride rate.

c. Calculate the coefficient of determination for the "stride rate on speed" regression of part (a) and for the "speed on stride rate" regression of part (b). How are these related?

65. The paper "Photocharge Effects in Dye Sensitized Ag[Br,I] Emulsions at Millisecond Range Exposures" (*Photographic Sci. and Engr.*, 1981, pp. 138–144) gives the accompanying data on x = % light absorption at 5800A and y = peak photovoltage:

x	4.0	8.7	12.7	19.1	21.4
y	.12	.28	.55	.68	.85
x	24.6	28.9	29.8	30.5	
y	1.02	1.15	1.34	1.29	

a. Construct a scatter plot of this data. What does it suggest?

b. Assuming that the simple linear regression model is appropriate, obtain the equation of the estimated regression line.

c. What proportion of the observed variation in peak photovoltage can be explained by the model relationship?

d. Predict peak photovoltage when % absorption is 19.1 and compute the value of the corresponding residual.

e. The paper's authors claim that there is a useful linear relationship between % absorption and peak photovoltage. Do you agree? Carry out a formal test.

f. Give an estimate of the change in expected peak photovoltage associated with a 1% increase in light absorption. Your estimate should convey information about the precision of estimation.

g. Repeat part (f) for the expected value of peak photovoltage when % light absorption is 20.

66. In Section 12.4 we presented a formula for $V(\hat{\beta}_0 + \hat{\beta}_1 x^*)$ and a CI for $\beta_0 + \beta_1 x^*$. Taking $x^* = 0$ gives $\sigma_{\hat{\beta}_0}^2$ and a CI for β_0. Use the data of Example 12.7 to calculate the estimated standard deviation of $\hat{\beta}_0$ and a 95% CI for the y-intercept of the true regression line.

67. Let s_x and s_y denote the sample standard deviations of the observed x's and y's, respectively [so $s_x^2 = \Sigma(x_i - \bar{x})^2/(n - 1)$ and similarly for s_y^2].

a. Show that an alternative expression for the estimated regression line $y = \hat{\beta}_0 + \hat{\beta}_1 x$ is

$$y = \bar{y} + r \cdot \frac{s_y}{s_x} (x - \bar{x})$$

b. This expression for the regression line can be interpreted as follows. Suppose $r = .5$. What then is the predicted y for an x that lies 1 SD (s_x units) above the mean of the x_i's? If r were 1, the prediction would be for y to lie 1 SD above its mean $\bar{y}$, but since $r = .5$, we predict a y that is only .5 SD ($.5s_y$ units) above $\bar{y}$. Using the data in Exercise 54 for a patient whose age is 1 SD below the average age in the sample, by how many standard deviations is the patient's predicted ΔCBG above or below the average ΔCBG for the sample?

68. Verify that the t statistic for testing $H_0: \beta_1 = 0$ in Section 12.3 is identical to the t statistic in Section 12.5 for testing $H_0: \rho = 0$.

69. Use the formula for computing SSE to verify that $r^2 = 1 - SSE/SST$.

70. The paper "Increased Oxygen Consumption During the Uptake of Water by the Eversible Vesicles of *Petrobius Brevistylis*" (*J. Insect Physiology*, 1977, pp. 1285–1294) presents the results of a regression of $y =$ increased oxygen uptake (in μL) above the mean resting rate on $x =$ weight increase (mg) when dehydrated insects were allowed access to distilled water. A sample size of $n = 20$ was used, and the computed summary statistics were (approximately, based on numbers read from a graph) $\Sigma x_i = 63.5$, $\Sigma y_i = 17.26$, $\Sigma x_i^2 = 311.74$, $\Sigma x_i y_i = 71.51$, and $\Sigma y_i^2 = 19.9625$.

a. Compute the equation of the estimated regression line.

b. There was only one observation made for an x value larger than 7: for $x_{20} = 9.8$, $y_{20} = 1.9$. The investigator would like to know whether the exclusion of this point greatly alters the estimated regression relationship. Compute the estimated regression line based just on the 19 pairs with (9.8, 1.9) deleted from the sample. What y would you predict using this new line when $x = 9.8$? [HINT: First recompute the summary statistics; for example, new $\Sigma x_i =$ old $\Sigma x_i - 9.8$.]

71. Reconsider the situation of Exercise 58, in which $x = \%$ RDF heat input and $y = \%$ efficiency for a particular boiler were related via the simple linear regression model $Y = \beta_0 + \beta_1 x + \epsilon$. Suppose that for a second boiler, these variables are also related via the simple linear regression model $Y = \gamma_0 + \gamma_1 x + \epsilon$ and that $V(\epsilon) = \sigma^2$ for both boilers. If the data set consists of n_1 observations on the first boiler and n_2 on the second and SSE_1 and SSE_2 denote the two error sums of squares, then a pooled estimate of σ^2 is $\hat{\sigma}^2 = (SSE_1 + SSE_2)/(n_1 + n_2 - 4)$. Let SS_{x1} and SS_{x2} denote $\Sigma(x_i - \bar{x})^2$ for the data on the first and second boilers, respectively. A test of $H_0: \beta_1 - \gamma_1 = 0$ (equal slopes) is based on the statistic

$$T = \frac{\hat{\beta}_1 - \hat{\gamma}_1}{\hat{\sigma}\sqrt{\dfrac{1}{SS_{x_1}} + \dfrac{1}{SS_{x_2}}}}$$

When H_0 is true, T has a t distribution with $n_1 + n_2 - 4$ df. Suppose the six observations on the second boiler are (0, 81.3), (10, 78.4), (20, 78.2), (25, 79.1), (30, 77.6), and (40, 77.4). Using this along with the data of Exercise 58, carry out a test at level .05 to see whether expected change in % efficiency associated with a 1% increase in RDF heat input is identical for the two boilers.

Bibliography

Draper, Norman, and Harry Smith, *Applied Regression Analysis* (2nd ed.), Wiley, New York, 1981. The most comprehensive and authoritative book on regression analysis currently in print.

Neter, John, William Wasserman, and Michael Kutner, *Applied Linear Statistical Models* (3rd ed.), Irwin, Homewood, IL, 1990. The first 15 chapters constitute an extremely readable and informative survey of regression analysis.

Younger, Mary Sue, *A Handbook for Linear Regression* (2nd ed.), Duxbury Press, Boston, 1985. A readable presentation of regression models and methodology; its discussion of the capabilities of the BMD, SAS, and SPSS "canned" statistical computer packages is especially enlightening.

13 Nonlinear and Multiple Regression

$$\Sigma\Sigma\Sigma\Sigma\Sigma\Sigma\Sigma\Sigma\Sigma\Sigma\Sigma\Sigma\Sigma\Sigma\Sigma\Sigma\Sigma\Sigma\Sigma$$

Introduction

The probabilistic model studied in Chapter 12 specified that the observed value of the dependent variable Y deviated from the linear regression function $\mu_{Y \cdot x} = \beta_0 + \beta_1 x$ by a random amount. Here we consider two ways of generalizing the simple linear regression model. The first way is to replace $\beta_0 + \beta_1 x$ by a nonlinear function of x, while the second is to use a regression function involving more than a single independent variable. After fitting a regression function of the chosen form to the given data, it is of course important to have methods available for making inferences about the parameters of the chosen model. Before these methods are used, though, the data analyst should first assess the validity of the chosen model. In Section 13.1, we discuss methods, based primarily on a graphical analysis of the residuals (observed minus predicted y's), for checking the aptness of a fitted model.

In Section 13.2, we consider nonlinear regression functions of a single independent variable x that are "intrinsically linear." By this we mean that it is possible to transform one or both of the variables so that the relationship between the new variables is linear. Another class of nonlinear relations is obtained by using polynomial regression functions of the form $\mu_{Y \cdot x} = \beta_0 + \beta_1 x + \beta_2 x^2 + \cdots + \beta_k x^k$; these polynomial models are the subject of Section 13.3. Multiple regression analysis involves building models for relating y to two or more independent variables. The focus in Section 13.4 is on interpretation of the parameters of various multiple regression models and on understanding and using the regression output of widely available statistical computer packages. The last section of the chapter surveys some extensions and pitfalls of multiple regression modeling.

13.1   Aptness of the Model and Model Checking

A plot of the observed pairs (x_i, y_i) is a necessary first step in deciding on the form of a mathematical relationship between x and y. It is possible to fit many functions other than a linear one ($y = b_0 + b_1 x$) to the data, using either the principle of least squares or another fitting method. Once a function of the chosen form has been fitted, it is important to check the fit of the model to see whether it is in fact appropriate. One way to study the fit is to superimpose a graph of the best-fit function on the scatter plot of the data. However, any tilt or curvature of the best-fit function may obscure some aspects of the fit that should be investigated. Furthermore, the scale on the vertical axis may make it difficult to assess the extent to which observed values deviate from the best-fit functions.

Residuals and Fitted Values

A more effective approach to assessment of model adequacy is to compute the fitted or predicted values $\hat{y}_i$ and the residuals $e_i = y_i - \hat{y}_i$ and then plot various functions of these computed quantities. We then examine the plots either to confirm our choice of model or for indications that the model is not appropriate. Suppose the simple linear regression model is correct and let $y = \hat{\beta}_0 + \hat{\beta}_1 x$ be the equation of the estimated regression line. Then the ith residual is $e_i = y_i - (\hat{\beta}_0 + \hat{\beta}_1 x)$. To derive properties of the residuals, let $e_i = Y_i - \hat{Y}_i$ represent the ith residual as a random variable (rv) (before observations are actually made). Then

$$E(Y_i - \hat{Y}_i) = E(Y_i) - E(\hat{\beta}_0 + \hat{\beta}_1 x_i) = \beta_0 + \beta_1 x_i - (\beta_0 + \beta_1 x_i) = 0 \qquad (13.1)$$

so each residual has expected value 0. Because $\hat{Y}_i \ (= \hat{\beta}_0 + \hat{\beta}_1 x_i)$ is a linear function of the Y_j's, so is $Y_i - \hat{Y}_i$ (where the coefficients depend on the x_j's). Thus, the normality of the Y_j's implies that each residual is normally distributed. It can also be shown that

$$V(Y_i - \hat{Y}_i) = \sigma^2 \cdot \left[1 - \frac{1}{n} - \frac{(x_i - \bar{x})^2}{\Sigma(x_j - \bar{x})^2} \right] \qquad (13.2)$$

Because we will want to know whether any given residual is unusually large, the **standardized residuals**

$$e_i^* = \frac{y_i - \hat{y}_i}{s\sqrt{1 - \frac{1}{n} - \frac{(x_i - \bar{x})^2}{\Sigma(x_j - \bar{x})^2}}} \qquad i = 1, \ldots n \qquad (13.3)$$

will be useful. Notice that the variances of the residuals differ from one another. If n is reasonably large, though, the bracketed term in (13.2) will be approximately 1, so some sources use e_i/s as the standardized residual. Computation of the e_i^*'s can be tedious, but several of the most widely used statistical computer packages (such as MINITAB) automatically provide them and (upon request) can construct various plots involving them.

Example 13.1   Exercise 9 in Chapter 12 presented data on $x =$ burner area liberation rate and $y = \mathrm{NO}_x$ emissions. Here we reproduce the data and give the fitted values, residuals, and standardized residuals. The estimated regression line is $y = -45.55 + 1.71x$, and $r^2 = .961$. Notice that the standardized residuals are not a constant multiple of the residuals (i.e., $e_i^* \neq e_i/s$):

x_i	y_i	$\hat{y}_i$	e_i	e_i^*
100	150	125.6	24.4	.75
125	140	168.4	−28.4	−.84
125	180	168.4	11.6	.35
150	210	211.1	−1.1	−.03
150	190	211.1	−21.1	−.62
200	320	296.7	23.3	.66
200	280	296.7	−16.7	−.47
250	400	382.3	17.7	.50
250	430	382.3	47.7	1.35
300	440	467.9	−27.9	−.80
300	390	467.9	−77.9	−2.24
350	600	553.4	46.6	1.39
400	610	639.0	−29.0	−.92
400	670	639.0	31.0	.99

∎

Diagnostic Plots

The basic plots that many statisticians recommend for an assessment of model validity and usefulness are the following:

1. e_i^* (or e_i) on the vertical axis versus x_i on the horizontal axis
2. e_i^* (or e_i) on the vertical axis versus $\hat{y}_i$ on the horizontal axis
3. $\hat{y}_i$ on the vertical axis versus y_i on the horizontal axis
4. A normal probability plot of the standardized residuals.

Plots 1 and 2 are called **residual plots** (against the independent variable and fitted values, respectively), while Plot 3 is fitted against observed values.

If Plot 3 yields points close to the 45° line [slope + 1 through (0, 0)], then the estimated regression function gives accurate predictions of the values actually observed. Thus, Plot 3 provides a visual assessment of model effectiveness in making predictions. Provided that the model is correct, neither residual plot should exhibit distinct patterns. The residuals should be randomly distributed about 0 according to a normal distribution, so all but a very few standardized residuals should lie between −2 and +2 (i.e., all but a few residuals within 2 SDs of their expected value 0). The plot of standardized residuals versus $\hat{y}$ is really a combination of the two other plots, showing implicitly both how residuals vary with x and how fitted values compare with observed values. This latter plot is the single one most often recommended for multiple regression analysis. Plot 4 allows the analyst to assess the plausibility of the assumption that ϵ has a normal distribution.

Example 13.2
(Example 13.1 continued)

Figure 13.1 presents a scatter plot of the data and the four plots recommended in 1–4 above. The plot of $\hat{y}$ versus y confirms the impression given by r^2 that x is effective in predicting y and also indicates that there is no observed y for which the predicted

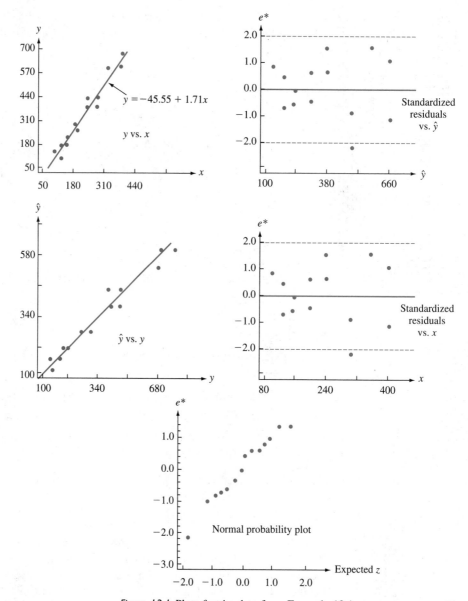

Figure 13.1 Plots for the data from Example 13.1

value is terribly far off the mark. Both residual plots show no unusual pattern or discrepant values. There is one standardized residual slightly outside the interval $(-2, 2)$, but this is not surprising in a sample of size 14. The normal probability plot of the standardized residuals is reasonably straight (in Chapter 14, we present a formal test for normality based on the points in a normal probability plot; when applied to the data here, it appears quite plausible that the ϵ distribution is normal). In summary, the plots leave us with no qualms about either the appropriateness of a simple linear relationship or the fit to the given data. ∎

Difficulties and Remedies

While we hope that our analysis will yield plots like those of Figure 13.1, quite frequently the plots will suggest one or more of the following difficulties:

1. A nonlinear probabilistic relationship between x and y is appropriate.
2. The variance of ϵ (and of Y) is not a constant σ^2 but depends on x.
3. The selected model fits the data well except for a very few discrepant or outlying data values, which may have greatly influenced the choice of the best-fit function.
4. The error term ϵ does not have a normal distribution.
5. When the subscript i indicates the time order of the observations, the ϵ_i's exhibit dependence over time.
6. One or more relevant independent variables have been omitted from the model.

Figure 13.2 presents residual plots corresponding to items 1–3, 5, and 6. In Chapter 4, we discussed patterns in normal probability plots that cast doubt on the assumption of an underlying normal distribution. Notice that the residuals from the data in Figure 13.2d with the circled point included would not by themselves necessarily suggest further analysis, yet when a new line is fit with that point deleted, the new line differs considerably from the original line. This type of behavior is more difficult to identify in multiple regression. It is most likely to arise when there is a single (or very few) data point(s) with independent variable value(s) far removed from the remainder of the data.

We now indicate briefly what remedies are available for the types of difficulties. For a more comprehensive discussion, one or more of the references on regression analysis should be consulted. If the residual plot looks something like that of Figure 13.2a, exhibiting a curved pattern, then a nonlinear function of x may be fit; in Section 13.2, we show how transformations can be used to this end, while in Section 13.3 we discuss polynomial models.

The residual plot of Figure 13.2b suggests that while a straight-line relationship may be reasonable, the assumption that $V(Y_i) = \sigma^2$ for each i is of doubtful validity. When the assumptions of Chapter 12 are valid, it can be shown that among all unbiased estimators of β_0 and β_1, the ordinary least-squares estimators have minimum variance. These estimators give equal weight to each (x_i, Y_i). If the variance of Y increases with x, then Y_i's for large x_i should be given less weight than those with small x_i. This suggests that β_0 and β_1 should be estimated by minimizing

$$f_w(b_0, b_1) = \Sigma w_i[y_i - (b_0 + b_1 x_i)]^2 \tag{13.4}$$

where the w_i's are weights that decrease with increasing x_i. Minimization of (13.4) yields **weighted least-squares** estimates. For example, if the standard deviation of Y is proportional to x (for $x > 0$)—that is, $V(Y) = kx^2$—then it can be shown that the weights $w_i = 1/x_i^2$ yield best estimators of β_0 and β_1. The books by John Neter et al.

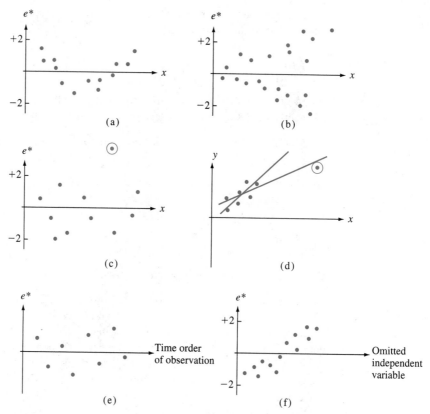

Figure 13.2 Plots that indicate abnormality in data: (a) nonlinear relationship; (b) nonconstant variance; (c) discrepant observation; (d) observation with large influence; (e) dependence in errors; (f) variable omitted

and by S. Chatterjee and Bertram Price contain more detail (see chapter bibliography). Weighted least squares is used quite frequently by econometricians (economists who use statistical methods) to estimate parameters.

When plots or other evidence suggest that the data set contains outliers or points having large influence on the resulting fit, one possible approach is to omit these outlying points and recompute the estimated regression equation. This would certainly be correct if it was found that the outliers resulted from errors in recording data values or experimental errors. If no assignable cause can be found for the outliers, it is still desirable to report the estimated equation both with and without outliers omitted. Yet another approach is to retain possible outliers but to use an estimation principle that puts relatively less weight on outlying values than does the principle of least squares. One such principle is MAD (minimize absolute deviations), which selects $\hat{\beta}_0$ and $\hat{\beta}_1$ to minimize $\Sigma |y_i - (b_0 + b_1 x_i)|$. Unlike the estimates of least squares, there are no nice formulas for the MAD estimates; their values must be found by using an iterative computational procedure. Such procedures are also used when it is suspected that the ϵ_i's have a distribution that is not normal but instead has "heavy tails" (making it much more likely than for the normal distribution that discrepant values will enter the sample); robust regression procedures are those that produce reliable estimates for a wide variety of underlying error distributions. Least-squares estimators are not robust in the same way that the sample mean $\bar{X}$ is not a robust estimator for μ.

When a plot suggests time dependence in the error terms, an appropriate analysis may involve a transformation of the y's or else a model explicitly including a time variable. Lastly, a plot such as that of Figure 13.2f, which shows a pattern in the residuals when plotted against an omitted variable, suggests that a multiple regression model that includes the previously omitted variable should be considered. We discuss such models in the last two sections of the chapter.

Exercises   Section 13.1 (1–14)

1. Suppose the variables x = commuting distance and y = commuting time are related according to the simple linear regression model with $\sigma = 10$.

  a. If $n = 5$ observations are made at the x values $x_1 = 5$, $x_2 = 10$, $x_3 = 15$, $x_4 = 20$, and $x_5 = 25$, calculate the standard deviations of the five corresponding residuals.

  b. Repeat part (a) for $x_1 = 5$, $x_2 = 10$, $x_3 = 15$, $x_4 = 20$, and $x_5 = 50$.

  c. What do the results of parts (a) and (b) imply about the deviation of the estimated line from the observation made at the largest sampled x value?

2. The x values and standardized residuals for the chlorine flow/etch rate data of Example 12.3 are displayed in the accompanying table. Construct a standardized residual plot and comment on its appearance.

x	1.50	1.50	2.00	2.50	2.50
e^*	.31	1.02	−1.15	−1.23	.23
x	3.00	3.50	3.50	4.00	
e^*	.73	−1.36	1.53	.07	

3. Example 12.5 presented the residuals from a simple linear regression of bark lead content y on traffic flow x.

  a. Plot the residuals against x. Does the resulting plot suggest that a straight-line regression function is a reasonable choice of model? Explain your reasoning.

  b. Using $s = 92.19$, compute the values of the standardized residuals. Is $e_i^* \approx e_i/s$ for $i = 1, \ldots, n$,

or are the e_i^*'s not close to being proportional to the e_i's?

  c. Plot the standardized residuals against x. Does the plot differ significantly in general appearance from the plot of part (a)?

4. Wear resistance of certain nuclear reactor components made of Zircaloy-2 is partly determined by properties of the oxide layer. The following data appears in a paper that proposed a new nondestructive testing method to monitor thickness of the layer ("Monitoring of Oxide Layer Thickness on Zircaloy-2 by the Eddy Current Test Method," *J. of Testing and Eval.*, 1987, pp. 333–336). The variables are x = oxide-layer thickness (μm) and y = eddy current response (arbitrary units).

x	0	7	17	114	133
y	20.3	19.8	19.5	15.9	15.1
x	142	190	218	237	285
y	14.7	11.9	11.5	8.3	6.6

The authors summarized the relationship by giving the equation of the least-squares line as $y = 20.6 - .047x$. Calculate and plot the residuals against x and then comment on the appropriateness of the simple linear regression model.

5. Refer to Exercise 4 and use $s = .7921$ to calculate the standardized residuals from a simple linear regression. Construct a standardized residual plot and comment.

6. Exercise 41 in Chapter 12 presented data on y = magnetic relaxation time and x = strength of the external magnetic field. A simple linear regression yields the estimated regression function $y = -8.78 + 18.87x$, with $s = 16.61$.

a. Compute the values of the residuals and plot the residuals against x. Does the plot suggest that a linear regression function is inappropriate?

b. Compute the values of the standardized residuals and plot them against x. Are there any unusually large (positive or negative) standardized residuals? Does this plot give the same message as the plot of part (a) regarding the appropriateness of a linear regression function?

c. Plot the predicted against the observed values. Does the plot indicate that the model gives reliable predictions?

d. Construct a normal probability plot of the standardized residuals and comment on its appearance.

7. The article "Effects of Gamma Radiation on Juvenile and Mature Cuttings of Quaking Aspen" (*Forest Science*, 1967, pp. 240–245) reports the following data on exposure time to radiation (x, in kr/16 hr) and dry weight of roots (y, in mg $\times 10^{-1}$).

x	0	2	4	6	8
y	110	123	119	86	62

a. Construct a scatter plot. Does the plot suggest that a linear probabilistic relationship is appropriate?

b. A linear regression results in the least-squares line $y = 127 - 6.65x$, with $s = 16.94$. Compute the residuals and standardized residuals and then construct residual plots. What do these plots suggest? What type of function should provide a better fit to the data than a straight line does?

8. The article "Pedunculate Oak Woodland in a Severe Environment" (*J. Ecology*, 1978, pp. 707–740) reports the following data on x = age (years) and y = annual trunk diameter growth increment (mm) for a sample of trees in a certain region:

x	17	23	30	37.5	40
y	2.20	1.25	.85	1.30	1.70
x	46.5	50	54	55	93
y	.75	.75	.50	1.00	.70

The estimated regression line is $y = 1.78 - .0154x$, with $s = .434$. Does a scatter plot suggest that a particular observation has excessively influenced the choice of a

best fit line? Is this impression confirmed by the standardized residuals?

9. Consider the following four (x, y) data sets; the first three have the same x values, so these values are listed only once (from Frank Anscombe, "Graphs in Statistical Analysis," *Amer. Statistician*, 1973, pp. 17–21):

Data Set	1–3	1	2	3	4	4
Variable	x	y	y	y	x	y
	10.0	8.04	9.14	7.46	8.0	6.58
	8.0	6.95	8.14	6.77	8.0	5.76
	13.0	7.58	8.74	12.74	8.0	7.71
	9.0	8.81	8.77	7.11	8.0	8.84
	11.0	8.33	9.26	7.81	8.0	8.47
	14.0	9.96	8.10	8.84	8.0	7.04
	6.0	7.24	6.13	6.08	8.0	5.25
	4.0	4.26	3.10	5.39	19.0	12.50
	12.0	10.84	9.13	8.15	8.0	5.56
	7.0	4.82	7.26	6.42	8.0	7.91
	5.0	5.68	4.74	5.73	8.0	6.89

For each of these four data sets, the values of the summary statistics Σx_i, Σx_i^2, Σy_i, Σy_i^2, and $\Sigma x_i y_i$ are identical, so all quantities computed from these five will be identical for the four sets—the least-squares line ($y = 3 + .5x$), *SSE*, s^2, r^2, t intervals, t statistics, and so on. The summary statistics provide no way of distinguishing among the four data sets. Based on a scatter plot and a residual plot for each set, comment on the appropriateness or inappropriateness of fitting a straight-line model; include in your comments any specific suggestions for how a "straight-line analysis" might be modified or qualified.

10. a. Show that $\Sigma_{i=1}^{n} e_i = 0$ when the e_i's are the residuals from a simple linear regression.

b. Are the residuals from a simple linear regression independent of one another, positively correlated, or negatively correlated? Explain.

c. Show that $\Sigma_{i=1}^{n} x_i e_i = 0$ for the residuals from a simple linear regression. [This result along with part (a) shows that there are two linear restrictions on the e_i's, resulting in a loss of 2 df when the squared residuals are used to estimate σ^2.]

d. Is it true that $\Sigma_{i=1}^{n} e_i^* = 0$? Give a proof or counterexample.

11. a. Express the ith residual $Y_i - \hat{Y}_i$ (where $\hat{Y}_i = \hat{\beta}_0 + \hat{\beta}_1 x_i$) in the form $\Sigma c_j Y_j$, a linear function of the Y_j's.

Then use rules of variance to verify that $V(Y_i - \hat{Y}_i)$ is given by expression (13.2).

b. It can be shown that $\hat{Y}_i$ and $Y_i - \hat{Y}_i$ (the ith predicted value and residual) are independent of one another. Use this fact, the relation $Y_i = \hat{Y}_i + (Y_i - \hat{Y}_i)$, and the expression for $V(\hat{Y})$ from Section 12.4 to again verify (13.2).

c. As x_i moves further away from $\bar{x}$, what happens to $V(\hat{Y}_i)$ and to $V(Y_i - \hat{Y}_i)$?

12. a. Could a linear regression result in residuals 23, –27, 5, 17, –8, 9, and 15? Why, or why not?

b. Could a linear regression result in residuals 23, –27, 5, 17, –8, –12, and 2 corresponding to x values 3, –4, 8, 12, –14, –20, and 25? Why, or why not? (HINT: See Exercise 10.)

13. Recall that $\hat{\beta}_0 + \hat{\beta}_1 x$ has a normal distribution with expected value $\beta_0 + \beta_1 x$ and variance

$$\sigma^2 \left\{ \frac{1}{n} + \frac{(x - \bar{x})^2}{\Sigma(x_i - \bar{x})^2} \right\}$$

so that

$$Z = \frac{\hat{\beta}_0 + \hat{\beta}_1 x - (\beta_0 + \beta_1 x)}{\sigma \left(\frac{1}{n} + \frac{(x - \bar{x})^2}{\Sigma(x_i - \bar{x})^2} \right)^{1/2}}$$

has a standard normal distribution. If $S = \sqrt{SSE/(n - 2)}$ is substituted for σ, the resulting variable has a t distribution with $n - 2$ df. By analogy, what is the distribution of any particular standardized residual? If $n = 25$, what is the probability that a particular standardized residual falls outside the interval $(-2.50, 2.50)$?

14. If there is at least one x value at which more than one observation has been made, there is a formal test procedure for testing

H_0: $\mu_{Y \cdot x} = \beta_0 + \beta_1 x$ for some values β_0, β_1 (the true regression function is linear)

versus

H_a: H_0 is not true (the true regression function is not linear)

Suppose observations are made at $x_1, x_2, \ldots, x_c$ and let $Y_{11}, Y_{12}, \ldots, Y_{1n_1}$ denote the n_1 observations when $x = x_1; \ldots; Y_{c1}, Y_{c2}, \ldots, Y_{cn_c}$ denote the n_c observations when $x = x_c$. With $n = \Sigma n_i$ (the total number of observations), SSE has $n - 2$ df. We break SSE into two pieces, $SSPE$ (pure error) and $SSLF$ (lack of fit), as follows:

$$SSPE = \Sigma_i \Sigma_j (Y_{ij} - \bar{Y}_{i \cdot})^2$$
$$= \Sigma \Sigma Y_{ij}^2 - \Sigma n_i \bar{Y}_{i \cdot}^2$$

$$SSLF = SSE - SSPE$$

The n_i observations at x_i contribute $n_i - 1$ df to $SSPE$, so the number of degrees of freedom for $SSPE$ is $\Sigma_i(n_i - 1) = n - c$ and the degrees of freedom for $SSLF$ is $n - 2 - (n - c) = c - 2$. Let $MSPE = SSPE/(n - c)$ and $MSLF = SSLF/(c - 2)$. Then it can be shown that while $E(MSPE) = \sigma^2$ whether or not H_0 is true, $E(MSLF) = \sigma^2$ if H_0 is true and $E(MSLF) > \sigma^2$ if H_0 is false.

Test statistic: $F = \dfrac{MSLF}{MSPE}$

Rejection region: $f \geq F_{\alpha, c-2, n-c}$

The following data comes from the paper "Changes in Growth Hormone Status Related to Body Weight of Growing Cattle" (*Growth*, 1977, pp. 241–247), with x = body weight and y = metabolic clearance rate/body weight.

x	110	110	110	230	230	230	360
y	235	198	173	174	149	124	115
x	360	360	360	505	505	505	505
y	130	102	95	122	112	98	96

(So $c = 4$, $n_1 = n_2 = 3$, $n_3 = n_4 = 4$.)

a. Test H_0 versus H_a at level .05 using the above lack-of-fit test.

b. Does a scatter plot of the data suggest that the relationship between x and y is linear? How does this compare with the result of part (a)? (A nonlinear regression function was used in the paper.)

13.2   Regression with Transformed Variables

The necessity for an alternative model to the linear probabilistic model $Y = \beta_0 + \beta_1 x + \epsilon$ may be suggested either by a theoretical argument or else by examining diagnostic plots from a linear regression analysis. In either case, settling on a model whose parameters can be easily estimated is desirable. An important class of such models is specified by means of functions that are "intrinsically linear."

Definition

> A function relating y to x is **intrinsically linear** if by means of a transformation on x and/or y, the function can be expressed as $y' = \beta_0 + \beta_1 x'$, where $x' = $ the transformed independent variable and $y' = $ the transformed dependent variable.

Four of the most useful intrinsically linear functions are given in Table 13.1. In each case, the appropriate transformation is either a log transformation—either base 10 or natural logarithm (base e)—or a reciprocal transformation. Representative graphs of the four functions appear in Figure 13.3.

Table 13.1 Useful intrinsically linear functions*

Function	Transformation(s) to Linearize	Linear Form
a. Exponential: $y = \alpha e^{\beta x}$	$y' = \ln(y)$	$y' = \ln(\alpha) + \beta x$
b. Power: $y = \alpha x^{\beta}$	$y' = \log(y),\ x' = \log(x)$	$y' = \log(\alpha) + \beta x'$
c. $y = \alpha + \beta \cdot \log(x)$	$x' = \log(x)$	$y = \alpha + \beta x'$
d. Reciprocal: $y = \alpha + \beta \cdot \dfrac{1}{x}$	$x' = \dfrac{1}{x}$	$y = \alpha + \beta x'$

*When $\log(\cdot)$ appears, either a base-10 or a base-e logarithm can be used.

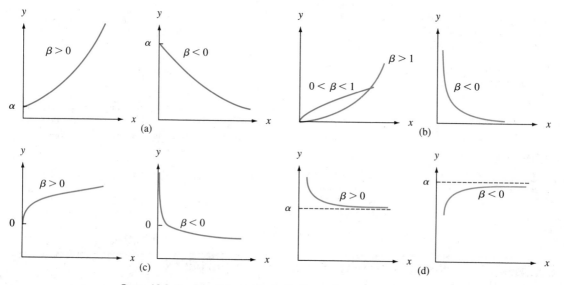

Figure 13.3 Graphs of the intrinsically linear functions given in Table 13.1

Thus, for an exponential function relationship, only y is transformed to achieve linearity, while for a power function relationship, both x and y are transformed. Because the variable x is in the exponent in an exponential relationship, y increases (if $\beta > 0$) or decreases (if $\beta < 0$) much more rapidly as x increases than is the case for the power function, though over a short interval of x values it can be difficult to differentiate between the two functions. Examples of functions that are not intrinsically linear are $y = \alpha + \gamma e^{\beta x}$ and $y = \alpha + \gamma x^{\beta}$.

Intrinsically linear functions lead directly to probabilistic models which, though not linear in x as a function, have parameters whose values are easily estimated using ordinary least squares.

Definition

> A probabilistic model relating Y to x is **intrinsically linear** if, by means of a transformation on Y and/or x, it can be reduced to a linear probabilistic model $Y' = \beta_0 + \beta_1 x' + \epsilon'$.

Corresponding to the four functions of Table 13.1, the intrinsically linear probabilistic models are

a. $Y = \alpha e^{\beta x} \cdot \epsilon$, a multiplicative exponential model, so that $\ln(Y) = Y' = \beta_0 + \beta_1 x' + \epsilon'$ with $x' = x$, $\beta_0 = \ln(\alpha)$, $\beta_1 = \beta$, and $\epsilon' = \ln(\epsilon)$.

b. $Y = \alpha x^{\beta} \cdot \epsilon$, a multiplicative power model, so that $\log(Y) = Y' = \beta_0 + \beta_1 x' + \epsilon'$ with $x' = \log(x)$, $\beta_0 = \log(\alpha)$, $\beta_1 = \beta$, and $\epsilon' = \log(\epsilon)$.

c. $Y = \alpha + \beta \log(x) + \epsilon$, so that $x' = \log(x)$ immediately linearizes the model.

d. $Y = \alpha + \beta \cdot 1/x + \epsilon$, so that $x' = 1/x$ yields a linear model.

The additive exponential and power models, $Y = \alpha e^{\beta x} + \epsilon$ and $Y = \alpha x^{\beta} + \epsilon$, are not intrinsically linear. Notice that both (a) and (b) require a transformation on Y and, as a result, a transformation on the error variable ϵ. In fact, if ϵ has a lognormal distribution (see Chapter 4) with $E(\epsilon) = e^{\sigma^2/2}$ and $V(\epsilon) = \tau^2$ independent of x, then the transformed models for both (a) and (b) will satisfy all the assumptions of Chapter 12 regarding the linear probabilistic model; this in turn implies that all inferences for the parameters of the transformed model based on these assumptions will be valid. If σ^2 is small, $\mu_{Y \cdot x} \approx \alpha e^{\beta x}$ in (a) or αx^{β} in (b).

The major advantage of an intrinsically linear model is that the parameters β_0 and β_1 of the transformed model can be immediately estimated using the principle of least squares simply by substituting x' and y' into the estimating formulas:

$$\hat{\beta}_1 = \frac{n \, \Sigma x_i' y_i' - \Sigma x_i' \, \Sigma y_i'}{n \, \Sigma (x_i')^2 - (\Sigma x_i')^2}$$

$$\hat{\beta}_0 = \frac{\Sigma y_i' - \hat{\beta}_1 \, \Sigma x_i'}{n}$$

(13.5)

Parameters of the original nonlinear model can then be estimated by transforming back $\hat{\beta}_0$ and/or $\hat{\beta}_1$ if necessary. In cases (a) and (b), when σ^2 is small an approximate confidence interval for $\mu_{Y \cdot x}$ results from taking antilogs in the interval for $\beta_0 + \beta_1 x$.*

Example 13.3

Taylor's equation for tool life y as a function of cutting time x states that $xy^c = k$ or, equivalently, that $y = \alpha x^\beta$. The paper "The Effect of Experimental Error on the Determination of Optimum Metal Cutting Conditions" (*J. Eng. for Industry*, 1967, pp. 315–322) observes that the relationship is not exact (deterministic) and that the parameters α and β must be estimated from data. Thus, an appropriate model is the multiplicative power model $Y = \alpha \cdot x^\beta \cdot \epsilon$, which the author fit to the accompanying data consisting of 12 carbide tool life observations (Table 13.2). In addition to the x, y, x', and y' values, the predicted transformed values ($\hat{y}'$) and the predicted values on the original scale ($\hat{y}$, after transforming back) are given.

Table 13.2 Data for Example 13.3

	x	y	$x' = \ln(x)$	$y' = \ln(y)$	$\hat{y}'$	$\hat{y} = e^{\hat{y}'}$
1	600.	2.3500	6.39693	0.85442	1.12754	3.0881
2	600.	2.6500	6.39693	0.97456	1.12754	3.0881
3	600.	3.0000	6.39693	1.09861	1.12754	3.0881
4	600.	3.6000	6.39693	1.28093	1.12754	3.0881
5	500.	6.4000	6.21461	1.85630	2.11203	8.2650
6	500.	7.8000	6.21461	2.05412	2.11203	8.2650
7	500.	9.8000	6.21461	2.28238	2.11203	8.2650
8	500.	16.5000	6.21461	2.80336	2.11203	8.2650
9	400.	21.5000	5.99146	3.06805	3.31694	27.5760
10	400.	24.5000	5.99146	3.19867	3.31694	27.5760
11	400.	26.0000	5.99146	3.25810	3.31694	27.5760
12	400.	33.0000	5.99146	3.49651	3.31694	27.5760

The summary statistics for fitting a straight line to the transformed data are $\Sigma x_i' = 74.41200$, $\Sigma y_i' = 26.22601$, $\Sigma x_i'^2 = 461.75874$, $\Sigma y_i'^2 = 67.74609$, and $\Sigma x_i' y_i' = 160.84601$, so

$$\hat{\beta}_1 = \frac{12(160.84601) - (74.41200)(26.22601)}{12(461.75874) - (74.41200)^2} = -5.3996$$

$$\hat{\beta}_0 = \frac{26.22601 - (-5.3996)(74.41200)}{12} = 35.6684$$

The estimated values of α and β, the parameters of the power function model, are $\hat{\beta} = \hat{\beta}_1 = -5.3996$ and $\hat{\alpha} = e^{\hat{\beta}_0} = 3.094491530 \cdot 10^{15}$. Thus, the estimated regression function is $\hat{\mu}_{Y \cdot x} = 3.094491530 \cdot 10^{15} \cdot x^{-5.3996}$. To recapture Taylor's (estimated) equation, set $y = 3.094491530 \cdot 10^{15} \cdot x^{-5.3996}$, whence $xy^{.185} = 740$.

---

*Strictly speaking, taking antilogs gives a confidence interval for the median of the Y distribution for the fixed x value—that is, for $\tilde{\mu}_{Y \cdot x}$. Because the lognormal distribution is positively skewed, $\mu_{Y \cdot z} > \tilde{\mu}_{Y \cdot x}$; the two are approximately equal if σ^2 is close to 0.

Figure 13.4a gives a plot of the standardized residuals from the linear regression using transformed variables (for which $r^2 = .922$); there is no apparent pattern in the plot, though one standardized residual is a bit large, and the residuals look as they should for a simple linear regression. Figure 13.4b pictures a plot of $\hat{y}$ versus y, which indicates satisfactory predictions on the original scale.

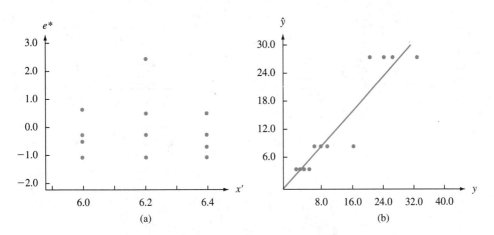

Figure 13.4 (a) Standardized residuals versus x' from Example 13.3; (b) $\hat{y}$ versus y from Example 13.3

To obtain a confidence interval for median tool life when cutting time is 500, we transform $x = 500$ to $x' = 6.21461$. Then $\hat{\beta}_0 + \hat{\beta}_1 x' = 2.1120$, and a 95% CI for $\beta_0 + \beta_1(6.21461)$ is (from Section 12.4) $2.1120 \pm (2.228)(.0824) = (1.928, 2.296)$. The 95% CI for $\tilde{\mu}_{Y \cdot 500}$ is then obtained by taking antilogs: $(e^{1.928}, e^{2.296}) = (6.876, 9.930)$. It is easily checked that for the transformed data $s^2 = \hat{\sigma}^2 \approx .081$. Because this is quite small, $(6.876, 9.930)$ is an approximate interval for $\mu_{Y \cdot 500}$. ∎

Example 13.4
In the article "Ethylene Synthesis in Lettuce Seeds: Its Physiological Significance" (*Plant Physiology*, 1972, pp. 719–722), ethylene content of lettuce seeds (y, in nL/g dry wt) was studied as a function of exposure time (x, in min) to an ethylene absorbent. Figure 13.5 presents both a scatter plot of the data and a plot of the residuals generated from a linear regression of y on x. Both plots show a strong curved pattern, suggesting that a transformation to achieve linearity is appropriate. In addition, a linear regression gives negative predictions for $x = 90$ and $x = 100$.

The author did not give any argument for a theoretical model, but his plot of $y' = \ln(y)$ versus x shows a strong linear relationship, suggesting that an exponential function will provide a good fit to the data. Table 13.3 shows the data values and other information from a linear regression of y' on x. The estimates of parameters of the linear model are $\hat{\beta}_1 = -.0323$ and $\hat{\beta}_0 = 5.941$, with $r^2 = .995$. The estimated regression function for the exponential model is $\hat{\mu}_{Y \cdot x} \approx e^{\beta_0} \cdot e^{\beta_1 x} = 380.32 e^{-.0323x}$. The predicted values $\hat{y}_i$ can then be obtained by substitution of x_i ($i = 1, \ldots, n$) into $\hat{\mu}_{Y \cdot x}$ or else by computing $\hat{y}_i = e^{\hat{y}'_i}$ where the $\hat{y}'_i$'s are the predictions from the

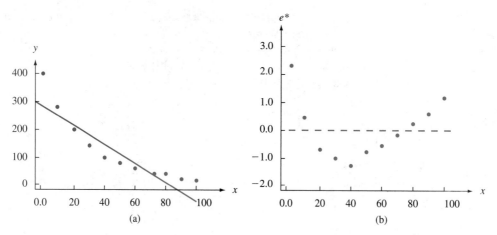

Figure 13.5 (a) Scatter plot; (b) residual plot from linear regression for the data in Example 13.4

Table 13.3 Data for Example 13.4

x	y	$y' = \ln(y)$	$\hat{y}'$	$\hat{y} = e^{\hat{y}'}$
2	408	6.01	5.876	353.32
10	274	5.61	5.617	275.12
20	196	5.28	5.294	199.12
30	137	4.92	4.971	144.18
40	90	4.50	4.647	104.31
50	78	4.36	4.324	75.50
60	51	3.93	4.001	54.64
70	40	3.69	3.677	39.55
80	30	3.40	3.354	28.62
90	22	3.09	3.031	20.72
100	15	2.71	2.708	15.00

transformed straight-line model. Figure 13.6 presents both a plot of e'^{*} versus x (the standardized residuals from a linear regression) and a plot of $\hat{y}$ versus y. These plots support the choice of an exponential model. ∎

In analyzing transformed data, one should keep in mind the following points:

1. Estimating β_1 and β_0 as in (13.5) above and then transforming back to obtain estimates of the original parameters is not equivalent to using the principle of least squares directly on the original model. Thus, for the exponential model, we could estimate α and β by minimizing $\Sigma(y_i - \alpha e^{\beta x_i})^2$. The resulting estimates would not be equal: $\hat{\alpha} \neq e^{\hat{\beta}_0}$ and $\hat{\beta} \neq \hat{\beta}_1$.

2. If the chosen model is not intrinsically linear, the approach summarized in (13.5) cannot be used. Instead, least squares (or some other fitting procedure) would

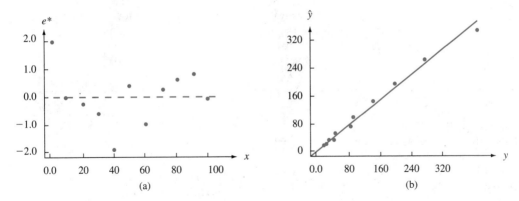

Figure 13.6 Plot of (a) standardized residuals (after transforming) versus x; (b) $\hat{y}$ versus y for data in Example 13.4

have to be applied to the untransformed model. Thus, for the additive exponential model $Y = \alpha e^{\beta x} + \epsilon$, least squares would involve minimizing $\Sigma(y_i - \alpha e^{\beta x_i})^2$. Taking partial derivatives with respect to α and β results in two nonlinear normal equations in α and β; these equations must then be solved using an iterative procedure. The book by Norman Draper and Harry Smith gives more details (see the chapter bibliography).

3. When the transformed linear model satisfies all the assumptions listed in Chapter 12, least squares yields best estimates of the transformed parameters. However, estimates of the original parameters may not be best in any sense, though they will be reasonable. For example, in the exponential model, the estimator $\hat{\alpha} = e^{\hat{\beta}_0}$ will not be unbiased, though it will be the maximum likelihood estimator of α if the error variable ϵ' is normally distributed. It is conceivable that using least squares directly (without transforming) could yield better estimates, though the computations would be quite burdensome.

4. If a transformation on Y has been made and one wishes to use the standard formulas to test hypotheses or construct CIs, ϵ' should be at least approximately normally distributed. To check this, the residuals from the transformed regression should be examined.

5. When Y is transformed, the r^2 value from the resulting regression refers to variation in the y_i''s explained by the transformed regression model. While a high value of r^2 here indicates a good fit of the estimated original nonlinear model to the observed y_i's, r^2 does not refer to these original observations. Perhaps the best way to assess the quality of the fit is to compute the predicted values $\hat{y}_i'$ using the transformed model, transform them back to the original y scale to obtain $\hat{y}_i$, and then plot $\hat{y}$ versus y. A good fit is then evidenced by points close to the 45° line. One could compute $SSE = \Sigma(y_i - \hat{y}_i)^2$ as a numerical measure of the goodness of fit. When the model was linear, we compared this to $SST = \Sigma(y_i - \bar{y})^2$, the total variation about the horizontal line at height $\bar{y}$; this led to r^2. In the nonlinear case,

though, it is not necessarily informative to measure total variation in this way, so an r^2 value is not as useful as in the linear case.

Exercises   Section 13.2 (15–23)

15. The accompanying data resulted from an experiment to investigate the relationship between applied stress x and time to rupture y for notched cold-rolled brass specimens exposed to a cracking solution ("On Initiation and Growth of Stress Corrosion Cracks in Tarnished Brass," *J. Electrochemical Soc.*, 1965, pp. 131–138):

x	22.5	25.0	28.0	30.5	38.0	40.5
y	44.0	42.0	33.5	28.0	18.0	13.6
x	42.5	48.0	54.5	55.0	70.0	
y	15.0	10.3	9.0	6.3	4.0	

A plot on log–log paper shows a pronounced linear pattern, suggesting a power model.

a. Show that fitting $Y' = \beta_0 + \beta_1 x' + \epsilon'$, where $Y' = \ln(Y)$ and $x' = \ln(x)$, results in $\hat{\beta}_1 = -2.16$ and $\hat{\beta}_0 = 10.67$. What are the estimates of the parameters of the power model?

b. Does a residual plot based on the transformed data and analysis confirm the choice of a power model?

c. Does a plot of $\hat{y}$ versus y confirm the efficacy of the power model for prediction of y?

d. Assuming that the power model is correct, test H_0: $\tilde{\mu}_{Y \cdot x} = \alpha x^{-2}$ ($SSE = .128$ for the transformed data).

e. Obtain a 95% CI for the exponent β in the power model.

f. Obtain a 95% CI for the median time to rupture when applied stress = 40.

16. The paper "The Luminosity–Spectral Index Relationship for Radio Galaxies" (*Nature*, 1972, pp. 88–89) suggested that for class S galaxies, $\ln(L_{178})$ is linearly related to spectral index, where L_{178} denotes luminosity at 178 MHz. Representative data appears below:

Spectral Index	.59	.67	.72	.80	.85	.90
$\ln(L_{178})$	23.6	25.6	26.4	25.7	26.8	26.7
Spectral Index	.94	.66	1.00	.86	1.03	.70
$\ln(L_{178})$	27.0	24.9	27.1	27.2	26.9	25.2

a. Estimate the parameters of the exponential model implied by the linear relationship between $\ln(L_{178})$ and spectral index.

b. What value of L_{178} would you predict for a spectral index of .75?

c. Compute a 95% prediction interval for luminosity when the spectral index is .95.

17. The following data on mass rate of burning x and flame length y is representative of that which appeared in the article "Some Burning Characteristics of Filter Paper" (*Combustion Science and Technology*, 1971, pp. 103–120):

x	1.7	2.2	2.3	2.6	2.7	3.0	3.2
y	1.3	1.8	1.6	2.0	2.1	2.2	3.0
x	3.3	4.1	4.3	4.6	5.7	6.1	
y	2.6	4.1	3.7	5.0	5.8	5.3	

a. Estimate the parameters of a power function model.

b. Construct diagnostic plots to check whether a power function is an appropriate model choice.

c. Test H_0: $\beta = \frac{4}{3}$ versus H_a: $\beta < \frac{4}{3}$, using a level .05 test.

d. Test the null hypothesis that states that the median flame length when burning rate is 5.0 is twice the median flame length when burning rate is 2.5 against the alternative that this is not the case.

18. An investigation of the influence of sodium benzoate concentration on the critical minimum pH necessary for the inhibition of Fe ("Mechanism of the Corrosion Inhibition of Fe by Sodium Benzoate," *Corrosion Science*, 1971, pp. 675–682) yielded the accompanying data, which suggests that expected critical minimum pH is linearly related to the natural logarithm of concentration:

Concentration	.01	.025	.1	.95
pH	5.1	5.5	6.1	7.3

a. What is the implied probabilistic model, and what are the estimates of the model parameters?

b. What critical minimum pH would you predict for a concentration of 1.0? Obtain a 95% PI for critical minimum pH when concentration is 1.0.

19. Thermal endurance tests were performed to study the relationship between temperature and lifetime of polyester enameled wire ("Thermal Endurance of Polyester Enameled Wires Using Twisted Wire Specimens," *IEEE Trans. Insulation*, 1965, pp. 38–44), resulting in the following data.

Temp.	200	200	200	200	200	200
Lifetime	5933	5404	4947	4963	3358	3878
Temp.	220	220	220	220	220	220
Lifetime	1561	1494	747	768	609	777
Temp.	240	240	240	240	240	240
Lifetime	258	299	209	144	180	184

a. Does a scatter plot of the data suggest a linear probabilistic relationship between lifetime and temperature?

b. What model is implied by a linear relationship between expected ln(lifetime) and 1/temperature? Does a scatter plot of the transformed data appear consistent with this relationship?

c. Estimate the parameters of the model suggested in part (b). What lifetime would you predict for a temperature of 220?

d. Because there are multiple observations at each x value, the method in Exercise 14 can be used to test the null hypothesis that states the model suggested in part (b) is correct. Carry out the test at level .01.

20. Exercise 14 presented data on body weight x and MCR/body weight y. Consider the following intrinsically linear functions for specifying the relationship between the two variables: (a) $\ln(y)$ versus x, (b) $\ln(y)$ versus $\ln(x)$, (c) y versus $\ln(x)$, (d) y versus $1/x$, and (e) $\ln(y)$ versus $1/x$. Use any appropriate diagnostic plots and analyses to decide which of these functions you would select to specify a probabilistic model. Explain your reasoning.

21. A plot appearing in the article "Thermal Conductivity of Polyethylene: The Effects of Crystal Size, Density, and Orientation on the Thermal Conductivity" (*Polymer Eng. and Science*, 1972, pp. 204–208) suggests that the expected value of thermal conductivity y is a linear function of $10^4 \cdot 1/x$ where x is lamellar thickness.

x	240	410	460	490	520	590	745	8300
y	12.0	14.7	14.7	15.2	15.2	15.6	16.0	18.1

a. Estimate the parameters of the regression function and the regression function itself.

b. Predict the value of thermal conductivity when lamellar thickness is 500 angstroms.

22. In each of the following cases, decide whether the given function is intrinsically linear. If so, identify x' and y' and then explain how a random error term ϵ can be introduced to yield an intrinsically linear probabilistic model.

a. $y = 1/(\alpha + \beta x)$ **b.** $y = 1/(1 + e^{\alpha + \beta x})$

c. $y = e^{e^{\alpha + \beta x}}$ (a Gompertz curve)

d. $y = \alpha + \beta e^{\lambda x}$

23. Suppose x and y are related according to a probabilistic exponential model $Y = \alpha e^{\beta x} \cdot \epsilon$ with $V(\epsilon)$ a constant independent of x (as was the case in the simple linear model $Y = \beta_0 + \beta_1 x + \epsilon$). Is $V(Y)$ a constant independent of x [as was the case for $Y = \beta_0 + \beta_1 x + \epsilon$, where $V(Y) = \sigma^2$]? Explain your reasoning. Draw a picture of a prototype scatter plot resulting from this model. Answer the same questions for the power model $Y = \alpha x^\beta \cdot \epsilon$.

Polynomial Regression

The nonlinear yet intrinsically linear models of Section 13.2 involved functions of the independent variable x that were either strictly increasing or strictly decreasing. In many situations, either theoretical reasoning or else a scatter plot of the data suggests that the true regression function $\mu_{Y \cdot x}$ has one or more peaks or valleys—that is, at least one relative minimum or maximum. In such cases, a polynomial function

$y = \beta_0 + \beta_1 x + \cdots + \beta_k x^k$ may provide a satisfactory approximation to the true regression function.

Definition

> The **kth degree polynomial regression model equation** is
>
> $$Y = \beta_0 + \beta_1 x + \beta_2 x^2 + \cdots + \beta_k x^k + \epsilon \tag{13.6}$$
>
> where ϵ is a normally distributed random variable with
>
> $$\mu_\epsilon = 0, \quad \sigma_\epsilon^2 = \sigma^2 \tag{13.7}$$

From (13.6) and (13.7) it follows immediately that

$$\mu_{Y \cdot x} = \beta_0 + \beta_1 x + \cdots + \beta_k x^k, \quad \sigma_{Y \cdot x}^2 = \sigma^2 \tag{13.8}$$

In words, the expected value of Y is a kth-degree polynomial function of x, while the variance of Y, which controls the spread of observed values about the regression function, is the same for each value of x. The observed pairs $(x_1, y_1), \ldots, (x_n, y_n)$ are assumed to have been generated independently from the model (13.6). Figure 13.7 illustrates both a quadratic and cubic model.

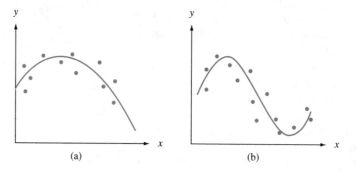

Figure 13.7 (a) Quadratic regression model; (b) Cubic regression model

Estimation of Parameters Using Least Squares

To estimate $\beta_0, \beta_1, \ldots, \beta_k$, consider a trial regression function $y = b_0 + b_1 x + \cdots + b_k x^k$. Then the goodness of fit of this function to the observed data can be measured by computing the sum of squared deviations

$$f(b_0, b_1, \ldots, b_k) = \sum_{i=1}^{n} [y_i - (b_0 + b_1 x_i + b_2 x_i^2 + \cdots + b_k x_i^k)]^2 \tag{13.9}$$

According to the principle of least squares, the estimates $\hat{\beta}_0, \hat{\beta}_1, \ldots, \hat{\beta}_k$ are those values of $b_0, b_1, \ldots, b_k$ that minimize (13.9). It should be noted that when x_1, $x_2, \ldots, x_n$ are all different, there is a polynomial of degree $n - 1$ that fits the data perfectly, so that the minimizing value of (13.9) is zero when $k = n - 1$. However, in virtually all applications the polynomial model (13.6) with k large is quite unrealistic, and in most applications $k = 2$ (quadratic) or $k = 3$ (cubic) is appropriate.

To find the minimizing values in (13.9), we take the $k + 1$ partial derivatives $\partial f/\partial b_0, \partial f/\partial b_1, \dots, \partial f/\partial b_k$ and equate them to 0, resulting in the system of normal equations for the estimates. Because the trial function $b_0 + b_1 x + \cdots + b_k x^k$ is linear in $b_0, \dots, b_k$ (though not in x), the $k + 1$ normal equations are linear in these unknowns. The system is

$$
\begin{aligned}
& b_0 n + b_1 \Sigma x_i + b_2 \Sigma x_i^2 + \cdots + b_k \Sigma x_i^k = \Sigma y_i \\
& b_0 \Sigma x_i + b_1 \Sigma x_i^2 + b_2 \Sigma x_i^3 + \cdots + b_k \Sigma x_i^{k+1} = \Sigma x_i y_i \\
& \qquad \cdot \qquad\qquad \cdot \qquad\qquad \cdot \\
& \qquad \cdot \qquad\qquad \cdot \qquad\qquad \cdot \\
& \qquad \cdot \qquad\qquad \cdot \qquad\qquad \cdot \\
& b_0 \Sigma x_i^k + b_1 \Sigma x_i^{k+1} + \cdots + b_k \Sigma x_i^{2k} = \Sigma x_i^k y_i
\end{aligned}
\tag{13.10}
$$

All standard statistical computer packages will automatically solve the system (13.10) and output the estimates as well as much other information.*

Example 13.5   The article "Determination of Biological Maturity and Effect of Harvesting and Drying Conditions on Milling Quality of Paddy" (*J. Agricultural Eng. Research*, 1975, pp. 353–361) reports the following data on date x of harvesting (number of days after flowering) and yield y (kg/ha) of paddy, a grain farmed in India:

x	16	18	20	22	24	26	28	30
y	2508	2518	3304	3423	3057	3190	3500	3883
x	32	34	36	38	40	42	44	46
y	3823	3646	3708	3333	3517	3241	3103	2776

The scatter plot appearing in Figure 13.8 suggests that the author's choice of a quadratic model is reasonable. This figure also gives MINITAB output from a quadratic regression, from which

$$\hat{\beta}_0 = -1070.4, \quad \hat{\beta}_1 = 293.48, \quad \hat{\beta}_2 = -4.5358$$

and the estimated regression function is

$$y = -1070.4 + 293.48x - 4.5358x^2$$

The estimated regression functions can be used to compute predicted values (substitute $x_1 = 16$, then $x_2 = 18$, etc., to obtain $\hat{y}_1, \hat{y}_2, \dots$) and then residuals $(y_i - \hat{y}_i)$. A residual plot indicates that a quadratic regression function is appropriate, though there is somewhat more spread in the residuals for small x than for large x.

---

*We will see in Section 13.4 that polynomial regression is a special case of multiple regression, so a command appropriate for this latter task is used.

```
YIELD
4200.+
     -
     -
     -
     -                                  *   *
3600.+                                          *
     -                              *       *
     -                 *    *            *
     -             *
     -                                          *   *
     -                                              *
3000.+                                                  *
     -                       *                              *
     -
     -
     -                                                          *
     -  *   *
2400.+
        +--------+--------+--------+--------+--------+ DAYS
       16.0     24.0     32.0     40.0     48.0     56.0
```

THE REGRESSION EQUATION is
YIELD = -1070 + 293 DAYS - 4.54 DAYSSQD

Predictor	Coef	Stdev	t-ratio	p
Constant	-1070.4	617.3	-1.73	0.107
DAYS	293.48	42.18	6.96	0.000
DAYSSQD	-4.5358	0.6744	-6.73	0.000

s = 203.9   R-sq = 79.4%   R-sq(adj) = 76.2%

Analysis of Variance

SOURCE	DF	SS	MS	F	p
Regression	2	2084779	1042389	25.08	0.000
Error	13	540388	41568		
Total	15	2625167			

Obs.	DAYS	YIELD	Fit	Stdev.Fit	Residual	St.Resid
1	16.0	2508.0	2464.2	135.6	43.8	0.29
2	18.0	2518.0	2742.7	104.8	-224.7	-1.28
3	20.0	3304.0	2984.9	83.0	319.1	1.71
4	22.0	3423.0	3190.9	71.3	232.1	1.22
5	24.0	3057.0	3360.6	68.4	-303.6	-1.58
6	26.0	3190.0	3494.0	70.7	-304.0	-1.59
7	28.0	3500.0	3591.1	74.2	-91.1	-0.48
8	30.0	3883.0	3651.9	76.4	231.1	1.22
9	32.0	3823.0	3676.4	76.4	146.6	0.78
10	34.0	3646.0	3664.6	74.2	-18.6	-0.10
11	36.0	3708.0	3616.6	70.7	91.4	0.48
12	38.0	3333.0	3532.3	68.4	-199.3	-1.04
13	40.0	3517.0	3411.6	71.3	105.4	0.55
14	42.0	3241.0	3254.7	83.0	-13.7	-0.07
15	44.0	3103.0	3061.5	104.8	41.5	0.24
16	46.0	2776.0	2832.1	135.6	-56.1	-0.37

Fit	Stdev.Fit	95% C.I.	95% P.I.
3431.8	69.2	(3282.2, 3581.4)	(2966.5, 3897.1)

Figure 13.8 Scatter plot and MINITAB output for Example 13.5

Weighted least squares can be used to obtain a more satisfactory fit, but the authors did not consider this. ∎

$\hat{\sigma}^2$ and R^2

To make further inferences about the parameters of the regression function, the error variance σ^2 must be estimated. With $\hat{y}_i = \hat{\beta}_0 + \hat{\beta}_1 x_i + \cdots + \hat{\beta}_k x_i^k$, the ith residual is $y_i - \hat{y}_i$, and the sum of squared residuals (error sum of squares) is $SSE = \Sigma(y_i - \hat{y}_i)^2$. The estimate of σ^2 is then

$$\hat{\sigma}^2 = s^2 = \frac{SSE}{n - (k + 1)} = MSE \tag{13.11}$$

where the denominator $n - (k + 1)$ is used because $k + 1$ df are lost in estimating $\beta_0, \beta_1, \ldots, \beta_k$.

If we again let $SST = \Sigma(y_i - \bar{y})^2$, then SSE/SST is the proportion of the total variation in the observed y_i's not explained by the polynomial model. The quantity $1 - SSE/SST$, the proportion of variation explained by the model, is called the **coefficient of multiple determination** and is denoted by R^2.

Suppose we consider fitting a cubic model to the data in Example 13.5. Because the cubic model includes the quadratic as a special case, the fit to a cubic will be at least as good as the fit to a quadratic. More generally, with $SSE_k = $ the error sum of squares from a kth-degree polynomial, $SSE_{k'} \leq SSE_k$ and $R_{k'}^2 \geq R_k^2$ whenever $k' > k$. Because the objective of regression analysis is to find a model that is both simple (relatively few parameters) and provides a good fit to the data, a higher-degree polynomial may not specify a better model than a lower-degree model despite its higher R^2 value. To balance the cost of using more parameters against the gain in R^2, many statisticians use the **adjusted coefficient of multiple determination**

$$\text{Adjusted } R^2 = 1 - \frac{n - 1}{n - (k + 1)} \cdot \frac{SSE}{SST} = \frac{(n - 1)R^2 - k}{n - 1 - k} \tag{13.12}$$

Adjusted R^2 adjusts the proportion of unexplained variation upward [since $(n - 1)/(n - k - 1) > 1$], which results in adjusted $R^2 < R^2$. Thus if $R_2^2 = .66$, $R_3^2 = .70$, and $n = 10$, then

$$\text{Adjusted } R_2^2 = \frac{9(.66) - 2}{10 - 3} = .563, \quad \text{Adjusted } R_3^2 = \frac{9(.70) - 3}{10 - 4} = .550$$

so the small gain in R^2 in going from a quadratic to a cubic model is not enough to offset the cost of adding an extra parameter to the model.

Example 13.6
(Example 13.5 continued)

SSE and SST are typically found on computer output in an ANOVA table. Figure 13.8 gives $SSE = 540{,}388$ and $SST = 2{,}625{,}167$ for the paddy yield data, from which

$$R^2 = 1 - \frac{SSE}{SST} = 1 - \frac{540{,}388}{2{,}625{,}167} = 1 - .206 = .794$$

Thus, 79.4% of the observed variation in yield can be attributed to the model relationship. Adjusted $R^2 = .762$, only a small downward change in R^2. The estimates of σ^2 and σ are

$$s^2 = \frac{SSE}{n - (k+1)} = \frac{540{,}388}{16 - (2+1)} = 41{,}568.3$$

$$s = \sqrt{s^2} = 203.9 \qquad\blacksquare$$

Besides computing R^2 and adjusted R^2, one should examine the usual diagnostic plots to determine whether model assumptions are valid or whether modification may be appropriate.

Confidence Intervals and Test Procedures

Because the y_i's appear in the normal equations (13.10) only on the right-hand side and in a linear fashion, the resulting estimates $\hat{\beta}_0, \ldots, \hat{\beta}_k$ are themselves linear functions of the y_i's. Thus, the estimators are linear functions of the Y_i's, so each $\hat{\beta}_i$ has a normal distribution. It can also be shown that each $\hat{\beta}_i$ is an unbiased estimator of β_i.

Let $\sigma_{\hat{\beta}_i}$ denote the standard deviation of the estimator $\hat{\beta}_i$. This standard deviation has the form

$$\sigma_{\hat{\beta}_i} = \sigma \cdot \left\{ \begin{array}{l} \text{a complicated expression involving } all \\ x_j\text{'s},\, x_j^2\text{'s}, \ldots, \text{and } x_j^k\text{'s} \end{array} \right\}$$

Fortunately, the expression in braces has been programmed into all of the most frequently used statistical computer packages—MINITAB, BMD, SAS, and SPSS. The estimated standard deviation of $\hat{\beta}_i$, $s_{\hat{\beta}_i}$, results from substituting s in place of σ in the above expression for $\sigma_{\hat{\beta}_i}$. These estimated standard deviations $s_{\hat{\beta}_0}, s_{\hat{\beta}_1}, \ldots,$ and $s_{\hat{\beta}_k}$ appear on output from all the aforementioned statistical packages. Let $S_{\hat{\beta}_i}$ denote the estimator of $\sigma_{\hat{\beta}_i}$—that is, the random variable whose observed value is $s_{\hat{\beta}_i}$. Then it can be shown that the standardized variable

$$T = \frac{\hat{\beta}_i - \beta_i}{S_{\hat{\beta}_i}} \tag{13.13}$$

has a t distribution based on $n - (k+1)$ df. This leads to the following inferential procedures.

A $100(1 - \alpha)\%$ CI for β_i, the coefficient of x^i in the polynomial regression function, is

$$\hat{\beta}_i \pm t_{\alpha/2, n-(k+1)} \cdot s_{\hat{\beta}_i}$$

A test of H_0: $\beta_i = \beta_{i0}$ is based on the t statistic value

$$t = \frac{\hat{\beta}_i - \beta_{i0}}{s_{\hat{\beta}_i}}$$

The test utilizes a t critical value with $n - (k + 1)$ df and is upper-, lower-, or two-tailed according to whether the inequality in H_a is >, <, or ≠.

A point estimate of $\mu_{Y \cdot x}$—that is, of $\beta_0 + \beta_1 x + \cdots + \beta_k x^k$—is $\hat{\mu}_{Y \cdot x} = \hat{\beta}_0 + \hat{\beta}_1 x + \cdots + \hat{\beta}_k x^k$. The estimated standard deviation of the corresponding estimator is rather complicated. Many computer packages will give this estimated standard deviation for *any* x value when requested to do so by a user. This, along with an appropriate standardized t variable, can be used to justify the following procedures.

Let x^* denote a specified value of x. A $100(1 - \alpha)\%$ CI for $\mu_{Y \cdot x^*}$ is

$$\hat{\mu}_{Y \cdot x^*} \pm t_{\alpha/2, n-(k+1)} \cdot \left\{ \begin{array}{c} \text{estimated SD of} \\ \hat{\mu}_{Y \cdot x^*} \end{array} \right\}$$

A $100(1 - \alpha)\%$ PI for a future y value to be observed when $x = x^*$ is

$$\hat{\mu}_{Y \cdot x^*} \pm t_{\alpha/2, n-(k+1)} \cdot \left\{ s^2 + \left(\begin{array}{c} \text{estimated SD} \\ \text{of } \hat{\mu}_{Y \cdot x^*} \end{array} \right)^2 \right\}^{1/2}$$

Example 13.7

Looking back to Figure 13.8, we see that $\hat{\beta}_2 = -4.5358$ and $s_{\hat{\beta}_2} = .6744$ (from the Stdev column at the top of the output). For testing H_0: $\beta_2 = 0$ versus H_a: $\beta_2 \neq 0$ (H_0 here says that the quadratic term in the model is unnecessary), the test statistic is $T = \hat{\beta}_2 / S_{\hat{\beta}_2}$, with computed value $-4.5358/.6744 = -6.73$. The test is based on $n - (k + 1) = 16 - 3 = 13$ df; with $t_{.025,13} = 2.160$, H_0 is rejected at level .05 if either $t \geq 2.160$ or $t \leq -2.160$. Since $-6.73 \leq -2.160$, H_0 is rejected at level .05, validating the inclusion of the quadratic term.

The last line of output in Figure 13.8 contains estimation and prediction information for $x = 25$:

$$\hat{\beta}_0 + \hat{\beta}_1(25) + \hat{\beta}_2(25)^2 = \text{Fit} = 3431.8$$

$$\left(\begin{array}{c} \text{estimated SD of} \\ \hat{\beta}_0 + \hat{\beta}_1(25) + \hat{\beta}_2(25)^2 \end{array} \right) = \text{Stdev. Fit} = 69.2$$

Thus, a 95% CI for mean yield when date of harvesting = 25 is

$$3431.8 \pm (2.160)(69.2) = 3431.8 \pm 149.5 = (3282.3, 3581.3)$$

whereas a 95% PI for yield when $x = 25$ is

$$3431.8 \pm (2.160)[41{,}568.3 + (69.2)^2]^{1/2} = 3431.8 \pm 465.1$$
$$= (2966.7, 3896.9)$$

The slight discrepancies between these interval limits and those appearing on output are attributable to rounding. Notice again that the PI is much wider than the corresponding CI. This is because s^2 is quite large compared to the Stdev. Fit.   ∎

Centering x Values

For the quadratic model with regression function $\mu_{Y \cdot x} = \beta_0 + \beta_1 x + \beta_2 x^2$, the parameters β_0, β_1, and β_2 characterize the behavior of the function near $x = 0$. For example, β_0 is the height at which the regression function crosses the vertical axis $x = 0$, while β_1 is the first derivative of the function at $x = 0$ (instantaneous rate of change of $\mu_{Y \cdot x}$ at $x = 0$). If the x_i's all lie far from 0, we may not have precise information about the values of these parameters. Let $\bar{x}$ = the average of the x_i's for which observations are to be taken and consider the model

$$Y = \beta_0^* + \beta_1^*(x - \bar{x}) + \beta_2^*(x - \bar{x})^2 + \epsilon \tag{13.14}$$

In the model (13.14), $\mu_{Y \cdot x} = \beta_0^* + \beta_1^*(x - \bar{x}) + \beta_2^*(x - \bar{x})^2$, and the parameters now describe the behavior of the regression function near the center $\bar{x}$ of the data.

To estimate the parameters of (13.14), we simply subtract $\bar{x}$ from each x_i to obtain $x_i' = x_i - \bar{x}$, and then use the x_i''s in place of the x_i's. An important benefit of this is that the coefficients of $b_0, \ldots, b_k$ in the normal equations (13.10) will be of much smaller magnitude than would be the case were the original x_i's used. When the system is solved by computer, this centering protects against any round-off error that may result.

Example 13.8   The paper "A Method for Improving the Accuracy of Polynomial Regression Analysis" (*J. Quality Technology*, 1971, pp. 149–155) reports the following data on x = cure temperature (°F) and y = ultimate shear strength of a rubber compound (psi), with $\bar{x} = 297.13$:

x	280	284	292	295	298	305	308	315
x'	−17.13	−13.13	−5.13	−2.13	.87	7.87	10.87	17.87
y	770	800	840	810	735	640	590	560

A computer analysis yielded the results shown in Table 13.4.

Table 13.4 Estimated cofficients and standard deviations for Example 13.8

Parameter	Estimate	Estimated SD	Parameter	Estimate	Estimated SD
β_0	−26,219.64	11,912.78	β_0^*	759.36	23.20
β_1	189.21	80.25	β_1^*	−7.61	1.43
β_2	−.3312	.1350	β_2^*	−.3312	.1350

The estimated regression function using the original model is $y = -26,219.64 + 189.21x - .3312x^2$, while for the centered model the function is $y = 759.36 -$

$7.61(x - 297.13) - .3312(x - 297.13)^2$. These estimated functions are identical; the only difference is that different parameters have been estimated for the two models. The estimated standard deviations indicate clearly that β_0^* and β_1^* have been more accurately estimated than β_0 and β_1. The quadratic parameters are identical ($\beta_2 = \beta_2^*$), as can be seen by comparing the x^2 term in (13.14) with the original model. We emphasize again that a major benefit of centering is the gain in computational accuracy, not only in quadratic but also in higher-degree models. ∎

The book by Neter, et al. is again a good source for more information about polynomial regression.

Exercises   Section 13.3 (24–32)

24. The following data on y = glucose concentration (g/L) and x = fermentation time (days) for a particular blend of malt liquor was read from a scatter plot appearing in the paper "Improving Fermentation Productivity with Reverse Osmosis" (*Food Tech.*, 1984, pp. 92–96):

x	1	2	3	4	5	6	7	8
y	74	54	52	51	52	53	58	71

a. Verify that a scatter plot of the data is consistent with the choice of a quadratic regression model.

b. The estimated quadratic regression equation is $y = 84.482 - 15.875x + 1.7679x^2$. Predict the value of glucose concentration for a fermentation time of 6 days and compute the corresponding residual.

c. Using $SSE = 61.77$, what proportion of observed variation can be attributed to the quadratic regression relationship?

d. The $n = 8$ standardized residuals based on the quadratic model are 1.91, −1.95, −.25, .58, .90, .04, −.66, and .20. Construct a plot of the standardized residuals versus x and a normal probability plot. Do the plots exhibit any troublesome features?

e. The estimated standard deviation of $\hat{\mu}_{Y \cdot 6}$—that is, $\hat{\beta}_0 + \hat{\beta}_1(6) + \hat{\beta}_2(36)$—is 1.69. Compute a 95% CI for $\mu_{Y \cdot 6}$.

f. Compute a 95% PI for a glucose concentration observation made after 6 days of fermentation time.

25. The viscosity (y) of an oil was measured by a cone and plate viscometer at six different cone speeds (x). It was assumed that a quadratic regression model was appro-priate, and the estimated regression function resulting from the $n = 6$ observations was

$$y = -113.0937 + 3.3684x - .01780x^2$$

a. Estimate $\mu_{Y \cdot 75}$, the expected viscosity when speed is 75 rpm.

b. What viscosity would you predict for a cone speed of 60 rpm?

c. If $\Sigma y_i^2 = 8386.43$, $\Sigma y_i = 210.70$, $\Sigma x_i y_i = 17{,}002.00$, and $\Sigma x_i^2 y_i = 1{,}419{,}780$, compute SSE, s^2, and s.

d. From part (c), $SST = 8386.43 - (210.70)^2/6 = 987.35$. Using SSE computed in part (c), what is the computed value of the coefficient of multiple determination R^2?

e. If the estimated standard deviation of $\hat{\beta}_2$ is $s_{\hat{\beta}_2} = .00226$, test $H_0: \beta_2 = 0$ versus $H_a: \beta_2 \neq 0$ at level .01.

26. Exercise 7 presented the following data on exposure time to radiation x and dry weight of roots y:

x	0	2	4	6	8
y	110	123	119	86	62

The estimated quadratic regression function is $y = 111.89 + 8.06x - 1.84x^2$.

a. Compute the predicted values and residuals. Then compute SSE and s^2.

b. Compute the coefficient of multiple determination R^2.

c. The estimated standard deviation of $\hat{\beta}_2$, the estimator of the quadratic coefficient β_2, is $s_{\hat{\beta}_2} = .480$.

Does the quadratic term belong in the model? State and test the appropriate hypotheses at level .05.

d. The estimated standard deviation of $\hat{\beta}_1$ is $s_{\hat{\beta}_1} = 4.01$. Use this and the information in part (c) to obtain joint CIs for β_1 and β_2 with joint confidence level (at least) 95%.

e. The estimated standard deviation of $\hat{\mu}_{Y\cdot4}$ ($= \hat{\beta}_0 + 4\hat{\beta}_1 + 16\hat{\beta}_2$) is 5.01. Compute a 90% CI for $\mu_{Y\cdot4}$.

f. Estimate the exposure time that maximizes expected dry weight of roots.

27. The article "A Simulation-Based Evaluation of Three Cropping Systems on Cracking-Clay Soils in a Summer Rainfall Environment" (*Agricultural Meteorology*, 1976, pp. 211–229) proposes a quadratic model for the relationship between water supply index (x) and farm wheat yield (y). Representative data and the resulting MINITAB output appear below:

x	1.2	1.3	1.5	1.8	2.1	2.3	2.5
y	790	950	740	1230	1000	1465	1370
x	2.9	3.1	3.2	3.3	3.9	4.0	4.3
y	1420	1625	1600	1720	1500	1550	1560

```
The regression equation is
yield = - 252 + 1000 index - 135 indexsqd

Predictor       Coef    Stdev  t-ratio        p
Constant      -251.6    285.1    -0.88    0.396
Index         1000.1    229.5     4.36    0.001
Indexsqd     -135.44    41.97    -3.23    0.008

s = 135.6           R-sq = 85.3%     R-sq(adj) = 82.6%

Analysis of Variance

SOURCE          DF        SS       MS        F        p
Regression       2   1170208   585104    31.83    0.000
Error           11    202228    18384
Total           13   1372435
```

a. Interpret the value of the coefficient of multiple determination.

b. Calculate a 95% CI for the coefficient of the quadratic predictor.

c. The estimated standard deviation of $\hat{\beta}_0 + \hat{\beta}_1x + \hat{\beta}_2x^2$ when $x = 2.5$ is 53.5. Test H_0: $\mu_{Y\cdot2.5} = 1500$ versus H_a: $\mu_{Y\cdot2.5} < 1500$ using $\alpha = .01$.

d. Obtain a 95% PI for wheat yield when the water supply index is 2.5 by using the information given in part (c).

28. The accompanying data was obtained from a study of a certain method for preparing pure alcohol from refinery streams ("Direct Hydration of Olefins," *Industrial and Eng. Chemistry*, 1961, pp. 209–211). The independent variable x is volume hourly space velocity, and the dependent variable y is the amount of conversion of isobutylene.

x	1	1	2	4	4	4	6
y	23.0	24.5	28.0	30.9	32.0	33.6	20.0

a. Assuming that a quadratic probabilistic model is appropriate, estimate the regression function.

b. Compute the predicted values and residuals and construct a residual plot. Does the plot look roughly as expected when the quadratic model is correct? Does the plot indicate that any observation has had a great influence on the fit? Does a scatter plot identify a point having large influence? If so, which point?

c. Compute s^2 and R^2. Does the quadratic model provide a good fit to the data?

d. In Exercise 11, it was noted that the predicted value $\hat{Y}_j$ and the residual $Y_j - \hat{Y}_j$ are independent of one another, so that $\sigma^2 = V(Y_j) = V(\hat{Y}_j) + V(Y_j - \hat{Y}_j)$. A computer printout gives the estimated standard deviations of the predicted values as .955, .955, .712, .777, .777, .777, and 1.407. Use these values along with s^2 to compute the estimated standard deviation of each residual. Then compute the standardized residuals and plot them against x. Does the plot look much like the plot of part (b)? Suppose you had standardized the residuals using just s in the denominator. Would the resulting values differ much from the correct values?

e. Using information given in part (d), compute a 90% PI for isobutylene conversion when volume hourly space velocity is 4.

29. The following data is a subset of data obtained in an experiment to study the relationship between soil pH x and $y = A1$. Concentration/EC ("Root Responses of Three *Gramineae* Species to Soil Acidity in an Oxisol and an Ultisol," *Soil Science*, 1973, pp. 295–302):

x	4.01	4.07	4.08	4.10	4.18
y	1.20	.78	.83	.98	.65

x	4.20	4.23	4.27	4.30	4.41
y	.76	.40	.45	.39	.30

x	4.45	4.50	4.58	4.68	4.70	4.77
y	.20	.24	.10	.13	.07	.04

A cubic model was proposed in the paper, but the version of MINITAB used by the author of the present text refused to include the x^3 term in the model, stating that "x^3 is highly correlated with other predictor variables." To remedy this, $\bar{x} = 4.3456$ was subtracted from each x value to yield $x' = x - \bar{x}$. A cubic regression was then requested to fit the model having regression function.

$$y = \beta_0^* + \beta_1^* x' + \beta_2^*(x')^2 + \beta_3^*(x')^3$$

The following computer output resulted:

Parameter	Estimate	Estimated SD
β_0^*	.3463	.0366
β_1^*	−1.2933	.2535
β_2^*	2.3964	.5699
β_3^*	−2.3968	2.4590

a. What is the estimated regression function for the "centered" model?

b. What is the estimated value of the coefficient β_3 in the "uncentered" model with regression function $y = \beta_0 + \beta_1 x + \beta_2 x^2 + \beta_3 x^3$? What is the estimate of β_2?

c. Using the cubic model, what value of y would you predict when soil pH is 4.5?

d. Carry out a test to decide whether the cubic term should be retained in the model.

30. In many polynomial regression problems, rather than fitting a "centered" regression function using $x' = x - \bar{x}$, computational accuracy can be improved by using a function of the standardized independent variable $x' = (x - \bar{x})/s_x$, where s_x is the standard deviation of the x_i's. Consider fitting the cubic regression function $y = \beta_0^* + \beta_1^* x' + \beta_2^*(x')^2 + \beta_3^*(x')^3$ to the following data resulting from a study of the relation between thrust efficiency y of supersonic propelling rockets and the half-divergence angle x of the rocket nozzle ("More on Correlating Data," *CHEMTECH*, 1976, pp. 266–270):

x	5	10	15	20	25	30	35
y	.985	.996	.988	.962	.940	.915	.878

Parameter	Estimate	Estimated SD
β_0^*	.9671	.0026
β_1^*	−.0502	.0051
β_2^*	−.0176	.0023
β_3^*	.0062	.0031

a. What value of y would you predict when the half-divergence angle is 20? When $x = 25$?

b. What is the estimated regression function $\hat{\beta}_0 + \hat{\beta}_1 x + \hat{\beta}_2 x^2 + \hat{\beta}_3 x^3$ for the "unstandardized" model?

c. Use a level .05 test to decide whether the cubic term should be deleted from the model.

d. What can you say about the relationship between SSE's and R^2's for the standardized and unstandardized models? Explain.

e. SSE for the cubic model is .00006300, while for a quadratic model SSE is .00014367. Compute R^2 for each model. Does the difference between the two suggest that the cubic term can be deleted?

31. The following data resulted from an experiment to assess the potential of unburnt colliery spoil as a medium for plant growth. The variables are $x = $ acid extractable cations and $y = $ exchangeable acidity/total cation exchange capacity ("Exchangeable Acidity in Unburnt Colliery Spoil," *Nature*, 1969, p. 161):

x	−23	−5	16	26	30	38
y	1.50	1.46	1.32	1.17	.96	.78

x	52	58	67	81	96	100	113
y	.77	.91	.78	.69	.52	.48	.55

Standardizing the independent variable x to obtain $x' = (x - \bar{x})/s_x$ and fitting the regression function $y = \beta_0^* + \beta_1^* x' + \beta_2^* (x')^2$ yielded the accompanying computer output:

Parameter	Estimate	Estimated SD
β_0^*	.8733	.0421
β_1^*	−.3255	.0316
β_2^*	.0448	.0319

a. Estimate $\mu_{Y \cdot 50}$.

b. Compute the value of the coefficient of multiple determination.

c. What is the estimated regression function $\hat{\beta}_0 + \hat{\beta}_1 x + \hat{\beta}_2 x^2$ using the unstandardized variable x?

d. What is the estimated standard deviation of $\hat{\beta}_2$ computed in part (c)?

e. Carry out a test using the standardized estimates to decide whether the quadratic term should be retained in the model. Repeat using the unstandardized estimates. Do your conclusions differ?

32. The paper "The Respiration in Air and in Water of the Limpets *Patella Caerulea* and *Patella Lusitanica*" (*Comp. Biochemistry and Physiology*, 1975, pp. 407–411) proposed a simple power model for the relationship between respiration rate y and temperature x for *P. caerulea* in air. However, a plot of $\ln(y)$ versus x exhibits a curved pattern. Fit the quadratic power model $Y = \alpha e^{\beta x + \gamma x^2} \cdot \epsilon$ to the accompanying data:

x	10	15	20	25	30
y	37.1	70.1	109.7	177.2	222.6

13.4   Multiple Regression Analysis

In multiple regression, the objective is to build a probabilistic model that relates a dependent variable Y to more than one independent or predictor variable. Let k represent the number of predictor variables ($k \geq 2$) and denote these predictors by $x_1, x_2, \ldots, x_k$. For example, in attempting to predict the selling price of a house, we might have $k = 3$ with $x_1 = $ size (ft^2.), $x_2 = $ age (years), and $x_3 = $ number of rooms.

Definition

> The **general additive multiple regression model equation** is
>
> $$Y = \beta_0 + \beta_1 x_1 + \beta_2 x_2 + \cdots + \beta_k x_k + \epsilon \qquad (13.15)$$
>
> where $E(\epsilon) = 0$ and $V(\epsilon) = \sigma^2$. In addition, for purposes of testing hypotheses and calculating CIs or PIs, it is assumed that ϵ is normally distributed.

Let $x_1^*, x_2^*, \ldots, x_k^*$ be particular values of $x_1, \ldots, x_k$. Then (13.15) implies that

$$\mu_{Y \cdot x_1^*, \ldots, x_k^*} = \beta_0 + \beta_1 x_1^* + \cdots + \beta_k x_k^* \qquad (13.16)$$

Thus, just as $\beta_0 + \beta_1 x$ describes the mean Y value as a function of x in simple linear regression, the **true regression function** $\beta_0 + \beta_1 x_1 + \cdots + \beta_k x_k$ gives the expected value of Y as a function of $x_1, \ldots, x_k$.

If an investigator has obtained observations on x_1 and x_2, one possible model is $Y = \beta_0 + \beta_1 x_1 + \beta_2 x_2 + \epsilon$. However, it is possible to construct other models by forming predictors that are mathematical functions of x_1 and/or x_2. For example, with $x_3 = x_1^2$ and $x_4 = x_1 x_2$, the model

$$Y = \beta_0 + \beta_1 x_1 + \beta_2 x_2 + \beta_3 x_3 + \beta_4 x_4 + \epsilon$$

has the general form of (13.15). In general, it is not only permissible for some predictors to be mathematical functions of others but also often highly desirable in the sense that the resulting model may be much more successful in explaining variation in y than any model without such predictors. This discussion also shows that polynomial regression is indeed a special case of multiple regression. For example, the quadratic model $Y = \beta_0 + \beta_1 x + \beta_2 x^2 + \epsilon$ has the form of (13.15) with $k = 2$, $x_1 = x$, and $x_2 = x^2$.

Model Interpretation*

For the case of two independent variables, x_1 and x_2, four useful multiple regression models are the following:

1. The first-order model, with

$$Y = \beta_0 + \beta_1 x_1 + \beta_2 x_2 + \epsilon$$

2. The second-order no-interaction model, with

$$Y = \beta_0 + \beta_1 x_1 + \beta_2 x_2 + \beta_3 x_1^2 + \beta_4 x_2^2 + \epsilon$$

3. The first-order interaction model, with

$$Y = \beta_0 + \beta_1 x_1 + \beta_2 x_2 + \beta_3 x_1 x_2 + \epsilon$$

4. The complete second order or full quadratic model, specified by

$$Y = \beta_0 + \beta_1 x_1 + \beta_2 x_2 + \beta_3 x_1^2 + \beta_4 x_2^2 + \beta_5 x_1 x_2 + \epsilon$$

Understanding the differences among these models is an important first step in building realistic regression models from the independent variables under study.

The first-order model is the most straightforward generalization of simple linear regression. It states that, for a fixed value of either variable, the expected value of Y is a linear function of the other variable and that the expected change in Y for a unit increase in x_1 (x_2) is β_1 (β_2) independent of the level of x_2 (x_1). Thus, if we graph the regression function as a function of x_1 for several different values of x_2, we obtain as contours of the regression function a collection of parallel lines as pictured in Figure 13.9a. The function $y = \beta_0 + \beta_1 x_1 + \beta_2 x_2$ specifies a plane in three-dimensional space; the first-order model says that each observed value of the dependent variable deviates from this plane by a random amount ϵ.

According to the second-order no-interaction model, if x_2 is fixed, the expected change in Y for a one-unit increase in x_1 is

$$\beta_0 + \beta_1(x_1 + 1) + \beta_2 x_2 + \beta_3(x_1 + 1)^2 + \beta_4 x_2^2$$
$$- (\beta_0 + \beta_1 x_1 + \beta_2 x_2 + \beta_3 x_1^2 + \beta_4 x_2^2) = \beta_1 + \beta_3 + 2\beta_3 x_1$$

---

*This subsection may be postponed until after the material on inferences has been covered.

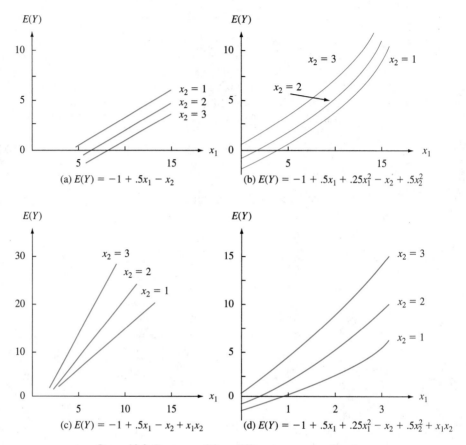

Figure 13.9 Contours of four different regression functions

Because this expected change does not depend on x_2, the contours of the regression function for different values of x_2 are still parallel to one another. However, the dependence of the expected change on the value of x_1 means that the contours are now curves rather than straight lines. This is pictured in Figure 13.9b. In this case, the regression surface is no longer a plane in three-dimensional space but is instead a curved surface.

The contours of the regression function for the first-order interaction model are nonparallel straight lines. This is because the expected change in Y when x_1 is increased by 1 is

$$\beta_0 + \beta_1(x_1 + 1) + \beta_2 x_2 + \beta_3(x_1 + 1)x_2$$
$$- (\beta_0 + \beta_1 x_1 + \beta_2 x_2 + \beta_3 x_1 x_2) = \beta_1 + \beta_3 x_2$$

This expected change depends on the value of x_2, so each contour line must have a different slope as in Figure 13.9c. The word *interaction* reflects the fact that an expected change in Y when one variable increases depends on the value of the other variable.

Finally, for the complete second order model, the expected change in Y when x_2 is held fixed while x_1 is increased by 1 unit is $\beta_1 + \beta_3 + 2\beta_3 x_1 + \beta_5 x_2$, which is a function of both x_1 and x_2. This implies that the contours of the regression function are both curved and not parallel to one another, as illustrated in Figure 13.9d.

Similar considerations apply to models constructed from more than two independent variables. In general, the presence of interaction terms in the model implies that the expected change in Y depends not only on the variable being increased or decreased but also on the values of some of the fixed variables. As in ANOVA, it is possible to have higher-way interaction terms (e.g., $x_1 x_2 x_3$), making model interpretation more difficult.

Implicit in the discussion thus far is the assumption that x_1 and x_2 are quantitative variables. It is also possible to build models involving one or more qualitative variables. Suppose x_1 is a quantitative variable and the other variable of interest is qualitative with three different levels (say, manufacturers 1, 2, and 3). Let

$$x_2 = \begin{cases} 1 & \text{if the second variable is at level 2 (e.g., manufacturer 2)} \\ 0 & \text{otherwise} \end{cases}$$

$$x_3 = \begin{cases} 1 & \text{if the second variable is at level 3 (e.g., manufacturer 3)} \\ 0 & \text{otherwise} \end{cases}$$

The values $x_2 = x_3 = 0$ correspond to the second variable being at level 1 ($x_2 = x_3 = 1$ is not allowed). The predictors x_2 and x_3 are often called *indicator* or *dummy* variables. Then two possible models are

$$Y = \beta_0 + \beta_1 x_1 + \beta_2 x_2 + \beta_3 x_3 + \epsilon$$

and

$$Y = \beta_0 + \beta_1 x_1 + \beta_2 x_2 + \beta_3 x_3 + \beta_4 x_1 x_2 + \beta_5 x_1 x_3 + \epsilon$$

The contour of the regression function for each of the three levels of the qualitative variable is pictured in Figure 13.10 for both these models. Chapter 9 of the book by Neter et al. contains an excellent discussion of models containing qualitative variables.

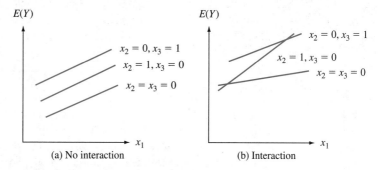

Figure 13.10 Regression function contours for a model with one quantitative variable (x_1) and one qualitative variable having three levels

Estimating Parameters

The data in simple linear regression consists of n pairs $(x_1, y_1), \ldots, (x_n, y_n)$. Suppose that a multiple regression model contains two predictor variables x_1 and x_2. Then the data set will consist of n triples $(x_{11}, x_{21}, y_1), (x_{12}, x_{22}, y_2), \ldots, (x_{1n}, x_{2n}, y_n)$. Here the first subscript on x refers to the predictor and the second to the observation number. More generally, the data now consists of n $(k + 1)$-tuples $(x_{11}, x_{21}, \ldots, x_{k1}, y_1)$, $(x_{12}, x_{22}, \ldots, x_{k2}, y_2), \ldots, (x_{1n}, x_{2n}, \ldots, x_{kn}, y_n)$, where x_{ij} is the value of the ith predictor x_i associated with the observed value y_j. The y_j's are assumed to have been observed independently of one another according to the model (13.15). To estimate the parameters $\beta_0, \beta_1, \ldots, \beta_k$ using the principle of least squares, form the sum of squared deviations of the observed y_j's from a trial function $y = b_0 + b_1 x_1 + \cdots + b_k x_k$:

$$f(b_0, b_1, \ldots, b_k) = \sum_j [y_j - (b_0 + b_1 x_{1j} + b_2 x_{2j} + \cdots + b_k x_{kj})]^2 \qquad (13.17)$$

The least-squares estimates are those values of $b_0, b_1, \ldots, b_k$ that minimize $f(b_0, \ldots, b_k)$. Taking the partial derivative of f with respect to each b_i $(i = 0, 1, \ldots, k)$ and equating all partials to zero yields the following system of **normal equations:**

$$b_0 n + b_1 \Sigma x_{1j} + b_2 \Sigma x_{2j} + \cdots + b_k \Sigma x_{kj} = \Sigma y_j$$
$$b_0 \Sigma x_{1j} + b_1 \Sigma x_{1j}^2 + b_2 \Sigma x_{1j} x_{2j} + \cdots + b_k \Sigma x_{1j} x_{kj} = \Sigma x_{1j} y_j$$

$$\qquad (13.18)$$

$$b_0 \Sigma x_{kj} + b_1 \Sigma x_{1j} x_{kj} + \cdots + b_{k-1} \Sigma x_{k-1, j} x_{kj} + b_k \Sigma x_{kj}^2 = \Sigma x_{kj} y_j$$

That these equations are linear in the unknowns $b_0, b_1, \ldots, b_k$ is a consequence of the regression function being linear in the parameters. Solving (13.18) yields the least-squares estimates $\hat{\beta}_0, \hat{\beta}_1, \ldots, \hat{\beta}_k$. In general, the system (13.18) can be solved by hand by first computing all coefficients of the b_i's and then using a technique such as Gaussian elimination. This is quite tedious, but fortunately any of the standard statistical regression packages will automatically solve for and print out $\hat{\beta}_0, \ldots, \hat{\beta}_k$.

Example 13.9   In the paper "An Ultracentrifuge Flour Absorption Method" (*Cereal Chemistry*, 1978, pp. 96–101), the authors studied the relationship between water absorption for wheat flour and various characteristics of the flour. In particular, the authors used a first-order multiple linear regression model to relate absorption y (%) to flour protein x_1 (%) and starch damage x_2 (Farrand units). The data is shown in Table 13.5.

Table 13.5 Data for Example 13.9

x_1	x_2	y	x_1	x_2	y
8.5	2	30.9	12.9	24	47.0
8.9	3	32.7	12.0	25	46.8
10.6	3	36.7	12.9	28	45.9
10.2	20	41.9	13.1	28	48.8
9.8	22	40.9	11.4	32	46.2
10.8	20	42.9	13.2	28	47.8
11.6	31	46.3	11.6	35	49.2
12.0	32	47.6	12.1	34	48.3
12.5	31	47.2	11.3	35	48.6
10.9	28	44.0	11.1	40	50.2
12.2	36	47.7	11.5	45	49.6
11.9	28	43.9	11.6	50	53.2
11.3	30	46.8	11.7	55	54.3
13.0	27	46.2	11.7	57	55.8

Figure 13.11 shows partial SPSS output from fitting the model $Y = \beta_0 + \beta_1 x_1 + \beta_2 x_2 + \epsilon$. In particular, from the B column at the bottom,

$$\hat{\beta}_0 = 19.43976, \quad \hat{\beta}_1 = 1.44228, \quad \hat{\beta}_2 = .33563$$

giving the estimated regression function

$$y = 19.43976 + 1.44228x_1 + .33563x_2$$

```
Multiple R              .98207
R Square                .96447
Adjusted R Square       .96163
Standard Error         1.09412

Analysis of Variance
                        DF    Sum of Squares    Mean Square
Regression               2         812.37959      406.18980
Residual                25          29.92755        1.19710

F =    339.31092     Signif F =    .0000

-------------- Variables in the .quation --------------

Variable             B        SE B     95% Confdnce     Intrvl B

STARCH            .33563     .01814         .29828         .37298
FLOUR            1.44228     .20764        1.01465        1.86991
(Constant)      19.43976    2.18829       14.93290       23.94662
```

Figure 13.11 SPSS output for Example 13.9

A point estimate of mean absorption when protein = 10 and damage = 25 is then

$$\hat{\mu}_{Y \cdot 10,25} = 19.43976 + (1.44228)(10) + (.33563)(25) \approx 42.3$$

This is also a point prediction for a single observation on absorption to be made when $x_1 = 10$ and $x_2 = 25$.   ■

$\hat{\sigma}^2$ and R^2

As for both simple linear regression and polynomial regression, the estimate of σ^2 is based on the sum of squared residuals

$$SSE = \Sigma[y_j - (\hat{\beta}_0 + \hat{\beta}_1 x_{1j} + \cdots + \hat{\beta}_k x_{kj})]^2$$

An efficient calculation formula for *SSE* is employed by most statistical computer packages. Because $k + 1$ parameters $(\beta_0, \beta_1, \ldots, \beta_k)$ have been estimated, $k + 1$ df are lost, so $n - (k + 1)$ df is associated with *SSE*, and

$$\hat{\sigma}^2 = s^2 = \frac{SSE}{n - (k + 1)} = MSE$$

With $SST = \Sigma(y_i - \bar{y})^2$, the proportion of total variation explained by the multiple regression model is $R^2 = 1 - SSE/SST$, the **coefficient of multiple determination**. As in polynomial regression, R^2 is often adjusted for the number of parameters in the model by the formula adjusted

$$R^2 = [(n - 1)R^2 - k]/[n - (k + 1)]$$

The positive square root of the coefficient of multiple determination is called the multiple correlation coefficient R. It can be shown that R is the sample correlation coefficient r between the observed y_j's and the predicted $\hat{y}_j$'s (i.e., using $x_j = \hat{y}_j$ in the formula for r results in $r = R$).

Example 13.10
(Example 13.9 continued)
The *SPSS* output of Figure 13.11 shows that $SSE = 29.92755 \approx 29.928$, so $\hat{\sigma}^2 = 29.928/[28 - (2 + 1)] = 1.197$, and $\hat{\sigma} = s = 1.094$. This is reported on the output as Standard Error = 1.09412. Also, $SST = \Sigma y_i^2 - (\Sigma y_i)^2/n = 60{,}035.12 - (1287.4)^2/28 = 842.31$ (the sum of the two *SS*s appearing in the ANOVA table), from which

$$R^2 = 1 - 29.928/842.31 = 1 - .036 = .964$$

Thus, 96.4% of observed variation in the observed y's can be explained by the chosen model. In addition, adjusted $R^2 = .962$.   ■

A Model Utility Test

A scatter plot of bivariate data [(x, y) pairs] will almost always indicate whether the simple linear regression model or a higher-degree polynomial will specify a useful relationship between the two variables. Unfortunately, with multivariate data, no analogous picture suggests whether a particular multiple regression model will be judged useful. The value of R^2 certainly communicates a preliminary message, but this value is sometimes deceptive because it can be greatly inflated by using a large number of predictors (large k) relative to the sample size n (this is the rationale behind adjusting R^2).

The model utility test in simple linear regression involved the null hypothesis H_0: $\beta_1 = 0$, according to which there is no useful relation between y and the single predictor x. Here we consider the assertion that $\beta_1 = 0, \beta_2 = 0, \ldots, \beta_k = 0$, which says that there is no useful relationship between y and *any* of the k predictors. If at least one of these β's is not 0, the corresponding predictor(s) is/are useful. The test is based on a statistic which has a particular F distribution when H_0 is true.

Null hypothesis: H_0: $\beta_1 = \beta_2 = \cdots = \beta_k = 0$

Alternative hypothesis: H_a: at least one $\beta_i \neq 0$ $(i = 1, \ldots, k)$

Test statistic value*: $f = \dfrac{R^2/k}{(1 - R^2)/[n - (k + 1)]}$ (13.19)

Rejection region for a level α test: $f \geq F_{\alpha,k,n-(k+1)}$

Except for a constant multiple, the test statistic here is $R^2/(1 - R^2)$, the ratio of explained to unexplained variation. If the proportion of explained variation is high relative to unexplained, we would naturally want to reject H_0 and confirm the utility of the model. However, if k is large relative to n, the factor $[(n - (k + 1))/k]$ will decrease f considerably.

Example 13.11   The ability of ecologists to identify regions of greatest species richness could have an impact on the preservation of genetic diversity, a major objective of the World Conservation Strategy. The paper "Prediction of Rarities from Habitat Variables: Coastal Plain Plants on Nova Scotian Lakeshores" (*Ecology*, 1992, pp. 1852–1859) used a sample of $n = 37$ lakes to obtain the estimated regression equation

$$y = 3.89 + .033x_1 + .024x_2 + .023x_3 - .0080x_4 - .13x_5 - .72x_6$$

where y = species richness, x_1 = watershed area, x_2 = shore width, x_3 = poor drainage (%), x_4 = water color (total color units), x_5 = sand (%), and x_6 = alkalinity. The coefficient of multiple determination was reported as $R^2 = .83$. Let's use the model utility F test with a significance level of .01. The relevant hypotheses are

$$H_0: \beta_1 = \beta_2 = \beta_3 = \beta_4 = \beta_5 = \beta_6 = 0$$

versus

$$H_a: \text{at least one among } \beta_1, \ldots, \beta_6 \text{ is not } 0$$

H_0 will be rejected if $f \geq F_{.01,6,30} = 3.47$

Since

$$f = \frac{.83/6}{.17/30} = 24.4 > 3.47$$

---

*f can also be written as *MSRegr/MSE*, where *MS = SS*/df, *SSRegr = SST − SSE*, and df for *SSRegr = k*

H_0 should be rejected. There does appear to be a useful linear relationship between y and at least one of the six predictors. ∎

Inferences About Model Parameters

Before testing hypotheses, constructing CIs, and making predictions, one should first examine diagnostic plots to see whether the model needs modification or whether there are outliers in the data. The recommended plots are (standardized) residuals versus each independent variable, residuals versus $\hat{y}$, $\hat{y}$ versus y, and a normal probability plot of the standardized residuals. Potential problems are suggested by the same patterns discussed in Section 13.1. Of particular importance is the identification of observations that have a large influence on the fit. In the next section, we describe several diagnostic tools suitable for this task.

Because each $\hat{\beta}_i$ is a linear function of the y_i's, the standard deviation of each $\hat{\beta}_i$ is the product of σ and a function of the x_{ij}'s, so an estimate $s_{\hat{\beta}_i}$ is obtained by substituting s for σ. Unfortunately, the function of the x_{ij}'s is quite complicated, but all standard regression computer packages compute and print the $s_{\hat{\beta}_i}$'s. Inferences concerning a single β_i are based on the standardized variable

$$T = \frac{\hat{\beta}_i - \beta_i}{s_{\hat{\beta}_i}}$$

which has a t distribution with $n - (k + 1)$ df.

The point estimate of $\mu_{Y \cdot x_1^*, \ldots, x_k^*}$, the expected value of Y when $x_1 = x_1^*, \ldots, x_k = x_k^*$, is $\hat{\mu}_{Y \cdot x_1^*, \ldots, x_k^*} = \hat{\beta}_0 + \hat{\beta}_1 x_1^* + \cdots + \hat{\beta}_k x_k^*$. The estimated standard deviation of the corresponding estimator is again a complicated expression involving the sample x_{ij}'s. However, the better statistical computer packages will calculate it on request. Inferences about $\mu_{Y \cdot x_1^*, \ldots, x_k^*}$ are based on standardizing its estimator to obtain a t variable having $n - (k + 1)$ df.

Inferences Based on the Model $Y = \beta_0 + \beta_1 x_1 + \cdots + \beta_k x_k + \epsilon$

1. A $100(1 - \alpha)\%$ CI for β_i, the coefficient of x_i in the regression function, is

   $$\hat{\beta}_i \pm t_{\alpha/2, n-(k+1)} \cdot s_{\hat{\beta}_i}$$

   Simultaneous confidence intervals for several β_i's for which the simultaneous confidence level is controlled can be obtained by applying the Bonferroni technique.

2. A test for H_0: $\beta_i = \beta_{i0}$ uses the t statistic value $t = (\hat{\beta}_i - \beta_{i0})/s_{\hat{\beta}_i}$ along with a t critical value with $n - (k + 1)$ df. The test is upper-, lower-, or two-tailed according to whether H_a contains the inequality >, <, or ≠.

3. A $100(1 - \alpha)\%$ CI for $\mu_{Y \cdot x_1^*, \ldots, x_k^*}$ is

   $$\hat{\mu}_{Y \cdot x_1^*, \ldots, x_k^*} \pm t_{\alpha/2, n-(k+1)} \cdot \{\text{estimated SD of } \hat{\mu}_{Y \cdot x_1^*, \ldots, x_k^*}\}$$

4. A $100(1 - \alpha)\%$ PI for a future y value is

   $$\hat{\mu}_{Y \cdot x_1^*, \ldots, x_k^*} \pm t_{\alpha/2, n-(k+1)} \cdot \{s^2 + (\text{estimated SD of } \hat{\mu}_{Y \cdot x_1^*, \ldots, x_k^*})^2\}^{1/2}$$

Example 13.12 Soil and sediment adsorption, the extent to which chemicals collect in a condensed form on the surface, is an important characteristic influencing the effectiveness of pesticides and various agricultural chemicals. The paper "Adsorption of Phosphate, Arsenate, Methanearsonate, and Cacodylate by Lake and Stream Sediments: Comparisons with Soils" (*J. of Environ. Qual.*, 1984, pp. 499–504) gives the accompanying data (Table 13.6) on y = phosphate adsorption index, x_1 = amount of extractable iron, and x_2 = amount of extractable aluminum.

Table 13.6 Data for Example 13.12

Observation	$x_1 = $ Extractable Iron	$x_2 = $ Extractable Aluminum	$y = $ Adsorption Index
1	61	13	4
2	175	21	18
3	111	24	14
4	124	23	18
5	130	64	26
6	173	38	26
7	169	33	21
8	169	61	30
9	160	39	28
10	244	71	36
11	257	112	65
12	333	88	62
13	199	54	40

The paper proposed the model

$$Y = \beta_0 + \beta_1 x_1 + \beta_2 x_2 + \epsilon$$

A computer analysis yielded the following information:

Parameter β_i	Estimate $\hat{\beta}_i$	Estimated SD $s_{\hat{\beta}_i}$
β_0	−7.351	3.485
β_1	.11273	.02969
β_2	.34900	.07131

$R^2 = .948$, adjusted $R^2 = .938$, $s = 4.379$

$\hat{\mu}_{Y \cdot 160,39} = -7.351 + (.11273)(160) + (.34900)(39) = 24.30$

Estimated SD of $\hat{\mu}_{Y \cdot 160,39} = 1.30$

A 99% CI for β_1, the change in expected adsorption associated with a 1-unit increase in extractable iron while extractable aluminum is held fixed, requires $t_{.005,13-(2+1)} = t_{.005,10} = 3.169$. The CI is

$$.11273 \pm (3.169)(.02969) = .11273 \pm .09409 \approx (.019, .207)$$

Similarly, a 99% interval for β_2 is

$$.34900 \pm (3.169)(.07131) = .34900 \pm .22598 \approx (.123, .575)$$

The Bonferroni technique implies that the simultaneous confidence level for both intervals is at least 98%.

A 95% CI for $\mu_{Y \cdot 160, 39}$, expected adsorption when extractable iron = 160 and extractable aluminum = 39, is

$$24.30 \pm (2.228)(1.30) = 24.30 \pm 2.90 = (21.40, 27.20)$$

A 95% PI for a future value of adsorption to be observed when $x_1 = 160$ and $x_2 = 39$ is

$$24.30 \pm (2.228)\{(4.379)^2 + (1.30)^2\}^{1/2} = 24.30 \pm 10.18 = (14.12, 34.48) \quad \blacksquare$$

Frequently, the hypothesis of interest has the form $H_0: \beta_i = 0$ for a particular i. For example, after fitting the six-predictor model in Example 13.11, the investigator might wish to test $H_0: \beta_4 = 0$. According to H_0, as long as the predictors $x_1, x_2, x_3, x_5,$ and x_6 remain in the model, x_4 contains no useful information about y. The test statistic value is the t ratio $\hat{\beta}_i / s_{\hat{\beta}_i}$. Many statistical computer packages report the t ratio and corresponding P-value for each predictor included in the model.

An F Test for a Group of Predictors

The model utility F test was appropriate for testing whether there is useful information about the dependent variable in *any* of the k predictors (i.e., whether $\beta_1 = \cdots = \beta_k = 0$). In many situations, one first builds a model containing k predictors and then wishes to know whether any of the predictors in a particular subset provide useful information about Y. For example, a model to be used to predict students' test scores might include a group of background variables such as family income and education levels and also some school characteristic variables such as class size and spending per pupil. One interesting hypothesis is that the school characteristic predictors can be dropped from the model.

Let's label the predictors as $x_1, x_2, \ldots, x_l, x_{l+1}, \ldots, x_k$, so that it is the last $k - l$ that we are considering deleting from the model. We then wish to test

$H_0: \beta_{l+1} = \beta_{l+2} = \cdots = \beta_k = 0$

(so the "reduced" model $Y = \beta_0 + \beta_1 x_1 + \cdots + \beta_l x_l + \epsilon$ is correct)

versus

$H_a:$ at least one among $\beta_{l+1}, \ldots, \beta_k$ is not 0

(so in the "full" model $Y = \beta_0 + \beta_1 x_1 + \cdots + \beta_k x_k + \epsilon$, at least one of the last $k - l$ predictors provides useful information).

The test is carried out by fitting both the full and reduced models. Because the full model contains not only the parameters of the reduced model but also some extra parameters, it should fit the data at least as well as the reduced model. That is, if we let SSE_k be the sum of squared residuals for the full model and SSE_l be the corresponding sum for the reduced model, then $SSE_k \le SSE_l$.* Intuitively, if SSE_k is a great deal smaller than SSE_l, the full model provides a much better fit than the reduced model; the appropriate test statistic should then depend on the reduction $SSE_l - SSE_k$ in unexplained variation. The formal procedure is

SSE_k = unexplained variation for the full model

SSE_l = unexplained variation for the reduced model

Test statistic value: $$f = \frac{(SSE_l - SSE_k)/(k - l)}{SSE_k/[n - (k + 1)]}$$ (13.20)

Rejection region: $f \ge F_{\alpha, k-l, n-(k+1)}$

Example 13.13
The data in Table 13.7 was taken from the paper "Applying Stepwise Multiple Regression Analysis to the Reaction of Formaldehyde with Cotton Cellulose" (*Textile Research J.*, 1984, pp. 157–165). The dependent variable y is durable press rating, a quantitative measure of wrinkle resistance. The four independent variables used in the model building process are x_1 = HCHO (formaldehyde) concentration, x_2 = catalyst ratio, x_3 = curing temperature, and x_4 = curing time.

Table 13.7 Data for Example 13.13

Obser-vation	x_1	x_2	x_3	x_4	y	Obser-vation	x_1	x_2	x_3	x_4	y
1	8	4	100	1	1.4	16	4	10	160	5	4.6
2	2	4	180	7	2.2	17	4	13	100	7	4.3
3	7	4	180	1	4.6	18	10	10	120	7	4.9
4	10	7	120	5	4.9	19	5	4	100	1	1.7
5	7	4	180	5	4.6	20	8	13	140	1	4.6
6	7	7	180	1	4.7	21	10	1	180	1	2.6
7	7	13	140	1	4.6	22	2	13	140	1	3.1
8	5	4	160	7	4.5	23	6	13	180	7	4.7
9	4	7	140	3	4.8	24	7	1	120	7	2.5
10	5	1	100	7	1.4	25	5	13	140	1	4.5
11	8	10	140	3	4.7	26	8	1	160	7	2.1
12	2	4	100	3	1.6	27	4	1	180	7	1.8
13	4	10	180	3	4.5	28	6	1	160	1	1.5
14	6	7	120	7	4.7	29	4	1	100	1	1.3
15	10	13	180	3	4.8	30	7	10	100	7	4.6

*The estimates $\hat{\beta}_0, \hat{\beta}_1, \ldots, \hat{\beta}_l$ will in general be different for the full and reduced models, so in general two different multiple regressions must be run to obtain SSE_l and SSE_k. If the variables are listed in the above order, though, most computer packages provide an ANOVA table for the full model that can be used to avoid fitting the reduced model.

Consider the full model consisting of $k = 14$ predictors: x_1, x_2, x_3, x_4, $x_5 = x_1^2$, ..., $x_8 = x_4^2$, $x_9 = x_1x_2$, ..., $x_{14} = x_3x_4$ (all first- and second-order predictors). Is the inclusion of the second-order predictors justified? That is, should the reduced model consisting of just the predictors x_1, x_2, x_3, and x_4 ($l = 4$) be used? Output resulting from fitting the two models follows:

Parameter	Estimate for Reduced Model	Estimate for Full Model
β_0	−.9122	−8.807
β_1	.16073	.1768
β_2	.21978	.7580
β_3	.011226	.10400
β_4	.10197	.5052
β_5	—	−.04393
β_6	—	−.035887
β_7	—	−.00003271
β_8	—	−.01646
β_9	—	.00588
β_{10}	—	.002702
β_{11}	—	.01178
β_{12}	—	−.0006547
β_{13}	—	.00242
β_{14}	—	.002526
R^2	.692	.921
SSE	17.4951	4.4782

The hypotheses to be tested are

$$H_0: \beta_5 = \beta_6 = \cdots = \beta_{14} = 0$$

versus

H_a: at least one among $\beta_5, \ldots, \beta_{14}$ is not 0

With $k = 14$ and $l = 4$, the F critical value for a test with $\alpha = .01$ is $F_{.01,10,15} = 3.80$. The test statistic value is

$$f = \frac{(17.4951 - 4.4782)/10}{4.4782/15} = \frac{1.3017}{.2985} = 4.36$$

Since $4.36 \geq 3.80$, H_0 is rejected. We conclude that the appropriate model should include at least one of the second-order predictors (Example 13.17 contains more information about selecting a model for this situation). ∎

Exercises   Section 13.4 (33–46)

33. The paper "The Value of Information for Selected Appliances" (*J. of Marketing Research*, 1980, pp. 14–25) suggests the plausibility of the general multiple regression model for relating the dependent variable y = the price of an air conditioner to $k = 3$ independent variables: x_1 = Btu-per-hour rating, x_2 = energy-efficiency ratio, and x_3 = number of settings. Suppose the model equation is

$$Y = -70 + .025x_1 + 20x_2 + 7.5x_3 + \epsilon$$

and $\sigma = 20$.

a. What is the expected value of price when Btu rating = 6000, energy-efficiency ratio = 8.0, and number of settings = 5?

b. What is the expected change in price associated with a 1-unit increase in energy efficiency ratio when the values of x_1 and x_3 are held fixed?

c. Assuming that ϵ has a normal distribution, what is the probability that a single observation on price is between 240 and 300 when x_1, x_2, and x_3 have the values specified in part (a)?

34. A chemist is studying how the yield (y) of a product from a certain chemical reaction depends on reaction temperature (x_1) and pressure (x_2). The following two models have been proposed for temperatures between 80 and 110 in combination with pressure values ranging between 50 and 70.

(i) $Y = 1200 + 15x_1 - 35x_2 + \epsilon$

(ii) $Y = -4500 + 75x_1 + 60x_2 - x_1x_2 + \epsilon$

a. Give the expected value of yield for each model when the temperature is 90. Repeat for a temperature of 95 and again for a temperature of 100. Graph the three mean values for each of the two models.

b. Referring to part (a), what is the expected change in yield associated with a 1-unit increase in pressure for a temperature of 90? Answer this for a temperature of 95 and again for a temperature of 100. How do these expected changes differ for the two models?

35. The accompanying output from a multiple regression analysis appeared in the paper "How to Optimize and Control the Wire Bonding Process: Part II" (*Solid State Technology*, 1991, pp. 67–70). The dependent variable y is ball bond shear strength (g), and the predictors are x_1 = force (g), x_2 = power (mw), x_3 = temperature (°C), and x_4 = time (ms).

1484XQ Ball Shear Strength

Predictor	Coeff	SD	t-ratio
Constant	39.3467	0.2722	144.55
Force	1.0958	0.3043	3.60
Power	6.7958	0.3043	22.33
Temperature	3.0625	0.3043	10.06
Time	1.8542	0.3043	6.09

The regression equation is:
Shear = 39.3 + 1.10 Force + 6.80 Power
                    + 3.06 Temperature + 1.85 Time.

s = 4.715

Analysis of Variance

Source	DF	SS	MS	F
Regression	4	14448.2	3612.1	162.71
Error	295	6557.7	22.2	
Total	299	21005.9		

a. Calculate and interpret the value of the coefficient of multiple determination.

b. Carry out a test of hypotheses at level .01 to see if the model specifies a useful relationship.

c. Suppose that the estimated standard deviation of $\hat{\beta}_0 + \hat{\beta}_1(35) + \hat{\beta}_2(75) + \hat{\beta}_3(200) + \hat{\beta}_4(20)$ is .950. Calculate a confidence interval for expected strength when the predictors have the indicated values using a confidence level of 95%.

d. Calculate and interpret a 95% CI for β_4.

36. The paper "Readability of Liquid Crystal Displays: A Response Surface" (*Human Factors*, 1983, pp. 185–190) used a multiple regression model with four independent variables to study accuracy in reading liquid crystal displays. The variables were

y = error percentage for subjects reading a four-digit liquid crystal display

x_1 = level of backlight (ranging from 0 to 122 cd/m²)

x_2 = character subtense (ranging from .025° to 1.34°)

x_3 = viewing angle (ranging from 0° to 60°)

x_4 = level of ambient light (ranging from 20 to 1500 lux)

The model fit to data was $Y = \beta_0 + \beta_1 x_1 + \beta_2 x_2 + \beta_3 x_3 + \beta_4 x_4 + \epsilon$. The resulting estimated coefficients were $\hat{\beta}_0 = 1.52$, $\hat{\beta}_1 = .02$, $\hat{\beta}_2 = -1.40$, $\hat{\beta}_3 = .02$, and $\hat{\beta}_4 = -.0006$.

a. Calculate an estimate of expected error percentage when $x_1 = 10$, $x_2 = .5$, $x_3 = 50$, and $x_4 = 100$.

b. Estimate the mean error percentage associated with a backlight level of 20, character subtense of .5, viewing angle of 10, and ambient light level of 30.

c. What is the estimated expected change in error percentage when the level of ambient light is increased by 1 unit while all other variables are fixed at the values given in part (a)? Answer for a 100-unit increase in ambient light level.

d. Explain why the answers in part (c) do not depend on the fixed values of x_1, x_2, and x_3. Under what conditions would there be such a dependence?

37. The estimated model discussed in Exercise 36 was based on $n = 30$ observations. The paper also reported the values $SST = 39.2$ and $SSE = 20.0$. Calculate the coefficient of multiple determination and then carry out the model utility test using $\alpha = .05$.

38. The paper "Development of a Model for Use in Maize Replant Decisions" (*Agronomy J.*, 1980, pp. 459–464) reports a summary of regression analyses using as variables y = % expected maize yield, x_1 = planting date (days after April 20), and x_2 = plant density [.0004047 × (plants/ha)]. The five planting dates were April 20, May 6, May 22, May 31, and June 10; the four plant densities were 30,890, 41,180, 51,480, and 61,780 plants/ha; and n was 180. The model

$$Y = \beta_0 + \beta_1 x_1 + \beta_2 x_2 + \beta_3 x_1^2 + \beta_4 x_2^2 + \epsilon$$

with $R^2 = .820$ gave a good fit to the data. Parameter estimates and estimated standard deviations were

Parameter	Estimate	Estimated SD
β_0	21.09	—
β_1	.653	.14
β_2	5.488	1.407
β_3	-.02059	.0027
β_4	-.10155	.0373

a. Graph the contours of the regression function for each of the given values of x_1 (estimated expected Y versus x_2 for each of the five x_1 values).

b. Assuming that the true model is given as shown above, what is the expected change in yield resulting from an increase of one in x_2? Estimate this expected change when plant density is 41,180 plants/ha.

39. Refer to the maize yield regression discussed in Exercise 38.

a. Test the hypothesis $H_0: \beta_1 = \beta_2 = \beta_3 = \beta_4 = 0$ (no linear relationship between Y and any of the predictors) against the alternative $H_a: \beta_i \neq 0$ for at least one i.

b. Compute joint CIs for β_3 and β_4 such that the joint confidence level is at least 95%.

c. Suppose $SSE = 1275.75$ (not given in the paper). Compute $\hat{\sigma}$ and also SST.

d. The authors also fit the regression function $y = \beta_0 + \beta_1 x_1 + \beta_2 x_2 + \beta_3 x_1^2 + \beta_4 x_2^2 + \beta_5 x_1^3 + \beta_6 x_2^3$. Supposing that the resulting SSE was 1247.30 and using SSE of part (c) for the model in Exercise 38, test $H_0: \beta_5 = \beta_6 = 0$ at level .05.

40. An experiment to investigate the effects of a new technique for degumming of silk yarn was described in the paper "Some Studies in Degumming of Silk with Organic Acids" (*J. Society of Dyers and Colourists*,

1992, pp. 79–86). One response variable of interest was y = weight loss (%). The experimenters made observations on weight loss for various values of three independent variables: x_1 = temperature (°C) = 90, 100, 110; x_2 = time of treatment (min) = 30, 75, 120; x_3 = tartaric acid concentration (g/L) = 0, 8, 16. In the regression analyses, the three values of each variable were coded as -1, 0, and 1, respectively, giving the accompanying data (the value y_8 = 19.3 was reported, but our value y_8 = 20.3 results in regression output identical to that appearing in the paper.

Obs.	1	2	3	4	5	6	7	8
x_1	-1	-1	1	1	-1	-1	1	1
x_2	-1	1	-1	1	0	0	0	0
x_3	0	0	0	0	-1	1	-1	1
y	18.3	22.2	23.0	23.0	3.3	19.3	19.3	20.3

Obs.	9	10	11	12	13	14	15
x_1	0	0	0	0	0	0	0
x_2	-1	-1	1	1	0	0	0
x_3	-1	1	-1	1	0	0	0
y	13.1	23.0	20.9	21.5	22.0	21.3	22.6

A multiple regression model with $k = 9$ predictors—x_1, $x_2, x_3, x_4 = x_1^2, x_5 = x_2^2, x_6 = x_3^2, x_7 = x_1 x_2, x_8 = x_1 x_3$, and $x_9 = x_2 x_3$—was fit to the data, resulting in $\hat{\beta}_0 = 21.967$, $\hat{\beta}_1 = 2.8125$, $\hat{\beta}_2 = 1.2750$, $\hat{\beta}_3 = 3.4375$, $\hat{\beta}_4 =$

-2.208, $\hat{\beta}_5 = 1.867$, $\hat{\beta}_6 = -4.208$, $\hat{\beta}_7 = -.975$, $\hat{\beta}_8 = -3.750$, $\hat{\beta}_9 = -2.325$, $SSE = 23.379$, and $R^2 = .938$.

a. Does this model specify a useful relationship? State and test the appropriate hypotheses using a significance level of .01.

b. The estimated standard deviation of $\hat{\mu}_Y$ when $x_1 = \cdots = x_9 = 0$ (i.e., when temperature = 100, time = 75, and concentration = 8) is 1.248. Calculate a 95% CI for expected weight loss when

temperature, time, and concentration have the specified values.

c. Calculate a 95% PI for a single weight-loss value to be observed when temperature, time, and concentration have values 100, 75, and 8, respectively.

41. Returning to Exercise 40, fitting the model with only x_1, x_2, and x_3 as predictors gave $R^2 = .456$ and $SSE = 203.82$. Is there a useful relationship between weight loss and at least one of the second-order predictors $x_4, x_5, \ldots, x_9$? State and test the appropriate hypotheses.

42. The paper "The Influence of Temperature and Sunshine on the Alpha-Acid Contents of Hops (*Agricultural Meteorology*, 1974, pp. 375–382) reports the following data on yield (y), mean temperature over the period between date of coming into hops and date of picking (x_1), and mean percentage of sunshine during the same period (x_2), for the fuggle variety of hop.

x_1	16.7	17.4	18.4	16.8	18.9	17.1
x_2	30	42	47	47	43	41
y	210	110	103	103	91	76
x_1	17.3	18.2	21.3	21.2	20.7	18.5
x_2	48	44	43	50	56	60
y	73	70	68	53	45	31

Here is partial MINITAB output from fitting the first-order model $Y = \beta_0 + \beta_1 x_1 + \beta_2 x_2 + \epsilon$ used in the paper:

```
Predictor    Coef     Stdev    t-ratio       p
Constant    415.11    82.52     5.03     0.000
Temp         -6.593    4.859    -1.36     0.208
Sunshine     -4.504    1.071    -4.20     0.002
s = 24.45    R-sq = 76.8%     R-sq(adj) = 71.6%
```

a. What is $\hat{\mu}_{Y \cdot 18.9, 43}$, and what is the corresponding residual?

b. Test $H_0: \beta_1 = \beta_2 = 0$ versus H_a: either β_1 or $\beta_2 \neq 0$ at level .05.

c. The estimated standard deviation of $\hat{\beta}_0 + \hat{\beta}_1 x_1 + \hat{\beta}_2 x_2$ when $x_1 = 18.9$ and $x_2 = 43$ is 8.20. Use this to obtain a 95% CI for $\mu_{Y \cdot 18.9, 43}$.

d. Use the information in part (c) to obtain a 95% PI

for yield in a future experiment when $x_1 = 18.9$ and $x_2 = 43$.

e. MINITAB reported that a 95% PI for yield when $x_1 = 18$ and $x_2 = 45$ is (35.94, 151.63). What is a 90% PI in this situation?

f. Given that x_2 is in the model, would you retain x_1?

g. When the model $Y = \beta_0 + \beta_2 x_2 + \epsilon$ is fit, the resulting value of R^2 is .721. Verify that the F statistic for testing H_0: $Y = \beta_0 + \beta_2 x_2 + \epsilon$ versus H_a: $Y = \beta_0 + \beta_1 x_1 + \beta_2 x_2 + \epsilon$ satisfies $t^2 = f$, where t is the value of the t statistic from part (f).

43. a. When the model $Y = \beta_0 + \beta_1 x_1 + \beta_2 x_2 + \beta_3 x_1^2 + \beta_4 x_2^2 + \beta_5 x_1 x_2 + \epsilon$ is fit to the hops data of Exercise 42, the estimate of β_5 is $\hat{\beta}_5 = .557$ with estimated standard deviation $s_{\hat{\beta}_5} = .94$. Test H_0: $\beta_5 = 0$ versus H_a: $\beta_5 \neq 0$.

b. Each t ratio $\hat{\beta}_i / s_{\hat{\beta}_i}$ $(i = 1, 2, 3, 4, 5)$ for the model of part (a) is less than 2 in absolute value, yet $R^2 = .861$ for this model. Would it be correct to drop each term from the model because of its small t ratio? Explain.

c. Using $R^2 = .861$ for the model of part (a), test H_0: $\beta_3 = \beta_4 = \beta_5 = 0$ (which says that all second-order terms can be deleted).

44. The article "The Undrained Strength of Some Thawed Permafrost Soils" (*Canadian Geotechnical J.*, 1979, pp. 420–427) contains the following data on undrained shear strength of sandy soil (y, in kPa), depth (x_1, in m),

and water content (x_2, in %). The predicted values and residuals were computed by fitting a full quadratic model, which resulted in the estimated regression function

$$y = -151.36 - 16.22x_1 + 13.48x_2 + .094x_1^2 - .253x_2^2 + .492x_1 x_2$$

a. Do plots of e^* versus x_1, e^* versus x_2, and e^* versus $\hat{y}$ suggest that the full quadratic model should be modified? Explain your answer.

b. The value of R^2 for the full quadratic model is .759. Test at level .05 the null hypothesis stating that there is no linear relationship between the dependent variable and any of the five predictors.

c. It can be shown that $V(Y) = \sigma^2 = V(\hat{Y}) + V(Y - \hat{Y})$. The estimate of σ is $\hat{\sigma} = s = 6.99$ (from the full quadratic model). First obtain the estimated standard deviation of $Y - \hat{Y}$ and then estimate the standard deviation of $\hat{Y}$ (i.e., $\hat{\beta}_0 + \hat{\beta}_1 x_1 + \hat{\beta}_2 x_2 + \hat{\beta}_3 x_1^2 + \hat{\beta}_4 x_2^2 + \hat{\beta}_5 x_1 x_2$) when $x_1 = 8.0$ and $x_2 = 33.1$. Finally, compute a 95% CI for $\mu_{Y \cdot 8.0, 33.1}$. [HINT: What is $(y - \hat{y})/e^*$?]

d. Fitting the first-order model with regression function $\mu_{Y \cdot x_1, x_2} = \beta_0 + \beta_1 x_1 + \beta_2 x_2$ results in $SSE = 894.95$. Test at level .05 the null hypothesis that states that all quadratic terms can be deleted from the model.

45. In an experiment to study factors influencing the wood specific gravity of slash pines ("Anatomical Factors

	y	x_1	x_2	$\hat{y}$	$y - \hat{y}$	Standardized Residual e^*
1	14.7	8.9	31.5	23.35	-8.65	-1.50
2	48.0	36.6	27.0	46.38	1.62	.54
3	25.6	36.8	25.9	27.13	-1.53	-.53
4	10.0	6.1	39.1	10.99	-.99	-.17
5	16.0	6.9	39.2	14.10	1.90	.33
6	16.8	6.9	38.3	16.54	.26	.04
7	20.7	7.3	33.9	23.34	-2.64	-.42
8	38.8	8.4	33.8	25.43	13.37	2.17
9	16.9	6.5	27.9	15.63	1.27	.23
10	27.0	8.0	33.1	24.29	2.71	.44
11	16.0	4.5	26.3	15.36	.64	.20
12	24.9	9.9	37.8	29.61	-4.71	-.91
13	7.3	2.9	34.6	15.38	-8.08	-1.53
14	12.8	2.0	36.4	7.96	4.84	1.02

Influencing Wood Specific Gravity of Slash Pines and the Implications for the Development of a High-Quality Pulpwood," *TAPPI*, 1964, pp. 401–404), a sample of 20 mature wood samples was obtained, and measurements were taken on number of fibers/mm^2 in springwood (x_1), number of fibers/mm^2 in summerwood (x_2), % springwood (x_3), light absorption in springwood (x_4), and light absorption in summerwood (x_5).

a. Fitting the regression function $\mu_{Y \cdot x_1, x_2, x_3, x_4, x_5} = \beta_0 + \beta_1 x_1 + \cdots + \beta_5 x_5$ resulted in $R^2 = .769$. Does the data indicate that there is a linear relationship between specific gravity and at least one of the predictors? Test using $\alpha = .01$.

b. When x_2 is dropped from the model, the value of R^2 remains at .769. Compute adjusted R^2 for both the full model and the model with x_2 deleted.

c. When x_1, x_2, and x_4 are all deleted, the resulting value of R^2 is .654. The total sum of squares is $SST = .0196610$. Does the data suggest that all of x_1, x_2, and x_4 have zero coefficients in the true regression model? Test the relevant hypotheses at level .05.

46. The accompanying data resulted from a study of the relationship between brightness of finished paper (y) and the variables H_2O_2% by weight (x_1), NaOH% by weight (x_2), silicate % by weight (x_3), and process temperature (x_4) ("Advantages of CEHDP Bleaching for High Brightness Kraft Pulp Production," *TAPPI*, 1964, pp. 170A–173A). Each independent variable was allowed to assume five different values, and these values were coded for regression analysis as -2, -1, 0, 1, and 2.

Test No.	H_2O_2 (x_1)	NaOH Conc (x_2)	Silicate Conc. (x_3)	Temp. (x_4)	Bright. (y)
1	-1	-1	-1	-1	83.9
2	$+1$	-1	-1	-1	84.9
3	-1	$+1$	-1	-1	83.4
4	$+1$	$+1$	-1	-1	84.2
5	-1	-1	$+1$	-1	83.8
6	$+1$	-1	$+1$	-1	84.7
7	-1	$+1$	$+1$	-1	84.0
8	$+1$	$+1$	$+1$	-1	84.8
9	-1	-1	-1	$+1$	84.5
10	$+1$	-1	-1	$+1$	86.0
11	-1	$+1$	-1	$+1$	82.6
12	$+1$	$+1$	-1	$+1$	85.1

Test No.	H_2O_2 (x_1)	NaOH Conc (x_2)	Silicate Conc. (x_3)	Temp. (x_4)	Bright. (y)
13	-1	-1	$+1$	$+1$	84.5
14	$+1$	-1	$+1$	$+1$	86.0
15	-1	$+1$	$+1$	$+1$	84.0
16	$+1$	$+1$	$+1$	$+1$	85.4
17	-2	0	0	0	82.9
18	$+2$	0	0	0	85.5
19	0	-2	0	0	85.2
20	0	$+2$	0	0	84.5
21	0	0	-2	0	84.7
22	0	0	$+2$	0	85.0
23	0	0	0	-2	84.9
24	0	0	0	$+2$	84.0
25	0	0	0	0	84.5
26	0	0	0	0	84.7
27	0	0	0	0	84.6
28	0	0	0	0	84.9
29	0	0	0	0	84.9
30	0	0	0	0	84.5
31	0	0	0	0	84.6

Variables	-2	-1	0	$+1$	$+2$
x_1 Hydrogen peroxide (100%), %wt	.1	.2	.3	.4	.5
x_2 NaOH, %wt	.1	.2	.3	.4	.5
x_3 Silicate (41°Bé), %wt	.5	1.5	2.5	3.5	4.5
x_4 Process temp., °F	130	145	160	175	190

a. When a (coded) model involving all linear terms, all quadratic terms, and all cross-product terms was fit, the estimated regression function was

$$y = 84.67 + .650x_1 - .258x_2 + .133x_3$$
$$+ .108x_4 - .135x_1^2 + .028x_2^2 + .028x_3^2$$
$$- .072x_4^2 + .038x_1 x_2 - .075x_1 x_3$$
$$+ .213x_1 x_4 + .200x_2 x_3 - .188x_2 x_4$$
$$+ .050x_3 x_4$$

Use this estimated model to predict brightness when H_2O_2 is .4%, NaOH is .4%, silicate is 3.5%, and temperature is 175. What are the values of the residuals for these values of the variables?

b. Express the estimated regression function in uncoded form.

c. $SST = 17.2567$, and R^2 for the model of part (a) is .885. When a model that includes only the four linear terms is fit, the resulting value of R^2 is .721. State and test at level .05 the null hypothesis that specifies that the coefficients of all quadratic and cross-product terms in the regression function are zero.

d. The estimated (coded) regression function when only linear terms are included is $\hat{\mu}_{Y \cdot x_1, x_2, x_3, x_4} =$

$85.5548 + .6500x_1 - .2583x_2 + .1333x_3 + .1083x_4$. When $x_1 = x_2 = x_3 = x_4 = 0$, the estimated standard deviation of $\hat{\mu}_{Y \cdot 0,0,0,0}$ is .0772. Suppose it had been believed that expected brightness for these values of the x_i's was at least 85.0. Does the given information contradict this belief? State and test the appropriate hypotheses.

13.5   Other Issues in Multiple Regression

In this section, we touch upon a number of issues that may arise when a multiple regression analysis is carried out. Consult the chapter references for a more extensive treatment of any particular topic.

Transformations in Multiple Regression

Often theoretical considerations suggest a nonlinear relation between a dependent variable and two or more independent variables, while on other occasions diagnostic plots indicate that some type of nonlinear function should be used. Frequently a transformation will linearize the model.

Example 13.14   An article in *Lubrication Eng.* ("Accelerated Testing of Solid Film Lubricants," 1972, pp. 365–372) reports on an investigation of wear life for solid film lubricant. Three sets of journal bearing tests were run on a Mil-L-8937-type film at each combination of three loads (3000, 6000, and 10,000 psi) and three speeds (20, 60, and 100 rpm), and the wear life (hours) was recorded for each run, as shown in Table 13.8.

Table 13.8 Wear-life data for Example 13.14

s	$l(1000s)$	w	s	$l(1000s)$	w
20	3	300.2	60	6	65.9
20	3	310.8	60	10	10.7
20	3	333.0	60	10	34.1
20	6	99.6	60	10	39.1
20	6	136.2	100	3	26.5
20	6	142.4	100	3	22.3
20	10	20.2	100	3	34.8
20	10	28.2	100	6	32.8
20	10	102.7	100	6	25.6
60	3	67.3	100	6	32.7
60	3	77.9	100	10	2.3
60	3	93.9	100	10	4.4
60	6	43.0	100	10	5.8
60	6	44.5			

The article contains the comment that a lognormal distribution is appropriate for W, since $\ln(W)$ is known to follow a normal law (recall from Chapter 4 that this is what defines a lognormal distribution). The model that appears is $W = (c/s^a l^b) \cdot \epsilon$, from which $\ln(W) = \ln(c) - a\ln(s) - b\ln(l) + \ln(\epsilon)$; so with $Y = \ln(W)$, $x_1 = \ln(s)$, $x_2 = \ln(l)$, $\beta_0 = \ln(c)$, $\beta_1 = -a$, and $\beta_2 = -b$, we have a multiple linear regression model. After computing $\ln(w_i)$, $\ln(s_i)$, and $\ln(l_i)$ for the above data, a first-order model in the transformed variables yielded the results shown in Table 13.9.

Table 13.9 Estimated coefficients and t ratios for Example 13.14

Parameter β_i	Estimate $\hat{\beta}_i$	Estimated SD $s_{\hat{\beta}_i}$	$t = \hat{\beta}_i / s_{\hat{\beta}_i}$
β_0	10.8719	.7871	13.81
β_1	−1.2054	.1710	−7.05
β_2	−1.3979	.2327	−6.01

The multiple coefficient of determination (for the transformed observations) has value $R^2 = .781$. The estimated regression function for the transformed variables is

$$\ln(w) = 10.87 - 1.21\ln(s) - 1.40\ln(l)$$

so that the original regression function is estimated as

$$w = e^{10.87} \cdot s^{-1.21} \cdot l^{-1.40}$$

The Bonferroni approach can be used to obtain simultaneous CIs for β_1 and β_2, and because $\beta_1 = -a$ and $\beta_2 = -b$, intervals for a and b are then immediately available. ∎

Standardizing Variables

In Section 13.3, we considered transforming x to $x' = x - \bar{x}$ before fitting a polynomial. For multiple regression, especially when values of variables are large in magnitude, it is advantageous to carry this coding one step further. Let $\bar{x}_j$ and s_j be the sample average and sample standard deviation of the x_{ij}'s $(j = 1, \ldots, n)$. We now code each variable x_j by $x'_j = (x_j - \bar{x}_j)/s_j$. The coded variable x'_j simply reexpresses any x_j value in units of standard deviation above or below the mean. Thus, if $\bar{x}_i = 100$ and $s_i = 20$, $x_i = 130$ becomes $x'_i = 1.5$ because 130 is 1.5 SDs above the mean of the values of x_i. For example, the coded full second-order model with two independent variables has regression function

$$E(Y) = \beta_0 + \beta_1 \left(\frac{x_1 - \bar{x}_1}{s_1} \right) + \beta_2 \left(\frac{x_2 - \bar{x}_2}{s_2} \right) + \beta_3 \left(\frac{x_1 - \bar{x}_1}{s_1} \right)^2$$

$$+ \beta_4 \left(\frac{x_2 - \bar{x}_2}{s_2} \right)^2 + \beta_5 \left(\frac{x_1 - \bar{x}_1}{s_1} \right) \left(\frac{x_2 - \bar{x}_2}{s_2} \right)$$

$$= \beta_0 + \beta_1 x'_1 + \beta_2 x'_2 + \beta_3 x'_3 + \beta_4 x'_4 + \beta_5 x'_5$$

The benefits of coding are (1) increased numerical accuracy in all computations (through less computer round-off error) and (2) more accurate estimation than for the parameters of the uncoded model because the individual parameters of the coded model characterize the behavior of the regression function near the center of the data rather than near the origin.

Example 13.15   The paper "The Value and the Limitations of High-Speed Turbo-Exhausters for the Removal of Tar-Fog from Carburetted Water-Gas" (*Soc. Chemical Industry J.*, 1946, pp. 166–168) presents the data (in Table 13.10) on y = tar content (grains/100 ft^3) of a gas stream as a function of x_1 = rotor speed (rpm) and x_2 = gas inlet temperature (°F). The data is also considered in the paper "Some Aspects of Nonorthogonal Data Analysis" (*J. Quality Technology*, 1973, pp. 67–79), which suggests using the coded model described above. The means and standard deviations are $\bar{x}_1 = 2991.13$,

Table 13.10 Data for Example 13.15

Run	y	x_1	x_2	x_1'	x_2'
1	60.0	2400	54.5	−1.52428	−.57145
2	61.0	2450	56.0	−1.39535	−.35543
3	65.0	2450	58.5	−1.39535	.00461
4	30.5	2500	43.0	−1.26642	−2.22763
5	63.5	2500	58.0	−1.26642	−.06740
6	65.0	2500	59.0	−1.26642	.07662
7	44.0	2700	52.5	−.75070	−.85948
8	52.0	2700	65.5	−.75070	1.01272
9	54.5	2700	68.0	−.75070	1.37276
10	30.0	2750	45.0	−.62177	−1.93960
11	26.0	2775	45.5	−.55731	−1.86759
12	23.0	2800	48.0	−.49284	−1.50755
13	54.0	2800	63.0	−.49284	.65268
14	36.0	2900	58.5	−.23499	.00461
15	53.5	2900	64.5	−.23499	.86870
16	57.0	3000	66.0	.02287	1.08472
17	33.5	3075	57.0	.21627	−.21141
18	34.0	3100	57.5	.28073	−.13941
19	44.0	3150	64.0	.40966	.79669
20	33.0	3200	57.0	.53859	−.21141
21	39.0	3200	64.0	.53859	.79669
22	53.0	3200	69.0	.53859	1.51677
23	38.5	3225	68.0	.60305	1.37276
24	39.5	3250	62.0	.66752	.50866
25	36.0	3250	64.5	.66752	.86870
26	8.5	3250	48.0	.66752	−1.50755
27	30.0	3500	60.0	1.31216	.22063
28	29.0	3500	59.0	1.31216	.07662
29	26.5	3500	58.0	1.31216	−.06740
30	24.5	3600	58.0	1.57002	−.06740
31	26.5	3900	61.0	2.34360	.36465

$s_1 = 387.81$, $\bar{x}_2 = 58.468$, and $s_2 = 6.944$, so $x_1' = (x_1 - 2991.13)/387.81$ and $x_2' = (x_2 - 58.468)/6.944$. With $x_3' = (x_1')^2$, $x_4' = (x_2')^2$, and $x_5' = x_1' \cdot x_2'$, fitting the full second-order model requires solving the system of six normal equations in six unknowns. A computer analysis yielded $\hat{\beta}_0 = 40.2660$, $\hat{\beta}_1 = -13.4041$, $\hat{\beta}_2 = 10.2553$, $\hat{\beta}_3 = 2.3313$, $\hat{\beta}_4 = -2.3405$, and $\hat{\beta}_5 = 2.5978$. The estimated regression equation is then

$$\hat{y} = 40.27 - 13.40x_1' + 10.26x_2' + 2.33x_3' - 2.34x_4' + 2.60x_5'$$

Thus, if $x_1 = 3200$ and $x_2 = 57.0$, $x_1' = .539$, $x_2' = -.211$, $x_3' = (.539)^2 = .2901$, $x_4' = (-.211)^2 = .0447$, and $x_5' = (.539)(-.211) = -.1139$, so

$$\hat{y} = 40.27 - (13.40)(.539) + (10.26)(-.211) + (2.33)(.2901)$$
$$- (2.34)(.0447) + (2.60)(-.1139) = 31.16$$
∎

Variable Selection

Often an experimenter will have data on a large number of predictors and then wish to build a regression model involving a subset of the predictors. The use of the subset will make the resulting model more manageable, especially if more data is to be subsequently collected, and also result in a model that is easier to interpret and understand than one with many more predictors. Two fundamental questions in connection with variable selection are the following:

1. If we can examine regressions involving all possible subsets of the predictors for which data is available, what criteria should be used to select a model?
2. If the number of predictors is too large to permit all regressions to be examined, is there a way of examining a reduced number of subsets among which a good model (or models) will be found?

To address Question (1) first, if the number of predictors is small (≤ 5, say), then it would not be too tedious to examine all possible regressions using any one of the readily available statistical computer packages (MINITAB, SAS, BMDP, SPSS). If data on at least six predictors is available, all possible regressions involve at least 64 ($= 2^6$) different models. Several packages will, for any specified m between 1 and 10, give the m best one-predictor models, the m best two-predictor models, and so on ("best" here means smallest SSE or, equivalently, largest R^2). MINITAB will do this for as many as 20 predictors, whereas BMDP will handle up to 27. The corresponding SSEs (or functions of them) can then be compared according to any of the criteria described next. The reason for specifying an m greater than 1 is to see if the best models have similar SSEs or R^2's.

Criteria for Variable Selection

As before, we use a subscript k to denote a quantity (say, SSE_k) computed from a model with k predictors (and thus $k + 1$ β_i's, because β_0 will always be included). For a fixed value of k, it is reasonable to identify the best model as the one having minimum SSE_k. The more difficult issue concerns comparison of SSE_k's for different values of k. Three different criteria, each one a simple function of SSE_k, are widely used.

1. R_k^2, the coefficient of multiple determination for a k-predictor model. Because R_k^2 will virtually always increase as k does (and can never decrease), it is not the k which maximizes R_k^2 that interests us. Instead we wish to identify a small k for which R_k^2 is nearly as large as R^2 for all predictors in the model.

2. $MSE_k = SSE/(n - k - 1)$, the mean squared error for a k-predictor model. This is often used in place of R_k^2, since while R_k^2 never decreases with increasing k, a small decrease in SSE_k obtained with one extra predictor can be more than offset by a decrease of one in the denominator of MSE_k. The objective is then to find the model having minimum MSE_k. Since adjusted $R_k^2 = 1 - MSE_k/MST$ where $MST = SST/(n - 1)$ is constant in k, examination of adjusted R_k^2 is equivalent to consideration of MSE_k.

3. The rationale for the third criterion, C_k, is more difficult to understand, but the criterion is gaining increasing acceptance among data analysts. Suppose the true regression model is specified by m predictors—that is,

$$Y = \beta_0 + \beta_1 x_1 + \cdots + \beta_m x_m + \epsilon, \qquad V(\epsilon) = \sigma^2$$

so that

$$E(Y) = \beta_0 + \beta_1 x_1 + \cdots + \beta_m x_m$$

Consider fitting a model by using a subset of k of these m predictors; for simplicity of notation, suppose we use $x_1, x_2, \ldots, x_k$. Then by solving the system of normal equations, estimates $\hat{\beta}_0, \hat{\beta}_1, \ldots, \hat{\beta}_k$ are obtained (but not, of course, estimates of any β's corresponding to carriers not in the fitted model). The true expected value $E(Y)$ can then be estimated by $\hat{Y} = \hat{\beta}_0 + \hat{\beta}_1 x_1 + \cdots + \hat{\beta}_k x_k$. Now consider the **normalized expected total error of estimation**

$$\Gamma_k = \frac{E\left(\sum_{i=1}^{n} [\hat{Y}_i - E(Y_i)]^2 \right)}{\sigma^2} = \frac{E(SSE_k)}{\sigma^2} + 2(k + 1) - n \qquad (13.21)$$

The second equality in (13.21) must be taken on faith because it requires a tricky expected-value argument. A particular subset is then appealing if its Γ_k value is small. Unfortunately, though, $E(SSE_k)$ and σ^2 are not known. To remedy this, let s^2 denote the estimate of σ^2 based on the model that includes all predictors for which data is available and define

$$C_k = \frac{SSE_k}{s^2} + 2(k + 1) - n$$

A desirable model is then specified by a subset of predictors for which C_k is small.

Example 13.16   The review article by Ron Hocking listed in the chapter bibliography reports on an analysis of data taken from the 1974 issues of *Motor Trend* magazine. The dependent variable y was gas mileage, there were $n = 32$ observations, and the predictors for which data was obtained were x_1 = engine shape (1 = straight and 0 = V), x_2 = number of cylinders, x_3 = transmission type (1 = manual and 0 = auto), x_4 = number of transmission speeds, x_5 = engine size, x_6 = horsepower, x_7 = number of carburetor barrels, x_8 = final drive ratio, x_9 = weight, and x_{10} = quarter-mile time. In Table 13.11, we present summary information from the analysis. The table describes for each k the subset having minimum SSE_k; reading down the variables column indicates which variable is added in going from k to $k + 1$ (in going from $k = 2$ to $k = 3$, both x_3 and x_{10} are added, and x_2 is deleted). Figure 13.12 contains plots of R_k^2, adjusted R_k^2, and C_k against k; these plots are an important visual aid in selecting a subset. The estimate of σ^2 is $s^2 = 6.24$, which is MSE_{10}. A simple model that rates highly according to all criteria is the one containing predictors x_3, x_9, and x_{10}.

Table 13.11 Best subsets for gas mileage data of Example 13.16

k = Number of Predictors	Variables	SSE_k	R_k^2	Adjusted R_k^2	C_k
1	9	247.2	.756	.748	11.6
2	2	169.7	.833	.821	1.2
3	3, 10, −2	150.4	.852	.836	.1
4	6	142.3	.860	.839	.8
5	5	136.2	.866	.840	1.8
6	8	133.3	.869	.837	3.4
7	4	132.0	.870	.832	5.2
8	7	131.3	.871	.826	7.1
9	1	131.1	.871	.818	9.0
10	2	131.0	.871	.809	11.0

Figure 13.12 R_k^2 and C_k plots for the gas mileage data

Generally speaking, when a subset of k predictors ($k < m$) is used to fit a model, the estimators $\hat{\beta}_0, \hat{\beta}_1, \ldots, \hat{\beta}_k$ will be biased for $\beta_0, \beta_1, \ldots, \beta_k$ and $\hat{Y}$ will also be a biased estimator for the true $E(Y)$ (all this because $m - k$ predictors are missing from the fitted model). However, as measured by the total normalized expected error Γ_k, estimates based on a subset can provide more precision than would be obtained using all possible predictors; essentially this greater precision is obtained at the price of introducing a bias in the estimators. A value of k for which $C_k \approx k + 1$ indicates that the bias associated with this k-predictor model would be small.

Example 13.17
(Example 13.13 continued)

Reconsider the durable press rating data introduced at the end of Section 13.4. The data set consisted of observations on four independent variables, from which we created a model containing 14 predictors: the four x_i's, the four x_i^2's, and the six $x_i x_j$'s. This "full" model was judged superior to the "reduced" model that involved only the predictors x_1, x_2, x_3, and x_4. The R^2 value for the full model was .921. However, computer output shows that several t ratios $\hat{\beta}_i / s_{\hat{\beta}_i}$ are quite small in magnitude, suggesting that there may be several among the 14 predictors that can be deleted.

Figure 13.13 presents selected output resulting from use of the BMDP package to identify the best three models of each given size. Only information on the best three six-predictor and the best three seven-predictor models is given. The best six-predictor model seemed superior to models including fewer predictors, whereas very little appeared to be gained by considering models with more than seven predictors. The choice between competing models is not exactly clear-cut. It would be nice if there were a good model that did not include any predictors based on one or more of the primary variables (e.g., the model with only x_2, x_4, x_2^2, x_4^2, and $x_2 x_4$). Unfortunately, this does not happen (though if we had looked at only the best single

			SUBSETS WITH		6 VARIABLES
R-SQUARED	ADJUSTED R-SQUARED	CK	VARIABLE	COEFFICIENT	T-STATISTIC
.884952	.854939	5.92	X2	.959914	6.66
			X1SQ	-.0372760	-3.82
			X2SQ	-.0389469	-5.82
			X1X3	.00368402	4.91
			X1X4	.0192505	3.23
			X2X3	-.00128271	-1.91
			INTERCEPT	-1.21835	
.881662	.850791	6.54	VARIABLE	COEFFICIENT	T-STATISTIC
			X2	.779668	7.93
			X4	.408872	2.98
			X1SQ	-.0313804	-3.40
			X2SQ	-.0394407	-5.81
			X1X3	.00356717	5.12
			X3X4	-.00208494	-2.27
			INTERCEPT	-1.34754	
.880382	.849177	6.79	VARIABLE	COEFFICIENT	T-STATISTIC
			X2	.760472	7.71
			X1SQ	-.0380752	-3.46
			X2SQ	-.0383770	-5.64
			X1X3	.00319522	5.13
			X1X4	.0384039	2.92
			X3X4	-.00090822	-1.62
			INTERCEPT	-.632344	

```
                                          SUBSETS WITH          7 VARIABLES
R-SQUARED      ADJUSTED
               R-SQUARED    CK    VARIABLE     COEFFICIENT      T-STATISTIC
 .899208        .867138    5.20      X2           .800294           8.40
                                     X3           .0966644          2.41
                                     X4           .129477           3.49
                                     X1SQ        -.0392306         -3.13
                                     X2SQ        -.0416758         -6.35
                                     X3SQ        -.000388272       -2.70
                                     X1X3         .00435437         4.08
                                     INTERCEPT  -7.27515

 .898832        .866643    5.27    VARIABLE     COEFFICIENT      T-STATISTIC
                                     X2           .825845           8.60
                                     X3           .0974167          2.42
                                     X1SQ        -.0392516         -3.12
                                     X2SQ        -.0433512         -6.57
                                     X3SQ        -.000390110       -2.71
                                     X4SQ         .0158501          3.47
                                     X1X3         .00436212         4.08
                                     INTERCEPT  -7.25606

 .896670        .863792    5.68    VARIABLE     COEFFICIENT      T-STATISTIC
                                     X2           .786050           8.15
                                     X3           .0820370          2.02
                                     X1SQ        -.0413732         -3.21
                                     X2SQ        -.0407510         -6.13
                                     X3SQ        -.000325208       -2.24
                                     X1X3         .00401885         3.79
                                     X1X4         .0195025          3.36
                                     INTERCEPT  -6.06342
```

Figure 13.13 Partial BMDP output for Example 13.17

model of any given size, such a model might have been overlooked). The best six-predictor model seems a reasonable choice. ∎

Stepwise Regression

When the number of predictors is too large to allow for explicit or implicit examination of all possible subsets, several alternative selection procedures generally will identify good models. The simplest such procedure is the **backward elimination** (BE) method. This method starts with the model in which all predictors under consideration are used. Let the set of all such predictors be $x_1, \ldots, x_m$. Then each t ratio $\hat{\beta}_i / s_{\hat{\beta}_i}$ $(i = 1, \ldots, m)$ appropriate for testing $H_0: \beta_i = 0$ versus $H_a: \beta_i \neq 0$ is examined. If the t ratio with the smallest absolute value is less than a prespecified constant t_{out}, that is, if

$$\min_{i=1,\ldots,m} \left| \frac{\hat{\beta}_i}{s_{\hat{\beta}_i}} \right| < t_{\text{out}}$$

then the predictor corresponding to the smallest ratio is eliminated from the model. The reduced model is now fit, the $m - 1$ t ratios are again examined, and another predictor is eliminated if it corresponds to the smallest absolute t ratio smaller than

t_{out}. In this way the algorithm continues until at some stage, all absolute t ratios are at least t_{out}. The model used is the one containing all predictors that were not eliminated. The value $t_{out} = 2$ is often recommended since most $t_{.05}$ values are near 2.

Example 13.18

(Example 13.15 continued)

For the coded full quadratic model in which y = tar content, the five potential predictors are $x'_1, x'_2, x'_3 = x'^2_1, x'_4 = x'^2_2$, and $x'_5 = x'_1 x'_2$ (so $m = 5$). Without specifying t_{out}, the predictor with the smallest absolute t ratio (asterisked) was eliminated at each stage, resulting in the sequence of models shown in Table 13.12.

Table 13.12 Data for Example 13.18

Step	Predictors	$\|t \text{ Ratio}\|$ 1	2	3	4	5
1	1, 2, 3, 4, 5	16.0	10.8	2.9	2.8	1.8*
2	1, 2, 3, 4	15.4	10.2	3.7	2.0*	—
3	1, 2, 3	14.5	12.2	4.3*	—	—
4	1, 2	10.9	9.1*	—	—	—
5	1	4.4*	—	—	—	—

Using $t_{out} = 2$, the resulting model would be based on x'_1, x'_2, and x'_3, since at Step 3 no predictor could be eliminated. It can be verified that each subset above is actually the best subset of its size, though this is by no means always the case. ∎

An alternative to the BE procedure is **forward selection** (FS). FS starts with no predictors in the model and considers fitting in turn the model with only x_1, only $x_2, \ldots$, and finally only x_m. The variable that, when fit, yields the largest absolute t ratio enters the model provided that the ratio exceeds the specified constant t_{in}. Suppose x_1 enters the model. Then models with $(x_1, x_2), (x_1, x_3), \ldots, (x_1, x_m)$ are considered in turn. The largest $|\hat{\beta}_j/s_{\beta_j}|$ $(j = 2, \ldots, m)$ then specifies the entering predictor provided that this maximum also exceeds t_{in}. This continues until at some step no absolute t ratios exceed t_{in}. The entered predictors then specify the model. The value $t_{in} = 2$ is often used for the same reason that $t_{out} = 2$ is used in BE. For the tar-content data, FS resulted in the sequence of models given at steps 5, 4, . . . , 1 above, so agreed with BE. This will not always be the case.

The stepwise procedure most widely used is a combination of FS and BE, denoted by FB. This procedure starts as does forward selection, by adding variables to the model, but after each addition examines those variables previously entered to see if any is a candidate for elimination. For example, if there are eight predictors under consideration and the current set consists of x_2, x_3, x_5, and x_6 with x_5 having just been added, the t ratios $\hat{\beta}_2/s_{\beta_2}$, $\hat{\beta}_3/s_{\beta_3}$, and $\hat{\beta}_6/s_{\beta_6}$ are examined. If the smallest absolute ratio is less than t_{out}, then the corresponding variable is eliminated from the model. The idea behind FB is that with forward selection, a single variable may be more strongly related to y than either of two or more other variables individually, but the combination of these variables may make the single variable subsequently redundant.

This actually happened with the gas mileage data of Example 13.16, with x_2 entering and subsequently leaving the model.

The FB procedure is part of several standard computer packages. The BMDP package specifies $t_{in} = 2$ and $t_{out} = \sqrt{3.9}$ (most packages actually use $f = t^2$ rather than t itself).

While in most situations these automatic selection procedures will identify a good model, there is no guarantee that the best or even a nearly best model will result. Close scrutiny should be given to data sets for which there appear to be strong relationships among some of the potential predictors; we will say more about this shortly.

Identification of Influential Observations

In simple linear regression, it is easy to spot an observation whose x value is much larger or much smaller than other x values in the sample. Such an observation may have a great impact on the estimated regression equation (whether it actually does depends on how consistent the corresponding y value is with the remainder of the data). In multiple regression, it is also desirable to know whether the values of the predictors for a particular observation are such that it has the potential for exerting great influence on the estimated equation. One method for identifying potentially influential observations relies on the fact that because each $\hat{\beta}_i$ is a linear function of $y_1, y_2, \ldots, y_n$, each predicted y value of the form $\hat{y} = \hat{\beta}_0 + \hat{\beta}_1 x_1 + \cdots + \hat{\beta}_k x_k$ is also a linear function of the y_j's. In particular, the predicted values corresponding to sample observations can be written as follows:

$$\hat{y}_1 = h_{11}y_1 + h_{12}y_2 + \cdots + h_{1n}y_n$$
$$\hat{y}_2 = h_{21}y_1 + h_{22}y_2 + \cdots + h_{2n}y_n$$
$$\vdots$$
$$\hat{y}_n = h_{n1}y_1 + h_{n2}y_2 + \cdots + h_{nn}y_n$$

Each coefficient h_{ij} is a function only of the x_{ij}'s in the sample and not of the y_j's. It can be shown that $h_{ij} = h_{ji}$ and that $0 \le h_{jj} \le 1$.

Let's focus on the "diagonal" coefficients $h_{11}, h_{22}, \ldots, h_{nn}$. The coefficient h_{jj} is the weight given to y_j in computing the corresponding predicted value $\hat{y}_j$. This quantity can also be expressed as a measure of the distance between the point $(x_{1j}, \ldots, x_{kj})$ in k-dimensional space and the center of the data $(\bar{x}_1, \ldots, \bar{x}_k)$. It is therefore natural to characterize an observation whose h_{jj} is relatively large as one with potentially large influence. Unless there is a perfect linear relationship among the k predictors, $\sum_{j=1}^{n} h_{jj} = k + 1$, so the average of the h_{jj}'s is $(k + 1)/n$. Some statisticians suggest that if $h_{jj} > 2(k + 1)/n$, the jth observation be cited as having potentially large influence; others use $3(k + 1)/n$ as the dividing line.

Example 13.19   The accompanying data appeared in the paper "Testing for the Inclusion of Variables in Linear Regression by a Randomization Technique" (*Technometrics*, 1966,

Beam Number	Specific Gravity (x_1)	Moisture Content (x_2)	Strength (y)
1	.499	11.1	11.14
2	.558	8.9	12.74
3	.604	8.8	13.13
4	.441	8.9	11.51
5	.550	8.8	12.38
6	.528	9.9	12.60
7	.418	10.7	11.13
8	.480	10.5	11.70
9	.406	10.5	11.02
10	.467	10.7	11.41

pp. 695–699), and was reanalyzed in Hoaglin and Welsch, "The Hat Matrix in Regression and ANOVA" (*Amer. Statistician*, 1978, pp. 17–23). The h_{ij}'s (with elements below the diagonal omitted by symmetry) are as follows:

	1	2	3	4	5	6	7	8	9	10
1	.418	−.002	.079	−.274	−.046	.181	.128	.222	.050	.242
2		.242	.292	.136	.243	.128	−.041	.033	−.035	.004
3			.417	−.019	.273	.187	−.126	.044	−.153	.004
4				.604	.197	−.038	.168	−.022	.275	−.028
5					.252	.111	−.030	.019	−.010	−.010
6						.148	.042	.117	.012	.111
7							.262	.145	.277	.174
8								.154	.120	.168
9									.315	.148
10										.187

Here $k = 2$ so $(k + 1)/n = \frac{3}{10} = .3$; since $h_{44} = .604 > 2(.3)$, the fourth data point is identified as potentially influential.   ∎

Another technique for assessing the influence of the jth observation that takes into account y_j as well as the predictor values involves deleting the jth observation from the data set and performing a regression based on the remaining observations. If the estimated coefficients from the "deleted observation" regression differ greatly from the estimates based on the full data, the jth observation has clearly had a substantial impact on the fit. One way to judge whether estimated coefficients change greatly is to express each change relative to the estimated standard deviation of the coefficient:

$$\frac{(\hat{\beta}_i \text{ before deletion}) - (\hat{\beta}_i \text{ after deletion})}{s_{\hat{\beta}_i}} = \frac{\text{change in } \hat{\beta}_i}{s_{\hat{\beta}_i}}$$

There exist efficient computational formulas that allow all this information to be obtained from the "no-deletion" regression, so that the additional n regressions are unnecessary.

Example 13.20
(Example 13.19 continued)

Consider separately deleting observations 1 and 6, whose residuals are the largest, and observation 4, where h_{jj} is large. Table 13.13 contains the relevant information.

Table 13.13 Changes in estimated coefficients for Example 13.20

Parameter	No-deletions Estimates	Estimated SD	Change When Point j Is Deleted		
			$j = 1$	$j = 4$	$j = 6$
β_0	10.302	1.896	2.710	−2.109	−.642
β_1	8.495	1.784	−1.772	1.695	.748
β_2	.2663	.1273	−.1932	.1242	.0329
e_j:			−3.25	−.96	2.20
h_{jj}:			.418	.604	.148

For deletion of both point 1 and point 4, the change in each estimate is in the range 1–1.5 SDs, which is reasonably substantial (this does not tell us what would happen if both points were simultaneously omitted). For point 6, however, the change is roughly .25 SD. Thus, points 1 and 4, but not 6, might well be omitted in calculating a regression equation. ∎

Multicollinearity

In many multiple regression data sets, the predictors $x_1, x_2, \ldots, x_k$ are highly interdependent. Suppose we consider the usual model

$$Y = \beta_0 + \beta_1 x_1 + \cdots + \beta_k x_k + \epsilon$$

with data $(x_{1j}, \ldots, x_{kj}, y_j)$ $(j = 1, \ldots, n)$ available for fitting. If we use the principle of least squares to regress x_i on the other predictors $x_1, \ldots, x_{i-1}, x_{i+1}, \ldots, x_k$, obtaining

$$\hat{x}_i = a_0 + a_1 x_1 + \cdots + a_{i-1} x_{i-1} + a_{i+1} x_{i+1} + \cdots + a_k x_k$$

it can be shown that

$$V(\hat{\beta}_i) = \frac{\sigma^2}{\sum_{j=1}^{n} (x_{ij} - \hat{x}_{ij})^2} \tag{13.22}$$

When the sample x_i values can be predicted very well from the other predictor values, the denominator of (13.22) will be small, so $V(\hat{\beta}_i)$ will be quite large. If this is the case for at least one predictor, the data is said to exhibit **multicollinearity**. Multicollinearity is often suggested by a regression computer output in which R^2 is large but some of the t ratios $\hat{\beta}_i / s_{\hat{\beta}_i}$ are small for predictors that, based on prior information and intuition, seem important. Another clue to the presence of multicollinearity lies in a $\hat{\beta}_i$ value that has the opposite sign from that which intuition would suggest, indicating that another predictor or collection of predictors is serving as a "proxy" for x_i.

An assessment of the extent of multicollinearity can be obtained by regressing each predictor in turn on the remaining $k - 1$ predictors. Let R_i^2 denote the value of R^2 in the regression with dependent variable x_i and predictors $x_1, \ldots, x_{i-1}$, $x_{i+1}, \ldots, x_k$. It has been suggested that severe multicollinearity is present if $R_i^2 > .9$ for any i. MINITAB will refuse to include a predictor in the model when its R_i^2 value is close to 1.

There is unfortunately no consensus among statisticians as to what remedies are appropriate when severe multicollinearity is present. One possibility involves continuing to use a model that includes all the predictors but estimating parameters by using something other than least squares. Consult a chapter reference for more details.

Exercises   Section 13.5 (47–52)

47. The paper "Bank Full Discharge of Rivers" (*Water Resources J.*, 1978, pp. 1141–1154) reports data on discharge amount (q, in m^3/sec), flow area (a, in m^2), and slope of the water surface (b, in m/m) obtained at a number of floodplain stations. A subset of the data appears below. The paper proposed a multiplicative power model $Q = \alpha a^\beta b^\gamma \epsilon$.

q	17.6	23.8	5.7	3.0	7.5
a	8.4	31.6	5.7	1.0	3.3
b	.0048	.0073	.0037	.0412	.0416
q	89.2	60.9	27.5	13.2	12.2
a	41.1	26.2	16.4	6.7	9.7
b	.0063	.0061	.0036	.0039	.0025

a. Use an appropriate transformation to make the model linear and then estimate the regression parameters for the transformed model. Finally, estimate α, β, and γ (the parameters of the original model). What would be your prediction of discharge amount when flow area is 10 and slope is .01?

b. Without actually doing any analysis, how would you fit a multiplicative exponential model $Q = \alpha e^{\beta a} e^{\gamma b} \epsilon$?

c. After the transformation to linearity in part (a), a 95% CI for the value of the transformed regression function when $a = 3.3$ and $b = .0046$ was obtained from computer output as (.217, 1.755). Obtain a 95% CI for $\alpha a^\beta b^\gamma$ when $a = 3.3$, $b = .0046$.

48. a. Referring to Exercise 45, the mean and standard deviation of x_3 were 52.540 and 5.4447, respectively, while those of x_5 were 89.195 and 3.6660, respectively. When the model involving these two standardized variables was fit, the estimated regression equation was $y = .5255 - .0236x_3' + .0097x_5'$. What value of specific gravity would you predict for a wood sample with % springwood = 50 and % light absorption in summerwood = 90?

b. The estimated standard deviation of the estimated coefficient $\hat{\beta}_3$ of x_3' (i.e., for $\hat{\beta}_3$ of the standardized model) was .0046. Obtain a 95% CI for β_3.

c. Using the information in parts (a) and (b), what is the estimated coefficient of x_3 in the unstandardized model (using only predictors x_3 and x_5), and what is the estimated standard deviation of the coefficient estimator (i.e., $s_{\hat{\beta}_3}$ for $\hat{\beta}_3$ in the unstandardized model)?

d. The estimate of σ for the two-predictor model is $s = .02001$, while the estimated standard deviation of $\hat{\beta}_0 + \hat{\beta}_3 x_3' + \hat{\beta}_5 x_5'$ when $x_3' = -.3747$ and $x_5' = -.2769$ (i.e., when $x_3 = 50.5$ and $x_5 = 88.9$) is .00482. Compute a 95% PI for specific gravity when % springwood = 50.5 and % light absorption in summerwood = 88.9.

49. In the accompanying table, we give the smallest *SSE* for each number of predictors $k (k = 1, 2, 3, 4)$ for a regression problem in which y = cumulative heat of hardening in cement, x_1 = % tricalcium aluminate, x_2 = % tricalcium silicate, x_3 = % aluminum ferrate, and x_4 = % dicalcium silicate.

Number of Predictors k	Carrier(s)	SSE
1	x_4	880.85
2	x_1, x_2	58.01
3	x_1, x_2, x_3	49.20
4	x_1, x_2, x_3, x_4,	47.86

In addition, $n = 13$, and $SST = 2715.76$.

a. Use the criteria discussed in the text to recommend the use of a particular regression model.

b. Would forward selection result in the best two-predictor model? Explain.

50. The paper "Creep and Fatigue Characteristics of Ferrocement Slabs" (*J. Ferrocement*, 1984, pp. 309–322) reported data on y = tensile strength (MPa), x_1 = slab thickness (cm), x_2 = load (kg), x_3 = age at loading (days), and x_4 = time under test (days) resulting from stress tests of $n = 9$ reinforced concrete slabs. The results of applying the BE elimination method of variable selection are summarized in the accompanying tabular format. Explain what occurred at each step of the procedure.

Step	1	2	3
Constant	8.496	12.670	12.989
x_1	−.29	−.42	−.49
T-RATIO	−1.33	−2.89	−3.14
x_2	.0104	.0110	.0116
T-RATIO	6.30	7.40	7.33
x_3	.0059		
T-RATIO	.83		
x_4	−.023	−.023	
T-RATIO	−1.48	−1.53	
S	.533	.516	.570
R-SQ	95.81	95.10	92.82

51. An analysis of the hops yield data of Exercise 42 yields the accompanying h_{ii}'s for the regression model $Y = \beta_0 + \beta_1 x_1 + \beta_2 x_2 + \epsilon$.

i	1	2	3	4	5	6
h_{ii}	.486	.131	.087	.219	.112	.159

i	7	8	9	10	11	12
h_{ii}	.172	.090	.460	.314	.301	.468

According to the rule of thumb given in the text, do any of the data points have a potentially large influence on the fit?

52. Refer to the water-discharge data given in Exercise 47 and let $y = \ln(q)$, $x_1 = \ln(a)$, and $x_2 = \ln(b)$. Consider fitting the model $Y = \beta_0 + \beta_1 x_1 + \beta_2 x_2 + \epsilon$.

a. The resulting h_{ii}'s are .138, .302, .266, .604, .464, .360, .215, .153, .214, and .284. Does any observation appear to be influential?

b. The estimated coefficients are $\hat{\beta}_0 = 1.5652$, $\hat{\beta}_1 = .9450$, $\hat{\beta}_2 = .1815$, and the corresponding estimated standard deviations are $s_{\hat{\beta}_0} = .7328$, $s_{\hat{\beta}_1} = .1528$, and $s_{\hat{\beta}_2} = .1752$. The second standardized residual is $e_2^* = 2.19$. When the second observation is omitted from the data set, the resulting estimated coefficients are $\hat{\beta}_0 = 1.8982$, $\hat{\beta}_1 = 1.025$, and $\hat{\beta}_2 = .3085$. Do any of these changes indicate that the second observation is influential?

c. Deletion of the fourth observation (why?) yields $\hat{\beta}_0 = 1.4592$, $\hat{\beta}_1 = .9850$, and $\hat{\beta}_2 = .1515$. Is this observation influential?

Supplementary Exercises /Chapter 13 (53–63)

53. The accompanying data on x = frequency (MHz) and y = output power (W) for a certain laser configuration was read from a graph in the paper "Frequency Dependence in RF Discharge Excited Waveguide CO_2 Lasers" (*IEEE J. Quantum Electronics*, 1984, pp. 509–514):

x	60	63	77	100	125	157	186	222
y	16	17	19	21	22	20	15	5

A computer analysis yielded the following information for a quadratic regression model: $\hat{\beta}_0 = -1.5127$, $\hat{\beta}_1 = .391901$, $\hat{\beta}_2 = -.00163141$, $s_{\hat{\beta}_2} = .00003391$, $SSE = .29$, $SST = 202.88$, and (estimated standard deviation of $\hat{\mu}_{Y \cdot 100}$) $= .1141$.

a. Does the quadratic model appear to be suitable for explaining observed variation in output power by relating it to frequency?

b. Would the simple linear regression model be nearly

as satisfactory as the quadratic model?

c. Do you think it would be worth considering a cubic model?

d. Compute a 95% CI for expected power output when frequency is 100.

e. Use a 95% PI to predict the power from a single experimental run when frequency is 100.

54. Conductivity is one important characteristic of glass. The paper "Structure and Properties of Rapidly Quenched $Li_2O-Al_2O-Nb_2O_5$ Glasses" (*J. Amer. Ceramic Soc.*, 1983, pp. 890–892) reports the accompanying data on $x = Li_2O$ content of a certain type of glass and $y =$ conductivity at 500 K:

x	19	20	24	27	29	30
y	$10^{-8.0}$	$10^{-7.1}$	$10^{-7.2}$	$10^{-6.7}$	$10^{-6.2}$	$10^{-6.8}$

x	31	39	40	43	45	50
y	$10^{-5.8}$	$10^{-5.3}$	$10^{-6.0}$	$10^{-4.7}$	$10^{-5.4}$	$10^{-5.1}$

(This is a subset of the data that appeared in the paper.) Propose a suitable model for relating y to x, estimate the model parameters, and predict conductivity when Li_2O content is 35.

55. The effect of manganese (Mn) on wheat growth is examined in the article "Manganese Deficiency and Toxicity Effects on Growth, Development and Nutrient Composition in Wheat" (*Agronomy J.*, 1984, pp. 213–217). A quadratic regression model was used to relate $y =$ plant height (cm) to $x = \log_{10}$(added Mn), with μM as the units for added Mn. The accompanying data was read from a scatter diagram appearing in the paper:

x	−1.0	−.4	0	.2	1.0
y	32	37	44	45	46

x	2.0	2.8	3.2	3.4	4.0
y	42	42	40	37	30

In addition, $\hat{\beta}_0 = 41.7422$, $\hat{\beta}_1 = 6.581$, $\hat{\beta}_2 = -2.3621$, $s_{\hat{\beta}_0} = .8522$, $s_{\hat{\beta}_1} = 1.002$, $s_{\hat{\beta}_2} = .3073$, and $SSE = 26.98$.

a. Is the quadratic model useful for describing the relationship between x and y? [HINT: Quadratic regression is a special case of multiple regression with $k = 2$, $x_1 = x$, and $x_2 = x^2$.] Apply an appropriate procedure.

b. Should the quadratic predictor be eliminated?

c. Estimate expected height for wheat treated with 10 μM of Mn using a 90% CI. [HINT: The estimated standard deviation of $\hat{\beta}_0 + \hat{\beta}_1 + \hat{\beta}_2$ is 1.031.]

56. The article "Chemithermomechanical Pulp from Mixed High Density Hardwoods" (*TAPPI*, July 1988, pp. 145–146) reports on a study in which the accompanying data was obtained in order to relate $y =$ specific surface area (cm^3/g) to $x_1 = \%$ NaOH used as a pretreatment chemical and $x_2 =$ treatment time (min) for a batch of pulp:

x_1	x_2	y
3	30	5.95
3	60	5.60
3	90	5.44
9	30	6.22
9	60	5.85
9	90	5.61
15	30	8.36
15	60	7.30
15	90	6.43

The accompanying MINITAB output resulted from a request to fit the model $Y = \beta_0 + \beta_1 x_1 + \beta_2 x_2 + \epsilon$.

```
The regression equation is
AREA = 6.05 + 0.142 NAOH - 0.0169 TIME

Predictor        Coef        Stdev     t-ratio          p
Constant       6.0483       0.5208       11.61      0.000
NAOH          0.14167      0.03301        4.29      0.005
TIME         -0.016944     0.006601       -2.57      0.043

s = 0.4851            R-sq = 80.7%   R-sq (adj) 74.2%
```

Analysis of Variance

SOURCE	DF	SS	MS	F	p
Regression	2	5.8854	2.9427	12.51	0.007
Error	6	1.4118	0.2353		
Total	8	7.2972			

a. What proportion of observed variation in specific surface area can be explained by the model relationship?

b. Does the chosen model appear to specify a useful relationship between the dependent variable and the predictors?

c. Provided that % NaOH remains in the model, would you suggest that the predictor *treatment time* be eliminated?

d. Calculate a 95% CI for the expected change in specific surface area associated with an increase of 1% in NaOH when treatment time is held fixed.

e. MINITAB reported that the estimated standard deviation of $\hat{\beta}_0 + \hat{\beta}_1(9) + \hat{\beta}_2(60)$ is .162. Calculate a prediction interval for the value of specific surface area to be observed when % NaOH = 9 and treatment time = 60.

57. A multiple regression analysis was carried out to relate y = tensile strength of a synthetic-fiber specimen to the variables x_1 = percent cotton and x_2 = drying time. The data set consisted of $n = 12$ observations.

a. The estimated coefficients were $\hat{\beta}_0 = 180.00$, $\hat{\beta}_1 = 1.000$, and $\hat{\beta}_2 = 10.500$. Calculate a point estimate of the expected tensile strength when percent cotton = 15 and drying time = 3.5.

b. Sums of squares were $SST = 1210.30$, and $SSE = 117.40$. What proportion of observed variation in tensile strength can be attributed to the model relationship?

c. Use the information in part (b) to decide whether the model with variables x_1 and x_2 specifies a useful relationship.

d. The estimated standard deviation of $\hat{\beta}_0 + \hat{\beta}_1 x_1 + \hat{\beta}_2 x_2$ when $x_1 = 18$ and $x_2 = 3.0$ was 1.20. Calculate a 95% PI for tensile strength of a fabric specimen for which $x_1 = 18$ and $x_2 = 3.0$.

58. A study was carried out to relate time to failure (y) of a certain machine component to the variables operating

voltage (x_1), motor speed (x_2), and operating temperature (x_3). The resulting data set consisted of $n = 20$ observations. When the model with the three variables x_1, x_2, and x_3 was fit to the data, the value of error sum of squares was 8212.5. Fitting the second-order interaction model (with quadratic predictors and all products of pairs of variables) gave an error sum of squares of 5027.1. Should at least one of the quadratic or interaction predictors be retained in the model? State and test the relevant hypotheses.

59. The article "A Statistical Analysis of the Notch Toughness of 9% Nickel Steels Obtained from Production Heats" (*J. of Testing and Eval.*, 1987, pp. 355–363) reports on the results of a multiple regression analysis relating Charpy v-notch toughness y (Joules) to the following variables: x_1 = plate thickness (mm), x_2 = carbon content (%), x_3 = manganese content (%), x_4 = phosphorus content (%), x_5 = sulphur content (%), x_6 = silicon content (%), x_7 = nickel content (%), x_8 = yield strength (Pa), and x_9 = tensile strength (Pa).

a. The best possible subsets involved adding variables in the order x_5, x_8, x_6, x_3, x_2, x_7, x_9, x_1, and x_4. The values of R_k^2, MSE_k, and C_k are as follows:

No. of Predictors	1	2	3	4
R_k^2	.354	.453	.511	.550
MSE_k	2295	1948	1742	1607
C_k	314	173	89.6	35.7

No. of Predictors	5	6	7	8	9
R_k^2	.562	.570	.572	.575	.575
MSE_k	1566	1541	1535	1530	1532
C_k	19.9	11.0	9.4	8.2	10.0

Which model would you recommend? Explain the rationale for your choice.

b. The authors also considered second-order models involving predictors x_i^2 and $x_i x_j$. Information on the best such models starting with the variables x_2, x_3, x_5, x_6, x_7, and x_8 is as follows (in going from the best four-predictor model to the best five-predictor model, x_8 was deleted and both $x_2 x_6$ and $x_7 x_8$ were entered; x_8 reentered at a later stage):

No. of Predictors	1	2	3	4	5
R_k^2	.415	.541	.600	.629	.650
MSE_k	2079	1636	1427	1324	1251
C_k	433	109	104	52.4	16.5

No. of Predictors	6	7	8	9	10
R_k^2	.652	.655	.658	.659	.659
MSE_k	1246	1237	1229	1229	1230
C_k	14.9	11.2	8.5	9.2	11.0

Which of these models would you recommend, and why? (NOTE: Models based on eight of the original variables did not yield marked improvement on those under consideration here.)

60. Reconsider the situation of Exercise 50 and test the utility of the model identified by the BE procedure.

61. A sample of $n = 20$ companies was selected, and the values of $y = $ stock price and $k = 15$ predictor variables (such as quarterly dividend, previous year's earnings, and debt ratio) were determined. When the multiple regression model using these 15 predictors was fit to the data, $R^2 = .90$ resulted.

a. Does the model appear to specify a useful relationship between y and the predictor variables? Carry out a test using significance level .05. [HINT: The F critical value for 15 numerator and 4 denominator df is 5.86.]

b. Based on the result of part (a), does a high R^2 value by itself imply that a model is useful? Under what circumstances might you be suspicious of a model with a high R^2 value?

c. With n and k as given above, how large would R^2 have to be for the model to be judged useful at the .05 level of significance?

62. Does exposure to air pollution result in decreased life expectancy? This question was examined in the paper "Does Air Pollution Shorten Lives?" (*Statistics and Public Policy*, Reading, Mass., Addison-Wesley, 1977). Data on

$y = $ total mortality rate (deaths per 10,000)
$x_1 = $ mean suspended particle reading ($\mu g/m^3$)
$x_2 = $ smallest sulfate reading ($[\mu g/m^3] \times 10$)
$x_3 = $ population density (people/mi^2)
$x_4 = $ (percent nonwhite) $\times$ 10
$x_5 = $ (percent over 65) $\times$ 10

for the year 1960 was recorded for $n = 117$ randomly selected standard metropolitan statistical areas. The estimated regression equation was

$$y = 19.607 + .041x_1 + .071x_2 + .001x_3 + .041x_4 + .687x_5$$

a. For this model, $R^2 = .827$. Using a .05 significance level, perform a model utility test.

b. The estimated standard deviation of $\hat{\beta}_1$ was .016. Calculate and interpret a 90% CI for β_1.

c. Given that the estimated standard deviation of $\hat{\beta}_4$ is .007, determine if percent nonwhite is an important variable in the model. Use a .01 significance level.

d. In 1960 the values of x_1, x_2, x_3, x_4, and x_5 for Pittsburgh were 166, 60, 788, 68, and 95, respectively. Use the given regression equation to predict Pittsburgh's mortality rate. How does your prediction compare with the actual 1960 value of 103 deaths per 10,000?

63. Given that $R^2 = .723$ for the model containing predictors x_1, x_4, x_5, and x_8 and $R^2 = .689$ for the model with predictors x_1, x_3, x_5, and x_6, what can you say about R^2 for the model containing predictors

a. x_1, x_3, x_4, x_5, x_6, and x_8? Explain.

b. x_1 and x_4? Explain.

Bibliography

Chatterjee, S., and Bertram Price, *Regression Analysis by Example* (2nd ed.), Wiley, New York, 1990. A brief but informative discussion of selected topics, especially multicollinearity and the use of biased estimation methods.

Daniel, Cuthbert, and Fred Wood, *Fitting Equations to Data*

(2nd ed.), Wiley, New York, 1980. Contains many insights and methods that evolved from the authors' extensive consulting experience.

Draper, Norman, and Harry Smith, *Applied Regression Analysis* (2nd ed.), Wiley, New York, 1981. See Chapter 12 bibliography.

Hoaglin, David, and Roy Welsch, "The Hat Matrix in Regression and ANOVA," *American Statistician*, 1978, pp. 17–23. Describes methods for detecting influential observations in a regression data set.

Hocking, Ron, "The Analysis and Selection of Variables in Linear Regression," *Biometrics*, 1976, pp. 1–49. An excellent survey of this topic.

Neter, John, William Wasserman, and Michael Kutner, *Applied Linear Statistical Models* (3rd ed.), Irwin, Homewood, IL, 1990. See Chapter 12 bibliography.

14 The Analysis of Categorical Data

$$\sum\sum\sum\sum\sum\sum\sum\sum\sum\sum\sum\sum\sum\sum\sum$$

Introduction

In the simplest type of situation considered in this chapter, each observation in a sample is classified as belonging to one of a finite number of categories (e.g., blood type could be one of the four categories O, A, B, or AB). With p_i denoting the probability that any particular observation belongs in category i (or the proportion of the population belonging to category i), we wish to test a null hypothesis that completely specifies the values of all the p_i's (such as H_0: $p_1 = .45$, $p_2 = .35$, $p_3 = .15$, $p_4 = .05$, when there are four categories). The test statistic will be a measure of the discrepancy between the observed numbers in the categories and the expected numbers when H_0 is true. Because a decision will be reached by comparing the computed value of the test statistic to a critical value of the chi-squared distribution, the procedure is called a chi-squared goodness-of-fit test.

Sometimes the null hypothesis specifies that the p_i depend on some smaller number of parameters without specifying the values of these parameters. For example, with three categories the null hypothesis might state that $p_1 = \theta^2$, $p_2 = 2\theta(1 - \theta)$, and $p_3 = (1 - \theta)^2$. For a chi-squared test to be performed, the values of any unspecified parameters must be estimated from the sample data. These problems are discussed in Section 14.2. The methods are then applied to test a null hypothesis that states that the sample comes from a particular family of distributions, such as the Poisson family (with λ estimated from the sample) or the normal family (with μ and σ estimated).

Chi-squared tests for two different situations are presented in Section 14.3. In the first, the null hypothesis states that the p_i's are the same for several different populations. The second type of situation involves taking a sample from a single population and classifying each individual with respect to two different categorical factors (such as religious preference and political party registration). The null hypothesis in this situation is that the two factors are independent within the population.

Goodness-of-Fit Tests When Category Probabilities Are Completely Specified

A binomial experiment consists of a sequence of independent trials in which each trial can result in one of two possible outcomes (the same two possibilities for each trial). The two possibilities are labeled S (for success) and F (for failure). The probability of success, denoted by p, is assumed to be constant from trial to trial, and the number n of trials is fixed at the outset of the experiment. In Chapter 8, we presented a large-sample z test for testing H_0: $p = p_0$. Notice that this null hypothesis specifies both $P(S)$ and $P(F)$, since if $P(S) = p_0$, then $P(F) = 1 - p_0$. Denoting $P(F)$ by q and $1 - p_0$ by q_0, the null hypothesis can alternatively be written as H_0: $p = p_0, q = q_0$. The z test is two-tailed when the alternative of interest is $p \neq p_0$.

A **multinomial experiment** generalizes a binomial experiment by allowing each trial to result in one of k possible outcomes, where k is an integer greater than 2. As an example, suppose a store accepts three different types of credit cards. A multinomial experiment would result from observing the type of credit card used—type 1, type 2, or type 3—by each of the next n customers who pay with a credit card. In general, we will refer to the k possible outcomes on any given trial as categories, and p_i will denote the probability that a trial results in category i. If the experiment consists of selecting n individuals or objects from a population and categorizing each one, then p_i is the proportion of the population falling in the ith category (such an experiment will be approximately multinomial provided that n is much smaller than the population size).

The null hypothesis of interest will specify the value of each p_i. For example, in the case $k = 3$, we might have H_0: $p_1 = .5$, $p_2 = .3$, $p_3 = .2$. The alternative hypothesis will state that H_0 is not true—that is, that at least one of the p_i's has a value different from that asserted by H_0 (in which case at least two must be different, since they sum to 1). The symbol p_{i0} will represent the value of p_i claimed by the null hypothesis. In the example just given, $p_{10} = .5$, $p_{20} = .3$, and $p_{30} = .2$.

Before the multinomial experiment is performed, the number of trials that will result in category i ($i = 1, 2, \ldots$, or k) is a random variable—just as the number of successes and the number of failures in a binomial experiment are random variables. This random variable will be denoted by N_i and its observed value by n_i. Since each trial results in exactly one of the k categories, $\Sigma N_i = n$, and the same is true of the n_i's. As an example, an experiment with $n = 100$ and $k = 3$ might yield $N_1 = 46$, $N_2 = 35$, and $N_3 = 19$.

The expected number of successes and expected number of failures in a binomial experiment are np and nq, respectively. When H_0: $p = p_0, q = q_0$ is true, the expected numbers of successes and failures are np_0 and nq_0, respectively. Similarly, in a multinomial experiment the expected number of trials resulting in category i is $E(N_i) = np_i$ ($i = 1, \ldots, k$). When H_0: $p_1 = p_{10}, \ldots, p_k = p_{k0}$ is true, these expected values become $E(N_1) = np_{10}$, $E(N_2) = np_{20}$, $\ldots$, $E(N_k) = np_{k0}$. For the case $k = 3$, H_0: $p_1 = .5$, $p_2 = .3$, $p_3 = .2$, and $n = 100$, $E(N_1) = 100(.5) = 50$, $E(N_2) = 30$, and $E(N_3) = 20$ when H_0 is true. The n_i's are often displayed in a tabular format consisting of a row of k cells, one for each category, as illustrated in Figure

14.1. The expected values when H_0 is true are displayed just below the observed values. The N_i's and n_i's are usually referred to as *observed cell counts* (or *observed cell frequencies*), and np_{10}, np_{20}, . . . , np_{k0} are the corresponding *expected cell counts* under H_0.

	$i = 1$	$i = 2$	. . .	$i = k$	Row total
Observed	n_1	n_2	. . .	n_k	n
Expected	np_{10}	np_{20}	. . .	np_{k0}	n

Figure 14.1 Observed and expected cell counts

The n_i's should all be reasonably close to the corresponding np_{i0}'s when H_0 is true. On the other hand, several of the observed counts should differ substantially from these expected counts when the actual values of the p_i's differ markedly from what the null hypothesis asserts. The test procedure involves measuring the discrepancy between the n_i's and the np_{i0}'s, with H_0 being rejected when the measured discrepancy is sufficiently large. It is natural to base a measure of discrepancy on the squared deviations $(n_1 - np_{10})^2$, $(n_2 - np_{20})^2$, . . . , $(n_k - np_{k0})^2$. An obvious way to combine these into an overall measure is to add them together to obtain $\Sigma(n_i - np_{i0})^2$. However, suppose $np_{10} = 100$ and $np_{20} = 10$. Then if $n_1 = 95$ and $n_2 = 5$, the two categories contribute the same squared deviations to the proposed measure. Yet n_1 is only 5% less than what would be expected when H_0 is true, whereas n_2 is 50% less.

To take relative magnitudes of the deviations into account, we will divide each squared deviation by the corresponding expected count and then combine. Before giving a more detailed description, we must discuss a type of probability distribution called the *chi-squared distribution*. This distribution was first introduced in Section 4.4 and was used in Chapter 7 to obtain a confidence interval for the variance σ^2 of a normal population. The chi-squared distribution has a single parameter ν, called the number of degrees of freedom (df) of the distribution, with possible values 1, 2, 3, ... Analogous to the critical value $t_{\alpha,\nu}$ for the t distribution, $\chi^2_{\alpha,\nu}$ is the value such that α of the area under the χ^2 curve in Figure 14.2 with ν df lies to the right of $\chi^2_{\alpha,\nu}$. Selected values of $\chi^2_{\alpha,\nu}$ are given in Appendix Table A.6.

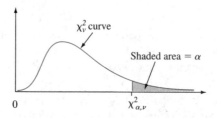

Figure 14.2 A critical value for a chi-squared distribution

Theorem

4/9/97

Provided that $np_i \geq 5$ for every i ($i = 1, 2, \ldots, k$), the random variable

$$\chi^2 = \sum_{i=1}^{k} \frac{(N_i - np_i)^2}{np_i} = \sum_{\text{all cells}} \frac{(\text{observed} - \text{expected})^2}{\text{expected}}$$

has approximately a chi-squared distribution with $k - 1$ df.

The fact that df $= k - 1$ is a consequence of the restriction $\Sigma N_i = n$. Although there are k observed cell counts, once any $k - 1$ are known, the remaining one is uniquely determined. That is, there are only $k - 1$ "freely determined" cell counts, and thus $k - 1$ df.

If np_{i0} is substituted for np_i in χ^2, the resulting test statistic has a chi-squared distribution when H_0 is true. Rejection of H_0 is appropriate when $\chi^2 \geq c$ (because large discrepancies between observed and expected counts lead to a large value of χ^2), and the choice $c = \chi^2_{\alpha,k-1}$ yields a test with significance level α.

Null hypothesis: $H_0: p_1 = p_{10}, p_2 = p_{20}, \ldots, p_k = p_{k0}$

Alternative hypothesis: H_a: at least one p_i does not equal p_{i0}

Test statistic value:

$$\chi^2 = \sum_{\text{all cells}} \frac{(\text{observed} - \text{expected})^2}{\text{expected}} = \sum_{i=1}^{k} \frac{(n_i - np_{i0})^2}{np_{i0}}$$

Rejection region: $\chi^2 \geq \chi^2_{\alpha,k-1}$

Example 14.1

4/9/97

If we focus on two different characteristics of an organism, each controlled by a single gene, and cross a pure strain having genotype AABB with a pure strain having genotype aabb (capital letters denoting dominant alleles and small letters recessive alleles), the resulting genotype will be AaBb. If these first-generation organisms are then crossed among themselves (a dihybrid cross), there will be four phenotypes depending on whether a dominant allele of either type is present. Mendel's laws of inheritance imply that these four phenotypes should have probabilities $\frac{9}{16}, \frac{3}{16}, \frac{3}{16}$, and $\frac{1}{16}$ of arising in any given dihybrid cross.

The paper "Linkage Studies of the Tomato" (*Trans. Royal Canadian Institute*, 1931, pp. 1–19) reports the following data on phenotypes from a dihybrid cross of tall cut-leaf tomatoes with dwarf potato-leaf tomatoes. There are $k = 4$ categories corresponding to the four possible phenotypes, with the null hypothesis being

$$H_0: p_1 = \frac{9}{16}, p_2 = \frac{3}{16}, p_3 = \frac{3}{16}, p_4 = \frac{1}{16}$$

The expected cell counts are $9n/16$, $3n/16$, $3n/16$, and $n/16$, and the test is based on $k - 1 = 3$ df. The total sample size was $n = 1611$. Observed and expected counts appear in Figure 14.3.

	$i = 1$ Tall, cut-leaf	$i = 2$ Tall, potato-leaf	$i = 3$ Dwarf, cut-leaf	$i = 4$ Dwarf, potato-leaf
n_i	926	288	293	104
np_{i0}	906.2	302.1	302.1	100.7

Figure 14.3 Observed and expected cell counts for Example 14.1

The contribution to χ^2 from the first cell is

$$\frac{(n_1 - np_{10})^2}{np_{10}} = \frac{(926 - 906.2)^2}{906.2} = .433$$

Cells 2, 3, and 4 contribute .658, .274, and .108, respectively, so $\chi^2 = .433 + .658 + .274 + .108 = 1.473$. A test with significance level .10 requires $\chi^2_{.10,3}$, the number in the 3 df row and .10 column of Appendix Table A.6. This critical value is 6.251. Since 1.473 is not at least 6.251, H_0 cannot be rejected even at this rather large level of significance. The data is quite consistent with Mendel's laws. ∎

Although we have developed the chi-squared test for situations in which $k > 2$, it can also be used when $k = 2$. The null hypothesis in this case can be stated as H_0: $p_1 = p_{10}$, since the relations $p_2 = 1 - p_1$ and $p_{20} = 1 - p_{10}$ make the inclusion of $p_2 = p_{20}$ in H_0 redundant. The alternative hypothesis is H_a: $p_1 \neq p_{10}$. These hypotheses can also be tested using a two-tailed z test with test statistic

$$Z = \frac{(N_1/n) - p_{10}}{\sqrt{\dfrac{p_{10}(1 - p_{10})}{n}}} = \frac{\hat{p}_1 - p_{10}}{\sqrt{\dfrac{p_{10}p_{20}}{n}}}$$

Surprisingly, the two test procedures are completely equivalent. This is because it can be shown that $Z^2 = \chi^2$ and $(z_{\alpha/2})^2 = \chi^2_{1,\alpha}$, so that $\chi^2 \geq \chi^2_{1,\alpha}$ if and only if (iff) $|Z| \geq z_{\alpha/2}$.*

Example 14.2

The developers of an English proficiency exam to be used in a large university system believe that 60% of all incoming freshmen will be able to pass the exam. In a random sample of 200 incoming freshmen, 105 pass the exam. Does this contradict the claim of the developers?

We wish to test H_0: $p_1 = .6$, $p_2 = .4$ versus H_a: H_0 is not true. The expected cell counts are $np_{10} = 200(.6) = 120$ and $np_{20} = 80$, so the computed χ^2 is

$$\chi^2 = \frac{(105 - 120)^2}{120} + \frac{(95 - 80)^2}{80} = 1.88 + 2.81 = 4.69$$

---

*The fact is that $(z_{\alpha/2})^2 = \chi^2_{1,\alpha}$ is a consequence of the relationship between the standard normal distribution and the chi-squared distribution with 1 df; if $Z \sim N(0, 1)$, then Z^2 has a chi-squared distribution with $\nu = 1$.

Using $\alpha = .05$, $\chi^2_{.05,1} = 3.843$. Since $4.69 \geq 3.843$, H_0 is rejected at level .05. Notice that $\chi^2_{.025,1} = 5.025$, so the P-value satisfies $.025 < P < .05$. With $\hat{p}_1 = 105/200 = .525$, the value of the z statistic is

$$z = \frac{.525 - .6}{\sqrt{\dfrac{(.6)(.4)}{200}}} = -2.165$$

and the z critical value for a level .05 test is $z_{.025} = 1.96$. Thus, $z^2 = (-2.165)^2 = 4.69 = \chi^2$ and $(z_{\alpha/2})^2 = (1.96)^2 = 3.84 = \chi^2_{1,.05}$, as must be the case. ∎

If the alternative hypothesis is either $H_a: p_1 > p_{10}$ or $H_a: p_1 < p_{10}$, the chi-squared test cannot be used. One must then revert to an upper- or lower-tailed z test.

As is the case with all test procedures, one must be careful not to confuse statistical significance with practical significance. A computed χ^2 that exceeds $\chi^2_{\alpha,k-1}$ may be a result of a very large sample size rather than any practical differences between the hypothesized p_{i0}'s and true p_i's. Thus, if $p_{10} = p_{20} = p_{30} = \frac{1}{3}$, but the true p_i's have values .330, .340, and .330, a large value of χ^2 is sure to arise with a sufficiently large n. Before rejecting H_0, the $\hat{p}_i$'s should be examined to see if they suggest a model different from that of H_0 from a practical point of view.

χ^2 When the p_i's Are Functions of Other Parameters

Frequently the p_i's are hypothesized to depend on a smaller number of parameters $\theta_1, \ldots, \theta_m$ ($m < k$). Then a specific hypothesis involving the θ_i's yields specific p_{i0}'s, which are then used in the χ^2 test.

Example 14.3

4/11/97

In a well-known genetics paper ("The Progeny in Generations F_{12} to F_{17} of a Cross Between a Yellow-Wrinkled and a Green-Round Seeded Pea," *J. Genetics*, 1923, pp. 255–331), the early statistician G. U. Yule analyzed data resulting from crossing garden peas. The dominant alleles in the experiment were Y = yellow color and R = round shape, resulting in the double dominant YR. Yule examined 269 four-seed pods resulting from a dihybrid cross and counted the number of YR seeds in each pod. Letting X denote the number of YR's in a randomly selected pod, possible X values are 0, 1, 2, 3, 4, which we identify with cells 1, 2, 3, 4, and 5 of a rectangular table (so, e.g., a pod with $X = 4$ yields an observed count in cell 5).

The hypothesis that the Mendelian laws are operative and that genotypes of individual seeds within a pod are independent of one another implies that X has a binomial distribution with $n = 4$ and $\theta = \frac{9}{16}$. We thus wish to test $H_0: p_1 = p_{10}, \ldots, p_5 = p_{50}$, where

$$p_{i0} = P(i - 1 \text{ YR's among 4 seeds when } H_0 \text{ is true})$$

$$= \binom{4}{i-1} \theta^{i-1}(1-\theta)^{4-(i-1)} \qquad i = 1, 2, 3, 4, 5; \ \theta = \frac{9}{16}$$

Yule's data and the computations appear in Figure 14.4 with expected cell counts $np_{i0} = 269p_{i0}$.

Cell i	1	2	3	4	5
YR peas/pods	0	1	2	3	4
Observed	16	45	100	82	26
Expected	9.86	50.68	97.75	83.78	26.93
$\dfrac{(observed - expected)^2}{expected}$	3.823	.637	.052	.038	.032

Figure 14.4 Observed and expected cell counts for Example 14.3

Thus, $\chi^2 = 3.823 + \cdots + .032 = 4.582$. Since $\chi^2_{.01,k-1} = \chi^2_{.01,4} = 13.277$, H_0 is not rejected at level .01.   ■

In Section 14.2, we will discuss goodness-of-fit tests when the values of parameters $\theta_1, \ldots, \theta_m$ are unspecified by H_0, so that we must estimate expected cell counts by first estimating the θ_i's.

χ^2 When the Underlying Distribution Is Continuous

We have so far assumed that the k categories are naturally defined in the context of the experiment under consideration. The χ^2 test can also be used to test whether a sample comes from a specific underlying continuous distribution. Let X denote the variable being sampled and suppose the hypothesized pdf of X is $f_0(x)$. As in the construction of a frequency distribution in Chapter 1, subdivide the measurement scale of X into k intervals $[a_0, a_1), [a_1, a_2), \ldots, [a_{k-1}, a_k)$, where the interval $[a_{i-1}, a_i)$ includes the value a_{i-1} but not a_i. The cell probabilities specified by H_0 are then

$$p_{i0} = P(a_{i-1} \le X < a_i) = \int_{a_{i-1}}^{a_i} f_0(x) \, dx$$

The cells should be chosen so that $np_{i0} \ge 5$ for $i = 1, \ldots, k$. Often they are selected so that the np_{i0}'s are equal.

Example 14.4   To see whether the time of onset of labor among expectant mothers is uniformly distributed throughout a 24-hour day, we can divide a day into k periods, each of length $24/k$. The null hypothesis states that $f(x)$ is the uniform pdf on the interval $[0, 24]$, so that $p_{i0} = 1/k$. The article "The Hour of Birth" (*British J. Preventive and Social Medicine*, 1953, pp. 43–59) reports on 1186 onset times, which were categorized into $k = 24$ 1-hour intervals beginning at midnight, resulting in cell counts of 52, 73, 89, 88, 68, 47, 58, 47, 48, 53, 47, 34, 21, 31, 40, 24, 37, 31, 47, 34, 36, 44, 78, and 59. Each expected cell count is $1186 \cdot \frac{1}{24} = 49.42$, and the resulting value of χ^2 is 162.77. Since $\chi^2_{.01,23} = 41.637$, the computed value is highly significant, and the null hypothesis is resoundingly rejected. Generally speaking, it appears that labor is much more likely to commence very late at night than during normal waking hours.   ■

For testing whether a sample comes from a specific normal distribution, the fundamental parameters are $\theta_1 = \mu$ and $\theta_2 = \sigma$, and each p_{i0} will be a function of these parameters.

Example 14.5

4|9|97

At a certain university, final exams are supposed to last 2 hours. The psychology department constructed a departmental final for an elementary course that was believed to satisfy the following criteria: (1) actual time taken to complete the exam is normally distributed, (2) $\mu = 100$ min, and (3) exactly 90% of all students will finish within the 2-hour period. To see whether this is actually the case, 120 students were randomly selected, and their completion times recorded. It was decided that $k = 8$ intervals should be used. The criteria imply that the 90th percentile of the completion time distribution is $\mu + 1.28\sigma = 120$. Since $\mu = 100$, this implies that $\sigma = 15.63$.

The eight intervals that divide the standard normal scale into eight equally likely segments are $[0, .32)$, $[.32, .675)$, $[.675, 1.15)$, $[1.15, \infty)$, and their four counterparts on the other side of 0. For $\mu = 100$ and $\sigma = 15.63$, these intervals become $[100, 105)$, $[105, 110.55)$, $[110.55, 117.97)$, and $[117.97, \infty)$. Thus, $p_{i0} = \frac{1}{8} = .125$ ($i = 1, \ldots, 8$), so each expected cell count is $np_{i0} = 120(.125) = 15$. The observed cell counts were 21, 17, 12, 16, 10, 15, 19, and 10, resulting in a χ^2 of 7.73. Since $\chi^2_{.10,7} = 12.017$ and 7.73 is not ≥ 12.017, there is no evidence for concluding that the criteria have not been met. ∎

Exercises   Section 14.1 (1–9)

1. Student consultants at a particular computer center face questions about programs written in FORTRAN (1), BASIC (2), Pascal (3), and Ada (4). The consultants have been hired based on the assumption that 40% of all questions concern FORTRAN programs, 25% concern BASIC programs, 25% concern Pascal programs, and 10% concern Ada programs. Let p_i denote the probability that a randomly selected question concerns a program written in language i ($i = 1, 2, 3, 4$). Use the accompanying data to test

H_0: $p_1 = .40$, $p_2 = .25$, $p_3 = .25$, $p_4 = .10$

versus

H_a: H_0 is not true

using a level .05 chi-squared test.

Cell	1	2	3	4
Frequency	52	38	21	9

2. It is hypothesized that when homing pigeons are disoriented in a certain manner, they will exhibit no preference for any direction of flight after takeoff (so

that the direction X should be uniformly distributed on the interval from 0° to 360°). To test this, 120 pigeons are disoriented, let loose, and the direction of flight of each is recorded; the resulting data appears below. Use the chi-squared test at level .10 to see if the data supports the hypothesis.

Direction	0–<45°	45–<90°	90–<135°
Frequency	12	16	17

Direction	135–<180°	180–<225°	225–<270°
Frequency	15	13	20

Direction	270–<315°	315–<360°
Frequency	17	10

3. Information has been loaded onto a single-disk storage device with ten concentric tracks in such a way that the probability of the access arm next being required to access track i is given by $p_i = (5.5 - |i - 5.5|)/30$ for $i = 1, \ldots, 10$ (i.e., the p_i's are $\frac{1}{30}, \frac{2}{30}, \frac{3}{30}, \frac{4}{30}, \frac{5}{30}, \frac{5}{30}, \frac{4}{30}, \frac{3}{30}, \frac{2}{30}, \frac{1}{30}$). A sample of 200 track numbers accessed resulted in the following data. Use the chi-squared test at level .10 to decide whether the data is consistent with the above access probabilities.

Track Number	1	2	3	4	5	6	7	8	9	10
Frequency	4	15	23	25	38	31	32	14	10	8

4. Sorghum is an important cereal crop whose quality and appearance could be affected by the presence of pigments in the pericarp (the walls of the plant ovary). The paper "A Genetic and Biochemical Study on Pericarp Pigments in a Cross Between Two Cultivars of Grain Sorghum, Sorghum Bicolor" (*Heredity*, 1976, pp. 413–416) reports on an experiment that involved an initial cross between CK60 sorghum (an American variety with white seeds) and Abu Taima (an Ethiopian variety with yellow seeds) to produce plants with red seeds and then a self-cross of the red-seeded plants. According to genetic theory, this F_2 cross should produce plants with red, yellow, or white seeds in the ratio $9:3:4$. The data from the experiment appears below; does the data confirm or contradict the genetic theory? Test at level .05.

Seed Color	Red	Yellow	White
Observed Frequency	195	73	100

5. Criminologists have long debated whether there is a relationship between weather conditions and the incidence of violent crime. The author of the paper "Is There a Season for Homicide?" (*Criminology*, 1988, pp. 287–296) classified 1361 homicides according to season, resulting in the accompanying data. Test the null hypothesis of equal proportions using $\alpha = .01$ by using the chi-squared table to say as much as possible about the *P*-value.

Winter	Spring	Summer	Fall
328	334	372	327

6. The paper "Psychiatric and Alcoholic Admissions Do Not Occur Disproportionately Close to Patients' Birthdays" (*Psychological Reports*, 1992, pp. 944–946) focuses on the existence of any relationship between date of patient admission for treatment of alcoholism and patient's birthday. Assuming a 365-day year (i.e., excluding leap year), in the absence of any relation, a patient's admission date is equally likely to be any one of the 365 possible days. The investigators established four different admission categories: (1) within 7 days of birthday, (2) between 8 and 30 days, inclusive, from

the birthday, (3) between 31 and 90 days, inclusive, from the birthday, and (4) more than 90 days from the birthday. A sample of 200 patients gave observed frequencies of 11, 24, 69, and 96 for categories 1, 2, 3, and 4, respectively. State and test the relevant hypotheses using a significance level of .01.

7. The response time of a computer system to a request for a certain type of information is hypothesized to have an exponential distribution with parameter $\lambda = 1$ sec (so if X = response time, the pdf of X under H_0 is $f_0(x) = e^{-x}$ for $x \geq 0$).

   a. If you had observed $X_1, X_2, \ldots, X_n$ and wanted to use the chi-squared test with five class intervals having equal probability under H_0, what would be the resulting class intervals?

   b. Carry out the chi-squared test using the following data resulting from a random sample of 40 response times:

   .10, .99, 1.14, 1.26, 3.24, .12, .26, .80, .79, 1.16, 1.76, .41, .59, .27, 2.22, .66, .71, 2.21, .68, .43, .11, .46, .69, .38, .91, .55, .81, 2.51, 2.77, .16, 1.11, .02, 2.13, .19, 1.21, 1.13, 2.93, 2.14, .34, .44

8. **a.** Show that another expression for the chi-squared statistic is

$$\chi^2 = \sum_{i=1}^{k} \frac{N_i^2}{np_{i0}} - n$$

   Why is it more efficient to compute χ^2 using this formula?

   b. When the null hypothesis is $H_0: p_1 = p_2 = \cdots = p_k = 1/k$ (i.e., $p_{i0} = 1/k$ for all i), how does the formula of part (a) simplify? Use the simplified expression to calculate χ^2 for the pigeon/direction data in Exercise 2.

9. **a.** Having obtained a random sample from a population, you wish to use a chi-squared test to decide whether the population distribution is standard normal. If you base the test on six class intervals having equal probability under H_0, what should be the class intervals?

   b. If you wish to use a chi-squared test to test H_0: the population distribution is normal with $\mu = .5$, $\sigma = .002$ and the test is to be based on six equiprobable (under H_0) class intervals, what should be these intervals?

   c. Use the chi-squared test with the intervals of part (b) to decide, based on the following 45 bolt diameters, whether bolt diameter is a normally distributed variable with $\mu = .5$ in., $\sigma = .002$ in.

.4974 .4976 .4991 .5014 .5008 .4993
.4994 .5010 .4997 .4993 .5013 .5000
.5017 .4984 .4967 .5028 .4975 .5013

.4972 .5047 .5069 .4977 .4961 .4987
.4990 .4974 .5008 .5000 .4967 .4977
.4992 .5007 .4975 .4998 .5000 .5008
.5021 .4959 .5015 .5012 .5056 .4991
.5006 .4987 .4968

14.2   Goodness of Fit for Composite Hypotheses

In the previous section, we presented a goodness-of-fit test based on a χ^2 statistic for deciding between H_0: $p_1 = p_{10}, \ldots, p_k = p_{k0}$ and the alternative H_a stating that H_0 is not true. The null hypothesis was a **simple hypothesis** in the sense that each p_{i0} was a specified number, so that the expected cell counts when H_0 was true were uniquely determined numbers.

In many situations, there are k naturally occurring categories, but H_0 states only that the p_i's are functions of other parameters $\theta_1, \ldots, \theta_m$ without specifying the values of these θ's. For example, a population may be in equilibrium with respect to proportions of the three genotypes AA, Aa, and aa. With p_1, p_2, and p_3 denoting these proportions (probabilities), one may wish to test

$$H_0: p_1 = \theta^2, p_2 = 2\theta(1 - \theta), p_3 = (1 - \theta)^2 \tag{14.1}$$

where θ represents the proportion of gene A in the population. This hypothesis is **composite** because knowing that H_0 is true does not uniquely determine the cell probabilities and expected cell counts but only their general form. To carry out a χ^2 test, the unknown θ_i's must first be estimated.

Similarly, we may be interested in testing to see whether a sample came from a particular family of distributions without specifying any particular member of the family. To use the χ^2 test to see whether the distribution is Poisson, for example, the parameter λ must be estimated. In addition, because there are actually an infinite number of possible values of a Poisson variable, these values must be grouped so that there are a finite number of cells. If H_0 states that the underlying distribution is normal, use of a χ^2 test must be preceded by a choice of cells and estimation of μ and σ.

χ^2 When Parameters Are Estimated

As before, k will denote the number of categories or cells and p_i will denote the probability of an observation falling in the ith cell. The null hypothesis now states that each p_i is a function of a small number of parameters $\theta_1, \ldots, \theta_m$ with the θ_i's otherwise unspecified:

$$H_0: p_1 = \pi_1(\boldsymbol{\theta}), \ldots, p_k = \pi_k(\boldsymbol{\theta}), \qquad \text{where } \boldsymbol{\theta} = (\theta_1, \ldots, \theta_m)$$
$$H_a: \text{the hypothesis } H_0 \text{ is not true} \tag{14.2}$$

For example, for H_0 of (14.1), $m = 1$ (there is only one θ), $\pi_1(\theta) = \theta^2$, $\pi_2(\theta) = 2\theta(1 - \theta)$, and $\pi_3(\theta) = (1 - \theta)^2$.

In the case $k = 2$, there is really only a single rv, N_1 (since $N_1 + N_2 = n$), which has a binomial distribution. The joint probability that $N_1 = n_1$ and $N_2 = n_2$ is then

$$P(N_1 = n_1, N_2 = n_2) = \binom{n}{n_1}p_1^{n_1} \cdot p_2^{n_2} \propto p_1^{n_1} \cdot p_2^{n_2}$$

where $p_1 + p_2 = 1$ and $n_1 + n_2 = n$. For general k, the joint distribution of $N_1, \ldots, N_k$ is the multinomial distribution (Section 5.1) with

$$P(N_1 = n_1, \ldots, N_k = n_k) \propto p_1^{n_1} \cdot p_2^{n_2} \cdots p_k^{n_k} \tag{14.3}$$

When H_0 is true, (14.3) becomes

$$P(N_1 = n_1, \ldots, N_k = n_k) \propto [\pi_1(\boldsymbol{\theta})]^{n_1} \cdots [\pi_k(\boldsymbol{\theta})]^{n_k} \tag{14.4}$$

To apply a chi-squared test, $\boldsymbol{\theta} = (\theta_1, \ldots, \theta_m)$ must be estimated.

Method of Estimation

Let $n_1, n_2, \ldots, n_k$ denote the observed values of $N_1, \ldots, N_k$. Then $\hat{\theta}_1, \ldots, \hat{\theta}_m$ are those values of the θ_i's that maximize (14.4).

The resulting estimators $\hat{\theta}_1, \ldots, \hat{\theta}_m$ are the **maximum likelihood estimators** of $\theta_1, \ldots, \theta_m$; this principle of estimation was discussed in Section 6.2.

Example 14.6

In humans there is a blood group, the MN group, that is composed of individuals having one of the three blood types M, MN, and N. Type is determined by two alleles, and there is no dominance, so the three possible genotypes give rise to three phenotypes. A population consisting of individuals in the MN group is in equilibrium if

$$P(M) = p_1 = \theta^2$$
$$P(MN) = p_2 = 2\theta(1 - \theta)$$
$$P(N) = p_3 = (1 - \theta)^2$$

for some θ. Suppose a sample from such a population yielded the results shown in Figure 14.5.

Type	M	MN	N	
Observed	125	225	150	$n = 500$

Figure 14.5 Observed counts for Example 14.6

Then

$$[\pi_1(\theta)]^{n_1}[\pi_2(\theta)]^{n_2}[\pi_3(\theta)]^{n_3} = [(\theta^2)]^{n_1}[2\theta(1 - \theta)]^{n_2}[(1 - \theta)^2]^{n_3}$$
$$= 2^{n_2} \cdot \theta^{2n_1 + n_2} \cdot (1 - \theta)^{n_2 + 2n_3}$$

Maximizing this with respect to θ (or, equivalently, maximizing the natural logarithm of this quantity, which is easier to differentiate) yields

$$\hat{\theta} = \frac{2n_1 + n_2}{[(2n_1 + n_2) + (n_2 + 2n_3)]} = \frac{2n_1 + n_2}{2n}$$

With $n_1 = 125$ and $n_2 = 225$, $\hat{\theta} = 475/1000 = .475$.   ■

Once $\boldsymbol{\theta} = (\theta_1, \ldots, \theta_m)$ has been estimated by $\hat{\boldsymbol{\theta}} = (\hat{\theta}_1, \ldots, \hat{\theta}_m)$, the estimated expected cell counts are the $n\pi_i(\hat{\boldsymbol{\theta}})$'s. These are now used in place of the np_{i0}'s of Section 14.1 to specify a χ^2 statistic.

Theorem

> Under general "regularity" conditions on $\theta_1, \ldots, \theta_m$ and the $\pi_i(\boldsymbol{\theta})$'s, if $\theta_1, \ldots, \theta_m$ are estimated by the method of maximum likelihood as described above and n is large,
>
> $$\chi^2 = \sum_{\text{all cells}} \frac{(\text{observed} - \text{estimated expected})^2}{\text{estimated expected}} = \sum_{i=i}^{k} \frac{[N_i - n\pi_i(\hat{\boldsymbol{\theta}})]^2}{n\pi_i(\hat{\boldsymbol{\theta}})}$$
>
> has approximately a chi-squared distribution with $k - 1 - m$ df when H_0 of (14.2) is true. An approximately level α test of H_0 versus H_a is then to reject H_0 if $\chi^2 \geq \chi^2_{\alpha,k-1-m}$. In practice, the test can be used if $n\pi_i(\hat{\boldsymbol{\theta}}) \geq 5$ for every i.

Notice that *the number of degrees of freedom is reduced by the number of θ_i's estimated.*

Example 14.7
(Example 14.6 continued)

With $\hat{\theta} = .475$ and $n = 500$, the estimated expected cell counts are $n\pi_1(\hat{\theta}) = 500(\hat{\theta})^2 = 112.81$, $n\pi_2(\hat{\theta}) = (500)(2)(.475)(1 - .475) = 249.38$, and $n\pi_3(\hat{\theta}) = 500 - 112.81 - 249.38 = 137.81$. Then

$$\chi^2 = \frac{(125 - 112.81)^2}{112.81} + \frac{(225 - 249.38)^2}{249.38} + \frac{(150 - 137.81)^2}{137.81} = 4.78$$

Since $\chi^2_{.05,k-1-m} = \chi^2_{.05,3-1-1} = \chi^2_{.05,1} = 3.843$ and $4.78 \geq 3.843$, H_0 is rejected.   ■

Example 14.8

Consider a series of games between two teams, I and II, that terminates as soon as one team has won four games (with no possibility of a tie). A simple probability model for such a series assumes that outcomes of successive games are independent and that the probability of team I winning any particular game is a constant θ. We arbitrarily designate I the better team, so that $\theta \geq .5$. Any particular series can then terminate after 4, 5, 6, or 7 games. Let $\pi_1(\theta)$, $\pi_2(\theta)$, $\pi_3(\theta)$, $\pi_4(\theta)$ denote the probability of termination in 4, 5, 6, and 7 games, respectively. Then

$$\pi_1(\theta) = P(\text{I wins in 4 games}) + P(\text{II wins in 4 games})$$
$$= \theta^4 + (1 - \theta)^4$$

$\pi_2(\theta) = P(\text{I wins 3 of the first 4 and the fifth})$

$\quad\quad + P(\text{I loses 3 of the first 4 and the fifth})$

$\quad\quad = \binom{4}{3} \theta^3 (1 - \theta) \cdot \theta + \binom{4}{1} \theta(1 - \theta)^3 \cdot (1 - \theta)$

$\quad\quad = 4\theta(1 - \theta)[\theta^3 + (1 - \theta)^3]$

$\pi_3(\theta) = 10\theta^2(1 - \theta)^2[\theta^2 + (1 - \theta)^2]$

$\pi_4(\theta) = 20\theta^3(1 - \theta)^3$

The paper "Seven-Game Series in Sports" by Groeneveld and Meeden (*Mathematics Magazine*, 1975, pp. 187–192) tested the fit of this model to results of National Hockey League playoffs during the period 1943–1967 (when league membership was stable). The data appears in Figure 14.6.

Cell	1	2	3	4	
Number of games played	4	5	6	7	
Observed frequency	15	26	24	18	$n = 83$
Estimated expected frequency	16.351	24.153	23.240	19.256	

Figure 14.6 Observed and expected counts for the simple model

The estimated expected cell counts are $83\pi_i(\hat{\theta})$, where $\hat{\theta}$ is the value of θ that maximizes

$$\{\theta^4 + (1 - \theta)^4\}^{15} \cdot \{4\theta(1 - \theta)[\theta^3 + (1 - \theta)^3]\}^{26}$$
$$\cdot \{10\theta^2(1 - \theta)^2[\theta^2 + (1 - \theta)^2]\}^{24} \cdot \{20\theta^3(1 - \theta)^3\}^{18} \quad (14.5)$$

Standard calculus methods fail to yield a nice formula for the maximizing value $\hat{\theta}$, so it must be computed using numerical methods. The result is $\hat{\theta} = .654$, from which $\pi_i(\hat{\theta})$ and the estimated expected cell counts are computed. The computed value of χ^2 is .360, while (since $k - 1 - m = 4 - 1 - 1 = 2$) $\chi^2_{.10,2} = 4.605$. There is thus no reason to reject the simple model as applied to NHL playoff series.

The cited paper also considered World Series data for the period 1903–1973. For the simple model, $\chi^2 = 5.97$, so the model does not seem appropriate. The suggested reason for this is that for the simple model

$$P(\text{series lasts six games} \mid \text{series lasts at least six games}) \geq .5 \quad (14.6)$$

while of the 38 series that actually lasted at least six games, only 13 lasted exactly six. The following alternative model is then introduced:

$\pi_1(\theta_1, \theta_2) = \theta_1^4 + (1 - \theta_1)^4,$

$\pi_2(\theta_1, \theta_2) = 4\theta_1(1 - \theta_1)[\theta_1^3 + (1 - \theta_1)^3],$

$\pi_3(\theta_1, \theta_2) = 10\theta_1^2(1 - \theta_1)^2\theta_2,$

$\pi_4(\theta_1, \theta_2) = 10\theta_1^2(1 - \theta_1)^2(1 - \theta_2)$

The first two π_i's are identical to the simple model, while θ_2 is the conditional probability of (14.6) (which can now be any number between 0 and 1). The values of $\hat{\theta}_1$ and $\hat{\theta}_2$ that maximize the expression analogous to (14.5) are determined numerically as $\hat{\theta}_1 = .614$, $\hat{\theta}_2 = .342$. A summary appears in Figure 14.7 with $\chi^2 = .384$. Since two parameters are estimated, df $= k - 1 - m = 1$ with $\chi^2_{.10,1} = 2.706$, indicating a good fit of the data to this new model.

Number of games played	4	5	6	7
Observed frequency	12	16	13	25
Estimated expected frequency	10.85	18.08	12.68	24.39

Figure 14.7 Observed and expected counts for the more complex model ■

One of the regularity conditions on the θ_i's in the theorem is that they be functionally independent of one another. That is, no single θ_i can be determined from the values of other θ_i's, so that m is the number of functionally independent parameters estimated. A general rule of thumb for degrees of freedom in a chi-squared test is the following.

$$\chi^2 \text{ df} = \begin{pmatrix} \text{number of freely} \\ \text{determined cell counts} \end{pmatrix} - \begin{pmatrix} \text{number of independent} \\ \text{parameters estimated} \end{pmatrix}$$

This rule will be used in connection with several different chi-squared tests in the next section.

Goodness of Fit for Discrete Distributions

Many experiments involve observing a random sample $X_1, X_2, \ldots, X_n$ from some discrete distribution. One may then wish to investigate whether the underlying distribution is a member of a particular family, such as the Poisson or negative binomial family. In the case of both a Poisson and a negative binomial distribution, the set of possible values is infinite, so the values must be grouped into k subsets before a chi-squared test can be used. The groupings should be done so that the expected frequency in each cell (group) is at least five. The last cell will then correspond to X values of $c, c + 1, c + 2, \ldots$ for some value c.

This grouping can considerably complicate the computation of the $\hat{\theta}_i$'s and estimated expected cell counts. This is because the theorem requires that the $\hat{\theta}_i$'s be obtained from the cell counts $N_1, \ldots, N_k$ rather than the sample values $X_1, \ldots, X_n$.

Example 14.9   Figure 14.8 presents count data on the number of *Larrea divaricata* plants found in each of 48 sampling quadrats, as reported in the article "Some Sampling Characteristics of Plants and Arthropods of the Arizona Desert" (*Ecology*, 1962, pp. 567–571).

Cell	1	2	3	4	5
Number of plants	0	1	2	3	≥ 4
Frequency	9	9	10	14	6

Figure 14.8 Observed counts for Example 14.9

The author fit a Poisson distribution to the data. Let λ denote the Poisson parameter and suppose for the moment that the six counts in cell 5 were actually 4, 4, 5, 5, 6, 6. Then denoting sample values by $x_1, \ldots, x_{48}$, nine of the x_i's were 0, nine were 1, and so on. The likelihood of the observed sample is

$$\frac{e^{-\lambda}\lambda^{x_1}}{x_1!} \cdot \ldots \cdot \frac{e^{-\lambda}\lambda^{x_{48}}}{x_{48}!} = \frac{e^{-48\lambda}\lambda^{\Sigma x_i}}{x_1! \cdots \cdot x_{48}!} = \frac{e^{-48\lambda}\lambda^{101}}{x_1! \cdots \cdot x_{48}!}$$

The value of λ for which this is maximized is $\hat{\lambda} = \Sigma x_i/n = 101/48 = 2.10$ (the value reported in the paper).

However, the $\hat{\lambda}$ required for χ^2 is obtained by maximizing expression (14.4) rather than the likelihood of the full sample. The cell probabilities are

$$\pi_i(\lambda) = \frac{e^{-\lambda}\lambda^{i-1}}{(i-1)!} \qquad i = 1, 2, 3, 4$$

$$\pi_5(\lambda) = 1 - \sum_{i=0}^{3} \frac{e^{-\lambda}\lambda^i}{i!}$$

so the right-hand side of (14.4) becomes

$$\left[\frac{e^{-\lambda}\lambda^0}{0!}\right]^9 \left[\frac{e^{-\lambda}\lambda^1}{1!}\right]^9 \left[\frac{e^{-\lambda}\lambda^2}{2!}\right]^{10} \left[\frac{e^{-\lambda}\lambda^3}{3!}\right]^{14} \left[1 - \sum_{i=0}^{3} \frac{e^{-\lambda}\lambda^i}{i!}\right]^6$$

There is no nice formula for $\hat{\lambda}$, the maximizing value of λ, in this latter expression, so it must be obtained numerically. ∎

Because the parameter estimates are usually much more difficult to compute from the grouped data than from the full sample, they are virtually always computed using this latter method. When these "full" estimators are used in the chi-squared statistic, the distribution of the statistic is altered and a level α test is no longer specified by the critical value $\chi^2_{\alpha, k-1-m}$.

Theorem

> Let $\hat{\theta}_1, \ldots, \hat{\theta}_m$ be the maximum likelihood estimators of $\theta_1, \ldots, \theta_m$ based on the full sample $X_1, \ldots, X_n$ and let χ^2 denote the statistic based on these estimators. Then the critical value c_α that specifies a level α upper-tailed test satisfies
>
> $$\chi^2_{\alpha, k-1-m} \leq c_\alpha \leq \chi^2_{\alpha, k-1} \qquad (14.7)$$

The test procedure implied by this theorem is the following:

$$
\begin{aligned}
&\text{If } \chi^2 \ge \chi^2_{\alpha,k-1}, \text{ reject } H_0 \\
&\text{If } \chi^2 \le \chi^2_{\alpha,k-1-m}, \text{ do not reject } H_0 \\
&\text{If } \chi^2_{\alpha,k-1-m} < \chi^2 < \chi^2_{\alpha,k-1}, \text{ withhold judgment}
\end{aligned}
\tag{14.8}
$$

Example 14.10
(Example 14.9 continued)

Using $\hat{\lambda} = 2.10$, the estimated expected cell counts are computed from $n\pi_i(\hat{\lambda})$ where $n = 48$. For example,

$$
n\pi_1(\hat{\lambda}) = 48 \cdot \frac{e^{-2.1}(2.1)^0}{0!} = (48)(e^{-2.1}) = 5.88
$$

Similarly, $n\pi_2(\hat{\lambda}) = 12.34$, $n\pi_3(\hat{\lambda}) = 12.96$, $n\pi_4(\hat{\lambda}) = 9.07$, and $n\pi_5(\hat{\lambda}) = 48 - 5.88 - \cdots - 9.07 = 7.75$. Then

$$
\chi^2 = \frac{(9 - 5.88)^2}{5.88} + \cdots + \frac{(6 - 7.75)^2}{7.75} = 6.31
$$

Since $m = 1$ and $k = 5$, at level .05 we need $\chi^2_{.05,3} = 7.815$ and $\chi^2_{.05,4} = 9.488$. Because $6.31 \le 7.815$, we do not reject H_0; at the 5% level, the Poisson distribution provides a reasonable fit to the data. Notice that $\chi^2_{.10,3} = 6.251$ and $\chi^2_{.10,4} = 7.779$, so at level .10 we would have to withhold judgment on whether the Poisson distribution was appropriate. ∎

Sometimes even the maximum likelihood estimates based on the full sample are quite difficult to compute. This is the case, for example, for the two-parameter (generalized) negative binomial distribution. In such situations, method-of-moments estimates are often used, and the resulting χ^2 compared to $\chi^2_{\alpha,k-1-m}$, though it is not known to what extent the use of moments estimators affects the true critical value.

Goodness of Fit for Continuous Distributions

The chi-squared test can also be used to test whether the sample comes from a specified family of continuous distributions, such as the exponential family or the normal family. The choice of cells (class intervals) is even more arbitrary in the continuous case than in the discrete case. To ensure that the chi-squared test is valid, the cells should be chosen independently of the sample observations. Once the cells are chosen, it is almost always quite difficult to estimate unspecified parameters (such as μ and σ in the normal case) from the observed cell counts, so instead mle's based on the full sample are computed. The critical value c_α again satisfies (14.7), and the test procedure is given by (14.8).

Example 14.11

The Institute of Nutrition of Central America and Panama (INCAP) has carried out extensive dietary studies and research projects in Central America. In one study reported in the November 1964 issue of the *American Journal of Clinical Nutrition*

("The Blood Viscosity of Various Socioeconomic Groups in Guatemala"), serum total cholesterol measurements for a sample of 49 low-income rural Indians were reported as follows (in mg/L):

```
204 108 140 152 158 129 175 146 157 174 192 194 144 152 135 223 145
231 115 131 129 142 114 173 226 155 166 220 180 172 143 148 171 143
124 158 144 108 189 136 136 197 131  95 139 181 165 142 162
```

Is it plausible that serum cholesterol level is normally distributed for this population? Suppose that prior to sampling, it was believed that plausible values for μ and σ were 150 and 30, respectively. The seven equiprobable class intervals for the standard normal distribution are $(-\infty, -1.07)$, $(-1.07, -.57)$, $(-.57, -.18)$, $(-.18, .18)$, $(.18, .57)$, $(.57, 1.07)$, and $(1.07, \infty)$, with each endpoint also giving the distance in standard deviations from the mean for any other normal distribution. For $\mu = 150$ and $\sigma = 30$, these intervals become $(-\infty, 117.9)$, $(117.9, 132.9)$, $(132.9, 144.6)$, $(144.6, 155.4)$, $(155.4, 167.1)$, $(167.1, 182.1)$, and $(182.1, \infty)$.

To obtain the estimated cell probabilities $\pi_1(\hat{\mu}, \hat{\sigma}), \ldots, \pi_7(\hat{\mu}, \hat{\sigma})$, we first need the mle's $\hat{\mu}$ and $\hat{\sigma}$. In Chapter 6, the mle of σ was shown to be $[\Sigma(x_i - \bar{x})^2/n]^{1/2}$ (rather than s), so with $s = 31.75$

$$\hat{\mu} = \bar{x} = 157.02, \quad \hat{\sigma} = \left[\frac{\Sigma(x_i - \bar{x})^2}{n}\right]^{1/2} = \left[\frac{(n-1)s^2}{n}\right]^{1/2} = 31.42$$

Each $\pi_i(\hat{\mu}, \hat{\sigma})$ is then the probability that a normal rv X with mean 157.02 and standard deviation 31.42 falls in the ith class interval. For example,

$$\pi_2(\hat{\mu}, \hat{\sigma}) = P(117.9 \leq X \leq 132.9) = P(-1.25 \leq Z \leq -.77) = .1150$$

so $n\pi_2(\hat{\mu}, \hat{\sigma}) = 49(.1150) = 5.64$. Observed and estimated expected cell counts are shown in Figure 14.9.

The computed χ^2 is 4.60. With $k = 7$ cells and $m = 2$ parameters estimated, $\chi^2_{.05, k-1} = \chi^2_{.05, 6} = 12.592$ and $\chi^2_{.05, k-1-m} = \chi^2_{.05, 4} = 9.488$. Since $4.60 \leq 9.488$, a normal distribution provides quite a good fit to the data.

Cell	$(-\infty, 117.9)$	$(117.9, 132.9)$	$(132.9, 144.6)$	$(144.6, 155.4)$
Observed	5	5	11	6
Estimated expected	5.17	5.64	6.08	6.64

Cell	$(155.4, 167.1)$	$(167.1, 182.1)$	$(182.1, \infty)$
Observed	6	7	9
Estimated expected	7.12	7.97	10.38

Figure 14.9 Observed and expected counts for Example 14.11   ∎

Example 14.12 The paper "Some Studies on Tuft Weight Distribution in the Opening Room" (*Textile Research J.*, 1976, pp. 567–573) reports the accompanying data on the distribution of output tuft weight X (mg) of cotton fibers for the input weight $x_0 = 70$.

Interval	0–8	8–16	16–24	24–32	32–40	40–48	48–56	56–64	64–70
Observed frequency	20	8	7	1	2	1	0	1	0
Expected frequency	18.0	9.9	5.5	3.0	1.8	.9	.5	.3	.1

The authors postulated a truncated exponential distribution:

$$H_0: f(x) = \frac{\lambda e^{-\lambda x}}{1 - e^{-\lambda x_0}}, \qquad 0 \le x \le x_0$$

The mean of this distribution is

$$\mu = \int_0^{x_0} x f(x) dx = \frac{1}{\lambda} - \frac{x_0 e^{-\lambda x_0}}{1 - e^{-\lambda x_0}}$$

The parameter λ was estimated by replacing μ by $\bar{x} = 13.086$ and solving the resulting equation to obtain $\hat{\lambda} = .0742$ (so $\hat{\lambda}$ is a method-of-moments estimate and not an mle). Then with $\hat{\lambda}$ replacing λ in $f(x)$, the estimated expected cell frequencies as displayed above are computed as

$$40\pi_i(\hat{\lambda}) = 40P(a_{i-1} \le X < a_i) = 40 \int_{a_{i-1}}^{a_i} f(x) \, dx = \frac{40(e^{-\hat{\lambda} a_{i-1}} - e^{-\hat{\lambda} a_i})}{1 - e^{-\hat{\lambda} x_0}}$$

where $[a_{i-1}, a_i)$ is the ith class interval. To obtain expected cell counts of at least 5, the last six cells are combined to yield observed counts of 20, 8, 7, 5 and expected counts of 18.0, 9.9, 5.5, 6.6. The computed value of chi-squared is then $\chi^2 = 1.34$. Because $\chi^2_{.05,2} = 5.992$, H_0 is not rejected, so the truncated exponential model provides a good fit. ■

A Special Test for Normality

Probability plots were introduced in Section 4.6 as an informal method for assessing the plausibility of any specified population distribution as the one from which the given sample was selected. The straighter the probability plot, the more plausible is the distribution on which the plot is based. A normal probability plot is used for checking whether *any* member of the normal distribution family is plausible. Let's denote the sample x_i's when ordered from smallest to largest by $x_{(1)}, x_{(2)}, \ldots, x_{(n)}$. Then the plot suggested for checking normality was a plot of the points $(x_{(i)}, y_i)$, where $y_i = \Phi^{-1}((i - .5)/n)$.

A quantitative measure of the extent to which points cluster about a straight line is the sample correlation coefficient r introduced in Chapter 12. Consider calculating r for the n pairs $(x_{(1)}, y_1), \ldots, (x_{(n)}, y_n)$. The y_i's here are not observed values in a random sample from a y population, so properties of this r are quite different from those described in Section 12.5. However, it is true that the more the r deviates from 1, the less the probability plot resembles a straight line (remember that a probability plot must slope upward). This idea can be extended to yield a formal test procedure: Reject the hypothesis of population normality if $r \leq c_\alpha$, where c_α is a critical value chosen to yield the desired significance level α. That is, the critical value is chosen so that when the population distribution is actually normal, the probability of obtaining an r value that is at most c_α (and thus incorrectly rejecting H_0) is the desired α. The developers of the MINITAB statistical computer package give critical values for $\alpha = .10, .05$, and $.01$ in combination with different sample sizes. These critical values are based on a slightly different definition of the y_i's than that given earlier.

MINITAB will also construct a normal probability plot based on these y_i's. The plot will be almost identical in appearance to that based on the earlier y_i's. When there are several tied $x_{(i)}$'s, MINITAB computes r by using the average of the corresponding y_i's as the second number in each pair.

Let $y_i = \Phi^{-1}[(i - .375)/(n + .25)]$ and compute the sample correlation coefficient r for the n pairs $(x_{(1)}, y_1), \ldots, (x_{(n)}, y_n)$. A test of

H_0: the population distribution is normal

versus

H_a: the population distribution is not normal

consists of rejecting H_0 when $r \leq c_\alpha$. Critical values c_α are given in Appendix Table A.14 for various significance levels α and sample sizes n.

Example 14.13   Consider the accompanying random sample of $n = 20$ observations, each the value of the width-to-length ratio of a beaded rectangle used by the Shoshoni to decorate leather handicrafts (*Lowie's Selected Papers in Anthropology*, 1960, pp. 137–142). Let's carry out a test at level .01 to see whether it is plausible that the population distribution of width-to-length ratios is normal. The critical value for $n = 20$ is .929, and the calculated value of r is .905. Since $.905 \leq .929$, H_0 is rejected. A normal probability plot exhibits curvature of the sort associated with a positively skewed distribution having a long upper tail.

	$x_{(i)}$	y_i	$x_{(i)}^2$	y_i^2	$x_{(i)}y_i$
1	.553	−1.87129	.305809	3.50172	−1.03482
2	.570	−1.40377	.324900	1.97057	−.80015
3	.576	−1.12690	.331776	1.26990	−.64909
4	.601	−.91718	.361201	.84121	−.55122
5	.606	−.66267	.367236	.43913	−.40158
6	.606	−.66267	.367236	.43913	−.40158
7	.609	−.44602	.370881	.19893	−.27162
8	.611	−.31325	.373321	.09812	−.19139
9	.615	−.18593	.378225	.03457	−.11435

	$x_{(i)}$	y_i	$x_{(i)}^2$	y_i^2	$x_{(i)}y_i$
10	.628	-.06165	.394384	.00380	-.03872
11	.654	.06165	.427716	.00380	.04032
12	.662	.18593	.438244	.03457	.12309
13	.668	.31325	.446224	.09812	.20925
14	.670	.44602	.448900	.19893	.29883
15	.672	.58740	.451584	.34504	.39473
16	.690	.74198	.476100	.55054	.51197
17	.693	.91718	.480249	.84121	.63560
18	.749	1.12690	.561001	1.26990	.84405
19	.844	1.40377	.712336	1.97057	1.18478
20	.933	1.87129	.870489	3.50172	1.74591
Sum:	13.210	.00404	8.887812	17.61148	1.53401

Exercises Section 14.2 (10–20)

10. Consider a large population of families in which each family has exactly three children. If the genders of the three children in any family are independent of one another, the number of male children in a randomly selected family will have a binomial distribution based on three trials.

a. Suppose a random sample of 160 families yields the following results. Test the relevant hypotheses by proceeding as in Example 14.6.

Number of male children	0	1	2	3
Frequency	14	66	64	16

b. Suppose a random sample of families in a nonhuman population resulted in observed frequencies of 15, 20, 12, and 3, respectively. Would the chi-squared test be based on the same number of degrees of freedom as the test in part (a)? Explain.

11. A study of sterility in the fruit fly ("Hybrid Dysgenesis in *Drosophila Melanogaster:* The Biology of Female and Male Sterility," *Genetics*, 1979, pp. 161–174) reports the following data on the number of ovaries developed for each female fly in a sample of size 1388. One model for unilateral sterility states that each ovary develops with some probability p independently of the other ovary. Test the fit of this model using χ^2.

x = Number of ovaries developed	0	1	2
Observed count	1212	118	58

12. The article "Feeding Ecology of the Red-Eyed Vireo and Associated Foliage-Gleaning Birds" (*Ecological Monographs*, 1971, pp. 129–152) presents the accompanying data on the variable X = the number of hops before the first flight and preceded by a flight. The author then proposed and fit a geometric probability distribution $[p(x) = P(X = x) = p^{x-1} \cdot q$ for $x = 1, 2, \ldots$ where $q = 1 - p]$ to the data. The total sample size was $n = 130$.

x	1	2	3	4	5	6	7	8	9	10	11	12
Number of times x observed	48	31	20	9	6	5	4	2	1	1	2	1

a. The likelihood is $(p^{x_1-1} \cdot q) \cdot \cdots \cdot (p^{x_n-1} \cdot q) = p^{\Sigma x_i - n} \cdot q^n$. Show that the mle of p is given by $\hat{p} = (\Sigma x_i - n)/\Sigma x_i$ and compute $\hat{p}$ for the given data.

b. Estimate the expected cell counts using $\hat{p}$ of part (a) [expected cell counts = $n \cdot (\hat{p})^{x-1} \cdot \hat{q}$ for $x = 1, 2, \ldots$] and test the fit of the model using a χ^2 test by combining the counts for $x = 7, 8, \ldots$, and 12 into one cell ($x \geq 7$).

13. A certain type of flashlight is sold with the four batteries included. A random sample of 150 flashlights is obtained, and the number of defective batteries in each is determined, resulting in the following data:

Number defective	0	1	2	3	4
Frequency	26	51	47	16	10

Let X be the number of defective batteries in a randomly selected flashlight. Test the null hypothesis that the distribution of X is Bin(4, θ). That is, with $p_i = P(i$ defectives), test

$$H_0: p_i = \binom{4}{i} \theta^i (1 - \theta)^{4-i} \qquad i = 0, 1, 2, 3, 4$$

[HINT: To obtain the mle of θ, write the likelihood (the function to be maximized) as $\theta^u (1 - \theta)^v$ where the exponents u and v are linear functions of the cell counts. Then take the natural log, differentiate with respect to θ, equate the result to 0, and solve for $\hat{\theta}$.]

14. In a genetics experiment, investigators looked at 300 chromosomes of a particular type and counted the number of sister-chromatid exchanges on each ("On the Nature of Sister-Chromatid Exchanges in 5-Bromo-deoxyuridine-Substituted Chromosomes," *Genetics*, 1979, pp. 1251–1264). A Poisson model was hypothesized for the distribution of the number of exchanges. Test the fit of a Poisson distribution to the data by first estimating λ and then combining the counts for $x = 8$ and $x = 9$ into one cell.

X = Number of exchanges	0	1	2	3	4	5	6	7	8	9
Observed counts	6	24	42	59	62	44	41	14	6	2

15. A paper in *Annals of Mathematical Statistics* reports the following data on the number of borers in each of 120 groups of borers. Does the Poisson pmf provide a plausible model for the distribution of the number of borers in a group? (HINT: Add the frequencies for 7, 8, . . . , 12 to establish a single category "≥ 7.")

Number of borers	0	1	2	3	4	5	6	7	8	9	10	11	12
Frequency	24	16	16	18	15	9	6	5	3	4	3	0	1

16. The article "A Probabilistic Analysis of Dissolved Oxygen–Biochemical Oxygen Demand Relationship in Streams" (*J. Water Resources Control Fed.*, 1969, pp. 73–90) reports data on the rate of oxygenation in streams at 20°C in a certain region. The sample mean and standard deviation were computed as $\bar{x} = .173$ and $s = .066$, respectively. Based on the accompanying frequency distribution, can it be concluded that oxygenation rate is a normally distributed variable? Use the chi-squared test with $\alpha = .05$.

Rate (per day)	Frequency
Below .100	12
.100–below .150	20
.150–below .200	23
.200–below .250	15
.250 or more	13

17. Each headlight on an automobile undergoing an annual vehicle inspection can be focused either too high (H), too low (L), or properly (N). Checking the two headlights simultaneously (and not distinguishing between left and right) results in the six possible outcomes HH, LL, NN, HL, HN, and LN. If the probabilities (population proportions) for the single headlight focus direction are $P(H) = \theta_1$, $P(L) = \theta_2$, and $P(N) = 1 - \theta_1 - \theta_2$ and the two headlights are focused independently of one another, the probabilities of the six outcomes for a randomly selected car are the following:

$$p_1 = \theta_1^2, \quad p_2 = \theta_2^2, \quad p_3 = (1 - \theta_1 - \theta_2)^2$$
$$p_4 = 2\theta_1\theta_2, \quad p_5 = 2\theta_1(1 - \theta_1 - \theta_2)$$
$$p_6 = 2\theta_2(1 - \theta_1 - \theta_2)$$

Use the data given below to test the null hypothesis

$$H_0: p_1 = \pi_1(\theta_1, \theta_2), \ldots, p_6 = \pi_6(\theta_1, \theta_2)$$

where the $\pi_i(\theta_1, \theta_2)$'s are given above.

Outcome	HH	LL	NN	HL	HN	LN
Frequency	49	26	14	20	53	38

(HINT: Write the likelihood as a function of θ_1 and θ_2, take the natural log, then compute $\partial/\partial\theta_1$ and $\partial/\partial\theta_2$, equate them to 0, and solve for $\hat{\theta}_1$, $\hat{\theta}_2$.)

18. The following data set consists of 25 observations on fracture toughness of base plate of 18% nickel maraging steel (from "Fracture Testing of Weldments," *ASTM Special Publ. No. 381*, 1965, pp. 328–356). Use the normal probability plot correlation coefficient test to decide whether a normal distribution provides a plausible model for fracture toughness.

69.5 71.9 72.6 73.1 73.3 73.5 74.1 74.2 75.3
75.5 75.7 75.8 76.1 76.2 76.9 77.0 77.9 78.1
79.6 79.7 79.9 80.1 82.2 83.7 93.7

19. The paper "Nonbloated Burned Clay Aggregate Concrete" (*J. Materials*, 1972, pp. 555–563) reports the following data on 7-day flexural strength of nonbloated burned clay aggregate concrete samples (psi):

257 327 317 300 340 340 343 374 377 386
383 393 407 407 434 427 440 407 450 440
456 460 456 476 480 490 497 526 546 700

Test at level .10 to decide whether flexural strength is a normally distributed variable.

20. Use the normal probability plot correlation coefficient test on the standardized residuals of Example 13.1 to assess the plausibility of assuming that the random deviation in the regression model is normally distributed.

Two-Way Contingency Tables

In the previous two sections, we discussed inferential problems in which the count data was displayed in a rectangular table of cells. Each table consisted of one row and a specified number of columns, where the columns corresponded to categories into which the population had been divided. We now study problems in which the data also consists of counts or frequencies, but the data table will now have I rows ($I \geq 2$) and J columns, so IJ cells. There are two commonly encountered situations in which such data arises:

1. There are I populations of interest, each corresponding to a different row of the table, and each population is divided into the same J categories. A sample is taken from the ith population ($i = 1, \ldots, I$), and the counts are entered in the cells in the ith row of the table.

2. There is a single population of interest, with each individual in the population categorized with respect to two different factors. There are I categories associated with the first factor, and J categories associated with the second factor. A single sample is taken, and the number of individuals belonging in both category i of factor 1 and category j of factor 2 is entered in the cell in row i, column j ($i = 1, \ldots, I; j = 1, \ldots, J$).

Let n_{ij} denote the number of individuals in the sample(s) falling in the (i, j)th cell (row i, column j) of the table—that is, the (i, j)th cell count. The table displaying the n_{ij}'s is called a **two-way contingency table**; a prototype is shown in Figure 14.10.

	1	2	...	j	...	J
1	n_{11}	n_{12}	...	n_{1j}	...	n_{1J}
2	n_{21}					⋮
⋮	⋮					
i	n_{i1}	...		n_{ij}	...	
⋮	⋮					
I	n_{I1}	...				n_{IJ}

Figure 14.10 A two-way contingency table

In situations of type 1, we want to investigate whether the proportions in the different categories are the same for all populations. The null hypothesis states that the populations are **homogeneous** with respect to these categories. In type 2 situations, we investigate whether the categories of the two factors occur independently of one another in the population.

In Chapter 8 we developed a large-sample z test for testing the null hypothesis that the proportion of successes in one population is identical to the proportion in a second population. Instead of using S and F to label the two sides of the dichotomy, let's label them as category 1 and category 2. The symbols p_{11} and p_{12} will represent the proportions of population 1 belonging to categories 1 and 2, respectively (in place of p_1 and q_1); p_{21} and p_{22} will denote the analogous proportions for the second population. In our use of doubly subscripted quantities, the first subscript will refer to the population or sample from that population, and the second subscript will identify the category. The null hypothesis that the two dichotomous populations are homogeneous with respect to category proportions, previously stated as $H_0: p_1 = p_2$, now becomes $H_0: p_{11} = p_{21}$ (in which case $p_{12} = p_{22}$ also).

Homogeneity for I Dichotomous Populations

Suppose now that there are I different dichotomous populations ($I > 2$), each consisting of the same two categories. Let p_{i1} (p_{i2}) denote the proportion of population i falling in category 1 (2). For example, we might focus on $I = 4$ different brands of television sets, with category 1 consisting of all sets that need no warranty repair work and category 2 consisting of all sets that do need such work. Then p_{11}, p_{21}, p_{31}, and p_{41} denote the proportions of the four brands falling in the first category. The null hypothesis of homogeneity states that $p_{11} = p_{21} = p_{31} = p_{41}$ (implying that $p_{12} = p_{22} = p_{32} = p_{42}$). More generally, we wish to test $H_0: p_{11} = p_{21} = \cdots = p_{I1}$ against the alternative that these category 1 proportions are not all the same.

To do so, we need sample data. Consider taking a random sample of size n_i from the ith population ($i = 1, \ldots, I$). The I samples are assumed to have been selected independently of one another. The total sample size is $n = \Sigma n_i$. Let $n_{i1} =$ the number among the n_i sampled from the ith population that fall in category 1, and $n_{i2} = n_i - n_{i1}$. Before the samples are selected, the number of individuals falling into categories 1 and 2 are random variables—$N_{11}, N_{21}, \ldots, N_{I1}$ for category 1 and N_{12}, $N_{22}, \ldots, N_{I2}$ for category 2. The n_{ij}'s are the observed values of the N_{ij}'s. These are customarily displayed in a tabular format as illustrated in Figure 14.11 and are called *observed cell counts*. $N_{.1}$ and $n_{.1}$ refer to the total number among the n sampled that fall in category 1—the total of the observed counts from the first column; $N_{.2}$ and $n_{.2}$ refer to the total for category 2. Notice that the row totals in the table are fixed in advance of the experiment, whereas the column totals are not.

The expected number of observations in the sample from population i that fall in the first category is $E(N_{i1}) = $ (sample size)(category proportion) $= n_i p_{i1}$ ($i = 1, \ldots, I$). When the null hypothesis is true, let p_1 denote the common value of $p_{11}, p_{21}, \ldots,$ and p_{I1}. In this case, the expected counts for category 1 are $n_1 p_1$, $n_2 p_1, \ldots,$ and $n_I p_1$, and they are $n_1(1 - p_1), \ldots,$ and $n_I(1 - p_1)$ for the second category. However, whereas H_0 asserts that there is a common proportion p_1, it does

	Category 1	Category 2	Sample size
Sample from population 1	n_{11}	n_{12}	n_1
Sample from population 2	n_{21}	n_{22}	n_2
$\vdots$	$\vdots$	$\vdots$	$\vdots$
Sample from population I	n_{I1}	n_{I2}	n_I
Category total	$n_{\cdot 1}$	$n_{\cdot 2}$	n

Figure 14.11 An $I \times 2$ contingency table

not specify what that value is. Construction of a test procedure requires that p_1 be estimated from the data. Since p_1 represents the proportion in category 1 for every population, the natural estimate is the fraction among all those sampled that fall in category 1:

$$\hat{p}_1 = \frac{n_{11} + n_{21} + \cdots + n_{I1}}{n_1 + n_2 + \cdots + n_I} = \frac{n_{\cdot 1}}{n}$$

Substituting this estimate into the expressions for expected counts when H_0 is true yields *estimated expected cell counts*.

Let $\hat{e}_{ij}$ denote the point estimate of $E(N_{ij})$, the expected number from the ith sample falling in category j, computed under the assumption that H_0 is true. Then

$$\hat{e}_{ij} = \begin{cases} n_i \hat{p}_1 = \dfrac{n_i \cdot n_{\cdot 1}}{n} & j = 1 \\[2ex] n_i(1 - \hat{p}_1) = \dfrac{n_i \cdot n_{\cdot 2}}{n} & j = 2 \end{cases}$$

That is, the estimated expected cell counts are given by

$$\hat{e}_{ij} = \frac{(i\text{th row total})(j\text{th column total})}{n} \tag{14.9}$$

Example 14.14 Three different sites off the coast of California are being considered as possible locations for a liquid natural gas terminal. To get information on public opinion regarding the LNG project, a separate random sample of individuals in each region is obtained. Each individual is then asked whether he/she favors building an LNG terminal at the proposed location in the region in which the individual resides. The observed cell counts appear in Figure 14.12.

Observed

	Favor	Oppose	
Region 1	198	202	$n_1 = 400$
Region 2	140	210	$n_2 = 350$
Region 3	133	217	$n_3 = 350$

$n = 1100$

Estimated expected

	Favor	Oppose	
Region 1	$\dfrac{(400)(471)}{1100} = 171.27$	$\dfrac{(400)(629)}{1100} = 228.73$	400
Region 2	$\dfrac{(350)(471)}{1100} = 149.86$	$\dfrac{(350)(629)}{1100} = 200.14$	350
Region 3	$\dfrac{(350)(471)}{1100} = 149.86$	$\dfrac{(350)(629)}{1100} = 200.14$	350

| 471 | 629 |

Figure 14.12 Observed and estimated expected cell counts for Example 14.14

The hypothesis of interest is $H_0: p_{11} = p_{21} = p_{31}$, where p_{i1} is the true proportion of individuals in region i who favor an LNG facility in that region. To compute the estimated expected cell counts, the column totals are computed (the row totals n_1, n_2, and n_3 are fixed in advance, since three different regions are sampled). The row and column totals are then entered on the margins of a table having the same dimensions as the observed count table, and formula (14.9) is employed. Calculations for this example are displayed in Figure 14.12. ■

Replacing $n._1$ by $N._1$ and $n._2$ by $N._2$ in $\hat{e}_{ij}$ gives the estimators (rv's) $\hat{E}_{ij}$ ($i = 1, \ldots,$ $I; j = 1, 2$). These estimators can be used to specify a chi-squared goodness-of-fit statistic that has the same general form as those in Sections 14.1 and 14.2:

$$\chi^2 = \sum_{\text{all cells}} \frac{(\text{observed} - \text{estimated expected})^2}{\text{estimated expected}} = \sum_{i=1}^{I} \sum_{j=1}^{2} \frac{(N_{ij} - \hat{E}_{ij})^2}{\hat{E}_{ij}}$$

A theorem from mathematical statistics states that when $H_0: p_{11} = p_{21} = \cdots = p_{I1}$ is true and all I sample sizes are large, χ^2 has approximately a chi-squared distribution with $I - 1$ df. This leads to the following test for homogeneity in I dichotomous populations.

Null hypothesis: H_0: $p_{11} = p_{21} = \cdots = p_{I1}$

Alternative hypothesis: H_a: at least two of the p_{i1}'s are unequal

Test statistic value:

$$\chi^2 = \sum_{\text{all cells}} \frac{(\text{observed} - \text{estimated expected})^2}{\text{estimated expected}} = \sum_{i=1}^{I} \sum_{j=1}^{2} \frac{(n_{ij} - \hat{e}_{ij})^2}{\hat{e}_{ij}}$$

Rejection region: $\chi^2 \geq \chi^2_{\alpha, I-1}$

In practice, the test can safely be applied as long as $\hat{e}_{ij} \geq 5$ for every i, j (all cells).

Example 14.15 The accompanying data resulted from an experiment in which seeds of five different types were planted and the number that germinated within 5 weeks of planting was recorded for each seed type ("Nondestructive Optical Methods of Food Quality Evaluation," *Food Science and Nutr.,* 1984, pp. 232–279). Let p_{i1} = the true proportion of seeds of type i that germinate within 5 weeks of planting. The hypotheses are

H_0: $p_{11} = p_{21} = p_{31} = p_{41} = p_{51}$

H_a: at least two p_{i1}'s are unequal

The test is based on $I - 1 = 4$ df, and because $\chi^2_{.01,4} = 13.277$, H_0 will be rejected at significance level .01 if $\chi^2 \geq 13.277$. The computed value of the test statistic is

$$\chi^2 = \frac{(31 - 22.5)^2}{22.5} + \frac{(57 - 53.3)^2}{53.3} + \cdots + \frac{(19 - 11.8)^2}{11.8}$$

$$= 3.2 + .25 + \cdots + 4.39 = 16.95$$

		Observed				Estimated		
		Germinated	Failed to germinate			Germinated	Failed to germinate	
	1	31	7	$n_1 = 38$		22.5	15.5	38
	2	57	33	$n_2 = 90$		53.3	36.7	90
Seed type	3	87	60	$n_3 = 147$		87.1	59.9	147
	4	52	44	$n_4 = 96$		56.9	39.1	96
	5	10	19	$n_5 = 29$		17.2	11.8	29
		$n_{.1} = 237$	$n_{.2} = 163$	$n = 400$		237	163	400

Figure 14.13 Observed and estimated expected counts for Example 14.15

Because $16.95 \geq 13.277$, H_0 is rejected at significance level .01. The true proportion of seeds that germinate appears to depend on seed type.   ■

The number of degrees of freedom for the test comes from our general rule of thumb: There are I freely determined cell counts (one in each row, since row totals are fixed), and the single parameter p_1 is estimated, giving df $= I - 1$. The test procedure can be used even when $I = 2$ (a comparison of two different population proportions). In this case, df $= 1$, and the chi-squared test is equivalent to the two-tailed z test of Section 9.4 because $z^2 = \chi^2$ and $(z_{\alpha/2})^2 = \chi^2_{\alpha,1}$. If H_0 is rejected in favor of the conclusion that the I populations are not homogeneous, there are multiple comparison methods available for identifying significant differences among the p_{i1}'s.

Homogeneity for I Populations, J Categories

We now assume that each individual in every one of the I populations belongs in exactly one of J categories. A sample of n_i individuals is taken from the ith population; let $n = \Sigma n_i$ and

n_{ij} = the number of individuals in the ith sample who fall into category j

$$n_{\cdot j} = \sum_{i=1}^{I} n_{ij} = \begin{array}{l} \text{the total number of individuals among} \\ \text{the } n \text{ sampled who fall into category } j \end{array}$$

The n_{ij}'s are recorded in a two-way contingency table with I rows and J columns. The sum of the n_{ij}'s in the ith row is n_i, while the sum of entries in the jth column is $n_{\cdot j}$.

With $J > 2$, the notation necessary to state the hypotheses is more cumbersome than in the case $J = 2$ just considered. Let

$$p_{ij} = \begin{array}{l} \text{the proportion of the individuals in} \\ \text{population } i \text{ who fall into category } j \end{array}$$

Thus, for population 1, the J proportions are $p_{11}, p_{12}, \ldots, p_{1J}$ (which sum to 1) and similarly for the other populations. The **null hypothesis of homogeneity** states that the proportion of individuals in category j is the same for each population, and that this is true for every category; that is, for every j, $p_{1j} = p_{2j} = \cdots = p_{Ij}$.

When H_0 is true, we can use $p_1, p_2, \ldots, p_J$ to denote the population proportions in the J different categories; these proportions are common to all I populations. The expected number of individuals in the ith sample who fall in the jth category when H_0 is true is then $E(N_{ij}) = n_i \cdot p_j$. To estimate $E(N_{ij})$, we must first estimate p_j, the proportion in category j. Among the total sample of n individuals, $N_{\cdot j}$ fall into category j, so we use $\hat{p}_j = N_{\cdot j}/n$ as the estimator (this can be shown to be the maximum likelihood estimator of p_j). Substitution of the estimate $\hat{p}_j$ for p_j in $n_i p_j$ yields the same formula for estimated expected counts under H_0 as in the dichotomous case:

$$\hat{e}_{ij} = \text{estimated expected count in cell } (i, j) = n_i \cdot \frac{n_{\cdot j}}{n}$$

$$= \frac{(i\text{th row total})(j\text{th column total})}{n} \qquad (14.10)$$

The test statistic also has the same form as in earlier problem situations. The number of degrees of freedom comes from the general rule of thumb. In each row of the table there are $J - 1$ freely determined cell counts (the sample size n_i is fixed), so there are a total of $I(J - 1)$ freely determined cells. Parameters $p_1, \ldots, p_J$ are estimated, but because $\Sigma p_i = 1$, only $J - 1$ of these are independent. Therefore $\text{df} = I(J - 1) - (J - 1) = (J - 1)(I - 1)$.

Null hypothesis: $H_0: p_{1j} = p_{2j} = \cdots = p_{Ij}$ $\qquad j = 1, 2, \ldots, J$

Alternative hypothesis: $H_a: H_0$ is not true

Test statistic value:

$$\chi^2 = \sum_{\text{all cells}} \frac{(\text{observed} - \text{estimated expected})^2}{\text{estimated expected}} = \sum_{i=1}^{I} \sum_{j=1}^{J} \frac{(n_{ij} - \hat{e}_{ij})^2}{\hat{e}_{ij}}$$

Rejection region: $\chi^2 \geq \chi^2_{\alpha,(I-1)(J-1)}$

The test can safely be applied as long as $\hat{e}_{ij} \geq 5$ for all cells.

Example 14.16   In a study to investigate the extent to which individuals are aware of industrial odors in a certain region ("Annoyance and Health Reactions to Odor from Refineries and Other Industries in Carson, California," *Environmental Research*, 1978, pp. 119–132), a sample of individuals was obtained from each of three different areas near industrial facilities. Each individual was asked whether he/she noticed odors (1) every day, (2) at least once/week, (3) at least once/month, (4) less often than once/month, or (5) not at all, resulting in the data given in Figure 14.14, with the estimated expected cell counts computed from (14.10).

Then

$$\chi^2 = \frac{(20 - 12.67)^2}{12.67} + \cdots + \frac{(53 - 26.20)^2}{26.20} = 70.63$$

With $(I - 1)(J - 1) = (2)(4) = 8$, $\chi^2_{.005,8} = 21.954$. Because $70.63 \geq 21.954$, the hypothesis of homogeneity is rejected at level .005 in favor of the conclusion that there are differences in perception of odors among the three areas.

Category

Observed	1	2	3	4	5	n_i
Area I	20	28	23	14	12	97
Area II	14	34	21	14	12	95
Area III	4	12	10	20	53	99
$n_{\cdot j}$	38	74	54	48	77	291

Expected	1	2	3	4	5	
Area I	12.67	24.67	18.00	16.00	25.67	97
Area II	12.41	24.16	17.63	15.67	25.14	95
Area III	12.93	25.18	18.37	16.33	26.20	99
	38	74	54	48	77	

Figure 14.14 Observed and estimated expected counts for Example 14.16

Figure 14.15 displays SPSS output; it shows that no cell has an expected frequency (E.F.) smaller than 5 and gives the P-value as .0000 to four decimal places. ∎

One way to extend the chi-squared analysis described here is to partition the overall χ^2 statistic into independent components. In Example 14.16, χ^2 can be broken down into χ^2 for area I versus area II with df = 4 and areas I and II versus area III with df = 4. The conclusion is that areas I and II do not differ significantly from one another while in combination they differ significantly from area III. This type of partitioning should be planned before the data is actually obtained. For unplanned comparisons, there are methods of multiple comparisons that can be employed. The book by B. S. Everitt contains a good exposition of these topics (see the chapter bibliography).

Independence of Factors in a Single Population

We focus now on the relationship between two different factors in a single population. The number of categories of the first factor will be denoted by I and the number of categories of the second factor by J. Each individual in the population is assumed to belong in exactly one of the I categories associated with the first factor and exactly one of the J categories associated with the second factor. For example, the population of interest might consist of all individuals who regularly watch the national news on television, with the first factor being preferred network (ABC, CBS, NBC, or PBS, so $I = 4$) and the second factor political philosophy (liberal, moderate, or conservative, giving $J = 3$).

Crosstabulation: AREA
By CATEGORY

CATEGORY → AREA	Count Exp Val Row Pct Col Pct	1.00	2.00	3.00	4.00	5.00	Row Total
	1.00	20 12.7 20.6% 52.6%	28 24.7 28.9% 37.8%	23 18.0 23.7% 42.6%	14 16.0 14.4% 29.2%	12 25.7 12.4% 15.6%	97 33.3%
	2.00	14 12.4 14.7% 36.8%	34 24.2 35.8% 45.9%	21 17.6 22.1% 38.9%	14 15.7 14.7% 29.2%	12 25.1 12.6% 15.6%	95 32.6%
	3.00	4 12.9 4.0% 10.5%	12 25.2 12.1% 16.2%	10 18.4 10.1% 18.5%	20 16.3 20.2% 41.7%	53 26.2 53.5% 68.8%	99 34.0%
	Column Total	38 13.1%	74 25.4%	54 18.6%	48 16.5%	77 26.5%	291 100.0%

Chi-Square	D.F.	Significance	Min E.F.	Cells with E.F.<5
70.64156	8	.0000	12.405	None

Figure 14.15 SPSS output for Example 14.16

For a sample of n individuals taken from the population, let n_{ij} denote the number among the n who fall both in category i of the first factor and category j of the second factor. The n_{ij}'s can be displayed in a two-way contingency table with I rows and J columns. In the case of homogeneity for I populations, the row totals were fixed in advance, and only the J column totals were random. Now only the total sample size is fixed, and both the $n_{i\cdot}$'s and $n_{\cdot j}$'s are observed values of random variables. To state the hypotheses of interest, let

p_{ij} = the proportion of individuals in the population who belong in category i of factor 1 and category j of factor 2

= P(a randomly selected individual falls in both category i of factor 1 and category j of factor 2)

Then

$$p_{i\cdot} = \sum_{j} p_{ij} = P\text{(a randomly selected individual falls in category } i \text{ of factor 1)}$$

$$p_{\cdot j} = \sum_{i} p_{ij} = P\text{(a randomly selected individual falls in category } j \text{ of factor 2)}$$

Recall from probability that two events A and B are independent if $P(A \cap B) = P(A) \cdot P(B)$. *The null hypothesis here says that an individual's category with respect to factor* 1 *is independent of the category with respect to factor* 2. In symbols, this becomes $p_{ij} = p_{i\cdot} \cdot p_{\cdot j}$ for every pair (i, j).

The expected count in cell (i, j) is $n \cdot p_{ij}$, so when the null hypothesis is true, $E(N_{ij}) = n \cdot p_{i\cdot} \cdot p_{\cdot j}$. To obtain a chi-squared statistic, we must therefore estimate the $p_{i\cdot}$'s $(i = 1, \ldots, I)$ and $p_{\cdot j}$'s $(j = 1, \ldots, J)$. The (maximum likelihood) estimates are

$$\hat{p}_{i\cdot} = \frac{n_{i\cdot}}{n} = \text{sample proportion for category } i \text{ of factor } 1$$

and

$$\hat{p}_{\cdot j} = \frac{n_{\cdot j}}{n} = \text{sample proportion for category } j \text{ of factor } 2$$

This gives estimated expected cell counts identical to those in the case of homogeneity.

$$\hat{e}_{ij} = n \cdot \hat{p}_{i\cdot} \cdot \hat{p}_{\cdot j} = n \cdot \frac{n_{i\cdot}}{n} \cdot \frac{n_{\cdot j}}{n} = \frac{n_{i\cdot} \cdot n_{\cdot j}}{n}$$

$$= \frac{(i \text{th row total})(j \text{th column total})}{n}$$

The test statistic is also identical to that used in testing for homogeneity, as is the number of degrees of freedom. This is because the number of freely determined cell counts is $IJ - 1$, since only the total n is fixed in advance. There are I estimated $p_{i\cdot}$'s, but only $I - 1$ are independently estimated since $\Sigma p_{i\cdot} = 1$, and similarly $J - 1$ $p_{\cdot j}$'s are independently estimated, so $I + J - 2$ parameters are independently estimated. The rule of thumb now yields df $= IJ - 1 - (I + J - 2) = IJ - I - J + 1 = (I - 1) \cdot (J - 1)$.

Null hypothesis: $H_0: p_{ij} = p_{i\cdot} \cdot p_{\cdot j}$ $i = 1, \ldots, I; j = 1, \ldots, J$

Alternative hypothesis: $H_a: H_0$ is not true

Test statistic value:

$$\chi^2 = \sum_{\text{all cells}} \frac{(\text{observed} - \text{estimated expected})^2}{\text{estimated expected}} = \sum_{i=1}^{I} \sum_{j=1}^{J} \frac{(n_{ij} - \hat{e}_{ij})^2}{\hat{e}_{ij}}$$

Rejection region: $\chi^2 \geq \chi^2_{\alpha, (I-1)(J-1)}$

The test can safely be applied as long as $\hat{e}_{ij} \geq 5$ for all cells.

Example 14.17   A study of the relationship between facility conditions at gasoline stations and aggressiveness in the pricing of gasoline ("An Analysis of Price Aggressiveness in

Gasoline Marketing," *J. Marketing Research,* 1970, pp. 36–42) reports the accompanying data based on a sample of $n = 441$ stations. At level .01, does the data suggest that facility conditions and pricing policy are independent of one another? Observed and estimated expected counts appear in Figure 14.16.

		Observed pricing policy				Expected pricing policy			
		Aggressive	Neutral	Nonaggressive	$n_{i\cdot}$	Aggressive	Neutral	Nonaggressive	
	Substandard	24	15	17	56	17.02	22.10	16.89	56
Condition	Standard	52	73	80	205	62.29	80.88	61.83	205
	Modern	58	86	36	180	54.69	71.02	54.29	180
	$n_{\cdot j}$	134	174	133	441	134	174	133	441

Figure 14.16 Observed and estimated expected counts for Example 14.17

Thus,

$$\chi^2 = \frac{(24 - 17.02)^2}{17.02} + \cdots + \frac{(36 - 54.29)^2}{54.29} = 22.47$$

and because $\chi^2_{.01,4} = 13.277$, the hypothesis of independence is rejected.

We conclude that knowledge of a station's pricing policy does give information about the condition of facilities at the station. In particular, stations with an aggressive pricing policy appear more likely to have substandard facilities than stations with a neutral or nonaggressive policy. ∎

Models and methods for analyzing data in which each individual is categorized with respect to three or more factors (multidimensional contingency tables) are discussed in several of the chapter references.

Exercises Section 14.3 (21–33)

21. The accompanying two-way table was constructed using data in the paper "Television Viewing and Physical Fitness in Adults" (*Research Quarterly for Exercise and Sport,* 1990, pp. 315–320). The author hoped to determine whether time spent watching television is associated with cardiovascular fitness. Subjects were asked about their television-viewing habits and were classified as physically fit if they scored in the excellent or very good category on a step test. We include

MINITAB output from a chi-squared analysis. The four TV groups corresponded to different amounts of time per day spent watching TV (0, 1–2, 3–4, or 5 or more hours). The 168 individuals represented in the first column were those judged physically fit. Expected counts appear below observed counts, and MINITAB displays the contribution to χ^2 from each cell. State and test the appropriate hypotheses using $\alpha = .05$.

	1	2	Total
1	35	147	182
	25.48	156.52	
2	101	629	730
	102.20	627.80	
3	28	222	250
	35.00	215.00	
4	4	34	38
	5.32	32.68	
Total	168	1032	1200

ChiSq = 3.557 + 0.579 +
0.014 + 0.002 +
1.400 + 0.228 +
0.328 + 0.053   = 6.161

df = 3

22. The accompanying data refers to leaf marks found on white clover samples selected from both long-grass areas and short-grass areas ("The Biology of the Leaf Mark Polymorphism in *Trifolium Repens* L., " *Heredity*, 1976, pp. 306–325). Use a χ^2 test to decide whether the true proportions of different marks are identical for the two types of regions.

	L	LL	Y + YL	O	Others	Sample size
Long-grass areas	409	11	22	7	277	726
Short-grass areas	512	4	14	11	220	761

Type of mark spans columns L, LL, Y + YL, O, Others.

23. The following data resulted from an experiment to study the effects of leaf removal on the ability of fruit of a certain type to mature ("Fruit Set, Herbivory, Fruit Reproduction, and the Fruiting Strategy of *Catalpa Speciosa*," *Ecology*, 1980, pp. 57–64):

Treatment	Number of Fruits Matured	Number of Fruits Aborted
Control	141	206
Two leaves removed	28	69
Four leaves removed	25	73
Six leaves removed	24	78
Eight leaves removed	20	82

Does the data suggest that the chance of a fruit maturing is affected by the number of leaves removed? State and test the appropriate hypotheses at level .01.

24. The article "Human Lateralization from Head to Foot: Sex-Related Factors" (*Science*, 1978, pp. 1291–1292) reports for both a sample of right-handed men and a sample of right-handed women the number of individuals whose feet were the same size, had a bigger left than right foot (a difference of half a shoe size or more), or had a bigger right than left foot.

	L > R	L = R	L < R	Sample size
Men	2	10	28	40
Women	55	18	14	87

Does the data indicate that gender has a strong effect on the development of foot asymmetry? State the appropriate null and alternative hypotheses, compute the value of χ^2, and place a bound or bounds on the *P*-value.

25. The paper "Susceptibility of Mice to Audiogenic Seizure Is Increased by Handling Their Dams During Gestation" (*Science*, 1976, pp. 427–428) reports on research into the effect of different injection treatments on the frequencies of audiogenic seizures.

Treatment	No response	Wild running	Clonic seizure	Tonic seizure
Thienylalanine	21	7	24	44
Solvent	15	14	20	54
Sham	23	10	23	48
Unhandled	47	13	28	32

Does the data suggest that the true percentages in the different response categories depend on the nature of the injection treatment? State and test the appropriate hypotheses using $\alpha = .005$.

26. The accompanying data on sex combinations of two recombinants resulting from six different male genotypes appears in the article "A New Method for Distinguishing Between Meiotic and Premeiotic Recombinational Events in *Drosophila Melonogaster*" (*Genetics*, 1979, pp. 543–554). Does the data support the hypothesis that the frequency distribution among the three sex combinations is homogeneous with respect to the different genotypes? Define the parameters of interest, state the appropriate H_0 and H_a, and perform the analysis.

		M/M	M/F	F/F
	1.	35	80	39
	2.	41	84	45
Male	3.	33	87	31
genotype	4.	8	26	8
	5.	5	11	6
	6.	30	65	20

Sex combination

27. Each of 325 individuals participating in a certain drug program was categorized both with respect to the presence or absence of hypoglycemia and with respect to mean daily dosage of insulin ("Relation of Body Weight and Insulin Dose to the Frequency of Hypoglycemia," *J. Amer. Medical Assn.*, 1974, pp. 192–194). Does the accompanying data support the claim that the presence/absence of hypoglycemia is independent of insulin dosage? Test using $\alpha = .05$.

Mean daily insulin dose

		<.25	.25–.49	.50–.74	.75–.99	≥1.0
Hypo-glycemia condition	Present	4	21	28	15	12
	Absent	40	74	59	26	46

28. A random sample of individuals who drive alone to work in a large metropolitan area was obtained, and each individual was categorized with respect to both size of car and commuting distance. Does the accompanying data suggest that commuting distance and size of car are related in the population sampled? State the appropriate hypotheses and use a level .05 chi-squared test.

Commuting distance

		0–<10	10–<20	≥20
	Subcompact	6	27	19
Size of car	Compact	8	36	17
	Midsize	21	45	33
	Full-size	14	18	6

29. Each individual in a random sample of high school and college students was cross-classified with respect to both political views and marijuana usage, resulting in the data displayed in the accompanying two-way table ("Attitudes About Marijuana and Political Views," *Psychological Reports*, 1973, pp. 1051–1054). Does the data support the hypothesis that political views and marijuana usage level are independent within the population?

Test the appropriate hypotheses using level of significance .01.

Usage level

		Never	Rarely	Frequently
	Liberal	479	173	119
Political views	Conservative	214	47	15
	Other	172	45	85

30. Show that the chi-squared statistic for the test of independence can be written in the form

$$\chi^2 = \sum_{i=1}^{I} \sum_{j=1}^{J} \left(\frac{N_{ij}^2}{\hat{E}_{ij}} \right) - n$$

Why is this formula more efficient computationally than the defining formula for χ^2?

31. Suppose that in Exercise 29 each student had been categorized with respect to political views, marijuana usage, and religious preferénce, with the categories of this latter factor being Protestant, Catholic, and other. The data could be displayed in three different two-way tables, one corresponding to each category of the third factor. With $p_{ijk} = P$(political category i, marijuana category j, and religious category k), the null hypothesis of independence of all three factors states that $p_{ijk} = p_{i\cdot\cdot}p_{\cdot j\cdot}p_{\cdot\cdot k}$. Let n_{ijk} denote the observed frequency in cell (i, j, k). Show how to estimate the expected cell counts assuming that H_0 is true ($\hat{e}_{ijk} = n\hat{p}_{ijk}$, so the $\hat{p}_{ijk}$'s must be determined). Then use the general rule of thumb to determine the number of degrees of freedom for the chi-squared statistic.

32. Suppose that in a particular state consisting of four distinct regions, a random sample of n_k voters is obtained from the kth region for $k = 1, 2, 3, 4$. Each voter is then classified according to which of candidates 1, 2, or 3 he/she prefers and according to voter registration (1 = Dem., 2 = Rep., 3 = Indep.). Let p_{ijk} denote the proportion of voters in region k who belong in candidate category i and registration category j. The null hypothesis of homogeneous regions is H_0: $p_{ij1} = p_{ij2} = p_{ij3} = p_{ij4}$ for all i, j (i.e., the proportion within each candidate/registration combination is the same for all four regions). Assuming that H_0 is true, determine $\hat{p}_{ijk}$ and $\hat{e}_{ijk}$ as functions of the observed n_{ijk}'s and use the general rule of thumb to obtain the number of degrees of freedom for the chi-squared test.

33. Consider the accompanying 2×3 table displaying the sample proportions that fell in the various combinations of categories (e.g., 13% of those in the sample were in the first category of both factors).

	1	2	3
1	.13	.19	.28
2	.07	.11	.22

a. Suppose the sample consisted of $n = 100$ people. Use the chi-squared test for independence with significance level .10.

b. Repeat part (a) assuming that the sample size was $n = 1000$.

c. What is the smallest sample size n for which these observed proportions would result in rejection of the independence hypothesis?

Supplementary Exercises Chapter 14 (34–42)

34. The paper "Birth Order and Political Success" (*Psych. Reports,* 1971, pp. 1239–1242) reports that among 31 randomly selected candidates for political office who came from families with four children, 12 were firstborn, 11 were middleborn, and 8 were lastborn. Use this data to test the null hypothesis that a political candidate from such a family is equally likely to be in any one of the four ordinal positions.

35. The results of an experiment to assess the effect of crude oil on fish parasites are described in the paper "Effects of Crude Oils on the Gastrointestinal Parasites of Two Species of Marine Fish" (*J. Wildlife Diseases,* 1983, pp. 253–258). Three treatments (corresponding to populations in the procedure described) were compared: (1) no contamination, (2) contamination by 1-year-old weathered oil, and (3) contamination by new oil. For each treatment condition, a sample of fish was taken, and then each fish was classified as either parasitized or not parasitized. Data compatible with that in the paper is given. Does the data indicate that the three treatments differ with respect to the true proportion of parasitized and nonparasitized fish? Test using $\alpha = .01$.

Treatment	Parasitized	Nonparasitized
Control	30	3
Old oil	16	8
New oil	16	16

36. Qualifications of male and female head and assistant college athletic coaches were compared in the paper "Sex Bias and the Validity of Believed Differences Between Male and Female Interscholastic Athletic Coaches" (*Research Quarterly for Exercise and Sport,* 1990, pp. 259–267). Each person in random samples of 2225 male coaches and 1141 female coaches was classified according to number of years of coaching experience to obtain the accompanying two-way table. Is there enough evidence to conclude that the proportions falling into the experience categories are different for men and women? Use $\alpha = .01$.

	Years of Experience				
Gender	1–3	4–6	7–9	10–12	13+
Male	202	369	482	361	811
Female	230	251	238	164	258

37. The authors of the article "Predicting Professional Sports Game Outcomes from Intermediate Game Scores" (*Chance,* 1992, pp. 18–22) used a chi-squared test to determine whether there was any merit to the idea that basketball games are not settled until the last quarter, whereas baseball games are over by the seventh inning. They also considered football and hockey. Data was collected for 189 basketball games, 92 baseball games, 80 hockey games, and 93 football games. The games analyzed were sampled randomly from all games played during the 1990 season for baseball and football and for the 1990–1991 season for basketball and hockey. For each game, the late-game leader was determined, and then it was noted whether the late-game leader actually ended up winning the game. The resulting data is summarized in the accompanying table.

Sport	Late-game Leader Wins	Late-game Leader Loses
Basketball	150	39
Baseball	86	6
Hockey	65	15
Football	72	21

The authors state that "*Late-game leader* is defined as the team that is ahead after three quarters in basketball and football, two periods in hockey, and seven innings in baseball. The chi-square value on three degrees of freedom is 10.52 ($P < .015$)."

a. State the relevant hypotheses and reach a conclusion using $\alpha = .05$.

b. Do you think that your conclusion in part (a) can be attributed to a single sport being an anomaly?

38. The accompanying two-way frequency table appears in the paper "Marijuana Use in College" (*Youth and Society,* 1979, pp. 323–334). Four hundred and forty-five college students were classified according to both frequency of marijuana use and parental use of alcohol and psychoactive drugs. Does the data suggest that parental usage and student usage are independent in the population from which the sample was drawn? Use the *P*-value method to reach a conclusion.

		Student level of marijuana use	
		Never	Occasional	Regular
Parental	Neither	141	54	40
use of alcohol	One	68	44	51
and drugs	Both	17	11	19

39. In a study of 2989 cancer deaths, the location of death (home, acute-care hospital, or chronic-care facility) and age at death were recorded, resulting in the given two-way frequency table ("Where Cancer Patients Die," *Public Health Reports,* 1983, p. 173). Using a .01 significance level, test the null hypothesis that age at death and location of death are independent.

		Location		
		Home	Acute-care	Chronic-care
	15–54	94	418	23
	55–64	116	524	34
Age	65–74	156	581	109
	Over 74	138	558	238

40. Refer to the data on fatigue crack propagation life given in Exercise 31 in Chapter 1. Carry out a test for normality based on the correlation coefficient for points in a normal probability plot. Use a significance level of .01.

41. Many shoppers have expressed unhappiness over plans by grocery stores to stop putting prices on individual grocery items. The paper "The Impact of Item Price Removal on Grocery Shopping Behavior" (*J. Marketing,* 1980, pp. 73–93) reports on a study in which each shopper in a sample was classified by age and by whether he or she felt the need for item pricing. Based on the accompanying data, does the need for item pricing appear to be independent of age?

	Age				
	<30	30–39	40–49	50–59	≥60
Number in sample	150	141	82	63	49
Number who want item pricing	127	118	77	61	41

42. Let p_1 denote the proportion of successes in a particular population. The test statistic value in Chapter 8 for testing $H_0: p_1 = p_{10}$ was $z = (\hat{p}_1 - p_{10})/\sqrt{p_{10}p_{20}/n}$, where $p_{20} = 1 - p_{10}$. Show that for the case $k = 2$, the chi-squared test statistic value of Section 14.1 satisfies $\chi^2 = z^2$. [HINT: First show that $(n_1 - np_{10})^2 = (n_2 - np_{20})^2$.]

Bibliography

Everitt, B. S., *The Analysis of Contingency Tables,* 2nd ed., Halsted Press, New York, 1992. A compact but informative survey of methods for analyzing categorical data, exposited with a minimum of mathematics.

Fienberg, Stephen. *The Analysis of Cross-Classified Categorical Data,* MIT Press, Cambridge, MA, 1977. A good introduction to the analysis of multidimensional contingency tables.

Mosteller, Frederick, and Richard Rourke, *Sturdy Statistics,* Addison-Wesley, Reading, MA, 1973. Contains several very readable chapters on the varied uses of chi-square.

Introduction

When the underlying population or populations are nonnormal, the t and F tests and t confidence intervals of Chapters 7–13 will in general have actual levels of significance or confidence levels that differ from the nominal levels (those prescribed by the experimenter through the choice of, say, $t_{.025}$, $F_{.01}$, etc.) α and $100(1 - \alpha)\%$, although the difference between actual and nominal levels may not be large when the departure from normality is not too severe. Because the t and F procedures require the distributional assumption of normality, they are not "distribution-free" procedures—alternatively, because they are based on a particular parametric family of distributions (normal), they are not "nonparametric" procedures.

In this chapter, we describe procedures that are valid [actual level α or confidence level $100(1 - \alpha)\%$] simultaneously for many different types of underlying distributions. Such procedures are called **distribution-free** or **nonparametric**. In Sections 15.1 and 15.2, we discuss two test procedures for analyzing a single sample of data; Section 15.3 presents a test procedure for use in two-sample problems. In Section 15.4, we develop distribution-free confidence intervals for μ and $\mu_1 - \mu_2$, while Section 15.5 describes distribution-free ANOVA procedures. These procedures are all competitors of the parametric (t and F) procedures described in earlier chapters, so it is important to compare the performance of the two types of procedures under both normal and nonnormal population models. Generally speaking, the distribution-free procedures perform almost as well as their t and F counterparts on the "home ground" of the normal distribution and will often yield a considerable improvement under nonnormal conditions.

The Sign Test

The sign test is a procedure for testing hypotheses about the median of a continuous distribution. Recall from Chapter 4 that the median $\tilde{\mu}$ of such a distribution is a number such that half the area under the density curve lies to the left of $\tilde{\mu}$ and half lies to the right of $\tilde{\mu}$. If X denotes the random variable (rv) whose distribution is under investigation, then $P(X \le \tilde{\mu}) = P(X \ge \tilde{\mu}) = .5$. The general null hypothesis will have the form H_0: $\tilde{\mu} = \tilde{\mu}_0$. To test H_0 against one of the three commonly encountered alternative hypotheses, we will have available a random sample $X_1, \ldots, X_n$ from the distribution of interest.

In Chapter 8, the t test was used to test hypotheses about the mean of a normal population. Because any normal distribution is symmetric, $\tilde{\mu} = \mu$, so the sign test can also be used to test hypotheses about a normal mean. We will comment shortly on the relative merits of the two test procedures. For now, we note that the sign test is valid whenever the distribution is continuous, whereas the t test was designed specifically for the normal distribution. Because the sign test can be used for a wide variety of underlying distributions, it is often referred to as a distribution-free test.

Testing H_0: $\tilde{\mu} = 0$

When $\tilde{\mu} = 0$, any X_i is equally likely to be positive or negative. If, however, the true value of $\tilde{\mu}$ is much larger than 0, X_i is much more likely to be positive than negative (see Figure 15.1), so we would expect most of the observed X_i's to be positive in this case. When most x_i's are positive, we would suspect that $\tilde{\mu} > 0$ rather than $\tilde{\mu} = 0$.

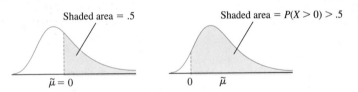

Figure 15.1 (a) $\tilde{\mu} = 0$, (b) $\tilde{\mu} > 0$

Define a test statistic Y by

Y = the number of X_i's for which $X_i > 0$

For testing H_0: $\tilde{\mu} = 0$ versus H_a: $\tilde{\mu} > 0$, the sign test rejects H_0 when $y \ge c$ (when a large number of the x_i's are positive). The constant c should now be chosen to control the probability of a type I error (rejecting H_0 when H_0 is true). Regard the observation of each X_i as constituting a trial, so that the experiment consists of n identical trials. If we now identify a positive X_i with a success and a nonpositive X_i with a failure, then when H_0 is true,

$$p = P(\text{success}) = P(X_i > 0) = P(X_i > \tilde{\mu}) = .5$$

Since the trials are independent, we have the following proposition.

Proposition

> When H_0: $\tilde{\mu} = 0$ is true, the statistic Y has a binomial distribution with parameters n and $p = .5$.

According to this proposition, a critical value c that ensures that P(reject H_0 when H_0 is true) = α for a specified α can be found from the binomial distribution with appropriate n and $p = .5$.

Example 15.1

4/21/97

The use of a particular cloud-seeding technique for inducing rain in a series of 15 independent experiments performed under similar climatic conditions yielded the following observations on the net gain in inches of rain after the seeding: $-.30$, 1.50, $.41$, $-.11$, 2.03, $.28$, $.76$, $-.29$, -1.14, $.09$, $-.92$, $.71$, $.36$, $.25$, $-.50$. Does this data indicate that the cloud-seeding technique is successful in its objective under the given conditions?

Let X represent the net gain in inches resulting from one particular seeding experiment and let $\tilde{\mu}$ denote the median of the distribution of X (assumed continuous). Then we will say that seeding is successful if $\tilde{\mu} > 0$, so that more than 50% of all seeding experiments result in a positive gain in rainfall. Because seeding is expensive, we will be convinced of its validity only if the data strongly indicates that $\tilde{\mu} > 0$, so we test H_0: $\tilde{\mu} = 0$ versus H_a: $\tilde{\mu} > 0$.

The test statistic Y has a binomial distribution with $n = 15$ and $p = .5$ when H_0 is true. From the binomial tables, $P(Y \geq 12) = 1 - B(11;\ 15,\ .5) = .018$ while $P(Y \geq 13) = .004$. Thus, a test with level of significance approximately .02 rejects H_0 if $y \geq 12$. Since only 9 of the 15 x_i's in the sample are positive, the observed value of Y is $y = 9$, which is not in the rejection region. At the chosen level of significance, H_0 cannot be rejected, so we cannot conclude that seeding is successful. The P-value is $P(Y \geq 9) = 1 - B(8;\ 15,\ .5) = .304$. ∎

The test statistic Y has a discrete distribution (binomial), so for any given level α there will not in general be a critical value that yields a test with exactly the desired level. In Example 15.1, there was no value of c yielding exactly level .01, so we used a test with a slightly higher level of significance. When the distribution of a test statistic is discrete, as is usually the case with a distribution-free procedure, the rejection region is chosen to yield an α as close to the desired level as possible.

The Sign Test for H_0: $\tilde{\mu} = \tilde{\mu}_0$

A minor modification of the test procedure for $\tilde{\mu}_0 = 0$ yields a procedure appropriate for any other null value. The new test statistic is the number of positive $(X_i - \tilde{\mu}_0)$'s, or, equivalently, the number of sample observations that exceed $\tilde{\mu}_0$.

4/21/97

> Null hypothesis: H_0: $\tilde{\mu} = \tilde{\mu}_0$
>
> Test statistic value: $y =$ the number of x_i's that exceed $\tilde{\mu}_0$

> | Alternative Hypothesis | Rejection Region |
> |---|---|
> | H_a: $\tilde{\mu} > \tilde{\mu}_0$ | $y \geq c_1$ (large y) |
> | H_a: $\tilde{\mu} < \tilde{\mu}_0$ | $y \leq c_2$ (small y) |
> | H_a: $\tilde{\mu} \neq \tilde{\mu}_0$ | Either $y \geq c$ or $y \leq n - c$ |
> | | (small or large y) |
>
> where the critical values c_1, c_2, and c are obtained from the binomial distribution Bin(n, .5) to yield the desired α.

Example 15.2

An electronics company will shortly begin to market a kit for a CB radio. The company wishes to claim that the kit can be assembled quite easily even by someone with no previous electronics experience. Let X be the assembly time for a randomly selected person of this type, with $f(x)$ denoting the distribution of X and $\tilde{\mu}$ the median of this distribution. It is decided that the claim will be validated only if experimental evidence strongly suggests that at least half of all such individuals can assemble the kit in less than 20 hours. The null hypothesis is then H_0: $\tilde{\mu} = 20$, while the appropriate alternative is H_a: $\tilde{\mu} < 20$; this choice of H_a ensures that the burden of proof lies with the claim of easy assembly.

With a sample size of $n = 15$, the test that rejects H_0 when Y (the number among the 15 assembly times that exceed 20) is 0, 1, 2, or 3 has $\alpha = .018$. For the sample $x_1 = 18.5$, $x_2 = 19.1$, $x_3 = 22.2$, $x_4 = 16.9$, $x_5 = 19.6$, $x_6 = 18.8$, $x_7 = 19.8$, $x_8 = 19.0$, $x_9 = 18.3$, $x_{10} = 19.2$, $x_{11} = 21.6$, $x_{12} = 17.5$, $x_{13} = 18.9$, $x_{14} = 18.0$, and $x_{15} = 19.4$, the test statistic has value $y = 2$. Since 2 falls in the rejection region for a level .018 test, H_0 is rejected at this level in favor of the conclusion that $\tilde{\mu} < 20$. ∎

The Normal Approximation

When $p = .5$, the binomial distribution can be approximated by a normal distribution for n as small as 10. Since $\mu_Y = np = .5n$ and $\sigma_Y^2 = np(1 - p) = .25n$, the test statistic

$$Z = \frac{Y - .5n}{.5\sqrt{n}}$$

has approximately a standard normal distribution when H_0 is true and $n \geq 10$. The upper-tailed test then rejects H_0 in favor of H_a: $\tilde{\mu} > \tilde{\mu}_0$ when $z \geq z_\alpha$, and rejection regions for the other two standard alternatives are identical to those of previous lower- and two-tailed z tests.

Example 15.3

Each of 20 randomly selected homeowners in a particular city was asked to reveal the amount by which his or her property tax bill had increased as a result of the most recent reassessment. The sample (dollar) amounts were

342 176 517 296 312 143 279 412 228 209
195 241 211 285 329 137 188 260 233 357

The county assessor has stated that the median increase for all county homeowners was \$300. Does the sample data strongly indicate that the true median increase for residents of this city was less than that for the county as a whole? Use the sign test with $\alpha = .01$.

The null hypothesis here is H_0: $\tilde{\mu} = 300$, and the alternative is H_a: $\tilde{\mu} < 300$, where $\tilde{\mu}$ is the median increase for city homeowners. The number of sample values that exceed 300 is $y = 6$, so

$$z = \frac{y - .5n}{.5\sqrt{n}} = \frac{6 - (.5)(20)}{.5\sqrt{20}} = -1.79$$

The P-value for a lower-tailed z test is $\Phi(z) = \Phi(-1.79) = .0367$. Since $.01 < .0367$, H_0 cannot be rejected at significance level $.01$. ■

β for the Sign Test

For testing H_0: $\tilde{\mu} = \tilde{\mu}_0$, as long as the underlying distribution is continuous, the sign test will control the probability of a type I error at the desired level α. To assess the ability of the test procedure to draw correct conclusions, we should also examine probabilities of type II errors. The test should be able to detect departures from the null hypothesis as evidenced by small β for alternatives of interest.

The difficulty with β calculations for the sign test is that we must assume not only a particular alternative value of $\tilde{\mu}$ but also a specific shape for the underlying density function. For example, if $\tilde{\mu} = \tilde{\mu}_0 + 1$ rather than $\tilde{\mu}_0$, β will also depend on whether the distribution assumed for the calculation is normal, uniform, exponential, or whatever. However, *for any particular underlying distribution, the test statistic Y is a binomial rv*, but if $\tilde{\mu} \neq \tilde{\mu}_0$ then p will be something other than $.5$. Thus, all β calculations reduce to binomial probability computations.

Example 15.4
(Example 15.1 continued)

Suppose H_0 is false because $\tilde{\mu} = 1$, and consider the two possible distributions of Figure 15.2.

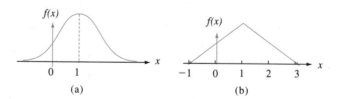

Figure 15.2 Two distributions for which H_0: $\tilde{\mu} = 0$ is false; (a) normal, $\tilde{\mu} = 1$, $\sigma = 1$; (b) symmetric triangular, $\tilde{\mu} = 1$

For the normal distribution, the probability of a positive net change is

$$p = P(X_i > 0) = P(Z > -1) = 1 - \Phi(-1) = .8413$$

Thus, Y has a binomial distribution with $n = 15$ and $p = .8413$. Using the rejection region $\{y \geq 12\} = \{12, 13, 14, 15\}$, the probability of not rejecting H_0 is

$$\beta(1) = P(Y \leq 11 \text{ when } Y \sim \text{Bin}(15, .8413))$$

$$= \sum_{y=0}^{11} \binom{15}{y} (.8413)^y (.1587)^{15-y} = .2050$$

If the underlying distribution is triangular as in Figure 15.2b, then a geometric argument yields $p = .875$, so that

$$\beta(1) = P(Y \leq 11 \text{ when } Y \sim \text{Bin}(15, .875)) = .1079 \qquad \blacksquare$$

These calculations show that β for the sign test depends not just on the alternative value of $\tilde{\mu}$ but also on the area under the density curve to the right of $\tilde{\mu}_0$. This area in turn depends on the particular shape of the density function.

Comparison of the Sign and t Tests

If it is assumed that the underlying distribution is normal, then either the sign test or the t test can be used to test H_0: $\tilde{\mu} = \tilde{\mu}_0$ against one of the usual alternative hypotheses. In this case, the t test is known to have minimum type II error probabilities among all level α tests, so is certainly a better test than the sign test. When the population distribution is symmetric but not normal and μ is finite (so $\tilde{\mu} = \mu$), β for the sign test will be larger than for the t test unless the distribution has very heavy tails compared to the normal distribution. The test procedure to be presented in the next section is better than the sign test and compares very favorably to the t test in symmetric situations. The sign test is thus better regarded as a procedure for drawing conclusions about a population median than as a serious competitor to the t test.

Exercises   Section 15.1 (1–7)

1. Determine the median of
   a. A uniform distribution on the interval $[a, b]$.
   b. An exponential distribution with parameter λ.

2. The *Los Angeles Times* (July 3, 1978) reported that under conditions of the desegregation plan to be implemented, any students who travel more than 45 minutes one way on a bus are required to have only 3 years of integrated schooling, while those traveling at most 45 minutes face a 5-year requirement. Let $\tilde{\mu}$ be the median of the distribution of travel time between two particular locations. Consider the following random sample of 20 travel times.

47.3 44.1 44.5 43.7 46.2 45.3 47.0
42.5 43.9 45.6 46.0 44.8 45.7 47.8
46.6 44.2 47.2 46.5 43.6 45.4

Use the sign test at level .10, with $Y =$ the number of times that exceed 45 minutes, to test H_0: $\tilde{\mu} = 45$ versus H_a: $\tilde{\mu} > 45$.

3. Ten soil samples taken from a particular region were subjected to chemical analysis to determine the pH of each sample. Here are the resulting sample pH's.

5.93 6.08 5.86 5.91 6.12
5.90 5.95 5.89 5.98 5.96

It had previously been believed that the median soil pH in this region was 6.0. Does this sample data strongly indicate that the true median pH is something other than 6.0? Let $Y =$ the number of pH's in the sample that exceed 6.0 and use the sign test at level .05 to test the appropriate hypotheses.

4. Suppose the distribution of times between successive alarms at a fire station is exponential, and we wish to test $H_0: \tilde{\mu} \leq 3$ hours versus $H_a: \tilde{\mu} > 3$ hours, where $\tilde{\mu}$ is the median time between successive alarms. Suppose a sample of ten such times is obtained, let $Y =$ the number of times that exceed 3 hours, and consider using the sign test with rejection region $\{8, 9, 10\}$.

a. What is the probability of a type I error when $\lambda = .231$? [λ is the parameter of the exponential distribution, which is related to $\tilde{\mu}$ by the result of Exercise 1b.]

b. What is the probability of a type II error if $\lambda = .1189$?

5. In the manufacture of a certain type of cylindrical rod, a lathe is periodically reset to the desired diameter of the rod. A production supervisor has recently become concerned over a possible tendency for the lathes to produce rods with larger and larger diameters immediately subsequent to setting (an upward trend in observed diameters). A sample of 12 rods is obtained, and the measured diameters are the following:

$x_1 = 4.935, \quad x_2 = 4.967, \quad x_3 = 4.989$
$x_4 = 5.012, \quad x_5 = 5.020, \quad x_6 = 4.997$
$x_7 = 5.025, \quad x_8 = 5.041, \quad x_9 = 5.008$
$x_{10} = 5.035, \quad x_{11} = 5.016, \quad x_{12} = 5.022$

a. Let $p =$ the probability that the $(i + 6)$th diameter X_{i+6} exceeds the ith diameter X_i. If there is actually no upward or downward trend in diameters, so that X_{i+6} is no more or less likely to exceed X_i than it is to fall below X_i, what is the value of p?

b. If $Y =$ the number of values of i for which $X_{i+6} > X_i$, and there is no upward or downward trend in diameters, what probability distribution does Y have?

c. Use the result of part (b) to test H_0: there is no trend versus H_a: there is an upward trend. Your test statistic should be Y, and you should use a level .1 (approximately) test. This simple test for trends, a modification of the sign test, is due to Cox and Stuart.

6. Suppose the underlying population is normal with $\sigma = 1$. For testing $H_0: \mu = 0$ versus $H_a: \mu > 0$ based on a sample of size $n = 15$, consider the z test that rejects H_0 if $z \geq 2.10$ (where $Z = (\bar{X} - \mu_0)/(\sigma/\sqrt{n}) = \sqrt{n}\bar{X}$).

a. What is α for this test?

b. When $\mu = 1$, how does β for this test compare to β for the sign test that rejects H_0 if $y \geq 12$?

7. A process for producing stainless steel bars was originally designed so that the nickel content of the bars was 10% by weight. A potential customer believes that the bars will be suitable for a particular application only if more than 75% of all bars produced contain at least 10% nickel (i.e., that fewer than 25% of all bars contain less than 10% nickel). The customer will not agree to purchase the bars unless experimental evidence strongly indicates that this requirement is satisfied.

If $f(x)$ denotes the density function of nickel content in a randomly selected bar, then the requirement is met only if the area under the graph of $f(x)$ to the right of 10 is more than .75. Equivalently, if θ denotes the 25th percentile of the nickel-content distribution, then we wish to test $H_0: \theta = 10$ versus $H_a: \theta > 10$ (so H_a: states that more than 75% of the area under the density curve lies to the right of 10). Suppose 20 bars are sampled and the observed values of nickel content are as follows:

10.23 10.95 11.02 10.14 9.83 10.30 12.05
11.26 10.58 10.18 10.38 11.49 9.52 10.86
11.00 10.19 10.94 11.13 10.25 11.02

Develop a test procedure for H_0 versus H_a and carry out the test for this data. Your α should be as close to .10 as possible. [HINT: Let $Y =$ the number of X_i's that exceed 10. If H_0 is true, what is the distribution of Y?]

15.2   The Wilcoxon Signed-Rank Test

Consider again the problem of testing a hypothesis about the median $\tilde{\mu}$ of a continuous distribution (i.e., population). If we wish to test $H_0: \tilde{\mu} = 0$ versus H_a: $\tilde{\mu} > 0$ using the sign test on the basis of ten observations from the population, then the rejection region $R = \{8, 9, 10\}$ has type I error probability $\alpha = .055$. Suppose our sample consists of the ordered observations $-.57$, $-.19$, $-.05$, $.76$, 1.30, 2.02, 2.17, 2.46, 2.68, and 3.02. Then the observed value of the test statistic Y, the number of positive observations, is $y = 7$, so at level $\alpha = .055$ we would not reject H_0 in favor of H_a.

A closer look at the data reveals that, although there are indeed three negative observations, these three observations are quite small in magnitude compared with the seven positive observations. The nonrejection of H_0 for this data is due to the way in which the sign test extracts information from the sample; magnitudes of the observations are discarded, and only the sign of each observation is retained. In a sense, this is the price that must be paid for having a test that is valid for a very large class of populations (*all* continuous populations).

Symmetric Distributions and a New Test Statistic

Frequently, an experimenter may be willing to assume more about the population of interest than just that it is continuous and, in particular, may be willing to assume that the population distribution has a symmetric shape. A population distribution is **symmetric** if the graph of the pdf to the left of $\tilde{\mu}$ is a reflection or mirror image of the graph to the right of $\tilde{\mu}$. The graphs of several symmetric pdf's are pictured in Figure 15.3.

Figure 15.3 Symmetric pdf's

If we assume that the population yielding the above sample is symmetric, then the conclusion that $\tilde{\mu} = 0$ (actually ≤ 0) seems quite suspect, for the data suggests strongly that the point of symmetry lies to the right of 0 ($\tilde{x} = 1.66$, which is much larger in magnitude than any of the negative observations). We now assume that *our sample comes from a continuous symmetric probability distribution*, and develop a test for making an inference about $\tilde{\mu}$ that is different from the sign test. When this new test is applied to the above sample, the data will argue strongly for the rejection of H_0 in favor of H_a. This is because the new test will take into account the magnitudes of the observations as well as the signs.

Figure 15.4 shows two different symmetric pdf's, one for which H_0 is true and one for which H_a is true. When H_0 is true, we expect the magnitudes of the negative

(a)   (b)

Figure 15.4

observations in the sample to be comparable to the magnitudes of the positive observations. If, however, H_0 is "grossly" untrue (Figure 15.4b), then observations of large absolute magnitude will tend to be positive rather than negative.

For the ten observations in our sample, suppose we proceed as follows:

1. Disregarding the signs of the observations, rank the collection of ten numbers in order of increasing magnitude.

2. Use as a test statistic S_+ = sum of the ranks associated with the positive observations.

Absolute magnitude	.05	.19	.57	.76	1.30	2.02	2.17	2.46	2.68	3.02
Rank	1	2	3	4	5	6	7	8	9	10
Signed rank	−1	−2	−3	4	5	6	7	8	9	10

The observed value of S_+ is $s_+ = 4 + 5 + 6 + 7 + 8 + 9 + 10 = 49$. H_0 is now rejected in favor of H_a: $\tilde{\mu} > 0$ when s_+ is too large because a large value of s_+ indicates that most of the observations with large absolute magnitude are positive, which in turn indicates a median greater than 0. If, on the other hand, observations with negative signs are intermingled with positive observations when signs are disregarded, then the observed s_+ will not be very large.

The Distribution of S_+

To decide between H_0 and H_a, we must know whether 49 is a "surprisingly large" value of S_+ when the null hypothesis is true. This entails obtaining the probability distribution of S_+ when H_0 is true and selecting as the rejection region a set of large s_+ values with small probability under H_0 (small α). If our observed s_+ is in the rejection region, we can then reject H_0 at the desired level α, and otherwise not reject H_0.

The probability distribution of S_+ is more difficult to obtain than was the distribution for the sign test. To illustrate, consider the case $n = 5$. The key observation in the derivation is this: When H_0 is true, *any* collection of signed ranks is *just as likely* as any other collection. Worded differently, the smallest observation in absolute magnitude is just as likely to be positive as negative, the second smallest observation in absolute magnitude is just as likely to be positive as negative, and similarly for the other observations. Thus, for $n = 5$, the sequence −1, +2, +3, −4, +5 of signed ranks is just as likely to be observed (when H_0 is true) as is the sequence −1, −2, −3, −4, +5, or any other sequence. Table 15.1 lists the possible signed rank sequences for $n = 5$ and the s_+ value associated with each sequence. Since there are 32 such sequences (two possible signs for the smallest observation, two for the second smallest, and so on, or $2 \cdot 2 \cdot 2 \cdot 2 \cdot 2 = 32$ sequences), each sequence has probability 1/32 of being observed when H_0 is true. From this the pmf of S_+ under H_0 can immediately be obtained as in Table 15.2. Note that for $n = 10$ observations

Table 15.1 Possible signed rank sequences for $n = 5$

Sequence					s_+	Sequence					s_+
−1	−2	−3	−4	−5	0	−1	−2	−3	+4	−5	4
+1	−2	−3	−4	−5	1	+1	−2	−3	+4	−5	5
−1	+2	−3	−4	−5	2	−1	+2	−3	+4	−5	6
−1	−2	+3	−4	−5	3	−1	−2	+3	+4	−5	7
+1	+2	−3	−4	−5	3	+1	+2	−3	+4	−5	7
+1	−2	+3	−4	−5	4	+1	−2	+3	+4	−5	8
−1	+2	+3	−4	−5	5	−1	+2	+3	+4	−5	9
+1	+2	+3	−4	−5	6	+1	+2	+3	+4	−5	10
−1	−2	−3	−4	+5	5	−1	−2	−3	+4	+5	9
+1	−2	−3	−4	+5	6	+1	−2	−3	+4	+5	10
−1	+2	−3	−4	+5	7	−1	+2	−3	+4	+5	11
−1	−2	+3	−4	+5	8	−1	−2	+3	+4	+5	12
+1	+2	−3	−4	+5	8	+1	+2	−3	+4	+5	12
+1	−2	+3	−4	+5	9	+1	−2	+3	+4	+5	13
−1	+2	+3	−4	+5	10	−1	+2	+3	+4	+5	14
+1	+2	+3	−4	+5	11	+1	+2	+3	+4	+5	15

Table 15.2 Null distribution of S_+ when $n = 5$

s_+	0	1	2	3	4	5	6	7
$p(s_+)$	1/32	1/32	1/32	2/32	2/32	3/32	3/32	3/32
s_+	8	9	10	11	12	13	14	15
$p(s_+)$	3/32	3/32	3/32	2/32	2/32	1/32	1/32	1/32

there are $2^{10} = 1024$ possible signed rank sequences, and to list these sequences would be very tedious. Each sequence, though, would receive probability 1/1024 under H_0, from which the null distribution (distribution when H_0 is true) of S_+ can be easily obtained.

As an example of how the probabilities in Table 15.2 were obtained, consider the S_+ value 7; there are three signed rank sequences for which $S_+ = 7$, and each receives probability 1/32 under H_0, so that $P(S_+ = 7$ when H_0 is true$) = 3/32$.

Table 15.2 can now be used to specify a rejection region for H_0: $\tilde{\mu} = 0$ versus H_a: $\tilde{\mu} > 0$ for which α can be calculated. Consider the rejection region $R = \{13, 14, 15\}$. Then

$$\alpha = P(\text{reject } H_0 \text{ when } H_0 \text{ is true})$$
$$= P(S_+ = 13, 14, \text{ or } 15 \text{ when } H_0 \text{ is true})$$
$$= 1/32 + 1/32 + 1/32 = 3/32$$
$$= .094$$

so that $R = \{13, 14, 15\}$ specifies a test with approximate level .1. For the rejection region $\{14, 15\}$, $\alpha = 2/32 = .063$. For the data set $x_1 = .58$, $x_2 = 2.50$, $x_3 = −.21$,

$x_4 = 1.23$, $x_5 = .97$, the signed rank sequence is $-1, +2, +3, +4, +5$, so $s_+ = 14$ and at level .063 H_0 would be rejected.

A General Description of the Wilcoxon Signed-Rank Test

Because the underlying distribution is assumed symmetric, $\mu = \tilde{\mu}$, so we will state the hypotheses of interest in terms of μ rather than $\tilde{\mu}$.*

Assumption

> $X_1, X_2, \ldots, X_n$ is a random sample from a continuous and symmetric probability distribution with mean (and median) μ.

When the hypothesized value of μ is μ_0, the absolute differences $|x_1 - \mu_0|, \ldots, |x_n - \mu_0|$ must be ranked from smallest to largest.

> Null hypothesis: H_0: $\mu = \mu_0$
>
> Test statistic value: $s_+ =$ the sum of the ranks associated with positive $(x_i - \mu_0)$'s
>
> | Alternative Hypothesis | Rejection Region for Level α Test |
> |---|---|
> | H_a: $\mu > \mu_0$ | $s_+ \geq c_1$ |
> | H_a: $\mu < \mu_0$ | $s_+ \leq c_2$ [where $c_2 = n(n+1)/2 - c_1$] |
> | H_a: $\mu \neq \mu_0$ | Either $s_+ \geq c$ or $s_+ \leq n(n+1)/2 - c$ |
>
> where the critical values c_1 and c are obtained from Appendix Table A.9 and satisfy $P(S_+ \geq c_1) \approx \alpha$ and $P(S_+ \geq c) \approx \alpha/2$ when H_0 is true.

Example 15.5

A manufacturer of electric irons, wishing to test the accuracy of the thermostat control at the 500°F setting, instructs a test engineer to obtain actual temperatures at that setting for 15 irons using a thermocouple. The resulting measurements are as follows:

494.6 510.8 487.5 493.2 502.6 485.0 495.9 498.2
501.6 497.3 492.0 504.3 499.2 493.5 505.8

The engineer believes it is reasonable to assume that a temperature deviation from 500° of any particular magnitude is just as likely to be positive as negative (the assumption of symmetry) but wants to protect against possible nonnormality of the actual temperature distribution, so she decides to use the Wilcoxon signed-rank test to see whether the data strongly suggests incorrect calibration of the iron.

The hypotheses are H_0: $\mu = 500$ versus H_a: $\mu \neq 500$, where $\mu =$ the true average actual temperature at the 500°F setting. Subtracting 500 from each x_i gives

-5.6, 10.8, -12.5, -6.8, 2.6, -15.0, -4.1, -1.8, 1.6, -2.7,
-8.0, 4.3, $-.8$, -6.5, 5.8

---

*If the tails of the distribution are "too heavy," as was the case with the Cauchy distribution mentioned in Chapter 6, then μ will not exist. In such cases, the Wilcoxon test will still be valid for tests concerning $\tilde{\mu}$.

The ranks are obtained by ordering these from smallest to largest without regard to sign:

Absolute magnitude	.8	1.6	1.8	2.6	2.7	4.1	4.3	5.6	5.8	6.5	6.8	8.0	10.8	12.5	15.0
Rank	1	2	3	4	5	6	7	8	9	10	11	12	13	14	15
Sign	−	+	−	+	−	−	+	−	+	−	−	−	+	−	−

Thus, $s_+ = 2 + 4 + 7 + 9 + 13 = 35$. From Appendix Table A.9, $P(S_+ \geq 95) = P(S_+ \leq 25) = .024$ when H_0 is true, so the two-tailed test with approximate level .05 rejects H_0 when either $s_+ \geq 95$ or ≤ 25 [the exact α is $2(.024) = .048$]. Since $s_+ = 35$ is not in the rejection region, it cannot be concluded at level .05 that μ is anything other than 500. Even at level .094 (approximately .1), H_0 is not rejected, since $P(S_+ \leq 30) = .047$ implies that s_+ values between 30 and 90 are not significant at that level. The P-value of the data is thus greater than .1. ■

Paired Observations

When the data consisted of pairs $(X_1, Y_1), \ldots, (X_n, Y_n)$ and the differences $D_1 = X_1 - Y_1, \ldots, D_n = X_n - Y_n$ were normally distributed, in Chapter 9 we used a paired t test to test hypotheses about the expected difference μ_D. If normality is not assumed, hypotheses about μ_D can be tested by using the Wilcoxon signed-rank test on the D_i's provided that the distribution of the differences is continuous and symmetric. *If X_i and Y_i both have continuous distributions that differ only with respect to their means* (so the Y distribution is the X distribution shifted by $\mu_1 - \mu_2 = \mu_D$), *then D_i will have a continuous symmetric distribution* (it is *not* necessary for the X and Y distributions to be symmetric individually). The null hypothesis is $H_0: \mu_D = \Delta_0$, and the test statistic S_+ is the sum of the ranks associated with the positive $(D_i - \Delta_0)$'s.

Example 15.6   An article in a 1949 issue of *Journal of Applied Physiology* ("Interval of Useful Consciousness at Various Altitudes") reported on the effect of a transfusion of a large amount of blood on the ability to perform a simple task at high altitude. The high-altitude condition (35,000 ft) was achieved by using a gas mask. Suppose it is

Table 15.3 Data for Example 15.6

Subject	1	2	3	4	5
Seconds of useful consciousness before transfusion	38	65	68	59	67
Seconds of useful consciousness after transfusion	45	77	81	70	73
d_i	−7	−12	−13	−11	−6
$d_i - (-5)$	−2	−7	−8	−6	−1
Rank of $\lvert d_i - (-5) \rvert$	2	4	5	3	1
Sign	−	−	−	−	−

believed that more blood would increase an individual's interval of useful consciousness following interruption of normal oxygen supply by an average of at most 5 seconds. Does the (slightly modified) data in Table 15.3 contradict this belief?

Because a 5-second increase corresponds to $\mu_D = -5$, we wish to test H_0: $\mu_D = -5$ versus H_a: $\mu_D < -5$. Since all signs are negative, the computed value of S_+ is $s_+ = 0$. From Appendix Table A.9 or Table 15.2, the test that rejects H_0 when $s_+ = 0$ has level .031, so the value 0 is significant at level .05. H_0 is thus rejected in favor of the conclusion that a transfusion increases time of useful consciousness by more than 5 seconds on average. ∎

Ties in the Wilcoxon Signal-Rank Test

Although a theoretical implication of the continuity of the underlying distribution is that ties will not occur, in practice they often do because of the discreteness of measuring instruments. If there are several data values with the same absolute magnitude, then they would be assigned the average of the ranks they would receive if they differed very slightly from one another. For example, if in Example 15.5 $x_8 = 498.2$ is changed to 498.4, then two different $(x_i - 500)$'s would have absolute magnitude 1.6. The ranks to be averaged would be 2 and 3, so each would be assigned rank 2.5.

A Large-Sample Approximation

Appendix Table A.9 provides critical values for level α tests only when $n \leq 20$. For $n > 20$, it can be shown that S_+ has approximately a normal distribution with

$$\mu_{S_+} = \frac{n(n+1)}{4}, \quad \sigma^2_{S_+} = \frac{n(n+1)(2n+1)}{24}$$

when H_0 is true.

The mean and variance result from noting that when H_0 is true (the symmetric distribution is centered at μ_0), then the rank i is just as likely to receive a + sign as it is to receive a − sign. Thus,

$$S_+ = W_1 + W_2 + W_3 + \cdots + W_n$$

where

$$W_1 = \begin{cases} 1 & \text{with probability .5} \\ 0 & \text{with probability .5} \end{cases} \cdots W_n = \begin{cases} n & \text{with probability .5} \\ 0 & \text{with probability .5} \end{cases}$$

($W_i = 0$ is equivalent to rank i being associated with a −, so i does not contribute to S_+.)

S_+ is then a sum of random variables, and when H_0 is true, these W_i's can be shown to be independent. Application of the rules of expected value and variance gives the mean and variance of S_+. Because the W_i's are not identically distributed, our version of the Central Limit Theorem cannot be applied, but there is a more general version of the theorem that can be used to justify the normality conclusion.

The large-sample test statistic is now given by

$$Z = \frac{S_+ - n(n + 1)/4}{\sqrt{n(n + 1)(2n + 1)/24}} \qquad (15.1)$$

For the three standard alternatives, the critical values for level α tests are the usual standard normal values z_α, $-z_\alpha$, and $\pm z_{\alpha/2}$.

Example 15.7   A particular type of steel beam has been designed to have a compressive strength (lb/in.^{2}) of at least 50,000. For each beam in a sample of 25 beams, the compressive strength was determined and is given in Table 15.4. Assuming that actual compressive strength is distributed symmetrically about the true average value, use the Wilcoxon test to decide if the true average compressive strength is less than the specified value. That is, test H_0: $\mu = 50,000$ versus H_a: $\mu < 50,000$ (favoring the claim that average compressive strength is at least 50,000).

Table 15.4 Data for Example 15.7

$x_i - 50,000$	Signed Rank	$x_i - 50,000$	Signed Rank	$x_i - 50,000$	Signed Rank
−10	−1	−99	−10	165	+18
−27	−2	113	+11	−178	−19
36	+3	−127	−12	−183	−20
−55	−4	−129	−13	−192	−21
73	+5	136	+14	−199	−22
−77	−6	−150	−15	−212	−23
−81	−7	−155	−16	−217	−24
90	+8	−159	−17	−229	−25
−95	−9				

The sum of the positively signed ranks is $3 + 5 + 8 + 11 + 14 + 18 = 59$, $n(n + 1)/4 = 162.5$, and $n(n + 1)(2n + 1)/24 = 1381.25$, so

$$z = \frac{59 - 162.5}{\sqrt{1381.25}} = -2.78$$

The lower-tailed level .01 test rejects H_0 if $z \le -2.33$. Since $-2.78 \le -2.33$, H_0 is rejected in favor of the conclusion that true average compressive strength is less than 50,000. ∎

When there are ties in the absolute magnitudes, so that average ranks must be used, it is still correct to standardize S_+ by subtracting $n(n + 1)/4$, but the following corrected formula for variance should be used:

$$\sigma_{S_+}^2 = \frac{1}{24}n(n + 1)(2n + 1) - \frac{1}{48}\Sigma(\tau_i - 1)(\tau_i)(\tau_i + 1) \tag{15.2}$$

where τ_i is the number of ties in the ith set of tied values and the sum is over all sets of tied values. If, for example, $n = 10$ and the signed ranks are 1, 2, −4, −4, 4, 6, 7, 8.5, 8.5, and 10, then there are two tied sets with $\tau_1 = 3$ and $\tau_2 = 2$, so the summation is $(2)(3)(4) + (1)(2)(3) = 30$ and $\sigma_+^2 = 96.25 - 30/48 = 95.62$. The denominator in (15.1) should be replaced by the square root of (15.2), though as this example shows, the correction is usually insignificant.

Efficiency of the Wilcoxon Signed-Rank Test

When the underlying distribution being sampled is normal, either the t test or the signed-rank test can be used to test a hypothesis about μ. The t test is the best test in such a situation because among all level α tests it is the one having minimum β. Since it is generally agreed that there are many experimental situations in which normality can be reasonably assumed, as well as some in which it should not be, there are two questions that must be addressed in an attempt to compare the two tests:

1. When the underlying distribution is normal (the "home ground" of the t test), how much is lost by using the signed-rank test?

2. When the underlying distribution is not normal, can a significant improvement be achieved by using the signed-rank test?

If the Wilcoxon test does not suffer much with respect to the t test on the "home ground" of the latter, and performs significantly better than the t test for a large number of other distributions, then there will be a strong case for using the Wilcoxon test.

Unfortunately, there are no simple answers to the two questions. Upon reflection, it is not surprising that the t test can perform poorly when the underlying distribution has "heavy tails" (i.e., when observed values lying far from μ are relatively more likely than they are when the distribution is normal). This is because the t-test behavior depends on the sample mean, which can be very unstable in the presence of heavy tails. The difficulty in producing answers to the two questions is that β for the Wilcoxon test is very difficult to obtain and study for *any* underlying distribution, and the same can be said for the t test when the distribution is not normal. Even if β were easily obtained, any measure of efficiency would clearly depend on which underlying distribution was postulated. A number of different efficiency measures have been proposed by statisticians; one that many statisticians regard as credible is called **asymptotic relative efficiency** (ARE). The ARE of one test with respect to another is essentially the limiting ratio of sample sizes necessary to obtain identical error probabilities for the two tests. Thus, if the ARE of one test with respect to a second equals .5, then when sample sizes are large, twice as large a sample size will be required of the first test in order that it perform as well as the second test. Although the ARE does not characterize test performance for small sample sizes, the following results can be shown to hold:

1. When the underlying distribution is normal, the ARE of the Wilcoxon test with respect to the *t* test is approximately .95.

2. For any distribution, the ARE will be at least .86 and for many distributions will be much greater than 1.

We can summarize these results by saying that in large-sample problems, the Wilcoxon test is never very much less efficient than the *t* test and may be much more efficient if the underlying distribution is far from normal. Though the issue is far from resolved in the case of sample sizes obtained in most practical problems, studies have shown that the Wilcoxon test performs reasonably and is thus a viable alternative to the *t* test.

Exercises   Section 15.2 (8–14)

8. Reconsider the problem described in Example 15.2 and use the Wilcoxon test with $\alpha = .01$ to test the given hypotheses.

9. Use the Wilcoxon test to analyze the data given in Exercise 3.

10. The accompanying data is a subset of the data reported in the paper "Synovial Fluid pH, Lactate, Oxygen and Carbon Dioxide Partial Pressure in Various Joint Diseases" (*Arthritis and Rheumatism,* 1971, pp. 476–477). The observations are pH values of synovial fluid (which lubricates joints and tendons) taken from the knees of individuals suffering from arthritis. Assuming that true average pH for nonarthritic individuals is 7.39, test at level .05 to see whether the data indicates a difference between average pH values for arthritic and nonarthritic individuals.

7.02 7.35 7.34 7.17 7.28 7.77 7.09
7.22 7.45 6.95 7.40 7.10 7.32 7.14

11. Both a gravimetric and a spectrophotometric method are under consideration for determining phosphate content of a particular material. Twelve samples of the material are obtained, each is split in half, and a determination is made on each half using one of the two methods, resulting in the following data.

Sample	1	2	3	4
Gravimetric	54.7	58.5	66.8	46.1
Spectrophotometric	55.0	55.7	62.9	45.5

Sample	5	6	7	8
Gravimetric	52.3	74.3	92.5	40.2
Spectrophotometric	51.1	75.4	89.6	38.4

Sample	9	10	11	12
Gravimetric	87.3	74.8	63.2	68.5
Spectrophotometric	86.8	72.5	62.3	66.0

Use the Wilcoxon test to decide whether one technique gives on average a different value than the other technique for this type of material.

12. Analyze the data in Exercise 2 using the large-sample version of the Wilcoxon procedure and compare the result with an analysis using a critical value from Appendix Table A.9.

13. The accompanying 25 observations on fracture toughness of base plate of 18% nickel maraging steel were reported in the paper "Fracture Testing of Weldments" (*ASTM Special Publ. No. 381,* 1965, pp. 328–356). Suppose a company will agree to purchase this steel for a particular application only if it can be strongly demonstrated from experimental evidence that true average toughness exceeds 75. Assuming that the fracture toughness distribution is symmetric, state and test the appropriate hypotheses at level .05 and compute a *P*-value.

69.5 71.9 72.6 73.1 73.3 73.5 74.1 74.2 75.3
75.5 75.7 75.8 76.1 76.2 76.2 76.9 77.0 77.9
78.1 79.6 79.7 80.1 82.2 83.7 93.7

14. Suppose that observations $X_1, X_2, \ldots, X_n$ are made on a process at times $1, 2, \ldots, n$. On the basis of this data, we wish to test

H_0: the X_i's constitute an independent and identically distributed sequence

versus

H_a: X_{i+1} tends to be larger than X_i for $i = 1, \ldots, n$ (an increasing trend)

Suppose the X_i's are ranked from 1 to n. Then when H_a is true, larger ranks tend to occur later in the sequence,

while if H_0 is true, large and small ranks tend to be mixed together. Let R_i be the rank of X_i and consider the test statistic $D = \sum_{i=1}^{n} (R_i - i)^2$. Then small values of D give support to H_a (e.g., the smallest value is 0 for $R_1 = 1, R_2 = 2, \ldots, R_n = n$), so H_0 should be rejected in favor of H_a if $d \le c$. When H_0 is true, any sequence of ranks has probability $1/n!$. Use this to find c for which the test has level as close to .10 as possible in the case $n = 4$. [HINT: List the 4! rank sequences, compute d for each one, and then obtain the null distribution of D. See the Lehmann book (in the chapter bibliography), p. 290, for more information.]

The Wilcoxon Rank-Sum Test

When at least one of the sample sizes in a two-sample problem is small, the t test requires the assumption of normality (at least approximately). While the test is the best test in the sense of minimizing β for fixed α, there are situations in which an investigator would want to use a test that is valid even if the underlying distributions are quite nonnormal. We now describe such a test, called the Wilcoxon rank-sum test. An alternative name for the procedure is the Mann–Whitney test, though the Mann–Whitney test statistic is sometimes expressed in a slightly different form from that of the Wilcoxon test. The Wilcoxon test procedure is distribution-free because it will have the desired level of significance for a very large class of underlying distributions.

Assumptions

$X_1, \ldots, X_m$ and $Y_1, \ldots, Y_n$ are two independent random samples from continuous distributions with means μ_1 and μ_2, respectively. The X and Y distributions have exactly the same shape and spread, the only possible difference between the two being in the values of μ_1 and μ_2.

When H_0: $\mu_1 - \mu_2 = \Delta_0$ is true, the X distribution is shifted by the amount Δ_0 to the right of the Y distribution, while when H_0 is false, the shift is by an amount other than Δ_0.

Development of the Test When $m = 3, n = 4$

Consider first testing H_0: $\mu_1 - \mu_2 = 0$. If μ_1 is actually much larger than μ_2, then most of the observed x's will fall to the right of the observed y's, while if H_0 is true then the observed values from the two samples should be intermingled. The test statistic will provide a quantification of how much intermingling there is in the two samples.

For concreteness, let $m = 3$ and $n = 4$. Then if all three observed x's were to the right of all four observed y's, this would provide strong evidence for rejecting H_0 in favor of H_a: $\mu_1 - \mu_2 \neq 0$, with a similar conclusion being appropriate if all three x's fall below all four of the y's. Suppose we pool the X's and Y's into a combined sample of size $m + n = 7$ and rank these observations from smallest to largest, with the smallest receiving rank 1 and the largest, rank 7. If either most of the largest ranks or most of the smallest ranks were associated with X observations, we would begin to suspect H_0. This suggests the test statistic

$$W = \text{the sum of the ranks in the combined sample} \tag{15.3}$$
$$\text{associated with } X \text{ observations}$$

For the values of m and n under consideration, the smallest possible value of W is $w = 1 + 2 + 3 = 6$ (if all three x's are smaller than all four y's), and the largest possible value is $w = 5 + 6 + 7 = 18$ (if all three x's are larger than all four y's).

As an example of the calculation of the observed W, suppose $x_1 = -3.10$, $x_2 = 1.67$, $x_3 = 2.01$, $y_1 = 5.27$, $y_2 = 1.89$, $y_3 = 3.86$, and $y_4 = .19$. Then the pooled ordered sample is -3.10, $.19$, 1.67, 1.89, 2.01, 3.86, and 5.27. The X ranks for this sample are 1 (for -3.10), 3 (for 1.67), and 5 (for 2.01), so the computed value of W is $w = 1 + 3 + 5 = 9$.

The test procedure based on the statistic (15.3) is to reject H_0 if the computed value w is "too extreme"—that is, $\geq c$ for an upper-tailed test, $\leq c$ for a lower-tailed test, and either $\geq c_1$ or $\leq c_2$ for a two-tailed test. The critical constant(s) c (c_1, c_2) should be chosen so that the test has the desired level of significance α. To see how this should be done, recall that when H_0 is true, all seven observations come from the same population. This means that under H_0, any possible triple of ranks associated with the three x's—such as $(1, 4, 5)$, $(3, 5, 6)$, or $(5, 6, 7)$—has the same probability as any other possible rank triple. Since there are $\binom{7}{3} = 35$ possible rank triples, under H_0 each rank triple has probability $\frac{1}{35}$. From a list of all 35 rank triples and the w value associated with each, the probability distribution of W can immediately be determined. For example, there are four rank triples that have w value 11—$(1, 3, 7)$, $(1, 4, 6)$, $(2, 3, 6)$, and $(2, 4, 5)$—so $P(W = 11) = \frac{4}{35}$. The summary of the listing and computations appears in Table 15.5.

Table 15.5 Probability distribution of W ($m = 3$, $n = 4$) when H_0 is true

w	6	7	8	9	10	11	12	13	14	15	16	17	18
$P(W = w)$	$\frac{1}{35}$	$\frac{1}{35}$	$\frac{2}{35}$	$\frac{3}{35}$	$\frac{4}{35}$	$\frac{4}{35}$	$\frac{5}{35}$	$\frac{4}{35}$	$\frac{4}{35}$	$\frac{3}{35}$	$\frac{2}{35}$	$\frac{1}{35}$	$\frac{1}{35}$

The distribution of Table 15.5 is symmetric about the value $w = (6 + 18)/2 = 12$, which is the middle value in the ordered list of possible W values. This is because the two rank triples (r, s, t) (with $r < s < t$) and $(8 - t, 8 - s, 8 - r)$ have values of w symmetric about 12, so for each triple with w value below 12, there is a triple with w value above 12 by the same amount.

If the alternative hypothesis is H_a: $\mu_1 - \mu_2 > 0$, then H_0 should be rejected in favor of H_a for large W values. Choosing as the rejection region the set of W values $\{17, 18\}$, $\alpha = P(\text{type I error}) = P(\text{reject } H_0 \text{ when } H_0 \text{ is true}) = P(W = 17 \text{ or } 18 \text{ when } H_0 \text{ is true}) = \frac{1}{35} + \frac{1}{35} = \frac{2}{35} = .057$; the region $\{17, 18\}$ therefore specifies a test with level of significance approximately .05. Similarly, the region $\{6, 7\}$, which is appropriate for H_a: $\mu_1 - \mu_2 < 0$, has $\alpha = .057 \approx .05$. The region $\{6, 7, 17, 18\}$, which is appropriate for the two-sided alternative, has $\alpha = \frac{4}{35} = .114$. The W value for the data given several paragraphs earlier was $w = 9$, which is rather close to the middle value 12, so H_0 would not be rejected at any reasonable level α for any one of the three H_a's.

General Description of the Wilcoxon Rank-Sum Test

The null hypothesis H_0: $\mu_1 - \mu_2 = \Delta_0$ is handled by subtracting Δ_0 from each X_i and using the $(X_i - \Delta_0)$'s as the X_i's were previously used. Recalling that for any positive integer K, the sum of the first K integers is $K(K + 1)/2$, the smallest possible value of the statistic W is $m(m + 1)/2$, which occurs when the $(X_i - \Delta_0)$'s are all to the left of the Y sample. The largest possible value of W occurs when the $(X_i - \Delta_0)$'s lie entirely to the right of the Y's; in this case, $W = (n + 1) + \ldots + (m + n) = (\text{sum of first } m + n \text{ integers}) - (\text{sum of first } n \text{ integers})$, which gives $m(m + 2n + 1)/2$. As with the special case $m = 3, n = 4$, the distribution of W is symmetric about the value that is halfway between the smallest and largest values; this middle value is $m(m + n + 1)/2$. Because of this symmetry, probabilities involving lower-tail critical values can be obtained from corresponding upper-tail values.

Null hypothesis: H_0: $\mu_1 - \mu_2 = \Delta_0$

Test statistic value: $w = \displaystyle\sum_{i=1}^{m} r_i$ where $r_1 = $ rank of $(x_i - \Delta_0)$ in the combined sample of $(m + n)$ x's and y's

Alternative Hypothesis	Rejection Region
H_a: $\mu_1 - \mu_2 > \Delta_0$	$w \geq c_1$
H_a: $\mu_1 - \mu_2 < \Delta_0$	$w \leq m(m + n + 1) - c_1$
H_a: $\mu_1 - \mu_2 \neq \Delta_0$	Either $w \geq c$ or
	$w \leq m(m + n + 1) - c$

where $P(W \geq c_1 \text{ when } H_0 \text{ is true}) = \alpha$, $P(W \geq c \text{ when } H_0 \text{ is true}) = \alpha/2$.

Because W has a discrete probability distribution, there will not always exist a critical value corresponding exactly to one of the usual levels of significance. Appendix Table A.10 gives upper-tail critical values for probabilities closest to .05, .025, .01, and .005, from which level .05 or .01 one- and two-tailed tests can be obtained. The table gives information only for $m = 3, 4, \ldots, 8$ and $n = m$, $m + 1, \ldots, 8$ (that is, $3 \leq m \leq n \leq 8$). For values of m and n that exceed 8, a normal approximation can be used. To use the table for small m and n, though, *the X and Y samples should be labeled so that* $m \leq n$.

Example 15.8   The urinary fluoride concentration (parts per million) was measured both for a sample of livestock grazing in an area previously exposed to fluoride pollution and for a similar sample grazing in an unpolluted region

Polluted	21.3	18.7	23.0	17.1	16.8	20.9	19.7
Unpolluted	14.2	18.3	17.2	18.4	20.0		

Does the data indicate strongly that the true average fluoride concentration for livestock grazing in the polluted region is larger than for the unpolluted region? Use the Wilcoxon rank-sum test at level $\alpha = .01$.

The sample sizes here are 7 and 5. To obtain $m \leq n$, label the unpolluted observations as the x's ($x_1 = 14.2, \ldots, x_5 = 20.0$) and the polluted observations as the y's. Thus, μ_1 is the true average fluoride concentration without pollution and μ_2 the true average concentration with pollution. The alternative hypothesis is H_a: $\mu_1 - \mu_2 < 0$ (pollution causes an increase in concentration), so a lower-tailed test is appropriate. Entering Appendix Table A.10 with $m = 5$ and $n = 7$, $P(W \geq 47$ when H_0 is true) $\approx .01$. The critical value for the lower-tailed test is therefore $m(m + n + 1) - 47 = 5(13) - 47 = 18$; H_0 will now be rejected if $w \leq 18$. The pooled ordered sample appears below; the computed W is $w = r_1 + r_2 + \cdots + r_5$ (where r_i is the rank of x_i) $= 1 + 5 + 4 + 6 + 9 = 25$. Since 25 is not ≤ 18, H_0 is not rejected at (approximately) level .01.

x	y	y	x	x	x	y	y	x	y	y	y
14.2	16.8	17.1	17.2	18.3	18.4	18.7	19.7	20.0	20.9	21.3	23.0
1	2	3	4	5	6	7	8	9	10	11	12

■

A Normal Approximation for W

When both m and n exceed 8, the distribution of W can be approximated by an appropriate normal curve, and this approximation can be used in place of Appendix Table A.10. To obtain the approximation, we need μ_W and σ_W^2 when H_0 is true. In this case, the rank R_i of $X_i - \Delta_0$ is equally likely to be any one of the possible values $1, 2, 3, \ldots, m + n$ (R_i has a discrete uniform distribution on the first $m + n$ positive integers), so $\mu_{R_i} = (m + n + 1)/2$. This gives, since $W = \Sigma R_i$,

$$\mu_W = \mu_{R_1} + \mu_{R_2} + \cdots + \mu_{R_m} = \frac{m(m + n + 1)}{2} \tag{15.4}$$

The variance of R_i is also easily computed to be $(m + n + 1)(m + n - 1)/12$. However, because the R_i's are not independent, $V(W) \neq mV(R_i)$. Using the fact that for any two distinct integers a and b between 1 and $m + n$ inclusive, $P(R_i = a,$

$R_j = b) = 1/[(m + n)(m + n - 1)]$ (two integers are being sampled without replacement), $\text{Cov}(R_i, R_j) = -(m + n + 1)/12$, yielding

$$\sigma_W^2 = \sum_{i=1}^{m} V(R_i) + \sum\sum_{i \neq j} \text{Cov}(R_i, R_j) = \frac{mn(m + n + 1)}{12} \qquad (15.5)$$

A CLT can then be used to conclude that when H_0 is true, the *test statistic*

$$Z = \frac{W - m(m + n + 1)/2}{\sqrt{mn(m + n + 1)/12}}$$

has approximately a standard normal distribution. This statistic is used in conjunction with the critical values z_α, $-z_\alpha$, and $\pm z_{\alpha/2}$ for upper-, lower-, and two-tailed tests, respectively.

Example 15.9

A paper in the *Journal of Applied Physiology* ("Histamine Content in Sputum from Allergic and Non-Allergic Individuals," 1969, pp. 535–539) reports the following data on sputum histamine level ($\mu g/g$ dry weight of sputum) for a sample of 9 individuals classified as allergics and another sample of 13 individuals classified as nonallergics.

Allergics   67.6, 39.6, 1651.0, 100.0, 65.9, 1112.0, 31.0, 102.4, 64.7
Nonallergics   34.3, 27.3, 35.4, 48.1, 5.2, 29.1, 4.7, 41.7, 48.0, 6.6, 18.9, 32.4, 45.5

Does the data indicate that there is a difference in true average sputum histamine level between allergics and nonallergics?

Since both sample sizes exceed 8, we use the normal approximation. The null hypothesis is $H_0: \mu_1 - \mu_2 = 0$, and observed ranks of the x_i's are $r_1 = 18$, $r_2 = 11$, $r_3 = 22$, $r_4 = 19$, $r_5 = 17$, $r_6 = 21$, $r_7 = 7$, $r_8 = 20$, and $r_9 = 16$, so $w = \Sigma r_i = 151$. The mean and variance of W are given by $\mu_W = 9(23)/2 = 103.5$ and $\sigma_W^2 = 9(13)(23)/12 = 224.25$. Thus,

$$z = \frac{151 - 103.5}{\sqrt{224.25}} = 3.17$$

The alternative hypothesis is $H_a: \mu_1 - \mu_2 \neq 0$, so at level .01 H_0 is rejected if either $z \geq 2.58$ or $z \leq -2.58$. Because $3.17 \geq 2.58$, H_0 is rejected, and we conclude that there is a difference in true average sputum histamine levels (the paper also used the Wilcoxon test).   ∎

Ties in the Wilcoxon Rank-Sum Test

Theoretically, the assumption of continuity of the two distributions ensures that all $m + n$ observed x's and y's will have different values. In practice, though, there will often be ties in the observed values. As with the Wilcoxon signed-rank test, the common practice in dealing with ties is to assign each of the tied observations in a particular set of ties the average of the ranks they would receive if they differed very slightly from one another.

If the sample sizes both exceed 8, the numerator of Z is still appropriate, but the denominator should be replaced by the square root of the adjusted variance

$$\sigma_W^2 = \frac{mn(m + n + 1)}{12}$$

$$- \frac{mn}{12(m + n)(m + n - 1)} \Sigma (\tau_i - 1)(\tau_i)(\tau_i + 1)$$

(15.6)

where τ_i is the number of tied observations in the ith set of ties and the sum is over all sets of ties. Unless there are a great many ties, there is little difference between (15.6) and (15.5).

Efficiency of the Wilcoxon Rank-Sum Test

When the distributions being sampled are both normal with $\sigma_1^2 = \sigma_2^2$, both the two-sample t test and the Wilcoxon rank-sum test can be used, but the t test is known to be the best among all possible tests in the sense of minimizing β for a fixed α. Because the Wilcoxon test is valid for a much broader class of underlying distributions than just the normal, it might be used if the validity of the normality assumption is open to question. One aspect of efficiency relates to how much is lost by using W instead of T when normality is correct. The other aspect of efficiency is concerned with the performance of W relative to T in nonnormal situations.

In the previous section, the notion of efficiency of test procedures was discussed in connection with the one-sample t test and Wilcoxon signed-rank test. The results for two-sample tests are the same as for the one-sample tests. When the normality assumption is correct, the rank-sum test is approximately 95% as efficient as the t test in large samples—meaning that the t test can give the same error probabilities as the Wilcoxon test using slightly smaller sample sizes. On the other hand, the Wilcoxon test will always be at least 86% as efficient as the t test and may be many times as efficient if the underlying distributions are very nonnormal, especially with heavy tails.

In summary, the Wilcoxon test never gives up very much to the t test and may be a distinct improvement on it, so if there is much doubt about the assumption of normality, the Wilcoxon test provides a good alternative means of analysis.

Lastly, we note that β calculations for the Wilcoxon test are quite difficult because the distribution of W when H_0 is false depends not only on $\mu_1 - \mu_2$ but also on the shapes of the two distributions. For most underlying distributions, the nonnull distribution of W is virtually intractable. This is why statisticians have developed large-sample (asymptotic relative) efficiency as a means of comparing tests.

Exercises   Section 15.3 (15–20)

15. In an experiment to compare the bond strength of two different adhesives, each adhesive was used in five bondings of two surfaces, and the force necessary to separate the surfaces was determined for each bonding. For adhesive 1 the resulting values were 229, 286, 245, 299, and 250, while the adhesive 2 observations were

213, 179, 163, 247, and 225. Let μ_i denote the true average bond strength of adhesive type i. Use the Wilcoxon rank-sum test at level .05 to test $H_0: \mu_1 = \mu_2$ versus $H_a: \mu_1 > \mu_2$.

16. The article "A Study of Wood Stove Particulate Emissions" (*J. Air Pollution Control Assn.*, 1979, pp. 724–728) reports the following data on burn time (hours) for samples of oak and pine. Test at level .05 to see whether there is any difference in true average burn time for the two types of wood.

Oak   1.72 .67 1.55 1.56 1.42 1.23 1.77 .48
Pine   .98 1.40 1.33 1.52 .73 1.20

17. A modification has been made to the process for producing a certain type of "time-zero" film (film that begins to develop as soon as a picture is taken). Because the modification involves extra cost, it will be incorporated only if sample data strongly indicates that the modification has decreased true average developing time by more than 1 second. Assuming that the developing-time distributions differ only with respect to location if at all, use the Wilcoxon rank-sum test at level .05 on the accompanying data to test the appropriate hypotheses.

Original
process   8.6 5.1 4.5 5.4 6.3 6.6 5.7 8.5
Modified
process   5.5 4.0 3.8 6.0 5.8 4.9 7.0 5.7

18. The accompanying data resulted from an experiment to compare the effects of vitamin C in orange juice and in synthetic ascorbic acid on the length of odontoblasts in guinea pigs over a 6-week period ("The Growth of the Odontoblasts of the Incisor Tooth as a Criterion of the Vitamin C Intake of the Guinea Pig," *J. Nutrition,* 1947, pp. 491–504). Use the Wilcoxon rank-sum test at level .01 to decide whether true average length differs for the two types of vitamin C intake. Compute also an approximate *P*-value.

Orange juice	8.2 9.4 9.6 9.7 10.0 14.5
	15.2 16.1 17.6 21.5
Ascorbic acid	4.2 5.2 5.8 6.4 7.0 7.3
	10.1 11.2 11.3 11.5

19. Test the hypotheses suggested in Exercise 18 using the following data:

Orange juice	8.2 9.5 9.5 9.7 10.0 14.5
	15.2 16.1 17.6 21.5
Ascorbic acid	4.2 5.2 5.8 6.4 7.0 7.3
	9.5 10.0 11.5 11.5

20. The paper "Measuring the Exposure of Infants to Tobacco Smoke" (*N. Engl. J. Med,*. 1984, pp. 1075–1078) reports on a study in which various measurements were taken both from a random sample of infants who had been exposed to household smoke and from a sample of unexposed infants. The accompanying data consists of observations on urinary concentration of cotanine, a major metabolite of nicotine (the values constitute a subset of the original data and were read from a plot that appeared in the paper). Does the data suggest that true average cotanine level is higher in exposed infants than in unexposed infants by more than 25? Carry out a test at significance level .05.

Unexposed   8 11 12 14   20   43 111
Exposed   35 56 83 92 128 150 176 208

15.4   Distribution-Free Confidence Intervals

The method we have used so far to construct a confidence interval (CI) can be described as follows: Start with a random variable (Z, T, χ^2, F, or the like) that depends on the parameter of interest and a probability statement involving the variable, manipulate the inequalities of the statement to isolate the parameter between random endpoints, and finally substitute computed values for random variables. Another general method for obtaining CIs takes advantage of a relationship between test procedures and CIs; a $100(1 - \alpha)\%$ CI for a parameter θ can be obtained from a level α test for $H_0: \theta = \theta_0$ versus $H_a: \theta \neq \theta_0$. This method will be used to derive intervals associated with the sign test, the Wilcoxon signed-rank test, and the Wilcoxon rank-sum test.

Before using the method to derive new intervals, reconsider the t test and the t interval. Suppose a random sample of $n = 25$ observations from a normal population yields summary statistics $\bar{x} = 100$, $s = 20$. Then a 90% CI for μ is

$$\left(\bar{x} - t_{.05,24} \cdot \frac{s}{\sqrt{25}}, \quad \bar{x} + t_{.05,24} \cdot \frac{s}{\sqrt{25}} \right) = (93.16, \ 106.84) \tag{15.7}$$

Suppose that instead of a CI, we had wished to test a hypothesis about μ. For H_0: $\mu = \mu_0$ versus H_a: $\mu \neq \mu_0$, the t test at level .10 specifies that H_0 should be rejected if t is either ≥ 1.711 or ≤ -1.711, where

$$t = \frac{\bar{x} - \mu_0}{s/\sqrt{25}} = \frac{100 - \mu_0}{20/\sqrt{25}} = \frac{100 - \mu_0}{4} \tag{15.8}$$

Consider now the null value $\mu_0 = 95$. Then $t = 1.25$ so H_0 is not rejected. Similarly, if $\mu_0 = 104$, then $t = -1$, so again H_0 is not rejected. However, if $\mu_0 = 90$, then $t = 2.5$, so H_0 is rejected, and if $\mu_0 = 108$, then $t = -2$, so H_0 is again rejected. By considering other values of μ_0 and the decision resulting from each one, the following general fact emerges: *Every number inside the interval (15.7) specifies a value of μ_0 for which t of (15.8) leads to nonrejection of H_0, while every number outside (15.7) corresponds to a t for which H_0 is rejected.* That is, for the fixed values of n, $\bar{x}$, and s, the set of all μ_0 values for which testing H_0: $\mu = \mu_0$ versus H_a: $\mu \neq \mu_0$ results in nonrejection of H_0 is precisely the interval (15.7).

Proposition

> Suppose we have a level α test procedure for testing H_0: $\theta = \theta_0$ versus H_a: $\theta \neq \theta_0$. For fixed sample values, let A denote the set of all values θ_0 for which H_0 is not rejected. Then A is a $100(1 - \alpha)\%$ CI for θ.

There are actually pathological examples in which the set A defined in the proposition is not an interval of θ values, but instead the complement of an interval or something even stranger. To be more precise, we should really replace the notion of a CI with that of a confidence set. In the three cases of interest here, the set A does turn out to be an interval.

The Sign Interval for $\tilde{\mu}$

Let the sample be denoted by $X_1, \ldots, X_n$. Then for testing H_0: $\tilde{\mu} = \tilde{\mu}_0$ versus H_a: $\tilde{\mu} \neq \tilde{\mu}_0$, the sign test statistic is $Y = $ number of X_i's that are $> \tilde{\mu}_0$ [the number of $(X_i - \tilde{\mu}_0)$'s that have positive signs]. H_0 is rejected if $y \geq c$ or $\leq n - c$; under H_0, Y has a binomial distribution with $p = .5$, and n, and c is determined by $P(Y \geq c) + P(Y \leq n - c) = \alpha$.

Example 15.10   The daily dietetic intake of nicotinic acid-equivalents (mg) was determined for a sample of 12 healthy individuals ("Assessment of Nicotinic Acid Status of

Population Groups," *Amer. J. Clinical Nutrition,* 1964, pp. 169–174). The sample observations, listed in order of increasing magnitude, were

16.6, 19.4, 21.7, 23.4, 24.2, 26.9, 27.3, 27.4, 28.0, 29.5, 30.7, 58.8

To obtain a $100(1 - \alpha)\%$ CI for $\tilde{\mu}$ = the true median daily nicotinic acid-equivalent intake, we must use the sign test at level α to test $H_0: \tilde{\mu} = \tilde{\mu}_0$ versus $H_a: \tilde{\mu} \neq \tilde{\mu}_0$ for various values of $\tilde{\mu}_0$. The CI is then the set $A = \{$all $\tilde{\mu}_0: H_0$ is not rejected$\}$.

For $n = 12$ and $p = .5$, $P(Y \leq 2) + P(Y \geq 10) = .0193 + .0193 = .0386 \approx .04$. If the rejection region for the sign test is $R = \{0, 1, 2, 10, 11, 12\}$, the test has (approximately) level .04, so the confidence level of the resulting interval will be 96%. The CI is most easily identified from the plot of the data in Figure 15.5.

Figure 15.5 Plot of the data for Example 15.10

If $\tilde{\mu}_0 < 21.7$, then y = number of x_i's $> \tilde{\mu}_0$ is at least 10 (10, 11, or 12), so H_0 would be rejected. Similarly, if $\tilde{\mu}_0 > 29.5$, then y is at most 2 (0, 1, or 2), so again H_0 is rejected. However, as Figure 15.5 indicates, if $\tilde{\mu}_0$ is between 21.7 and 29.5, then $3 \leq y \leq 9$ so H_0 is not rejected. The set A is then $\{\tilde{\mu}_0: 21.7 < \tilde{\mu}_0 < 29.5\}$, so our 96% CI for $\tilde{\mu}$ is (21.7, 29.5). ∎

Example 15.10 shows that after the data values are ordered from smallest to largest, the interval for $\tilde{\mu}$ includes all numbers between two of these ordered values.

Proposition

> Denote the ordered sample values by $x_{(1)}, x_{(2)}, \ldots, x_{(n)}$, where $x_{(1)}$ is the smallest, $x_{(2)}$ the next smallest, and so on. Suppose the level α sign test rejects $H_0: \tilde{\mu} = \tilde{\mu}_0$ if either $y \geq c$ or $\leq n - c$. Then a $100(1 - \alpha)\%$ CI for $\tilde{\mu}$ is $(x_{(n-c+1)}, x_{(c)})$.

Example 15.11
(Example 15.10 continued)

For $n = 12$ and $c = 10$, the level of significance is .04. Therefore, the $100(1 - .04)\% = 96\%$ CI is $(x_{(12-10+1)}, x_{(10)}) = (x_{(3)}, x_{(10)}) = (21.7, 29.5)$. The sample mean and sample standard deviation are easily computed as $\bar{x} = 27.83$ and $s = 10.61$, from which the 95% t interval for μ, assuming the distribution to be normal, is (21.09, 34.57). That this t interval is much longer than the sign interval is due primarily to the single outlying value 58.8. ∎

Notice that in contrast to the t interval, the sign interval is relatively insensitive to a few outlying values. In Example 15.10, the two extreme values on each end of the sample can be moved farther from the middle without affecting the interval.

When the population is normal, $\tilde{\mu} = \mu$, so the sign interval is a competitor to the t interval. In this case, the sign interval will usually be longer than the t interval, since outlying values in a normal sample are very unusual. The Wilcoxon interval, to be described next, is a better competitor of the t interval.

Finally, because of the discreteness of the binomial distribution, the usual confidence levels are not achievable. In Example 15.11, we settled for 96% rather than 95% confidence.

The Wilcoxon Signed-Rank Interval

To test H_0: $\mu = \mu_0$ versus H_a: $\mu \neq \mu_0$ using the Wilcoxon signed-rank test, where μ is the mean of a continuous symmetric distribution, the absolute values $|x_1 - \mu_0|, \ldots, |x_n - \mu_0|$ are ordered from smallest to largest, with the smallest receiving rank 1 and the largest rank n. Each rank is then given the sign of its associated $x_i - \mu_0$, and the test statistic is the sum of the positively signed ranks. The two-tailed test rejects H_0 if s_+ is either $\geq c$ or $\leq n(n + 1)/2 - c$, where c is obtained from Appendix Table A.9 once the desired level of significance α is specified.

Just as was the case for the sign interval, for fixed $x_1, \ldots, x_n$ the $100(1 - \alpha)\%$ signed-rank interval will consist of all μ_0 for which H_0: $\mu = \mu_0$ is not rejected at level α. To identify this interval, it is convenient to express the test statistic S_+ in another form:

$$S_+ = \text{the number of pairwise averages } (X_i + X_j)/2 \text{ with } i \leq j \text{ that} \qquad (15.9)$$
$$\text{are} \geq \mu_0$$

That is, if we average each x_j in the list with each x_i to its left, including in addition $(x_j + x_j)/2$ (which is just x_j) and count the number of these averages that are $\geq \mu_0$, s_+ results. In moving from left to right in the list of sample values, we are simply averaging every pair of observations in the sample [again including $(x_j + x_j)/2$] exactly once, so the order in which the observations are listed before averaging is not important. The equivalence of the two methods for computing s_+ is not difficult to verify. The number of pairwise averages is $\binom{n}{2} + n$ (the first term due to averaging of different observations and the second due to averaging each x_i with itself), which equals $n(n + 1)/2$. If either too many or too few of these pairwise averages are $\geq \mu_0$, H_0 is rejected.

Example 15.12   The following observations are values of cerebral metabolic rate for rhesus monkeys: $x_1 = 4.51, x_2 = 4.59, x_3 = 4.90, x_4 = 4.93, x_5 = 6.80, x_6 = 5.08, x_7 = 5.67$. There are 28 pairwise averages which are, in increasing order,

4.51, 4.55, 4.59, 4.705, 4.72, 4.745, 4.76, 4.795, 4.835, 4.90,
4.915, 4.93, 4.99, 5.005, 5.08, 5.09, 5.13, 5.285, 5.30, 5.375,
5.655, 5.67, 5.695, 5.85, 5.865, 5.94, 6.235, 6.80

The first few and the last few of these are pictured on a measurement axis in Figure 15.6.

Figure 15.6 Plot of the data for Example 15.12

Because of the discreteness of the distribution of S_+, $\alpha = .05$ cannot be obtained exactly. The rejection region $\{0, 1, 2, 26, 27, 28\}$ has $\alpha = .046$, which is as close as possible to .05, so the level is approximately .05. Thus, if the number of pairwise averages $\geq \mu_0$ is between 3 and 25, inclusive, H_0 is not rejected. From Figure 15.6 the (approximate) 95% CI for μ is (4.59, 5.94). ∎

In general, once the pairwise averages are ordered from smallest to largest, the endpoints of the Wilcoxon interval are two of the "extreme" averages. To express this precisely, let the smallest pairwise average be denoted by $\bar{x}_{(1)}$, the next smallest by $\bar{x}_{(2)}, \ldots$, and the largest by $\bar{x}_{(n(n+1)/2)}$.

Proposition

> If the level α Wilcoxon signed-rank test for $H_0: \mu = \mu_0$ versus $H_a: \mu \neq \mu_0$ is to reject H_0 if either $s_+ \geq c$ or $s_+ \leq n(n + 1)/2 - c$, then a $100(1 - \alpha)\%$ CI for μ is
>
> $$(\bar{x}_{(n(n+1)/2-c+1)}, \quad \bar{x}_{(c)}) \qquad (15.10)$$

In words, the interval extends from the dth smallest pairwise average to the dth largest average, where $d = n(n + 1)/2 - c + 1$. Appendix Table A.11 gives for $n = 5$, $6, \ldots, 25$ the values of c that correspond to the usual confidence levels.

Example 15.13
(Example 15.12 continued)

For $n = 7$, an 89.1% interval (approximately 90%) is obtained by using $c = 24$ (since the rejection region $\{0, 1, 2, 3, 4, 24, 25, 26, 27, 28\}$ has $\alpha = .109$). The interval is $(\bar{x}_{(28-24+1)}, \bar{x}_{(24)}) = (\bar{x}_{(5)}, \bar{x}_{(24)}) = (4.72, 5.85)$, which extends from the fifth smallest to the fifth largest pairwise average. ∎

Other Uses and Properties of the Wilcoxon Signed-Rank Interval

While the derivation of the interval depended on having a single sample from a continuous symmetric distribution with mean (median) μ, when the data is paired, the interval constructed from the differences $d_1, d_2, \ldots, d_n$ is a CI for the mean (median) difference μ_D. In this case, the symmetry of X and Y distributions need not be assumed; as long as the X and Y distributions have the same shape, the $X - Y$ distribution will be symmetric, so only continuity is required.

For $n > 20$, the large-sample approximation to the Wilcoxon test based on standardizing S_+ gives an approximation to c in (15.10). The result [for a $100(1 - \alpha)\%$ interval] is

$$c \approx \frac{n(n + 1)}{4} + z_{\alpha/2} \sqrt{\frac{n(n + 1)(2n + 1)}{24}}$$

The efficiency of the Wilcoxon interval relative to the t interval is roughly the same as that for the Wilcoxon test relative to the t test. In particular, for large samples when the underlying population is normal, the Wilcoxon interval will tend to be slightly longer than the t interval, but if the population is quite nonnormal (symmetric but with heavy tails), then the Wilcoxon interval will tend to be much shorter than the t interval.

The Wilcoxon Rank-Sum Interval

The Wilcoxon rank-sum test for testing $H_0: \mu_1 - \mu_2 = \Delta_0$ is carried out by first combining the $(X_i - \Delta_0)$'s and Y_j's into one sample of size $m + n$ and ranking them from smallest (rank 1) to largest (rank $m + n$). The test statistic W is then the sum of the ranks of the $(X_i - \Delta_0)$'s. For the two-sided alternative, H_0 is rejected if w is either too small or too large.

To obtain the associated CI for fixed x_i's and y_j's we must determine the set of all Δ_0 values for which H_0 is not rejected. This is easiest to do if we first express the test statistic in a slightly different form. The smallest possible value of W is $m(m + 1)/2$, corresponding to every $(X_i - \Delta_0)$ less than every Y_j, and there are mn differences of the form $(X_i - \Delta_0) - Y_j$. A bit of manipulation gives

$$W = [\text{number of } (X_i - Y_j - \Delta_0)\text{'s} \geq 0] + \frac{m(m + 1)}{2}$$

$$= [\text{number of } (X_i - Y_j)\text{'s} \geq \Delta_0] + \frac{m(m + 1)}{2}$$

(15.11)

Thus, rejecting H_0 if the number of $(x_i - y_j)$'s $\geq \Delta_0$ is either too small or too large is equivalent to rejecting H_0 for small or large w.

Expression (15.11) suggests that we compute $x_i - y_j$ for each i and j and order these mn differences from smallest to largest. Then if the null value Δ_0 is not either smaller than most of the differences or larger than most, $H_0: \mu_1 - \mu_2 = \Delta_0$ is not rejected. Varying Δ_0 now shows that a CI for $\mu_1 - \mu_2$ will have as its lower endpoint one of the ordered $(x_i - y_j)$'s, and similarly for the upper endpoint.

Proposition

> Let $x_1, \ldots, x_m$ and $y_1, \ldots, y_n$ be the observed values in two independent samples from continuous distributions that differ only in location (and not in shape). With $d_{ij} = x_i - y_j$ and the ordered differences denoted by $d_{ij(1)}$,

$d_{ij(2)}, \ldots, d_{ij(mn)}$, the general form of a $100(1 - \alpha)\%$ CI for $\mu_1 - \mu_2$ is

$$(d_{ij(mn-c+1)}, \quad d_{ij(c)}) \tag{15.12}$$

where c is the critical constant for the two-tailed level α Wilcoxon rank-sum test.

Notice that the form of the Wilcoxon rank-sum interval (15.12) is very similar to the Wilcoxon signed-rank interval (15.10); (15.10) uses pairwise averages from a single sample, while (15.12) uses pairwise differences from two samples. Appendix Table A.12 gives values of c for selected values of m and n.

Example 15.14   The paper "Some Mechanical Properties of Impregnated Bark Board" (*Forest Products J.*, 1977, pp. 31–38) reports the following data on maximum crushing strength (psi) for a sample of epoxy-impregnated bark board and for a sample of bark board impregnated with another polymer:

Epoxy (x's)   10,860 11,120 11,340 12,130 14,380 13,070
Other (y's)   4,590   4,850   6,510   5,640   6,390

Obtain a 95% CI for the true average difference in crushing strength between the epoxy-impregnated board and the other type of board.

From Appendix Table A.12, since the smaller sample size is 5 and the larger sample size is 6, $c = 26$ for a confidence level of approximately 95%. The d_{ij}'s appear in Table 15.6. The five smallest d_{ij}'s [$d_{ij(1)}, \ldots, d_{ij(5)}$] are 4350, 4470, 4610, 4730, and 4830, while the five largest d_{ij}'s are (in descending order) 9790, 9530, 8740, 8480, and 8220. Thus, the CI is $(d_{ij(5)}, d_{ij(26)}) = (4830, 8220)$. ∎

Table 15.6 Differences for the rank-sum interval in Example 15.14

	d_{ij}	4590	4850	5640	6390	6510
				y_j		
	10,860	6270	6010	5220	4470	4350
	11,120	6530	6270	5480	4730	4610
x_i	11,340	6750	6490	5700	4950	4830
	12,130	7540	7280	6490	5740	5620
	13,070	8480	8220	7430	6680	6560
	14,380	9790	9530	8740	7990	7870

When m and n are both large, the Wilcoxon test statistic has approximately a normal distribution. This can be used to derive a large-sample approximation for the value c in (15.12). The result is

$$c \approx \frac{mn}{2} + z_{\alpha/2} \sqrt{\frac{mn(m + n + 1)}{12}} \tag{15.13}$$

As with the signed-rank interval, the rank-sum interval (15.12) is quite efficient with respect to the t interval; in large samples, (15.12) will tend to be only a bit longer than the t interval when the underlying populations are normal and may be considerably shorter than the t interval if the underlying populations have heavier tails than do normal populations.

Exercises/Section 15.4 (21–28)

21. Use the interval associated with the sign test to compute a 99% CI for the true median travel time $\tilde{\mu}$ using the data in Exercise 2.

22. Compute a 98% CI for the true median soil pH $\tilde{\mu}$ using the data in Exercise 3. For the given sample size, can a 95% CI be obtained? Explain.

23. The paper "The Lead Content and Acidity of Christchurch Precipitation" (*New Zealand J. Science,* 1980, pp. 311–312) reports the accompanying data on lead concentration ($\mu g/L$) in samples gathered during eight different summer rainfalls: 17.0, 21.4, 30.6, 5.0, 12.2, 11.8, 17.3, and 18.8. Assuming that the lead-content distribution is symmetric, use the Wilcoxon signed-rank interval to obtain a 95% CI for μ.

24. Compute the 99% signed-rank interval for true average pH μ (assuming symmetry) using the data in Exercise 10. [HINT: Try to compute only those pairwise averages having relatively small or large values (rather than all 105 averages).]

25. Compute a 94% CI for μ_D of Example 15.6 using the data given there.

26. The following observations are amounts of hydrocarbon emissions resulting from road wear of bias-belted tires under a 522-kg load inflated at 228 kPa and driven at 64 km/hr for 6 hours ("Characterization of Tire Emissions Using an Indoor Test Facility," *Rubber Chemistry and Technology,* 1978, pp. 7–25): .045, .117, .062, and .072. What confidence levels are achievable for this sample size using the signed-rank interval? Select an appropriate confidence level and compute the interval.

27. Compute the 90% rank-sum CI for $\mu_1 - \mu_2$ using the data in Exercise 15.

28. Compute a 99% CI for $\mu_1 - \mu_2$ using the data in Exercise 16.

15.5   Distribution-Free ANOVA

The single-factor ANOVA model of Chapter 10 for comparing I population or treatment means assumed that for $i = 1, 2, \ldots, I$, a random sample of size J_i was drawn from a normal population with mean μ_i and variance σ^2. This can be written as

$$X_{ij} = \mu_i + \epsilon_{ij}, \qquad j = 1, \ldots, J_i;\ i = 1, \ldots, I \qquad (15.14)$$

where the ϵ_{ij}'s are independent and normally distributed with mean zero and variance σ^2. While the normality assumption was required for the validity of the F test described in Chapter 10, the validity of the Kruskal–Wallis test for testing equality of the μ_i's depends only on the ϵ_{ij}'s having the same continuous distribution.

The Kruskal–Wallis Test

Let $N = \Sigma J_i$, the total number of observations in the data set, and suppose we rank all N observations from 1 (the smallest X_{ij}) to N (the largest X_{ij}). When H_0: $\mu_1 = \mu_2 = \cdots = \mu_I$ is true, the N observations all come from the same distribution, in which case all possible assignments of the ranks $1, 2, \ldots, N$ to the I samples are equally likely and we expect ranks to be intermingled in these samples. If, however, H_0 is false, then some samples will consist mostly of observations having small ranks in the combined sample while others will consist mostly of observations having large ranks. More specifically, if R_{ij} denotes the rank of X_{ij} among the N observations, and $R_{i\cdot}$ and $\bar{R}_{i\cdot}$ denote, respectively, the total and average of the ranks in the ith sample, then when H_0 is true

$$E(R_{ij}) = \frac{N+1}{2}, \quad E(\bar{R}_{i\cdot}) = \frac{1}{J_i} \sum_j E(R_{ij}) = \frac{N+1}{2}$$

The K–W test statistic is a measure of the extent to which the $\bar{R}_{i\cdot}$'s deviate from their common expected value $(N + 1)/2$, and H_0 is rejected if the computed value of the statistic indicates too great a discrepancy between observed and expected rank averages.

Test Statistic

$$K = \frac{12}{N(N+1)} \sum_{j=1}^{I} J_i \left(\bar{R}_{i\cdot} - \frac{N+1}{2} \right)^2$$

$$= \frac{12}{N(N+1)} \sum_{i=1}^{I} \frac{R_{i\cdot}^2}{J_i} - 3(N+1) \tag{15.15}$$

The second expression for K is the computational formula; it involves the rank totals ($R_{i\cdot}$'s) rather than the averages and requires only one subtraction.

If H_0 is rejected when $k \geq c$, then c should be chosen so that the test has level α. That is, c should be the upper-tail critical value of the distribution of K when H_0 is true. Under H_0, each possible assignment of the ranks to the I samples is equally likely, so in theory all such assignments can be enumerated, the value of K determined for each one, and the null distribution obtained by counting the number of times each value of K occurs. Clearly this computation is tedious, so while there are tables of the exact null distribution and critical values for small values of the J_i's, we will use the following "large-sample" approximation.

Proposition

When H_0 is true and either

$$I = 3, \quad J_i \geq 6 \qquad (i = 1, 2, 3)$$

or

$$I > 3, \quad J_i \geq 5 \qquad (i = 1, \ldots, I)$$

then K has approximately a chi-squared distribution with $I - 1$ df.

> This implies that a test with approximate significance level α rejects H_0 if $k \geq \chi^2_{\alpha, I-1}$.

Example 15.15 The accompanying observations (Table 15.7) on axial stiffness index resulted from a study of metal plate–connected trusses in which five different plate lengths—4 in., 6 in., 8 in., 10 in., and 12 in.—were used ("Modeling Joints Made with Light-Gauge Metal Connector Plates," *Forest Products J.,* 1979, pp. 39–44).

Table 15.7 Data and ranks for Example 15.15

	$i = 1$ (4"):	309.2	309.7	311.0	316.8	326.5	349.8	409.5		
	$i = 2$ (6"):	331.0	347.2	348.9	361.0	381.7	402.1	404.5		
	$i = 3$ (8"):	351.0	357.1	366.2	367.3	382.0	392.4	409.9		
	$i = 4$ (10"):	346.7	362.6	384.2	410.6	433.1	452.9	461.4		
	$i = 5$ (12"):	407.4	410.7	419.9	441.2	441.8	465.8	473.4		
									$r_{i.}$	$\bar{r}_{i.}$
	$i = 1$:	1	2	3	4	5	10	24	49	7.00
	$i = 2$:	6	8	9	13	17	21	22	96	13.71
Ranks	$i = 3$:	11	12	15	16	18	20	25	117	16.71
	$i = 4$:	7	14	19	26	29	32	33	160	22.86
	$i = 5$:	23	27	28	30	31	34	35	208	29.71

The computed value of K is

$$k = \frac{12}{35(36)} \left[\frac{(49)^2}{7} + \frac{(96)^2}{7} + \frac{(117)^2}{7} + \frac{(160)^2}{7} + \frac{(208)^2}{7} \right] - 3(36)$$

$$= 20.12$$

At level .01, $\chi^2_{.01,4} = 13.277$, and since $20.12 \geq 13.277$, H_0 is rejected and we conclude that expected axial stiffness does depend on plate length. ■

Friedman's Test for a Randomized Block Experiment

Suppose $X_{ij} = \mu + \alpha_i + \beta_j + \epsilon_{ij}$, where α_i is the ith treatment effect, β_j is the jth block effect, and the ϵ_{ij}'s are drawn independently from the same continuous (but not necessarily normal) distribution. Then to test $H_0: \alpha_1 = \alpha_2 = \cdots = \alpha_I = 0$, the null hypothesis of no treatment effects, the observations are first ranked separately from 1 to I within each block, and then the rank average $\bar{r}_{i.}$ is computed for each of the I treatments. When H_0 is true, the $\bar{r}_{i.}$'s should be close to one another, since within each block all $I!$ assignments of ranks to treatments are equally likely. Friedman's test statistic measures the discrepancy between the expected value $(I + 1)/2$ of each rank average and the $\bar{r}_{i.}$'s.

Test Statistic

$$F_r = \frac{12J}{I(I+1)} \sum_{i=1}^{I} \left(\bar{R}_{i\cdot} - \frac{I+1}{2} \right)^2 = \frac{12}{IJ(I+1)} \sum R_{i\cdot}^2 - 3J(I+1)$$

As with the K–W test, Friedman's test rejects H_0 when the computed value of the test statistic is too large. For the cases $I = 3$, $J = 2, \ldots, 15$ and $I = 4$, $J = 2, \ldots, 8$, Lehmann's book gives the upper-tail critical values for the test. Alternatively, for even moderate values of J, the test statistic F_r has approximately a chi-squared distribution with $I - 1$ df when H_0 is true, so H_0 can be rejected if $f_r \geq \chi^2_{\alpha, I-1}$.

Example 15.16

The paper "Physiological Effects During Hypnotically Requested Emotions" (*Psychosomatic Med.*, 1963, pp. 334–343) reports the following data (Table 15.8) on skin potential (mV) when the emotions of fear, happiness, depression, and calmness were requested from each of eight subjects.

Table 15.8 Data and ranks for Example 15.16

x_{ij}	Blocks (subjects)									
	1	2	3	4	5	6	7	8		
Fear	23.1	57.6	10.5	23.6	11.9	54.6	21.0	20.3		
Happiness	22.7	53.2	9.7	19.6	13.8	47.1	13.6	23.6		
Depression	22.5	53.7	10.8	21.1	13.7	39.2	13.7	16.3		
Calmness	22.6	53.1	8.3	21.6	13.3	37.0	14.8	14.8		
Ranks	1	2	3	4	5	6	7	8	$r_{i\cdot}$	$r_{i\cdot}^2$
Fear	4	4	3	4	1	4	4	3	27	729
Happiness	3	2	2	1	4	3	1	4	20	400
Depression	1	3	4	2	3	2	2	2	19	361
Calmness	2	1	1	3	2	1	3	1	14	196
										1686

Thus,

$$f_r = \frac{12}{4(8)(5)} (1686) - 3(8)(5) = 6.45$$

At level .05, $\chi^2_{.05,3} = 7.815$, and because 6.45 is not ≥ 7.815, H_0 is not rejected. There is no evidence that average skin potential depends on which emotion is requested. ■

The book by Myles Hollander and Douglas Wolfe (see chapter bibliography) discusses multiple comparisons procedures associated with the Kruskal–Wallis and Friedman tests, as well as other aspects of distribution-free ANOVA.

Exercises   Section 15.5 (29–33)

29. The accompanying data refers to concentration of the radioactive isotope strontium-90 in milk samples obtained from five randomly selected dairies in each of four different regions.

	1	6.4	5.8	6.5	7.7	6.1
Region	2	7.1	9.9	11.2	10.5	8.8
	3	5.7	5.9	8.2	6.6	5.1
	4	9.5	12.1	10.3	12.4	11.7

Test at level .10 to see whether true average strontium-90 concentration differs for at least two of the regions.

30. The paper "Production of Gaseous Nitrogen in Human Steady-State Conditions" (*J. Applied Physiology,* 1972, pp. 155–159) reports the following observations on the amount of nitrogen expired (in liters) under four dietary regimens: fasting (1), 23% protein (2), 32% protein (3), and 67% protein (4). Use the K–W test at level .05 to test equality of the corresponding μ_i's.

1. 4.079 4.859 3.540 5.047 3.298
2. 4.368 5.668 3.752 5.848 3.802
3. 4.169 5.709 4.416 5.666 4.123
4. 4.928 5.608 4.940 5.291 4.674

1. 4.679 2.870 4.648 3.847
2. 4.844 3.578 5.393 4.374
3. 5.059 4.403 4.496 4.688
4. 5.038 4.905 5.208 4.806

31. The accompanying data on cortisol level was reported in the paper "Cortisol, Cortisone, and 11-Deoxycortisol Levels in Human Umbilical and Maternal Plasma in Relation to the Onset of Labor" (*J. Obstetric Gynaecology British Commonwealth,* 1974, pp. 737–745). Experimental subjects were pregnant women who were delivered between 38 and 42 weeks gestation. Group 1 individuals elected to deliver by Caesarean section before labor onset, group 2 delivered by emergency Caesarean during induced labor, and group 3 individuals experienced spontaneous labor. Use the K–W test at level .05 to test for equality of the three population means.

Group 1   262 307 211   323 454 339
          304 154 287   356
Group 2   465 501 455   355 468 362
Group 3   343 772 207 1048 838 687

32. In a test to determine if soil pretreated with small amounts of Basic-H makes the soil more permeable to water, soil samples were divided into blocks, and each block received each of the four treatments under study. The treatments were (A) water with .001% Basic-H flooded on control soil, (B) water without Basic-H on control soil, (C) water with Basic-H flooded on soil pretreated with Basic-H, and (D) water without Basic-H on soil pretreated with Basic-H. Test at level .01 to see if there are any effects due to the different treatments.

			Blocks		
	1	2	3	4	5
A	37.1	31.8	28.0	25.9	25.5
B	33.2	25.3	20.2	20.3	18.3
C	58.9	54.2	49.2	47.9	38.2
D	56.7	49.6	46.4	40.9	39.4

	6	7	8	9	10
A	25.3	23.7	24.4	21.7	26.2
B	19.3	17.3	17.0	16.7	18.3
C	48.8	47.8	40.2	44.0	46.4
D	37.1	37.5	39.6	35.1	36.5

33. In an experiment to study the way in which different anesthetics affect plasma epinephrine concentration, ten dogs were selected and concentration was measured while under the influence of the anesthetics isoflurane, halothane, and cyclopropane ("Sympathoadrenal and Hemodynamic Effects of Isoflurane, Halothane, and Cyclopropane in Dogs," *Anesthesiology,* 1974, pp. 465–470). Test at level .05 to see whether there is an anesthetic effect on concentration.

			Dog		
	1	2	3	4	5
Isoflurane	.28	.51	1.00	.39	.29
Halothane	.30	.39	.63	.38	.21
Cyclopropane	1.07	1.35	.69	.28	1.24

	6	7	8	9	10
Isoflurane	.36	.32	.69	.17	.33
Halothane	.88	.39	.51	.32	.42
Cyclopropane	1.53	.49	.56	1.02	.30

Supplementary Exercises Chapter 15 (34–40)

34. The paper "Effects of a Rice-Rich Versus Potato-Rich Diet on Glucose, Lipoprotein, and Cholesterol Metabolism in Noninsulin-Dependent Diabetics" (*Amer. J. Clinical Nutr.,* 1984, pp. 598–606) gives the accompanying data on cholesterol-synthesis rate for eight diabetic subjects. Subjects were fed a standardized diet with potato or rice as the major carbohydrate source. Participants received both diets for specified periods of time, with cholesterol synthesis rate (mmol/day) measured at the end of each dietary period. The analysis presented in this paper used a distribution-free test. Use such a test with significance level .05 to determine whether the true mean cholesterol-synthesis rate differs significantly for the two sources of carbohydrates.

Cholesterol-synthesis rate

Subject	1	2	3	4	5	6	7	8
Potato	1.88	2.60	1.38	4.41	1.87	2.89	3.96	2.31
Rice	1.70	3.84	1.13	4.97	.86	1.93	3.36	2.15

35. High-pressure sales tactics or door-to-door salespeople can be quite offensive. Many people succumb to such tactics, sign a purchase agreement, and later regret their actions. In the mid-1970s, the Federal Trade Commission implemented regulations clarifying and extending rights of purchasers to cancel such agreements. The accompanying data is a subset of that given in the paper "Evaluating the FTC Cooling-Off Rule" (*J. Consumer Affairs,* 1977, pp. 101–106). Individual observations are cancellation rates for each of nine salespeople during each of 4 years. Use an appropriate test at level .05 to see if true average cancellation rate depends on the year.

Salesperson

	1	2	3	4	5	6	7	8	9
1973	2.8	5.9	3.3	4.4	1.7	3.8	6.6	3.1	0.0
1974	3.6	1.7	5.1	2.2	2.1	4.1	4.7	2.7	1.3
1975	1.4	.9	1.1	3.2	.8	1.5	2.8	1.4	.5
1976	2.0	2.2	.9	1.1	.5	1.2	1.4	3.5	1.2

36. The given data on phosphorus concentration in topsoil for four different soil treatments appeared in the article "Fertilisers for Lotus and Clover Establishment on a Sequence of Acid Soils on the East Otago Uplands" (*N. Zeal. J. Exper. Ag.,* 1984, pp. 119–129). Use a distribution-free procedure to test the null hypothesis of no difference in true mean phosphorus concentration (mg/g) for the four soil treatments.

Treatment					
I	8.1	5.9	7.0	8.0	9.0
II	11.5	10.9	12.1	10.3	11.9
III	15.3	17.4	16.4	15.8	16.0
IV	23.0	33.0	28.4	24.6	27.7

37. Refer to the data of Exercise 36 and compute a 95% CI for the difference between true average concentrations for treatments II and III.

38. The study reported in "Gait Patterns During Free Choice Ladder Ascents" (*Human Movement Sci.,* 1983, pp. 187–195) was motivated by publicity concerning the increased accident rate for individuals climbing ladders. A number of different gait patterns were used by subjects climbing a portable straight ladder according to specified instructions. The ascent times for seven subjects who used a lateral gait and six subjects who used a four-beat diagonal gait are given.

Lateral	.86	1.31	1.64	1.51	1.53	1.39	1.09
Diagonal	1.27	1.82	1.66	.85	1.45	1.24	

a. Carry out a test using $\alpha = .05$ to see if the data suggests any difference in the true average ascent times for the two gaits.

b. Compute a 95% CI for the difference between the true average gait times.

39. Suppose we wish to test

H_0: the X and Y distributions are identical

versus

H_a: the X distribution is less spread out than the Y distribution

The accompanying figure pictures X and Y distributions for which H_a is true. The Wilcoxon rank-sum test is not appropriate in this situation because when H_a is true as pictured, the Y's will tend to be at the extreme ends of the combined sample (resulting in small and large Y ranks), so the sum of X ranks will result in a W value that is neither large nor small.

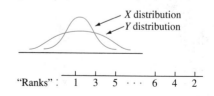

Consider modifying the procedure for assigning ranks as follows: After the combined sample of $m + n$ observations is ordered, the smallest observation is given rank 1, the largest observation is given rank 2, the second smallest is given rank 3, the second largest is given rank 4, and so on. Then if H_a is true as pictured, the X values will tend to be in the middle of the sample and thus receive large ranks. Let W' denote the sum of the X ranks and consider rejecting H_0 in favor of H_a when $w' \geq c$. When H_0 is true, every possible set of X ranks has the same probability, so W' *has the same distribution as does W when H_0 is true.* Thus, c can be chosen from Appendix Table A.10 to yield a level α test. The data below refers to medial muscle thickness for arterioles from the lungs of both children who died from sudden infant death syndrome (x's) and a control group of children (y's). Carry out the test of H_0 versus H_a at level .05.

SIDS   4.0   4.4   4.8   4.9
Control   3.7   4.1   4.3   5.1   5.6

See Lehmann for more information on this, called the Siegel–Tukey test.

40. The ranking procedure described in Exercise 39 is somewhat asymmetric, because the smallest observation receives rank 1 while the largest receives rank 2, and so on. Suppose both the smallest and the largest receive rank 1, the second smallest and second largest receive rank 2, and so on, and let W'' be the sum of the X ranks. The null distribution of W'' is not identical to the null distribution of W, so different tables are needed. Consider the case $m = 3$, $n = 4$. List all 35 possible orderings of the three X values among the seven observations (e.g., 1, 3, 7 or 4, 5, 6), assign ranks in the manner described, compute the value of W'' for each possibility, and then tabulate the null distribution of W''. For the test that rejects if $w'' \geq c$, what value of c prescribes approximately a level .10 test? This is the Ansari–Bradley test; see the book by Hollander and Wolfe.

Bibliography

Hollander, Myles, and Douglas Wolfe, *Nonparametric Statistical Methods,* Wiley, New York, 1973. A very good reference on distribution-free methods with an excellent collection of tables.

Lehmann, Erich, *Nonparametrics: Statistical Methods Based on Ranks,* Holden-Day, San Francisco, 1975. An excellent discussion of the most important distribution-free methods, presented with a great deal of insightful commentary.

Marascuilo, Leonard, and Maryellen McSweeney, *Nonparametric and Distribution-Free Methods for the Social Sciences,* Brooks/Cole, Pacific Grove, CA, 1977. A good survey, with particular attention given to multiple comparison methods.

16 Quality Control Methods

Introduction

Quality characteristics of manufactured products have received much attention from design engineers and production personnel as well as those concerned with financial management. An article of faith over the years was that very high quality levels and economic well-being were incompatible goals. Recently, however, it has become increasingly apparent that raising quality levels can lead to decreased costs, a greater degree of consumer satisfaction, and thus increased profitability. This has resulted in renewed emphasis on statistical techniques for designing quality into products and for identifying quality problems at various stages of production and distribution.

Control charting is now used extensively in industry as a diagnostic technique for monitoring production processes to identify instability and unusual circumstances. After an introduction to basic ideas in Section 16.1, a number of different control charts are presented in the next four sections. The basis for most of these lies in our earlier work concerning probability distributions of various statistics such as a sample mean $\bar{X}$ and sample proportion $\hat{p} = X/n$.

Another commonly encountered situation in industrial settings involves a decision by a customer as to whether a batch of items offered by a supplier is of acceptable quality. In the last section of the chapter, we briefly survey some acceptance sampling methods for deciding, based on sample data, on the disposition of a batch.

Besides control charts and acceptance sampling plans, which were first developed in the 1920s and 1930s, statisticians and engineers have recently introduced many new statistical methods for identifying types and levels of production inputs that will ensure high-quality output. Japanese investigators, and in particular the engineer/statistician G. Taguchi and his disciples, have been very influential in this respect, and there is now a large body of material known as "Taguchi methods." The ideas of experimental design, and in particular fractional factorial experiments, are key ingredients. There is still much controversy in the statistical community as to which designs and methods of analysis are best suited to the task at hand. A recent critique is contained in the expository paper by George Box et al., cited in the chapter bibliography, and the book by Thomas Ryan listed there is also a good source of information.

16.1 General Comments on Control Charts

A central message throughout this book has been the pervasiveness of naturally occurring variation associated with any characteristic or attribute of different individuals or objects. In a manufacturing context, no matter how carefully machines are calibrated, environmental factors are controlled, materials and other inputs are monitored, and workers are trained, diameter will vary from bolt to bolt, some plastic sheets will be stronger than others, some fuses will be defective while others are not, and so on. We might think of such natural random variation as uncontrollable background noise.

There are, however, other sources of variation that may have a pernicious impact on the quality of items produced by some process. Such variation may be attributable to contaminated material, incorrect machine settings, unusual tool wear, and the like. These sources of variation have been termed *assignable causes* in the quality control literature. **Control charts** provide a mechanism for recognizing situations where assignable causes may be adversely affecting product quality. Once a chart indicates an out-of-control situation, an investigation can be launched to identify causes and take corrective action.

A basic element of control charting is that samples have been selected from the process of interest at a sequence of time points. Depending on the aspect of the process under investigation, some statistic, such as the sample mean or sample proportion of defective items, is chosen. The value of this statistic is then calculated for each sample in turn. A traditional control chart then results from plotting these calculated values over time, as illustrated in Figure 16.1.

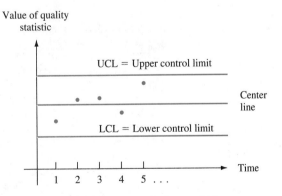

Figure 16.1 A prototypical control chart

Notice that in addition to the plotted points themselves, the chart has a center line and two control limits. The basis for the choice of a center line is sometimes a target value or design specification, for example, a desired value of the bearing diameter. In other cases, the height of the center line is estimated from the data. If the points on the chart all lie between the two control limits, the process is deemed to be in control. That is, the process is believed to be operating in a stable fashion reflecting only natural random variation. An out-of-control "signal" occurs whenever a plotted point falls outside the limits. This is assumed to be attributable to some assignable cause, and a search for such causes commences. The limits are designed so that an in-control process generates very few false alarms, whereas a process not in control quickly gives rise to a point outside the limits.

There is a strong analogy between the logic of control charting and our earlier work in hypothesis testing. The null hypothesis here is that the process is in control. When an in-control process yields a point *outside* the control limits (an out-of-control signal), a type I error has occurred. On the other hand, a type II error results when an out-of-control process produces a point *inside* the control limits. Appropriate choice of sample size and control limits (the latter corresponding to specifying a rejection region in hypothesis testing) will make the associated error probabilities suitably small.

We emphasize that "in control" is not synonymous with "meets design specifications or tolerance." The extent of natural variation may be such that the percentage of items not conforming to specification is much higher than can be tolerated. In such cases, a major restructuring of the process will be necessary to improve process capability. An in-control process is simply one whose behavior with respect to variation is stable over time, showing no indications of unusual extraneous causes.

As a final introductory comment, software for control charting is now widely available. The journal *Quality Progress* contains many advertisements for statistical quality control computer packages. In addition, SAS and the newest version of MINITAB, among other general-purpose packages, have attractive quality control capabilities.

Exercises   Section 16.1 (1–3)

1. A control chart for thickness of rolled-steel sheets is based on an upper control limit of .0520 in. and a lower limit of .0475 in. The first ten values of the quality statistic (in this case $\bar{x}$, the sample mean thickness of $n = 5$ sample sheets) are .0506, .0493, .0502, .0501, .0512, .0498, .0485, .0500, .0505, and .0483. Construct the initial part of the quality control chart and comment on its appearance.

2. Refer to Exercise 1 and suppose the ten most recent values of the quality statistic are .0493, .0485, .0490, .0503, .0492, .0486, .0495, .0494, .0493, and .0488. Construct the relevant portion of the corresponding control chart, and comment on its appearance.

3. Suppose a control chart is constructed so that the probability of a point falling outside the control limits when the process is actually in control is .002. What is the probability that ten successive points (based on independently selected samples) will be within the control limits? What is the probability that 25 successive points will all lie within the control limits? What is the smallest number of successive points plotted for which the probability of observing at least one outside the control limits exceeds .10?

Control Charts for Process Location

Suppose the quality characteristic of interest is associated with a variable whose observed values result from making measurements. For example, the characteristic might be resistance of electrical wire (ohms), internal diameter of molded rubber expansion joints (cm), or hardness of a certain alloy (Brinell units). One important use of control charts is to see whether some measure of location of the variable's distribution remains stable over time. The most popular chart for this purpose is the $\bar{X}$ chart.

The $\bar{X}$ Chart Based on Known Parameter Values

Because there is uncertainty as to the value of the variable for any particular item or specimen, we denote such a *random* variable (rv) by X. Assume that for an in-control process, X has a normal distribution with mean value μ and standard deviation σ. Then if $\bar{X}$ denotes the sample mean for a random sample of size n selected at a particular time point, we know that

1. $E(\bar{X}) = \mu$
2. $\sigma_{\bar{X}} = \sigma/\sqrt{n}$
3. $\bar{X}$ has a normal distribution.

It follows that

$$P(\mu - 3\sigma_{\bar{X}} \leq \bar{X} \leq \mu + 3\sigma_{\bar{X}}) = P(-3.00 \leq Z \leq 3.00) = .9974$$

where Z is a standard normal rv.* It is thus highly likely that for an in-control process, the sample mean will fall within 3 standard deviations ($3\sigma_{\bar{X}}$) of the process mean μ.

Consider first the case in which the values of both μ and σ are known. Suppose that at each of the time points 1, 2, 3, . . . , a random sample of size n is available and let $\bar{x}_1, \bar{x}_2, \bar{x}_3, \ldots$ denote the calculated values of the corresponding sample means. An $\bar{X}$ chart results from plotting these $\bar{x}_i$'s over time—that is, plotting the points $(1, \bar{x}_1)$, $(2, \bar{x}_2)$, $(3, \bar{x}_3)$, and so on—and then drawing horizontal lines across the plot at

$$\text{LCL} = \text{lower control limit} = \mu - 3 \cdot \frac{\sigma}{\sqrt{n}}$$

$$\text{UCL} = \text{upper control limit} = \mu + 3 \cdot \frac{\sigma}{\sqrt{n}}$$

Such a plot is often called a 3-sigma chart. Any point outside the control limits suggests that the process may have been out of control at that time, so a search for assignable causes should be initiated.

---

*The use of charts based on 3 SD limits is traditional, but tradition is certainly not inviolable.

Example 16.1   Once each day, three specimens of motor oil are randomly selected from the production process, and each is analyzed to determine viscosity. The accompanying data (Table 16.1) is for a 25-day period. Extensive experience with this process suggests that when the process is in control, viscosity of a specimen is normally distributed with mean 10.5 and standard deviation .18. Thus, $\sigma_{\bar{x}} = \sigma/\sqrt{n} = .18/\sqrt{3} = .104$, so the 3 SD control limits are

$$\text{LCL} = \mu - 3 \cdot \frac{\sigma}{\sqrt{n}} = 10.5 - 3(.104) = 10.188$$

$$\text{UCL} = \mu + 3 \cdot \frac{\sigma}{\sqrt{n}} = 10.5 + 3(.104) = 10.812$$

All points on the control chart shown in Figure 16.2 are between the control limits, indicating stable behavior of the process mean over this time period (the standard deviation and range for each sample will be used in the next subsection).

Table 16.1 Viscosity data for Example 16.1

Day	Viscosity Observations			$\bar{x}$	s	Range
1	10.37	10.19	10.36	10.307	.101	.18
2	10.48	10.24	10.58	10.433	.175	.34
3	10.77	10.22	10.54	10.510	.276	.55
4	10.47	10.26	10.31	10.347	.110	.21
5	10.84	10.75	10.53	10.707	.159	.31
6	10.48	10.53	10.50	10.503	.025	.05
7	10.41	10.52	10.46	10.463	.055	.11
8	10.40	10.38	10.69	10.490	.173	.31
9	10.33	10.35	10.49	10.390	.087	.16
10	10.73	10.45	10.30	10.493	.218	.43
11	10.41	10.68	10.25	10.447	.217	.43
12	10.00	10.60	10.71	10.437	.382	.71
13	10.37	10.50	10.34	10.403	.085	.16
14	10.47	10.60	10.75	10.607	.140	.28
15	10.46	10.46	10.56	10.493	.058	.10
16	10.44	10.68	10.32	10.480	.183	.36
17	10.65	10.42	10.26	10.443	.196	.39
18	10.73	10.72	10.83	10.760	.061	.11
19	10.39	10.75	10.27	10.470	.250	.48
20	10.59	10.23	10.35	10.390	.183	.36
21	10.47	10.67	10.64	10.593	.108	.20
22	10.40	10.55	10.38	10.443	.093	.17
23	10.24	10.71	10.27	10.407	.263	.47
24	10.37	10.69	10.40	10.487	.177	.32
25	10.46	10.35	10.37	10.393	.059	.11

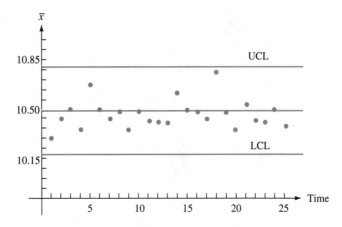

Figure 16.2 $\bar{X}$ chart for the viscosity data of Example 16.1

$\bar{X}$ Charts Based on Estimated Parameters

In practice it frequently happens that values of μ and σ are unknown, so must be estimated from sample data prior to determining control limits. This is especially true when a process is first subjected to a quality control analysis. Again denote the number of observations in each sample by n and let k represent the number of samples available. Typical values of n are 3, 4, 5, or 6, and it is recommended that k be at least 20. We assume that the k samples were gathered during a period when the process was believed to be in control. More will be said about this assumption shortly.

With $\bar{x}_1, \bar{x}_2, \ldots , \bar{x}_k$ denoting the k calculated sample means, the usual estimate of μ is simply the average of these means:

$$\hat{\mu} = \bar{\bar{x}} = \frac{\displaystyle\sum_{i=1}^{k} \bar{x}_i}{k}$$

We present two different commonly used methods for estimating σ, one based on the k sample standard deviations and the other on the k sample ranges (recall that the sample range is the difference between the largest and smallest sample observations). Prior to the wide availability of good calculators and statistical computer software, ease of hand calculation was of paramount consideration, so the range method predominated. However, in the case of a normal population distribution, the unbiased estimator of σ based on S is known to have smaller variance than that based on the sample range. Statisticians say that the former estimator is more *efficient* than the latter. The loss in efficiency for the estimator is slight when n is very small but becomes important for $n > 4$.

Recall that the sample standard deviation is not an unbiased estimator for σ. When $X_1, \ldots , X_n$ is a random sample from a normal distribution, it can be shown (cf. Exercise 6.37) that

$$E(S) = a_n \cdot \sigma$$

where

$$a_n = \frac{\sqrt{2}\,\Gamma(n/2)}{\sqrt{n-1}\,\Gamma[(n-1)/2]}$$

and $\Gamma(\cdot)$ denotes the gamma function (see Section 4.4). A tabulation of a_n for selected n follows:

n	3	4	5	6	7	8
a_n	.886	.921	.940	.952	.959	.965

Let

$$\bar{S} = \frac{\sum_{i=1}^{k} S_i}{k}$$

where $S_1, S_2, \ldots, S_k$ are the sample standard deviations for the k samples. Then

$$E(\bar{S}) = \frac{1}{k} E\left(\sum_{i=1}^{k} S_i\right) = \frac{1}{k} \sum_{i=1}^{k} E(S_i) = \frac{1}{k} \sum_{i=1}^{k} a_n \cdot \sigma = a_n \cdot \sigma$$

Thus,

$$E\left(\frac{\bar{S}}{a_n}\right) = \frac{1}{a_n} E(\bar{S}) = \frac{1}{a_n} \cdot a_n \cdot \sigma = \sigma$$

so $\hat{\sigma} = \bar{S}/a_n$ is an unbiased estimator of σ.

Control Limits Based on the Sample Standard Deviations

$$\text{LCL} = \bar{\bar{x}} - 3 \cdot \frac{\bar{s}}{a_n \sqrt{n}}$$

$$\text{UCL} = \bar{\bar{x}} + 3 \cdot \frac{\bar{s}}{a_n \sqrt{n}}$$

where

$$\bar{\bar{x}} = \frac{\sum_{i=1}^{k} \bar{x}_i}{k}, \quad \bar{s} = \frac{\sum_{i=1}^{k} s_i}{k}$$

Example 16.2   Referring to the viscosity data of Example 16.1, we had $n = 3$ and $k = 25$. The values of $\bar{x}_i$ and s_i ($i = 1, \ldots, 25$) appear in Table 16.1, from which it follows that $\bar{\bar{x}} = 261.896/25 = 10.476$ and $\bar{s} = 3.834/25 = .153$. With $a_3 = .886$, we have

$$\text{LCL} = 10.476 - 3 \cdot \frac{.153}{.886\sqrt{3}} = 10.476 - .299 = 10.177$$

$$\text{UCL} = 10.476 + 3 \cdot \frac{.153}{.886\sqrt{3}} = 10.476 + .299 = 10.775$$

These limits differ a bit from the earlier ones based on $\mu = 10.5$ and $\sigma = .18$ because now $\hat{\mu} = 10.476$ and $\hat{\sigma} = \bar{s}/a_3 = .173$. Inspection of Table 16.1 shows that every $\bar{x}_i$ is between these new limits, so again no out-of-control situation is evident.   ∎

To obtain an estimate of σ based on the sample range, note that if $X_1, \ldots, X_n$ form a random sample from a normal distribution then

$$R = \text{range}(X_1, \ldots, X_n)$$
$$= \max(X_1, \ldots, X_n) - \min(X_1, \ldots, X_n)$$
$$= \max(X_1 - \mu, \ldots, X_n - \mu) - \min(X_1 - \mu, \ldots, X_n - \mu)$$
$$= \sigma \left\{ \max\left(\frac{X_1 - \mu}{\sigma}, \ldots, \frac{X_n - \mu}{\sigma} \right) - \min\left(\frac{X_1 - \mu}{\sigma}, \ldots, \frac{X_n - \mu}{\sigma} \right) \right\}$$
$$= \sigma \cdot \{ \max(Z_1, \ldots, Z_n) - \min(Z_1, \ldots, Z_n) \}$$

where $Z_1, \ldots, Z_n$ are independent standard normal rv's. Thus,

$$E(R) = \sigma \cdot E(\text{range of a standard normal sample})$$
$$= \sigma \cdot b_n$$

so that R/b_n is an unbiased estimator of σ.

Now denote the ranges for the k samples in the quality control data set by r_1, $r_2, \ldots, r_k$. The argument just given implies that the estimate

$$\hat{\sigma} = \frac{\frac{1}{k} \sum_{i=1}^{k} r_i}{b_n} = \frac{\bar{r}}{b_n}$$

comes from an unbiased estimator for σ. Selected values of b_n appear in the accompanying table [their computation is based on using statistical theory and numerical integration to determine $E(\min(Z_1, \ldots, Z_n))$ and $E(\max(Z_1, \ldots, Z_n))$].

n	3	4	5	6	7	8
b_n	1.693	2.058	2.325	2.536	2.706	2.844

Control Limits
Based on the
Sample Ranges

$$LCL = \bar{\bar{x}} - 3 \cdot \frac{\bar{r}}{b_n \sqrt{n}}$$

$$UCL = \bar{\bar{x}} + 3 \cdot \frac{\bar{r}}{b_n \sqrt{n}}$$

where $\bar{r} = \Sigma_{i=1}^{k} r_i / k$ and $r_1, \ldots, r_k$ are the k individual sample ranges.

Example 16.3
(Example 16.2
continued)

It is easily verified from Table 16.1 that $\bar{r} = .292$, so $\hat{\sigma} = .292/b_3 = .292/1.693 = .172$ and

$$LCL = 10.476 - 3 \cdot \frac{.292}{1.693\sqrt{3}} = 10.476 - .299 = 10.177$$

$$UCL = 10.476 + 3 \cdot \frac{.292}{1.693\sqrt{3}} = 10.476 + .299 = 10.775$$

These limits are identical to those based on $\bar{s}$, and again every $\bar{x}_i$ lies between the limits.   ■

Recomputing Control Limits

We have assumed that the sample data used for estimating μ and σ was obtained from an in-control process. Suppose, though, that one of the points on the resulting control chart falls outside the control limits. Then if an assignable cause for this out-of-control situation can be found and verified, it is recommended that new control limits be calculated after deleting the corresponding sample from the data set. Similarly, if more than one point falls outside the original limits, new limits should be determined after eliminating any such point for which an assignable cause can be identified and dealt with. It may even happen that one or more points fall outside the new limits, in which case the deletion/recomputation process must be repeated.

Performance Characteristics of Control Charts

Generally speaking, a control chart will be effective if it gives very few out-of-control signals when the process is in control, but shows a point outside the control limits almost as soon as the process goes out of control. One assessment of a chart's effectiveness is based on the notion of "error probabilities." Suppose the variable of interest is normally distributed with known σ (the same value for an in-control or out-of-control process). In addition, consider a 3-sigma chart based on the target value μ_0, with $\mu = \mu_0$ when the process is in control. One error probability is

$\alpha = P$(a single sample gives a point outside the control limits when $\mu = \mu_0$)

$= P(\bar{X} > \mu_0 + 3\sigma/\sqrt{n}$ or $\bar{X} < \mu_0 - 3\sigma/\sqrt{n}$ when $\mu = \mu_0$)

$$= P\left(\frac{\bar{X} - \mu_0}{\sigma/\sqrt{n}} > 3 \text{ or } \frac{\bar{X} - \mu_0}{\sigma/\sqrt{n}} < -3 \text{ when } \mu = \mu_0\right)$$

The standardized variable $Z = (\bar{X} - \mu_0)/(\sigma/\sqrt{n})$ has a standard normal distribution when $\mu = \mu_0$, so

$$\alpha = P(Z > 3 \text{ or } Z < -3) = \Phi(-3.00) + 1 - \Phi(3.00) = .0026$$

If 3.09 rather than 3 had been used to determine the control limits (this is customary in Great Britain), then

$$\alpha = P(Z > 3.09 \text{ or } Z < -3.09) = .0020$$

The use of 3-sigma limits makes it highly unlikely that an out-of-control signal will result from an in-control process.

Now suppose the process goes out of control because μ has shifted to $\mu + \Delta\sigma$ (Δ might be positive or negative); Δ is the number of standard deviations by which μ has changed. A second error probability is

$$\beta = P\left(\begin{array}{c}\text{a single sample gives a point inside} \\ \text{the control limits when } \mu = \mu_0 + \Delta\sigma\end{array}\right)$$
$$= P(\mu_0 - 3\sigma/\sqrt{n} < \bar{X} < \mu_0 + 3\sigma/\sqrt{n} \text{ when } \mu = \mu_0 + \Delta\sigma)$$

We now standardize by first subtracting $\mu_0 + \Delta\sigma$ from each term inside the parentheses and then dividing by $\sigma/\sqrt{n}$. This gives

$$\beta = P(-3 - \sqrt{n}\Delta < \text{standard normal rv} < 3 - \sqrt{n}\Delta)$$
$$= \Phi(3 - \sqrt{n}\Delta) - \Phi(-3 - \sqrt{n}\Delta)$$

This error probability depends on Δ, which determines the size of the shift, and on the sample size n. In particular, for fixed Δ, β will decrease as n increases (the larger the sample size, the more likely it is that an out-of-control signal will result), and for fixed n, β decreases as $|\Delta|$ increases (the larger the magnitude of a shift, the more likely it is that an out-of-control signal will result). The accompanying table gives β for selected values of Δ when $n = 4$:

Δ	.25	.50	.75	1.00	1.50	2.00	2.50	3.00
β when $n = 4$	.9936	.9772	.9332	.8413	.5000	.1587	.0668	.0013

It is clear that a small shift is quite likely to go undetected in a single sample.

If 3 is replaced by 3.09 in the control limits, then α decreases from .0026 to .002, but for any fixed n and σ, β will increase. This is just a manifestation of the inverse relationship between the two types of error probabilities in hypothesis testing. For example, changing 3 to 2.5 will increase α and decrease β.

The error probabilities discussed thus far are computed under the assumption that the variable of interest is normally distributed. If the distribution is only slightly nonnormal, the Central Limit Theorem effect implies that $\bar{X}$ will have approximately a normal distribution even when n is small, in which case the stated error probabilities will be approximately correct. This is, of course, no longer the case when the variable's distribution deviates considerably from normality.

A second performance assessment involves expected or average run length needed to observe an out-of-control signal. When the process is in control, we should expect to observe many samples before seeing one whose $\bar{x}$ lies outside the control limits. On the other hand, if a process goes out of control, the expected number of samples in order to detect this should be small.

Let p denote the probability that a single sample yields an $\bar{x}$ value outside the control limits, that is,

$$p = P(\bar{X} < \mu_0 - 3\sigma/\sqrt{n} \text{ or } \bar{X} > \mu_0 + 3\sigma/\sqrt{n})$$

Consider first an in-control process, so that $\bar{X}_1, \bar{X}_2, \bar{X}_3, \ldots$ are all normally distributed with mean value μ_0 and standard deviation $\sigma/\sqrt{n}$. Define an rv Y by

Y = the first i for which $\bar{X}_i$ falls outside the control limits

Thinking of each sample number as a trial and an out-of-control sample as a success, Y is the number of (independent) trials necessary to observe a success. This Y has a geometric distribution, and we showed in Example 3.18 that $E(Y) = 1/p$. The acronym ARL (for *average run length*) is often used in place of $E(Y)$. Because $p = \alpha$ for an in-control process, we have

$$\text{ARL} = E(Y) = \frac{1}{p} = \frac{1}{\alpha} = \frac{1}{.0026} = 384.62$$

Replacing 3 in the control limits by 3.09 gives ARL = 1/.002 = 500.

Now suppose that at a particular time point the process mean shifts to $\mu = \mu_0 + \Delta\sigma$. If we define Y to be the first i subsequent to the shift for which a sample generates an out-of-control signal, it is again true that ARL = $E(Y) = 1/p$, but now $p = 1 - \beta$. The accompanying table gives selected ARLs for a 3-sigma chart when $n = 4$. These results again show the chart's effectiveness in detecting large shifts but also its inability to quickly identify small shifts. When sampling is done rather infrequently, a great many items are likely to be produced before a small shift in μ is detected. The CUSUM procedures discussed in Section 16.5 were developed to address this deficiency.

Δ	.25	.50	.75	1.00	1.50	2.00	2.50	3.00
ARL when $n = 4$	156.25	43.86	14.97	6.30	2.00	1.19	1.07	1.0013

Supplemental Rules for $\bar{X}$ Charts

The inability of $\bar{X}$ charts with 3-sigma limits to quickly detect small shifts in the process mean has prompted investigators to develop procedures that provide improved behavior in this respect. One approach involves introducing additional conditions that cause an out-of-control signal to be generated. The following conditions were recommended by Western Electric (then a subsidiary of AT&T). An intervention to take corrective action is appropriate whenever one of these conditions is satisfied:

1. Two out of three successive points fall outside 2-sigma limits on the same side of the center line.
2. Four out of five successive points fall outside 1-sigma limits on the same side of the center line.
3. Eight successive points fall on the same side of the center line.

A quality control text should be consulted for a discussion of these and other supplemental rules. In Section 16.5, we present a different type of procedure to overcome the $\bar{X}$ chart's deficiency.

Robust Control Charts

The presence of outliers in the sample data tends to reduce the sensitivity of control-charting procedures when parameters must be estimated. This is because the control limits are moved outward from the center line, making the identification of unusual points more difficult. We do *not* want the statistic whose values are plotted to be resistant to outliers because that would mask any out-of-control signal. For example, plotting sample medians would be less effective than plotting $\bar{x}_1, \bar{x}_2, \ldots$ as is done on an $\bar{X}$ chart.

The paper "Robust Control Charts" by David M. Rocke (*Technometrics*, 1989, pp. 173–184) presents a study of procedures for which control limits are based on statistics resistant to the effects of outliers. Rocke recommends control limits calculated from the *interquartile range* (IQR), which is very similar to the fourth spread introduced in Chapter 1. In particular,

$$\text{IQR} = \begin{cases} (\text{2nd largest } x_i) - (\text{2nd smallest } x_i) & n = 4, 5, 6, 7 \\ (\text{3rd largest } x_i) - (\text{3rd smallest } x_i) & n = 8, 9, 10, 11 \end{cases}$$

For a random sample from a normal distribution, $E(\text{IQR}) = k_n \sigma$; the values of k_n are given in the accompanying table:

n	4	5	6	7	8
k_n	.596	.990	1.282	1.512	.942

The suggested control limits are

$$\text{LCL} = \bar{\bar{x}} - 3 \cdot \frac{\text{IQR}}{k_n \sqrt{n}}, \quad \text{UCL} = \bar{\bar{x}} + 3 \cdot \frac{\text{IQR}}{k_n \sqrt{n}}$$

The values of $\bar{x}_1, \bar{x}_2, \bar{x}_3, \ldots$ are plotted. Simulations reported in the paper indicated that the performance of the chart with these limits is superior to that of the traditional $\bar{X}$ chart.

Exercises   Section 16.2 (4–13)

4. In the case of known μ and σ, what control limits would be such that the probability of a single point being outside the limits for an in-control process is .005?

5. Consider a 3-sigma control chart with center line at μ_0 and based on $n = 5$. Assuming normality, calculate the probability that a single point will fall outside the control limits when the actual process mean is

    a. $\mu_0 + .5\sigma$

    b. $\mu_0 - \sigma$

    c. $\mu_0 + 2\sigma$

6. The accompanying table gives data on moisture content for specimens of a certain type of fabric. Determine control limits for a chart with center line at height 13.00 based on $\sigma = .600$, construct the control chart, and comment on its appearance.

7. Refer to the data given in Exercise 6 and construct a control chart with an estimated center line and limits based on using the sample standard deviations to estimate σ. Is there any evidence that the process is out of control?

8. Refer to Exercises 6 and 7 and now employ control limits based on using the sample ranges to estimate σ. Does the process appear to be in control?

9. The accompanying table gives sample means and standard deviations, each based on $n = 6$ observations of the refractive index of fiber-optic cable. Construct a control chart and comment on its appearance. [HINT: $\Sigma \bar{x}_i = 2317.07$ and $\Sigma s_i = 30.34$.]

Sample No.	Moisture-Content Observations					$\bar{x}$	s	Range
1	12.2	12.1	13.3	13.0	13.0	12.72	.536	1.2
2	12.4	13.3	12.8	12.6	12.9	12.80	.339	.9
3	12.9	12.7	14.2	12.5	12.9	13.04	.669	1.7
4	13.2	13.0	13.0	12.6	13.9	13.14	.477	1.3
5	12.8	12.3	12.2	13.3	12.0	12.52	.526	1.3
6	13.9	13.4	13.1	12.4	13.2	13.20	.543	1.5
7	12.2	14.4	12.4	12.4	12.5	12.78	.912	2.2
8	12.6	12.8	13.5	13.9	13.1	13.18	.526	1.3
9	14.6	13.4	12.2	13.7	12.5	13.28	.963	2.4
10	12.8	12.3	12.6	13.2	12.8	12.74	.329	.9
11	12.6	13.1	12.7	13.2	12.3	12.78	.370	.9
12	13.5	12.3	12.8	13.1	12.9	12.92	.438	1.2
13	13.4	13.3	12.0	12.9	13.1	12.94	.559	1.4
14	13.5	12.4	13.0	13.6	13.4	13.18	.492	1.2
15	12.3	12.8	13.0	12.8	13.5	12.88	.432	1.2
16	12.6	13.4	12.1	13.2	13.3	12.92	.554	1.3
17	12.1	12.7	13.4	13.0	13.9	13.02	.683	1.8
18	13.0	12.8	13.0	13.3	13.1	13.04	.182	.5
19	12.4	13.2	13.0	14.0	13.1	13.14	.573	1.6
20	12.7	12.4	12.4	13.9	12.8	12.84	.619	1.5
21	12.6	12.8	12.7	13.4	13.0	12.90	.316	.8
22	12.7	13.4	12.1	13.2	13.3	12.94	.541	1.3

Day	$\bar{x}$	s	Day	$\bar{x}$	s
1	95.47	1.30	13	97.02	1.28
2	97.38	.88	14	95.55	1.14
3	96.85	1.43	15	96.29	1.37
4	96.64	1.59	16	96.80	1.40
5	96.87	1.52	17	96.01	1.58
6	96.52	1.27	18	95.39	.98
7	96.08	1.16	19	96.58	1.21
8	96.48	.79	20	96.43	.75
9	96.63	1.48	21	97.06	1.34
10	96.50	.80	22	98.34	1.60
11	97.22	1.42	23	96.42	1.22
12	96.55	1.65	24	95.99	1.18

10. Refer to Exercise 9. An assignable cause was found for the unusually high sample average refractive index on day 22. Recompute control limits after deleting the data from this day. What do you conclude?

11. Consider the control chart based on control limits $\mu_0 \pm 2.81 \; \sigma/\sqrt{n}$.

   a. What is the ARL when the process is in control?

   b. What is the ARL when $n = 4$ and the process mean has shifted to $\mu = \mu_0 + \sigma$?

   c. How do the values of parts (a) and (b) compare to the corresponding values for a 3-sigma chart?

12. Apply the supplemental rules suggested in the text to the data of Exercise 6. Are there any out-of-control signals?

13. Calculate control limits for the data of Exercise 6, using the robust procedure presented in this section.

16.3 Control Charts for Process Variation

The control charts discussed in the previous section were designed to control location (equivalently, central tendency) of a process, with particular attention to the mean as a measure of location. It is equally important to ensure that a process is under control with respect to variation. In fact, most practitioners recommend that control be established on variation *prior to* constructing an $\bar{X}$ chart or any other chart for controlling location. In this section, we consider charts for variation based on the sample standard deviation S and also charts based on the sample range R. The former are generally preferred because the standard deviation gives a more efficient assessment of variation than does the range, but R charts were used first and tradition dies hard.

The S Chart

We again suppose that k independently selected samples are available, each one consisting of n observations on a normally distributed variable. Denote the sample standard deviations by $s_1, s_2, \ldots, s_k$, with $\bar{s} = \Sigma s_i/k$. The values $s_1, s_2, s_3, \ldots$ are plotted in sequence on an S chart. The center line of the chart will be at height $\bar{s}$, and the 3-sigma limits necessitate determining $3\sigma_S$ (just as 3-sigma limits of an $\bar{X}$ chart required $3\sigma_{\bar{X}} = 3\sigma/\sqrt{n}$, with σ then estimated from the data).

Recall that for any rv Y, $V(Y) = E(Y^2) - [E(Y)]^2$, and that a sample variance S^2 is an unbiased estimator of σ^2, that is, $E(S^2) = \sigma^2$. Thus,

$$V(S) = E(S^2) - [E(S)]^2 = \sigma^2 - (a_n\sigma)^2 = \sigma^2(1 - a_n^2)$$

where values of a_n for $n = 3, \ldots, 8$ are tabulated in the previous section. The standard deviation of S is then

$$\sigma_S = \sqrt{V(S)} = \sigma\sqrt{1 - a_n^2}$$

It is natural to estimate σ using $s_1, \ldots, s_k$ as was done in the previous section, namely, $\hat{\sigma} = \bar{s}/a_n$. Substituting $\hat{\sigma}$ for σ in the expression for σ_S gives the quantity used to calculate 3-sigma limits.

> The 3-sigma control limits for an S control chart are
>
> $$LCL = \bar{s} - 3\bar{s}\sqrt{1 - a_n^2}/a_n$$
>
> $$UCL = \bar{s} + 3\bar{s}\sqrt{1 - a_n^2}/a_n$$
>
> The expression for LCL will be negative if $n \leq 5$, in which case it is customary to use LCL = 0.

Example 16.4   Table 16.2 displays observations on stress resistance of plastic sheets (the force, in psi, necessary to crack a sheet). There are $k = 22$ samples, obtained at equally spaced time points, and $n = 4$ observations in each sample. It is easily verified that $\Sigma s_i = 51.10$ and $\bar{s} = 2.32$, so the center of the S chart will be at 2.32 (though because $n = 4$, LCL = 0 and the center line will not be midway between the control limits). From the previous section, $a_4 = .921$, from which the UCL is

Table 16.2 Stress-resistance data for Example 16.4

Sample No.	Observations				SD	Range
1	29.7	29.0	28.8	30.2	.64	1.4
2	32.2	29.3	32.2	32.9	1.60	3.6
3	35.9	29.1	32.1	31.3	2.83	6.8
4	28.8	27.2	28.5	35.7	3.83	8.5
5	30.9	32.6	28.3	28.3	2.11	4.3
6	30.6	34.3	34.8	26.3	3.94	8.5
7	32.3	27.7	30.9	27.8	2.30	4.6
8	32.0	27.9	31.0	30.8	1.76	4.1
9	24.2	27.5	28.5	31.1	2.85	6.9
10	33.7	24.4	34.3	31.0	4.53	9.9
11	35.3	33.2	31.4	28.0	3.09	7.3
12	28.1	34.0	31.0	30.8	2.41	5.9
13	28.7	28.9	25.8	29.7	1.71	3.9
14	29.0	33.0	30.2	30.1	1.71	4.0
15	33.5	32.6	33.6	29.2	2.07	4.4
16	26.9	27.3	32.1	28.5	2.37	5.2
17	30.4	29.6	31.0	33.8	1.83	4.2
18	29.0	28.9	31.8	26.7	2.09	5.1
19	33.8	30.9	31.7	28.2	2.32	5.6
20	29.7	27.9	29.1	30.1	.96	2.2
21	27.9	27.7	30.2	32.9	2.43	5.2
22	30.0	31.4	27.7	28.1	1.72	3.7

$$UCL = 2.32 + 3(2.32)(\sqrt{1 - (.921)^2})/.921$$
$$= 2.32 + 2.94$$
$$= 5.26$$

The resulting control chart is shown in Figure 16.3. All plotted points are well within the control limits, suggesting stable process behavior with respect to variation.

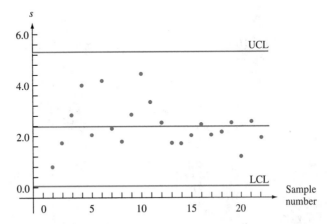

Figure 16.3 *S* chart for stress-resistance data of Example 16.4   ■

The R Chart

Let $r_1, r_2, \ldots, r_k$ denote the k sample ranges and $\bar{r} = \Sigma r_i/k$. The center line of an R chart will be at height $\bar{r}$. Determination of the control limits requires σ_R, where R denotes the range (prior to making observations—as a random variable) of a random sample of size n from a normal distribution with mean value μ and standard deviation σ. Because

$$R = \max(X_1, \ldots, X_n) - \min(X_1, \ldots, X_n)$$
$$= \sigma\{\max(Z_1, \ldots, Z_n) - \min(Z_1, \ldots, Z_n)\}$$

where $Z_i = (X_i - \mu)/\sigma$, and the Z_i's are standard normal rv's, it follows that

$$\sigma_R = \sigma \cdot \begin{pmatrix} \text{standard deviation of the range of a random sample} \\ \text{of size } n \text{ from a standard normal distribution} \end{pmatrix}$$
$$= \sigma \cdot c_n$$

The values of c_n for $n = 3, \ldots, 8$ appear in the accompanying table:

n	3	4	5	6	7	8
c_n	.888	.880	.864	.848	.833	.820

It is customary to estimate σ by $\hat{\sigma} = \bar{r}/b_n$ as discussed in the previous section. This gives $\hat{\sigma}_R = c_n\bar{r}/b_n$ as the estimated standard deviation of R.

> The 3-sigma limits for an R chart are
>
> $$\text{LCL} = \bar{r} - 3c_n\bar{r}/b_n$$
> $$\text{UCL} = \bar{r} + 3c_n\bar{r}/b_n$$
>
> The expression for LCL will be negative if $n \leq 6$, in which case LCL = 0 should be used.

Example 16.5
(Example 16.4 continued)

Table 16.2 yields $\Sigma r_i = 115.3$ and $\bar{r} = 5.24$. Since $n = 4$, LCL = 0. With $b_4 = 2.058$ and $c_4 = .880$,

$$\text{UCL} = 5.24 + 3(.880)(5.24)/2.058 = 11.96$$

The R chart appears as Figure 16.4. As with the S chart, all points are between the limits, indicating an in-control process as far as variation is concerned.

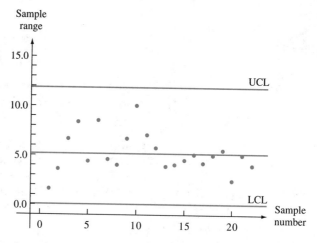

Figure 16.4 R chart for stress-resistance data of Example 16.5　　■

Charts Based on Probability Limits

Consider an $\bar{X}$ chart based on the in-control (target) value μ_0 and known σ. When the variable of interest is normally distributed and the process is in control,

$$P(\bar{X}_i > \mu_0 + 3\sigma/\sqrt{n}) = .0013 = P(\bar{X}_i < \mu_0 - 3\sigma/\sqrt{n})$$

That is, the probability that a point on the chart falls above the UCL is .0013, as is the probability that the point falls below the LCL (using 3.09 in place of 3 gives .001 for each probability). When control limits are based on estimates of μ and σ, these probabilities will be approximately correct provided that n is not too small and k is at least 20.

By contrast, it is *not* the case for a 3-sigma S chart that $P(S_i > \text{UCL}) = P(S_i < \text{LCL}) = .0013$, nor is it true for a 3-sigma R chart that

$P(R_i > \text{UCL}) = P(R_i < \text{LCL}) = .0013$. This is because neither the sample standard deviation S nor the sample range R has a normal distribution even when the population distribution is normal. Instead, both S and R have skewed distributions. The best that can be said for 3-sigma S and R charts is that an in-control process is quite unlikely to yield a point at any particular time that is outside the control limits. Some authors have advocated the use of control limits for which the "exceedance probability" for each limit is approximately .001. The book *Statistical Methods for Quality Improvement* (see the chapter bibliography) contains more information on this topic.

Exercises   Section 16.3 (14–18)

14. A manufacturer of dustless chalk instituted a quality control program to monitor chalk density. The sample standard deviations of densities for 24 different subgroups, each consisting of $n = 8$ chalk specimens, were as follows:

.204 .315 .096 .184 .230 .212 .322 .287
.145 .211 .053 .145 .272 .351 .159 .214
.388 .187 .150 .229 .276 .118 .091 .056

Calculate limits for an S chart, construct the chart, and check for out-of-control points. If there is an out-of-control point, delete it and repeat the process.

15. Subgroups of power supply units are selected once each hour from an assembly line and the high-voltage output of each unit is determined.

   a. Suppose the sum of the resulting sample ranges for 30 subgroups, each consisting of four units, is 85.2. Calculate control limits for an R chart.

   b. Repeat part (a) if each subgroup consists of eight units and the sum is 106.2.

16. Calculate control limits for both an S chart and an R chart using the moisture content data from Exercise 6. Then check for the presence of any out-of-control signals.

17. Calculate control limits for an S chart from the refractive index data of Exercise 9. Does the process appear to be in control with respect to variability? Why, or why not?

18. When S^2 is the sample variance of a normal random sample, $(n - 1)S^2/\sigma^2$ has a chi-squared distribution with $n - 1$ df, so

$$P\left(\chi^2_{.999,n-1} < \frac{(n - 1)S^2}{\sigma^2} < \chi^2_{.001,n-1}\right) = .998$$

from which

$$P\left(\frac{\sigma^2 \chi^2_{.999,n-1}}{n - 1} < S^2 < \frac{\sigma^2 \chi^2_{.001,n-1}}{n - 1}\right) = .998$$

This suggests that an alternative chart for controlling process variation involves plotting the sample variances and using the control limits

$$\text{LCL} = \bar{s^2}\chi^2_{.999,n-1}/(n - 1)$$

$$\text{UCL} = \bar{s^2}\chi^2_{.001,n-1}/(n - 1)$$

Construct the corresponding chart for the data of Exercise 9. [HINT: The lower- and upper- tailed chi-squared critical values for 5 df are .210 and 20.515, respectively.]

16.4   Control Charts for Attributes

The term *attribute data* is used in the quality control literature to describe two situations:

   1. Each item produced is either defective or nondefective (conforms to specifications or does not).

2. A single item may have one or more defects, and the number of defects is determined.

In the former case, a control chart is based on the binomial distribution; in the latter case, the Poisson distribution is the basis for a chart.

The p Chart for Fraction Defective

Suppose that when a process is in control, the probability that any particular item is defective is p (equivalently, p is the long-run proportion of defective items for an in-control process) and that different items are independent of one another with respect to their conditions. Consider a sample of n items obtained at a particular time, and let X be the number of defectives and $\hat{p} = X/n$. Because X has a binomial distribution, $E(X) = np$ and $V(X) = np(1 - p)$, so

$$E(\hat{p}) = p, \quad V(\hat{p}) = \frac{p(1 - p)}{n}$$

Also, if $np \geq 5$ and $n(1 - p) \geq 5$, $\hat{p}$ has approximately a normal distribution.

In the case of known p (or a chart based on target value), the control limits are

$$\text{LCL} = p - 3\sqrt{\frac{p(1 - p)}{n}}, \quad \text{UCL} = p + 3\sqrt{\frac{p(1 - p)}{n}}$$

If each sample consists of n items, the number of defective items in the ith sample is x_i, and $\hat{p}_i = x_i/n$, then $\hat{p}_1, \hat{p}_2, \hat{p}_3, \ldots$ are plotted on the control chart.

Usually the value of p must be estimated from the data. Suppose that k samples from what is believed to be an in-control process are available and let

$$\bar{p} = \frac{\sum\limits_{i=1}^{k} \hat{p}_i}{k}$$

The estimate $\bar{p}$ is then used in place of p in the aforementioned control limits.

The p chart for the fraction of defective items has its center line at height $\bar{p}$ and control limits

$$\text{LCL} = \bar{p} - 3\sqrt{\frac{\bar{p}(1 - \bar{p})}{n}}$$

$$\text{UCL} = \bar{p} + 3\sqrt{\frac{\bar{p}(1 - \bar{p})}{n}}$$

If LCL is negative, it is replaced by 0.

Example 16.6   A sample of 100 cups from a particular dinnerware pattern was selected on each of 25 successive days, and each was examined for defects. The resulting numbers of unacceptable cups and corresponding sample proportions are as follows:

Day (i)	1	2	3	4	5	6	7	8	9	10	11	12	13
x_i	7	4	3	6	4	9	6	7	5	3	7	8	4
$\hat{p}_i$	.07	.04	.03	.06	.04	.09	.06	.07	.05	.03	.07	.08	.04

Day (i)	14	15	16	17	18	19	20	21	22	23	24	25
x_i	6	2	9	7	6	7	11	6	7	4	8	6
$\hat{p}_i$	.06	.02	.09	.07	.06	.07	.11	.06	.07	.04	.08	.06

Assuming that the process was in control during this period, let's establish control limits and construct a p chart. We have that $\Sigma\hat{p}_i = 1.52$, giving $\bar{p} = 1.52/25 = .0608$ and

$$LCL = .0608 - 3\sqrt{(.0608)(.9392)/100} = .0608 - .0717 = -.0109$$
$$LCL = .0608 + 3\sqrt{(.0608)(.9392)/100} = .0608 + .0717 = .1325$$

The LCL is therefore set at 0. The control chart pictured in Figure 16.5 shows that all points are within the control limits. This is consistent with the assumption of an in-control process.

Figure 16.5 Control chart for fraction-defective data of Example 16.6

The c Chart for Number of Defectives

We now consider situations in which the observation at each time point is the number of defects in a unit of some sort. The unit may consist of a single item (e.g., one automobile) or a group of items (e.g., blemishes on a set of four tires). In the second case, the group size is assumed to be the same at each time point.

The control chart for number of defectives is based on the Poisson probability distribution. Recall that if Y is a Poisson random variable with parameter θ, then

$$E(Y) = \theta, \quad V(Y) = \theta, \quad \sigma_Y = \sqrt{\theta}$$

Also, Y has approximately a normal distribution when θ is large ($\theta \geq 10$ will suffice for most purposes). Furthermore, if $Y_1, Y_2, \ldots, Y_n$ are independent Poisson variables with parameters $\theta_1, \theta_2, \ldots, \theta_n$, it can be shown that $Y_1 + \cdots + Y_n$ has a Poisson distribution with parameter $\theta_1 + \cdots + \theta_n$. In particular, if $\theta_1 = \cdots = \theta_n = \theta$ (the distribution of the number of defects per item is the same for each item), then the Poisson parameter is $\lambda = n\theta$.

Let λ denote the Poisson parameter for the number of defects in a unit (it is the expected number of defects per unit). In the case of known λ (or a chart based on a target value),

$$\text{LCL} = \lambda - 3\sqrt{\lambda}, \quad \text{UCL} = \lambda + 3\sqrt{\lambda}$$

With x_i denoting the total number of defects in the ith unit ($i = 1, 2, 3, \ldots$), then points at heights $x_1, x_2, x_3, \ldots$ are plotted on the chart.

Usually the value of λ must be estimated from the data. Since $E(X_i) = \lambda$, it is natural to use the estimate $\hat{\lambda} = \bar{x}$ (based on $x_1, x_2, \ldots, x_k$).

The c chart for the number of defectives in a unit has center line at height $\bar{x}$ and

$$\text{LCL} = \bar{x} - 3\sqrt{\bar{x}}$$
$$\text{UCL} = \bar{x} + 3\sqrt{\bar{x}}$$

If LCL is negative, it is replaced by 0.

Example 16.7   A company manufactures metal panels that are baked after first being coated with a slurry of powdered ceramic. Flaws sometimes appear in the finish of these panels, and the company wishes to establish a control chart for the number of flaws. The numbers of flaws in each of the 24 panels sampled at regular time intervals are as follows:

```
 7 10   9 12 13 6 13   7 5 11   8 10
13   9 21 10   6 8   3 12 7 11  14 10
```

With $\Sigma x_i = 235$, $\hat{\lambda} = \bar{x} = 235/24 = 9.79$. The control limits are

$$\text{LCL} = 9.79 - 3\sqrt{9.79} = .40, \quad \text{UCL} = 9.79 + 3\sqrt{9.79} = 19.18$$

The control chart appears in Figure 16.6. The point corresponding to the fifteenth panel lies above the UCL. Upon investigation, the slurry used on that panel was discovered to be of unusually low viscosity (an assignable cause). Eliminating that observation from the data set gives $\bar{x} = 214/23 = 9.30$ and new control limits

$$\text{LCL} = 9.30 - 3\sqrt{9.30} = .15, \quad \text{UCL} = 9.30 + 3\sqrt{9.30} = 18.45$$

Figure 16.6 Control chart for number of flaws data of Example 16.7

The remaining 23 observations all lie between these limits, indicating an in-control process. ■

Control Charts Based on Transformed Data

The use of 3-sigma control limits is presumed to result in $P(\text{statistic} < \text{LCL}) \approx P(\text{statistic} > \text{UCL}) \approx .0013$ when the process is in control. However, when p is small, the normal approximation to the distribution of $\hat{p} = X/n$ will often not be very accurate in the extreme tails. Table 16.3 gives evidence of this behavior for selected values of p and n (the value of p is used to calculate the control limits). In many cases, the probability that a single point falls outside the control limits is very different from the nominal probability of .0026.

Table 16.3 In-control probabilities for a p chart

p	n	$P(\hat{p} < \text{LCL})$	$P(\hat{p} > \text{UCL})$	P (out-of-control point)
.10	100	.00003	.00198	.00201
.10	200	.00048	.00299	.00347
.10	400	.00044	.00171	.00215
.05	200	.00004	.00266	.00270
.05	400	.00020	.00207	.00227
.05	600	.00031	.00189	.00220
.02	600	.00007	.00275	.00282
.02	800	.00036	.00374	.00410
.02	1000	.00023	.00243	.00266

This problem can be remedied by applying a transformation to the data. Let $h(X)$ denote a function applied to transform the binomial variable X. Then $h(\cdot)$ should be chosen so that $h(X)$ has approximately a normal distribution *and* this approximation is accurate in the tails. A recommended transformation is based on the arcsin (i.e., $\sin^{-1}$) function:

$$Y = h(X) = \sin^{-1}(\sqrt{X/n})$$

Then Y is approximately normal with mean value $\sin^{-1}(\sqrt{p})$ and variance $1/4n$; note that the variance is independent of p. Let $y_i = \sin^{-1}(\sqrt{x_i/n})$. Then points on the control chart are plotted at heights $y_1, y_2, \ldots$. For known n the control limits are

$$\text{LCL} = \sin^{-1}(\sqrt{p}) - 3\sqrt{1/4n}, \quad \text{UCL} = \sin^{-1}(\sqrt{p}) + 3\sqrt{1/4n}$$

When p is not known, $\sin^{-1}(\sqrt{p})$ is replaced by $\bar{y}$.

Similar comments apply to the Poisson distribution when λ is small. The suggested transformation is $Y = h(X) = 2\sqrt{X}$, which has mean value $2\sqrt{\lambda}$ and variance 1. Resulting control limits are $2\sqrt{\lambda} \pm 3$ when λ is known and $\bar{y} \pm 3$ otherwise. The book *Statistical Methods for Quality Improvement* listed in the chapter bibliography discusses these issues in greater detail.

Exercises   Section 16.4 (19–26)

19. On each of the previous 25 days, 100 electronic devices of a certain type were randomly selected and subjected to a severe heat stress test. The total number of items that failed to pass the test was 578.

 a. Determine control limits for a 3-sigma p chart.

 b. The highest number of failed items on a given day was 39, and the lowest number was 13. Does either of these correspond to an out-of-control point? Explain.

20. A sample of 200 ROM computer chips was selected on each of 30 consecutive days, and the number of nonconforming chips on each day was as follows: 10, 18, 24, 17, 37, 19, 7, 25, 11, 24, 29, 15, 16, 21, 18, 17, 15, 22, 12, 20, 17, 18, 12, 24, 30, 16, 11, 20, 14, 28. Construct a p chart and examine it for any out-of-control points.

21. When $n = 150$, what is the smallest value of $\bar{p}$ for which the LCL in a p chart is positive?

22. Refer to the data of Exercise 20 and construct a control chart using the $\sin^{-1}$ transformation as suggested in the text.

23. The accompanying observations are numbers of defects in 25 1-square-yard specimens of woven fabric of a

certain type: 3, 7, 5, 3, 4, 2, 8, 4, 3, 3, 6, 7, 2, 3, 2, 4, 7, 3, 2, 4, 4, 1, 5, 4, 6. Construct a c chart for the number of defects.

24. For what $\bar{x}$ values will the LCL in a c chart be negative?

25. In some situations, the sizes of sampled specimens vary, and larger specimens are expected to have more defects than smaller ones. For example, sizes of fabric samples inspected for flaws might vary over time. Alternatively, the number of items inspected might change with time. Let

$$u_i = \frac{\text{the number of defects observed at time } i}{\text{size of entity inspected at time } i}$$

$$= \frac{x_i}{g_i}$$

where "size" might refer to area, length, volume, or simply the number of items inspected. Then a *u* **chart** plots $u_1, u_2, \ldots$, has center line $\bar{u}$, and the control limits for the ith observations are $\bar{u} \pm 3\sqrt{\bar{u}/g_i}$.

 Painted panels were examined in time sequence, and for each one, the number of blemishes in a specified sampling region was determined. The surface area (ft^2) of the region examined varied from panel to panel. Results were as follows:

Panel	Area Examined	No. of Blemishes	Panel	Area Examined	No. of Blemishes
1	.8	3	11	.6	1
2	.6	2	12	.8	3
3	.8	3	13	.8	5
4	.8	2	14	1.0	4
5	1.0	5	15	1.0	6
6	1.0	5	16	1.0	12
7	.8	10	17	.8	3
8	1.0	12	18	.6	3
9	.6	4	19	.6	5
10	.6	2	20	.6	1

Construct a u chart.

26. Construct a control chart for the data of Exercise 23 by using the transformation suggested in the text.

16.5 CUSUM Procedures

A defect of the traditional $\bar{X}$ chart is its inability to detect a relatively small change in a process mean. This is largely a consequence of the fact that whether a process is judged out of control at a particular time depends only on the sample at that time, and not on the past history of the process. **Cumulative sum (CUSUM)** control charts and procedures have been designed to remedy this defect.

There are two equivalent versions of a CUSUM procedure for a process mean, one graphical and the other computational. The computational version is used almost exclusively in practice, but the logic behind the procedure is most easily grasped by first considering the graphical form.

The V-Mask

Let μ_0 denote a target value or goal for the process mean and define *cumulative sums* by

$$S_1 = \bar{x}_1 - \mu_0$$

$$S_2 = (\bar{x}_1 - \mu_0) + (\bar{x}_2 - \mu_0) = \sum_{i=1}^{2} (\bar{x}_i - \mu_0)$$

.

.

.

$$S_l = (\bar{x}_1 - \mu_0) + \cdots + (\bar{x}_l - \mu_0) = \sum_{i=1}^{l} (\bar{x}_i - \mu_0)$$

(in the absence of a target value, $\bar{x}$ is used in place of μ_0). These cumulative sums are plotted over time. That is, at time l, we plot a point at height S_l. At the current time point r, the plotted points are $(1, S_1)$, $(2, S_2)$, $(3, S_3)$, . . . , (r, S_r).

Now a V-shaped "mask" is superimposed on the plot, as shown in Figure 16.7. The point 0, which lies a distance d behind the point at which the two arms of the mask intersect, is positioned at the current CUSUM point (r, S_r). At time r, the process is judged out of control if any of the plotted points lies outside the V-mask—either above the upper arm or below the lower arm. When the process is in control, the $\bar{x}_i$'s will vary around the target value μ_0, so successive S_i's should vary around 0. Suppose, however, that at a certain time, the process mean shifts to a value larger than the target. From that point on differences $\bar{x}_i - \mu_0$ will tend to be positive, so that successive S_i's will increase and plotted points will drift upward. If a shift has occurred prior to the current time point r, there is a good chance that (r, S_r) will be substantially higher than some other points in the plot, in which case these other points will be below the lower arm of the mask. Similarly, a shift to a value smaller than the target will subsequently result in points above the upper arm of the mask.

Any particular V-mask is determined by specifying the "lead distance" d and "half-angle" θ, or, equivalently, by specifying d and the length h of the vertical line

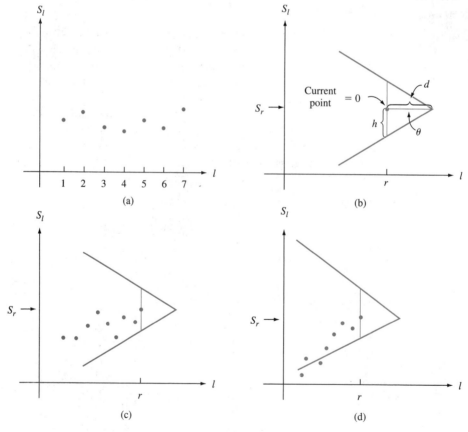

Figure 16.7 CUSUM plots: (a) Successive points (I, S_i), in a CUSUM plot; (b) a V-mask with $0 = (r, S_r)$; (c) an in-control process; (d) an out-of-control process

segment from 0 to the lower (or to the upper) arm of the mask. One method for deciding which mask to use involves specifying the size of a shift in the process mean that is of particular concern to an investigator. Then the parameters of the mask are chosen to give desired values of α and β, the false-alarm probability and probability of not detecting the specified shift, respectively. An alternative method involves selecting the mask that yields specified values of the ARL (average run length) both for an in-control process and for a process in which the mean has shifted by a designated amount. After developing the computational form of the CUSUM procedure, we will illustrate the second method of construction.

Example 16.8 A wood products company manufactures charcoal briquettes for barbecues. It packages these briquettes in bags of various sizes, the largest of which is supposed to contain 40 lb. Table 16.4 displays the weights of bags from 16 different samples, each of size $n = 4$. The first 10 of these were drawn from a normal distribution with $\mu = \mu_0 = 40$ and $\sigma = .5$. Starting with the eleventh sample, the mean has shifted upward to $\mu = 40.3$.

Table 16.4 Observations, $\bar{x}$'s, and cumulative sums for Example 16.8

Sample Number	Observations				$\bar{x}$	$\Sigma(\bar{x}_i - 40)$
1	40.77	39.95	40.86	39.21	40.20	.20
2	38.94	39.70	40.37	39.88	39.72	-.08
3	40.43	40.27	40.91	40.05	40.42	.34
4	39.55	40.10	39.39	40.89	39.98	.32
5	41.01	39.07	39.85	40.32	40.06	.38
6	39.06	39.90	39.84	40.22	39.76	.14
7	39.63	39.42	40.04	39.50	39.65	-.21
8	41.05	40.74	40.43	39.40	40.41	.20
9	40.28	40.89	39.61	40.48	40.32	.52
10	39.28	40.49	38.88	40.72	39.84	.36
11	40.57	40.04	40.85	40.51	40.49	.85
12	39.90	40.67	40.51	40.53	40.40	1.25
13	40.70	40.54	40.73	40.45	40.61	1.86
14	39.58	40.90	39.62	39.83	39.98	1.84
15	40.16	40.69	40.37	39.69	40.23	2.07
16	40.46	40.21	40.09	40.58	40.34	2.41

Figure 16.8 displays an $\bar{X}$ chart with center line at height 40 and control limits at

$$\mu_0 \pm 3\sigma_{\bar{x}} = 40 \pm 3 \cdot (.5/\sqrt{4}) = 40 \pm .75$$

No point on the chart lies outside the control limits. This chart suggests a stable process for which the mean has remained on target.

Figure 16.8 $\bar{X}$ control chart for the data of Example 16.8

Figure 16.9 shows CUSUM plots with a particular V-mask superimposed. The plot in Figure 16.9a is for current time $r = 12$. All points in this plot lie inside the arms of the mask. However, the plot for $r = 13$ displayed in Figure 16.9b gives an out-of-control signal. The point falling below the lower arm of the mask suggests an increase in the value of the process mean. The mask at $r = 16$ is even more emphatic in its out-of-control message. This is in marked contrast to the ordinary $\bar{X}$ chart.

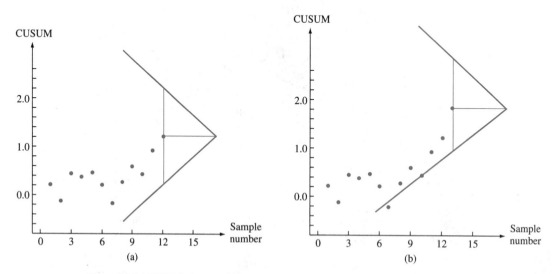

Figure 16.9 CUSUM plots and V-masks for data of Example 16.8. (a) V-mask at time $r = 12$; Process in control; (b) V-mask at time $r = 13$; Out-of-control signal

Computational Form of the CUSUM Procedure

We first describe the computational version of the CUSUM procedure, then show its equivalence to the graphical form, and finally discuss designing a procedure to meet specified criteria.

Let $d_0 = e_0 = 0$, and calculate $d_1, d_2, d_3, \ldots$ and $e_1, e_2, e_3, \ldots$ recursively using the relationships

$$d_l = \max[0, d_{l-1} + (\bar{x}_l - (\mu_0 + k))]$$
$$e_l = \max[0, e_{l-1} - (\bar{x}_l - (\mu_0 - k))] \qquad (l = 1, 2, 3, \ldots)$$

Here the symbol k denotes the slope of the lower arm of the V-mask, and its value is customarily taken as $\Delta/2$ (where Δ is the size of a shift in μ on which attention is focused).

If at current time r, either $d_r > h$ or $e_r > h$, the process is judged to be out of control. The first inequality suggests the process mean has shifted to a value greater than the target, whereas $e_r > h$ indicates a shift to a smaller value.

Example 16.9   Reconsider the charcoal briquette data displayed in Table 16.4 of Example 16.8. The target value is $\mu_0 = 40$, and the size of a shift to be quickly detected is $\Delta = .3$. Thus,

$$k = \frac{\Delta}{2} = .15, \quad \mu_0 + k = 40.15, \quad \mu_0 - k = 39.85$$

so

$$d_l = \max[0, d_{l-1} + (\bar{x}_l - 40.15)]$$
$$e_l = \max[0, e_{l-1} - (\bar{x}_l - 39.85)]$$

Calculations of the first few d_l's proceeds as follows:

$$d_0 = 0$$
$$d_1 = \max[0, d_0 + (\bar{x}_1 - 40.15)]$$
$$= \max[0, 0 + (40.20 - 40.15)]$$
$$= .05$$
$$d_2 = \max[0, d_1 + (\bar{x}_2 - 40.15)]$$
$$= \max[0, .05 + (39.72 - 40.15)]$$
$$= 0$$
$$d_3 = \max[0, d_2 + (\bar{x}_3 - 40.15)]$$
$$= \max[0, 0 + (40.42 - 40.15)]$$
$$= .27$$

The remaining calculations are summarized in Table 16.5.

Table 16.5 CUSUM calculations for Example 16.9

Sample Number	$\bar{x}_i$	$\bar{x}_i - 40.15$	d_i	$\bar{x}_i - 39.85$	e_l
1	40.20	.05	.05	.35	0
2	39.72	−.43	0	−.13	.13
3	40.42	.27	.27	.57	0
4	39.98	−.17	.10	.13	0
5	40.06	−.09	.01	.21	0
6	39.76	−.39	0	−.09	.09
7	39.65	−.50	0	−.20	.29
8	40.41	.26	.26	.56	0
9	40.32	.17	.43	.47	0
10	39.84	−.31	.12	−.01	.01
11	40.49	.34	.46	.64	0
12	40.40	.25	.71	.55	0
13	40.61	.46	1.17	.76	0
14	39.98	−.17	1.00	.13	0
15	40.23	.08	1.08	.38	0
16	40.34	.19	1.27	.49	0

The value $h = .95$ gives a CUSUM procedure with desirable properties—false alarms (incorrect out-of-control signals) rarely occur, yet a shift of $\Delta = .3$ will usually be detected rather quickly. With this value of h, the first out-of-control signal comes after the thirteenth sample is available. Since $d_{13} = 1.17 > .95$, it appears that the mean has shifted to a value larger than the target. This is the same message as the one given by the V-mask in Figure 16.9b. ∎

Equivalence of the V-Mask and Computational Form

Again let r denote the current time point, so that $\bar{x}_1, \bar{x}_2, \ldots, \bar{x}_r$ are available. Figure 16.10 displays a V-mask with the point labeled 0 at (r, S_r). The slope of the lower arm, which we denote by k, is h/d. Thus, the points on the lower arm above r, $r - 1$, $r - 2$, ... are at heights $S_r - h$, $S_r - h - k$, $S_r - h - 2k$, and so on.

The process is in control if all points are on or between the arms of the mask. We wish to describe this condition algebraically. To do so, let

$$T_l = \sum_{i=1}^{l} [\bar{x}_i - (\mu_0 + k)], \qquad l = 1, 2, 3, \ldots, r$$

The conditions under which all points are on or above the lower arm are

$$S_r - h \leq S_r \quad \text{(trivially satisfied) i.e., } S_r \leq S_r + h$$
$$S_r - h - k \leq S_{r-1} \qquad \text{i.e., } S_r \leq S_{r-1} + h + k$$
$$S_r - h - 2k \leq S_{r-2} \qquad \text{i.e., } S_r \leq S_{r-2} + h + 2k$$
$$\vdots \qquad\qquad\qquad \vdots$$

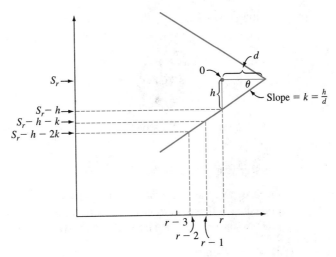

Figure 16.10 A V-mask with slope of lower arm = k

Now subtract rk from both sides of each inequality to obtain

$$S_r - rk \leq S_r - rk + h \qquad \text{i.e., } T_r \leq T_r + h$$
$$S_r - rk \leq S_{r-1} - (r-1)k + h \quad \text{i.e., } T_r \leq T_{r-1} + h$$
$$S_r - rk \leq S_{r-2} - (r-2)k + h \quad \text{i.e., } T_r \leq T_{r-2} + h$$

. . .

. . .

. . .

Thus, all plotted points lie on or above the lower arm if and only if (iff) $T_r - T_r \leq h$, $T_r - T_{r-1} \leq h$, $T_r - T_{r-2} \leq h$, and so on. This is equivalent to

$$T_r - \min(T_1, T_2, \ldots, T_r) \leq h$$

In a similar manner, if we let

$$V_r = \sum_{i=1}^{r} [\bar{x}_i - (\mu_0 - k)] = S_r + rk$$

it can be shown that all points lie on or below the upper arm iff

$$\max(V_1, \ldots, V_r) - V_r \leq h$$

If we now let

$$d_r = T_r - \min(T_1, \ldots, T_r)$$
$$e_r = \max(V_1, \ldots, V_r) - V_r$$

it is easily seen that $d_1, d_2, \ldots$ and $e_1, e_2, \ldots$ can be calculated recursively as illustrated earlier. For example, the expression for d_r follows from consideration of two cases:

1. $\min(T_1, \ldots, T_r) = T_r$,   whence $d_r = 0$
2. $\min(T_1, \ldots, T_r) = \min(T_1, \ldots, T_{r-1})$,   so that

$$d_r = T_r - \min(T_1, \ldots, T_{r-1})$$
$$= \bar{x}_r - (\mu_0 + k) + T_{r-1} - \min(T_1, \ldots, T_{r-1})$$
$$= \bar{x}_r - (\mu_0 + k) + d_{r-1}$$

Since d_r cannot be negative, it is the larger of these two quantities.

Designing a CUSUM Procedure

Let Δ denote the size of a shift in μ that is to be quickly detected using a CUSUM procedure.* It is common practice to let $k = \Delta/2$. Now suppose a quality control practitioner specifies desired values of two average run lengths:

1. ARL when the process is in control ($\mu = \mu_0$);
2. ARL when the process is out of control because the mean has shifted by Δ ($\mu = \mu_0 + \Delta$ or $\mu = \mu_0 - \Delta$)

A chart developed by Kenneth Kemp ("The Use of Cumulative Sums for Sampling Inspection Schemes," *Applied Statistics*, 1962, p. 23), called a *nomogram*, can then be used to determine values of h and n that achieve the specified ARLs.† This chart is shown as Figure 16.11. The method for using the chart is described in the accompanying box. Either the value of σ must be known or an estimate is used in its place.

Using the Kemp Nomogram	
	1. Locate the desired ARLs on the in-control and out-of-control scales. Connect these two points with a line.
	2. Note where the line crosses the k' scale, and solve for n using the equation

$$k' = \frac{\Delta/2}{\sigma/\sqrt{n}}$$

Then round n up to the nearest integer.
3. Connect the point on the k' scale with the point on the in-control ARL scale using a second line and note where this line crosses the h' scale. Then $h = (\sigma/\sqrt{n}) \cdot h'$.

The value $h = .95$ was employed in Example 16.9. In that situation, it follows that the in-control ARL is 500 and the out-of-control ARL (for $\Delta = .3$) is 7.

---

*This contrasts with earlier notation, where Δ represented the number of standard deviations by which μ changed.
†The word *nomogram* is not specific to this chart; nomograms are used for many other purposes.

Figure 16.11 The Kemp nomogram
SOURCE: Kemp, Kenneth W., "The Use of Cumulative Sums for Sampling Inspection Schemes," *Applied Statistics,* Vol.. XI, 1962, p. 23. With permission of the Royal Statistical Society.

Example 16.10   The target value for the diameter of the interior core of a hydraulic pump is 2.250 in. If the standard deviation of core diameter is $\sigma = .004$, what CUSUM procedure will yield an in-control ARL of 500 and an ARL of 5 when the mean core diameter shifts by the amount .003 in.?

Connecting the point 500 on the in-control ARL scale to the point 5 on the out-of-control ARL scale and extending the line to the k' scale on the far left in Figure 16.11 gives $k' = .74$. Thus,

$$k' = .74 = \frac{\Delta/2}{\sigma/\sqrt{n}} = \frac{.0015}{.004/\sqrt{n}} = .375\sqrt{n}$$

so

$$\sqrt{n} = \frac{.74}{.375} = 1.973, \quad n = (1.973)^2 = 3.894$$

The CUSUM procedure should therefore be based on the sample size $n = 4$. Now connecting .74 on the k' scale to 500 on the in-control ARL scale gives $h' = 3.2$, from which

$$h = (\sigma/\sqrt{n}) \cdot (3.2) = (.004/\sqrt{4})(3.2) = .0064$$

An out-of-control signal results as soon as either $d_r > .0064$ or $e_r > .0064$.   ■

We have discussed CUSUM procedures for controlling process location. There are also CUSUM procedures for controlling process variation and for attribute data. The chapter references should be consulted for information on these procedures.

Exercises   Section 16.5 (27–30)

27. Containers of a certain treatment for septic tanks are supposed to contain 16 oz of liquid. A sample of five containers is selected from the production line once each hour and the sample average content is determined. Consider the following results: 15.992, 16.051, 16.066, 15.912, 16.030, 16.060, 15.982, 15.899, 16.038, 16.074, 16.029, 15.935, 16.032, 15.960, 16.055. Using $\Delta = .10$ and $h = .20$, employ the computational form of the CUSUM procedure to investigate the behavior of this process.

28. The target value for the diameter of a certain type of drive shaft is .75 in. The size of the shift in the average diameter considered important to detect is .002 in. Sample average diameters for successive groups of $n = 4$ shafts are as follows: .7507, .7504, .7492, .7501, .7503, .7510, .7490, .7497, .7488, .7504, .7516, .7472,

.7489, .7483, .7471, .7498, .7460, .7482, .7470, .7493, .7462, .7481. Use the computational form of the CUSUM procedure with $h = .003$ to see if the process mean remained on target throughout the time of observation.

29. The standard deviation of a certain dimension on an aircraft part is .005 cm. What CUSUM procedure will give an in-control ARL of 600 and an out-of-control ARL of 4 when the mean value of the dimension shifts by .004 cm?

30. When the out-of-control ARL corresponds to a shift of one standard deviation in the process mean, what are the characteristics of the CUSUM procedure that has ARLs of 250 and 4.8 for the in-control and out-of-control conditions, respectively?

16.6   Acceptance Sampling

Items coming from a production process are often sent in groups to another company or commercial establishment. A group might consist of all units from a particular production run or shift, in a shipping container of some sort, sent in response to a particular order, and so on. The group of items is usually called a *lot,* the sender is referred to as a *producer,* and the recipient of the lot is the *consumer.* Our focus will be on situations in which each item is either defective or nondefective, with p denoting the proportion of defective units in the lot. The consumer would naturally want to accept the lot only if the value of p is suitably small. Acceptance sampling is that part of applied statistics dealing with methods for deciding whether the consumer should accept or reject a lot.

Until quite recently, control chart procedures and acceptance sampling techniques were regarded by practitioners as equally important parts of quality control methodology. This is no longer the case. The reason is that the use of control charts and other recently developed strategies offers the opportunity to design quality into a product, whereas acceptance sampling deals with what has already been produced and thus does not provide for any direct control over process quality. This led the late American quality control expert W. E. Deming, a major force in persuading the Japanese to make substantial use of quality control methodology, to argue strongly against the use of acceptance sampling in many situations. In a similar vein, the recent book by Ryan devotes several chapters to control charts and mentions acceptance sampling only in passing. As a reflection of this deemphasis, we content ourselves here with a brief introduction to basic concepts. More information can be found in several references of the chapter bibliography.

Single-Sampling Plans

The most straightforward type of acceptance sampling plan involves selecting a single random sample of size n and then rejecting the lot if the number of defectives in the sample exceeds a specified critical value c. Let the rv X denote the number of defective items in the lot and A denote the event that the lot is accepted. Then $P(A) = P(X \leq c)$ is a function of p; the larger the value of p, the smaller will be the probability of accepting the lot.

If the sample size n is large relative to N, $P(A)$ is calculated using the hypergeometric distribution (the number of defectives in the lot is Np):

$$P(X \leq c) = \sum_{x=0}^{c} h(x; n, Np, N) = \sum_{x=0}^{c} \frac{\binom{Np}{x} \cdot \binom{N(1-p)}{n-x}}{\binom{N}{n}}$$

When n is small relative to N (the rule of thumb suggested earlier was $n \leq .05N$, but some authors employ the less conservative rule $n \leq .10N$), the binomial distribution can be used:

$$P(X \leq c) = \sum_{x=0}^{c} b(x; n, p) = \sum_{x=0}^{c} \binom{n}{x} p^x (1-p)^{n-x}$$

Finally, if $P(A)$ is large only when p is small (this depends on the value of c), the Poisson approximation to the binomial distribution is justified:

$$P(X \leq c) \approx \sum_{x=0}^{c} p(x; np) = \sum_{x=0}^{c} \frac{e^{-np}(np)^x}{x!}$$

The behavior of a sampling plan can be nicely summarized by graphing $P(A)$ as a function of p. Such a graph is called the **operating characteristic (OC) curve** for the plan.

Example 16.11   Consider the sampling plan with critical value $c = 2$ and sample size $n = 50$, and suppose that the lot size N exceeds 1000, so the binomial distribution can be used. This gives

$$P(A) = P(X \le 2) = (1 - p)^{50} + 50p(1 - p)^{49} + 1255p^2(1 - p)^{48}$$

The accompanying table shows $P(A)$ for selected values of p, and the corresponding operating characteristic (OC) curve is shown in Figure 16.12.

p	.01	.02	.03	.04	.05	.06	.07	.08	.09	.10	.12	.15
$P(A)$	.986	.922	.811	.677	.541	.416	.311	.226	.161	.112	.051	.014

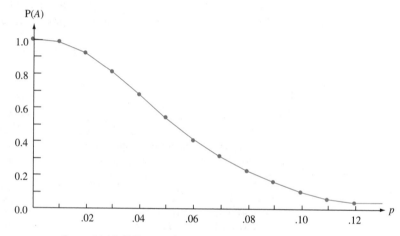

Figure 16.12 OC curve for sampling plan with $c = 2$, $n = 50$

The OC curve for the plan of Example 16.11 has $P(A)$ near 1 for p very close to 0. However, in many applications a defective rate of 8% [for which $P(A) = .226$] or even just 5% [$P(A) = .541$] would be considered excessive, in which case the acceptance probabilities are too high. Increasing the critical value c while holding n fixed gives a plan for which $P(A)$ increases at each p (except 0 and 1), so the new OC curve lies above the old one. This is desirable for p near 0 but not for larger values of p. Holding c constant while increasing n gives a lower OC curve, which is fine for larger p but not for p close to 0. We want an OC curve that is higher for very small p and lower for larger p. This can be achieved by both increasing n and adjusting c.

Designing a Single-Sample Plan

An effective sampling plan is one with the following characteristics:

1. It has a specified high probability of accepting lots that the producer considers to be of good quality.

2. It has a specified low probability of accepting lots that the consumer considers to be of poor quality.

A plan of this sort can be developed by proceeding as follows. Let's designate two different values of p, one for which $P(A)$ is a specified value close to 1 and the other for which $P(A)$ is a specified value near 0. These two values of p—say, p_1 and p_2—are often called the **acceptable quality level (AQL)** and the **lot tolerance percent defective (LTPD).** That is, we require a plan for which

1. $P(A) = 1 - \alpha$ when $p = p_1 = $ AQL (α small)
2. $P(A) = \beta$ when $p = p_2 = $ LTPD (β small)

This is analogous to seeking a test procedure with specified type I error probability α and specified type II error probability β when testing hypotheses. For example, we might have

$$\text{AQL} = .01, \quad \alpha = .05 \quad (P(A) = .95)$$
$$\text{LTPD} = .045, \quad \beta = .10 \quad (P(A) = .10)$$

Because X is discrete, we must typically be content with values of n and c that approximately satisfy these conditions.

Table 16.6 gives information from which n and c can be determined in the case $\alpha = .05$, $\beta = .10$.

Table 16.6 Factors for determining n and c for a single-sample plan with $\alpha = .05$, $\beta = .10$

c	np_1	np_2	p_2/p_1	c	np_i	np_2	p_2/p_1
0	.051	2.30	45.10	8	4.695	12.99	2.77
1	.355	3.89	10.96	9	5.425	14.21	2.62
2	.818	5.32	6.50	10	6.169	15.41	2.50
3	1.366	6.68	4.89	11	6.924	16.60	2.40
4	1.970	7.99	4.06	12	7.690	17.78	2.31
5	2.613	9.28	3.55	13	8.464	18.86	2.24
6	3.285	10.53	3.21	14	9.246	20.13	2.18
7	3.981	11.77	2.96	15	10.040	21.29	2.12

Example 16.12   Let's determine a plan for which AQL $= p_1 = .01$, LTPD $= p_2 = .045$. The ratio of p_2 to p_1 is

$$\frac{\text{LTPD}}{\text{AQL}} = \frac{p_2}{p_1} = \frac{.045}{.01} = 4.50$$

This value lies between the ratio 4.89 given in Table 16.6, for which $c = 3$, and 4.06, for which $c = 4$. Once one of these values of c is chosen, n can be determined either by dividing the np_1 value in Table 16.6 by p_1 or via np_2/p_2. Thus four different plans (two values of c, and for each two values of n) give approximately the specified value of α and β. Consider, for example, using $c = 3$ and

$$n = \frac{np_1}{p_1} = \frac{1.366}{.01} = 136.6 \approx 137$$

Then

$$\alpha = 1 - P(X \leq 3 \text{ when } p = p_1)$$

$$= 1 - \sum_{x=0}^{3} \binom{137}{x}(.01)^x(.99)^{137-x} = .050$$

(the Poisson approximation with $\lambda = 1.37$ also gives .050) and

$$\beta = P(X \leq 3 \text{ when } p = p_2) = .131$$

The plan with $c = 4$ and n determined from $np_2 = 7.99$ has $n = 178$, $\alpha = .034$, $\beta = .094$. The larger sample size results in a plan with both α and β smaller than the corresponding specified values.   ∎

The book by Douglas Montgomery cited in the chapter bibliography contains a chart from which c and n can be determined for *any* specified α and β.

It may happen that the number of defective items in the sample reaches $c + 1$ before all items have been examined. For example, in the case $c = 3$ and $n = 137$, it may be that the 125th item examined is the fourth defective item, so that the remaining 12 items need not be examined. However, it is generally recommended that all items be examined even when this does occur in order to provide a lot-by-lot quality history and estimates of p over time.

Double-Sampling Plans

In a double-sampling plan, the number of defective items x_1 in an initial sample of size n_1 is determined. There are then three possible courses of action: immediately accept the lot, immediately reject the lot, or take a second sample of n_2 items and reject or accept the lot depending on the total number of $x_1 + x_2$ of defective items in the two samples. Besides the two sample sizes, a specific plan is characterized by three further numbers c_1, r_1, and c_2 as follows:

1. Reject the lot if $x_1 \geq r_1$.

2. Accept the lot if $x_1 \leq c_1$.

3. If $c_1 < x_1 < r_1$, take a second sample; then accept the lot if $x_1 + x_2 \leq c_2$ and reject it otherwise.

Example 16.13   Consider the double-sampling plan with $n_1 = 80, n_2 = 80, c_1 = 2, r_1 = 5$, and $c_2 = 6$. Thus, the lot will be accepted if (1) $x_1 = 0, 1$, or 2; (2) $x_1 = 3$ and $x_2 = 0, 1, 2$, or 3; or (3) $x_1 = 4$ and $x_2 = 0, 1$, or 2.

Assuming that the lot size is large enough for the binomial approximation to apply, the probability $P(A)$ of accepting the lot is

$$P(A) = P(X_1 = 0, 1, \text{ or } 2) + P(X_1 = 3, X_2 = 0, 1, 2, \text{ or } 3)$$
$$+ P(X_1 = 4, X_2 = 0, 1, \text{ or } 2)$$

$$= \sum_{x_1=0}^{2} b(x_1; 80, p) + b(3; 80, p) \sum_{x_2=0}^{3} b(x_2; 80, p)$$

$$+ b(4; 80, p) \sum_{x_2 = 0}^{2} b(x_2; 80, p)$$

Again the graph of $P(A)$ versus p is the plan's OC curve. The OC curve for this plan appears in Figure 16.13.

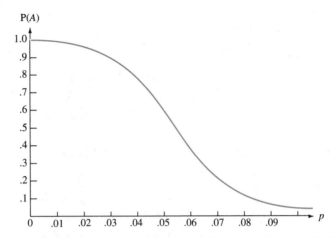

Figure 16.13 OC curve for the double-sampling plan of Example 16.13

One standard method for designing a double-sampling plan involves proceeding as suggested earlier for single-sample plans. Specify values p_1 and p_2 along with corresponding acceptance probabilities $1 - \alpha$ and β. Then find a plan that satisfies these conditions. The book by Montgomery provides tables similar to Table 16.6 for this purpose in the cases $n_2 = n_1$ and $n_2 = 2n_1$ with $1 - \alpha = .95, \beta = .10$. Much more extensive tabulations of plans are available in other sources.

Analogous to standard practice with single-sample plans, it is recommended that all items in the first sample be examined even when the $(r_1 + 1)$st defective is discovered prior to inspection of the n_1th item. However, it is customary to terminate inspection of the second sample if the number of defectives is sufficient to justify rejection before all items have been examined. This is referred to as *curtailment* in the second sample. Under curtailment, it can be shown that the expected number of items inspected in a double-sampling plan is smaller than the number of items examined in a single-sampling plan when the OC curves of the two plans are close to being identical. This is the major virtue of double-sampling plans. For more on these matters as well as a discussion of multiple and sequential sampling plans (these latter plans involve selecting items for inspection one by one rather than in groups), a book on quality control should be consulted.

Rectifying Inspection and Other Design Criteria

In some situations, sampling inspection is carried out using *rectification*. For single-sample plans, this means that each defective item in the sample is replaced with a satisfactory one, and if the number of defectives in the sample exceeds the acceptance cutoff *c, all* items in the lot are examined and good items are substituted for any defectives. Let N denote the lot size. One important characteristic of a sampling plan with rectifying inspection is **average outgoing quality,** denoted by **AOQ.** This is the long-run proportion of defective items among those sent on after the sampling plan is employed. Now defectives will occur only among the $N - n$ items not inspected in a lot judged acceptable on the basis of a sample. Suppose, for example, $P(A) = P(X \le c) = .985$ when $p = .01$. Then in the long run, 98.5% of the $N - n$ items not in the sample will not be inspected, of which we expect 1% to be defective. This implies that the expected number of defectives in a randomly selected batch is $(N - n) \cdot P(A) \cdot p = .00985(N - n)$. Dividing this by the number of items in a lot gives average outgoing quality:

$$AOQ = \frac{(N - n) \cdot P(A) \cdot p}{N}$$

$$\approx P(A) \cdot p \quad \text{if } N \gg n$$

Because $AOQ = 0$ when either $p = 0$ or $p = 1$ [$P(A) = 0$ in the latter case], it follows that there is a value of p between 0 and 1 for which AOQ is a maximum. The maximum value of AOQ is called the **average outgoing quality limit, AOQL.** For example, for the plan with $n = 137$ and $c = 3$ discussed earlier, AOQL = .0142, the value of AOQ at $p \approx .02$.

Proper choices of n and c will yield a sampling plan for which AOQL is a specified small number. Such a plan is not, however, unique, so another condition can be imposed. Frequently this second condition will involve the **average** (i.e., expected) **total number inspected,** denoted by **ATI.** The number of items inspected in a randomly chosen lot is a random variable that takes on the value n with probability $P(A)$ and N with probability $1 - P(A)$. Thus, the expected number of items inspected in a randomly selected lot is

$$ATI = n \cdot P(A) + N \cdot (1 - P(A))$$

It is common practice to select a sampling plan that has a specified AOQL and, in addition, minimum ATI at a particular quality level p.

Standard Sampling Plans

It may seem as though the determination of a sampling plan that simultaneously satisfies several criteria would be quite difficult. Fortunately, others have already laid the groundwork in the form of extensive tabulations of such plans. MIL STD 105D, developed by the military after World War II, is the most widely used set of plans. A civilian version, ANSI/ASQC Z1.4, is quite similar to the military version. A third set of plans that is quite popular was developed at Bell Laboratories prior to World War II by two applied statisticians named Dodge and Romig. The book by Montgomery contains a readable introduction to the use of these plans.

Exercises   Section 16.6 (31–38)

31. Consider the single-sample plan with $c = 2$ and $n = 50$, as discussed in Example 16.11, but now suppose that the lot size is $N = 500$. Calculate $P(A)$, the probability of accepting the lot, for $p = .01, .02, \ldots, .10$ using the hypergeometric distribution. Does the binomial approximation give satisfactory results in this case?

32. A sample of 50 items is to be selected from a batch consisting of 5000 items. The batch will be accepted if the sample contains at most one defective item. Calculate the probability of lot acceptance for $p = .01, .02, \ldots, .10$, and sketch the OC curve.

33. Refer to Exercise 32 and consider the plan with $n = 100$ and $c = 2$. Calculate $P(A)$ for $p = .01, .02, \ldots, .05$, and sketch the two OC curves on the same set of axes. Which of the two plans is preferable (leaving aside the cost of sampling), and why?

34. Develop a single-sample plan for which AQL $= .02$ and LTPD $= .07$ in the case $\alpha = .05$, $\beta = .10$. Once values of n and c have been determined, calculate the achieved values of α and β for the plan.

35. Consider the double-sampling plan for which both sample sizes are 50, the lot is accepted after the first sample if the number of defectives is at most one, rejected if the number of defectives is at least four, and is rejected after the second sample if the total number of defectives is six or more. Calculate the probability of accepting the lot when $p = .01, .05$, and .10.

36. Some sources advocate a somewhat more restrictive type of doubling-sampling plan in which $r_1 = c_2 + 1$, that is, the lot is rejected if at either stage the (total) number of defectives is at least r_1 (see the book by Montgomery). Consider this type of sampling plan with $n_1 = 50$, $n_2 = 100$, $c_1 = 1$, and $r_1 = 4$. Calculate the probability of lot acceptance when $p = .02, .05$, and .10.

37. Refer to Example 16.11, in which a single-sample plan with $n = 50$ and $c = 2$ was employed.
  a. Calculate AOQ for $p = .01, .02, \ldots, .10$. What does this suggest about the value of p for which AOQ is a maximum and the corresponding AOQL?
  b. Determine the value of p for which AOQ is a maximum and the corresponding value of AOQL. [HINT: Use calculus.]
  c. For $N = 2000$, calculate ATI for the values of p given in part (a).

38. Consider the single-sample plan that utilizes $n = 50$ and $c = 1$ when $N = 2000$. Determine the values of AOQ and ATI for selected values of p and graph each of these against p. Also determine the value of AOQL.

Supplementary Exercises Chapter 16 (39–44)

39. Observations on shear strength for 26 subgroups of test spot welds, each consisting of six welds, yield $\Sigma \bar{x}_i = 10,980$, $\Sigma s_i = 402$, and $\Sigma r_i = 1074$. Calculate control limits for any relevant control charts.

40. The number of scratches on the surface of each of 24 rectangular metal plates is determined, yielding the following data: 8, 1, 7, 5, 2, 0, 2, 3, 4, 3, 1, 2, 5, 7, 3, 4, 6, 5, 2, 4, 0, 10, 2, 6. Construct an appropriate control chart and comment.

41. The following numbers are observations on tensile strength of synthetic fabric specimens selected from a production process at equally spaced time intervals. Construct appropriate control charts and comment (assume an assignable cause is identifiable for any out-of-control observations).

1. 51.3 51.7 49.5	**12.** 49.6 48.4 50.0
2. 51.0 50.0 49.3	**13.** 49.8 51.2 49.7
3. 50.8 51.1 49.0	**14.** 50.4 49.9 50.7
4. 50.6 51.1 49.0	**15.** 49.4 49.5 49.0
5. 49.6 50.5 50.9	**16.** 50.7 49.0 50.0
6. 51.3 52.0 50.3	**17.** 50.8 49.5 50.9
7. 49.7 50.5 50.3	**18.** 48.5 50.3 49.3
8. 51.8 50.3 50.0	**19.** 49.6 50.6 49.4
9. 48.6 50.5 50.7	**20.** 50.9 49.4 49.7
10. 49.6 49.8 50.5	**21.** 54.1 49.8 48.5
11. 49.9 50.7 49.8	**22.** 50.2 49.6 51.5

42. An alternative to the p chart for the fraction defective is the *np chart for number defective*. This chart has UCL $= n\bar{p} + 3\sqrt{n\bar{p}(1 - \bar{p})}$, LCL $= n\bar{p} - 3\sqrt{n\bar{p}(1 - \bar{p})}$, and the *number* of defectives from each sample is plotted on the chart. Construct such a chart for the data of Example 16.6. Will the use of an *np* chart always give the same message as the use of a *p* chart (that is, are the two charts equivalent)?

43. Resistance observations (ohms) for subgroups of a certain type of register gave the following summary quantities:

i	n_i	$\bar{x}_i$	s_i	i	n_i	$\bar{x}_i$	s_i
1	4	430.0	22.5	11	4	445.2	27.3
2	4	418.2	20.6	12	4	430.1	22.2
3	3	435.5	25.1	13	4	427.2	24.0
4	4	427.6	22.3	14	4	439.6	23.3
5	4	444.0	21.5	15	3	415.9	31.2
6	3	431.4	28.9	16	4	419.8	27.5
7	4	420.8	25.4	17	3	447.0	19.8
8	4	431.4	24.0	18	4	434.4	23.7
9	4	428.7	21.2	19	4	422.2	25.1
10	4	440.1	25.8	20	4	425.7	24.4

Construct appropriate control limits. [HINT: Use $\bar{x} = \Sigma n_i \bar{x}_i / \Sigma n_i$ and $s^2 = \Sigma (n_i - 1)s_i^2 / \Sigma (n_i - 1)$.]

44. Let α be a number between 0 and 1 and define a sequence $W_1, W_2, W_3, \ldots$ by $W_0 = \mu$ and $W_t = \alpha \bar{X}_t + (1 - \alpha)W_{t-1}$ for $t = 1, 2, \ldots$. Substituting for W_{t-1} its representation in terms of $\bar{X}_{t-1}$ and W_{t-2}, then substituting for W_{t-2}, etc., results in

$$W_t = \alpha \bar{X}_t + \alpha(1 - \alpha)\bar{X}_{t-1} + \cdots$$
$$+ \alpha(1 - \alpha)^{t-1}\bar{X}_1 + (1 - \alpha)^t \mu$$

The fact that W_t depends not only on $\bar{X}_t$ but also on averages for past time points, albeit with (exponentially) decreasing weights, suggests that changes in the process mean will be more quickly reflected in the W_t's than in the individual $\bar{X}_t$'s.

a. Show that $E(W_t) = \mu$.

b. Let $\sigma_t^2 = V(W_t)$, and show that

$$\sigma_t^2 = \frac{\alpha[1 - (1 - \alpha)^{2t}]}{2 - \alpha} \cdot \frac{\sigma^2}{n}$$

c. An *exponentially weighted moving-average control chart* plots the W_t's and uses control limits $\mu_0 \pm 3\sigma_t$ (or $\bar{\bar{x}}$ in place of μ_0). Construct such a chart for the data of Example 16.9, using $\mu_0 = 40$.

Bibliography

Box, George, Soren Bisgaard, and Conrad Fung, "An Explanation and Critique of Taguchi's Contributions to Quality Engineering," *Quality and Reliability Engineering International*, 1988, pp. 123–131.

Montgomery, Douglas C., *Introduction to Statistical Quality Control* (2nd ed.), Wiley, New York, 1991. This is a comprehensive introduction to many aspects of quality control at roughly the same level as this book.

Ryan, Thomas P., *Statistical Methods for Quality Improvement*, Wiley, New York, 1989. Captures very nicely the modern flavor of quality control with minimal demands on the background of readers.

Wadsworth, Harrison, Kenneth Stephens, and A. Blanton Godfrey, *Modern Methods for Quality Control and Improvement*, Wiley, New York, 1986. Provides broad coverage of topics at a level in between the two aforementioned books.

Appendix
Tables

Appendix Tables

Table A.1 Cumulative Binomial Probabilities

$$B(x; n, p) = \sum_{y=0}^{x} b(y; n, p)$$

a. $n = 5$

							p								
	0.01	0.05	0.10	0.20	0.25	0.30	0.40	0.50	0.60	0.70	0.75	0.80	0.90	0.95	0.99
0	.951	.774	.590	.328	.237	.168	.078	.031	.010	.002	.001	.000	.000	.000	.000
1	.999	.977	.919	.737	.633	.528	.337	.188	.087	.031	.016	.007	.000	.000	.000
x 2	1.000	.999	.991	.942	.896	.837	.683	.500	.317	.163	.104	.058	.009	.001	.000
3	1.000	1.000	1.000	.993	.984	.969	.913	.812	.663	.472	.367	.263	.081	.023	.001
4	1.000	1.000	1.000	1.000	.999	.998	.990	.969	.922	.832	.763	.672	.410	.226	.049

b. $n = 10$

							p								
	0.01	0.05	0.10	0.20	0.25	0.30	0.40	0.50	0.60	0.70	0.75	0.80	0.90	0.95	0.99
0	.904	.599	.349	.107	.056	.028	.006	.001	.000	.000	.000	.000	.000	.000	.000
1	.996	.914	.736	.376	.244	.149	.046	.011	.002	.000	.000	.000	.000	.000	.000
2	1.000	.988	.930	.678	.526	.383	.167	.055	.012	.002	.000	.000	.000	.000	.000
3	1.000	.999	.987	.879	.776	.650	.382	.172	.055	.011	.004	.001	.000	.000	.000
4	1.000	1.000	.998	.967	.922	.850	.633	.377	.166	.047	.020	.006	.000	.000	.000
x 5	1.000	1.000	1.000	.994	.980	.953	.834	.623	.367	.150	.078	.033	.002	.000	.000
6	1.000	1.000	1.000	.999	.996	.989	.945	.828	.618	.350	.224	.121	.013	.001	.000
7	1.000	1.000	1.000	1.000	1.000	.998	.988	.945	.833	.617	.474	.322	.070	.012	.000
8	1.000	1.000	1.000	1.000	1.000	1.000	.998	.989	.954	.851	.756	.624	.264	.086	.004
9	1.000	1.000	1.000	1.000	1.000	1.000	1.000	.999	.994	.972	.944	.893	.651	.401	.096

c. $n = 15$

							p								
	0.01	0.05	0.10	0.20	0.25	0.30	0.40	0.50	0.60	0.70	0.75	0.80	0.90	0.95	0.99
0	.860	.463	.206	.035	.013	.005	.000	.000	.000	.000	.000	.000	.000	.000	.000
1	.990	.829	.549	.167	.080	.035	.005	.000	.000	.000	.000	.000	.000	.000	.000
2	1.000	.964	.816	.398	.236	.127	.027	.004	.000	.000	.000	.000	.000	.000	.000
3	1.000	.995	.944	.648	.461	.297	.091	.018	.002	.000	.000	.000	.000	.000	.000
4	1.000	.999	.987	.836	.686	.515	.217	.059	.009	.001	.000	.000	.000	.000	.000
5	1.000	1.000	.998	.939	.852	.722	.403	.151	.034	.004	.001	.000	.000	.000	.000
6	1.000	1.000	1.000	.982	.943	.869	.610	.304	.095	.015	.004	.001	.000	.000	.000
x 7	1.000	1.000	1.000	.996	.983	.950	.787	.500	.213	.050	.017	.004	.000	.000	.000
8	1.000	1.000	1.000	.999	.996	.985	.905	.696	.390	.131	.057	.018	.000	.000	.000
9	1.000	1.000	1.000	1.000	.999	.996	.966	.849	.597	.278	.148	.061	.002	.000	.000
10	1.000	1.000	1.000	1.000	1.000	.999	.991	.941	.783	.485	.314	.164	.013	.001	.000
11	1.000	1.000	1.000	1.000	1.000	1.000	.998	.982	.909	.703	.539	.352	.056	.005	.000
12	1.000	1.000	1.000	1.000	1.000	1.000	1.000	.996	.973	.873	.764	.602	.184	.036	.000
13	1.000	1.000	1.000	1.000	1.000	1.000	1.000	1.000	.995	.965	.920	.833	.451	.171	.010
14	1.000	1.000	1.000	1.000	1.000	1.000	1.000	1.000	1.000	.995	.987	.965	.794	.537	.140

SOURCE: Adapted from *Statistics for Management*, by Lincoln L. Chao. Copyright © 1980 by Wadsworth, Inc.

(continued)

Table A.1 Cumulative Binomial Probabilities *(cont.)*

$$B(x;\, n,\, p) = \sum_{y\,=\,0}^{x} b(y;\, n,\, p)$$

d. $n = 20$

							p									
	0.01	0.05	0.10	0.20	0.25	0.30	0.40	0.50	0.60	0.70	0.75	0.80	0.90	0.95	0.99
0	.818	.358	.122	.012	.003	.001	.000	.000	.000	.000	.000	.000	.000	.000	.000
1	.983	.736	.392	.069	.024	.008	.001	.000	.000	.000	.000	.000	.000	.000	.000
2	.999	.925	.677	.206	.091	.035	.004	.000	.000	.000	.000	.000	.000	.000	.000
3	1.000	.984	.867	.411	.225	.107	.016	.001	.000	.000	.000	.000	.000	.000	.000
4	1.000	.997	.957	.630	.415	.238	.051	.006	.000	.000	.000	.000	.000	.000	.000
5	1.000	1.000	.989	.804	.617	.416	.126	.021	.002	.000	.000	.000	.000	.000	.000
6	1.000	1.000	.998	.913	.786	.608	.250	.058	.006	.000	.000	.000	.000	.000	.000
7	1.000	1.000	1.000	.968	.898	.772	.416	.132	.021	.001	.000	.000	.000	.000	.000
8	1.000	1.000	1.000	.990	.959	.887	.596	.252	.057	.005	.001	.000	.000	.000	.000
9	1.000	1.000	1.000	.997	.986	.952	.755	.412	.128	.017	.004	.001	.000	.000	.000
10	1.000	1.000	1.000	.999	.996	.983	.872	.588	.245	.048	.014	.003	.000	.000	.000
11	1.000	1.000	1.000	1.000	.999	.995	.943	.748	.404	.113	.041	.010	.000	.000	.000
12	1.000	1.000	1.000	1.000	1.000	.999	.979	.868	.584	.228	.102	.032	.000	.000	.000
13	1.000	1.000	1.000	1.000	1.000	1.000	.994	.942	.750	.392	.214	.087	.002	.000	.000
14	1.000	1.000	1.000	1.000	1.000	1.000	.998	.979	.874	.584	.383	.196	.011	.000	.000
15	1.000	1.000	1.000	1.000	1.000	1.000	1.000	.994	.949	.762	.585	.370	.043	.003	.000
16	1.000	1.000	1.000	1.000	1.000	1.000	1.000	.999	.984	.893	.775	.589	.133	.016	.000
17	1.000	1.000	1.000	1.000	1.000	1.000	1.000	1.000	.996	.965	.909	.794	.323	.075	.001
18	1.000	1.000	1.000	1.000	1.000	1.000	1.000	1.000	.999	.992	.976	.931	.608	.264	.017
19	1.000	1.000	1.000	1.000	1.000	1.000	1.000	1.000	1.000	.999	.997	.988	.878	.642	.182

x (row label appears at left of rows 10–19)

(continued)

Table A.1 Cumulative Binomial Probabilities *(cont.)*

$$B(x; n, p) = \sum_{y=0}^{x} b(y; n, p)$$

e. $n = 25$

		0.01	0.05	0.10	0.20	0.25	0.30	0.40	0.50	0.60	0.70	0.75	0.80	0.90	0.95	0.99
	0	.778	.277	.072	.004	.001	.000	.000	.000	.000	.000	.000	.000	.000	.000	.000
	1	.974	.642	.271	.027	.007	.002	.000	.000	.000	.000	.000	.000	.000	.000	.000
	2	.998	.873	.537	.098	.032	.009	.000	.000	.000	.000	.000	.000	.000	.000	.000
	3	1.000	.966	.764	.234	.096	.033	.002	.000	.000	.000	.000	.000	.000	.000	.000
	4	1.000	.993	.902	.421	.214	.090	.009	.000	.000	.000	.000	.000	.000	.000	.000
	5	1.000	.999	.967	.617	.378	.193	.029	.002	.000	.000	.000	.000	.000	.000	.000
	6	1.000	1.000	.991	.780	.561	.341	.074	.007	.000	.000	.000	.000	.000	.000	.000
	7	1.000	1.000	.998	.891	.727	.512	.154	.022	.001	.000	.000	.000	.000	.000	.000
	8	1.000	1.000	1.000	.953	.851	.677	.274	.054	.004	.000	.000	.000	.000	.000	.000
	9	1.000	1.000	1.000	.983	.929	.811	.425	.115	.013	.000	.000	.000	.000	.000	.000
	10	1.000	1.000	1.000	.994	.970	.902	.586	.212	.034	.002	.000	.000	.000	.000	.000
	11	1.000	1.000	1.000	.998	.980	.956	.732	.345	.078	.006	.001	.000	.000	.000	.000
x	12	1.000	1.000	1.000	1.000	.997	.983	.846	.500	.154	.017	.003	.000	.000	.000	.000
	13	1.000	1.000	1.000	1.000	.999	.994	.922	.655	.268	.044	.020	.002	.000	.000	.000
	14	1.000	1.000	1.000	1.000	1.000	.998	.966	.788	.414	.098	.030	.006	.000	.000	.000
	15	1.000	1.000	1.000	1.000	1.000	1.000	.987	.885	.575	.189	.071	.017	.000	.000	.000
	16	1.000	1.000	1.000	1.000	1.000	1.000	.996	.946	.726	.323	.149	.047	.000	.000	.000
	17	1.000	1.000	1.000	1.000	1.000	1.000	.999	.978	.846	.488	.273	.109	.002	.000	.000
	18	1.000	1.000	1.000	1.000	1.000	1.000	1.000	.993	.926	.659	.439	.220	.009	.000	.000
	19	1.000	1.000	1.000	1.000	1.000	1.000	1.000	.998	.971	.807	.622	.383	.033	.001	.000
	20	1.000	1.000	1.000	1.000	1.000	1.000	1.000	1.000	.991	.910	.786	.579	.098	.007	.000
	21	1.000	1.000	1.000	1.000	1.000	1.000	1.000	1.000	.998	.967	.904	.766	.236	.034	.000
	22	1.000	1.000	1.000	1.000	1.000	1.000	1.000	1.000	1.000	.991	.968	.902	.463	.127	.002
	23	1.000	1.000	1.000	1.000	1.000	1.000	1.000	1.000	1.000	.998	.993	.973	.729	.358	.026
	24	1.000	1.000	1.000	1.000	1.000	1.000	1.000	1.000	1.000	1.000	.999	.996	.928	.723	.222

Table A.2 Cumulative Poisson Probabilities

$$F(x; \lambda) = \sum_{y=0}^{x} \frac{e^{-\lambda}\lambda^{y}}{y!}$$

		λ									
		.1	.2	.3	.4	.5	.6	.7	.8	.9	1.0
	0	.905	.819	.741	.670	.607	.549	.497	.449	.407	.368
	1	.995	.982	.963	.938	.910	.878	.844	.809	.772	.736
	2	1.000	.999	.996	.992	.986	.977	.966	.953	.937	.920
x	3		1.000	1.000	.999	.998	.997	.994	.991	.987	.981
	4				1.000	1.000	1.000	.999	.999	.998	.996
	5							1.000	1.000	1.000	.999
	6										1.000

(continued)

Table A.2 Cumulative Poisson Probabilities *(cont.)*

$$F(x;\ \lambda) = \sum_{y\,=\,0}^{x} \frac{e^{-\lambda}\lambda^{y}}{y!}$$

						λ						
	2.0	3.0	4.0	5.0	6.0	7.0	8.0	9.0	10.0	15.0	20.0
0	.135	.050	.018	.007	.002	.001	.000	.000	.000	.000	.000
1	.406	.199	.092	.040	.017	.007	.003	.001	.000	.000	.000
2	.677	.423	.238	.125	.062	.030	.014	.006	.003	.000	.000
3	.857	.647	.433	.265	.151	.082	.042	.021	.010	.000	.000
4	.947	.815	.629	.440	.285	.173	.100	.055	.029	.001	.000
5	.983	.916	.785	.616	.446	.301	.191	.116	.067	.003	.000
6	.995	.966	.889	.762	.606	.450	.313	.207	.130	.008	.000
7	.999	.988	.949	.867	.744	.599	.453	.324	.220	.018	.001
8	1.000	.996	.979	.932	.847	.729	.593	.456	.333	.037	.002
9		.999	.992	.968	.916	.830	.717	.587	.458	.070	.005
10		1.000	.997	.986	.957	.901	.816	.706	.583	.118	.011
11			.999	.995	.980	.947	.888	.803	.697	.185	.021
12			1.000	.998	.991	.973	.936	.876	.792	.268	.039
13				.999	.996	.987	.966	.926	.864	.363	.066
14				1.000	.999	.994	.983	.959	.917	.466	.105
15					.999	.998	.992	.978	.951	.568	.157
16					1.000	.999	.996	.989	.973	.664	.221
17						1.000	.998	.995	.986	.749	.297
18							.999	.998	.993	.819	.381
19							1.000	.999	.997	.875	.470
20								1.000	.998	.917	.559
21									.999	.947	.644
22									1.000	.967	.721
23										.981	.787
24										.989	.843
25										.994	.888
26										.997	.922
27										.998	.948
28										.999	.966
29										1.000	.978
30											.987
31											.992
32											.995
33											.997
34											.999
35											.999
36											1.000

x (row label for the leftmost index column, positioned near rows 18)

Table A.3 Standard Normal Curve Areas $\Phi(z) = P(Z \le z)$

z	0.00	0.01	0.02	0.03	0.04	0.05	0.06	0.07	0.08	0.09
−3.4	0.0003	0.0003	0.0003	0.0003	0.0003	0.0003	0.0003	0.0003	0.0003	0.0002
−3.3	0.0005	0.0005	0.0005	0.0004	0.0004	0.0004	0.0004	0.0004	0.0004	0.0003
−3.2	0.0007	0.0007	0.0006	0.0006	0.0006	0.0006	0.0006	0.0005	0.0005	0.0005
−3.1	0.0010	0.0009	0.0009	0.0009	0.0008	0.0008	0.0008	0.0008	0.0007	0.0007
−3.0	0.0013	0.0013	0.0013	0.0012	0.0012	0.0011	0.0011	0.0011	0.0010	0.0010
−2.9	0.0019	0.0018	0.0017	0.0017	0.0016	0.0016	0.0015	0.0015	0.0014	0.0014
−2.8	0.0026	0.0025	0.0024	0.0023	0.0023	0.0022	0.0021	0.0021	0.0020	0.0019
−2.7	0.0035	0.0034	0.0033	0.0032	0.0031	0.0030	0.0029	0.0028	0.0027	0.0026
−2.6	0.0047	0.0045	0.0044	0.0043	0.0041	0.0040	0.0039	0.0038	0.0037	0.0036
−2.5	0.0062	0.0060	0.0059	0.0057	0.0055	0.0054	0.0052	0.0051	0.0049	0.0048
−2.4	0.0082	0.0080	0.0078	0.0075	0.0073	0.0071	0.0069	0.0068	0.0066	0.0064
−2.3	0.0107	0.0104	0.0102	0.0099	0.0096	0.0094	0.0091	0.0089	0.0087	0.0084
−2.2	0.0139	0.0136	0.0132	0.0129	0.0125	0.0122	0.0119	0.0116	0.0113	0.0110
−2.1	0.0179	0.0174	0.0170	0.0166	0.0162	0.0158	0.0154	0.0150	0.0146	0.0143
−2.0	0.0228	0.0222	0.0217	0.0212	0.0207	0.0202	0.0197	0.0192	0.0188	0.0183
−1.9	0.0287	0.0281	0.0274	0.0268	0.0262	0.0256	0.0250	0.0244	0.0239	0.0233
−1.8	0.0359	0.0352	0.0344	0.0336	0.0329	0.0322	0.0314	0.0307	0.0301	0.0294
−1.7	0.0446	0.0436	0.0427	0.0418	0.0409	0.0401	0.0392	0.0384	0.0375	0.0367
−1.6	0.0548	0.0537	0.0526	0.0516	0.0505	0.0495	0.0485	0.0475	0.0465	0.0455
−1.5	0.0668	0.0655	0.0643	0.0630	0.0618	0.0606	0.0594	0.0582	0.0571	0.0559
−1.4	0.0808	0.0793	0.0778	0.0764	0.0749	0.0735	0.0722	0.0708	0.0694	0.0681
−1.3	0.0968	0.0951	0.0934	0.0918	0.0901	0.0885	0.0869	0.0853	0.0838	0.0823
−1.2	0.1151	0.1131	0.1112	0.1093	0.1075	0.1056	0.1038	0.1020	0.1003	0.0985
−1.1	0.1357	0.1335	0.1314	0.1292	0.1271	0.1251	0.1230	0.1210	0.1190	0.1170
−1.0	0.1587	0.1562	0.1539	0.1515	0.1492	0.1469	0.1446	0.1423	0.1401	0.1379
−0.9	0.1841	0.1814	0.1788	0.1762	0.1736	0.1711	0.1685	0.1660	0.1635	0.1611
−0.8	0.2119	0.2090	0.2061	0.2033	0.2005	0.1977	0.1949	0.1922	0.1894	0.1867
−0.7	0.2420	0.2389	0.2358	0.2327	0.2296	0.2266	0.2236	0.2206	0.2177	0.2148
−0.6	0.2743	0.2709	0.2676	0.2643	0.2611	0.2578	0.2546	0.2514	0.2483	0.2451
−0.5	0.3085	0.3050	0.3015	0.2981	0.2946	0.2912	0.2877	0.2843	0.2810	0.2776
−0.4	0.3446	0.3409	0.3372	0.3336	0.3300	0.3264	0.3228	0.3192	0.3156	0.3121
−0.3	0.3821	0.3783	0.3745	0.3707	0.3669	0.3632	0.3594	0.3557	0.3520	0.3483
−0.2	0.4207	0.4168	0.4129	0.4090	0.4052	0.4013	0.3974	0.3936	0.3897	0.3859
−0.1	0.4602	0.4562	0.4522	0.4483	0.4443	0.4404	0.4364	0.4325	0.4286	0.4247
−0.0	0.5000	0.4960	0.4920	0.4880	0.4840	0.4801	0.4761	0.4721	0.4681	0.4641

Table A.3 Standard Normal Curve Areas *(cont.)* $\Phi(z) = P(Z \leq z)$

z	0.00	0.01	0.02	0.03	0.04	0.05	0.06	0.07	0.08	0.09
0.0	0.5000	0.5040	0.5080	0.5120	0.5160	0.5199	0.5239	0.5279	0.5319	0.5359
0.1	0.5398	0.5438	0.5478	0.5517	0.5557	0.5596	0.5636	0.5675	0.5714	0.5753
0.2	0.5793	0.5832	0.5871	0.5910	0.5948	0.5987	0.6026	0.6064	0.6103	0.6141
0.3	0.6179	0.6217	0.6255	0.6293	0.6331	0.6368	0.6406	0.6443	0.6480	0.6517
0.4	0.6554	0.6591	0.6628	0.6664	0.6700	0.6736	0.6772	0.6808	0.6844	0.6879
0.5	0.6915	0.6950	0.6985	0.7019	0.7054	0.7088	0.7123	0.7157	0.7190	0.7224
0.6	0.7257	0.7291	0.7324	0.7357	0.7389	0.7422	0.7454	0.7486	0.7517	0.7549
0.7	0.7580	0.7611	0.7642	0.7673	0.7704	0.7734	0.7764	0.7794	0.7823	0.7852
0.8	0.7881	0.7910	0.7939	0.7967	0.7995	0.8023	0.8051	0.8078	0.8106	0.8133
0.9	0.8159	0.8186	0.8212	0.8238	0.8264	0.8289	0.8315	0.8340	0.8365	0.8389
1.0	0.8413	0.8438	0.8461	0.8485	0.8508	0.8531	0.8554	0.8577	0.8599	0.8621
1.1	0.8643	0.8665	0.8686	0.8708	0.8729	0.8749	0.8770	0.8790	0.8810	0.8830
1.2	0.8849	0.8869	0.8888	0.8907	0.8925	0.8944	0.8962	0.8980	0.8997	0.9015
1.3	0.9032	0.9049	0.9066	0.9082	0.9099	0.9115	0.9131	0.9147	0.9162	0.9177
1.4	0.9192	0.9207	0.9222	0.9236	0.9251	0.9265	0.9278	0.9292	0.9306	0.9319
1.5	0.9332	0.9345	0.9357	0.9370	0.9382	0.9394	0.9406	0.9418	0.9429	0.9441
1.6	0.9452	0.9463	0.9474	0.9484	0.9495	0.9505	0.9515	0.9525	0.9535	0.9545
1.7	0.9554	0.9564	0.9573	0.9582	0.9591	0.9599	0.9608	0.9616	0.9625	0.9633
1.8	0.9641	0.9649	0.9656	0.9664	0.9671	0.9678	0.9686	0.9693	0.9699	0.9706
1.9	0.9713	0.9719	0.9726	0.9732	0.9738	0.9744	0.9750	0.9756	0.9761	0.9767
2.0	0.9772	0.9778	0.9783	0.9788	0.9793	0.9798	0.9803	0.9808	0.9812	0.9817
2.1	0.9821	0.9826	0.9830	0.9834	0.9838	0.9842	0.9846	0.9850	0.9854	0.9857
2.2	0.9861	0.9864	0.9868	0.9871	0.9875	0.9878	0.9881	0.9884	0.9887	0.9890
2.3	0.9893	0.9896	0.9898	0.9901	0.9904	0.9906	0.9909	0.9911	0.9913	0.9916
2.4	0.9918	0.9920	0.9922	0.9925	0.9927	0.9929	0.9931	0.9932	0.9934	0.9936
2.5	0.9938	0.9940	0.9941	0.9943	0.9945	0.9946	0.9948	0.9949	0.9951	0.9952
2.6	0.9953	0.9955	0.9956	0.9957	0.9959	0.9960	0.9961	0.9962	0.9963	0.9964
2.7	0.9965	0.9966	0.9967	0.9968	0.9969	0.9970	0.9971	0.9972	0.9973	0.9974
2.8	0.9974	0.9975	0.9976	0.9977	0.9977	0.9978	0.9979	0.9979	0.9980	0.9981
2.9	0.9981	0.9982	0.9982	0.9983	0.9984	0.9984	0.9985	0.9985	0.9986	0.9986
3.0	0.9987	0.9987	0.9987	0.9988	0.9988	0.9989	0.9989	0.9989	0.9990	0.9990
3.1	0.9990	0.9991	0.9991	0.9991	0.9992	0.9992	0.9992	0.9992	0.9993	0.9993
3.2	0.9993	0.9993	0.9994	0.9994	0.9994	0.9994	0.9994	0.9995	0.9995	0.9995
3.3	0.9995	0.9995	0.9995	0.9996	0.9996	0.9996	0.9996	0.9996	0.9996	0.9997
3.4	0.9997	0.9997	0.9997	0.9997	0.9997	0.9997	0.9997	0.9997	0.9997	0.9998

Table A.4 The Incomplete Gamma Function $\qquad F(x;\ \alpha) = \int_0^x \frac{1}{\Gamma(\alpha)} y^{\alpha-1} e^{-y} dy$

x \ α	1	2	3	4	5	6	7	8	9	10
1	.632	.264	.080	.019	.004	.001	.000	.000	.000	.000
2	.865	.594	.323	.143	.053	.017	.005	.001	.000	.000
3	.950	.801	.577	.353	.185	.084	.034	.012	.004	.001
4	.982	.908	.762	.567	.371	.215	.111	.051	.021	.008
5	.993	.960	.875	.735	.560	.384	.238	.133	.068	.032
6	.998	.983	.938	.849	.715	.554	.394	.256	.153	.084
7	.999	.993	.970	.918	.827	.699	.550	.401	.271	.170
8	1.000	.997	.986	.958	.900	.809	.687	.547	.407	.283
9		.999	.994	.979	.945	.884	.793	.676	.544	.413
10		1.000	.997	.990	.971	.933	.870	.780	.667	.542
11			.999	.995	.985	.962	.921	.857	.768	.659
12			1.000	.998	.992	.980	.954	.911	.845	.758
13				.999	.996	.989	.974	.946	.900	.834
14				1.000	.998	.994	.986	.968	.938	.891
15					.999	.997	.992	.982	.963	.930

Table A.5 Critical Values $t_{\alpha,\nu}$ for the t Distribution

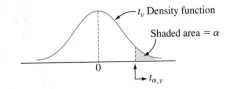

t_ν Density function

Shaded area = α

0

$t_{\alpha,\nu}$

| ν | α | | | | | | |
	.10	.05	.025	.01	.005	.001	.0005
1	3.078	6.314	12.706	31.821	63.657	318.31	636.62
2	1.886	2.920	4.303	6.965	9.925	22.326	31.598
3	1.638	2.353	3.182	4.541	5.841	10.213	12.924
4	1.533	2.132	2.776	3.747	4.604	7.173	8.610
5	1.476	2.015	2.571	3.365	4.032	5.893	6.869
6	1.440	1.943	2.447	3.143	3.707	5.208	5.959
7	1.415	1.895	2.365	2.998	3.499	4.785	5.408
8	1.397	1.860	2.306	2.896	3.355	4.501	5.041
9	1.383	1.833	2.262	2.821	3.250	4.297	4.781
10	1.372	1.812	2.228	2.764	3.169	4.144	4.587
11	1.363	1.796	2.201	2.718	3.106	4.025	4.437
12	1.356	1.782	2.179	2.681	3.055	3.930	4.318
13	1.350	1.771	2.160	2.650	3.012	3.852	4.221
14	1.345	1.761	2.145	2.624	2.977	3.787	4.140
15	1.341	1.753	2.131	2.602	2.947	3.733	4.073
16	1.337	1.746	2.120	2.583	2.921	3.686	4.015
17	1.333	1.740	2.110	2.567	2.898	3.646	3.965
18	1.330	1.734	2.101	2.552	2.878	3.610	3.922
19	1.328	1.729	2.093	2.539	2.861	3.579	3.883
20	1.325	1.725	2.086	2.528	2.845	3.552	3.850
21	1.323	1.721	2.080	2.518	2.831	3.527	3.819
22	1.321	1.717	2.074	2.508	2.819	3.505	3.792
23	1.319	1.714	2.069	2.500	2.807	3.485	3.767
24	1.318	1.711	2.064	2.492	2.797	3.467	3.745
25	1.316	1.708	2.060	2.485	2.787	3.450	3.725
26	1.315	1.706	2.056	2.479	2.779	3.435	3.707
27	1.314	1.703	2.052	2.473	2.771	3.421	3.690
28	1.313	1.701	2.048	2.467	2.763	3.408	3.674
29	1.311	1.699	2.045	2.462	2.756	3.396	3.659
30	1.310	1.697	2.042	2.457	2.750	3.385	3.646
40	1.303	1.684	2.021	2.423	2.704	3.307	3.551
60	1.296	1.671	2.000	2.390	2.660	3.232	3.460
120	1.289	1.658	1.980	2.358	2.617	3.160	3.373
∞	1.282	1.645	1.960	2.326	2.576	3.090	3.291

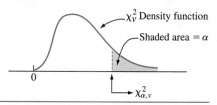

Table A.6 Critical Values $\chi^2_{\alpha,\nu}$ for the Chi-Squared Distribution

ν	.995	.99	.975	.95	.90	.10	.05	.025	.01	.005
1	0.000	0.000	0.001	0.004	0.016	2.706	3.843	5.025	6.637	7.882
2	0.010	0.020	0.051	0.103	0.211	4.605	5.992	7.378	9.210	10.597
3	0.072	0.115	0.216	0.352	0.584	6.251	7.815	9.348	11.344	12.837
4	0.207	0.297	0.484	0.711	1.064	7.779	9.488	11.143	13.277	14.860
5	0.412	0.554	0.831	1.145	1.610	9.236	11.070	12.832	15.085	16.748
6	0.676	0.872	1.237	1.635	2.204	10.645	12.592	14.440	16.812	18.548
7	0.989	1.239	1.690	2.167	2.833	12.017	14.067	16.012	18.474	20.276
8	1.344	1.646	2.180	2.733	3.490	13.362	15.507	17.534	20.090	21.954
9	1.735	2.088	2.700	3.325	4.168	14.684	16.919	19.022	21.665	23.587
10	2.156	2.558	3.247	3.940	4.865	15.987	18.307	20.483	23.209	25.188
11	2.603	3.053	3.816	4.575	5.578	17.275	19.675	21.920	24.724	26.755
12	3.074	3.571	4.404	5.226	6.304	18.549	21.026	23.337	26.217	28.300
13	3.565	4.107	5.009	5.892	7.041	19.812	22.362	24.735	27.687	29.817
14	4.075	4.660	5.629	6.571	7.790	21.064	23.685	26.119	29.141	31.319
15	4.600	5.229	6.262	7.261	8.547	22.307	24.996	27.488	30.577	32.799
16	5.142	5.812	6.908	7.962	9.312	23.542	26.296	28.845	32.000	34.267
17	5.697	6.407	7.564	8.682	10.085	24.769	27.587	30.190	33.408	35.716
18	6.265	7.015	8.231	9.390	10.865	25.989	28.869	31.526	34.805	37.156
19	6.843	7.632	8.906	10.117	11.651	27.203	30.143	32.852	36.190	38.580
20	7.434	8.260	9.591	10.851	12.443	28.412	31.410	34.170	37.566	39.997
21	8.033	8.897	10.283	11.591	13.240	29.615	32.670	35.478	38.930	41.399
22	8.643	9.542	10.982	12.338	14.042	30.813	33.924	36.781	40.289	42.796
23	9.260	10.195	11.688	13.090	14.848	32.007	35.172	38.075	41.637	44.179
24	9.886	10.856	12.401	13.848	15.659	33.196	36.415	39.364	42.980	45.558
25	10.519	11.523	13.120	14.611	16.473	34.381	37.652	40.646	44.313	46.925
26	11.160	12.198	13.844	15.379	17.292	35.563	38.885	41.923	45.642	48.290
27	11.807	12.878	14.573	16.151	18.114	36.741	40.113	43.194	46.962	49.642
28	12.461	13.565	15.308	16.928	18.939	37.916	41.337	44.461	48.278	50.993
29	13.120	14.256	16.147	17.708	19.768	39.087	42.557	45.772	49.586	52.333
30	13.787	14.954	16.791	18.493	20.599	40.256	43.773	46.979	50.892	53.672
31	14.457	15.655	17.538	19.280	21.433	41.422	44.985	48.231	52.190	55.000
32	15.134	16.362	18.291	20.072	22.271	42.585	46.194	49.480	53.486	56.328
33	15.814	17.073	19.046	20.866	23.110	43.745	47.400	50.724	54.774	57.646
34	16.501	17.789	19.806	21.664	23.952	44.903	48.602	51.966	56.061	58.964
35	17.191	18.508	20.569	22.465	24.796	46.059	49.802	53.203	57.340	60.272
36	17.887	19.233	21.336	23.269	25.643	47.212	50.998	54.437	58.619	61.581
37	18.584	19.960	22.105	24.075	26.492	48.363	52.192	55.667	59.891	62.880
38	19.289	20.691	22.878	24.884	27.343	49.513	53.384	56.896	61.162	64.181
39	19.994	21.425	23.654	25.695	28.196	50.660	54.572	58.119	62.426	65.473
40	20.706	22.164	24.433	26.509	29.050	51.805	55.758	59.342	63.691	66.766

For $\nu > 40$, $x^2_{\alpha,\nu} \approx \nu \left(1 - \dfrac{2}{9\nu} + z_\alpha \sqrt{\dfrac{2}{9\nu}}\right)^3$

Table A.7 Critical Values F_{α, ν_1, ν_2} for the F Distribution

$\alpha = .05$

ν_2 \ ν_1	1	2	3	4	5	6	7	8	9	10	12	15	20	24	30	40	60	120	∞
1	161.4	199.5	215.7	224.6	230.2	234.0	236.8	238.9	240.5	241.9	243.9	245.9	248.0	249.1	250.1	251.1	252.2	253.3	254.3
2	18.51	19.00	19.16	19.25	19.30	19.33	19.35	19.37	19.38	19.40	19.41	19.43	19.45	19.45	19.46	19.47	19.48	19.49	19.50
3	10.13	9.55	9.28	9.12	9.01	8.94	8.89	8.85	8.81	8.79	8.74	8.70	8.66	8.64	8.62	8.59	8.57	8.55	8.53
4	7.71	6.94	6.59	6.39	6.26	6.16	6.09	6.04	6.00	5.96	5.91	5.86	5.80	5.77	5.75	5.72	5.69	5.66	5.63
5	6.61	5.79	5.41	5.19	5.05	4.95	4.88	4.82	4.77	4.74	4.68	4.62	4.56	4.53	4.50	4.46	4.43	4.40	4.36
6	5.99	5.14	4.76	4.53	4.39	4.28	4.21	4.15	4.10	4.06	4.00	3.94	3.87	3.84	3.81	3.77	3.74	3.70	3.67
7	5.59	4.74	4.35	4.12	3.97	3.87	3.79	3.73	3.68	3.64	3.57	3.51	3.44	3.41	3.38	3.34	3.30	3.27	3.23
8	5.32	4.46	4.07	3.84	3.69	3.58	3.50	3.44	3.39	3.35	3.28	3.22	3.15	3.12	3.08	3.04	3.01	2.97	2.93
9	5.12	4.26	3.86	3.63	3.48	3.37	3.29	3.23	3.18	3.14	3.07	3.01	2.94	2.90	2.86	2.83	2.79	2.75	2.71
10	4.96	4.10	3.71	3.48	3.33	3.22	3.14	3.07	3.02	2.98	2.91	2.85	2.77	2.74	2.70	2.66	2.62	2.58	2.54
11	4.84	3.98	3.59	3.36	3.20	3.09	3.01	2.95	2.90	2.85	2.79	2.72	2.65	2.61	2.57	2.53	2.49	2.45	2.40
12	4.75	3.89	3.49	3.26	3.11	3.00	2.91	2.85	2.80	2.75	2.69	2.62	2.54	2.51	2.47	2.43	2.38	2.34	2.30
13	4.67	3.81	3.41	3.18	3.03	2.92	2.83	2.77	2.71	2.67	2.60	2.53	2.46	2.42	2.38	2.34	2.30	2.25	2.21
14	4.60	3.74	3.34	3.11	2.96	2.85	2.76	2.70	2.65	2.60	2.53	2.46	2.39	2.35	2.31	2.27	2.22	2.18	2.13
15	4.54	3.68	3.29	3.06	2.90	2.79	2.71	2.64	2.59	2.54	2.48	2.40	2.33	2.29	2.25	2.20	2.16	2.11	2.07
16	4.49	3.63	3.24	3.01	2.85	2.74	2.66	2.59	2.54	2.49	2.42	2.35	2.28	2.24	2.19	2.15	2.11	2.06	2.01
17	4.45	3.59	3.20	2.96	2.81	2.70	2.61	2.55	2.49	2.45	2.38	2.31	2.23	2.19	2.15	2.10	2.06	2.01	1.96
18	4.41	3.55	3.16	2.93	2.77	2.66	2.58	2.51	2.46	2.41	2.34	2.27	2.19	2.15	2.11	2.06	2.02	1.97	1.92
19	4.38	3.52	3.13	2.90	2.74	2.63	2.54	2.48	2.42	2.38	2.31	2.23	2.16	2.11	2.07	2.03	1.98	1.93	1.88
20	4.35	3.49	3.10	2.87	2.71	2.60	2.51	2.45	2.39	2.35	2.28	2.20	2.12	2.08	2.04	1.99	1.95	1.90	1.84
21	4.32	3.47	3.07	2.84	2.68	2.57	2.49	2.42	2.37	2.32	2.25	2.18	2.10	2.05	2.01	1.96	1.92	1.87	1.81
22	4.30	3.44	3.05	2.82	2.66	2.55	2.46	2.40	2.34	2.30	2.23	2.15	2.07	2.03	1.98	1.94	1.89	1.84	1.78
23	4.28	3.42	3.03	2.80	2.64	2.53	2.44	2.37	2.32	2.27	2.20	2.13	2.05	2.01	1.96	1.91	1.86	1.81	1.76
24	4.26	3.40	3.01	2.78	2.62	2.51	2.42	2.36	2.30	2.25	2.18	2.11	2.03	1.98	1.94	1.89	1.84	1.79	1.73
25	4.24	3.39	2.99	2.76	2.60	2.49	2.40	2.34	2.28	2.24	2.16	2.09	2.01	1.96	1.92	1.87	1.82	1.77	1.71
26	4.23	3.37	2.98	2.74	2.59	2.47	2.39	2.32	2.27	2.22	2.15	2.07	1.99	1.95	1.90	1.85	1.80	1.75	1.69
27	4.21	3.35	2.96	2.73	2.57	2.46	2.37	2.31	2.25	2.20	2.13	2.06	1.97	1.93	1.88	1.84	1.79	1.73	1.67
28	4.20	3.34	2.95	2.71	2.56	2.45	2.36	2.29	2.24	2.19	2.12	2.04	1.96	1.91	1.87	1.82	1.77	1.71	1.65
29	4.18	3.33	2.93	2.70	2.55	2.43	2.35	2.28	2.22	2.18	2.10	2.03	1.94	1.90	1.85	1.81	1.75	1.70	1.64
30	4.17	3.32	2.92	2.69	2.53	2.42	2.33	2.27	2.21	2.16	2.09	2.01	1.93	1.89	1.84	1.79	1.74	1.68	1.62
40	4.08	3.23	2.84	2.61	2.45	2.34	2.25	2.18	2.12	2.08	2.00	1.92	1.84	1.79	1.74	1.69	1.64	1.58	1.51
60	4.00	3.15	2.76	2.53	2.37	2.25	2.17	2.10	2.04	1.99	1.92	1.84	1.75	1.70	1.65	1.59	1.53	1.47	1.39
120	3.92	3.07	2.68	2.45	2.29	2.17	2.09	2.02	1.96	1.91	1.83	1.75	1.66	1.61	1.55	1.50	1.43	1.35	1.25
∞	3.84	3.00	2.60	2.37	2.21	2.10	2.01	1.94	1.88	1.83	1.75	1.67	1.57	1.52	1.46	1.39	1.32	1.22	1.00

(continued)

Table A.7 Critical Values F_{α,v_1,v_2} for the F Distribution (cont.)

$\alpha = .01$

$v_2 \backslash v_1$	1	2	3	4	5	6	7	8	9	10	12	15	20	24	30	40	60	120	∞
1	4052	4999.5	5403	5625	5764	5859	5928	5981	6022	6056	6106	6157	6209	6235	6261	6287	6313	6339	6366
2	98.50	99.00	99.17	99.25	99.30	99.33	99.36	99.37	99.39	99.40	99.42	99.43	99.45	99.46	99.47	99.47	99.48	99.49	99.50
3	34.12	30.82	29.46	28.71	28.24	27.91	27.67	27.49	27.35	27.23	27.05	26.87	26.69	26.60	26.50	26.41	26.32	26.22	26.13
4	21.20	18.00	16.69	15.98	15.52	15.21	14.98	14.80	14.66	14.55	14.37	14.20	14.02	13.93	13.84	13.75	13.65	13.56	13.46
5	16.26	13.27	12.06	11.39	10.97	10.67	10.46	10.29	10.16	10.05	9.89	9.72	9.55	9.47	9.38	9.29	9.20	9.11	9.02
6	13.75	10.92	9.78	9.15	8.75	8.47	8.26	8.10	7.98	7.87	7.72	7.56	7.40	7.31	7.23	7.14	7.06	6.97	6.88
7	12.25	9.55	8.45	7.85	7.46	7.19	6.99	6.84	6.72	6.62	6.47	6.31	6.16	6.07	5.99	5.91	5.82	5.74	5.65
8	11.26	8.65	7.59	7.01	6.63	6.37	6.18	6.03	5.91	5.81	5.67	5.52	5.36	5.28	5.20	5.12	5.03	4.95	4.86
9	10.56	8.02	6.99	6.42	6.06	5.80	5.61	5.47	5.35	5.26	5.11	4.96	4.81	4.73	4.65	4.57	4.48	4.40	4.31
10	10.04	7.56	6.55	5.99	5.64	5.39	5.20	5.06	4.94	4.85	4.71	4.56	4.41	4.33	4.25	4.17	4.08	4.00	3.91
11	9.65	7.21	6.22	5.67	5.32	5.07	4.89	4.74	4.63	4.54	4.40	4.25	4.10	4.02	3.94	3.86	3.78	3.69	3.60
12	9.33	6.93	5.95	5.41	5.06	4.82	4.64	4.50	4.39	4.30	4.16	4.01	3.86	3.78	3.70	3.62	3.54	3.45	3.36
13	9.07	6.70	5.74	5.21	4.86	4.62	4.44	4.30	4.19	4.10	3.96	3.82	3.66	3.59	3.51	3.43	3.34	3.25	3.17
14	8.86	6.51	5.56	5.04	4.69	4.46	4.28	4.14	4.03	3.94	3.80	3.66	3.51	3.43	3.35	3.27	3.18	3.09	3.00
15	8.68	6.36	5.42	4.89	4.56	4.32	4.14	4.00	3.89	3.80	3.67	3.52	3.37	3.29	3.21	3.13	3.05	2.96	2.87
16	8.53	6.23	5.29	4.77	4.44	4.20	4.03	3.89	3.78	3.69	3.55	3.41	3.26	3.18	3.10	3.02	2.93	2.84	2.75
17	8.40	6.11	5.18	4.67	4.34	4.10	3.93	3.79	3.68	3.59	3.46	3.31	3.16	3.08	3.00	2.92	2.83	2.75	2.65
18	8.29	6.01	5.09	4.58	4.25	4.01	3.84	3.71	3.60	3.51	3.37	3.23	3.08	3.00	2.92	2.84	2.75	2.66	2.57
19	8.18	5.93	5.01	4.50	4.17	3.94	3.77	3.63	3.52	3.43	3.30	3.15	3.00	2.92	2.84	2.76	2.67	2.58	2.49
20	8.10	5.85	4.94	4.43	4.10	3.87	3.70	3.56	3.46	3.37	3.23	3.09	2.94	2.86	2.78	2.69	2.61	2.52	2.42
21	8.02	5.78	4.87	4.37	4.04	3.81	3.64	3.51	3.40	3.31	3.17	3.03	2.88	2.80	2.72	2.64	2.55	2.46	2.36
22	7.95	5.72	4.82	4.31	3.99	3.76	3.59	3.45	3.35	3.26	3.12	2.98	2.83	2.75	2.67	2.58	2.50	2.40	2.31
23	7.88	5.66	4.76	4.26	3.94	3.71	3.54	3.41	3.30	3.21	3.07	2.93	2.78	2.70	2.62	2.54	2.45	2.35	2.26
24	7.82	5.61	4.72	4.22	3.90	3.67	3.50	3.36	3.26	3.17	3.03	2.89	2.74	2.66	2.58	2.49	2.40	2.31	2.21
25	7.77	5.57	4.68	4.18	3.85	3.63	3.46	3.32	3.22	3.13	2.99	2.85	2.70	2.62	2.54	2.45	2.36	2.27	2.17
26	7.72	5.53	4.64	4.14	3.82	3.59	3.42	3.29	3.18	3.09	2.96	2.81	2.66	2.58	2.50	2.42	2.33	2.23	2.13
27	7.68	5.49	4.60	4.11	3.78	3.56	3.39	3.26	3.15	3.06	2.93	2.78	2.63	2.55	2.47	2.38	2.29	2.20	2.10
28	7.64	5.45	4.57	4.07	3.75	3.53	3.36	3.23	3.12	3.03	2.90	2.75	2.60	2.52	2.44	2.35	2.26	2.17	2.06
29	7.60	5.42	4.54	4.04	3.73	3.50	3.33	3.20	3.09	3.00	2.87	2.73	2.57	2.49	2.41	2.33	2.23	2.14	2.03
30	7.56	5.39	4.51	4.02	3.70	3.47	3.30	3.17	3.07	2.98	2.84	2.70	2.55	2.47	2.39	2.30	2.21	2.11	2.01
40	7.31	5.18	4.31	3.83	3.51	3.29	3.12	2.99	2.89	2.80	2.66	2.52	2.37	2.29	2.20	2.11	2.02	1.92	1.80
60	7.08	4.98	4.13	3.65	3.34	3.12	2.95	2.82	2.72	2.63	2.50	2.35	2.20	2.12	2.03	1.94	1.84	1.73	1.60
120	6.85	4.79	3.95	3.48	3.17	2.96	2.79	2.66	2.56	2.47	2.34	2.19	2.03	1.95	1.86	1.76	1.66	1.53	1.38
∞	6.63	4.61	3.78	3.32	3.02	2.80	2.64	2.51	2.41	2.32	2.18	2.04	1.88	1.79	1.70	1.59	1.47	1.32	1.00

SOURCE: This table is reproduced with the kind permission of the Trustees of Biometrika from E. S. Pearson and H. O. Hartley (eds.), *The Biometrika Tables for Statisticians*, vol. 1, 3rd ed., *Biometrika*, 1966.

| | | \multicolumn{10}{c}{m} | | | | | | | | |
v	α	2	3	4	5	6	7	8	9	10	11
5	.05	3.64	4.60	5.22	5.67	6.03	6.33	6.58	6.80	6.99	7.17
	.01	5.70	6.98	7.80	8.42	8.91	9.32	9.67	9.97	10.24	10.48
6	.05	3.46	4.34	4.90	5.30	5.63	5.90	6.12	6.32	6.49	6.65
	.01	5.24	6.33	7.03	7.56	7.97	8.32	8.61	8.87	9.10	9.30
7	.05	3.34	4.16	4.68	5.06	5.36	5.61	5.82	6.00	6.16	6.30
	.01	4.95	5.92	6.54	7.01	7.37	7.68	7.94	8.17	8.37	8.55
8	.05	3.26	4.04	4.53	4.89	5.17	5.40	5.60	5.77	5.92	6.05
	.01	4.75	5.64	6.20	6.62	6.96	7.24	7.47	7.68	7.86	8.03
9	.05	3.20	3.95	4.41	4.76	5.02	5.24	5.43	5.59	5.74	5.87
	.01	4.60	5.43	5.96	6.35	6.66	6.91	7.13	7.33	7.49	7.65
10	.05	3.15	3.88	4.33	4.65	4.91	5.12	5.30	5.46	5.60	5.72
	.01	4.48	5.27	5.77	6.14	6.43	6.67	6.87	7.05	7.21	7.36
11	.05	3.11	3.82	4.26	4.57	4.82	5.03	5.20	5.35	5.49	5.61
	.01	4.39	5.15	5.62	5.97	6.25	6.48	6.67	6.84	6.99	7.13
12	.05	3.08	3.77	4.20	4.51	4.75	4.95	5.12	5.27	5.39	5.51
	.01	4.32	5.05	5.50	5.84	6.10	6.32	6.51	6.67	6.81	6.94
13	.05	3.06	3.73	4.15	4.45	4.69	4.88	5.05	5.19	5.32	5.43
	.01	4.26	4.96	5.40	5.73	5.98	6.19	6.37	6.53	6.67	6.79
14	.05	3.03	3.70	4.11	4.41	4.64	4.83	4.99	5.13	5.25	5.36
	.01	4.21	4.89	5.32	5.63	5.88	6.08	6.26	6.41	6.54	6.66
15	.05	3.01	3.67	4.08	4.37	4.59	4.78	4.94	5.08	5.20	5.31
	.01	4.17	4.84	5.25	5.56	5.80	5.99	6.16	6.31	6.44	6.55
16	.05	3.00	3.65	4.05	4.33	4.56	4.74	4.90	5.03	5.15	5.26
	.01	4.13	4.79	5.19	5.49	5.72	5.92	6.08	6.22	6.35	6.46
17	.05	2.98	3.63	4.02	4.30	4.52	4.70	4.86	4.99	5.11	5.21
	.01	4.10	4.74	5.14	5.43	5.66	5.85	6.01	6.15	6.27	6.38
18	.05	2.97	3.61	4.00	4.28	4.49	4.67	4.82	4.96	5.07	5.17
	.01	4.07	4.70	5.09	5.38	5.60	5.79	5.94	6.08	6.20	6.31
19	.05	2.96	3.59	3.98	4.25	4.47	4.65	4.79	4.92	5.04	5.14
	.01	4.05	4.67	5.05	5.33	5.55	5.73	5.89	6.02	6.14	6.25
20	.05	2.95	3.58	3.96	4.23	4.45	4.62	4.77	4.90	5.01	5.11
	.01	4.02	4.64	5.02	5.29	5.51	5.69	5.84	5.97	6.09	6.19
24	.05	2.92	3.53	3.90	4.17	4.37	4.54	4.68	4.81	4.92	5.01
	.01	3.96	4.55	4.91	5.17	5.37	5.54	5.69	5.81	5.92	6.02
30	.05	2.89	3.49	3.85	4.10	4.30	4.46	4.60	4.72	4.82	4.92
	.01	3.89	4.45	4.80	5.05	5.24	5.40	5.54	5.65	5.76	5.85
40	.05	2.86	3.44	3.79	4.04	4.23	4.39	4.52	4.63	4.73	4.82
	.01	3.82	4.37	4.70	4.93	5.11	5.26	5.39	5.50	5.60	5.69
60	.05	2.83	3.40	3.74	3.98	4.16	4.31	4.44	4.55	4.65	4.73
	.01	3.76	4.28	4.59	4.82	4.99	5.13	5.25	5.36	5.45	5.53
120	.05	2.80	3.36	3.68	3.92	4.10	4.24	4.36	4.47	4.56	4.64
	.01	3.70	4.20	4.50	4.71	4.87	5.01	5.12	5.21	5.30	5.37
∞	.05	2.77	3.31	3.63	3.86	4.03	4.17	4.29	4.39	4.47	4.55
	.01	3.64	4.12	4.40	4.60	4.76	4.88	4.99	5.08	5.16	5.23

(continued)

Table A.8 Critical Values $Q_{\alpha,m,v}$ for the Studentized Range Distribution (cont.)

12	13	14	15	16	17	18	19	20	α	v
7.32	7.47	7.60	7.72	7.83	7.93	8.03	8.12	8.21	.05	5
10.70	10.89	11.08	11.24	11.40	11.55	11.68	11.81	11.93	.01	
6.79	6.92	7.03	7.14	7.24	7.34	7.43	7.51	7.59	.05	6
9.48	9.65	9.81	9.95	10.08	10.21	10.32	10.43	10.54	.01	
6.43	6.55	6.66	6.76	6.85	6.94	7.02	7.10	7.17	.05	7
8.71	8.86	9.00	9.12	9.24	9.35	9.46	9.55	9.65	.01	
6.18	6.29	6.39	6.48	6.57	6.65	6.73	6.80	6.87	.05	8
8.18	8.31	8.44	8.55	8.66	8.76	8.85	8.94	9.03	.01	
5.98	6.09	6.19	6.28	6.36	6.44	6.51	6.58	6.64	.05	9
7.78	7.91	8.03	8.13	8.23	8.33	8.41	8.49	8.57	.01	
5.83	5.93	6.03	6.11	6.19	6.27	6.34	6.40	6.47	.05	10
7.49	7.60	7.71	7.81	7.91	7.99	8.08	8.15	8.23	.01	
5.71	5.81	5.90	5.98	6.06	6.13	6.20	6.27	6.33	.05	11
7.25	7.36	7.46	7.56	7.65	7.73	7.81	7.88	7.95	.01	
5.61	5.71	5.80	5.88	5.95	6.02	6.09	6.15	6.21	.05	12
7.06	7.17	7.26	7.36	7.44	7.52	7.59	7.66	7.73	.01	
5.53	5.63	5.71	5.79	5.86	5.93	5.99	6.05	6.11	.05	13
6.90	7.01	7.10	7.19	7.27	7.35	7.42	7.48	7.55	.01	
5.46	5.55	5.64	5.71	5.79	5.85	5.91	5.97	6.03	.05	14
6.77	6.87	6.96	7.05	7.13	7.20	7.27	7.33	7.39	.01	
5.40	5.49	5.57	5.65	5.72	5.78	5.85	5.90	5.96	.05	15
6.66	6.76	6.84	6.93	7.00	7.07	7.14	7.20	7.26	.01	
5.35	5.44	5.52	5.59	5.66	5.73	5.79	5.84	5.90	.05	16
6.56	6.66	6.74	6.82	6.90	6.97	7.03	7.09	7.15	.01	
5.31	5.39	5.47	5.54	5.61	5.67	5.73	5.79	5.84	.05	17
6.48	6.57	6.66	6.73	6.81	6.87	6.94	7.00	7.05	.01	
5.27	5.35	5.43	5.50	5.57	5.63	5.69	5.74	5.79	.05	18
6.41	6.50	6.58	6.65	6.73	6.79	6.85	6.91	6.97	.01	
5.23	5.31	5.39	5.46	5.53	5.59	5.65	5.70	5.75	.05	19
6.34	6.43	6.51	6.58	6.65	6.72	6.78	6.84	6.89	.01	
5.20	5.28	5.36	5.43	5.49	5.55	5.61	5.66	5.71	.05	20
6.28	6.37	6.45	6.52	6.59	6.65	6.71	6.77	6.82	.01	
5.10	5.18	5.25	5.32	5.38	5.44	5.49	5.55	5.59	.05	24
6.11	6.19	6.26	6.33	6.39	6.45	6.51	6.56	6.61	.01	
5.00	5.08	5.15	5.21	5.27	5.33	5.38	5.43	5.47	.05	30
5.93	6.01	6.08	6.14	6.20	6.26	6.31	6.36	6.41	.01	
4.90	4.98	5.04	5.11	5.16	5.22	5.27	5.31	5.36	.05	40
5.76	5.83	5.90	5.96	6.02	6.07	6.12	6.16	6.21	.01	
4.81	4.88	4.94	5.00	5.06	5.11	5.15	5.20	5.24	.05	60
5.60	5.67	5.73	5.78	5.84	5.89	5.93	5.97	6.01	.01	
4.71	4.78	4.84	4.90	4.95	5.00	5.04	5.09	5.13	.05	120
5.44	5.50	5.56	5.61	5.66	5.71	5.75	5.79	5.83	.01	
4.62	4.68	4.74	4.80	4.85	4.89	4.93	4.97	5.01	.05	∞
5.29	5.35	5.40	5.45	5.49	5.54	5.57	5.61	5.65	.01	

SOURCE: This table is abridged from Table 29 in *Biometrika Tables for Statisticians*, vol. 1, 3rd ed., by E. S. Pearson and H. O. Hartley (eds.). Reproduced with the kind permission of the Trustees of *Biometrika*, 1966.

Table A.9 Upper-Tail Critical Values and Probabilities for the Null Distribution of the Wilcoxon Signed-Rank Statistic S_+

$$P_0(S_+ \geq c_1) = P(S_+ \geq c_1 \text{ when } H_0 \text{ is true})$$

n	c_1	$P_0(S_+ \geq c_1)$	n	c_1	$P_0(S_+ \geq c_1)$
3	6	.125		78	.011
4	9	.125		79	.009
	10	.062		81	.005
5	13	.094	14	73	.108
	14	.062		74	.097
	15	.031		79	.052
6	17	.109		84	.025
	19	.047		89	.010
	20	.031		92	.005
	21	.016	15	83	.104
7	22	.109		84	.094
	24	.055		89	.053
	26	.023		90	.047
	28	.008		95	.024
8	28	.098		100	.011
	30	.055		101	.009
	32	.027		104	.005
	34	.012	16	93	.106
	35	.008		94	.096
	36	.004		100	.052
9	34	.102		106	.025
	37	.049		112	.011
	39	.027		113	.009
	42	.010		116	.005
	44	.004	17	104	.103
10	41	.097		105	.095
	44	.053		112	.049
	47	.024		118	.025
	50	.010		125	.010
	52	.005		129	.005
11	48	.103	18	116	.098
	52	.051		124	.049
	55	.027		131	.024
	59	.009		138	.010
	61	.005		143	.005
12	56	.102	19	128	.098
	60	.055		136	.052
	61	.046		137	.048
	64	.026		144	.025
	68	.010		152	.010
	71	.005		157	.005
13	64	.108	20	140	.101
	65	.095		150	.049
	69	.055		158	.024
	70	.047		167	.010
	74	.024		172	.005

SOURCE: Adapted from W. J. Dixon and F. J. Massey, Jr., *Introduction to Statistical Analysis* (3rd ed.), New York: McGraw-Hill Book Company, Copyright © 1969.

Table A.10 Upper-Tail Critical Values and Probabilities for the Null Distribution of the Wilcoxon Rank-Sum Statistic W

$$P_0(W \geq c) = P(W \geq c \text{ when } H_0 \text{ is true})$$

m	n	c	$P_0(W \geq c)$	m	n	c	$P_0(W \geq c)$
3	3	15	.05			40	.004
	4	17	.057		6	40	.041
		18	.029			41	.026
	5	20	.036			43	.009
		21	.018			44	.004
	6	22	.048		7	43	.053
		23	.024			45	.024
		24	.012			47	.009
	7	24	.058			48	.005
		26	.017		8	47	.047
		27	.008			49	.023
	8	27	.042			51	.009
		28	.024			52	.005
		29	.012	6	6	50	.047
		30	.006			52	.021
4	4	24	.057			54	.008
		25	.029			55	.004
		26	.014		7	54	.051
	5	27	.056			56	.026
		28	.032			58	.011
		29	.016			60	.004
		30	.008		8	58	.054
	6	30	.057			61	.021
		32	.019			63	.01
		33	.010			65	.004
		34	.005	7	7	66	.049
	7	33	.055			68	.027
		35	.021			71	.009
		36	.012			72	.006
		37	.006		8	71	.047
	8	36	.055			73	.027
		38	.024			76	.01
		40	.008			78	.005
		41	.004	8	8	84	.052
5	5	36	.048			87	.025
		37	.028			90	.01
		39	.008			92	.005

SOURCE: Adapted from W. J. Dixon and F. J. Massey, Jr., *Introduction to Statistical Analysis* (3rd ed.), New York: McGraw-Hill Book Company, Copyright © 1969.

Table A.11 Critical Constant *c* for the Wilcoxon Signed-Rank Interval

$$(\bar{x}_{(n(n+1)/2-c+1)}, \bar{x}_{(c)})$$

n	Confidence Level (%)	c	n	Confidence Level (%)	c	n	Confidence Level (%)	c
5	93.8	15	13	99.0	81	20	99.1	173
	87.5	14		95.2	74		95.2	158
6	96.9	21		90.6	70		90.3	150
	93.7	20	14	99.1	93	21	99.0	188
	90.6	19		95.1	84		95.0	172
7	98.4	28		89.6	79		89.7	163
	95.3	26	15	99.0	104	22	99.0	204
	89.1	24		95.2	95		95.0	187
8	99.2	36		90.5	90		90.2	178
	94.5	32	16	99.1	117	23	99.0	221
	89.1	30		94.9	106		95.2	203
9	99.2	44		89.5	100		90.2	193
	94.5	39	17	99.1	130	24	99.0	239
	90.2	37		94.9	118		95.1	219
10	99.0	52		90.2	112		89.9	208
	95.1	47	18	99.0	143	25	99.0	257
	89.5	44		95.2	131		95.2	236
11	99.0	61		90.1	124		89.9	224
	94.6	55	19	99.1	158			
	89.8	52		95.1	144			
12	99.1	71		90.4	137			
	94.8	64						
	90.8	61						

SOURCE: Derived from W. J. Dixon and F. J. Massey, Jr., *Introduction to Statistical Analysis* (3rd ed.), New York: McGraw-Hill Book Company, Copyright © 1969.

Table A.12 Critical Constant c for the Wilcoxon Rank-Sum Interval $(d_{ij(mn-c+1)}, d_{ij(c)})$

Smaller Sample Size

Larger Sample Size	5 Confidence Level (%)	c	6 Confidence Level (%)	c	7 Confidence Level (%)	c	8 Confidence Level (%)	c
5	99.2	25						
	94.4	22						
	90.5	21						
6	99.1	29	99.1	34				
	94.8	26	95.9	31				
	91.8	25	90.7	29				
7	99.0	33	99.2	39	98.9	44		
	95.2	30	94.9	35	94.7	40		
	89.4	28	89.9	33	90.3	38		
8	98.9	37	99.2	44	99.1	50	99.0	56
	95.5	34	95.7	40	94.6	45	95.0	51
	90.7	32	89.2	37	90.6	43	89.5	48
9	98.8	41	99.2	49	99.2	56	98.9	62
	95.8	38	95.0	44	94.5	50	95.4	57
	88.8	35	91.2	42	90.9	48	90.7	54
10	99.2	46	98.9	53	99.0	61	99.1	69
	94.5	41	94.4	48	94.5	55	94.5	62
	90.1	39	90.7	46	89.1	52	89.9	59
11	99.1	50	99.0	58	98.9	66	99.1	75
	94.8	45	95.2	53	95.6	61	94.9	68
	91.0	43	90.2	50	89.6	57	90.9	65
12	99.1	54	99.0	63	99.0	72	99.0	81
	95.2	49	94.7	57	95.5	66	95.3	74
	89.6	46	89.8	54	90.0	62	90.2	70

Smaller Sample Size

Larger Sample Size	9 Confidence Level (%)	c	10 Confidence Level (%)	c	11 Confidence Level (%)	c	12 Confidence Level (%)	c
9	98.9	69						
	95.0	63						
	90.6	60						
10	99.0	76	99.1	84				
	94.7	69	94.8	76				
	90.5	66	89.5	72				
11	99.0	83	99.0	91	98.9	99		
	95.4	76	94.9	83	95.3	91		
	90.5	72	90.1	79	89.9	86		
12	99.1	90	99.1	99	99.1	108	99.0	116
	95.1	82	95.0	90	94.9	98	94.8	106
	90.5	78	90.7	86	89.6	93	89.9	101

SOURCE: Derived from W. J. Dixon and F. J. Massey, Jr., *Introduction to Statistical Analysis* (3rd ed.), New York: McGraw-Hill Book Company, Copyright © 1969.

Table A.13 Curves of $\beta = P(\text{Type II Error})$ for t Tests

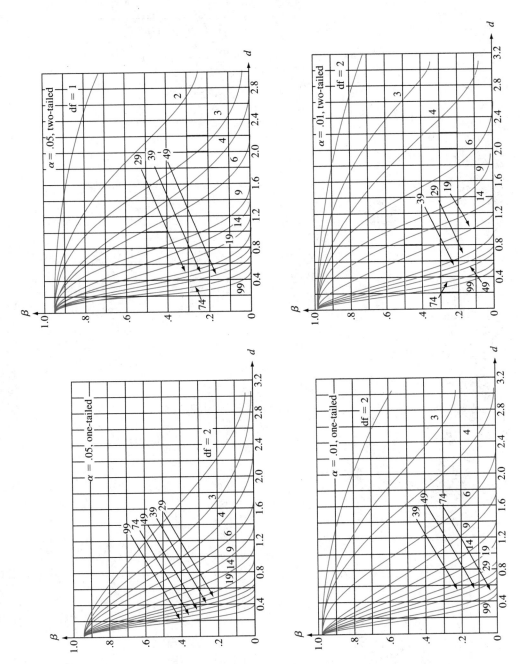

Table A.14 Critical Values c_α for the Test of Normality Based on r from a Normal Probability Plot

		α		
		.10	*.05*	*.01*
	5	.9033	.8804	.8320
	10	.9347	.9180	.8804
	15	.9506	.9383	.9110
	20	.9600	.9503	.9290
n	25	.9662	.9582	.9408
	30	.9707	.9639	.9490
	40	.9767	.9715	.9597
	50	.9807	.9764	.9664
	60	.9835	.9799	.9710
	75	.9865	.9835	.9757

SOURCE: MINITAB Reference Manual

Answers to Odd-Numbered Exercises

Chapter 1

1. Each of the following is one of many possible samples.
   a. *L.A. Times, Oberlin Tribune, Gainesville Sun, St. Louis Post-Dispatch*   **b.** Wadsworth Publishing Co., Dupont, Walt Disney, Proctor and Gamble   **c.** Heather Groves, Greg Ridgeway, Marc Tran, Paula Rodriguez   **d.** KDFC, WCLV, KFI, KKGO

3. Concrete: all registered voters in California, all mutual funds marketed in the United States, all books published in 1990
   Hypothetical: all grade point averages for University of California undergraduates during the next academic year, fuel efficiencies (mpg) for all Honda Accords during the next model year, diameters of all bolts to be manufactured by a particular company during the next week.

5. One could take a single random sample selected from the population of all CSU students or a separate random sample from students at each of the 20 campuses. This study is enumerative.

7. **a.** 8   **b.** An analytic study

9. **a.**

2	32
3	8473954672
4	60838159
5	11680404506165970
6	79426405320961072246989203
7	10556355622430580
8	08324432
9	268320537636381
10	5483425846
11	5293993
12	37
13	8
14	36
15	0350
16	
17	
18	9

Stem: Ones
Leaf: Tenths

b. An observation with a stem of 7   **c.** Quite spread out   **d.** Positively skewed   **e.** Yes, the largest observation 18.9

11.

Class	Frequency	Relative Frequency
4000 −< 4200	1	.01
4200 −< 4400	2	.02
4400 −< 4600	9	.09
4600 −< 4800	13	.13
4800 −< 5000	18	.18
5000 −< 5200	22	.22
5200 −< 5400	20	.20
5400 −< 5600	7	.07
5600 −< 5800	7	.07
5800 −< 6000	1	.01
	100	1.00

13. **a.**

Class	Frequency	Relative Frequency	Density
.15 −< .25	8	.02192	0.2192
.25 −< .35	14	.03836	0.3836
.35 −< .45	28	.07671	0.7671
.45 −< .50	24	.06575	1.3150
.50 −< .55	39	.10685	2.1370
.55 −< .60	51	.13973	2.7946
.60 −< .65	106	.29041	5.8082
.65 −< .70	84	.23014	4.6028
.70 −< .75	11	.03014	0.6028
	$n = 365$	1.00001	

b. 6%   **c.** 26%

15.

Category	Frequency	Relative Frequency
J	10	.1667
F	9	.1500
B	7	.1167
M	4	.0667
C	3	.0500
N	6	.1000
O	21	.3500
	$n = 60$	1.0001

17. b. .518   **c.** .635

21. No, to 1.01

23.

6L	430
6H	769689
7L	42014202
7H	
8L	011211410342
8H	9595578
9L	30
9H	58

The gap in the data–no scores in the high 70s

25. a. $\bar{x} \approx 192.57$, $\tilde{x} = 189.0$ **b.** New mean = 189.71, median unchanged

27. 191.0, 7.14%

29. 10% trimmed mean = 11.46

31. a. $\bar{x} = 1.0297$, $\tilde{x} = 1.009$ **b.** .383

33. a. .7   **b.** Also .7   **c.** 13

35. $\tilde{x} = 68.0$, 20% trimmed mean = 66.2, 30% trimmed mean = 67.5

37. a. $s^2 = 14.00$, $s = 3.74$ **b.** Identical to part (a)   **c.** Again, $s^2 = 14.00$ **d.** New $s^2 = 1400$

39. a. 56.80, 197.8040   **b.** .5016, .708

41. $\Sigma y_i = 31.2$, $\Sigma y_i^2 = 146.08$, $s_y^2 = 5.4151 = s_x^2$, nothing special

43. a. .0272, .1649   **b.** .222   **c.** There are no outliers.

45. –3.5; add any number to each of the five deviations.

47. The lower fourth, median, and upper fourth are 104, 132, and 152.5, respectively, from which $f_s = 48.5$. There are no outliers.

49. $s^2 = 541.51$. Divide by $(10)^2$.

51. The boxplots are similar in shape, but the seeded boxplot is shifted to the right of the control boxplot and stretches out more toward larger values.

53. a. 62   **b.** Most frequent category

55. 10.65

57. a. $\bar{y} = a\bar{x} + b$, $s_y^2 = a^2 s_x^2$ **b.** 189.14, 1.87

59. $\bar{x} = .9255$, $s = .0809$, $\tilde{x} = .93$, small amount of variability, slight bit of skewness

61.

	$\bar{x}$	$\tilde{x}$	Fourths	s
Smoking	693.364	693.000	609.5, 760.0	103.864
Nonsmoking	961.091	947.000	916.5, 977.0	113.810

Volumes for nonsmokers appear to exceed those for smokers; boxplots show several outliers in the nonsmoking sample and much less variability in the middle 50% than for smokers.

63. a. $s_y = s_x$ **b.** $s_z = 1$

65.

1H	96, 90, 88, 86	Stem = Tenths
2L	43, 36	Leaf = Hundredths and
2H	87, 99, 92, 91, 92	thousandths
3L	24, 00, 46, 05, 04, 34, 15, 35, 32, 49, 24, 01, 33, 03, 18, 06, 45, 45, 37, 20, 29	
3H	99, 93, 75, 99, 67, 53, 57, 54, 61	
4L	11, 48, 19, 47, 19, 08, 41, 28, 12	
4H	51, 98, 89, 68	
5L		
5H	85	

Class	Frequency	Relative frequency
.150 –< .200	4	.073
.200 –< .250	2	.036
.250 –< .300	5	.091
.300 –< .350	21	.382
.350 –< .400	9	.164
.400 –< .450	9	.164
.450 –< .500	4	.073
.500 –< .550	0	.000
.550 –< .600	1	.018
	55	1.001

$\bar{x} = .3489$, $s = .0800$, $\tilde{x} = .3370$, $f_s = .4035 - .3035 = .1000$, $.4035 + 1.5f_s = .5535$, so .585 is a mild outlier. The distribution of values is reasonably symmetric, with a high peak in the low .3's.

67. a. All points fall on a 45° line. Points fall below a 45° line.   **b.** Points fall well below a 45° line, indicating a substantial positive skew.

Chapter 2

1. a. $\{(A, A), (A, E), (A, J), (E, A), (E, E), (E, J), (J, A),$
$(J, E), (J, J)\}$ **b.** $\{(A, E), (A, J), (E, A), (J, A)\}$
Note: A = American, J = Asian. **c.** $\{(A, E),$
$(A, J), (E, A), (E, E), (E, J), (J, A), (J, E), (J. J)\}$;
complement = $\{(A, A)\}$, a simple event.

3. a. $\mathcal{S} = \{(1, 1, 1), (1, 1, 2), (1, 1, 3), (1, 2, 1), (1, 2, 2),$
$(1, 2, 3), (1, 3, 1), (1, 3, 2), (1, 3, 3), (2, 1, 1), (2, 1, 2),$
$(2, 1, 3), (2, 2, 1), (2, 2, 2), (2, 2, 3), (2, 3, 1), (2, 3, 2),$
$(2, 3, 3), (3, 1, 1), (3, 1, 2), (3, 1, 3), (3, 2, 1), (3, 2, 2),$
$(3, 2, 3), (3, 3, 1), (3, 3, 2), (3, 3, 3)\}$ **b.** $\{(1, 1, 1),$
$(2, 2, 2), (3, 3, 3)\}$ **c.** $\{(1, 2, 3), (1, 3, 2), (2, 1, 3),$
$(2, 3, 1), (3, 1, 2), (3, 2, 1)\}$ **d.** $\{(1, 1, 1), (1, 1, 3),$
$(1, 3, 1), (1, 3, 3), (3, 1, 1), (3, 1, 3), (3, 3, 1), (3, 3, 3)\}$

5. a. $\{11, 22, 33\}$ **b.** $\{1213, 1312, 1231, 1321,$
$2123, 2132, 2312, 2321, 3123, 3132, 3213, 3231\}$

7. a. There are 35 outcomes in $\mathcal{S}$. **b.** $\{AABABAB,$
$AABAABB, AAABBAB, AAABABB, AAAABBB\}$

11. a. .07 **b.** .30 **c.** .57

13. a. .36 **b.** .64 **c.** .53 **d.** .47 **e.** .17
f. .75

15. a. .913 **b.** .913

17. a. .4 **b.** .75 **c.** .25

19. a. $(M, 2), (M, 4), (Q, 2), (Q, 4)$ **b.** .32 **c.** .41

21. a. .10 **b.** .18, .19 **c.** .41 **d.** .59 **e.** .31
f. .69

23. .40

27. a. .98 **b.** .02 **c.** .13 **d.** .24

29. a. 1/9 **b.** 8/9 **c.** 2/9

31. a. 720 **b.** 120 **c.** 1/30

33. a. 2401 **b.** .050 **c.** .701 **d.** .408

35. a. 15,504 **b.** 840 **c.** .0542 **d.** .0578

37. .20

39. .0456

41. a. 369,600 **b.** .00006504

43. .067, .333, .667

47. $P(A \mid B)$

49. .027

51. a. .05 **b.** .12 **c.** .56, .44 **d.** .25 **e.** .533
f. .444, .556

53. .04

55. a. .5 **b.** .0455 **c.** .682 **d.** .0189

59. a. .21 **b.** .455 **c.** .264, .274

61. a. .578, .278, .144 **b.** 0, .457, .543

63. b. .54 **c.** .68 **d.** .74 **e.** .7941

65. $P(\text{Mean} \mid S) = .3922$, $P(\text{Median} \mid S) = .2941$, so Mean
and Median are most and least likely, respectively.

67. a. .126 **b.** .05 **c.** .1125 **d.** .2725
e. .5325 **f.** .2113

69. Only A_2 and A_3 are independent.

71. a. No **b.** .3, .5

73. a. .00601 **b.** .00421

75. .0059

77. a. .10, .20 **b.** 0

79. a. $p(2 - p)$ **b.** $1 - (1 - p)^n$ **c.** $(1 - p)^3$
d. $.9 + (1 - p)^3(.1)$ **e.** $.1(1 - p)^3/[.9 + .1(1 - p)^3]$
$= .0137$ for $p = .5$

81. .8588, .9949

83. .0625, .625

85. a. 1140 **b.** 969 **c.** 1020 **d.** .85

87. .0714, .0762

89. .0204, .0098

91. a. .0083 **b.** .2 **c.** .2

93. .1071, .5357

95. .905

97. a. .90416 **b.** .7656

99. .008

101. .362, .348, .290

103. a. Granite **b.** Basalt in both cases **c.** .175
d. $p > 14/17$

105. a. .0625 **b.** .15625 **c.** .375 **d.** .0198

107. a. .12, .88 **b.** .18, .38

109. $P(\text{at least } 1) = 1 - (1 - p_1)(1 - p_2)(1 - p_3)(1 - p_4)$
$P(\text{at least } 2) = p_1 p_2 p_3' p_4' + \cdots + p_1' p_2' p_3 p_4 +$
$p_1' p_2 p_3 p_4' + \cdots + p_1 p_2 p_3' p_4' + p_1 p_2 p_3 p_4'$

Chapter 3

1. $x = 0$ for *FFF*; $x = 1$ for *SFF, FSF,* and *FFS*; $x = 2$ for *SSF, SFS,* and *FSS,* and $x = 3$ for *SSS*

3. Z = average of the two numbers, with possible values $2/2, 3/2, \ldots , 12/2$; W = absolute value of the difference, with possible values 0, 1, 2, 3, 4, 5

5. No. In Example 3.4, let $Y = 1$ if at most three batteries are examined and let $Y = 0$ otherwise. Then Y has only two values.

7. a. $\{0, 1, \ldots , 12\}$; discrete   **c.** $\{1, 2, 3, \ldots \}$; discrete   **e.** $\{0, c, 2c, \ldots , 10,000c\}$ where c is the royalty per book; discrete   **g.** $\{x : m \leq x \leq M\}$ where $m(M)$ is the minimum (maximum) possible tension; continuous

9. a. $\{2, 4, 6, 8, \ldots \}$, that is, $\{2(1), 2(2), 2(3), 2(4), \ldots \}$, an infinte sequence; discrete   **b.** $\{2, 3, 4, 5, 6, \ldots \}$, that is, $\{1 + 1, 1 + 2, 1 + 3, 1 + 4, \ldots \}$, an infinite sequence; discrete

11. a. $p(4) = .45$, $p(6) = .40$, $p(8) = .15$, $p(x) = 0$ for $x \neq 4, 6,$ or 8

13. a. .70   **b.** .45   **c.** .55   **d.** .71   **e.** .65   **f.** .45

15. a. (1, 2), (1, 3), (1, 4), (1, 5), (2, 3), (2, 4), (2, 5), (3, 4), (3, 5), (4, 5)   **b.** $p(0) = .3, p(1) = .6, p(2) = .1$ **c.** $F(x) = 0$ for $x < 0$, = .3 for $0 \leq x < 1$, = .9 for $1 \leq x < 2$, and = 1 for $2 \leq x$

17. a. .64   **b.** .256   **c.** It is *A; AUUUA, UAUUA, UUAUA, UUUAA;* .02408

19. $p(1) = p(2) = p(3) = p(4) = .25$

21. $F(x) = 0$ for $x < 0$, = .10 for $0 \leq x < 1$, = .25 for $1 \leq x < 2$, = .45 for $2 \leq x < 3$, = .70 for $3 \leq x < 4$, = .90 for $4 \leq x < 5$, = .96 for $5 \leq x < 6$, and = 1 for $6 \leq x$

23. a. $p(1) = .30, p(3) = .10, p(4) = .05, p(6) = .15, p(12) = .40$ **b.** .30, .60

25. a. $p(x) = (\frac{2}{3})^{x-1} \cdot (\frac{1}{3})$ for $x = 1, 2, 3, \ldots$
b. $p(y) = (\frac{2}{3})^{y-2} \cdot (\frac{1}{3})$ for $y = 2, 3, 4, 5, \ldots$
c. $p(z) = (\frac{25}{54}) \cdot (\frac{2}{3})^{2z-2}$ for $z = 1, 2, 3, \ldots$

29. a. 3.24   **b.** 1.785

31. a. 16.38, 272.298, 3.9936   **b.** 401   **c.** 2496   **d.** 13.66

33. Yes, because $\sum_{x=1}^{\infty} 1/x^2 < \infty$.

35. $700

37. $E(h(X)) = .408$, so gamble.

39. $V(-X) = V(X)$

41. a. 32.5   **b.** 7.5
c. $V(X) = E[X(X - 1)] + E(X) - E(X^2)$

43. a. .25, .11, .06, .04, .01   **b.** 2.64, 1.54, .04, conservative   **c.** They are equal.   **d.** $p(-1) = .02 = p(1)$, $p(0) = .96$

45. a. .850   **b.** .200   **c.** .200   **d.** .701
e. .851   **f.** .000   **g.** .570

47. a. .354   **b.** .115   **c.** .918

49. a. .944   **b.** .449   **c.** .816

51. .1478

53. a. .017   **b.** .811, .425   **c.** .006, .902, .586

55. When $p = .9$, the probability is .99 for A and .9963 for B. If $p = .5$, these probabilities are .75 and .6875, respectively.

57. The tabulation for $p > .5$ is unnecessary.

59. a. 30, 21   **b.** 50, 25

61. $P(|X - \mu| \geq 2\sigma) = .042$ when $p = .5$ and = .065 when $p = .75$, compared to the upper bound of .25. Using $k = 3$ in place of $k = 2$, these probabilities are .002 and .004, respectively, whereas the upper bound is .11.

63. a. .242   **b.** .030   **c.** .939

65. $h(x; 15, 10, 20)$, .033

67. a. $h(x; 10, 10, 20)$ **b.** .033

69. a. $nb(x; 2, .5)$ **b.** .188   **c.** .688   **d.** 2, 4

71. $nb(x; 6, .5)$, 6

73. a. .932   **b.** .065   **c.** .068   **d.** .492   **e.** .251

75. a. .583   **b.** .493   **c.** 10, 3.16

77. Poisson with $\lambda = 5$ **a.** .602   **b.** .133

79. a. .091, .900, .283   **b.** 12, 3.464   **c.** .530, .011

81. a. .099   **b.** .135   **c.** 2

83. a. 4   **b.** .215   **c.** At least $-\ln(.1)/2 \approx 1.1513$ years

85. a. .221   **b.** 6,800,000   **c.** $p(x; 1608.5)$

89. b. 3.114, .405, .636

91. a. $b(x; 15, .75)$ **b.** .764   **c.** .707
d. 11.25, 2.81   **e.** .310

93. .991

95. a. $p(x; .5)$ **b.** .076   **c.** .090

97. 1.813, 3.05

99. $p(2) = p^2$, $p(3) = (1 - p)p^2$, $p(4) = (1 - p)p^2$, $p(x) = [1 - p(2) - \ldots - p(x - 3)](1 - p)p^2$ for $x = 5$, 6, 7, . . . ; .99950841

101. a. 16 **b.** 4 **c.** .804 **d.** $h(x; 5, 4, 12)$; 1.67, .707

103. a. .135 **b.** .00144 **c.** $\sum_{x=0}^{\infty} [p(x; 2)]^5$

105. 3.590

107. a. No **b.** .0273

109. b. $.6p(x; \lambda) + .4p(x; \mu)$ **c.** $(\lambda + \mu)/2$
d. $(\lambda - \mu)^2/4 + (\lambda + \mu)/2$

111. $\sum_{i=1}^{10} (p_{i+j+1} + p_{i-j-1})p_i$ where $p_k = 0$ if $k < 0$ or $k > 10$.

Chapter 4

1. a. .25 **b.** .50 **c.** .4375

3. b. .50 **c.** .6875 **d.** .6328

5. a. 3 **b.** .125 **c.** .109 **d.** .704

7. a. $f(x) = .1$ for $25 \le x \le 35$ and 0 otherwise
b. .20 **c.** .40 **d.** .20

9. a. .562 **b.** .438, .438 **c.** .071

11. a. .25 **b.** .1875 **c.** .9375 **d.** 1.4142
e. $f(x) = x/2$ for $0 < x < 2$

13. a. 1.33 **b.** .222, .471 **c.** 2

15. a. For $0 \le x \le 1$, $F(x) = 2[x - x^2/2]$. **b.** .75
c. .3125, .3125 **d.** .5 **e.** .293 **f.** .333, .0556, .2357

17. a. For $2 \le x \le 4$, $F(x) = .25[3x - 7 - (x - 3)^3]$ **b.** 3
c. 3, .2

19. a. .597 **b.** .369
c. $f(x) = .3466 - .25 \ln(x)$ for $0 < x < 4$

21. 314.79

23. 248, 3.60

25. b. 1.8(90th percentile for X) + 32
c. $a(X$ percentile) + b

27. a. 2.14 **b.** .81 **c.** 1.17 **d.** .97 **e.** 2.41

29. a. 2.54 **b.** 1.34 **c.** −.42

31. a. .9452 **b.** .9370 **c.** .9544

33. a. .1587 **b.** .6247 **c.** .8413

35. a. 31.7 **b.** 17.225 **c.** 3.192

37. .002

39. 10, .2

41. a. .9938 **b.** .9876, 1 **c.** .0052

43. 15.66

45. .3174 for $k = 1$, .0456 for $k = 2$, .0026 for $k = 3$, as compared to the bounds of 1, .25, and .111, respectively.

47. a. Exact: .212, .577, .573 Approximate: .211, .567, .596 **b.** Exact: .885, .575, .017 Approximate: .885, .579, .012 **c.** Exact: .002, .029, .617 Approximate: .003, .033, .599

49. a. .9666 **b.** .0099, 0

51. b. normal, $\mu = 239$, $\sigma^2 = 12.96$

53. a. 120 **b.** 1.329 **c.** .371 **d.** .735 **e.** 0

55. a. .594 **b.** .092 **c.** .537

57. a. .424 **b.** .567; yes **c.** 60 **d.** 66

59. a. .777 **b.** .368 **c.** .145 **d.** 13.863

61. a. n/λ; 20 **b.** .930 **c.** $1 - \sum_{k=0}^{n-1} \dfrac{e^{-\lambda t}(\lambda t)^k}{k!}$

63. $-\ln(1 - p)/\lambda$, .693/λ

67. a. .632, .632, .064 **b.** .470 **c.** 172.727

71. a. .8212 **b.** .1120

73. a. 149.157, 223.595 **b.** .9834 **c.** .0921
d. 148.41 **e.** 9.83 **f.** 125.90

75. $\alpha = \beta$

77. b. $[\Gamma(\alpha + \beta) \cdot \Gamma(m + \beta)]/[\Gamma(\alpha + \beta + m) \cdot \Gamma(\beta)]$, $\beta/(\alpha + \beta)$

79. The point corresponding to the largest observation deviates substantially from the linear pattern determined by the other points.

81. The plot is straight; population normality is plausible.

83. Plot $\ln(x)$ vs. z percentile. The pattern is straight, so a lognormal population distribution is plausible.

85. There is substantial curvature in the plot. λ is a scale parameter (as is σ for the normal family).

87. a. $F(y) = \frac{1}{48}(y^2 - y^3/18)$ for $0 \le y \le 12$ **b.** .259, .5, .241 **c.** 6, 43.2, 7.2 **d.** .518 **e.** 3.75

89. a. $f(x) = x^2$ for $0 \le x < 1$ and $= \frac{7}{4} - \frac{3}{4}x$ for $1 \le x \le 7/3$
b. .917 **c.** 1.213

91. a. .9162 **b.** .9549 **c.** 1.3374

93. b. $F(x) = 0$ for $x < -1$, $= (4x + x^3/3)/9 + 11/27$ for $-1 \le x \le 2$, and $= 1$ for $x > 2$
c. No. $F(0) < .5 \Rightarrow \tilde{\mu} > 0$ **d.** $Y \sim \text{Bin}(10, 5/27)$

95. a. .368, .828, .460   **b.** 352.53
  c. $1/\beta \cdot \exp[-\exp(-(x - \alpha)/\beta)] \cdot \exp(-(x - \alpha)/\beta)$
  d. α **e.** $\mu = 201.95$, mode = 150, $\tilde{\mu} = 182.99$
97. a. μ **b.** No.   **c.** 0   **d.** $(\alpha - 1)\beta$ **e.** $\nu - 2$
99. b. $p(1 - \exp(-\lambda_1 x)) + (1 - p)(1 - \exp(-\lambda_2 x))$ for $x \geq 0$
  c. $p/\lambda_1 + (1-p)/\lambda_2$ **d.** $V(X) = 2p/\lambda_1^2 + 2(1-p)/\lambda_2^2 - \mu^2$
  e. 1, $CV > 1$ **f.** $CV < 1$

101. a. Lognormal.   **b.** 1   **c.** 2.72, .0185
103. a. $Y \sim$ gamma with parameters α and 60β
  b. Gamma with parameters α and $c\beta$
105. a. λ **b.** $(\alpha/\beta^\alpha)x^{\alpha-1}$; $\alpha > 1$; $\alpha < 1$
  c. $F(x) = 1 - \exp[-\alpha(x - x^2/2\beta)]$
107. a. $g(\mu)$, $[g'(\mu)]^2 \cdot V(X)$ **b.** $v/20$, $v/800$

Chapter 5

1. a. .20   **b.** .42   **c.** At least one hose is in use at
  each pump; .70.   **d.** $p_X(x) = .16, .34, .50$ for $x = 0$,
  1, 2, respectively; $p_Y(y) = .24, .38, .38$ for $y = 0, 1, 2$,
  respectively; .50   **e.** No; $p(0, 0) \neq p_X(0) \cdot p_Y(0)$
3. a. .15   **b.** .40   **c.** .22   **d.** .17, .46
5. a. .054   **b.** .00018
7. a. .030   **b.** .120   **c.** .300   **d.** .380   **e.** Yes.
9. a. 3/380,000   **b.** .3024   **c.** .3593
  d. $10Kx^2 + .05$ for $20 \leq x \leq 30$ **e.** No
11. a. $e^{-\lambda - \mu} \cdot \lambda^x \cdot \mu^y/x!y!$ **b.** $e^{-\lambda - \mu} \cdot [1 + \lambda + \mu]$
  c. $e^{-(\lambda + \mu)} \cdot (\lambda + \mu)^m/m!$; Poisson $(\lambda + \mu)$
13. a. $e^{-x - y}$ for $x \geq 0$, $y \geq 0$ **b.** .400   **c.** .594
  d. .330
15. a. $F(y) = 1 - e^{-\lambda y} + (1 - e^{-\lambda y})^2 - (1 - e^{-\lambda y})^3$ for $y \geq 0$.
  b. $2/3\lambda$
17. a. .25   **b.** .318   **c.** .637
  d. $f_x(x) = 2\sqrt{R^2 - x^2}/\pi R^2$ for $-R \leq x \leq R$; no
19. a. $K(x^2 + y^2)/(10Kx^2 + .05)$; $K(x^2 + y^2)/(10Ky^2 + .05)$
  b. .556, .549   **c.** 25.37, 2.87
21. $= f_Y(y)$ for every x
23. .15
25. L^2
27. .25 hr
29. $-2/3$
31. a. $-.1082$ **b.** $-.0131$
37. a.

$\bar{x}$	160	180	200	205	225	250
$p(\bar{x})$	.25	.30	.09	.20	.12	.04

$E(\bar{X}) = 190 = \mu$

  b.

s^2	0	800	1250	4050
$p(s^2)$	.38	.30	.12	.20

$E(S^2) = 1200 = \sigma^2$

39.

Proportion	0	.1	.2	.3	.4	.5
Probability	.000	.000	.000	.001	.005	.027

Proportion	.6	.7	.8	.9	1.0
Probability	.088	.201	.302	.269	.107

41. a.

$\bar{x}$	1	1.5	2	2.5	3	3.5	4
$p(\bar{x})$	.16	.24	.25	.20	.10	.04	.01

  b. .85   **c.**

r	0	1	2	3
$p(r)$	.30	.40	.22	.08

45. a. .6826   **b.** .1056
47. a. .6026   **b.** .2981
49. .7720
51. a. 0   **b.** 0
53. a. .0968   **b.** .8882
55. .9616
57. a. .9332, .9544   **b.** .0062, .6826   **c.** .8719
  d. .8413, .6826
59. a. 3.5, 2.27, 1.51   **b.** 15.4, 75.94, 8.71
61. a. .695   **b.** 4.0675 > 2.6775
63. a. .7620   **b.** .8414
65. .1588
67. a. 2400   **b.** 1205; independence   **c.** 2400, 41.77
69. a. 158, 430.25   **b.** .9788
71. a. Both approximately normal (CLT)   **b.** Approximately normal, mean = 2, SD = 1.621   **c.** .2354
  d. .0068; yes
73. a. .2, .5, .3 for $x = 7, 9, 10$; .10, .35, .55 for $y = 7, 9, 10$
  b. .25   **c.** No   **d.** 18.25   **e.** 1.05
75. a. 3/81,250   **b.** $f_X(x) = k(250x - 10x^2)$ for $0 \leq x$
  ≤ 20 and $= k(450x - 30x^2 + \frac{1}{2}x^3)$ for $20 < x \leq 30$; $f_Y(y)$
  results from substituting y for x in $f_X(x)$. They are not
  independent.   **c.** .355   **d.** 25.969   **e.** 204.6154,
  $-.894$ **f.** 7.66
77. ≈ 1

79. a. 400 min   **b.** 70

81. 97

83. .9973

87. *b, c.* Chi-squared with $\nu = n$.

89. a. $\sigma_W^2/(\sigma_W^2 + \sigma_E^2)$ **b.** .9999

91. 26, 1.64

Chapter 6

1. a. 65.72,   4.53   **b.** 65.10   **c.** 66.05   **d.** .70

3. a. 76.818; $\bar{X}$ **b.** 76.88   **c.** .083   **d.** .216

5. a. 2.54   **b.** Variance $= (\sigma_1^2/m) + (\sigma_1^2/n)$ **c.** .286

7. a. 120.6   **b.** 1,206,000   **c.** .80   **d.** 120.0

9. a. 2.11   **b.** .119

11. b. $\left[\dfrac{p_1 q_1}{n_1} + \dfrac{p_2 q_2}{n_2}\right]^{1/2}$ **c.** Use $\hat{p}_i = x_i/n_i$ and $\hat{q}_i = 1 - \hat{p}_i$ in place of p_i and q_i in part (b) for $i = 1, 2$ **d.** $-.245$ **e.** .041

15. a. $\hat{\theta} = \Sigma X_i^2/2n$ **b.** 74.505

17. b. .444

19. a. $\hat{p} = 2\hat{\lambda} - .30 = .20$ **b.** $\hat{p} = (100\hat{\lambda} - 9)/70$

21. b. $\hat{\alpha} = 5$, $\hat{\beta} = 28.0/\Gamma(1.2)$

23. $\hat{\lambda}_1 = \bar{x}$, $\hat{\lambda}_2 = \bar{y}$, $(\lambda_1 \stackrel{.}{-} \lambda_2) = \bar{x} - \bar{y}$

25. a. 384.4, 18.86   **b.** 415.42

29. a. $\hat{\theta} = \min(X_i)$, $\hat{\lambda} = n/\Sigma[X_i - \min(X_i)]$ **b.** .64, .202

33. With $x_i =$ time between birth $i - 1$ and birth i, $\hat{\lambda} = 6/\Sigma_{i=1}^{6} ix_i = .0436$.

35. 29.5

37. 1.0132

Chapter 7

1. a. 99.5%   **b.** 85%   **c.** 2.96   **d.** 1.15

3. a. (4.52, 5.18)   **b.** (4.12, 5.00)   **c.** 55

5. By a factor of 4; the length is decreased by a factor of 5.

7. a. $(\bar{x} - 1.645\sigma/\sqrt{n}, \infty)$; $(4.57, \infty)$ **b.** $(\bar{x} - z_\alpha \cdot \sigma/\sqrt{n}, \infty)$ **c.** $(-\infty, \bar{x} + z_\alpha \cdot \sigma/\sqrt{n})$; $(-\infty, 59.7)$

9. 950, .8714

11. (7.80, 8.54); no assumptions

13. (150.2, 154.4)

15. (.182, .316)

17. a. 385   **b.** 342

19. $\bar{x} \pm z_{\alpha/2} \cdot \sqrt{\bar{x}/n}$; a 95% CI is (3.50, 4.62)

21. a. 1.321   **b.** -1.321 **c.** -1.746 **d.** .99   **e.** .95   **f.** .002

23. a. (8.19, 10.85)   **b.** $(-\infty, \bar{x} + t_{\alpha,n-1} \cdot s/\sqrt{n})$; $(-\infty, 10.54)$

25. (170.5, 205.5)

27. (33.53, 43.79)

29. (14.66, 29.14)

31. a. 18.307   **b.** 3.940   **c.** .95   **d.** .10

33. (3.6, 8.1); no

35. (490.0, 786.2)

37. a. (.626, .734)   **b.** 1083   **c.** (.636, .724)

39. $(-.84, -.16)$

41. 246

43. $(2t_r/\chi^2_{1-\alpha/2,2r}, 2t_r/\chi^2_{\alpha/2,2r}) = (65.3, 232.5)$

45. a. $(\max(x_i)/(1 - \alpha/2)^{1/n}, \max(x_i)/(\alpha/2)^{1/n})$ **b.** $(\max(x_i), \max(x_i)/\alpha^{1/n})$ **c.** (b); (4.2, 7.65)

47. (73.6, 78.8) versus (75.1, 79.6)

49. a. (1237.1, 3015.9)   **c.** It is much wider.   **d.** $k \to 1.645$; width $\to 3.29\sigma$; width $\to 0$

Chapter 8

1. a. Yes   **b.** No   **c.** No   **d.** Yes   **e.** No   **f.** Yes

5. H_0: $\sigma = .05$ versus H_a: $\sigma < .05$. I: conclude variability in thickness is satisfactory when it isn't. II: Conclude variability in thickness isn't satisfactory when in fact it is.

7. I: concluding that the plant isn't in compliance when it is; II: concluding that the plant is in compliance when it isn't.

9. **a.** H_0: $\mu = 1300$ versus H_a: $\mu > 1300$ **b.** Normal with mean 1300, SD. 13.416; .01 **c.** Normal with mean 1350, SD. 13.416; .0808 **d.** Replace 1331.26 by 1322.07; .0188 **e.** $z \geq 2.33$

11. **a.** $\mu =$ true average braking distance, H_0: $\mu = 120$, H_a: $\mu < 120$. **b.** R_2 **c.** .002; replace 115.20 by 114.87 **d.** .4522 **e.** .01, .002

13. **a.** .01 **b.** .5040, .5596; no.

15. **a.** .001 **b.** .01 **c.** .10

17. **a.** 1.5 **b.** Don't reject H_0. **c.** .002 **d.** .5398 **e.** 88 **f.** .0003

19. **a.** $z = 2.18 \geq 1.645$, so reject H_0 **b.** .0901 **c.** H_0: $\mu = 250$ versus H_a: $\mu < 250$.

21. $z = 3.33 \geq 1.645$, so reject H_0.

23. $z = -1.82 > -2.33$, so don't reject H_0.

25. $z = -1.13 > -1.28$, so don't reject H_0.

27. Yes, because $z = -6.17 \leq -z_\alpha$ for any reasonable α.

29. $t = -33.5 < -4.604$, so reject H_0.

33. Yes, because $z = 2.46 > 1.96$.

35. Reject H_0 because $z = 3.55 \geq 1.645$.

37. $-6.1 \leq -2.33$; yes

39. $n = 25$, $\{5, 6, \ldots, 25\}$, $\alpha = .098$, $\beta(.3) = .090$

41. $z = -.26 > -2.33$; no

43. **a.** Don't reject H_0. **b.** Don't reject H_0. **c.** Don't reject H_0. **d.** Reject H_0. **e.** Don't reject H_0. **f.** Don't reject H_0.

45. **a.** .0358 **b.** .0802 **c.** .5824 **d.** .1586 **e.** 0

47. **a.** $.05 < P$-value $< .10$ **b.** P-value $\approx .01$ **c.** P-value $< .001$ **d.** P-value $\approx .20$

49. Because P-value $> .01$, H_0 cannot be rejected.

51. $.025 < P$-value $< .05$, so reject H_0; normal.

53. P-value $\ll .0005$; the average level does appear to exceed the standard.

55. **a.** .9703, .8554, .4325, .0944 **b.** .4013, .1056, .0062, .0000 **c.** No

57. Yes

59. **a.** Yes, because $t = 12.9 \geq 2.228$. **b.** The population distribution is normal.

61. **a.** No; no **b.** No, because $z = .44$ and P-value $= .33 > .10$.

63. **a.** Approximately .75; approximately .2 (from Table A.13) **b.** $n = 28$

65. $t = 2.83 < 3.733$, so don't reject H_0.

67. **a.** $-2.060 < 1.70 < 2.060$, so don't reject H_0; prior research is not contradicted. **b.** P-value $\approx .10$

69. P-value $= .0005$, so reject H_0: $p = .75$ in favor of H_a: $p < .75$.

71. P-value ≈ 0, so conclude that $\mu > 15$.

73. $.01 < P$-value $< .025$, so H_0 cannot be rejected at level .01

75. **a.** Test statistic $= 2\Sigma X_i / \mu_0$; for H_a: $\mu > \mu_0$, reject H_0 if test statistic value $\geq \chi^2_{\alpha, 2n}$. Rejection regions for $<$ and $\neq$ are lower- and two-tailed, respectively. **b.** Test statistic value $= 19.65 > 8.260$, so H_0 is not rejected using $\alpha = .01$.

77. **a.** Yes

Chapter 9

1. **a.** $-.4$ h; it doesn't. **b.** .0724, .2691 **c.** No

3. Test statistic value $= 6.41$. Reject H_0.

5. **a.** $z = -2.90$, so reject H_0. **b.** .0019 **c.** .8212 **d.** 66

7. Yes, since $z = 1.83 > 1.645$.

9. **a.** $z = 2.89$, so don't use the high purity steel. **b.** .2981 **c.** s_1 and s_2 replace σ_1 and σ_2; same conclusion.

11. $(-11.23, -6.31)$

13. $\hat{\theta} = -.97$, $\hat{\sigma}_{\hat{\theta}} = s_{\hat{\theta}} = 1.05$, $z = -.92$, so don't reject H_0; no.

15. They increase.

17. $t = -2.00$, so don't reject H_0.

19. $(-20.3, -6.9)$

21. **a.** 9.66 **b.** P-value $= .0018 < .01$, so reject H_0 **c.** $(2.9, 34.9)$

23. **a.** Yes; because $-2.46 \leq -1.734$, reject H_0. **b.** $.01 < P$-value $< .025$ **c.** $\beta \approx .7$ **d.** $m = n = 15$

25. $(-17.58, -9.74)$

27. Assume normality and $\sigma_1 = \sigma_2$. Because $4.93 \geq 2.120$, conclude that $\mu_1 \neq \mu_2$.

29. $t' = -1.29$ and $\nu = 8$; because $-1.29 > -2.896$, don't reject H_0.

31. Do not reject H_0 because $t = 1.73 < 2.998$.

33. Because $t = .36 < 1.771$, do not reject H_0.

35. $(-.811, -.017)$

37. $(12.11, 164.55)$

39. $(-10.81, -4.39)$

41. H_0 is rejected because $-4.18 \leq -2.33$.

43. P-value $= .4247$, so H_0 cannot be rejected.

45. a. $z = 1.43 < 1.96$, so don't reject H_0.   **b.** $n = 6582$

47. Identifying p_1 with the placebo and p_2 with the aspirin treatment, the CI is $(1.39, 2.25)$.

49. $(-.35, .07)$

51. a. 3.69   **b.** 4.82   **c.** 2.07   **d.** 2.71   **e.** 4.30
   f. .212   **g.** .95   **h.** .94

53. $f = .585$, so H_0 is not rejected; no.

55. $f = 1.22$, so it is quite plausible that $\sigma_1 = \sigma_2$.

57. No; $t = 3.22 \geq 2.878$, so reject H_0 using $\alpha = .01$.

59. $(-.18, 5.28)$

61. Assuming that the distribution of differences is normal, the CI is $(-.110, .186)$.

63. A normal probability plot suggests that the good visibility distribution is not normal, so the pooled t test should not be used.

65. a. $m = 141, n = 47$ **b.** $m = 240, n = 160$

67. a. $-3.922 < 3.55 < 3.922$, so don't reject H_0.
   b. $.002 < P$-value $< .01$

69. .9015, .8264, .0294, .0000; true average IQs; no

71. $z = .62$, so there doesn't appear to be any difference.

73. $1.22 < 1.350$; H_0 is not rejected.

75. a. Yes, because $z = -6.4$.   **b.** $z = 1.1 < 1.645$, so true average weight does not appear to exceed 1400.

77. $z = -5.3$, so P-value ≈ 0; there does appear to be a difference.

79. $f = 4.57$, so H_0 cannot be rejected for either $\alpha = .02$ or $\alpha = .10$; it is plausible that the amount of variability is the same for the two types of children.

Chapter 10

1. $f = 1.85$, $F_{.05,4,15} = 3.06$, so don't reject H_0.

3. $f = 1.30$ and $F_{.05,2,21} = 3.47$, so there don't appear to be any differences.

5. $f = 1.73$ and $F_{.01,2,27} = 5.49$, so the three grades don't appear to differ.

7. $f = 1.70$ and $F_{.05,3,16} = 3.24$, so H_0 cannot be rejected.

9. $f = 3.96$ and $F_{.05,3,20} = 3.10$, so H_0 is rejected and there appear to be differences among the grains.

11. $w = 36.09$

3	1	4	2	5
437.5	462.0	469.3	512.8	532.1

Brands 2 and 5 don't appear to differ, nor does there appear to be any difference between brands 1, 3, and 4, but each brand in the first group appears to differ significantly from all brands in the second group.

13.

3	1	4	2	5
427.5	462.0	469.3	502.8	532.1

15. $w = 5.94$

2	1	3	4
24.69	26.08	29.95	33.84

The only significant differences are between 4 and both 1 and 2.

17. $(-.701, -.091)$

19. Any value of SSE between 422.16 and 431.88 will work.

21. a. $f = 22.6$ and $F_{.01,5,78} \approx 3.3$, so reject H_0.
   b. $(-99.16, -35.64)$, $(29.34, 94.16)$

23.

	1	2	3	4
1	–	2.88 ± 5.81	7.43 ± 5.81	12.78 ± 5.48
2		–	4.55 ± 6.13	9.90 ± 5.81
3			–	5.35 ± 5.81
4				–

4	3	2	1
		.016519	

25. a. 1.1724, .066076, .006337   **b.** .029309, .000419
   c. $f = 15.1$ and $F_{.05,4,70} \approx 2.51$, so there appear to be differences.
     39.4
   d.

1	2	3	4	5

27. $f = 3.75 \geq 3.10 = F_{.05,3,20}$, so brands appear to differ.

31. Approximately .62

33. $\arcsin(\sqrt{x/n})$

35. a. $3.68 < 4.94$, so H_0 is not rejected.
   b. $.029 > .01$, so again H_0 is not rejected.

37. $f = .39$, so the three brands do not appear to differ.

39. The CI is $(-.144, .474)$, which does include 0.

41. $f = 3.96 < 4.07$, so $H_0: \sigma_A^2 = 0$ cannot be rejected.

43. $f = 1123$, so H_0 is resoundingly rejected. Only μ_1 and μ_4 are judged to not differ significantly.

Chapter 11

1. a. $f_A = 1.55$, so don't reject H_{0A}.   **b.** $f_B = 2.98$, so don't reject H_{0B}.

3. a. $f_A = 12.987$, $F_{.01,3,9}$, so conclude that there is a gas rate effect; $f_B = 105.31$, so conclude that there is a liquid rate effect.   **b.** $w = 95.44$; 231.75 325.25 441.0 613.25, so only the lowest two rates do not differ significantly from one another.
c. 336.75 382.25 419.25 473 so only the lowest and highest rates appear to differ significantly from one another.

5. $f_A = 2.65$, $F_{.01,3,12} = 5.95$, so there appears to be no effect due to angle of pull.

7. a.

Source	df	SS	MS	f
Treatments	2	28.78	14.39	1.04
Blocks	17	2977.67	175.16	12.68
Error	34	469.55	13.81	
Total	53	3476.00		

True average adaptation score does not appear to depend on which treatment is given.   **b.** Yes; f_B is quite large, suggesting great variability between subjects.

9.

Source	df	SS	MS	f	$F_{.05}$
Treatments	3	81.19	27.06	22.4	3.01
Blocks	8	66.50	8.31		
Error	24	29.06	1.21		
Total	35	176.75			

1	4	3	2
8.55	9.22	10.77	12.44

11. b. Each SS is multiplied by c^2, but f_A and f_B are unchanged.

13. a. Approximately .20, .43   **b.** Approximately .30

15. a. $f_A = 3.76$, $f_B = 6.82$, $f_{AB} = .74$, and $F_{.05,2,9} = 4.26$, so the amount of carbon fiber addition appears significant.   **b.** $f_A = 6.54$, $f_B = 5.33$, $f_{AB} = .27$

17. a.

Source	df	SS	MS	f
Coal	2	1.00241	.50121	29.49
NaOH	2	.12431	.06216	3.66
Interaction	4	.01456	.00364	.21
Error	9	.15295	.01699	
Total	17	1.29423		

Type of coal does appear to affect total acidity.
b. Coals 1 and 3 don't differ significantly from one another, but both differ significantly from coal 2.

19. a, b.

Source	df	SS	MS	f
A	2	22941.80	11470.90	22.98
B	4	22765.53	5691.38	5.60
AB	8	3993.87	499.23	.49
Error	15	15253.50	1016.90	
Total	29	64954.70		

H_{0A} and H_{0B} are both rejected.

23. $(-1.39, -1.05)$

25. a.

Source	df	SS	MS	f	$F_{.05}$
A	2	14,144.44	7,072.22	61.06	3.35
B	2	5,511.27	2,755.64	23.79	3.35
C	2	244,696.39	122,348.20	1,056.27	3.35
AB	4	1,069.62	267.41	2.31	2.73
AC	4	62.67	15.67	.14	2.73
BC	4	331.67	82.92	.72	2.73
ABC	8	1,080.77	135.10	1.17	2.31
Error	27	3,127.50	115.83		
Total	53	270,024.33			

d. $Q_{.05,3,27} = 3.51$, $w = 8.90$, and all three of the levels differ significantly from one another.

27.

Source	df	SS	MS	f
A	2	12.896	6.448	1.04
B	1	100.041	100.041	16.10
C	3	393.416	131.139	21.10
AB	2	1.646	.823	<1
AC	6	71.021	11.837	1.905
BC	3	1.542	.514	<1
ABC	6	9.771	1.629	<1
Error	72	447.500	6.215	
Total	95	1037.833		

a. No interaction effects are significant.   **b.** Factor B and factor C main effects are significant.
c. $w = 1.89$; only machines 2 and 4 do not differ significantly from one another.

29. The P-value column shows that several interaction effects are significant at level .01.

31.

Source	df	SS	MS	f
A	6	67.32	11.02	
B	6	51.06	8.51	
C	6	5.43	.91	.61
Error	30	44.26	1.48	
Total	48	168.07		

$F_{.05,6,30} = 2.42$, $f_C = .61$, so H_{0C} is not rejected.

33.

Source	df	SS	MS	F
A	4	28.88	7.22	10.7
B	4	23.70	5.93	8.79
C	4	.62	.155	<1
Error	12	8.10	.675	
Total	24	61.30		

Since $F_{.05,4,12} = 3.26$, both A and B are significant.

35.

Source	df	MS	f
A	2	2207.329	2259*
B	1	47.255	48.4*
C	2	491.783	503*
D	1	.044	<1
AB	2	15.303	15.7*
AC	4	275.446	282*
AD	2	.470	<1
BC	2	2.141	2.19
BD	1	.273	<1
CD	2	.247	<1
ABC	4	3.714	3.80
ABD	2	4.072	4.17*
ACD	4	.767	<1
BCD	2	.280	<1
ABCD	4	.347	<1
Error	36	.977	
Total	71	93.621	

* Denotes a significant F ratio.

37. a. $\hat{\beta}_1 = 54.38$, $\hat{\gamma}_{11}^{AC} = -2.21$, $\hat{\gamma}_{21}^{AC} = 2.21$.

b.

Source	Effect Contrast	MS	f
A	1,307	71,177.04	436.7
B	1,305	70,959.34	435.4
C	529	11,660.04	71.54
AB	199	1,650.04	10.12
AC	−53	117.04	<1
BC	57	135.38	<1
ABC	27	30.38	<1
Error		162.98	

39.

Source	SS	f
A	136,640.02	1,007.6
B	139,644.19	1,029.8
C	24,616.02	181.5
D	20,377.52	150.3
AB	2,173.52	16.0
AC	2.52	<1
AD	58.52	<1
BC	165.02	1.2
BD	9.19	<1
CD	17.52	<1
ABC	42.19	<1
ABD	117.19	<1
ACD	188.02	1.4
BCD	13.02	<1
ABCD	204.19	1.5
Error	4339.33	
Total	328,607.98	

$F_{.05,1,32} \approx 4.15$, so only the four main effects and the AB interaction appear significant.

41.

Source	df	SS	f
A	1	.436	<1
B	1	.099	<1
C	1	.109	<1
D	1	414.12	851
AB	1	.003	<1
AC	1	.078	<1
AD	1	.017	<1
BC	1	1.404	3.62
BD	1	.456	<1
CD	1	2.190	4.50
Error	5	2.434	

$F_{.05,1,5} = 6.61$, so only the factor D main effect is judged significant.

43. a. 1: (1), *ab*, *cd*, *abcd*; 2: *a*, *b*, *acd*, *bcd*; 3: *c*, *d*, *abc*, *abd*; 4: *ac*, *bc*, *ad*, *bd*.

b.

Source	df	SS	f
A	1	14,028.13	53.89
B	1	92,235.13	345.33
C	1	3.13	<1
D	1	18.00	<1
AC	1	105.13	<1
AD	1	200.00	<1
BC	1	91.13	<1
BD	1	420.50	1.62
ABC	1	276.13	1.06
ABD	1	2.00	<1
ACD	1	450.00	1.73
BCD	1	2.00	<1
Blocks	7	898.88	<1
Error	12	3,123.72	
Total	31	111,853.88	

$F_{.01,1,12} = 9.33$, so only the A and B main effects are significant.

45. a. *ABFG;* (1), *ab, cd, ce, de, fg, acf, adf, adg, aef, acg, aeg, bcg, bcf, bdf, bdg, bef, beg, abcd, abce, abde, abfg, cdfg, cefg, defg, acdef, acdeg, bcdef, bcdeg, abcdfg, abcefg, abdefg.* {*A, BCDE, ACDEFG, BFG*}, {*B, ACDE, BCDEFG, AFG*}, {*C, ABDE, DEFG, ABCFG*}, {*D, ABCE, CEFG, ABDFG*}, {*E, ABCD, CDFG, ABEFG*}, {*F, ABCDEF, CDEG, ABG*}, {*G, ABCDEG, CDEF, ABF*}.   **b.** 1: (1), *aef, beg, abcd, abfg, cdfg, acdeg, bcdef;* 2: *ab, cd, fg, aeg, bef, acdef, bcdeg, abcdfg;* 3: *de, acg, adf, bcf, bdg, abce, cefg, abdefg;* 4: *ce, acf, adg, bcg, bdf, abde, defg, abcefg.*

47. *SSA* = 2.250, *SSB* = 7.840, *SSC* = .360, *SSD* = 52.563, *SSE* = 10.240, *SSAB* = 1.563, *SSAC* = 7.563, *SSAD* = .090, *SSAE* = 4.203, *SSBC* = 2.103, *SSBD* = .010, *SSBE* = .123, *SSCD* = .010, *SSCE* = .063, *SSDE* = 4.840. Error *SS* = sum of two-factor *SS*'s = 20.568, Error *MS* = 2.057, $F_{.01,1,10}$ = 10.04, so only the *D* main effect is significant.

49.

Source	df	SS	MS	f
A main effects	1	322.667	322.667	980.38
B main effects	3	35.623	11.874	36.08
Interaction	3	8.557	2.852	8.67
Error	16	5.266	0.329	
Total	23	372.113		

$F_{.05,3,16}$ = 3.24, so interactions appear to be present.

51. *MSA* = 6.940,   *MSB* = 1.870,   *MSC* = 6.165, *MSAB* = 1.350,   *MSAC* = 3.660,   *MSBC* = 2.633, *MSABC* = .728, *MSE* = .600.

53.

Source	df	SS	MS	f
Diet	2	18,138	9,069	28.9
Temp.	2	5,182	2,591	8.3
Interaction	4	1,737	434.3	1.4
Error	36	11,291	313.6	
Total	44	36,348		

H_{0AB} cannot be rejected, but both main effects are significant. Diet and temperature both appear to affect average energy intake.

Chapter 12

1. a. .5050   **b.** 1.3   **c.** 130   **d.** −130

3. a. .095   **b.** −.475   **c.** .830, 1.305   **d.** .4207, .3446   **e.** .0036

5. a. −.01,   −.10   **b.** 3.00,   2.50   **c.** .3627   **d.** .4641

7. a. $10b_0 + 1269b_1$ = 475, $1269b_0 + 172,809b_1$ = 62,631   **b.** $\hat{\beta}_1$ = .19990826 ≈ .200, $\hat{\beta}_0$ = 22.13164131 ≈ 22.13   **c.** .200   **d.** 2.40   **e.** 46.13

9. a. $y = -45.5519 + 1.7114x$ **b.** 339.51   **c.** −85.57   **d.** The $\hat{y}_i$'s are 125.6, 168.4, 168.4, 211.1, 211.1, 296.7, 296.7, 382.3, 382.3, 467.9, 467.9, 553.4, 639.0, 639.0; a 45° line through (0, 0).

11. b. $y = -2.182 + .660x$ **c.** 7.72   **d.** 7.72

13. a. *s* = .06155   **b.** r^2 = .801

15. a. The pattern in the plot shows some curvature, so the use of the simple linear regression model is questionable.   **b.** $y = 6.735 - .14202x$ **c.** *s* = 1.481   **d.** r^2 = .775

19. $\hat{\beta}_1 = \Sigma x_i Y_i / \Sigma x_i^2$

21.

Data set	r^2	s	
1	.43	4.03	Most effective: set 3
2	.99	4.03	Least effective: set 1
3	.99	1.90	

23. a. .001017   **b.** (−.00956, −.00516)

25. a. *t* = 5.29 > 3.106, so the model is judged useful.   **b.** *P*-value = .000 < .01, so again the model appears useful.   **c.** *t* = −2.92, so reject H_0: β_1 = 1.5 in favor of H_a: β_1 < 1.5; the data contradicts prior belief.

27. a. *t* = 3.40 ≥ 2.650, so reject H_0.   **b.** (1.08, 2.32)

29. $t = -5.56 \le -t_{.0005,9}$ = −4.781, so even at level .001 we reject H_0: β_1 = 0. According to this test, the model can be judged useful, even though the scatter plot shows marked curvature.

31. *f* = 96.0 and $F_{.01,1,9}$ = 10.56, so the model is judged useful.

35. *d* = 1.20, df = 14, and β ≈ .1.

37. a. (36.10, 40.42)   **b.** (31.87, 44.65)   **c.** Narrower, because $\bar{x}$ = 2.667   **d.** No; the danger of extrapolation

39. a. .40 is closer to $\bar{x} = .7495$ **b.** (.745, .875)
c. (.059, .523)

41. a. (318.70, 343.19) **b.** (291.39, 370.50)
d. (258.76, 289.89) and (357.34, 380.05)

43. Using a 98% confidence level for each interval separately (simultaneous confidence level $\geq$ 94%) gives (326.96, 373.64), (359.86, 404.74), and (390.27, 438.33).

45. A 95% PI is (1.06, 8.14).

47. a. $x_2 = x_3 = 12$, yet $y_2 \neq y_3$. **b.** Yes **c.** $y = -19.670 + 3.2847x$ **d.** (24.93, 67.12)

49. a. .7428 **b.** .552

51. $r = .9066$, $t = 7.75$, and P-value $< .0005$, so conclude that $\rho > 0$.

53. $r = .773$, yet $t = 2.44 < 2.776$; so H_0: $\rho = 0$ cannot be rejected.

55. a. The x plot is a bit curved but not disturbingly so in light of the small sample size. The y plot is quite straight.
b. $t = 6.3 \geq 3.355$, so there does appear to be a linear relationship.

57. $t = 2.20 \geq 1.96$, so it appears that $\rho \neq 0$. But $r = .022$ suggests that ρ is not much different from zero, indicating little practical significance.

59. a. $t = -1.24 > -2.201$, so H_0 cannot be rejected.
b. .970

61. a. .836 **b.** $t = 8.44$, so the relationship does appear to be useful. **c.** $t = -3.11 \leq -2.145$, so reject H_0 and conclude that $\mu_{Y \cdot 500} \neq 10$.

63. a. $y = 14.1904 - .14892x$ **b.** $t = -1.43$, so don't reject H_0: $\beta_1 = -.10$. **c.** No; $\Sigma(x - \bar{x})^2 = 143$ here and 182 for the given data. **d.** A 95% CI for $\mu_{Y \cdot 28}$ is (9.599, 10.443).

65. a. A substantial linear relationship **b.** $y = -.08259 + .044649x$ **c.** 98.3% **d.** .7702, $-.0902$ **e.** Yes; $t = 19.96$ **f.** (.0394, .0499) **g.** (.762, .858)

67. b. .573

71. $t = -1.14$, so it is plausible that $\beta_1 = \gamma_1$.

Chapter 13

1. a. 6.32, 8.37, 8.94, 8.37, and 6.32 **b.** 7.87, 8.49, 8.83, 8.94, and 2.83 **c.** The deviation is likely to be much smaller for the x values of part (b).

3. a. Yes. **b.** $-.75, .31, -.74, 1.13, .43, -.72, 1.43, .93, -1.51, -1.27, .90$ **c.** No.

5. $-.50, -.75, -.50, .79, .90, .93, .19, 1.46, -1.80, -1.12$; curvature.

7. a. No. **b.** e_i's are $-16.60, 9.70, 19.00, -.70, -11.40$; e_i^*'s are $-1.55, .68, 1.25, -.05, -1.06$; a quadratic function.

9. For set 1, simple linear regression is appropriate. A quadratic regression is reasonable for set 2. In set 3, (13, 12.74) appears very inconsistent with the remaining data. The estimated slope for set 4 depends largely on the single observation (19, 12.5), and evidence for a linear relationship is not compelling.

11. c. $V(\hat{Y}_i)$ increases, and $V(Y_i - \hat{Y}_i)$ decreases.

13. t with $n - 2$ df; .02

15. a. $\hat{\alpha} = 43,044.94$, $\hat{\beta} = -2.16$ **b.** Yes **c.** Yes **d.** $t = -1.53$, so don't reject H_0. **f.** $x = 40$ $\Rightarrow x' = 3.6889$; the C.I. for $\beta_0 + \beta_1 x'$ is (2.6203, 2.7833), so the desired interval is (13.74, 16.17).

17. a. $\Sigma x_i' = 15.501$, $\Sigma y_i' = 13.352$, $\Sigma(x_i')^2 = 20.228$, $\Sigma x_i' y_i' = 18.109$, $\Sigma(y_i')^2 = 16.572$, $\hat{\beta}_1 = 1.254$, $\hat{\beta}_0 = -.468$, $\hat{\alpha} = .626$, $\hat{\beta} = 1.254$ **c.** $t = -1.07$, so don't reject H_0. **d.** H_0: $\beta = 1$, $t = -4.30$, so reject H_0.

19. a. No **b.** $Y' = \beta_0 + \beta_1 \cdot (1/t) + \epsilon'$, where $Y' = \ln(Y)$, so $Y = \alpha e^{\beta/t} \cdot \epsilon$. **c.** $\hat{\beta} = \hat{\beta}_1 = 3735.45$, $\hat{\beta}_0 = -10.2045$, $\hat{\alpha} = (3.70034) \cdot (10^{-5})$. $\hat{y}' = 6.7748$, $\hat{y} = 875.5$ **d.** $SSE = 1.39587$, $SSPE = 1.36594$ (using transformed values), $f = .33 < 8.68 = F_{.01,1,15}$, so don't reject H_0.

21. a. $\hat{\mu}_{Y \cdot x} = 18.14 - 1485/x$ **b.** $\hat{y} = 15.17$

23. For the exponential model, $V(Y | x) = \alpha^2 e^{2\beta x} \sigma^2$, which does depend on x. A similar result holds for the power model.

25. a. 39.41 **b.** 24.93 **c.** 217.82; 72.61; 8.52 **d.** .779 **e.** $t = -7.88$, so reject H_0 and retain the quadratic predictor.

27. a. $R^2 = .853$, so 85.3% of the observed variation in yield can be attributed to the quadratic model relationship. **b.** $(-227.83, -43.07)$ **c.** $t = -1.83 > -2.718 = -t_{.01,11}$, so don't reject H_0. **d.** (1081.37, 1722.99)

29. a. $.3463 - 1.2933(x - \bar{x}) + 2.3964(x - \bar{x})^2 -$
$2.3968\,(x - \bar{x})^3$ **b.** -2.3968; 33.6430 **c.** $.1949$
d. $t = -.97$, so the cubic term can be deleted.

31. a. $.873$ **b.** $.919$ **c.** $1.200887 - .01048314x +$
$.00002618x^2$ **d.** $.00077118$ **e.** $t = 1.40$, so delete the quadratic term; no

33. a. \$227.50   **b.** \$20   **c.** $.8407$

35. a. $.688$ **b.** $f = 162.71 \geq 3.4$, so the model appears useful.   **c.** $(1235.11,\ 1238.83)$ **d.** $(1.26,\ 2.45)$

37. $f = 6.0 \geq 2.76$, so the model is judged useful.

39. a. $f = 199 \geq 3.85$, so the model is judged useful.
b. Using $t_{.01}$ gives simultaneous confidence level
$\geq 96\%$: $(-.02688, -.01430)$ and $(-.18846, -.01464)$.
c. 2.700, 7087.50 **d.** $f = 1.97 < 3.0$, so inclusion of x_1^3 and x_2^3 is not justified.

41. $f = 6.43 \geq 4.95 = F_{.05,6,5}$, so at least one of the second-order predictors should be retained (the conclusion changes if $\alpha = .01$ is used).

43. a. $t = .59 < 2.447$, so the interaction predictor should be deleted.   **b.** No   **c.** $f = 1.34 < 4.76$, so none of the second-order predictors should be included.

45. a. $f = 9.32 \geq 4.69$, indicating a useful relation between y and at least one of the predictors.   **b.** $.687$, $.707$
c. $f = 2.32 < 3.34$, so don't reject H_0.

47. a. The dependent variable is $\ln(q)$, and the predictors

are $x_1 = \ln(a)$ and $x_2 = \ln(b)$; $\hat{\beta} = \hat{\beta}_1 = .9450$,
$\hat{\gamma} = \hat{\beta}_2 = .1815$, $\hat{\alpha} = 4.7836$, $\hat{q} = 18.27$ **b.** Now regress $\ln(q)$ against $x_1 = a$ and $x_2 = b$.   **c.** $(1.24,\ 5.78)$

49.

k	R^2	adj. R^2	C_k
1	.676	.647	138.2
2	.979	.975	2.7
3	.9819	.976	3.2
4	.9824		4

a. The model with $k = 2$ **b.** No

51. No

53. a. $f = 1783$, so the model appears useful.
b. $t = -48.1 \leq -6.689$, so even at level .001 the quadratic predictor should be retained.   **c.** No
d. $(21.07, 21.65)$ **e.** $(20.67, 22.05)$

55. a. $f = 30.8 \geq 9.55 = F_{.01,2,7}$, so the model appears useful.   **b.** $t = -7.69$ and P-value $< .001$, so retain the quadratic predictor.   **c.** $(44.01, 47.91)$

57. a. 231.75 **b.** $.903$ **c.** $f = 41.9$, indicating a useful relationship.   **d.** $(220.9, 238.1)$

59. There are several reasonable choices in each case.

61. a. $f = 2.4$; no   **b.** No; k large relative to n **c.** At least $.9565$

63. a. At least $.723$ **b.** At most $.723$

Chapter 14

1. $\chi^2 = 5.92 < 7.815 = \chi^2_{.05,3}$, so don't reject H_0.

3. $\chi^2 = 6.61 < 14.684 = \chi^2_{.10,9}$, so don't reject H_0.

5. $\chi^2 = 4.03$ and P-value $> .10$, so don't reject H_0.

7. a. $[0, .2231)$, $[.2231, .5108)$, $[.5108, .9163)$, $[.9163, 1.0694)$, and $[1.0694, \infty)$ **b.** $\chi^2 = 1.25 < \chi^2_{\alpha,4}$ for any reasonable α, so the specified exponential distribution is quite plausible.

9. a. $(-\infty, -.97)$, $[-.97, -.43)$, $[-.43, 0)$, $[0, .43)$, $[.43, .97)$, and $[.97, \infty)$ **b.** $(-\infty, .49806)$, $[.49806, .49914)$, $[.49914, .5)$, $[.5, .50086)$, $[.50086, .50194)$, and $[.50194, \infty)$ **c.** $\chi^2 = 5.53$, $\chi^2_{.10,5} = 9.236$, so P-value $> .10$ and the specified normal distribution is plausible.

11. $\hat{p} = .0843$, $\chi^2 = 280.3 > \chi^2_{\alpha,1}$ for any tabulated α, so the model gives a poor fit.

13. The likelihood is proportional to $\theta^{233}(1 - \theta)^{367}$, from which $\hat{\theta} = .3883$. The estimated expected counts are 21.00, 53.33, 50.78, 21.50, and 3.41. Combining cells 4 and 5, $\chi^2 = 1.62$, so don't reject H_0.

15. $\hat{\lambda} = 3.167$, from which $\chi^2 = 103.98 \gg \chi^2_{\alpha,k-1} = \chi^2_{\alpha,7}$ for any tabulated α, so the Poisson distribution provides a very poor fit.

17. $\hat{\theta}_1 = (2n_1 + n_3 + n_5)/2n = .4125$, $\hat{\theta}_2 = .2750$, $\chi^2 = 30.58$, $\chi^2_{.01,3} = 11.344$, so reject H_0.

19. $r = 967$, $c_{.10} = .9707$ for $n = 30$, so at this level the hypothesis of normality must be rejected (a very close call, though, and normality would not be rejected for $\alpha = .05$ or $.01$).

21. $6.161 < 7.815$, so (surprisingly) there is no evidence of an association.

23. $\chi^2 = 24.82 \geq 13.277$, so number of leaves removed does appear to have an impact.

25. $\chi^2 = 27.66 \geq 23.587$, so reject H_0.

27. $\chi^2 = 12.37 \geq 9.488$, so condition and dosage level do not appear to be independent.

29. $\chi^2 = 64.65 \geq 13.277$, so independence is not plausible.

31. df $= 20$

33. a. $\chi^2 = .6806 < \chi^2_{.10,2} = 4.605$, so don't reject H_0.
  b. Now $\chi^2 = 6.806$, so H_0 is rejected. **c.** $n = 677$

35. $\chi^2 = 13.0 \geq 9.210$, so the treatments appear to differ.

37. a. P-value $< \alpha$, so reject H_0. **b.** Yes, baseball.

39. $\chi^2 = 197.62$, indicating strongly that age and location are dependent.

41. $\chi^2 = 11.65 \geq 9.488 = \chi^2_{.05,4}$, so the hypothesis of independence is rejected.

Chapter 15

1. a. $(a + b)/2$ **b.** $\ln(\lambda)/2$

3. At level .05, reject H_0 if either $y \geq 9$ or $y \leq 1$. Because $y = 2$, don't reject H_0.

5. a. $p = .50$ **b.** Bin(6, .50)   **c.** The rejection region $\{5, 6\}$ gives $\alpha = .1094 \approx .10$. Because $y = 5$, reject H_0.

7. When H_0 is true, Y has a Bin(20, .75) distribution, so $\{18, 19, 20\}$ gives $\alpha = .091 \approx .10$. Because $y = 18$, reject H_0.

9. Reject H_0 if either $s_+ \geq 47$ or ≤ 8. Because $s_+ = 14$, don't reject H_0.

11. Reject H_0 if either $s_+ \geq 64$ or ≤ 14. Only ranks 1 and 5 are negatively signed, so $s_+ = 72$ and H_0 is rejected.

13. Reject H_0 if $z \geq 1.645$. Because $z = (226.5 - 162.5)/37.16 = 1.72 \geq 1.645$, reject H_0. P-value $\approx 1 - \Phi(1.72) = .0427$.

15. $w = 38 \geq 36 = c_1$, so reject H_0.

17. $w = 65$, whereas $c_1 = 84$, so H_0 is not rejected.

19. $z = 2.54 < 2.58$, so don't reject H_0.

21. $(x_{(5)}, x_{(16)}) = (44.1, 46.6)$

23. $(\bar{x}_{(5)}, \bar{x}_{(32)}) = (11.15, 23.80)$

25. $(-13.0, -6.0)$

27. $(d_{ij(5)}, d_{ij(21)}) = (16, 87)$

29. $k = 14.06 \geq 6.251$, so reject H_0.

31. $k = 9.23 \geq 5.992$, so reject H_0.

33. $f_r = 2.60 < 5.992$, so don't reject H_0.

35. $f_r = 9.62 > 7.815 = \chi^2_{.05,3}$, so reject H_0.

37. $(-5.9, -3.8)$

39. $w' = 26 < 27$, so don't reject H_0.

Chapter 16

1. All points on the chart fall between the control limits.

3. .9802, .9512, 53

5. a. .0301   **b.** .2236   **c.** .6808

7. LCL = 12.20, UCL = 13.70. No.

9. LCL = 94.91, UCL = 98.17. There appears to be a problem on the 22nd day.

11. a. 200   **b.** 4.78   **c.** 384.62 (larger), 6.30 (smaller)

13. LCL = 12.37, UCL = 13.53

15. a. LCL = 0, UCL = 6.48   **b.** LCL = .48, UCL = 6.60

17. LCL = .045, UCL = 2.484. Yes, since all points are inside the control limits.

19. a. LCL = .105, UCL = .357   **b.** Yes, since .39 > UCL.

21. $\bar{p} > 3/53$

23. LCL = 0, UCL = 10.1

25. When area = .6, LCL = 0 and UCL = 14.6; when area = .8, LCL = 0 and UCL = 13.4; when area = 1.0, LCL = 0 and UCL = 12.6.

27.

l:	1	2	3	4	5	6	7	8
d_l:	0	.001	.017	0	0	.010	0	0
e_l:	0	0	0	.038	0	0	0	.054

l:	9	10	11	12	13	14	15
d_l:	0	.024	.003	0	0	0	.005
e_l:	0	0	0	.015	0	0	0

There are no out-of-control signals.

29. $n = 5$, $h = .00626$

31. Hypergeometric probabilities (calculated on an HP21S calculator) are .9919, .9317, .8182, .6775, .5343, .4047, .2964, .2110, .1464, and .0994, whereas the corresponding binomial probabilities are .9862, .9216, .8108, .6767, .5405, .4162, .3108, .2260, .1605, and .1117. The approximation is satisfactory.

33. .9206, .6767, .4198, .2321, .1183; the plan with $n = 100$, $c = 2$ is preferable.

35. .9981 .5968, and .0688

37. a. .010, .018, .024, .027, .027, .025, .022, .018, .014, .011 **b.** .0447, .0274 **c.** 77.3, 202.1, 418.6, 679.9, 945.1, 1188.8, 1393.6, 1559.3, 1686.1, 1781.6

39. $\bar{x}$ chart based on sample standard deviations: LCL = 402.42, UCL = 442.20. $\bar{x}$ chart based on sample ranges: LCL = 402.36, UCL = 442.26. S chart: LCL = .55, UCL = 30.37. R chart: LCL = 0, UCL = 82.75.

41. S chart: LCL = 0, UCL = 2.3020; because $s_{21} = 2.931 > $ UCL, the process appears to be out of control at this time. Because an assignable cause is identified, recalculate limits after deletion: for an S chart, LCL = 0 and UCL = 2.0529; for an $\bar{x}$ chart, LCL = 48.583 and UCL = 51.707. All points on both charts lie between the control limits.

43. $\bar{\bar{x}} = 430.65$, $s = 24.2905$; for an S chart, UCL = 62.43 when $n = 3$ and UCL = 55.11 when $n = 4$; for an $\bar{x}$ chart, LCL = 383.16 and UCL = 478.14 when $n = 3$, and LCL = 391.09 and UCL = 470.21 when $n = 4$.

Index